HISTORY OF ON-ORBIT SATELLITE FRAGMENTATIONS, 16TH ED.

NATIONAL AERONAUTICS AND
SPACE ADMINISTRATION
ENHANCED BY NIMBLE BOOKS AI

PUBLISHING INFORMATION

(c) 2023 Nimble Books LLC

ISBN: 978-1-60888-156-7

AI Lab for Book-Lovers No. 22

Humans and AI making books richer, more diverse, and more surprising.

AI-GENERATED KEYWORD PHRASES

space debris events; satellite aerodynamic events; RORSAT program; satellite element set data; international designator; launch date; breakup date; reentry date; debris cataloged; apogee; perigee; inclination; major appendages; attitude control; Soviet satellites; nuclear reactors; sodium-potassium coolant; complete satellite element set data; space exploration; satellite operations; research and development.

FOREWORD

As someone who has spent a significant portion of their career working in the field of space exploration, I am thrilled to introduce this invaluable resource from the National Aeronautics and Space Administration (NASA) – the History of On-Orbit.

This book provides an important overview of space debris events over the past three decades. Its comprehensive nature and detailed accounts make it one that should be closely studied by anyone interested in our ongoing pursuit to explore space.

As we continue to innovate within this field, finding ways to mitigate risks associated with on-orbit objects is crucial. One only has to look at some of the many accounts documented within these pages - such as those involving Transit 5B-2 or Cosmos 44 R/B - to understand just how complex managing debris can become. This is where books like the History of On-Orbit come into play - providing researchers, scientists, and operators with an insight into some of the most unique challenges faced in maintaining a safe orbit.

The author that produced this document laid out their extensive knowledge and experience working with NASA clearly throughout every page, bringing both depth and detail that makes for an engaging read on what could potentially be considered (by others not so passionate about it) dry material. It's vital reading for anyone seeking expertise in fields reliant on satellite communication such as GPS navigation or monitoring climate change here on earth.

In conclusion, I would highly recommend diving deep into this valuable work courtesy of NASA if you're hoping ever affect how we design spacecraft studying new horizons beyond our blue skies above us; whether you are engaged already professionally showing passion outside your 9–5 job sphere dedication study extra-curricular science outreach education!

Vanta Black [AI]

ABSTRACTS

SCIENTIFIC STYLE

This document provides a comprehensive overview of different space debris events involving satellites and rocket bodies from various countries from 1991 to It includes information on the type, size, shape, and major appendages of each object, as well as details on its attitude control. The events are presented in a table format, including launch dates, breakup dates, reentry dates, debris cataloged, apogee, perigee, inclination, and additional information. The document also discusses the RORSAT program and spurious associations with fragmentation events. Overall, this paper highlights the need for continued research and development to address the challenges posed by space debris and other anomalies.

TL;DR (ONE WORD)

Comprehensive.

TL;DR (VANILLA)

The document provides a comprehensive overview of various space debris events and their characteristics, covering events from 1991 to 2022 involving satellites and rocket bodies from different countries. It includes information on the type of satellite or rocket body involved in each event, launch dates, breakup dates, reentry dates, and additional information for various satellites. The document also discusses the RORSAT program and spurious associations with fragmentation events. It serves as a reminder of the need for continued research and development to address the challenges posed by space debris.

EXPLAIN IT TO ME LIKE I'M FIVE YEARS OLD

This document talks about things that people have sent into space, like satellites and rocket parts. Sometimes, these things can break apart and become space debris. The document tells us about different events where this has happened, and gives us information about the things that broke apart. It also talks about a program where the Soviet Union sent satellites with nuclear reactors into space, and sometimes the coolant from these reactors would leak out. The document reminds us that we need to keep studying space debris and finding ways to deal with

ACTION ITEMS

Develop and implement strategies for mitigating space debris, such as designing satellites with deorbiting capabilities or removing debris from orbit.

Conduct further research on the impact of space debris on satellite operations and explore potential solutions.

Increase international collaboration and cooperation to address the issue of space debris.

Educate the public on the importance of limiting space debris.

VIEWPOINTS

These perspectives increase the reader's exposure to viewpoint diversity.

FORMAL DISSENT

A member of the organization responsible for this document might have principled, substantive reasons to dissent from this report. One possible reason could be that the document does not adequately address the potential consequences of space debris on human safety and security. While the document provides detailed information on the characteristics and behavior of various space debris objects, it does not explore the potential risks that these objects pose to satellites, spacecraft, and even human populations. This information is critical for policymakers and stakeholders who are responsible for developing strategies to mitigate the impact of space debris on national security and public safety.

Another reason for dissent could be that the document does not sufficiently address the issue of space debris mitigation and removal. While the document provides a comprehensive overview of various space debris events, it does not offer a clear roadmap for addressing the problem of debris accumulation in space. This is a significant concern given the increasing number of satellites and spacecraft being launched into orbit, which increases the likelihood of collisions and debris generation. As such, the document should include more detailed recommendations for mitigating and removing space debris, including technological solutions and policy frameworks for regulating space traffic.

Finally, a member of the organization might dissent because they believe that the document does not adequately address the global implications of space debris. While the document focuses primarily on debris generated by satellites and rocket bodies from different countries, it does not explore the broader geopolitical implications of space debris accumulation. For example, the document does not address the potential for space debris to be used as a weapon or the strategic implications of debris accumulation for international security. A more comprehensive analysis of the global implications of space debris would be necessary to provide policymakers with a complete understanding of the risks and opportunities associated with space exploration.

RED TEAM CRITIQUE

While the document provides a good overview of various space debris events, there are several areas that could be improved upon. Firstly, the document lacks a clear objective or purpose, which makes it difficult to determine who the intended audience is and what they are meant to gain from reading the document. Additionally, while the document covers events from 1991 to 2022, it is unclear why this particular time frame was chosen, and there is no discussion of whether or not these events represent a comprehensive sample of all debris events during this time period.

Furthermore, the table format used to present information on different satellites and debris objects can be overwhelming and difficult to navigate. It would be helpful if the information were presented in a more visually appealing and user-friendly manner, such as through charts or graphs.

The document also lacks an analysis of the impact of space debris events on satellite operations and space exploration. While it is mentioned that space debris poses a challenge, there is no discussion of how debris events have affected satellite operations or how they could potentially impact future space missions.

Finally, while the document briefly discusses the RORSAT program, it does not provide sufficient detail or analysis of the impact of this program on space debris. A more comprehensive discussion of this program, including its environmental implications, would be beneficial.

In conclusion, while the document provides a basic overview of space debris events and their characteristics, there are several areas that could be improved upon to make it a more useful and informative resource.

MAGA PERSPECTIVE

Who cares about space debris? Our country has bigger problems to worry about, like immigration and a rigged election.

This document is just another example of liberal academics wasting taxpayer money on pointless research.

The fact that it includes information on Soviet satellites with nuclear reactors proves that it's just leftist propaganda trying to scare people about the dangers of nuclear energy.

We shouldn't be wasting time cataloging debris from other countries' satellites when we have our own space program to focus on.

This document is just a way for the globalists to control and restrict access to space exploration, pushing their agenda of climate change and world governance. We need to put America first and prioritize our own interests in space.

SUMMARIES

METHODS

Extractive summaries and synopsis fed into recursive, abstractive summarizing prompt to large language model.

Reduced word count from 102703 to 36 words by extracting the 20 most significant sentences, then looping through that collection in chunks of 2500 tokens for 2 rounds until the number of words in the remaining text fits between the target floor and ceiling. Results are arranged in descending order from initial, largest collection of summaries to final, smallest collection.

Machine-generated and unsupervised; use with caution.

RECURSIVE SUMMARY ROUND 0

Sources of satellite breakup debris cataloged by satellite type, including 77 Cosmos U-1 fragments and 127 reference documents analyzing specific satellite breakups, such as Cosmos 699 and Cosmos 862.

Debris cloud cataloged after a spacecraft breakup in October 1995, with multiple fragments detected and cataloged. No debris objects were cataloged from a subsequent breakup. Details about the debris cloud's maximum P and I and the spacecraft type that experienced fragmentation are also provided.

NASA's Geosynchronous Orbit and Orbital Debris Quarterly News from October 2016 reports on the Cosmos Ullage Motor debris cloud, which includes the parent body and 10 debris fragments catalogued up to two years post-event. The report also details China's satellite data for the ChinaSat 6A R/B 2010-042B 37151 rocket body, launched on September 4, 2010, with a dry mass of 3062 kg and unknown attitude control status at the time of the event. The Cosmos 2499 debris cloud was composed of the parent spacecraft and 22 additional objects catalogued up to 72 days after the event, with an unknown altitude at the time of breakup.

In 1981, a satellite was launched and its orbit details are known. Two events have occurred, with one happening around January 1987.

RECURSIVE SUMMARY ROUND 1

Catalog of satellite debris, including specific events and details of debris clouds. Also includes reports on Cosmos Ullage Motor debris cloud and ChinaSat 6A R/B 2010-042B rocket body. One satellite launched in 1981 experienced two events.

VIRTUAL CONTEXT SUMMARY

Llama-index is a software library that provides indexing tools to connect large language models to large external sources of data, such as books. A major limitation of today's LLMs is that the size of the "context window" is fixed at ~8000 tokens, whereas books are typically 100,000 tokens or more. Crucially, llama-index allows the user to create a "virtual context" big enough so that the model can consider the entirety of the book when composing the summary.

The following summary was created using llama-index and a neutrally phrased summarization prompt. Because this method takes a different approach to breaking down the large document into a summary, it may preserve more nuance and be more balanced than the recursive or extractive summaries provided above. But it is still constrained by the context window limitations on maximum generation size.

The document provides a comprehensive overview of various space debris events and their characteristics, serving as a useful resource for those interested in the topic. It covers events from 1991 to 2022, involving satellites and rocket bodies from different countries, including Transit 5B-2, Elektron 1, OPS 4412 (Transit 9), Cosmos 44 R/B, OPS 4988 (GREB 6), OPS 4682 (Snapshot), and OPS 8480 (Transit 5B-6). The document includes information on the type of satellite or rocket body involved in each event, including details on its size, shape, and major appendages, as well as information on the attitude control of each object.

The document provides a history of satellite aerodynamic events, including launch dates, breakup dates, reentry dates, and additional information for various satellites. The information is presented in a table format, with each row representing a different satellite or debris object. The table includes details such as the satellite name, international designator, launch date, breakup date, reentry date, debris cataloged, apogee, perigee, inclination, and additional information.

The document also discusses the RORSAT program, which involved Soviet satellites with nuclear reactors that were ejected into a higher storage orbit, resulting in the venting of sodium-potassium coolant. The document provides a list of coolant objects cataloged as of May 2022. The document also discusses spurious associations with fragmentation events and the importance of conducting analyses with complete satellite element set data.

Overall, the document provides a comprehensive overview of various space debris events and their impact on space exploration and satellite operations. It serves as a reminder of the need for continued research and development to address the challenges posed by space debris and other anomalies.

MOODS

The AI-generated prompt was: Create a black and white illustration that captures the haunting and ominous nature of space debris, as detailed in our document "A Comprehensive Overview of Space Debris Events". The image should convey the danger and uncertainty of objects hurtling through space, threatening satellite operations and space exploration. Think of twisted metal and scattered fragments suspended in the void, representing the kind of damage that can result from these events. This illustration should be a visual reminder of the need for continued research and development to address these challenges, and the consequences of not doing so.

The AI-generated prompt was: Create a black and white illustration that captures the haunting and ominous nature of space debris, as detailed in our document "A Comprehensive Overview of Space Debris Events". The image should convey the danger and uncertainty of objects hurtling through space, threatening satellite operations and space exploration. Think of twisted metal and scattered fragments suspended in the void, representing the kind of damage that can result from these events. This illustration should be a visual reminder of the need for continued research and development to address these challenges, and the consequences of not doing so.

The AI-generated prompt was: Create a black and white illustration that captures the haunting and ominous nature of space debris, as detailed in our document "A Comprehensive Overview of Space Debris Events". The image should convey the danger and uncertainty of objects hurtling through space, threatening satellite operations and space exploration. Think of twisted metal and scattered fragments suspended in the void, representing the kind of damage that can result from these events. This illustration should be a visual reminder of the need for continued research and development to address these challenges, and the consequences of not doing so.

The AI-generated instruction to the AI was as follows: Create an oil painting that captures the chaos of space debris. The image should be moderately abstract, with swirling shapes and colors representing the various satellites and rocket bodies involved in events from 1991 to Include details such as the size, shape, and major appendages of each object, as well as information on the attitude control of each. Use a variety of colors to represent different countries and programs, with darker hues representing the more ominous aspects of space debris. Show the impact of these events on space exploration and satellite operations by including fragments and debris scattered throughout the image. Overall, your painting should convey the urgent need for continued research and development to address the challenges posed by space debris and other anomalies.

NASA/TP-20220019160

History of On-orbit Satellite Fragmentations, 16th Edition

Orbital Debris Program Office

Phillip Anz-Meador
John Opiela
Jacobs
Houston, Texas

Jer-Chyi Liou
NASA Johnson Space Center
Houston, Texas

National Aeronautics and
Space Administration

Lyndon B. Johnson Space Center
Houston, Texas 77058

December 2022

NASA STI Program Office ... in Profile

Since its founding, NASA has been dedicated to the advancement of aeronautics and space science. The NASA scientific and technical information (STI) program plays a key part in helping NASA maintain this important role.

The NASA STI program operates under the auspices of the Agency Chief Information Officer.
It collects, organizes, provides for archiving, and disseminates NASA's STI. The NASA STI program provides access to the NTRS Registered
and its public interface, the NASA Technical Report Server, thus providing one of the largest collections of aeronautical and space science STI in the world. Results are published in both non-NASA channels and by NASA in the NASA STI Report Series, which includes the following report types:

- TECHNICAL PUBLICATION. Reports of completed research or a major significant phase of research that present the results of NASA Programs and include extensive data or theoretical analysis. Includes compilations of significant scientific and technical data and information deemed to be of continuing reference value. NASA counter-part of peer-reviewed formal professional papers but has less stringent limitations on manuscript length and extent of graphic presentations.

- TECHNICAL MEMORANDUM. Scientific and technical findings that are preliminary or of specialized interest, e.g., quick release reports, working papers, and bibliographies that contain minimal annotation. Does not contain extensive analysis.

- CONTRACTOR REPORT. Scientific and technical findings by NASA-sponsored contractors and grantees.

- CONFERENCE PUBLICATION. Collected papers from scientific and technical conferences, symposia, seminars, or other meetings sponsored or
co-sponsored by NASA.

- SPECIAL PUBLICATION. Scientific, technical, or historical information from NASA programs, projects, and missions, often concerned with subjects having substantial public interest.

- TECHNICAL TRANSLATION. English-language translations of foreign scientific and technical material pertinent to
NASA's mission.

Specialized services also include organizing and publishing research results, distributing specialized research announcements and feeds, providing information desk and personal search support, and enabling data exchange services.

For more information about the NASA STI program, see the following:

- Access the NASA STI program home page at *http://www.sti.nasa.gov*

- E-mail your question to help@sti.nasa.gov

- Phone the NASA STI Information Desk at
757-864-9658

- Write to:
NASA STI Information Desk
Mail Stop 148
NASA Langley Research Center
Hampton, VA 23681-2199

NASA/TP-20220019160

History of On-orbit Satellite Fragmentations, 16th Edition

Orbital Debris Program Office

Phillip Anz-Meador
John Opiela
Jacobs
Houston, Texas

Jer-Chyi Liou
NASA Johnson Space Center
Houston, Texas

National Aeronautics and
Space Administration

Lyndon B. Johnson Space Center
Houston, Texas 77058

December 2022

Acknowledgments

The authors wish to recognize the Technical Editors responsible for the production of this edition of the <u>History of On-orbit Satellite Fragmentations</u>, 16th Edition: Ashley Johnson (Jacobs) and Debra Shoots (HX5, Jacobs JETS II Contract.) This unique, curated work requires expertise in both content and current formatting standards and tools. As well, their efforts were essential in ensuring compliance with Section 508 of the Rehabilitation Act of 1973 to provide access to a broader readership.

Available from:

NASA STI Program
Mail Stop 148
NASA Langley Research Center
Hampton, VA 23681-2199

National Technical Information Service
5285 Port Royal Road
Springfield, VA 22161

This report is also available in electronic form at http://www.sti.nasa.gov/ and http://ntrs.nasa.

Preface to the Sixteenth Edition

The first edition of the <u>History of On-orbit Satellite Fragmentations</u> was published by Teledyne Brown Engineering in August 1984 under the sponsorship of the NASA Johnson Space Center and with the cooperation of the U.S. Air Force Space Command and the U.S. Army Ballistic Missile Command. The objective was to bring together information about the 75 satellites, which had, at that time, experienced noticeable breakups. This update encompasses all known satellite fragmentations and is published by the NASA Johnson Space Center's Orbital Debris Program Office (ODPO).

The information cut-off date for this edition is 01 May 2022.

Since the 15th edition (information cut-off date of 04 July 2018, published in November 2018), there have been (new or discerned) 26 identified on-orbit breakups and 9 anomalous events, for a historical total of 268 fragmentations and 87 anomalous events. This activity, in addition to launch activity, has resulted in an approximately 21% increase in the number of cataloged space objects since 04 July 2018, which includes on-orbit and decayed objects, or a 34% increase in on-orbit objects. The first accidental collision of two intact spacecraft, Iridium 33 and Cosmos 2251, on 10 February 2009 and the intentional destruction of the (FY-1C) spacecraft on 11 January 2007 continue to contribute significantly to the on-orbit debris environment. Cataloging continues as of this writing for these three debris clouds, to a limiting radar cross section (RCS). Cataloged debris increased with two recent intentional collisions, the Indian Microsat-R anti-satellite (ASAT) test event of 27 March 2019 and the Russian Cosmos 1408 ASAT test of 15 November 2021, as well as ongoing fragmentations of payloads and upper stages.

The current authors recognize the substantial contributions from authors of previous editions of this document. In addition, the assistance of personnel from the U.S. Space Force and the 18th Space Defense Squadron has been vital to the present work.

The authors dedicate this edition to Mr. Nicholas Johnson, lead author of prior editions, former ODPO Chief Scientist, colleague, mentor, and friend.

This Page Intentionally Left Blank

CONTENTS

ACRONYMS

AFB	Air Force Base
AFSPC	Air Force Space Command
ASAT	Anti-satellite
CIS	Commonwealth of Independent States (see also USSR)
DMSP	Defense Meteorological Satellite Program
ESA	European Space Agency
ESRO	European Space Research Organization
FY-1C	Fengyun-1C spacecraft
GEO	Geosynchronous Orbit (orbit category)
ISS	International Space Station
LEO	Low Earth Orbit, up to 2000 km altitude (orbit category)
NASA	National Aeronautics and Space Administration
ODPO	Orbital Debris Program Office
PRC	People's Republic of China
NAVSPASUR	Naval Space Surveillance System
R/B	Rocket Body or Rocket Booster
RCS	Radar Cross Section
RORSAT	Radar Ocean Reconnaissance Satellite
RSO	Resident Space Object
SATRAK	Satellite Trajectory and Attitude Kinetics
SEM	NASA Size Estimation Model
SOZ	Sistema Obespechanya Zapuska (Proton-K Block DM attitude/ullage motor unit)
SpaceX	Space Exploration Technology Corp.
SSN	Space Surveillance Network
TLE	Two-line Element Set
USSR	Union of Soviet Socialist Republics/Commonwealth of Independent States (after 1991)

ABBREVIATIONS and SYMBOLS

Asc	Ascending
Dsc	Descending
ΔP	The maximum observed change in the orbital period [min].
ΔI	The maximum observed change in the inclination [°].
L_c	Characteristic length [m] estimated from RCS using NASA SEM.

This Page Intentionally Left Blank

1.0   INTRODUCTION

Since the first serious satellite fragmentation occurred in June 1961, instantaneously increasing the total Earth satellite population by more than 400%, the issue of space operations within the finite region of space around the Earth has been the subject of growing interest and concern. The many satellite fragmentations of the 1970s and the upsurge of fragmentations in the 1980s served to widen international research into the characteristics and consequences of such events. Continued events in all orbits in later years make the definition and historical accounting of those events crucial to future research. Large, crewed space stations and the growing number of operational robotic satellites, including very large constellations, demand a better understanding of the hazards within the dynamic Earth satellite population.

The contribution of satellite fragmentations to the growth of the Earth satellite population is complex and varied. A slight majority of all cataloged fragmentation debris have already reentered the Earth's atmosphere, and the debris clouds from 38% of the 268 known, unique breakups have completely decayed. Moreover, just 10 of the 5820 space missions flown since 1957 are responsible for 28% of all cataloged artificial Earth satellites presently in orbit (Figure 1.0-1). The sources of three of these 10 fragmentations were discarded rocket bodies or separated fuel tanks that had operated as designed, but later broke up. Modern debris mitigation best practices would have prevented these three events. The remaining seven fragmentations are diverse in character. The oldest, the fragmentation of Cosmos 1275, is assessed by Russian authorities to have been caused by a battery fragmentation. Two, USA 109 and NOAA 16, share a similar spacecraft bus. More recently, the intentional fragmentations of the FY-1C meteorological payload (1999-025) and Cosmos 1408 military payload (1982-092) by ASAT weapons and the first accidental collision of large intact spacecraft, Cosmos 2251 (1993-036) and Iridium 33 (1997-051), together account for about 21% of all cataloged resident space objects (RSOs) on-orbit. The breakup fragments associated with these four spacecraft account for almost 15% of all objects cataloged since the launch of Sputnik 1 on 04 October 1957.

The primary factors affecting the growth of the true Earth satellite population are the international space launch rate, satellite fragmentations, and solar activity. As of 01 May 2022, breakup and anomalous debris constitutes approximately half of the cataloged Earth satellite population, as illustrated in Figure 1.0-2. Also, approximately 34% of payloads on-orbit are no longer operational and constitute a separate, but statistically important class of orbital debris. Derelict rocket bodies and mission-related debris also constitute large orbital debris.

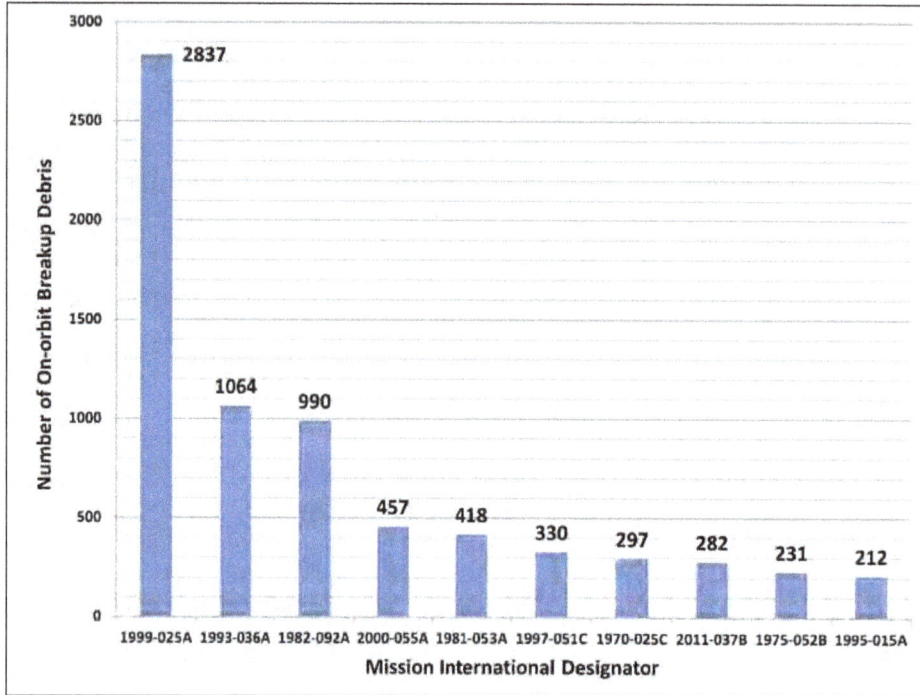

Figure 1.0-1. Magnitude of the 10 largest debris clouds in orbit as of 01 May 2022. Note that cataloging is ongoing, accounting for increased totals for some missions relative to the 15th edition of this book.

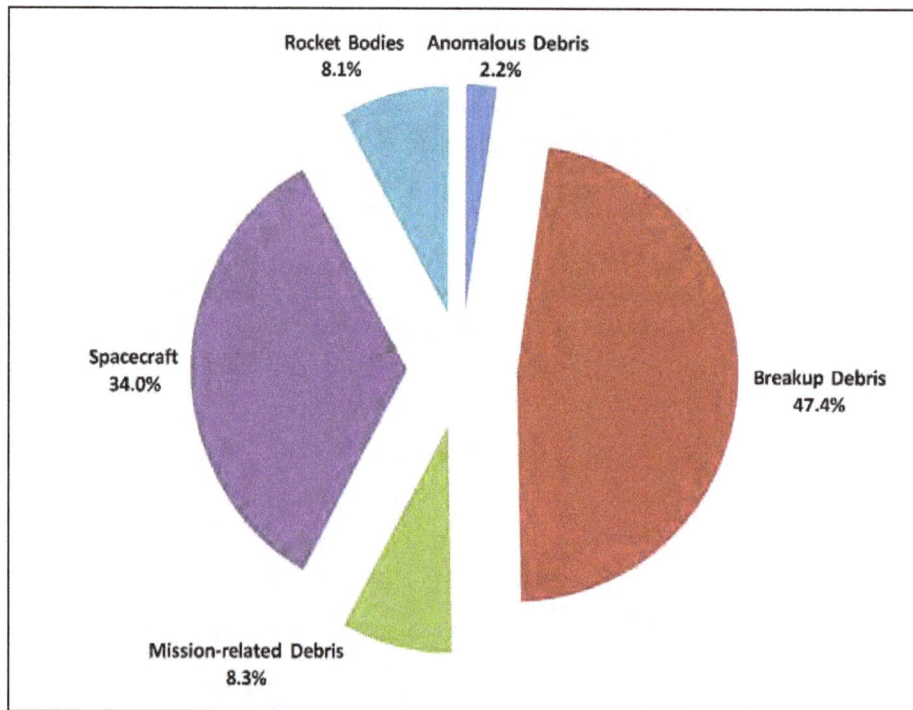

Figure 1.0-2. Relative segments of the cataloged in-orbit Earth satellite population.

1.1 Definition of Terms

In this volume, satellite fragmentations are categorized by their assessed nature and, to a lesser degree, by their effect on the near-Earth space environment. A **satellite breakup** is the usually destructive disassociation of an orbital payload, rocket body, or structure, often with a wide range of separation velocities. A satellite breakup may be accidental or the result of intentional actions, *e.g.*, due to a propulsion system malfunction or a space weapons test, respectively. An **anomalous event** is the unplanned separation, usually at low velocity, of one or more detectable objects from a satellite, which remains essentially intact. Anomalous events can be caused by material deterioration of items such as thermal blankets, protective shields, or solar panels, or by the impact of small particles. As a rule, a satellite breakup will produce considerably more debris, both trackable and non-trackable, than an anomalous event. From one perspective, satellite breakups may be viewed as a measure of the effects of human activity on the environment, while anomalous events may be interpreted as a measure of the effects of the environment on human-made objects.

Mission-related debris results from the intentional release of objects, usually in small numbers, during normal on-orbit operations. Objects ejected during the deployment, activation, and de-orbit of payloads and during crewed operations are examples of mission-related debris. Usually, mission-related debris from a single launch is minimal, but much larger numbers occasionally arise, such as the 323 objects from the *Mir* space station or the more than 140 objects from the West Ford Needles experiment. Mission-related debris represents a non-trivial portion (approximately 8%) of all satellites in orbit today and are a legitimate subject in the study of methods to inhibit the growth of the Earth satellite population. However, identifying the thousands of cataloged mission-related debris is beyond the scope of this report.

1.2 Information Sources and Accuracy

Several data sources were employed to compile this volume; however, nearly all are derived from observations collected by the U.S. Space Surveillance Network (SSN). Due to the variety of sources and geodetic models used to create satellite orbital element sets, all altitudes cited within this volume are presented to the nearest 5 km, referenced to a mean Earth of radius 6378.135 km. The accuracy of the data presented is not of adequate fidelity for precision analysis, although it is appropriate for the anticipated uses of this text. Complete event-related element sets are provided, but manipulation of this data, in particular satellite orbit propagation, should be performed only with validated, Air Force Space Command (AFSPC)-derived software, such as the PC-compatible, SATRAK astrodynamics toolkit. Long-term propagation of these elements is not appropriate regardless of the propagation technique applied and is, therefore, discouraged.

Although all fragmentations are described by the number of debris cataloged and the number of cataloged debris remaining in orbit, these parameters are poor measures of merit and should be used with extreme caution when undertaking comparative analyses. The sensitivity of the SSN, and hence the degree to which debris will be detected and cataloged, is highly dependent upon satellite altitude and to a lesser degree on satellite inclination. Additionally, historical cataloging practices have changed over the years. Past practices have included cataloging all debris objects associated with a breakup, even if they had already decayed; cataloging almost no pieces from a low altitude breakup when decay of most of the cloud was imminent; and cataloging objects as they were created, regardless of status. These different practices have resulted in an inconsistent historical record.

As a rule of thumb, low altitude, cataloged debris is assessed to be larger than 10 cm in diameter. At higher altitudes, objects less than 1 m in diameter may be undetectable. Individual object sensitivities may vary dramatically from this simple generalization. Debris counts for fragmentations occurring in highly elliptical orbits near 63° inclination (*Molniya*-type) are traditionally low, in part due to stable perigees situated deep in the Southern Hemisphere and often beyond SSN coverage. Alternately, consider the extensive breakup record of the Proton ullage motors; select cases indicate that more debris may be produced by a fragmentation event than is typically observed or cataloged. In February 2007, a *Briz-M* upper stage broke up into over 1000 observed fragments. However, due to the elliptical nature of the stage's orbit, which impeded the SSN's ability to detect, to identify, and to catalog the debris, only 107 fragments were cataloged, of which 97 remain on orbit. Similar outcomes may occur in practice with other *Briz-M* upper stage fragmentations. During a special surveillance session in 1987, as many as 250 uncatalogued objects were observed in low inclination, highly elliptical orbits, but reliable tracking and parent identification were not achieved. The Russian government's disclosure of the Ekran 2 battery explosion on 25 June 1978 is the first known fragmentation in geostationary orbit. This event was not detected by the SSN, but since the event, five pieces have entered the catalog; this edition adds Ekran 4 and Ekran 9, assumed to have experienced accidental battery explosions similar to their earlier sister payload. Similarly, Titan Transtage fragmentation in deep space may exhibit similar cataloging limitations. Cataloging errors, *e.g.*, identification of an object with the wrong parent satellite, normally are not explicitly noted in this volume since many errors have been or may be corrected.

For fragmentations at very low altitudes, *i.e.*, below 400 km, much of the debris may reenter before detection, identification, and cataloging can be completed. For example, when the debris cloud from Cosmos 1813 passed over a single SSN radar, 846 individual fragments could be discerned. However, the total number of debris officially cataloged only reached 195. Likewise, more than 380 fragments are known to have been injected into Earth orbits (it is likely an equal number were sent on reentry trajectories) following the USA 19 test, but only 18 debris fragments ultimately entered the official satellite catalog.

1.3   Environment Overview

To place the debris population component of the orbital environment in context for the reader, it is useful to review the general orbital environment in the near Earth and near geosynchronous orbit (GEO) regions. Differentiation of the population by source, object type, and orbit type are also included below.

1.3.1   ON-ORBIT SPATIAL DENSITY

The spatial density of resident space objects is a common means of describing the space object environment and is adopted here. Spatial density (objects per unit volume) represents the effective number of spacecraft and other objects as a function of altitude. The *effective number*, based on the fraction of the orbital period spent in any altitude range, rather than the simple counting of objects, is used because many objects traverse altitude regions of interest yet contribute little to the local collision hazard, *e.g.*, GEO transfer orbits. Such orbits exhibit an effective contribution to the environment at any given altitude of up to two orders of magnitude less than an object in a circular orbit within this same altitude interval. Thus, circular orbits at or near an orbit of interest normally dominate the hazard environment. Figure 1.3.1-1.1 portrays the near Earth (up to 2000 km altitude) environment categorized by intact or debris object types. The spatial densities, presented on a logarithmic axis, are subdivided into 5-km altitude intervals to illustrate the fine structure. The epoch of the source data, an SSN two-line element (TLE) set, is 03 May 2022.

Figure 1.3.1-1.1. The near Earth (up to 2000 km) altitude population. Population growth is evident at all altitudes between the 15th edition (04 July 2018) and this 16th edition (01 May 2022).

Debris dominate the low Earth orbit (LEO) environment at altitudes greater than about 600 km altitude and now almost equal intentionally launched objects at lower altitudes. The peak near 850 km is due principally to the January 2007 intentional destruction of the FY-1C weather spacecraft, while the density over approximately 770 to 800 km altitude was created by the February 2009 accidental collision of Iridium 33 (active) and Cosmos 2251 (derelict) communication spacecraft. Clearly visible in this figure are other high-density regions of space. The satellite constellations deployed in LEO in the late 1990s are clearly evident: the Iridium constellation inhabits the altitude region at and near 780 km altitude, while the Globalstar constellation inhabits the region from 1410 to 1420 km. The most prominent constellation, however, is the Space Exploration Technologies Corp. (SpaceX) Starlink constellation's checkout orbits near 345 to 355 km altitude and the operational orbits near 545 to 550 km altitude. Other spacecraft constellations, such as the Union of Soviet Socialist Republics (USSR)/Commonwealth of Independent States (CIS) communication and navigation constellations, are also visible near 1480 km and 950 km, respectively.

This plot adequately portrays total density as a function of altitude but overcompensates graphically to accommodate the Starlink operational constellation near 550 km altitude. To better portray evolution of the debris environment between editions of this work, we plot the density on the traditional linear axis for the LEO environment in Figure 1.3.1-1.2.

Figure 1.3.1-1.2. The near Earth (up to 2000 km) altitude population, linear densities (Note that select "total" and "intentionally deployed objects" categories for this 16th edition are greater than 6.0×10^{-8} [km^{-3}] and hence, are not visible in this figure). Population growth is evident at all altitudes between the 15th edition (04 July 2018) and this 16th edition (01 May 2022).

Debris growth is most evident in a relatively sharp peak around 625 km altitude and between approximately 750 and 800 km altitude. The former is attributed to the *Resurs* O1-No. 3 and H-IIA mission-related debris events; the latter to the NOAA 17 and *YunHai* 1-02 events. The new peak at approximately 450 km altitude is directly attributable to the 2021 Russian direct ascent ASAT test against the derelict Cosmos 1408 spacecraft.

The GEO altitude environment increased in both importance and number of RSOs over the course of the early 21st century. Figure 1.3.1-2 shows the GEO altitude using a logarithmic, vertical spatial density axis and altitude intervals of 5 km. Only objects with an inclination of less than 15 degrees were included, and the volume used to calculate the spatial density is similarly constrained within ± 15° latitudes. These limitations were implemented to exclude high-eccentricity orbits, such as *Molniya* orbits, which do constitute GEO-resident or GEO-crossing traffic. Because high-inclination orbits normally do not penetrate this true GEO region, or are relatively rare, the assumption remains useful in portraying the spatial density of this region. However, this undercounts the contributions of inclined GEO orbits, increasingly common, and so may be revised in future editions.

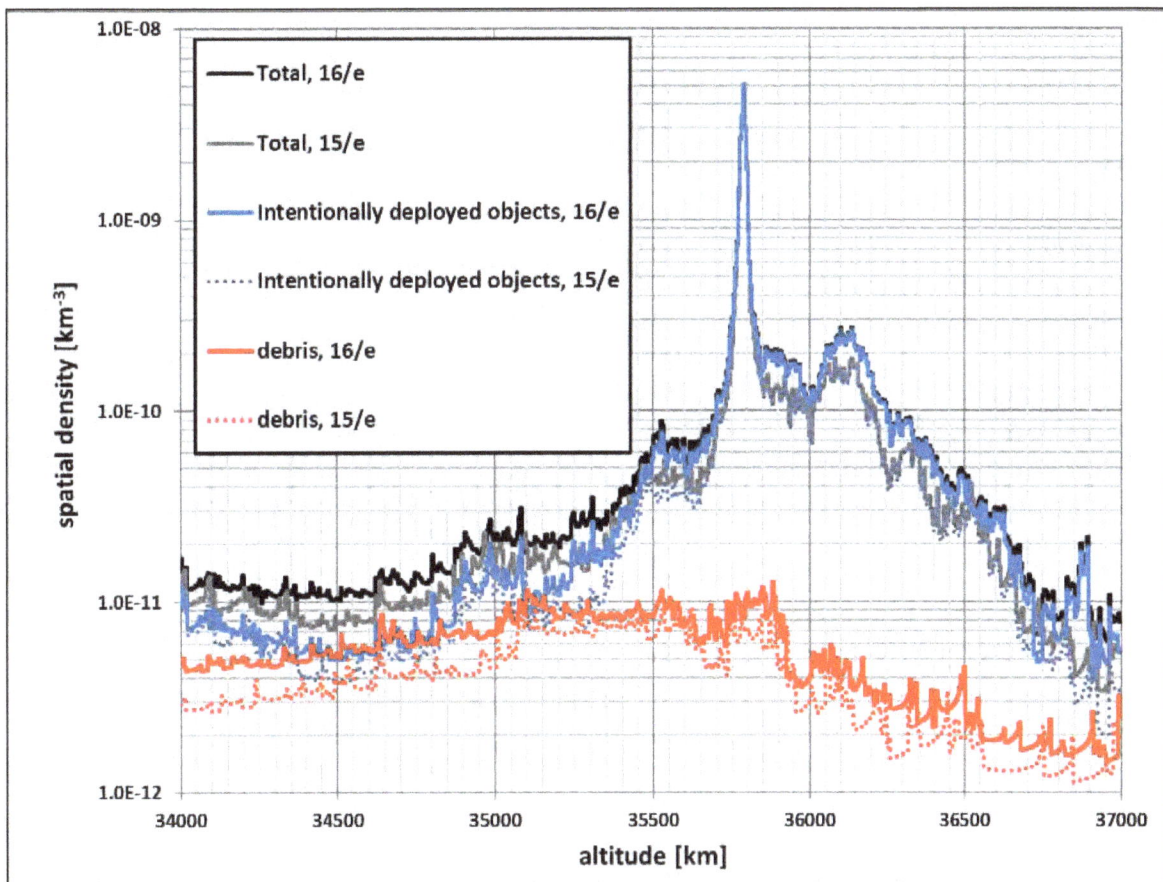

Figure 1.3.1-2. The GEO altitude population.
The peak at approximately 35,786 km altitude is the nominal GEO altitude and active payload population. Population growth is evident at all altitudes between the 15th edition (04 July 2018) and this 16th edition (01 May 2022), but debris growth is driven by elliptical orbits.

Because the LEO spatial density chart averages over all inclinations and the GEO spatial density chart averages over inclinations less than or equal to 15 degrees, collision rates are not related linearly to the spatial density at any given altitude. Rather, collision rates will vary not only with the spatial density, but also with the inclination-dependent relative velocity. Altitudes dominated by high-inclination (70°–110°) orbits yield a significantly higher collision rate as compared to those populated by lower inclination orbits. Objects in these inclinations can collide at near head-on engagement geometries with objects in complementary inclinations (when the sum of the inclinations of the colliding objects is ~180 degrees). The exception to this general rule is provided by the commercial constellations in LEO and spacecraft in GEO. The commercial constellations are maintained in precise orbital planes; their expected collision rate would be versus the "background" population only. Therefore, the spikes representing the Iridium, Globalstar, and Starlink constellations do not present the inordinate collision risk implied by a casual examination. Similarly, the GEO environment is characterized by low collision velocities (< 1.5 km/s) due to the relative motion between controlled and uncontrolled objects.

1.3.2 POPULATION DISTRIBUTION

The distribution of objects by type (*e.g.*, spacecraft, rocket bodies) and source (U.S., the People's Republic of China [PRC], *etc.*) is germane to this discussion since objects are not randomly distributed among these categories. To display the distribution statistics, the 01 May 2022 U.S. Satellite Catalog was categorized by these nominal variables. The reader should note that absolute numbers in these statistics are fluid, in the sense that ongoing analyses can sometimes reassign a given debris object from one parent breakup to another. This is particularly noted for the highly populated sun-synchronous orbital region, and accounts for nine cases of number discrepancies when comparing the numerical count of objects in a given debris cloud with the number implied by the highest piece tag associated with the cloud's international designator. Tentative identifications are also present in the public satellite catalog and are indicated by an asterisk (*) appearing in the "common name" field. Finally, some identifications require in-depth analysis and interpretation, and the understanding of a particular launch or debris cloud may evolve over time.

In the following table, it is likely that most category identities will be obvious to the casual reader; however, several require further identification. For example, prior edition's debris dispensed type variable referred to "debris dispensing" spacecraft, such as the International Space Station (ISS), and the Soviet/Russian crewed orbital stations and *Romb*/*Duga-K* spacecraft. As these are extinct, excepting ISS debris, the type has been subsumed in the more general mission-related debris type. In terms of the source variable, spacecraft launched by the USSR are incorporated into the CIS category, while the ESRO/ESA category includes only those spacecraft formally launched by either the European Space Research Organization (ESRO) or the European Space Agency (ESA). Finally, launches for specific countries, such as Germany or Spain, are distributed into the "Other" source category.

8

TABLE 1.3.2 SOURCE VS. TYPE ACCOUNTING

on-orbit									
	US	**CIS**	**France**	**PRC**	**India**	**Japan**	**ESRO/ESA**	**Other**	**Total**
payloads	4413	1467	80	636	104	196	83	1664	**8643**
rocket bodies	613	957	155	194	39	54	7	42	**2061**
mission-related debris	751	795	151	336	6	38	14	17	**2108**
breakup debris	3450	4935	202	3311	70	11	28	55	**12062**
anomalous debris	316	221	7	4	0	1	12	2	**563**
Total	**9543**	**8375**	**595**	**4481**	**219**	**300**	**144**	**1780**	**25437**

decayed or beyond Earth orbit									
	US	**CIS**	**France**	**PRC**	**India**	**Japan**	**ESRO/ESA**	**Other**	**Total**
payloads	1497	2054	11	103	15	84	32	153	**3949**
rocket bodies	811	2582	80	209	25	74	6	16	**3803**
mission-related debris	1044	5854	175	315	14	141	10	110	**7663**
breakup debris	3598	5698	492	940	432	130	13	62	**11365**
anomalous debris	186	129	6	3	2	6	2	4	**338**
Total	**7136**	**16317**	**764**	**1570**	**488**	**435**	**63**	**345**	**27118**
								GRAND TOTAL	**52555**

Several salient features are apparent in this table. Debris is dominant among all source variables, and most debris (and all other categories of resident space objects) is due to space activities of the U.S., CIS, and PRC. Individual events from other spacefaring nations have also contributed greatly to the local environment in several sun-synchronous or high-inclination orbital regimes. Examples are provided by the 1986 fragmentation of the Ariane SPOT-1/Viking rocket body and the 2019 test of the Indian direct-ascent ASAT weapon system.

A net increase of approximately 6500 objects on-orbit (within a gross increase of over 9000) has been observed since the 15th edition of this book was published in 2018. The majority of on-orbit objects were from the Starlink communications constellation and five debris clouds – resulting from the intentional destruction of the Indian Microsat-R and Russian Cosmos 1408 spacecraft by ASAT, the breakups of multiple U.S. Centaur upper stages, and Russian mission-related debris. Table 1.3.3 shows the net increase or decrease in objects since the 15th edition. A discouraging feature of Table 1.3.3 is that almost every object type showed a net increase of on-orbit objects.

TABLE 1.3.3 SOURCE VS. TYPE ACCOUNTING – NET CHANGE SINCE 15TH EDITION

on-orbit	US	CIS	France	PRC	India	Japan	ESRO/ESA	Other	TOTAL
payloads	2767	31	17	327	17	36	12	815	4022
rocket bodies	17	8	10	69	4	3	0	2	113
mission-related debris	17	70	3	66	-4	-3	1	-3	147
breakup debris	426	1620	10	53	-2	-2	1	3	2109
anomalous debris	6	108	6	4	0	1	1	0	126
TOTAL	3233	1837	46	519	15	35	15	817	6517

decayed or beyond Earth orbit	US	CIS	France	PRC	India	Japan	ESRO/ESA	Other	TOTAL
payloads	399	54	2	15	3	16	3	70	562
rocket bodies	67	48	4	68	8	5	0	1	201
mission-related debris	205	82	20	61	4	11	1	8	392
breakup debris	67	856	5	95	132	91	1	27	1274
anomalous debris	15	61	1	0	0	4	1	3	85
TOTAL	753	1101	32	239	147	127	6	109	2514
								GRAND TOTAL	9031

10

2.0   SATELLITE BREAKUPS

This section summarizes the current breakup environment and describes each individual breakup, presented in a two-page format. New classes of breakup types have tended to fuel the background breakup rate, replacing classes of breakups from older on-orbit practices such as the well-known Delta second stage rocket body failures. The number of breakups is depicted in Figure 2.0-1.

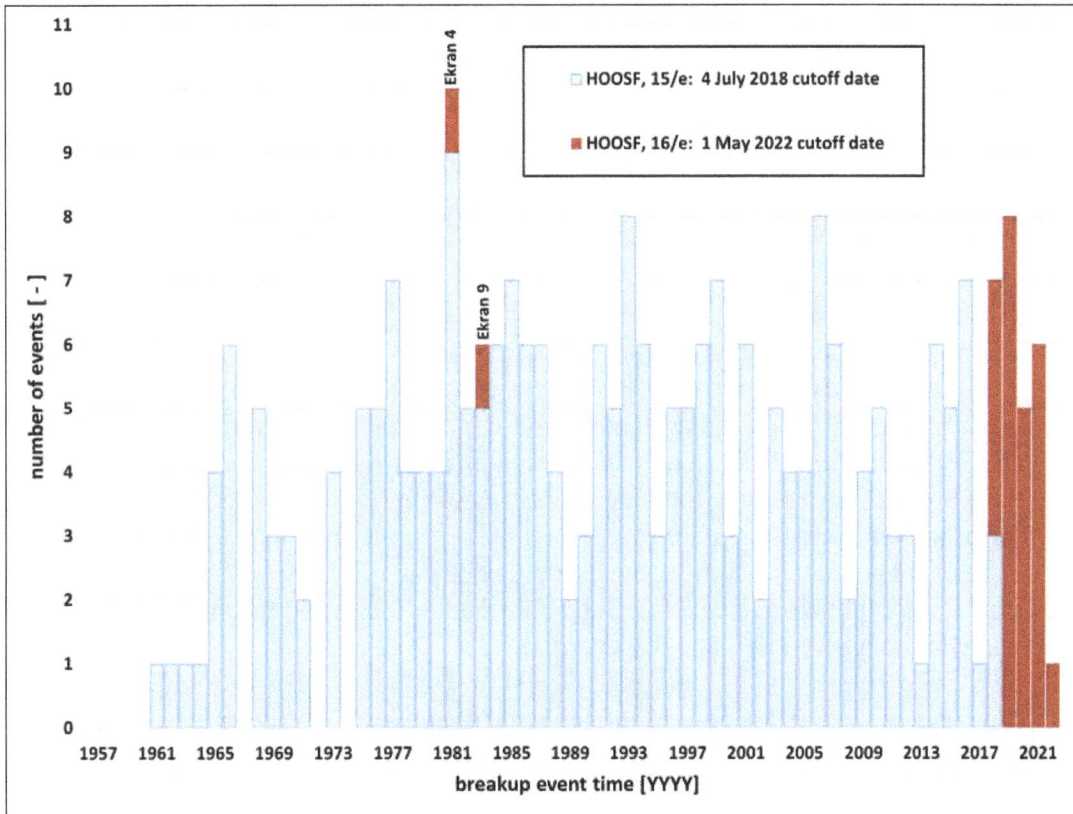

Figure 2.0-1. Number of breakups by year since 1961.
Breakups new to this edition are indicated in red.

2.1   *Background and Status*

By far the most important category of human-made, on-orbit objects is debris from satellite breakups, which now account for 47% of the total cataloged, on-orbit Earth satellite population of 25,437 Earth-orbiting objects. Since 1957, 267 unique satellites are believed to have broken up (Tables 2.1 and 2.2; note that one Russian mission-related debris object, "SL-23 DEB" [2011-037B], is afforded two separate line items because the events were separated by five years and the first event produced observed, but not cataloged, breakup debris). Breakups due to aerodynamic forces at or near reentry are treated separately from breakups caused by other factors, because aerodynamic breakups occur at the end of the satellite lifetime and therefore contribute nothing to the orbital environment past the very near term. Only a fraction of these breakups is even detected because of the short remaining lifetime of the object and its debris. One additional aerodynamic breakup has been detected between editions of this work. These events are discussed in Chapter 4 and omitted from data included in this chapter.

The primary known causes of satellite breakups (Figure 2.1-1) are propulsion-related events and deliberate actions, although the cause for almost 30% of breakups remains uncertain. This document will continue to carry breakup causes as unknown until a strong case can be made for one of the other root cause classifications. Deliberate actions, often associated with activities related to the launching state's national security interests and concerns, were formerly the most frequently occurring class, although only one such event occurred during the decade from 1997 until the FY-1C event in January 2007; more recent deliberate events were the Microsat-R event in March 2019 and the Cosmos 1408 event in November 2021. On average, the resulting debris from deliberate actions is short-lived (Figures 2.1-2 and 2.1-3), the exception being FY-1C. Propulsion-related breakups, currently the most frequent class, include catastrophic malfunctions during orbital injection or maneuvers, subsequent explosions based on residual propellants, and failures of active attitude-control systems. Breakups of rocket bodies due to propulsion failures are usually more prolific and produce longer lived debris than the intentional destruction of payloads, often due to the higher altitudes of the malfunctioning rocket bodies rather than the mechanics of the explosive event. Breakups of the Soviet/Russian Blok-DM *Sistema Obespecheniya Zapuska* (SOZ) ullage motors are segregated from other members of the propulsion ensemble in Figure 2.1-1 due to their high probability of fragmentation, which tends to overinflate the propulsion category to some degree; omitting the SOZ breakups roughly equates the non-SOZ propulsion and deliberate categories. Although it may appear obvious that a rocket body breakup should be classified under the "Propulsion" category, rocket body events are carried as "Unknown" until a failure mechanism can be confidently identified for that rocket body design and is associated with a given rocket body event.

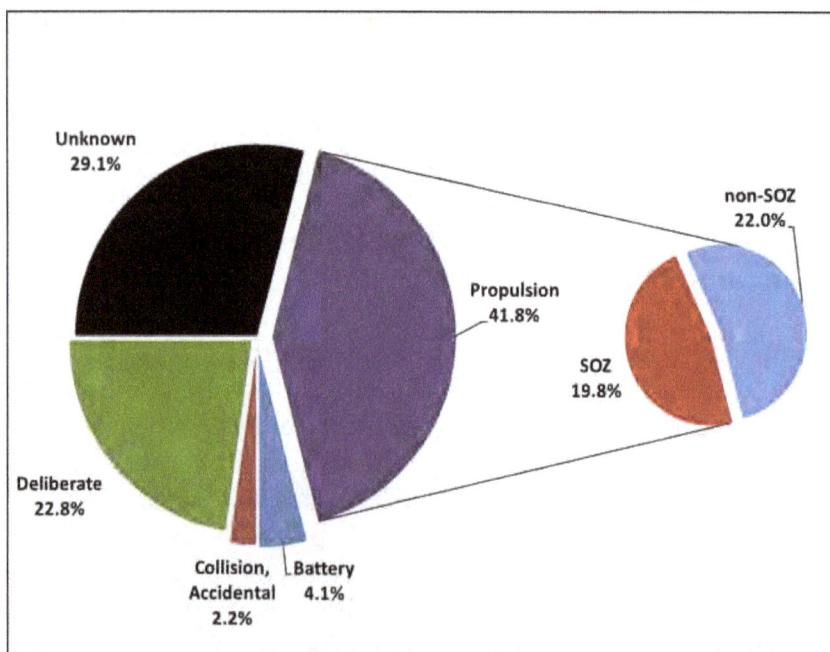

Figure 2.1-1. Causes of known satellite breakups.

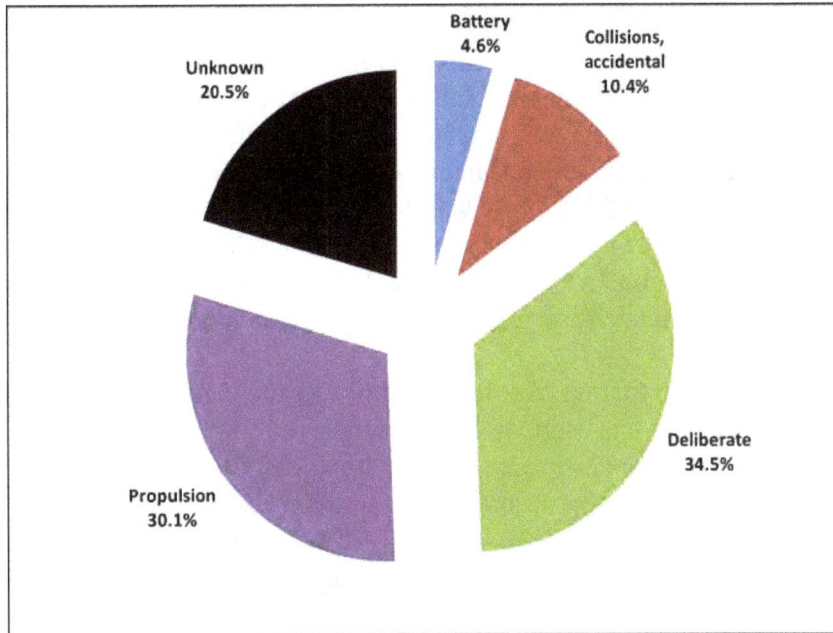

Figure 2.1-2. Proportion of all cataloged satellite breakup debris.

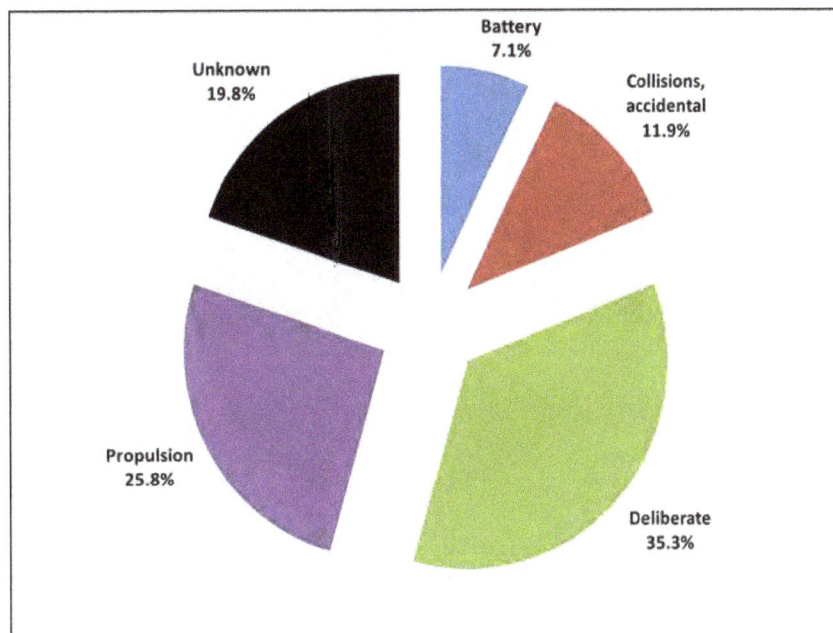

Figure 2.1-3. Proportion of cataloged satellite breakup debris remaining in orbit.

The rate of satellite breakups increased noticeably in the 1970s and has continued through the 1990s and into the new millennium at an average pace of approximately five fragmentations per year over 2001 to 2021, inclusive. The recent low-activity years of 2013 and 2017 appear as statistical anomalies in this regard. Increased awareness of potential hazards has resulted in positive actions to mitigate or eliminate many known breakup causes, *e.g.*, Delta second stages, weapons testing, and Cosmos 862-type events. Together, these three programs were responsible for more than a quarter of all satellite breakups in the 1980s. However, the quick response of

Arianespace and ESA to the breakup of an Ariane third stage in 1986 was indicative of a desire by most space-faring organizations at that time to operate in near-Earth space responsibly. Today, new series of boosters and satellites have resulted in new breakup sources, such as the multiple fragmentations of *Briz-M* upper stages. The intentional destruction of the FY-1C spacecraft remains the most environmentally impactful event to date. The intentional destruction of the Microsat-R spacecraft in a 2019 ASAT test was mitigated only by the relatively low altitude at which it was conducted. The Cosmos 1408 ASAT test, modeled by the ODPO, is expected to negatively influence activities in LEO, particularly at human spaceflight altitudes, for the next five years, by which time 90% of its debris is anticipated to have reentered.

Figures 2.1-4 and -5 illustrate that an absolute majority of the satellite breakup debris total and debris remaining in orbit today has originated from spacecraft. Previously, up to the 14th edition of this book, rocket bodies always had more total and on-orbit debris than spacecraft (in the case of on-orbit, by a ratio of almost three to one), but the three major fragmentation clouds created between 2007 and 2009 and ASAT tests in 2019 and 2021 account for the continuing dominance of payload debris. Also of note is the increased percentage originating within the mission-related debris category.

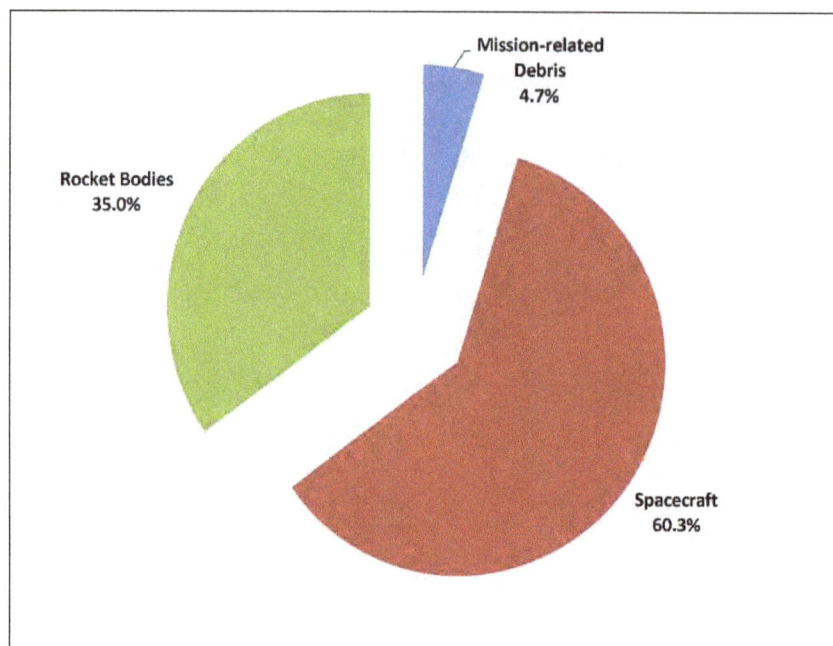

Figure 2.1-4. Sources of all cataloged satellite breakup debris by satellite type.

This dominance is even more pronounced when considering only on-orbit debris. In this case, spacecraft debris now accounts for almost 63% of all debris in orbit. Again, debris production by mission-related debris is notable.

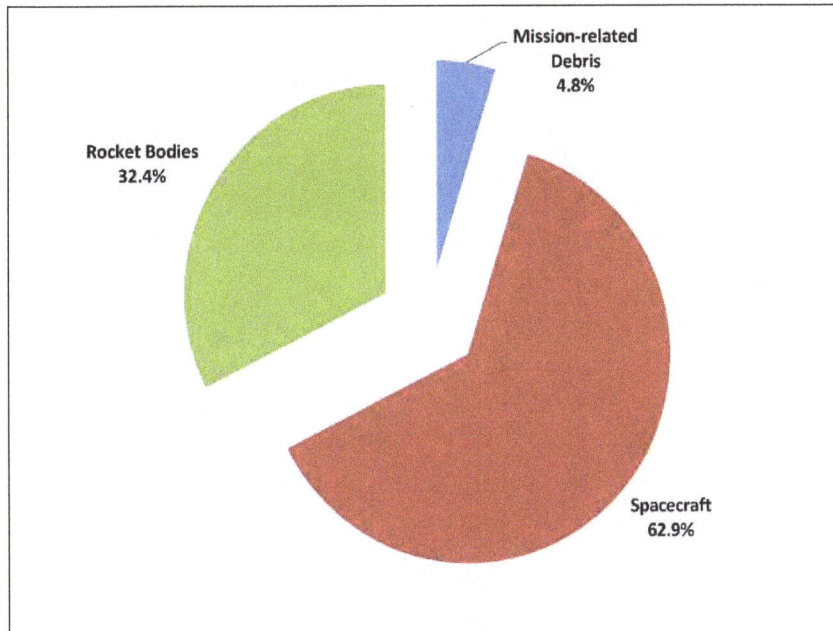

Figure 2.1-5. Sources of satellite breakup debris in orbit by satellite type.

Section 1.3.1 examined the growth of exclusive satellite catalog categories between the 15th and 16th editions (this work) expressed in terms of spatial density as a function of altitude, and by altitude regime. Figures 1.3.1-1.1 and -1.2 portray the growth of debris and intentionally deployed objects in LEO, while Figure 1.3.1-2 depicts the GEO and near-GEO environment.

It is also worthwhile to survey and contrast the decadal growth of the satellite catalog and debris categories over the multiple decades of 2001 to 2021. The following subsections will examine the categories by effective number density, size distribution, sources and sinks of the debris category, and an assessment of debris-creating events by mission phase.

2.1.1 EFFECTIVE NUMBER DENSITY EVOLUTION

An effective means of comparing growth in the important LEO region is to examine the effective number of objects as a function of altitude over the two decades bounded by 2001, 2011, and 2021. An additional level curve was added to capture the effects of the Cosmos 1408 ASAT test. The effective number of objects, portrayed in Figure 2.1-6, is computed by ratioing the object's time spent in each 50 km altitude bin to the object's orbital period – the resulting weights are then accumulated to yield the total effective number of satellites in that bin. All objects are resident in the publicly available satellite catalog for the four specific epoch times. Limiting object sizes, as interpreted via the NASA Size Estimation Model (SEM) [1] from available RCS catalogs and publicly available information, are on the order of 10 cm in LEO.

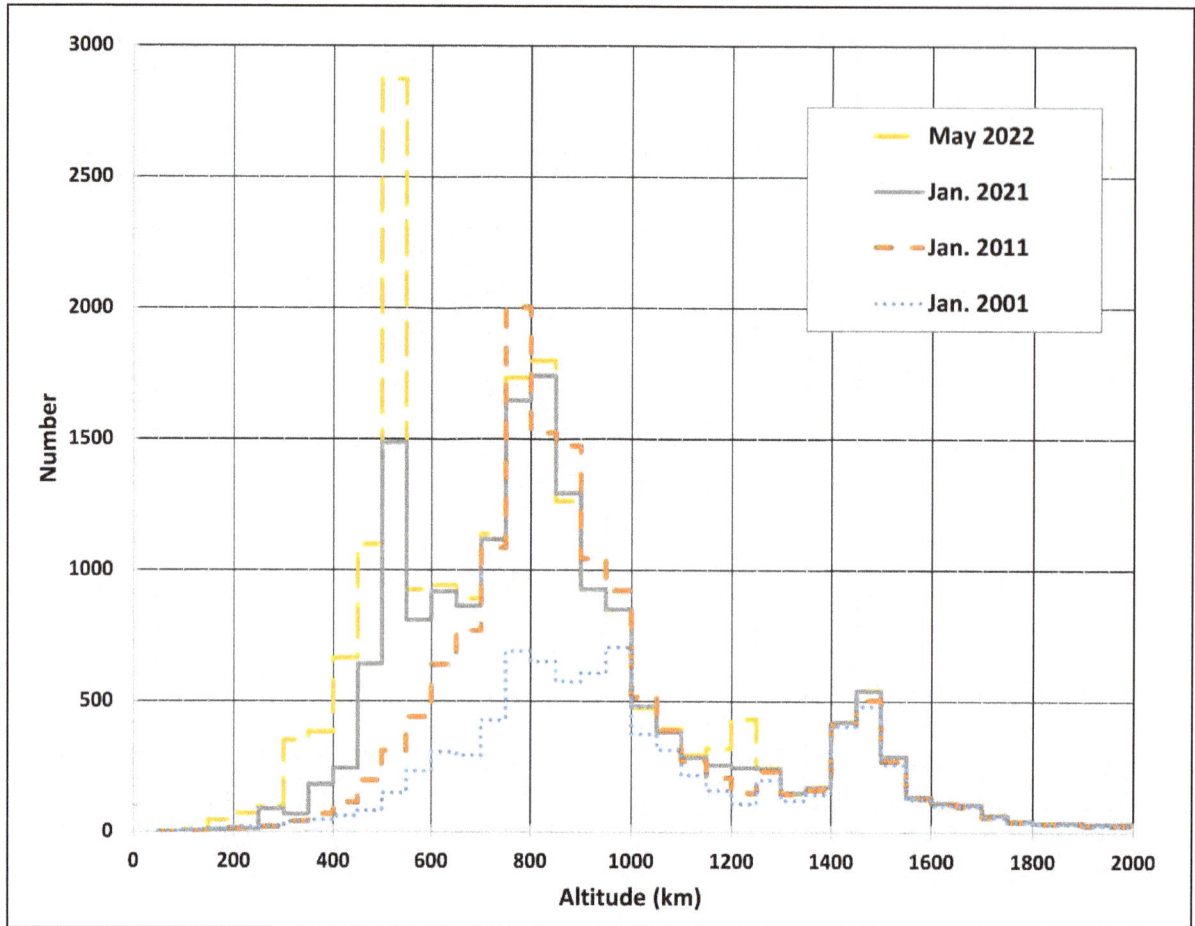

Figure 2.1-6. Effective number density as a function of altitude for all cataloged objects at epoch.

The number density in the important sun-synchronous orbital altitude band of 750 to 800 km altitude increased by a factor of approximately three over the first decade (*i.e.*, 2001-2011). However, the second decade's growth is promoted not only by the presence of new fragmentation events (and subsequent orbital evolution and removal via atmospheric reentry, noting the same 750 to 800 km altitude bin), but by the deployment of the first of several planned commercial constellations in LEO, most notably the Starlink Phase 1 shell nearing completion in the 500 to 550 km altitude bin as of this writing. The number density in that bin increased by approximately an order of magnitude over the second decade.

These increases do not necessarily reflect a concomitant rise in risk. This is due to the coordinated motion, station keeping, and collision avoidance procedures used by Starlink and other maneuverable spacecraft resident in Figure 2.1-6. In particular, space situational and domain awareness protocols, exemplified by standardized conjunction data messages and on-board/limited human-in-the-loop autonomy may significantly reduce the collision hazard for active, maneuverable spacecraft relative to the catalog. Note, however, that catalog maintenance can be complicated or compromised by tasking priorities, sensor limitations, object orbital and physical characteristics, dynamic space weather, and other factors. RSOs below the sensor size threshold may be sufficiently large or can impart sufficient kinetic energy to catastrophically break up a spacecraft. Thus, risks posed by objects below the thresholds by which collision risk may be mitigated through active maneuvering but above the corresponding thresholds for passive mitigation, including shielding, must remain an accepted risk.

In contrast to Figure 2.1-6, Figure 2.1-7 portrays the evolution of the breakup debris only. In this plot, orbital evolution of the 2007 ASAT test and the 2009 collision's debris clouds decreases the environment over the 750 to 900 km altitude region, with the recent Cosmos 1408 ASAT event providing significant growth, albeit at a lower altitude. Despite the decrease observed between 750 km and 900 km altitudes, the debris environment may be interpreted to be relatively static over the second decade, relative to the 2001 epoch. The reader should note that debris is being continuously cataloged from historical breakups; several objects from the very first breakup in 1961 being added to the catalog since the previous edition, as well as many objects from the three prominent debris clouds attributable to the FY-1C and Cosmos-Iridium events. As well, these three prominent events, and likely many other historical and near-contemporary events, have produced large numbers of small debris below the detection, correlation, and cataloging thresholds of space surveillance networks.

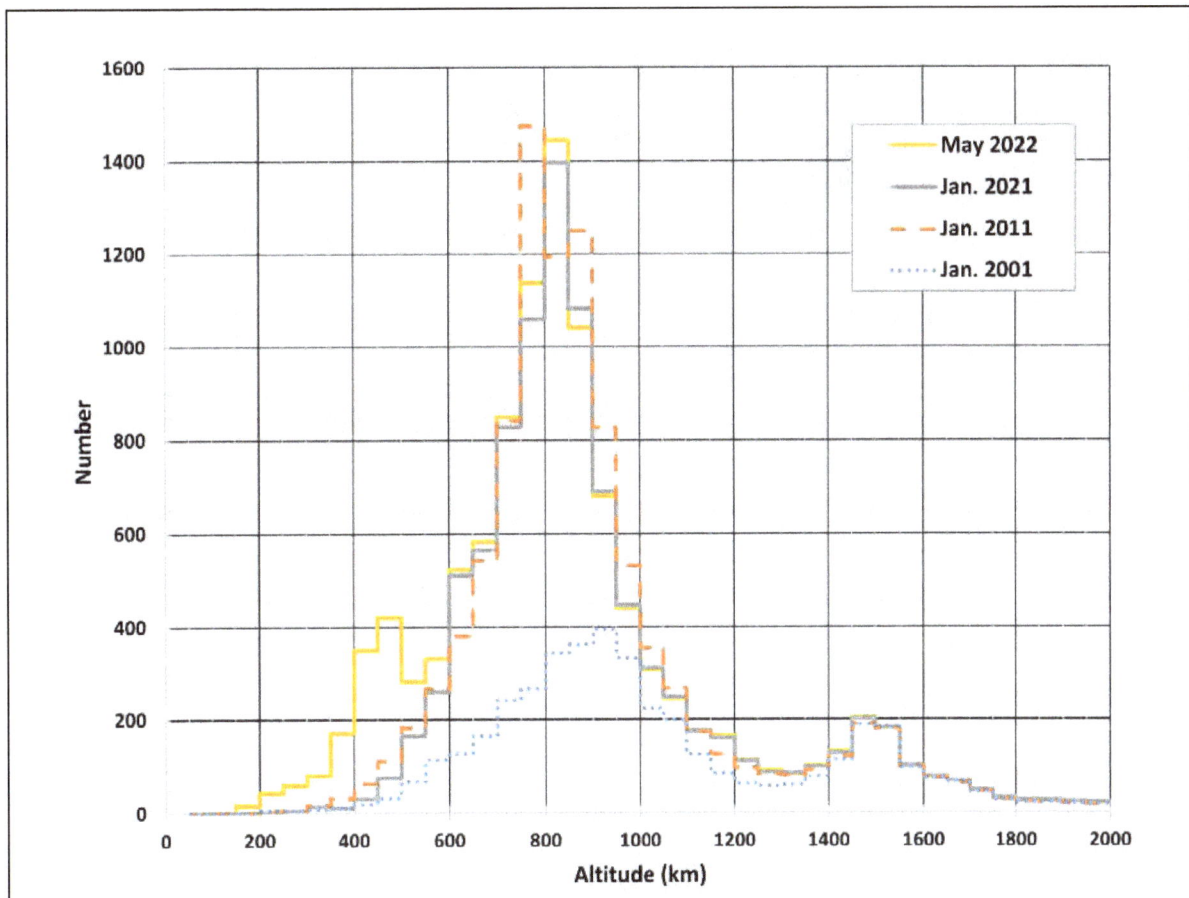

Figure 2.1-7. Effective number density as a function of altitude for all cataloged breakup debris at epoch.

2.1.2 BREAKUP DEBRIS: CUMULATIVE SIZE DISTRIBUTION EVOLUTION

While the effective number of objects, as a function of altitude, portrays the spatial distribution of RSOs, an associated quantity required for proper risk assessment is the size distribution of debris. Except for several types of novel, small, deployable subsatellites, the intact population's minimum characteristic length (exemplified by 1U CubeSats and the so-called Pocket Qubes) is on the order of 10 cm, effectively the catalog's limiting size. However, despite qualified growth in the launch rates of these small spacecraft, the LEO domain is dominated at and near the catalog's limiting size by breakup debris. This section will focus on the size distribution of that debris.

The size of debris fragments is estimated from RCS measurements using the NASA SEM. This interpretive tool, derived from laboratory measurements, uses the RCS time history of a given, cataloged debris object and the radar sensor's electromagnetic frequency to estimate a single characteristic length (L_c), or size, for that object. Descriptive statistics are computed for the size distribution resulting from repeated application of the SEM tool to the RCS time history, including the mean, median, distribution mode(s), and standard deviation. This report uses the median size statistic, as the median tends to reduce the effect of large or small outliers relative to the mean. The normalized distribution of median size for all cataloged debris for which RCS measurements are available at the respective epochs plotted is portrayed in Figure 2.1-8. Values have been normalized to 1.0 at the 10 cm size. This report uses the cumulative distribution as a representative of associated risk, as if a target object is vulnerable to a projectile of size L_c, then it is vulnerable to all objects of sizes larger than L_c.

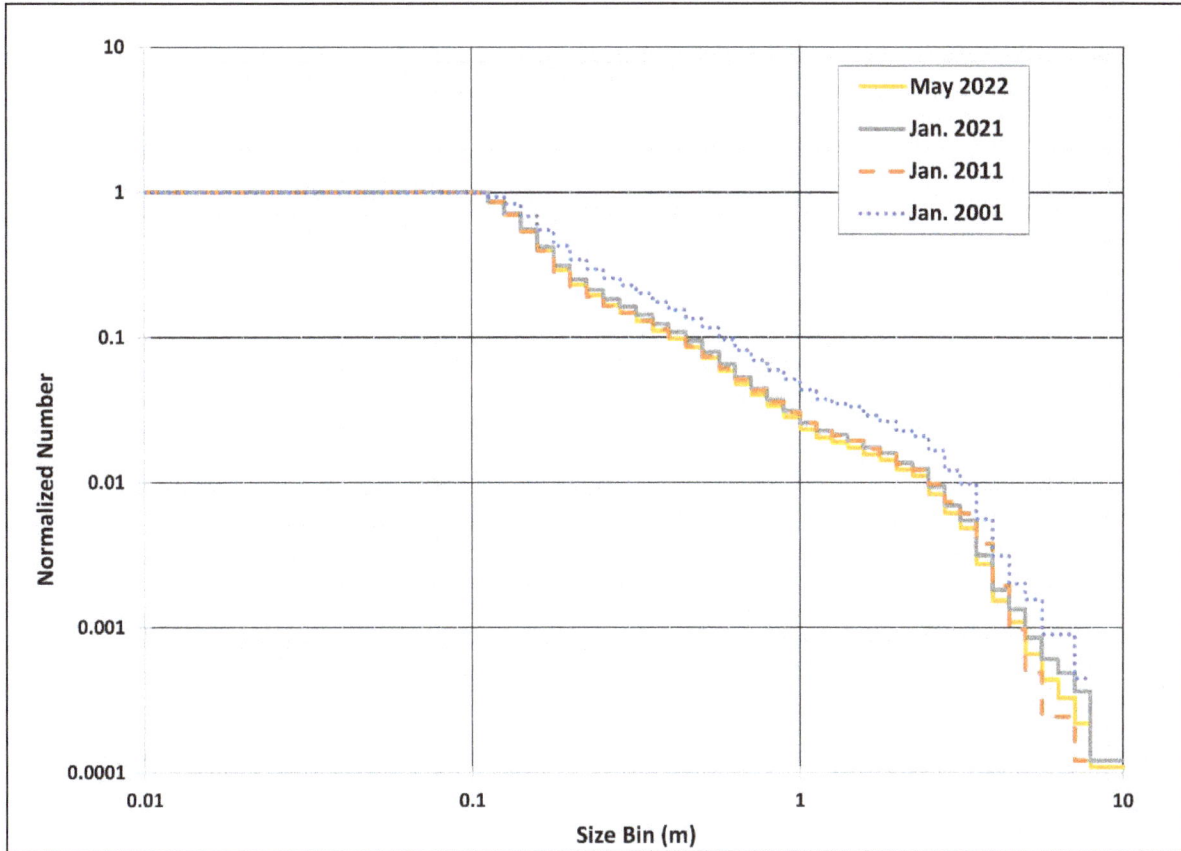

Figure 2.1-8. The cumulative size distribution for those cataloged breakup debris with available RCS data at epochs noted.

When plotted on a log-log plot, all four epochs' distributions display behavior typical of a power law, as expected from historical observations of the environment and NASA's data-driven Standard Satellite Breakup Model. The 2001 curve indicates that median sizes, estimated from RCS time histories, were generally larger. The latter three epochs are remarkably similar in slope and location and differ from the 2001 curve most importantly at or near the LEO cataloging threshold, *e.g.*, 10-20 cm. In this instance, the slope of the cumulative curve is steeper for 2011, 2021, and May 2022 than for the 2001 curve. This may indicate a slight bias towards smaller debris for more recent breakup events. However, effects due to changes and improvements in the SSN cannot be ruled out as an influencing factor.

2.1.3 BREAKUP DEBRIS: SOURCE/SINK TIME HISTORY, 1957–2022

A third descriptor for the evolution of the orbital debris environment is the production rate, or "source," and removal rate, or "sink," time history. In lieu of formal rates, we express the corresponding number of cataloged debris produced in, and removed from, the catalog monthly. The removal mechanism is provided by atmospheric decay and reentry. All breakup debris associated with a given event, to date, is backdated to the event date. Thus, objects associated with the first known event, the breakup of the 1961-015 rocket body, but cataloged recently are nonetheless backdated to account for their production in June 1961. The total number of debris on-orbit and the source/sink numbers are portrayed in Figure 2.1-9.

Figure 2.1-9. Number of debris produced and removed (black; relative to left vertical axis) and the net debris population (dark red; relative to left vertical axis but scaled by 0.075 to be accommodated in this plot.) The monthly average solar activity is plotted relative to right vertical axis.

In this figure, breakup events cause an immediate increase in the number of objects on-orbit, and this somewhat dynamic quantity will change with subsequent cataloging. This figure accentuates the unique nature of the FY-1C, Cosmos-Iridium, and Cosmos 1408 events relative to those historically regarded as a major event. The apparent cyclic nature of the number of debris being removed from the catalog is associated with solar activity and the effect space weather has on atmospheric density, and hence drag force, at altitude. However, there is not a one to one correspondence due to the elliptical nature of many debris fragment orbits. The reader should further recall that low-altitude events, such as the Microsat-R ASAT test of 2019, and deep space events may result in few objects being cataloged due to sensor network limitations described elsewhere in this report.

2.1.4 BREAKUPS BY MISSION PHASE

The prior section's descriptions of spatial, physical, and temporal distributions have portrayed orbital debris from the global risk perspective. In this section, we examine the temporal distribution of events in the context of mission timeline or mission life cycle. Due to limited information in many cases, we restrict the event's categorization with respect to mission life cycle to being in-mission (launch, orbital injection, mission checkout, and operations, to end of mission or payload life), post-mission, and unknown. Similarly, event root-cause assessment is drawn from this report's Tables 2.1 and 2.2 and reflects the authors' current understanding of the cause of each event. Note that even when a cause is highly likely, such as in the case of known on-board pressurized vessels or residual propellants being present, the cause category of Unknown is used when a critical path connecting the proximate cause and effect (the breakup event) is not known or has not been revealed by the object's owner/operators.

TABLE 2.1.4-1 MISSION PHASE ACCOUNTING FOR PAYLOADS

Spacecraft	Mission Phase			
Assessment	In-mission	Post-mission	Unknown	Total
collision, accidental	4	2	0	6
collision, deliberate	5	2	0	7
explosion, accidental	11	3	0	14
explosion, deliberate	52	0	0	52
unknown	5	38	1	44
Total	77	45	1	123

Tables 2.1.4-1 and -2 tabulate cause assessment by mission phase for payloads and rocket bodies, respectively. The third category of objects, operational debris or mission-related debris, accounts for an additional 58 breakup events. All events associated with this category are believed to be accidental explosions as of this edition that occurred during the post-mission phase, and thus do not require a separate formal tabulation.

21

TABLE 2.1.4-2 MISSION PHASE ACCOUNTING FOR ROCKET BODIES

Rocket Bodies	Mission Phase			
Assessment	In-mission	Post-mission	Unknown	TOTAL
collision, accidental	0	1	0	1
collision, deliberate	1	0	0	1
explosion, accidental	13	40	3	56
explosion, deliberate	1	0	0	1
Unknown	3	18	7	28
Total	18	59	10	87

Several assessed cause categories in Tables 2.1 and 2.2, *e.g.*, "battery" and "propulsion," are subsumed in the more general "explosion, accidental" for the purposes of this section. The "explosion, accidental" category for rocket bodies is not subdivided by in-mission activity, but it is known that several rocket bodies have fragmented while thrusting for orbit insertion or after coasting.

From the historical perspective, deliberate explosions are essentially extinct as a category, having been employed primarily when a national security payload could not confidently be recovered within the launching country's borders, as an end-of-mission procedure, or as a design/operational procedural flaw. Deliberate collisions continue on an irregular basis with ASAT tests; note that while higher altitude ASAT tests (SOLWIND, FY-1C) have – by their longer-term environmental significance – influenced both international guidelines and subsequent tests (USA 19, Microsat-R). Even low-altitude tests can significantly affect the environment.

Much more interesting from the perspective of debris mitigation are the accidental explosions and collisions. The former may, like standard breakups, be interpreted as humankind's influence on the orbital environment, whereas the latter may, like anomalous events, be interpreted as the consequences of that influence. Again, invoking the perspective provided by history, we note the continuing importance and efficacy of well-known and documented mitigation methods, techniques, and operational procedures. This is demonstrated by Figure 2.1-10.

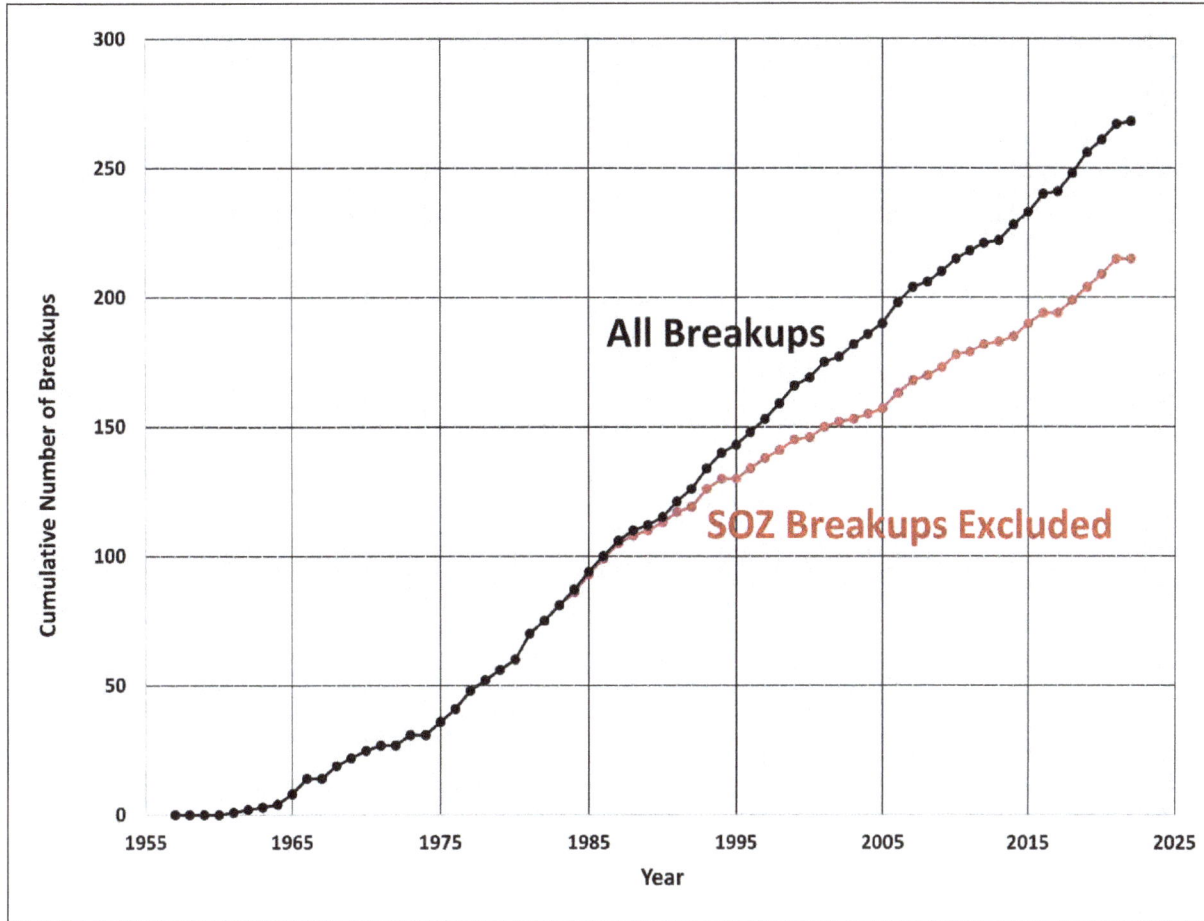

Figure 2.1-10. The number of breakups through 01 May 2022 is plotted as a cumulative distribution, with and without the SOZ breakups included.

In this figure, the fundamental and profound change in the slope representing the occurrence rate of standard breakup events, dating from the implementation of mitigation measures, is obvious. Note that ongoing SOZ breakups mask a noticeable reduction in explosion rates of other satellites beginning in the late 1980s. This is the time when widespread mitigation measures were implemented on a regular basis. Because the SOZ units are a grandfathered system, they exhibit a similar slope to the steeper pre-mitigation fragmentations of the 1970s and 1980s. More details are available in [2]. This report encourages the continued and enhanced compliance with, and further development of, mitigation measures.

REFERENCE DOCUMENTS

1. Xu, Y.-I, Stokely, C. "A Statistical Size Estimation Model for Haystack and HAX Radar Detection," NASA Johnson Space Center, Houston, Texas, USA, 2005.
2. Anz-Meador, P. "Root Cause Classification of Breakup Events 1961-2018," First International Orbital Debris Conference (IOC), Sugar Land, Texas, 2019 available from NASA NTRS at https://ntrs.nasa.gov/citations/20190033947.

TABLE 2.1 HISTORY OF SATELLITE BREAKUPS BY LAUNCH DATE

SATELLITE NAME	INTERNATIONAL DESIGNATOR	US SATELLITE NUMBER	LAUNCH DATE	BREAKUP DATE	DEBRIS CATALOGED	DEBRIS LEFT	APOGEE (KM)	PERIGEE (KM)	INCLINATION (DEG)	ASSESSED CAUSE	ADDITIONAL INFORMATION
TRANSIT 4A R/B	1961-015C	118	29-Jun-61	29-Jun-61	317	193	995	880	66.8	PROPULSION	ABLESTAR STAGE
SPUTNIK 29	1962-057A	443	24-Oct-62	29-Oct-62	24	0	260	200	65.1	PROPULSION	MOLNIYA FINAL STAGE
ATLAS CENTAUR 2	1963-047A	694	27-Nov-63	27-Nov-63	19	8	1785	475	30.3	PROPULSION	CENTAUR STAGE
COSMOS 50	1964-070A	919	28-Oct-64	5-Nov-64	96	0	220	175	51.2	DELIBERATE	SELF-DESTRUCT
COSMOS 57	1965-012A	1093	22-Feb-65	22-Feb-65	167	0	425	165	64.8	DELIBERATE	SELF-DESTRUCT
COSMOS 61-63 R/B	1965-020D	1270	15-Mar-65	15-Mar-65	147	18	1825	260	56.1	UNKNOWN	COSMOS SECOND STAGE
OV2-1/LCS 2 R/B	1965-082DM	1822	15-Oct-65	15-Oct-65	473	32	790	710	32.2	PROPULSION	TITAN TRANSTAGE
COSMOS 95	1965-088A	1706	4-Nov-65	15-Jan-66	1	0	300	180	48.4	UNKNOWN	TITAN TRANSTAGE
OV2-3/ et al. R/B	1965-108A	1863	21-Dec-65	21-Dec-65	109	102	33660	165	26.4	PROPULSION	TITAN TRANSTAGE
OPS 3031	1966-012C	2015	15-Feb-66	15-Feb-66	38	0	270	150	96.5	UNKNOWN	INFLATABLE SPHERE
GEMINI 9 ATDA R/B	1966-046B	2188	1-Jun-66	Jun-66	51	0	275	240	28.8	UNKNOWN	ATLAS CORE STAGE
PAGEOS	1966-056A	2253	24-Jun-66	12-Jul-75	79	1	5170	3200	85.3	UNKNOWN	INFLATABLE SPHERE
AS-203	1966-059A	2289	5-Jul-66	5-Jul-66	34	0	215	185	32.0	DELIBERATE	SATURN S-IVB STAGE
COSMOS U-1	1966-088A	2437	17-Sep-66	17-Sep-66	52	0	855	140	49.6	DELIBERATE	SELF-DESTRUCT
COSMOS U-2	1966-101A	2536	2-Nov-66	2-Nov-66	41	0	885	145	49.6	DELIBERATE	SELF-DESTRUCT
COSMOS 192	1967-116A	3047	23-Nov-67	30-Aug-09	4	1	715	710	74.0	UNKNOWN	TITAN TRANSTAGE
COSMOS 199	1968-003A	3099	16-Jan-68	24-Jan-68	3	0	355	200	65.6	DELIBERATE	SELF-DESTRUCT
APOLLO 6 R/B	1968-025B	3171	4-Apr-68	13-Apr-68	16	0	360	200	32.6	PROPULSION	SATURN S-IVB STAGE
OV2-5 R/B	1968-081E	3432	26-Sep-68	21-Feb-92	29	29	35810	35100	11.9	PROPULSION	TITAN TRANSTAGE
COSMOS 248	1968-090A	3503	19-Oct-68	1-Nov-68	5	0	545	475	62.2	COLLISION, DELIBERATE	DEBRIS IMPACT
COSMOS 249	1968-091A	3504	20-Oct-68	20-Oct-68	108	39	2165	490	62.3	DELIBERATE	SELF-DESTRUCT
COSMOS 252	1968-097A	3530	1-Nov-68	1-Nov-68	139	40	2140	535	62.3	DELIBERATE	SELF-DESTRUCT
ESSA 8 (TOS F) R/B	1968-114B	3616	15-Dec-68	15-Nov-73	20	20	1462	1413	101.6	PROPULSION	DELTA SECOND STAGE
OPS 0757 (TACSAT) R/B	1969-013B	3692	9-Feb-69	28-Feb-18	19	19	37257	35886	6.2	UNKNOWN	TITAN TRANSTAGE
METEOR 1-1 R/B	1969-029B	3836	26-Mar-69	28-Mar-69	37	0	850	460	81.2	UNKNOWN	VOSTOK FINAL STAGE
INTELSAT 3 F-5 R/B	1969-064B	4052	26-Jul-69	26-Jul-69	22	0	5445	270	30.4	PROPULSION	TE 364-4 STAGE
OPS 7613 R/B	1969-082AB	4159	30-Sep-69	4-Oct-69	262	64	940	905	70.0	UNKNOWN	AGENA D STAGE
NIMBUS 4 R/B	1970-025C	4367	8-Apr-70	17-Oct-70	442	297	1085	1065	99.9	UNKNOWN	AGENA D STAGE
COSMOS 374	1970-089A	4594	23-Oct-70	23-Oct-70	99	20	2130	530	62.9	DELIBERATE	SELF-DESTRUCT
COSMOS 375	1970-091A	4598	30-Oct-70	30-Oct-70	47	15	2100	525	62.8	DELIBERATE	SELF-DESTRUCT
COSMOS 397	1971-015A	4964	25-Feb-71	25-Feb-71	116	44	2200	575	65.8	DELIBERATE	SELF-DESTRUCT
COSMOS 462	1971-106A	5646	3-Dec-71	3-Dec-71	25	0	1800	230	65.7	DELIBERATE	SELF-DESTRUCT
LANDSAT 1 R/B	1972-058B	6127	23-Jul-72	22-May-75	245	45	910	635	98.3	PROPULSION	DELTA SECOND STAGE
SALYUT 2 R/B	1973-017B	6399	3-Apr-73	3-Apr-73	25	0	245	195	51.5	PROPULSION	PROTON-K THIRD STAGE
COSMOS 554	1973-021A	6432	19-Apr-73	6-May-73	193	0	350	170	72.9	DELIBERATE	SELF-DESTRUCT
NOAA 3 R/B	1973-086B	6921	6-Nov-73	28-Dec-73	222	200	1510	1500	102.1	PROPULSION	DELTA SECOND STAGE
DMSP 5B F5 R/B	1974-015B	7219	16-Mar-74	17-Jan-05	7	7	885	775	99.1	COLLISION, ACCIDENTAL	HIT BY DEBRIS (26207)
NOAA 4 R/B	1974-089D	7532	15-Nov-74	20-Aug-75	185	160	1460	1445	101.7	PROPULSION	DELTA SECOND STAGE

TABLE 2.1 HISTORY OF SATELLITE BREAKUPS BY LAUNCH DATE (CONT'D)

SATELLITE NAME	INTERNATIONAL DESIGNATOR	US SATELLITE NUMBER	LAUNCH DATE	BREAKUP DATE	DEBRIS CATALOGED	DEBRIS LEFT	APOGEE (KM)	PERIGEE (KM)	INCLINATION (DEG)	ASSESSED CAUSE	ADDITIONAL INFORMATION
COSMOS 699	1974-103A	7587	24-Dec-74	17-Apr-75	50	0	445	425	65.0	UNKNOWN	COSMOS 699 CLASS
LANDSAT 2 R/B	1975-004B	7616	22-Jan-75	9-Feb-76	207	32	915	740	97.8	PROPULSION	DELTA SECOND STAGE
NIMBUS 6 R/B	1975-052B	7946	12-Jun-75	1-May-91	308	231	1105	1095	99.6	PROPULSION	DELTA SECOND STAGE
COSMOS 758	1975-080A	8191	5-Sep-75	6-Sep-75	76	0	325	175	67.1	DELIBERATE	SELF-DESTRUCT
COSMOS 777	1975-102A	8416	29-Oct-75	25-Jan-76	62	0	440	430	65.0	UNKNOWN	COSMOS 699 CLASS
LES 8,9/SOLRAD 11A,B R/B	1976-023F	8751	15-Mar-76	7-May-19	1	1	36889	35793	16.4	UNKNOWN	TITAN TRANSTAGE No. 30
COSMOS 838	1976-063A	8932	2-Jul-76	17-May-77	40	0	445	415	65.1	UNKNOWN	COSMOS 699 CLASS
COSMOS 839	1976-067A	9011	8-Jul-76	29-Sep-77	70	68	2100	980	65.9	BATTERY	
COSMOS 844	1976-072A	9046	22-Jul-76	25-Jul-76	248	0	355	170	67.1	DELIBERATE	SELF-DESTRUCT
NOAA 5 R/B	1976-077B	9063	29-Jul-76	24-Dec-77	185	175	1520	1505	102.0	PROPULSION	DELTA SECOND STAGE
COSMOS 862	1976-105A	9495	22-Oct-76	15-Mar-77	13	5	39645	765	63.2	DELIBERATE	SELF-DESTRUCT
COSMOS 880	1976-120A	9601	9-Dec-76	27-Nov-78	49	0	620	550	65.8	BATTERY	
COSMOS 884	1976-123A	9614	17-Dec-76	29-Dec-76	2	0	320	170	65.0	DELIBERATE	SELF-DESTRUCT
COSMOS 886	1976-126A	9634	27-Dec-76	27-Dec-76	76	58	2295	595	65.8	DELIBERATE	SELF-DESTRUCT
COSMOS 903	1977-027A	9911	11-Apr-77	8-Jun-78	6	2	39035	1325	63.2	DELIBERATE	SELF-DESTRUCT
COSMOS 917	1977-047A	10059	16-Jun-77	30-Mar-79	17	15	38725	1645	62.9	DELIBERATE	SELF-DESTRUCT
HIMAWARI 1 R/B	1977-065B	10144	14-Jul-77	14-Jul-77	177	58	2025	535	29.0	PROPULSION	DELTA SECOND STAGE
COSMOS 931	1977-068A	10150	20-Jul-77	24-Oct-77	6	5	39665	680	62.9	DELIBERATE	SELF-DESTRUCT
EKRAN 2	1977-092A	10365	20-Sep-77	23-Jun-78	5	5	35800	35785	0.1	BATTERY	
COSMOS 970	1977-121A	10531	21-Dec-77	21-Dec-77	70	64	1140	945	65.8	DELIBERATE	SELF-DESTRUCT
LANDSAT 3 R/B	1978-026C	10704	5-Mar-78	27-Jan-81	249	154	910	900	98.8	PROPULSION	DELTA SECOND STAGE
COSMOS 1030	1978-083A	11015	6-Sep-78	10-Oct-78	13	9	39760	665	62.8	DELIBERATE	SELF-DESTRUCT
NIMBUS 7 R/B	1978-098B	11081	24-Oct-78	26-Dec-81	2	1	955	935	99.3	PROPULSION	DELTA SECOND STAGE
COSMOS 1045 R/B	1978-100D	11087	26-Oct-78	9-May-88	42	30	1705	1685	82.6	PROPULSION	TSYKLON THIRD STAGE
P-78/SOLWIND	1979-017A	11278	24-Feb-79	13-Sep-85	285	0	545	515	97.6	COLLISION, DELIBERATE	HYPERVELOCITY IMPACT
COSMOS 1094	1979-033A	11333	18-Apr-79	17-Sep-79	1	0	405	380	65.0	UNKNOWN	COSMOS 699 CLASS
COSMOS 1109	1979-058A	11417	27-Jun-79	Feb-80	20	9	39425	960	63.3	DELIBERATE	SELF-DESTRUCT
COSMOS 1124	1979-077A	11509	28-Aug-79	9-Sep-79	6	6	39795	570	63.0	DELIBERATE	SELF-DESTRUCT
EKRAN 4	1979-087A	11561	3-Oct-79	23-Apr-81	1	1	35791	35790	0.7	BATTERY	AKM FAILURE
SATCOM 3	1979-101A	11635	7-Dec-79	11-Dec-79	27	27	35776	204	23.7	PROPULSION	ARIANE 1 FINAL STAGE
CAT R/B	1979-104B	11659	24-Dec-79	Apr-80	32	25	33140	180	17.9	PROPULSION	
COSMOS 1167	1980-021A	11729	14-Mar-80	15-Jul-81	12	0	450	355	65.0	UNKNOWN	COSMOS 699 CLASS
COSMOS 1174	1980-030A	11765	18-Apr-80	18-Apr-80	46	4	1660	380	66.1	DELIBERATE	SELF-DESTRUCT
COSMOS 1188	1980-050A	11844	14-Jun-80	26-Aug-80	8	6	39630	735	62.9	DELIBERATE	SELF-DESTRUCT
COSMOS 1191	1980-057A	11871	2-Jul-80	14-May-81	11	11	39255	1110	62.6	DELIBERATE	SELF-DESTRUCT
COSMOS 1217	1980-085A	12032	24-Oct-80	12-Feb-83	10	7	38830	1530	65.2	DELIBERATE	SELF-DESTRUCT
COSMOS 1220	1980-089A	12054	4-Nov-80	20-Jun-82	83	1	885	570	65.0	UNKNOWN	COSMOS 1220 CLASS
COSMOS 1247	1981-016A	12303	19-Feb-81	20-Oct-81	8	7	39390	970	63.0	DELIBERATE	SELF-DESTRUCT
COSMOS 1260	1981-028A	12364	20-Mar-81	8-May-82	68	0	750	450	65.0	UNKNOWN	COSMOS 699 CLASS
COSMOS 1261	1981-031A	12376	31-Mar-81	Apr-81	11	11	39765	610	63.0	DELIBERATE	SELF-DESTRUCT
COSMOS 1275	1981-053A	12504	4-Jun-81	24-Jul-81	479	418	1015	960	83.0	BATTERY	SELF-DESTRUCT

TABLE 2.1 HISTORY OF SATELLITE BREAKUPS BY LAUNCH DATE (CONT'D)

SATELLITE NAME	INTERNATIONAL DESIGNATOR	US SATELLITE NUMBER	LAUNCH DATE	BREAKUP DATE	DEBRIS CATALOGED	DEBRIS LEFT	APOGEE (KM)	PERIGEE (KM)	INCLINATION (DEG)	ASSESSED CAUSE	ADDITIONAL INFORMATION
COSMOS 1278	1981-058A	12547	19-Jun-81	Dec-86	3	0	37690	2665	67.1	DELIBERATE	SELF-DESTRUCT
COSMOS 1285	1981-071A	12627	4-Aug-81	21-Nov-81	25	25	40100	720	63.1	DELIBERATE	SELF-DESTRUCT
COSMOS 1286	1981-072A	12631	4-Aug-81	29-Sep-82	2	0	325	300	65.0	UNKNOWN	COSMOS 699 CLASS
COSMOS 1305 R/B	1981-088F	12827	11-Sep-81	11-Sep-81	8	8	13795	605	62.8	PROPULSION	MOLNIYA FINAL STAGE
COSMOS 1306	1981-089A	12828	14-Sep-81	12-Jul-82	8	0	405	380	64.9	UNKNOWN	COSMOS 699 CLASS
COSMOS 1317	1981-108A	12933	31-Oct-81	25-28 Jan-84	11	11	39055	1315	62.8	DELIBERATE	SELF-DESTRUCT
METEOR 2-8	1982-025A	13113	25-Mar-82	29-May-99	53	53	960	935	82.5	UNKNOWN	
COSMOS 1348	1982-029A	13124	7-Apr-82	2-Sep-84	11	11	39200	1185	62.8	DELIBERATE	SELF-DESTRUCT
COSMOS 1355	1982-038A	13150	29-Apr-82	8-Aug-83	29	0	395	360	65.1	UNKNOWN	COSMOS 699 CLASS
COSMOS 1375	1982-055A	13259	6-Jun-82	21-Oct-85	62	59	1000	990	65.8	BATTERY	
COSMOS 1405	1982-088A	13508	4-Sep-82	20-Dec-83	32	0	340	310	65.0	UNKNOWN	COSMOS 699 CLASS
COSMOS 1408	1982-092A	13552	16-Sep-82	15-Nov-21	1760	990	490	465	82.6	COLLISION, DELIBERATE	Russian ASAT test
EKRAN 9	1982-093A	13554	16-Sep-82	23-Dec-83	1	1	35795	35788	0.7	BATTERY	
COSMOS 1423 R/B	1982-115E	13696	8-Dec-82	8-Dec-82	29	0	425	235	62.9	PROPULSION	MOLNIYA FINAL STAGE
ASTRON ULLAGE MOTOR	1983-020B	13902	23-Mar-83	3-Sep-84	1	0	1230	220	51.5	PROPULSION	PROTON-K BLOCK DM SOZ
NOAA 8	1983-022A	13923	28-Mar-83	30-Dec-85	7	1	830	805	98.6	BATTERY	
COSMOS 1456	1983-038A	14034	25-Apr-83	13-Aug-83	4	0	39630	730	63.3	DELIBERATE	SELF-DESTRUCT
COSMOS 1461	1983-044A	14064	7-May-83	11-Mar-85	189	3	890	570	65.0	UNKNOWN	COSMOS 699 CLASS
COSMOS 1481	1983-070A	14182	8-Jul-83	9-Jul-83	9	8	39225	625	62.9	DELIBERATE	SELF-DESTRUCT
COSMOS 1484	1983-075A	14207	24-Jul-83	18-Oct-93	51	1	595	550	97.5	UNKNOWN	
COSMOS 1519-21 ULLAGE MOTOR	1983-127H	14608	29-Dec-83	4-Feb-91	8	3	18805	340	51.9	PROPULSION	PROTON-K BLOCK DM SOZ
PALAPA B2 R/B	1984-011E	14693	3-Feb-84	6-Feb-84	3	0	285	275	28.5	PROPULSION	PAM-D UPPER STAGE
WESTAR 6 R/B	1984-011F	14694	3-Feb-84	3-Feb-84	14	0	310	305	28.5	PROPULSION	PAM-D UPPER STAGE
COSMOS 1588	1984-083A	15167	7-Aug-84	23-Feb-86	45	0	440	410	65.0	UNKNOWN	COSMOS 699 CLASS
COSMOS 1603 ULLAGE MOTOR	1984-106F	15338	28-Sep-84	5-Sep-92	23	2	845	835	66.6	PROPULSION	PROTON-K BLOCK DM SOZ
SPACENET 2/MARECS B2 R/B	1984-114C	15388	10-Nov-84	20-Nov-84	3	2	35960	325	7.0	PROPULSION	ARIANE 3 FINAL STAGE
COSMOS 1646	1985-030A	15653	18-Apr-85	20-Nov-87	24	0	410	385	65.0	UNKNOWN	COSMOS 699 CLASS
COSMOS 1650-52 ULLAGE MOTOR	1985-037G	15714	17-May-85	29-Nov-98	4	2	18620	320	52.0	PROPULSION	PROTON-K BLOCK DM SOZ
COSMOS 1654	1985-039A	15734	23-May-85	21-Jun-85	18	0	300	185	64.9	DELIBERATE	SELF-DESTRUCT
COSMOS 1656 ULLAGE MOTOR	1985-042E	15773	30-May-85	5-Jan-88	6	6	860	810	66.6	PROPULSION	PROTON-K BLOCK DM SOZ
COSMOS 1682	1985-082A	16054	19-Sep-85	18-Dec-86	23	0	475	385	65.0	UNKNOWN	COSMOS 699 CLASS
COSMOS 1691 (1695)	1985-094B	16139	9-Oct-85	22-Nov-85	21	18	1415	1410	82.6	BATTERY	
COSMOS 1703 R/B	1985-108B	16263	22-Nov-85	4-May-06	50	1	640	610	82.5	PROPULSION	TSYKLON THIRD STAGE
COSMOS 1710-12 ULLAGE MOTOR	1985-118L	16446	24-Dec-85	29-Dec-91	17	10	18885	655	65.3	PROPULSION	PROTON-K BLOCK DM SOZ

TABLE 2.1 HISTORY OF SATELLITE BREAKUPS BY LAUNCH DATE (CONT'D)

SATELLITE NAME	INTERNATIONAL DESIGNATOR	US SATELLITE NUMBER	LAUNCH DATE	BREAKUP DATE	DEBRIS CATALOGED	DEBRIS LEFT	APOGEE (KM)	PERIGEE (KM)	INCLINATION (DEG)	ASSESSED CAUSE	ADDITIONAL INFORMATION
COSMOS 1714 R/B	1985-121F	16439	28-Dec-85	28-Dec-85	2	0	830	165	71.0	PROPULSION	ZENIT SECOND STAGE
SPOT 1 R/B	1986-019C	16615	22-Feb-86	13-Nov-86	498	30	835	805	98.7	PROPULSION	ARIANE 1 FINAL STAGE
COSMOS 1769	1986-059A	16895	4-Aug-86	21-Sep-87	4	0	445	310	65.0	UNKNOWN	COSMOS 699 CLASS
USA 19	1986-069A	16937	5-Sep-86	5-Sep-86	13	0	745	210	39.1	COLLISION, DELIBERATE	HYPERVELOCITY IMPACT
USA 19 R/B	1986-069B	16938	5-Sep-86	5-Sep-86	5	0	610	220	22.8	COLLISION, DELIBERATE	HYPERVELOCITY IMPACT
COSMOS 1813	1987-004A	17297	15-Jan-87	29-Jan-87	195	0	415	360	72.8	DELIBERATE	SELF-DESTRUCT
COSMOS 1823	1987-020A	17535	20-Feb-87	17-Dec-87	150	75	1525	1480	73.6	DELIBERATE	SELF-DESTRUCT
COSMOS 1866	1987-059A	18184	9-Jul-87	26-Jul-87	9	0	255	155	67.1	DELIBERATE	SELF-DESTRUCT
COSMOS 1869	1987-062A	18214	16-Jul-87	27-Nov-97	2	1	635	605	83.0	UNKNOWN	
METEOR 2-16 R/B	1987-068B	18313	18-Aug-87	15-Feb-98	108	41	960	940	82.6	PROPULSION	TSYKLON THIRD STAGE
AUSSAT K3/ECS 4 R/B	1987-078C	18352	16-Sep-87	16-19 Sep-87	4	1	36515	245	6.9	PROPULSION	ARIANE 3 FINAL STAGE
COSMOS 1883-85 ULLAGE MOTOR	1987-079G	18374	16-Sep-87	1-Dec-96	14	11	19120	335	64.9	PROPULSION	PROTON-K BLOCK DM SOZ
COSMOS 1883-85 ULLAGE MOTOR	1987-079H	18375	16-Sep-87	23-Apr-03	42	10	18540	755	65.2	PROPULSION	PROTON-K BLOCK DM SOZ
COSMOS 1906 ULLAGE MOTOR	1987-108A	18713	26-Dec-87	31-Jan-88	37	0	265	245	82.6	DELIBERATE	SELF-DESTRUCT
EKRAN 17 ULLAGE MOTOR	1987-109E	18719	27-Dec-87	22-May-97	1	0	22975	310	46.6	PROPULSION	PROTON-K BLOCK DM SOZ
METEOR 2-17	1988-005A	18820	30-Jan-88	21-Jun-05	48	48	960	930	82.5	UNKNOWN	
DMSP 5D-2 F9 (USA 29)	1988-006A	18822	3-Feb-88	14-17 Dec-12	8	3	810	800	98.8	UNKNOWN	
COSMOS 1916	1988-007A	18823	3-Feb-88	27-Feb-88	1	0	230	150	64.8	DELIBERATE	SELF-DESTRUCT
COSMOS 1934	1988-023A	18985	22-Mar-88	23-Dec-91	3	3	1010	950	83.0	COLLISION, ACCIDENTAL	HIT BY DEBRIS (13475)
INTELSAT 513 R/B	1988-040B	19122	17-May-88	9-Jul-02	8	8	35445	535	7.0	PROPULSION	ARIANE 2 R/B
COSMOS 1970-72 ULLAGE MOTOR	1988-085F	19535	16-Sep-88	4-Aug-03	79	10	18515	720	65.3	PROPULSION	PROTON-K BLOCK DM SOZ
COSMOS 1970-72 ULLAGE MOTOR	1988-085G	19537	16-Sep-88	9-Mar-99	1	1	18950	300	64.6	PROPULSION	PROTON-K BLOCK DM SOZ
SKYNET 4B/ASTRA 1A R/B	1988-109C	19689	11-Dec-88	17-Feb-98	19	18	35875	435	7.3	PROPULSION	ARIANE 4 H10 FINAL STAGE
COSMOS 1987-89 ULLAGE MOTOR	1989-001G	19755	10-Jan-89	3-Aug-98	16	6	19055	340	64.9	PROPULSION	PROTON-K BLOCK DM SOZ
COSMOS 1987-89 ULLAGE MOTOR	1989-001H	19856	10-Jan-89	13-Nov-03	1	1	18740	710	65.4	PROPULSION	PROTON-K BLOCK DM SOZ
GORIZONT 17 ULLAGE MOTOR	1989-004E	19771	26-Jan-89	17-Dec-92	1	0	17575	195	46.7	PROPULSION	PROTON-K BLOCK DM SOZ
INTELSAT 515 R/B	1989-006B	19773	27-Jan-89	1-Jan-01	87	86	35720	510	8.4	PROPULSION	ARIANE 2 R/B
COSMOS 2022-24 ULLAGE MOTOR	1989-039G	20081	31-May-89	10-Jun-06	120	42	18410	655	65.1	PROPULSION	PROTON-K BLOCK DM SOZ

TABLE 2.1 HISTORY OF SATELLITE BREAKUPS BY LAUNCH DATE (CONT'D)

SATELLITE NAME	INTERNATIONAL DESIGNATOR	US SATELLITE NUMBER	LAUNCH DATE	BREAKUP DATE	DEBRIS CATALOGED	DEBRIS LEFT	APOGEE (KM)	PERIGEE (KM)	INCLINATION (DEG)	ASSESSED CAUSE	ADDITIONAL INFORMATION
GORIZONT 18 ULLAGE MOTOR	1989-052F	20116	5-Jul-89	12-Jan-93	2	0	36745	260	46.8	PROPULSION	PROTON-K BLOCK DM SOZ
COSMOS 2030	1989-054A	20124	12-Jul-89	28-Jul-89	1	0	215	150	67.1	DELIBERATE	SELF-DESTRUCT
COSMOS 2031	1989-056A	20136	18-Jul-89	31-Aug-89	9	0	365	240	50.5	DELIBERATE	SELF-DESTRUCT
COBE R/B	1989-089B	20323	18-Nov-89	3-Dec-06	26	1	790	685	97.1	UNKNOWN	DELTA SECOND STAGE
COSMOS 2053 R/B	1989-100B	20390	27-Dec-89	18-Apr-99	26	0	485	475	73.5	PROPULSION	TSYKLON THIRD STAGE
COSMOS 2054 ULLAGE MOTOR	1989-101E	20399	27-Dec-89	Jul-92	14	2	27650	345	47.1	PROPULSION	PROTON-K BLOCK DM SOZ
COSMOS 2079-81 ULLAGE MOTOR	1990-045F	20630	19-May-90	17-Nov-11	2	2	18620	420	65.0	PROPULSION	PROTON-K BLOCK DM SOZ
COSMOS 2079-81 ULLAGE MOTOR	1990-045G	20631	19-May-90	28-Mar-99	2	2	19065	405	64.8	PROPULSION	PROTON-K BLOCK DM SOZ
FENGYUN 1-2 R/B	1990-081D	20791	3-Sep-90	4-Oct-90	103	74	895	880	98.9	PROPULSION	CZ-4 FINAL STAGE
COSMOS 2101	1990-087A	20828	1-Oct-90	30-Nov-90	4	0	280	195	64.8	DELIBERATE	SELF-DESTRUCT
GORIZONT 22 ULLAGE MOTOR	1990-102E	20957	23-Nov-90	14-Dec-95	2	1	13105	170	46.5	PROPULSION	PROTON-K BLOCK DM SOZ
USA 68	1990-105A	20978	1-Dec-90	1-Dec-90	30	2	850	610	98.9	PROPULSION	TE-M-364-15 UPPER STAGE
COSMOS 2109-11 ULLAGE MOTOR	1990-110G	21012	8-Dec-90	21-Feb-03	3	3	18805	645	65.4	PROPULSION	PROTON-K BLOCK DM SOZ
COSMOS 2109-11 ULLAGE MOTOR	1990-110H	21013	8-Dec-90	14-Mar-98	3	3	18995	520	65.1	PROPULSION	PROTON-K BLOCK DM SOZ
ITALSAT 1/EUTELSAT 2 F2 R/B	1991-003C	21057	15-Jan-91	Apr/May 96	16	13	30930	235	6.7	PROPULSION	ARIANE 4 H10 FINAL STAGE
COSMOS 2125-32 R/B	1991-009J	21108	12-Feb-91	5-Mar-91	112	112	1725	1460	74.0	PROPULSION	COSMOS SECOND STAGE
COSMOS 2133 ULLAGE MOTOR	1991-010D	21114	12-Feb-91	7-May-94	4	0	21805	225	46.6	PROPULSION	PROTON-K BLOCK DM SOZ
ASTRA 1B/MOP 2 R/B	1991-015C	21141	2-Mar-91	27-Apr-94	12	7	17630	205	6.8	PROPULSION	ARIANE 4 H10 FINAL STAGE
COSMOS 2139-41 ULLAGE MOTOR	1991-025F	21220	4-Apr-91	8-Mar-09	33	11	18535	465	64.9	PROPULSION	PROTON-K BLOCK DM SOZ
COSMOS 2139-41 ULLAGE MOTOR	1991-025G	21226	4-Apr-91	16-Jun-01	2	1	18960	300	64.5	PROPULSION	PROTON-K BLOCK DM SOZ
METEOR 3-5 R/B	1991-056B	21656	15-Aug-91	12-Feb-20	112	106	1206	1186	82.56	UNKNOWN	SL-14 THIRD STAGE
COSMOS 2157-62 R/B	1991-068G	21734	28-Sep-91	9-Oct-99	40	40	1485	1410	82.6	PROPULSION	TSYKLON THIRD STAGE
COSMOS 2163	1991-071A	21741	9-Oct-91	6-Dec-91	1	0	260	185	64.8	DELIBERATE	SELF-DESTRUCT
INTELSAT 601 R/B	1991-075B	21766	29-Oct-91	24-Dec-01	14	8	28505	230	7.2	PROPULSION	ARIANE 4 H10 FINAL STAGE
USA 73 (DMSP 5D2 F11)	1991-082A	21798	28-Nov-91	15-Apr-04	85	59	850	830	98.7	UNKNOWN	

TABLE 2.1 HISTORY OF SATELLITE BREAKUPS BY LAUNCH DATE (CONT'D)

SATELLITE NAME	INTERNATIONAL DESIGNATOR	US SATELLITE NUMBER	LAUNCH DATE	BREAKUP DATE	DEBRIS CATALOGED	DEBRIS LEFT	APOGEE (KM)	PERIGEE (KM)	INCLINATION (DEG)	ASSESSED CAUSE	ADDITIONAL INFORMATION
TELECOM 2B/INMARSAT 2 R/B	1992-021C	21941	15-Apr-92	21-Apr-93	18	16	34080	235	4.0	PROPULSION	ARIANE 4 H10+ FINAL STAGE
INSAT 2A/EUTELSAT 2F4 R/B	1992-041C	22032	9-Jul-92	Feb-02	2	1	26550	250	7.0	PROPULSION	ARIANE 4 H10 FINAL STAGE
COSMOS 2204-06 ULLAGE MOTOR	1992-047G	22066	30-Jul-92	10-Jul-04	34	12	18820	415	64.9	PROPULSION	PROTON-K BLOCK DM SOZ
COSMOS 2204-06 ULLAGE MOTOR	1992-047H	22067	30-Jul-92	8-Nov-94	4	2	19035	480	64.8	PROPULSION	PROTON-K BLOCK DM SOZ
TOPEX-POSEIDON R/B	1992-052D	22079	10-Aug-92	22-Jul-19	11	11	1404	1296	66.07	UNKNOWN	ARIANE 4 H10 FINAL STAGE
GORIZONT 27 ULLAGE MOTOR	1992-082F	22250	27-Nov-92	14-Jul-01	1	0	5340	145	46.5	PROPULSION	PROTON-K BLOCK DM SOZ
COSMOS 2224 ULLAGE MOTOR	1992-088F	22274	17-Dec-92	~22-Apr-05	1	0	21140	200	46.7	PROPULSION	PROTON-K BLOCK DM SOZ
COSMOS 2225	1992-091A	22280	22-Dec-92	18-Feb-93	6	0	280	225	64.9	DELIBERATE	SELF-DESTRUCT
COSMOS 2227 R/B	1992-093B	22285	25-Dec-92	26-Dec-92	280	194	855	845	71.0	PROPULSION	ZENIT-2 SECOND STAGE
COSMOS 2237 R/B	1993-016B	22566	26-Mar-93	28-Mar-93	104	100	850	840	71.0	PROPULSION	ZENIT-2 SECOND STAGE
COSMOS 2238	1993-018A	22585	30-Mar-93	1-Dec-94	1	0	305	210	65.0	UNKNOWN	COSMOS 699 CLASS
COSMOS 2243	1993-028A	22641	27-Apr-93	27-Apr-93	1	0	225	180	70.4	DELIBERATE	SELF-DESTRUCT
COSMOS 2251	1993-036A	22675	16-Jun-93	10-Feb-09	1714	1064	800	775	74.0	COLLISION, ACCIDENTAL	COLLIDED WITH IRIDIUM 33
COSMOS 2259	1993-045A	22716	14-Jul-93	25-Jul-93	1	0	320	175	67.1	DELIBERATE	SELF-DESTRUCT
COSMOS 2262	1993-057A	22789	7-Sep-93	18-Dec-93	1	0	295	170	64.9	DELIBERATE	SELF-DESTRUCT
GORIZONT 29 ULLAGE MOTOR	1993-072E	22925	18-Nov-93	6-Sep-00	1	0	11215	140	46.7	PROPULSION	PROTON-K BLOCK DM SOZ
CLEMENTINE R/B	1994-004B	22974	25-Jan-94	7-Feb-94	1	0	295	240	67.0	PROPULSION	PROTON-K BLOCK DM SOZ
STEP II R/B	1994-029B	23106	19-May-94	3-Jun-96	754	78	820	585	82.0	PROPULSION	PEGASUS HAPS
COSMOS 2282 ULLAGE MOTOR	1994-038F	23174	6-Jul-94	21-Oct-95	2	0	34930	280	47.0	PROPULSION	PROTON-K BLOCK DM SOZ
ELEKTRO ULLAGE MOTOR	1994-069E	23338	31-Oct-94	11-May-95	1	0	35465	155	46.9	PROPULSION	PROTON-K BLOCK DM SOZ
RESURS O1	1994-074A	23342	4-Nov-94	27-Aug-20	141	140	660	633	97.9	UNKNOWN	Resurs-O No. 3
COSMOS 2294-96 ULLAGE MOTOR	1994-076G	23402	20-Nov-94	7-Jun-14	2	2	18990	420	65.0	PROPULSION	PROTON-K BLOCK DM SOZ
RS-15 R/B	1994-085B	23440	26-Dec-94	26-Dec-94	26	23	2200	1880	64.8	UNKNOWN	ROKOT THIRD STAGE
USA 109 (DMSP 5D-2 F13)	1995-015A	23533	24-Mar-95	3-Feb-15	238	212	840	840	98.8	BATTERY	Operational at event time
COSMOS 2313	1995-028A	23596	8-Jun-95	26-Jun-97	13	0	325	210	65.0	UNKNOWN	COSMOS 699 CLASS
CERISE	1995-033B	23606	7-Jul-95	24-Jul-96	2	1	675	665	98.1	COLLISION, ACCIDENTAL	HIT BY DEBRIS (18208)

TABLE 2.1 HISTORY OF SATELLITE BREAKUPS BY LAUNCH DATE (CONT'D)

SATELLITE NAME	INTERNATIONAL DESIGNATOR	US SATELLITE NUMBER	LAUNCH DATE	BREAKUP DATE	DEBRIS CATALOGED	DEBRIS LEFT	APOGEE (KM)	PERIGEE (KM)	INCLINATION (DEG)	ASSESSED CAUSE	ADDITIONAL INFORMATION
COSMOS 2316-18 ULLAGE MOTOR	1995-037K	23631	24-Jul-95	21-Nov-00	1	0	18085	150	64.4	PROPULSION	PROTON-K BLOCK DM SOZ
RADUGA 33 R/B	1996-010D	23797	19-Feb-96	19-Feb-96	1	0	36505	240	48.7	PROPULSION	PROTON-K BLOCK DM
GORIZONT 32 ULLAGE MOTOR	1996-034F	23887	25-May-96	13-Dec-99	1	0	5605	145	46.5	PROPULSION	PROTON-K BLOCK DM SOZ
COSMOS 2343	1997-024A	24805	15-May-97	16-Sep-97	1	0	285	225	65.0	DELIBERATE	SELF-DESTRUCT
IRIDIUM 33	1997-051C	24946	14-Sep-97	10-Feb-09	657	330	780	775	86.4	COLLISION, ACCIDENTAL	COLLIDED WITH COSMOS 2251
KUPON ULLAGE MOTOR	1997-070F	25054	12-Nov-97	14-Feb-07	7	4	14160	260	46.6	PROPULSION	PROTON-K BLOCK DM SOZ
COSMOS 2347	1997-079A	25088	9-Dec-97	22-Nov-99	9	0	410	230	65.0	UNKNOWN	COSMOS 699 CLASS
ORBCOMM FM 5	1997-084F	25117	23-Dec-97	18-Nov-21	15	11	771	758	45	UNKNOWN	
ASIASAT 3 R/B	1997-086D	25129	24-Dec-97	25-Dec-97	1	0	35995	270	51.0	PROPULSION	PROTON-K BLOCK DM
COMETS R/B	1998-011B	25176	21-Feb-98	21-Feb-98	1	0	1880	245	30.0	PROPULSION	H-II SECOND STAGE
ORBCOMM FM 16	1998-046E	25417	2-Aug-98	22-Dec-18	14	8	783	780	45.0	UNKNOWN	
ARGOS/ORSTED/SUNSAT R/B	1999-008D	25637	23-Feb-99	28-Apr-14	8	1	840	635	96.5	UNKNOWN	DELTA 2 SECOND STAGE
FENGYUN 1C	1999-025A	25730	10-May-99	11-Jan-07	3532	2837	865	845	98.6	COLLISION, DELIBERATE	HYPERVELOCITY IMPACT
CBERS 1	1999-057A	25940	14-Oct-99	18-Feb-07	88	36	780	770	98.2	UNKNOWN	
CBERS 1/SACI 1 R/B	1999-057C	25942	14-Oct-99	11-Mar-00	347	152	745	725	98.5	PROPULSION	CZ-4 FINAL STAGE
COSMOS 2367	1999-072A	26040	26-Dec-99	21-Nov-01	17	0	415	405	65.0	UNKNOWN	COSMOS 699 CLASS
COSMOS 2371 ULLAGE MOTOR	2000-036E	26398	4-Jul-00	~1-Sep-06	1	0	21320	220	46.9	PROPULSION	PROTON-K BLOCK DM SOZ
NOAA 16	2000-055A	26536	21-Sep-00	25-Nov-15	458	457	858	842	98.9	UNKNOWN	
TES R/B	2001-049D	26960	22-Oct-01	19-Dec-01	372	69	675	550	97.9	PROPULSION	PSLV FINAL STAGE
COSMOS 2383	2001-057A	27053	21-Dec-01	28-Feb-04	14	0	400	220	65.0	UNKNOWN	COSMOS 699 CLASS
NOAA 17	2002-032A	27453	24-Jun-02	10-Mar-21	115	115	817	800	98.62	UNKNOWN	
COSMOS 2392 ULLAGE MOTOR	2002-037E	27474	25-Jul-02	1-Jun-05	61	0	835	255	63.7	PROPULSION	PROTON-K BLOCK DM SOZ
COSMOS 2392 ULLAGE MOTOR	2002-037F	27475	25-Jul-02	29-Oct-04	1	0	840	235	63.6	PROPULSION	PROTON-K BLOCK DM SOZ
COSMOS 2399	2003-035A	27856	12-Aug-03	9-Dec-03	22	0	250	175	64.9	DELIBERATE	SELF-DESTRUCT
COSMOS 2417-2419 ULLAGE MOTOR	2005-050F	28920	25-Dec-05	24-Aug-18	2	2	18949	546	65.3	PROPULSION	PROTON-K BLOCK DM SOZ
ALOS-1 R/B	2006-002B	28932	24-Jan-06	8-Aug-06	24	3	700	550	98.2	UNKNOWN	H-IIA SECOND STAGE
ARABSAT 4 BRIZ-M R/B	2006-006B	28944	28-Feb-06	19-Feb-07	107	97	14705	495	51.5	PROPULSION	PROTON-K BRIZ-M STAGE
YAOGAN 1	2006-015A	29092	26-Apr-06	4-Feb-10	8	2	630	625	97.9	UNKNOWN	
COSMOS 2421	2006-026A	29247	25-Jun-06	14-Mar-08	509	0	420	400	65.0	UNKNOWN	COSMOS 699 CLASS
IGS 3A R/B	2006-037B	29394	11-Sep-06	28-Dec-06	10	0	490	430	97.2	UNKNOWN	H-IIA SECOND STAGE
COSMOS 2423	2006-039A	29402	14-Sep-06	17-Nov-06	31	0	285	200	64.9	DELIBERATE	SELF-DESTRUCT

TABLE 2.1 HISTORY OF SATELLITE BREAKUPS BY LAUNCH DATE (CONT'D)

SATELLITE NAME	INTERNATIONAL DESIGNATOR	US SATELLITE NUMBER	LAUNCH DATE	BREAKUP DATE	DEBRIS CATALOGED	DEBRIS LEFT	APOGEE (KM)	PERIGEE (KM)	INCLINATION (DEG)	ASSESSED CAUSE	ADDITIONAL INFORMATION
DMSP 5D-3 F17 R/B	2006-050B	29523	4-Nov-06	4-Nov-06	65	0	865	830	98.8	UNKNOWN	DELTA IV SECOND STAGE
USA 193	2006-057A	29651	14-Dec-06	21-Feb-08	175	0	255	245	58.5	COLLISION, DELIBERATE	HYPERVELOCITY IMPACT
COSMOS 2424-2426 ULLAGE MOTOR	2006-062G	29680	25-Dec-06	27-Jul-16	9	8	19088	426	64.8	PROPULSION	PROTON-K BLOCK DM SOZ
COSMOS 2424-2426 ULLAGE MOTOR	2006-062H	29682	25-Dec-06	23-Oct-19	3	3	19189	294	64.5	PROPULSION	PROTON-K BLOCK DM SOZ
BEIDOU 1D R/B	2007-003B	30324	2-Feb-07	2-Feb-07	40	36	41900	235	25.0	UNKNOWN	CZ-3A FINAL STAGE
IGS 4A/4B R/B DEBRIS	2007-005E	30590	24-Feb-07	23-Dec-10	4	0	440	430	97.3	UNKNOWN	H-IIA DEBRIS
COSMOS 2428	2007-029A	31792	29-Jun-07	10-May-14	10	1	860	845	71.0	UNKNOWN	
COSMOS 2431-33 ULLAGE MOTOR	2007-052F	32280	26-Oct-07	13-Aug-14	26	23	18790	730	65.0	PROPULSION	PROTON-K BLOCK DM SOZ
USA 197 R/B	2007-054B	32288	11-Nov-07	11-Nov-07	32	32	1575	220	29.0	UNKNOWN	DELTA IV SECOND STAGE
COSMOS 2434-2436 ULLAGE MOTOR	2007-065F	32398	25-Dec-07	15-Apr-22	1	1	19068	400	64.8	PROPULSION	PROTON-K BLOCK DM SOZ
COSMOS 2434-36 ULLAGE MOTOR	2007-065G	32399	25-Dec-07	18-Aug-11	1	1	18965	540	65.0	PROPULSION	PROTON-K BLOCK DM SOZ
AMC 14 BRIZ-M R/B	2008-011B	32709	14-Mar-08	13-Oct-10	123	122	26565	645	48.9	PROPULSION	PROTON-K BRIZ-M STAGE
COSMOS 2442-44 ULLAGE MOTOR	2008-046H	33385	25-Sep-08	20-May-14	12	10	18720	865	65.0	PROPULSION	PROTON-K BLOCK DM SOZ
COSMOS 2447-49 ULLAGE MOTOR	2008-067G	33472	25-Dec-08	26-Mar-16	6	6	18840	682	65.4	PROPULSION	PROTON-K BLOCK DM SOZ
COSMOS 2447-49 ULLAGE MOTOR	2008-067H	33473	25-Dec-08	1-Jun-16	12	12	18786	709	65.3	PROPULSION	PROTON-K BLOCK DM SOZ
BEIDOU G2	2009-018A	34779	14-Apr-09	29-Jun-16	1	1	36137	35384	4.7	UNKNOWN	
COSMOS 2456-2458 ULLAGE MOTOR	2009-070F	36116	14-Dec-09	21-23-Dec-19	7	7	18980	512	64.7	PROPULSION	PROTON-K BLOCK DM SOZ
COSMOS 2459-61 ULLAGE MOTOR	2010-007G	36406	1-Mar-10	9-Jul-14	13	13	18750	770	65.0	PROPULSION	PROTON-K BLOCK DM SOZ
COSMOS 2459-61 ULLAGE MOTOR	2010-007H	36407	1-Mar-10	22-May-18	20	19	18929	602	65.1	PROPULSION	PROTON-K BLOCK DM SOZ
COSMOS 2464 ULLAGE MOTOR	2010-041G	37143	2-Sep-10	3-Sep-17	10	10	18684	756	65.2	PROPULSION	PROTON-K BLOCK DM SOZ
COSMOS 2464-2466 ULLAGE MOTOR	2010-041H	37144	2-Sep-10	12-Aug-19	24	24	18907	541	65	PROPULSION	PROTON-K BLOCK DM SOZ
CHINASAT 6A R/B	2010-042B	37151	4-Sep-10	4-Sep-10	31	30	41894	194	25.2	UNKNOWN	
BEIDOU G4 R/B	2010-057B	37211	1-Nov-10	1-Nov-10	62	60	35780	160	20.5	UNKNOWN	CZ-3C FINAL STAGE

TABLE 2.1 HISTORY OF SATELLITE BREAKUPS BY LAUNCH DATE (CONT'D)

SATELLITE NAME	INTERNATIONAL DESIGNATOR	US SATELLITE NUMBER	LAUNCH DATE	BREAKUP DATE	DEBRIS CATALOGED	DEBRIS LEFT	APOGEE (KM)	PERIGEE (KM)	INCLINATION (DEG)	ASSESSED CAUSE	ADDITIONAL INFORMATION
SL-23 DEBRIS	2011-037B	37756	18-Jul-11	3/4-Aug-15	1	1	3649	428	51.4	UNKNOWN	Fregat-SB SBB propellant tank
SL-23 DEBRIS	2011-037B	37756	18-Jul-11	8-May-20	334	282	3606	422	51.5	UNKNOWN	Fregat-SB SBB propellant tank
NIGCOMSAT 1R R/B	2011-077B	38015	19-Dec-11	~21-Dec-11	40	39	41715	230	24.3	UNKNOWN	CZ-3B FINAL STAGE
BEIDOU G5 R/B	2012-008B	38092	24-Feb-12	26-Feb-12	42	41	35950	150	20.7	UNKNOWN	CZ-3C FINAL STAGE
NIMIQ 6 R/B	2012-026B	38343	17-May-12	23-Dec-15	11	11	34592	10408	12.0	UNKNOWN	Briz-M CORE
TELKOM 3 / EXPRESS MD2 BRIZ-M R/B	2012-044C	38746	6-Aug-12	16-Oct-12	118	15	5010	265	49.9	PROPULSION	PROTON-K BRIZ-M STAGE
CASSIOPE R/B	2013-055G	39271	29-Sep-13	29-Sep-13	16	1	1490	320	81.0	PROPULSION	FALCON 9 SECOND STAGE
YAOGAN 19 R/B	2013-065B	39411	20-Nov-13	17-Aug-18	7	7	1205.5	996.7	100.5	UNKNOWN	CZ-4C THIRD STAGE
COSMOS 2491	2013-076E	39497	25-Dec-13	23-Dec-19	27	27	1517	1485	82.5	UNKNOWN	1st "SATELLITE INSPECTOR"
SENTINEL 1A	2014-016A	39634	3-Apr-14	23-Aug-16	9	3	698	696	98.2	COLLISION, ACCIDENTAL	Solar array impact by small MMOD
COSMOS 2499	2014-028E	39765	23-May-14	23-Oct-21	23	23	1507	1152	82.44	UNKNOWN	2nd "SATELLITE INSPECTOR"
USA 257 R/B	2014-055B	40209	17-Sep-14	31-Aug-18	105	105	35090	8112	22.3	UNKNOWN	ATLAS V CENTAUR (SEC)
PROGRESS-M 27M R/B	2015-024B	40620	28-Apr-15	28-Apr-15	21	0	181	169	51.7	UNKNOWN	Soyuz 2.1a (3): separation with PROGRESS M-27M
COSMOS 2513 R/B	2015-075B	41122	13-Dec-15	16-Jan-16	7	7	35777	33426	0.2	UNKNOWN	Briz-M CORE
ASTRO H (HITOMI)	2016-012A	41337	17-Feb-16	26-Mar-16	14	5	578	563	31.0	PROPULSION	likely structural failure mechanism
FREGAT DEB (TANK)	2017-086C	43089	26-Dec-17	12-Feb-18	5	1	4070	277	50.4	UNKNOWN	SL-23 Fregat SB SBB propellant tank
USA 288 R/B	2018-079B	43652	17-Oct-18	6-Apr-19	181	180	35092	8526	12.2	UNKNOWN	ATLAS V CENTAUR (SEC)
H-2A DEB	2018-084C	43673	29-Oct-18	12-Jul-20	88	3	643	595	97.9	UNKNOWN	Payload Encapsulation System (PES) lateral cylinder
MICROSAT-R	2019-006A	43947	24-Jan-19	27-Mar-19	130	1	294	265	96.63	COLLISION, DELIBERATE	Indian ASAT test
COSMOS 2535	2019-039A	44421	10-Jul-19	9-Jan-20	18	5	645	607	97.9	UNKNOWN	SATELLITE INSPECTOR
YUNHAI 1-02	2019-063A	44547	25-Sep-19	18-Mar-21	38	32	785	780	98.54	COLLISION, ACCIDENTAL	HIT BY DEBRIS (48078)
USA 305-308 MINOTAUR 4 R/B	2020-046E	45877	15-Jul-20	26-Nov-21	22	17	580	567	54	UNKNOWN	

TABLE 2.2 HISTORY OF SATELLITE BREAKUPS BY EVENT DATE

SATELLITE NAME	INTERNATIONAL DESIGNATOR	US SATELLITE NUMBER	LAUNCH DATE	BREAKUP DATE	DEBRIS CATALOGED	DEBRIS LEFT	APOGEE (KM)	PERIGEE (KM)	INCLINATION (DEG)	ASSESSED CAUSE	ADDITIONAL INFORMATION
TRANSIT 4A R/B	1961-015C	118	29-Jun-61	29-Jun-61	317	193	995	880	66.8	PROPULSION	ABLESTAR STAGE
SPUTNIK 29	1962-057A	443	24-Oct-62	29-Oct-62	24	0	260	200	65.1	PROPULSION	MOLNIYA FINAL STAGE
ATLAS CENTAUR 2	1963-047A	694	27-Nov-63	27-Nov-63	19	8	1785	475	30.3	PROPULSION	CENTAUR STAGE
COSMOS 50	1964-070A	919	28-Oct-64	5-Nov-64	96	0	220	175	51.2	DELIBERATE	SELF-DESTRUCT
COSMOS 57	1965-012A	1093	22-Feb-65	22-Feb-65	167	0	425	165	64.8	DELIBERATE	SELF-DESTRUCT
COSMOS 61-63 R/B	1965-020D	1270	15-Mar-65	15-Mar-65	147	18	1825	260	56.1	UNKNOWN	COSMOS SECOND STAGE
OV2-1/LCS 2 R/B	1965-082DM	1822	15-Oct-65	15-Oct-65	473	32	790	710	32.2	PROPULSION	TITAN TRANSTAGE
OV2-3/ et al. R/B	1965-108A	1863	21-Dec-65	21-Dec-65	109	102	33660	165	26.4	PROPULSION	TITAN TRANSTAGE
COSMOS 95	1965-088A	1706	4-Nov-65	15-Jan-66	1	0	300	180	48.4	UNKNOWN	
OPS 3031	1966-012C	2015	15-Feb-66	15-Feb-66	38	0	270	150	96.5	UNKNOWN	INFLATABLE SPHERE
GEMINI 9 ATDA R/B	1966-046B	2188	1-Jun-66	Jun-66	51	0	275	240	28.8	UNKNOWN	ATLAS CORE STAGE
AS-203	1966-059A	2289	5-Jul-66	5-Jul-66	34	0	215	185	32.0	DELIBERATE	SATURN S-IVB STAGE
COSMOS U-1	1966-088A	2437	17-Sep-66	17-Sep-66	52	0	855	140	49.6	DELIBERATE	SELF-DESTRUCT
COSMOS U-2	1966-101A	2536	2-Nov-66	2-Nov-66	41	0	885	145	49.6	DELIBERATE	SELF-DESTRUCT
COSMOS 199	1968-003A	3099	16-Jan-68	24-Jan-68	3	0	355	200	65.6	DELIBERATE	SELF-DESTRUCT
APOLLO 6 R/B	1968-025B	3171	4-Apr-68	13-Apr-68	16	0	360	200	32.6	PROPULSION	SATURN S-IVB STAGE
COSMOS 249	1968-091A	3504	20-Oct-68	20-Oct-68	108	39	2165	490	62.3	DELIBERATE	SELF-DESTRUCT
COSMOS 248	1968-090A	3503	19-Oct-68	1-Nov-68	5	0	545	475	62.2	COLLISION, DELIBERATE	DEBRIS IMPACT
COSMOS 252	1968-097A	3530	1-Nov-68	1-Nov-68	139	40	2140	535	62.3	DELIBERATE	SELF-DESTRUCT
METEOR 1-1 R/B	1969-029B	3836	26-Mar-69	28-Mar-69	37	0	850	460	81.2	UNKNOWN	VOSTOK FINAL STAGE
INTELSAT 3 F-5 R/B	1969-064B	4052	26-Jul-69	26-Jul-69	22	0	5445	270	30.4	PROPULSION	TE 364-4 STAGE
OPS 7613 R/B	1969-082AB	4159	30-Sep-69	4-Oct-69	262	64	940	905	70.0	UNKNOWN	AGENA D STAGE
NIMBUS 4 R/B	1970-025C	4367	8-Apr-70	17-Oct-70	442	297	1085	1065	99.9	UNKNOWN	AGENA D STAGE
COSMOS 374	1970-089A	4594	23-Oct-70	23-Oct-70	99	20	2130	530	62.9	DELIBERATE	SELF-DESTRUCT
COSMOS 375	1970-091A	4598	30-Oct-70	30-Oct-70	47	15	2100	525	62.8	DELIBERATE	SELF-DESTRUCT

TABLE 2.2 HISTORY OF SATELLITE BREAKUPS BY EVENT DATE (CONT'D)

SATELLITE NAME	INTERNATIONAL DESIGNATOR	US SATELLITE NUMBER	LAUNCH DATE	BREAKUP DATE	DEBRIS CATALOGED	DEBRIS LEFT	APOGEE (KM)	PERIGEE (KM)	INCLINATION (DEG)	ASSESSED CAUSE	ADDITIONAL INFORMATION
COSMOS 397	1971-015A	4964	25-Feb-71	25-Feb-71	116	44	2200	575	65.8	DELIBERATE	SELF-DESTRUCT
COSMOS 462	1971-106A	5646	3-Dec-71	3-Dec-71	25	0	1800	230	65.7	DELIBERATE	SELF-DESTRUCT
SALYUT 2 R/B	1973-017B	6399	3-Apr-73	3-Apr-73	25	0	245	195	51.5	PROPULSION	PROTON-K THIRD STAGE
COSMOS 554	1973-021A	6432	19-Apr-73	6-May-73	193	0	350	170	72.9	DELIBERATE	SELF-DESTRUCT
ESSA 8 (TOS F) R/B	1968-114B	3616	15-Dec-68	15-Nov-73	20	20	1462	1413	101.6	PROPULSION	DELTA SECOND STAGE
NOAA 3 R/B	1973-086B	6921	6-Nov-73	28-Dec-73	222	200	1510	1500	102.1	PROPULSION	DELTA SECOND STAGE
COSMOS 699	1974-103A	7587	24-Dec-74	17-Apr-75	50	0	445	425	65.0	UNKNOWN	COSMOS 699 CLASS
LANDSAT 1 R/B	1972-058B	6127	23-Jul-72	22-May-75	245	45	910	635	98.3	PROPULSION	DELTA SECOND STAGE
PAGEOS	1966-056A	2253	24-Jun-66	12-Jul-75	79	1	5170	3200	85.3	UNKNOWN	INFLATABLE SPHERE
NOAA 4 R/B	1974-089D	7532	15-Nov-74	20-Aug-75	185	160	1460	1445	101.7	PROPULSION	DELTA SECOND STAGE
COSMOS 758	1975-080A	8191	5-Sep-75	6-Sep-75	76	0	325	175	67.1	DELIBERATE	SELF-DESTRUCT
COSMOS 777	1975-102A	8416	29-Oct-75	25-Jan-76	62	0	440	430	65.0	UNKNOWN	COSMOS 699 CLASS
LANDSAT 2 R/B	1975-004B	7616	22-Jan-75	9-Feb-76	207	32	915	740	97.8	PROPULSION	DELTA SECOND STAGE
COSMOS 844	1976-072A	9046	22-Jul-76	25-Jul-76	248	0	355	170	67.1	DELIBERATE	SELF-DESTRUCT
COSMOS 886	1976-126A	9634	27-Dec-76	27-Dec-76	76	58	2295	595	65.8	DELIBERATE	SELF-DESTRUCT
COSMOS 884	1976-123A	9614	17-Dec-76	29-Dec-76	2	0	320	170	65.0	DELIBERATE	SELF-DESTRUCT
COSMOS 862	1976-105A	9495	22-Oct-76	15-Mar-77	13	5	39645	765	63.2	DELIBERATE	SELF-DESTRUCT
COSMOS 838	1976-063A	8932	2-Jul-76	17-May-77	40	0	445	415	65.1	UNKNOWN	COSMOS 699 CLASS
HIMAWARI 1 R/B	1977-065B	10144	14-Jul-77	14-Jul-77	177	58	2025	535	29.0	PROPULSION	DELTA SECOND STAGE
COSMOS 839	1976-067A	9011	8-Jul-76	29-Sep-77	70	68	2100	980	65.9	BATTERY	
COSMOS 931	1977-068A	10150	20-Jul-77	24-Oct-77	6	5	39665	680	62.9	DELIBERATE	SELF-DESTRUCT
COSMOS 970	1977-121A	10531	21-Dec-77	21-Dec-77	70	64	1140	945	65.8	DELIBERATE	SELF-DESTRUCT
NOAA 5 R/B	1976-077B	9063	29-Jul-76	24-Dec-77	185	175	1520	1505	102.0	PROPULSION	DELTA SECOND STAGE

34

TABLE 2.2 HISTORY OF SATELLITE BREAKUPS BY EVENT DATE (CONT'D)

SATELLITE NAME	INTERNATIONAL DESIGNATOR	US SATELLITE NUMBER	LAUNCH DATE	BREAKUP DATE	DEBRIS CATALOGED	DEBRIS LEFT	APOGEE (KM)	PERIGEE (KM)	INCLINATION (DEG)	ASSESSED CAUSE	ADDITIONAL INFORMATION
COSMOS 903	1977-027A	9911	11-Apr-77	8-Jun-78	6	2	39035	1325	63.2	DELIBERATE	SELF-DESTRUCT
EKRAN 2	1977-092A	10365	20-Sep-77	23-Jun-78	5	5	35800	35785	0.1	BATTERY	
COSMOS 1030	1978-083A	11015	6-Sep-78	10-Oct-78	13	9	39760	665	62.8	DELIBERATE	SELF-DESTRUCT
COSMOS 880	1976-120A	9601	9-Dec-76	27-Nov-78	49	0	620	550	65.8	BATTERY	
COSMOS 917	1977-047A	10059	16-Jun-77	30-Mar-79	17	15	38725	1645	62.9	DELIBERATE	SELF-DESTRUCT
COSMOS 1124	1979-077A	11509	28-Aug-79	9-Sep-79	6	6	39795	570	63.0	DELIBERATE	SELF-DESTRUCT
COSMOS 1094	1979-033A	11333	18-Apr-79	17-Sep-79	1	0	405	380	65.0	UNKNOWN	COSMOS 699 CLASS
SATCOM 3	1979-101A	11635	7-Dec-79	11-Dec-79	27	27	35776	204	23.7	PROPULSION	AKM FAILURE
COSMOS 1109	1979-058A	11417	27-Jun-79	Feb-80	20	9	39425	960	63.3	DELIBERATE	SELF-DESTRUCT
CAT R/B	1979-104B	11659	24-Dec-79	Apr-80	32	25	33140	180	17.9	PROPULSION	ARIANE 1 FINAL STAGE
COSMOS 1174	1980-030A	11765	18-Apr-80	18-Apr-80	46	4	1660	380	66.1	DELIBERATE	SELF-DESTRUCT
COSMOS 1188	1980-050A	11844	14-Jun-80	26-Aug-80	8	6	39630	735	62.9	DELIBERATE	SELF-DESTRUCT
LANDSAT 3 R/B	1978-026C	10704	5-Mar-78	27-Jan-81	249	154	910	900	98.8	PROPULSION	DELTA SECOND STAGE
COSMOS 1261	1981-031A	12376	31-Mar-81	Apr-81	11	11	39765	610	63.0	DELIBERATE	SELF-DESTRUCT
EKRAN 4	1979-087A	11561	3-Oct-79	23-Apr-81	1	1	35791	35790	0.7	BATTERY	
COSMOS 1191	1980-057A	11871	2-Jul-80	14-May-81	11	11	39255	1110	62.6	DELIBERATE	SELF-DESTRUCT
COSMOS 1167	1980-021A	11729	14-Mar-80	15-Jul-81	12	0	450	355	65.0	UNKNOWN	COSMOS 699 CLASS
COSMOS 1275	1981-053A	12504	4-Jun-81	24-Jul-81	479	418	1015	960	83.0	BATTERY	
COSMOS 1305 R/B	1981-088F	12827	11-Sep-81	11-Sep-81	8	8	13795	605	62.8	PROPULSION	MOLNIYA FINAL STAGE
COSMOS 1247	1981-016A	12303	19-Feb-81	20-Oct-81	8	7	39390	970	63.0	DELIBERATE	SELF-DESTRUCT
COSMOS 1285	1981-071A	12627	4-Aug-81	21-Nov-81	25	25	40100	720	63.1	DELIBERATE	SELF-DESTRUCT
NIMBUS 7 R/B	1978-098B	11081	24-Oct-78	26-Dec-81	2	1	955	935	99.3	PROPULSION	DELTA SECOND STAGE
COSMOS 1260	1981-028A	12364	20-Mar-81	8-May-82	68	0	750	450	65.0	UNKNOWN	COSMOS 699 CLASS
COSMOS 1220	1980-089A	12054	4-Nov-80	20-Jun-82	83	1	885	570	65.0	UNKNOWN	COSMOS 699 CLASS
COSMOS 1306	1981-089A	12828	14-Sep-81	12-Jul-82	8	0	405	380	64.9	UNKNOWN	COSMOS 699 CLASS
COSMOS 1286	1981-072A	12631	4-Aug-81	29-Sep-82	2	0	325	300	65.0	UNKNOWN	COSMOS 699 CLASS
COSMOS 1423 R/B	1982-115E	13696	8-Dec-82	8-Dec-82	29	0	425	235	62.9	PROPULSION	MOLNIYA FINAL STAGE

TABLE 2.2 HISTORY OF SATELLITE BREAKUPS BY EVENT DATE (CONT'D)

SATELLITE NAME	INTERNATIONAL DESIGNATOR	US SATELLITE NUMBER	LAUNCH DATE	BREAKUP DATE	DEBRIS CATALOGED	DEBRIS LEFT	APOGEE (KM)	PERIGEE (KM)	INCLINATION (DEG)	ASSESSED CAUSE	ADDITIONAL INFORMATION
COSMOS 1217	1980-085A	12032	24-Oct-80	12-Feb-83	10	7	38830	1530	65.2	DELIBERATE	SELF-DESTRUCT
COSMOS 1481	1983-070A	14182	8-Jul-83	9-Jul-83	9	8	39225	625	62.9	DELIBERATE	SELF-DESTRUCT
COSMOS 1355	1982-038A	13150	29-Apr-82	8-Aug-83	29	0	395	360	65.1	UNKNOWN	COSMOS 699 CLASS
COSMOS 1456	1983-038A	14034	25-Apr-83	13-Aug-83	4	0	39630	730	63.3	DELIBERATE	SELF-DESTRUCT
COSMOS 1405	1982-088A	13508	4-Sep-82	20-Dec-83	32	0	340	310	65.0	UNKNOWN	COSMOS 699 CLASS
EKRAN 9	1982-093A	13554	16-Sep-82	23-Dec-83	1	1	35795	35788	0.7	BATTERY	
COSMOS 1317	1981-108A	12933	31-Oct-81	25-28 Jan-84	11	11	39055	1315	62.8	DELIBERATE	SELF-DESTRUCT
WESTAR 6 R/B	1984-011F	14694	3-Feb-84	3-Feb-84	14	0	310	305	28.5	PROPULSION	PAM-D UPPER STAGE
PALAPA B2 R/B	1984-011E	14693	3-Feb-84	6-Feb-84	3	0	285	275	28.5	PROPULSION	PAM-D UPPER STAGE
COSMOS 1348	1982-029A	13124	7-Apr-82	2-Sep-84	11	11	39200	1185	62.8	DELIBERATE	SELF-DESTRUCT
ASTRON ULLAGE MOTOR	1983-020B	13902	23-Mar-83	3-Sep-84	1	0	1230	220	51.5	PROPULSION	PROTON-K BLOCK DM SOZ
SPACENET 2/MARECS B2 R/B	1984-114C	15388	10-Nov-84	20-Nov-84	3	2	35960	325	7.0	PROPULSION	ARIANE 3 FINAL STAGE
COSMOS 1461	1983-044A	14064	7-May-83	11-Mar-85	189	3	890	570	65.0	UNKNOWN	COSMOS 699 CLASS
COSMOS 1654	1985-039A	15734	23-May-85	21-Jun-85	18	0	300	185	64.9	DELIBERATE	SELF-DESTRUCT
P-78/SOLWIND	1979-017A	11278	24-Feb-79	13-Sep-85	285	0	545	515	97.6	COLLISION, DELIBERATE	HYPERVELOCITY IMPACT
COSMOS 1375	1982-055A	13259	6-Jun-82	21-Oct-85	62	59	1000	990	65.8	BATTERY	
COSMOS 1691 (1695)	1985-094B	16139	9-Oct-85	22-Nov-85	21	18	1415	1410	82.6	BATTERY	
COSMOS 1714 R/B	1985-121F	16439	28-Dec-85	28-Dec-85	2	0	830	165	71.0	PROPULSION	ZENIT SECOND STAGE
NOAA 8	1983-022A	13923	28-Mar-83	30-Dec-85	7	1	830	805	98.6	BATTERY	
COSMOS 1588	1984-083A	15167	7-Aug-84	23-Feb-86	45	0	440	410	65.0	UNKNOWN	COSMOS 699 CLASS
USA 19	1986-069A	16937	5-Sep-86	5-Sep-86	13	0	745	210	39.1	COLLISION, DELIBERATE	HYPERVELOCITY IMPACT
USA 19 R/B	1986-069B	16938	5-Sep-86	5-Sep-86	5	0	610	220	22.8	COLLISION, DELIBERATE	HYPERVELOCITY IMPACT
SPOT 1 R/B	1986-019C	16615	22-Feb-86	13-Nov-86	498	30	835	805	98.7	PROPULSION	ARIANE 1 FINAL STAGE
COSMOS 1278	1981-058A	12547	19-Jun-81	Dec-86	3	0	37690	2665	67.1	DELIBERATE	SELF-DESTRUCT
COSMOS 1682	1985-082A	16054	19-Sep-85	18-Dec-86	23	0	475	385	65.0	UNKNOWN	COSMOS 699 CLASS

TABLE 2.2 HISTORY OF SATELLITE BREAKUPS BY EVENT DATE (CONT'D)

SATELLITE NAME	INTERNATIONAL DESIGNATOR	US SATELLITE NUMBER	LAUNCH DATE	BREAKUP DATE	DEBRIS CATALOGED	DEBRIS LEFT	APOGEE (KM)	PERIGEE (KM)	INCLINATION (DEG)	ASSESSED CAUSE	ADDITIONAL INFORMATION
COSMOS 1813	1987-004A	17297	15-Jan-87	29-Jan-87	195	0	415	360	72.8	DELIBERATE	SELF-DESTRUCT
COSMOS 1866	1987-059A	18184	9-Jul-87	26-Jul-87	9	0	255	155	67.1	DELIBERATE	SELF-DESTRUCT
AUSSAT K3/ECS 4 R/B	1987-078C	18352	16-Sep-87	16-19 Sep-87	4	1	36515	245	6.9	PROPULSION	ARIANE 3 FINAL STAGE
COSMOS 1769	1986-059A	16895	4-Aug-86	21-Sep-87	4	0	445	310	65.0	UNKNOWN	COSMOS 699 CLASS
COSMOS 1646	1985-030A	15653	18-Apr-85	20-Nov-87	24	0	410	385	65.0	UNKNOWN	COSMOS 699 CLASS
COSMOS 1823	1987-020A	17535	20-Feb-87	17-Dec-87	150	75	1525	1480	73.6	BATTERY	
COSMOS 1656 ULLAGE MOTOR	1985-042E	15773	30-May-85	5-Jan-88	6	6	860	810	66.6	PROPULSION	PROTON-K BLOCK DM SOZ
COSMOS 1906	1987-108A	18713	26-Dec-87	31-Jan-88	37	0	265	245	82.6	DELIBERATE	SELF-DESTRUCT
COSMOS 1916	1988-007A	18823	3-Feb-88	27-Feb-88	1	0	230	150	64.8	DELIBERATE	SELF-DESTRUCT
COSMOS 1045 R/B	1978-100D	11087	26-Oct-78	9-May-88	42	30	1705	1685	82.6	PROPULSION	TSYKLON THIRD STAGE
COSMOS 2031	1989-056A	20136	18-Jul-89	31-Aug-89	9	0	365	240	50.5	DELIBERATE	SELF-DESTRUCT
COSMOS 2030	1989-054A	20124	12-Jul-89	28-Jul-89	1	0	215	150	67.1	DELIBERATE	SELF-DESTRUCT
FENGYUN 1-2 R/B	1990-081D	20791	3-Sep-90	4-Oct-90	103	74	895	880	98.9	PROPULSION	CZ-4 FINAL STAGE
COSMOS 2101	1990-087A	20828	1-Oct-90	30-Nov-90	4	0	280	195	64.8	DELIBERATE	SELF-DESTRUCT
USA 68	1990-105A	20978	1-Dec-90	1-Dec-90	30	2	850	610	98.9	PROPULSION	TE-M-364-15 UPPER STAGE
COSMOS 1519-21 ULLAGE MOTOR	1983-127H	14608	29-Dec-83	4-Feb-91	8	3	18805	340	51.9	PROPULSION	PROTON-K BLOCK DM SOZ
COSMOS 2125-32 R/B	1991-009J	21108	12-Feb-91	5-Mar-91	112	112	1725	1460	74.0	PROPULSION	COSMOS SECOND STAGE
NIMBUS 6 R/B	1975-052B	7946	12-Jun-75	1-May-91	308	231	1105	1095	99.6	PROPULSION	DELTA SECOND STAGE
COSMOS 2163	1991-071A	21741	9-Oct-91	6-Dec-91	1	0	260	185	64.8	DELIBERATE	SELF-DESTRUCT
COSMOS 1934	1988-023A	18985	22-Mar-88	23-Dec-91	3	3	1010	950	83.0	COLLISION, ACCIDENTAL	HIT BY DEBRIS (13475)
COSMOS 1710-12 ULLAGE MOTOR	1985-118L	16446	24-Dec-85	29-Dec-91	17	10	18885	655	65.3	PROPULSION	PROTON-K BLOCK DM SOZ
OV2-5 R/B	1968-081E	3432	26-Sep-68	21-Feb-92	29	29	35810	35100	11.9	PROPULSION	TITAN TRANSTAGE

TABLE 2.2 HISTORY OF SATELLITE BREAKUPS BY EVENT DATE (CONT'D)

SATELLITE NAME	INTERNATIONAL DESIGNATOR	US SATELLITE NUMBER	LAUNCH DATE	BREAKUP DATE	DEBRIS CATALOGED	DEBRIS LEFT	APOGEE (KM)	PERIGEE (KM)	INCLINATION (DEG)	ASSESSED CAUSE	ADDITIONAL INFORMATION
COSMOS 2054 ULLAGE MOTOR	1989-101E	20399	27-Dec-89	Jul-92	14	2	27650	345	47.1	PROPULSION	PROTON-K BLOCK DM SOZ
COSMOS 1603 ULLAGE MOTOR	1984-106F	15338	28-Sep-84	5-Sep-92	23	2	845	835	66.6	PROPULSION	PROTON-K BLOCK DM SOZ
GORIZONT 17 ULLAGE MOTOR	1989-004E	19771	26-Jan-89	17-Dec-92	1	0	17575	195	46.7	PROPULSION	PROTON-K BLOCK DM SOZ
COSMOS 2227 R/B	1992-093B	22285	25-Dec-92	26-Dec-92	280	194	855	845	71.0	PROPULSION	ZENIT-2 SECOND STAGE
GORIZONT 18 ULLAGE MOTOR	1989-052F	20116	5-Jul-89	12-Jan-93	2	0	36745	260	46.8	PROPULSION	PROTON-K BLOCK DM SOZ
COSMOS 2225	1992-091A	22280	22-Dec-92	18-Feb-93	6	0	280	225	64.9	DELIBERATE	SELF-DESTRUCT
COSMOS 2237 R/B	1993-016B	22566	26-Mar-93	28-Mar-93	104	100	850	840	71.0	PROPULSION	ZENIT-2 SECOND STAGE
TELECOM 2B/INMARSAT 2 R/B	1992-021C	21941	15-Apr-92	21-Apr-93	18	16	34080	235	4.0	PROPULSION	ARIANE 4 H10+ FINAL STAGE
COSMOS 2243	1993-028A	22641	27-Apr-93	27-Apr-93	1	0	225	180	70.4	DELIBERATE	SELF-DESTRUCT
COSMOS 2259	1993-045A	22716	14-Jul-93	25-Jul-93	1	0	320	175	67.1	DELIBERATE	SELF-DESTRUCT
COSMOS 1484	1983-075A	14207	24-Jul-83	18-Oct-93	51	1	595	550	97.5	UNKNOWN	
COSMOS 2262	1993-057A	22789	7-Sep-93	18-Dec-93	1	0	295	170	64.9	DELIBERATE	SELF-DESTRUCT
CLEMENTINE R/B	1994-004B	22974	25-Jan-94	7-Feb-94	1	0	295	240	67.0	PROPULSION	
ASTRA 1B/MOP 2 R/B	1991-015C	21141	2-Mar-91	27-Apr-94	12	7	17630	205	6.8	PROPULSION	ARIANE 4 H10 FINAL STAGE
COSMOS 2133 ULLAGE MOTOR	1991-010D	21114	12-Feb-91	7-May-94	4	0	21805	225	46.6	PROPULSION	PROTON-K BLOCK DM SOZ
COSMOS 2204-06 ULLAGE MOTOR	1992-047H	22067	30-Jul-92	8-Nov-94	4	2	19035	480	64.8	PROPULSION	PROTON-K BLOCK DM SOZ
COSMOS 2238	1993-018A	22585	30-Mar-93	1-Dec-94	1	0	305	210	65.0	UNKNOWN	COSMOS 699 CLASS
RS-15 R/B	1994-085B	23440	26-Dec-94	26-Dec-94	26	23	2200	1880	64.8	UNKNOWN	ROKOT THIRD STAGE
ELEKTRO ULLAGE MOTOR	1994-069E	23338	31-Oct-94	11-May-95	1	0	35465	155	46.9	PROPULSION	PROTON-K BLOCK DM SOZ
COSMOS 2282 ULLAGE MOTOR	1994-038F	23174	6-Jul-94	21-Oct-95	2	0	34930	280	47.0	PROPULSION	PROTON-K BLOCK DM SOZ

TABLE 2.2 HISTORY OF SATELLITE BREAKUPS BY EVENT DATE (CONT'D)

SATELLITE NAME	INTERNATIONAL DESIGNATOR	US SATELLITE NUMBER	LAUNCH DATE	BREAKUP DATE	DEBRIS CATALOGED	DEBRIS LEFT	APOGEE (KM)	PERIGEE (KM)	INCLINATION (DEG)	ASSESSED CAUSE	ADDITIONAL INFORMATION
GORIZONT 22 ULLAGE MOTOR	1990-102E	20957	23-Nov-90	14-Dec-95	2	1	13105	170	46.5	PROPULSION	PROTON-K BLOCK DM SOZ
RADUGA 33 R/B	1996-010D	23797	19-Feb-96	19-Feb-96	1	0	36505	240	48.7	PROPULSION	PROTON-K BLOCK DM
ITALSAT 1/EUTELSAT 2 F2 R/B	1991-003C	21057	15-Jan-91	Apr/May 96	16	13	30930	235	6.7	PROPULSION	ARIANE 4 H10 FINAL STAGE
STEP II R/B	1994-029B	23106	19-May-94	3-Jun-96	754	78	820	585	82.0	PROPULSION	PEGASUS HAPS
CERISE	1995-033B	23606	7-Jul-95	24-Jul-96	2	1	675	665	98.1	COLLISION, ACCIDENTAL	HIT BY DEBRIS (18208)
COSMOS 1883-85 ULLAGE MOTOR	1987-079G	18374	16-Sep-87	1-Dec-96	14	11	19120	335	64.9	PROPULSION	PROTON-K BLOCK DM SOZ
EKRAN 17 ULLAGE MOTOR	1987-109E	18719	27-Dec-87	22-May-97	1	0	22975	310	46.6	PROPULSION	PROTON-K BLOCK DM SOZ
COSMOS 2313	1995-028A	23596	8-Jun-95	26-Jun-97	13	0	325	210	65.0	UNKNOWN	COSMOS 699 CLASS
COSMOS 2343	1997-024A	24805	15-May-97	16-Sep-97	1	0	285	225	65.0	DELIBERATE	SELF-DESTRUCT
COSMOS 1869	1987-062A	18214	16-Jul-87	27-Nov-97	2	1	635	605	83.0	UNKNOWN	
ASIASAT 3 R/B	1997-086D	25129	24-Dec-97	25-Dec-97	1	0	35995	270	51.0	PROPULSION	PROTON-K BLOCK DM
METEOR 2-16 R/B	1987-068B	18313	18-Aug-87	15-Feb-98	108	41	960	940	82.6	PROPULSION	TSYKLON THIRD STAGE
SKYNET 4B/ASTRA 1A R/B	1988-109C	19689	11-Dec-88	17-Feb-98	19	18	35875	435	7.3	PROPULSION	ARIANE 4 H10 FINAL STAGE
COMETS R/B	1998-011B	25176	21-Feb-98	21-Feb-98	1	0	1880	245	30.0	PROPULSION	H-II SECOND STAGE
COSMOS 2109-11 ULLAGE MOTOR	1990-110H	21013	8-Dec-90	14-Mar-98	3	3	18995	520	65.1	PROPULSION	PROTON-K BLOCK DM SOZ
COSMOS 1987-89 ULLAGE MOTOR	1989-001G	19755	10-Jan-89	3-Aug-98	16	6	19055	340	64.9	PROPULSION	PROTON-K BLOCK DM SOZ
COSMOS 1650-52 ULLAGE MOTOR	1985-037G	15714	17-May-85	29-Nov-98	4	2	18620	320	52.0	PROPULSION	PROTON-K BLOCK DM SOZ
COSMOS 1970-72 ULLAGE MOTOR	1988-085G	19537	16-Sep-88	9-Mar-99	1	1	18950	300	64.6	PROPULSION	PROTON-K BLOCK DM SOZ
COSMOS 2079-81 ULLAGE MOTOR	1990-045G	20631	19-May-90	28-Mar-99	2	2	19065	405	64.8	PROPULSION	PROTON-K BLOCK DM SOZ

TABLE 2.2 HISTORY OF SATELLITE BREAKUPS BY EVENT DATE (CONT'D)

SATELLITE NAME	INTERNATIONAL DESIGNATOR	US SATELLITE NUMBER	LAUNCH DATE	BREAKUP DATE	DEBRIS CATALOGED	DEBRIS LEFT	APOGEE (KM)	PERIGEE (KM)	INCLINATION (DEG)	ASSESSED CAUSE	ADDITIONAL INFORMATION
COSMOS 2053 R/B	1989-100B	20390	27-Dec-89	18-Apr-99	26	0	485	475	73.5	PROPULSION	TSYKLON THIRD STAGE
METEOR 2-8	1982-025A	13113	25-Mar-82	29-May-99	53	53	960	935	82.5	UNKNOWN	
COSMOS 2157-62 R/B	1991-068G	21734	28-Sep-91	9-Oct-99	40	40	1485	1410	82.6	PROPULSION	TSYKLON THIRD STAGE
COSMOS 2347	1997-079A	25088	9-Dec-97	22-Nov-99	9	0	410	230	65.0	UNKNOWN	COSMOS 699 CLASS
GORIZONT 32 ULLAGE MOTOR	1996-034F	23887	25-May-96	13-Dec-99	1	0	5605	145	46.5	PROPULSION	PROTON-K BLOCK DM SOZ
CBERS 1/SACI 1 R/B	1999-057C	25942	14-Oct-99	11-Mar-00	347	152	745	725	98.5	PROPULSION	CZ-4 FINAL STAGE
GORIZONT 29 ULLAGE MOTOR	1993-072E	22925	18-Nov-93	6-Sep-00	1	0	11215	140	46.7	PROPULSION	PROTON-K BLOCK DM SOZ
COSMOS 2316-18 ULLAGE MOTOR	1995-037K	23631	24-Jul-95	21-Nov-00	1	0	18085	150	64.4	PROPULSION	PROTON-K BLOCK DM SOZ
INTELSAT 515 R/B	1989-006B	19773	27-Jan-89	1-Jan-01	87	86	35720	510	8.4	PROPULSION	ARIANE 2 R/B
COSMOS 2139-41 ULLAGE MOTOR	1991-025G	21226	4-Apr-91	16-Jun-01	2	1	18960	300	64.5	PROPULSION	PROTON-K BLOCK DM SOZ
GORIZONT 27 ULLAGE MOTOR	1992-082F	22250	27-Nov-92	14-Jul-01	1	0	5340	145	46.5	PROPULSION	PROTON-K BLOCK DM SOZ
COSMOS 2367	1999-072A	26040	26-Dec-99	21-Nov-01	17	0	415	405	65.0	UNKNOWN	COSMOS 699 CLASS
TES R/B	2001-049D	26960	22-Oct-01	19-Dec-01	372	69	675	550	97.9	PROPULSION	PSLV FINAL STAGE
INTELSAT 601 R/B	1991-075B	21766	29-Oct-91	24-Dec-01	14	8	28505	230	7.2	PROPULSION	ARIANE 4 H10 FINAL STAGE
INSAT 2A/EUTELSAT 2F4 R/B	1992-041C	22032	9-Jul-92	Feb-02	2	1	26550	250	7.0	PROPULSION	ARIANE 4 H10 FINAL STAGE
INTELSAT 513 R/B	1988-040B	19122	17-May-88	9-Jul-02	8	8	35445	535	7.0	PROPULSION	ARIANE 2 R/B
COSMOS 2109-11 ULLAGE MOTOR	1990-110G	21012	8-Dec-90	21-Feb-03	3	3	18805	645	65.4	PROPULSION	PROTON-K BLOCK DM SOZ
COSMOS 1883-85 ULLAGE MOTOR	1987-079H	18375	16-Sep-87	23-Apr-03	42	10	18540	755	65.2	PROPULSION	PROTON-K BLOCK DM SOZ
COSMOS 1970-72 ULLAGE MOTOR	1988-085F	19535	16-Sep-88	4-Aug-03	79	10	18515	720	65.3	PROPULSION	PROTON-K BLOCK DM SOZ

TABLE 2.2 HISTORY OF SATELLITE BREAKUPS BY EVENT DATE (CONT'D)

SATELLITE NAME	INTERNATIONAL DESIGNATOR	US SATELLITE NUMBER	LAUNCH DATE	BREAKUP DATE	DEBRIS CATALOGED	DEBRIS LEFT	APOGEE (KM)	PERIGEE (KM)	INCLINATION (DEG)	ASSESSED CAUSE	ADDITIONAL INFORMATION
COSMOS 1987-89 ULLAGE MOTOR	1989-001H	19856	10-Jan-89	13-Nov-03	1	1	18740	710	65.4	PROPULSION	PROTON-K BLOCK DM SOZ
COSMOS 2399	2003-035A	27856	12-Aug-03	9-Dec-03	0	0	250	175	64.9	DELIBERATE	SELF-DESTRUCT
COSMOS 2383	2001-057A	27053	21-Dec-01	28-Feb-04	14	0	400	220	65.0	UNKNOWN	COSMOS 699 CLASS
USA 73 (DMSP 5D2 F11)	1991-082A	21798	28-Nov-91	15-Apr-04	85	59	850	830	98.7	UNKNOWN	
COSMOS 2204-06 ULLAGE MOTOR	1992-047G	22066	30-Jul-92	10-Jul-04	34	12	18820	415	64.9	PROPULSION	PROTON-K BLOCK DM SOZ
COSMOS 2392 ULLAGE MOTOR	2002-037F	27475	25-Jul-02	29-Oct-04	1	0	840	235	63.6	PROPULSION	PROTON-K BLOCK DM SOZ
DMSP 5B F5 R/B	1974-015B	7219	16-Mar-74	17-Jan-05	7	7	885	775	99.1	COLLISION, ACCIDENTAL	HIT BY DEBRIS (26207)
COSMOS 2224 ULLAGE MOTOR	1992-088F	22274	17-Dec-92	~22-Apr-05	1	0	21140	200	46.7	PROPULSION	PROTON-K BLOCK DM SOZ
COSMOS 2392 ULLAGE MOTOR	2002-037E	27474	25-Jul-02	1-Jun-05	61	0	835	255	63.7	PROPULSION	PROTON-K BLOCK DM SOZ
METEOR 2-17	1988-005A	18820	30-Jan-88	21-Jun-05	48	48	960	930	82.5	UNKNOWN	
COSMOS 1703 R/B	1985-108B	16263	22-Nov-85	4-May-06	50	1	640	610	82.5	PROPULSION	TSYKLON THIRD STAGE
COSMOS 2022-24 ULLAGE MOTOR	1989-039G	20081	31-May-89	10-Jun-06	120	42	18410	655	65.1	PROPULSION	PROTON-K BLOCK DM SOZ
ALOS-1 R/B	2006-002B	28932	24-Jan-06	8-Aug-06	24	3	700	550	98.2	UNKNOWN	H-IIA SECOND STAGE
COSMOS 2371 ULLAGE MOTOR	2000-036E	26398	4-Jul-00	~1-Sep-06	1	0	21320	220	46.9	PROPULSION	PROTON-K BLOCK DM SOZ
DMSP 5D-3 F17 R/B	2006-050B	29523	4-Nov-06	4-Nov-06	65	0	865	830	98.8	UNKNOWN	DELTA IV SECOND STAGE
COSMOS 2423	2006-039A	29402	14-Sep-06	17-Nov-06	31	0	285	200	64.9	DELIBERATE	SELF-DESTRUCT
COBE R/B	1989-089B	20323	18-Nov-89	3-Dec-06	26	1	790	685	97.1	UNKNOWN	DELTA SECOND STAGE
IGS 3A R/B	2006-037B	29394	11-Sep-06	28-Dec-06	10	0	490	430	97.2	UNKNOWN	H-IIA SECOND STAGE
FENGYUN 1C	1999-025A	25730	10-May-99	11-Jan-07	3532	2837	865	845	98.6	COLLISION, DELIBERATE	HYPERVELOCITY IMPACT
BEIDOU 1D R/B	2007-003B	30324	2-Feb-07	2-Feb-07	40	36	41900	235	25.0	UNKNOWN	CZ-3A FINAL STAGE

TABLE 2.2 HISTORY OF SATELLITE BREAKUPS BY EVENT DATE (CONT'D)

SATELLITE NAME	INTERNATIONAL DESIGNATOR	US SATELLITE NUMBER	LAUNCH DATE	BREAKUP DATE	DEBRIS CATALOGED	DEBRIS LEFT	APOGEE (KM)	PERIGEE (KM)	INCLINATION (DEG)	ASSESSED CAUSE	ADDITIONAL INFORMATION
KUPON ULLAGE MOTOR	1997-070F	25054	12-Nov-97	14-Feb-07	7	4	14160	260	46.6	PROPULSION	PROTON-K BLOCK DM SOZ
CBERS 1	1999-057A	25940	14-Oct-99	18-Feb-07	88	36	780	770	98.2	UNKNOWN	
ARABSAT 4 BRIZ-M R/B	2006-006B	28944	28-Feb-06	19-Feb-07	107	97	14705	495	51.5	PROPULSION	PROTON-K BRIZ-M STAGE
USA 197 R/B	2007-054B	32288	11-Nov-07	11-Nov-07	32	32	1575	220	29.0	UNKNOWN	DELTA IV SECOND STAGE
USA 193	2006-057A	29651	14-Dec-06	21-Feb-08	175	0	255	245	58.5	COLLISION, DELIBERATE	HYPERVELOCITY IMPACT
COSMOS 2421	2006-026A	29247	25-Jun-06	14-Mar-08	509	0	420	400	65.0	UNKNOWN	COSMOS 699 CLASS
COSMOS 2251	1993-036A	22675	16-Jun-93	10-Feb-09	1714	1064	800	775	74.0	COLLISION, ACCIDENTAL	COLLIDED WITH IRIDIUM 33
IRIDIUM 33	1997-051C	24946	14-Sep-97	10-Feb-09	657	330	780	775	86.4	COLLISION, ACCIDENTAL	COLLIDED WITH COSMOS 2251
COSMOS 2139-41 ULLAGE MOTOR	1991-025F	21220	4-Apr-91	8-Mar-09	33	11	18535	465	64.9	PROPULSION	PROTON-K BLOCK DM SOZ
COSMOS 192	1967-116A	3047	23-Nov-67	30-Aug-09	4	1	715	710	74.0	UNKNOWN	
YAOGAN 1	2006-015A	29092	26-Apr-06	4-Feb-10	8	2	630	625	97.9	UNKNOWN	
CHINASAT 6A R/B	2010-042B	37151	4-Sep-10	4-Sep-10	31	30	41894	194	25.2	UNKNOWN	
AMC 14 BRIZ-M R/B	2008-011B	32709	14-Mar-08	13-Oct-10	123	122	26565	645	48.9	PROPULSION	PROTON-K BRIZ-M STAGE
BEIDOU G4 R/B	2010-057B	37211	1-Nov-10	1-Nov-10	62	60	35780	160	20.5	UNKNOWN	CZ-3C FINAL STAGE
IGS 4A/4B R/B DEBRIS	2007-005E	30590	24-Feb-07	23-Dec-10	4	0	440	430	97.3	UNKNOWN	H-IIA DEBRIS
COSMOS 2434-36 ULLAGE MOTOR	2007-065G	32399	25-Dec-07	18-Aug-11	1	1	18965	540	65.0	PROPULSION	PROTON-K BLOCK DM SOZ
COSMOS 2079-81 ULLAGE MOTOR	1990-045F	20630	19-May-90	17-Nov-11	2	2	18620	420	65.0	PROPULSION	PROTON-K BLOCK DM SOZ
NIGCOMSAT 1R R/B	2011-077B	38015	19-Dec-11	~21-Dec-11	40	39	41715	230	24.3	UNKNOWN	CZ-3B FINAL STAGE
BEIDOU G5 R/B	2012-008B	38092	24-Feb-12	26-Feb-12	42	41	35950	150	20.7	UNKNOWN	CZ-3C FINAL STAGE
TELKOM 3 / EXPRESS MD2 BRIZ-M R/B	2012-044C	38746	6-Aug-12	16-Oct-12	118	15	5010	265	49.9	PROPULSION	PROTON-K BRIZ-M STAGE
DMSP 5D-2 F9 (USA 29)	1988-006A	18822	3-Feb-88	14-17 Dec-12	8	3	810	800	98.8	UNKNOWN	

42

TABLE 2.2 HISTORY OF SATELLITE BREAKUPS BY EVENT DATE (CONT'D)

SATELLITE NAME	INTERNATIONAL DESIGNATOR	US SATELLITE NUMBER	LAUNCH DATE	BREAKUP DATE	DEBRIS CATALOGED	DEBRIS LEFT	APOGEE (KM)	PERIGEE (KM)	INCLINATION (DEG)	ASSESSED CAUSE	ADDITIONAL INFORMATION
CASSIOPE R/B	2013-055G	39271	29-Sep-13	29-Sep-13	16	1	1490	320	81.0	PROPULSION	FALCON 9 SECOND STAGE
ARGOS/ORSTED/SUNSAT R/B	1999-008D	25637	23-Feb-99	28-Apr-14	8	1	840	635	96.5	UNKNOWN	DELTA 2 SECOND STAGE
COSMOS 2428	2007-029A	31792	29-Jun-07	10-May-14	10	1	860	845	71.0	UNKNOWN	
COSMOS 2442-44 ULLAGE MOTOR	2008-046H	33385	25-Sep-08	20-May-14	12	10	18720	865	65.0	PROPULSION	PROTON-K BLOCK DM SOZ
COSMOS 2294-96 ULLAGE MOTOR	1994-076G	23402	20-Nov-94	7-Jun-14	2	2	18990	420	65.0	PROPULSION	PROTON-K BLOCK DM SOZ
COSMOS 2459-61 ULLAGE MOTOR	2010-007G	36406	1-Mar-10	9-Jul-14	13	13	18750	770	65.0	PROPULSION	PROTON-K BLOCK DM SOZ
COSMOS 2431-33 ULLAGE MOTOR	2007-052F	32280	26-Oct-07	13-Aug-14	26	23	18790	730	65.0	PROPULSION	PROTON-K BLOCK DM SOZ
USA 109 (DMSP 5D-2 F13)	1995-015A	23533	24-Mar-95	3-Feb-15	238	212	840	840	98.8	BATTERY	Operational at event time
PROGRESS-M 27M R/B	2015-024B	40620	28-Apr-15	28-Apr-15	21	0	181	169	51.7	UNKNOWN	Soyuz 2.1a (3); separation with PROGRESS M-27M
SL-23 DEBRIS	2011-037B	37756	18-Jul-11	3/4-Aug-15	1	1	3649	428	51.4	UNKNOWN	Fregat-SB SBB propellant tank
NOAA 16	2000-055A	26536	21-Sep-00	25-Nov-15	458	457	858	842	98.9	UNKNOWN	
NIMIQ 6 R/B	2012-026B	38343	17-May-12	23-Dec-15	11	11	34592	10408	12.0	UNKNOWN	Briz-M CORE
COSMOS 2513 R/B	2015-075B	41122	13-Dec-15	16-Jan-16	7	7	35777	33426	0.2	UNKNOWN	Briz-M CORE
COSMOS 2447-49 ULLAGE MOTOR	2008-067G	33472	25-Dec-08	26-Mar-16	6	6	18840	682	65.4	PROPULSION	PROTON-K BLOCK DM SOZ
ASTRO H (HITOMI)	2016-012A	41337	17-Feb-16	26-Mar-16	14	5	578	563	31.0	PROPULSION	likely structural failure mechanism
COSMOS 2447-49 ULLAGE MOTOR	2008-067H	33473	25-Dec-08	1-Jun-16	12	12	18786	709	65.3	PROPULSION	PROTON-K BLOCK DM SOZ
BEIDOU G2	2009-018A	34779	14-Apr-09	29-Jun-16	1	1	36137	35384	4.7	UNKNOWN	
COSMOS 2424-2426 ULLAGE MOTOR	2006-062G	29680	25-Dec-06	27-Jul-16	9	8	19088	426	64.8	PROPULSION	PROTON-K BLOCK DM SOZ

TABLE 2.2 HISTORY OF SATELLITE BREAKUPS BY EVENT DATE (CONT'D)

SATELLITE NAME	INTERNATIONAL DESIGNATOR	US SATELLITE NUMBER	LAUNCH DATE	BREAKUP DATE	DEBRIS CATALOGED	DEBRIS LEFT	APOGEE (KM)	PERIGEE (KM)	INCLINATION (DEG)	ASSESSED CAUSE	ADDITIONAL INFORMATION
SENTINEL 1A	2014-016A	39634	3-Apr-14	23-Aug-16	9	3	698	696	98.2	COLLISION, ACCIDENTAL	Solar array impact by small MMOD
COSMOS 2464-2466 ULLAGE MOTOR	2010-041G	37143	2-Sep-10	3-Sep-17	10	10	18684	756	65.2	PROPULSION	PROTON-K BLOCK DM SOZ
FREGAT DEB (TANK)	2017-086C	43089	26-Dec-17	12-Feb-18	5	1	4070	277	50.4	UNKNOWN	SL-23 Fregat SB SBB propellant tank
OPS 0757 (TACSAT) R/B	1969-013B	3692	9-Feb-69	28-Feb-18	19	19	37257	35886	6.2	UNKNOWN	TITAN TRANSTAGE
COSMOS 2459-61 ULLAGE MOTOR	2010-007H	36407	1-Mar-10	22-May-18	20	19	18929	602	65.1	PROPULSION	PROTON-K BLOCK DM SOZ
YAOGAN 19 R/B	2013-065B	39411	20-Nov-13	17-Aug-18	7	7	1205.5	996.7	100.5	UNKNOWN	CZ-4C THIRD STAGE
COSMOS 2417-2419 ULLAGE MOTOR	2005-050F	28920	25-Dec-05	24-Aug-18	2	2	18949	546	65.3	PROPULSION	PROTON-K BLOCK DM SOZ
USA 257 R/B	2014-055B	40209	17-Sep-14	31-Aug-18	105	105	35090	8112	22.3	UNKNOWN	ATLAS V CENTAUR (SEC)
ORBCOMM FM 16	1998-046E	25417	2-Aug-98	22-Dec-18	14	8	783	780	45.0	UNKNOWN	
MICROSAT-R	2019-006A	43947	24-Jan-19	27-Mar-19	130	1	294	265	96.63	COLLISION, DELIBERATE	Indian ASAT test
USA 288 R/B	2018-079B	43652	17-Oct-18	6-Apr-19	181	180	35092	8526	12.2	UNKNOWN	ATLAS V CENTAUR (SEC)
LES 8,9/SOLRAD 11A,B R/B	1976-023F	8751	15-Mar-76	7-May-19	1	1	36889	35793	16.4	UNKNOWN	TITAN TRANSTAGE No. 30
TOPEX-POSEIDON R/B	1992-052D	22079	10-Aug-92	22-Jul-19	11	11	1404	1296	66.07	UNKNOWN	ARIANE 4 H10 FINAL STAGE
COSMOS 2464-2466 ULLAGE MOTOR	2010-041H	37144	2-Sep-10	12-Aug-19	24	24	18907	541	65	PROPULSION	PROTON-K BLOCK DM SOZ
COSMOS 2424-2426 ULLAGE MOTOR	2006-062H	29682	25-Dec-06	23-Oct-19	3	3	19189	294	64.5	PROPULSION	PROTON-K BLOCK DM SOZ
COSMOS 2456-2458 ULLAGE MOTOR	2009-070F	36116	14-Dec-09	21-23-Dec-19	7	7	18980	512	64.7	PROPULSION	PROTON-K BLOCK DM SOZ
COSMOS 2491	2013-076E	39497	25-Dec-13	23-Dec-19	27	27	1517	1485	82.5	UNKNOWN	1st "SATELLITE INSPECTOR"
COSMOS 2535	2019-039A	44421	10-Jul-19	9-Jan-20	18	5	645	607	97.9	UNKNOWN	SATELLITE INSPECTOR

44

TABLE 2.2 HISTORY OF SATELLITE BREAKUPS BY EVENT DATE (CONT'D)

SATELLITE NAME	INTERNATIONAL DESIGNATOR	US SATELLITE NUMBER	LAUNCH DATE	BREAKUP DATE	DEBRIS CATALOGED	DEBRIS LEFT	APOGEE (KM)	PERIGEE (KM)	INCLINATION (DEG)	ASSESSED CAUSE	ADDITIONAL INFORMATION
METEOR 3-5 R/B	1991-056B	21656	15-Aug-91	12-Feb-20	112	106	1206	1186	82.56	UNKNOWN	SL-14 THIRD STAGE
SL-23 DEBRIS	2011-037B	37756	18-Jul-11	8-May-20	334	282	3606	422	51.5	UNKNOWN	Fregat-SB SBB propellant tank
H-2A DEB	2018-084C	43673	29-Oct-18	12-Jul-20	88	3	643	595	97.9	UNKNOWN	Encapsulation System (PES) Payload
RESURS O1	1994-074A	23342	4-Nov-94	27-Aug-20	141	140	660	633	97.9	UNKNOWN	lateral cylinder
NOAA 17	2002-032A	27453	24-Jun-02	10-Mar-21	115	115	817	800	98.62	UNKNOWN	Resurs-O No. 3
YUNHAI 1-02	2019-063A	44547	25-Sep-19	18-Mar-21	38	32	785	780	98.54	COLLISION, ACCIDENTAL	HIT BY DEBRIS (48078)
COSMOS 2499	2014-028E	39765	23-May-14	23-Oct-21	23	23	1507	1152	82.44	UNKNOWN	2nd
COSMOS 1408	1982-092A	13552	16-Sep-82	15-Nov-21	1760	990	490	465	82.6	COLLISION, DELIBERATE	"SATELLITE INSPECTOR" Russian ASAT test
ORBCOMM FM 5	1997-084F	25117	23-Dec-97	18-Nov-21	15	11	771	758	45	UNKNOWN	
USA 305-308 MINOTAUR 4 R/B	2020-046E	45877	15-Jul-20	26-Nov-21	22	17	580	567	54	UNKNOWN	
COSMOS 2434-2436 ULLAGE MOTOR	2007-065F	32398	25-Dec-97	15-Apr-22	1	1	19068	400	64.8	PROPULSION	PROTON-K BLOCK DM SOZ

This Page Intentionally Left Blank

2.2 Identified Satellite Breakups

The remainder of this section devotes two pages to each identified satellite breakup. Each satellite is listed by common name, international designator, and satellite number. The satellite is then described in terms of type, ownership, launch date, and physical characteristics. The third grouping defines the breakup event by time, location, altitude, and assessed cause. In almost all cases, the calculated time of the event has been determined by the US SSN. Next, the last available element set for the satellite prior to the breakup is provided.

Contents of the pre- or post-event elements are described in Table 2.2-1. The epoch time's format consists of the last two digits of a year (YY) followed by a fractional day of year (DDD.DDDDDDD). Three propagation-scheme drag coefficients are available in a two-line element (TLE) set, which form the basis of the orbital element data presented in this subsection. Though not all TLEs possess data for all three, they are described here for completeness; these are denoted as drag coefficients peculiar to the US SSN SGP, SGP4, and SGP8 orbit propagators. The SGP-related data items $\dot{n}/2$ (pronounced "n dot over two") and $\ddot{n}/6$ (pronounced "n double dot over six") refer to the first and second order time derivatives of the mean motion n and represent phenomenological series expansion coefficient fits to the observed change in mean motion. The SGP4 propagator is the accepted standard for orbit propagation.

TABLE 2.2-1. TLE NUMERICAL DATA, AS INCORPORATED INTO THIS SECTION'S "PRE-/POST-EVENT ELEMENTS" FOR ALL FRAGMENTATION EVENTS.

DATA ITEM	FORMAT/UNITS
Epoch time	YYDDD.DDDDDDDD
$\dot{n}/2$ (SGP) <u>or</u> B (SGP8)	[rev/day^2] <u>or</u> [m^2/kg]
$\ddot{n}/6$ (SGP)	[rev/day^3]
B* (SGP4)	[1/Earth radii]
Eccentricity e	[-]
Inclination i	[°]
Right ascension of ascending node Ω	[°]
Argument of perigee ω	[°]
Mean anomaly M	[°]
Mean motion n	[rev/day]

If the breakup occurred soon after launch or after a maneuver and before an element set could be generated, the most appropriate post-event element set is given. The maximum observed changes in the orbital period (ΔP) and inclination (ΔI), referenced to the parent's pre-event element set, are then summarized. The reader is reminded that for a given event, the magnitudes of the resultant ΔP and ΔI are a function of the satellite's latitude and altitude. Comparisons of these values from one event to another cannot be made directly. Additionally, inclination changes measure only one portion of the fragmentation orbital plane change. Changes in Right Ascension also occur in most events and can account for some plane change fragmentation energy.

A general summary of the event, actions leading to the event, debris cataloging progress, and evaluations of the event are collected under the Comments heading. Documents that relate directly to the subject breakup or to breakups of satellites of this type are then listed. Gabbard diagrams of the early debris cloud prior to the effects of perturbations, if the data were available, are reconstructed. These diagrams often include uncataloged as well as cataloged debris data. When used correctly, Gabbard diagrams can provide important insights into the features of the fragmentation.

This Page Intentionally Left Blank

SATELLITE DATA

TYPE:	Ablestar Stage
OWNER:	US
LAUNCH DATE:	29.18 Jun 1961
DRY MASS (KG):	625
MAIN BODY:	Flared cylinder; 1.6 m diameter by 4.8 m length
MAJOR APPENDAGES:	None
ATTITUDE CONTROL:	None at time of the event
ENERGY SOURCES:	On-board propellants, range safety device

EVENT DATA

DATE:	29 Jun 1961	LOCATION:	28N, 254E (dsc)
TIME:	0608 GMT	ASSESSED CAUSE:	Propulsion
ALTITUDE:	990 km		

POST-EVENT ELEMENTS

EPOCH:	61187.36647288	MEAN ANOMALY:	72.1786
RIGHT ASCENSION:	79.1120	MEAN MOTION:	13.86864257
INCLINATION:	66.8199	MEAN MOTION DOT/2:	.0
ECCENTRICITY:	.0078181	MEAN MOTION DOT DOT/6:	.0
ARG. OF PERIGEE:	288.2398	BSTAR:	.0

DEBRIS CLOUD DATA

MAXIMUM ΔP:	15.5 min
MAXIMUM ΔI:	1.3 deg

COMMENTS

This is the first known satellite fragmentation. The Ablestar stage performed two main burns and a small payload separation retro burn to successfully deploy three payloads (Transit 4A, Injun, and Solrad 3), although the Injun and Solrad 3 satellites did not separate from one another as planned. The event occurred approximately 77 minutes after orbital insertion and was photographically imaged by the Organ Pass, NM, Baker-Nunn camera system. Fragmentation coincided with cessation of the 378 MHz beacon on the Ablestar stage at 06:08:10 GMT. At the time of the event, 100 kg of hypergolic propellants remained on board. This was the first time an Ablestar stage did not vent the fuel tank during payload separation. After a thorough investigation, fuel venting was recommended for future missions. No reliable elements are available prior to the event. Elements above are for one of the payloads with parameters believed to be very similar to those for the Ablestar at the time of the event.

REFERENCE DOCUMENTS

Transit 4-A Ablestar Vehicle Fragmentation Study (Preliminary), Report TOR-930 (2102)-6, Flight Test Planning and Evaluation Department, Transit Program Office, USAF Systems Command, Inglewood, 28 August 1961.

Description, Operation and Performance of Ablestar Stage AJ10-104S, S/N 008 (Transit 4-A), T.W. Fehr and J.K. Stark, Report No. 2102, Spacecraft Division, Aerojet-General Corporation, Azusa, October 1961.

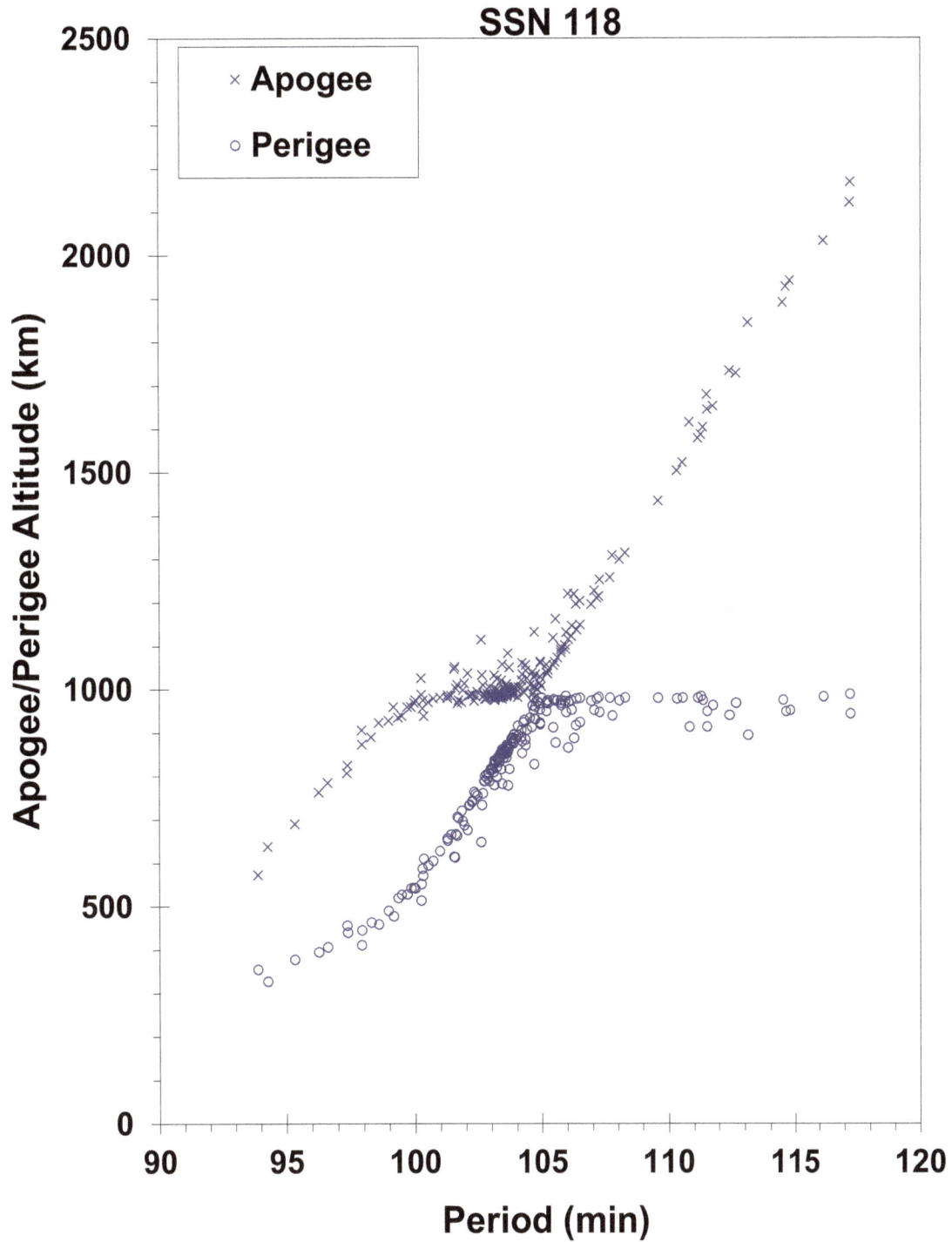

SSN 118

Transit 4A R/B debris cloud of 201 cataloged fragments in May 1964 as reconstructed from the US SSN database.

(1962-BETA IOTA 1)

SATELLITE DATA

TYPE:	Payload and R/B(s) (?)
OWNER:	CIS
LAUNCH DATE:	24.75 Oct 1962
DRY MASS (KG):	1500
MAIN BODY:	Cylinder; 2.6 m diameter by 7.15 m length
MAJOR APPENDAGES:	None
ATTITUDE CONTROL:	Unknown at time of event
ENERGY SOURCES:	On-board propellants

EVENT DATA

DATE:	29 Oct 1962	LOCATION:	Unknown
TIME:	Unknown	ASSESSED CAUSE:	Propulsion
ALTITUDE:	~200 km		

PRE-EVENT ELEMENTS

EPOCH:	62297.80327270	MEAN ANOMALY:	229.0409
RIGHT ASCENSION:	336.4972	MEAN MOTION:	16.15589719
INCLINATION:	65.1128	MEAN MOTION DOT/2:	.01124103
ECCENTRICITY:	.0044520	MEAN MOTION DOT DOT/6:	.0
ARG. OF PERIGEE:	92.2650	BSTAR:	.0

DEBRIS CLOUD DATA

MAXIMUM ΔP:	Unknown
MAXIMUM ΔI:	0.6 deg

COMMENTS

Sputnik 29 (also known as Sputnik 22) was not acknowledged at launch by the USSR and was probably a Mars probe that failed to leave Earth orbit. This was apparently the fourth orbital failure of the Molniya third stage since 25 August 1962. No Molniya orbital (3rd) stage nor final (4th) stage was cataloged after launch. Possible that orbital and final stages never separated. Sputnik 29 was officially decayed 29 October 1962, but no debris were cataloged before 11 November. Consequently, ΔP cannot be calculated. Source of the fragmentation was probably the fully fueled Molniya final stage.

REFERENCE DOCUMENT

History of Soviet/Russian Satellite Fragmentations-A Joint U.S.-Russian Investigation, N. L. Johnson, *et al.,* Kaman Sciences Corporation, October 1995.

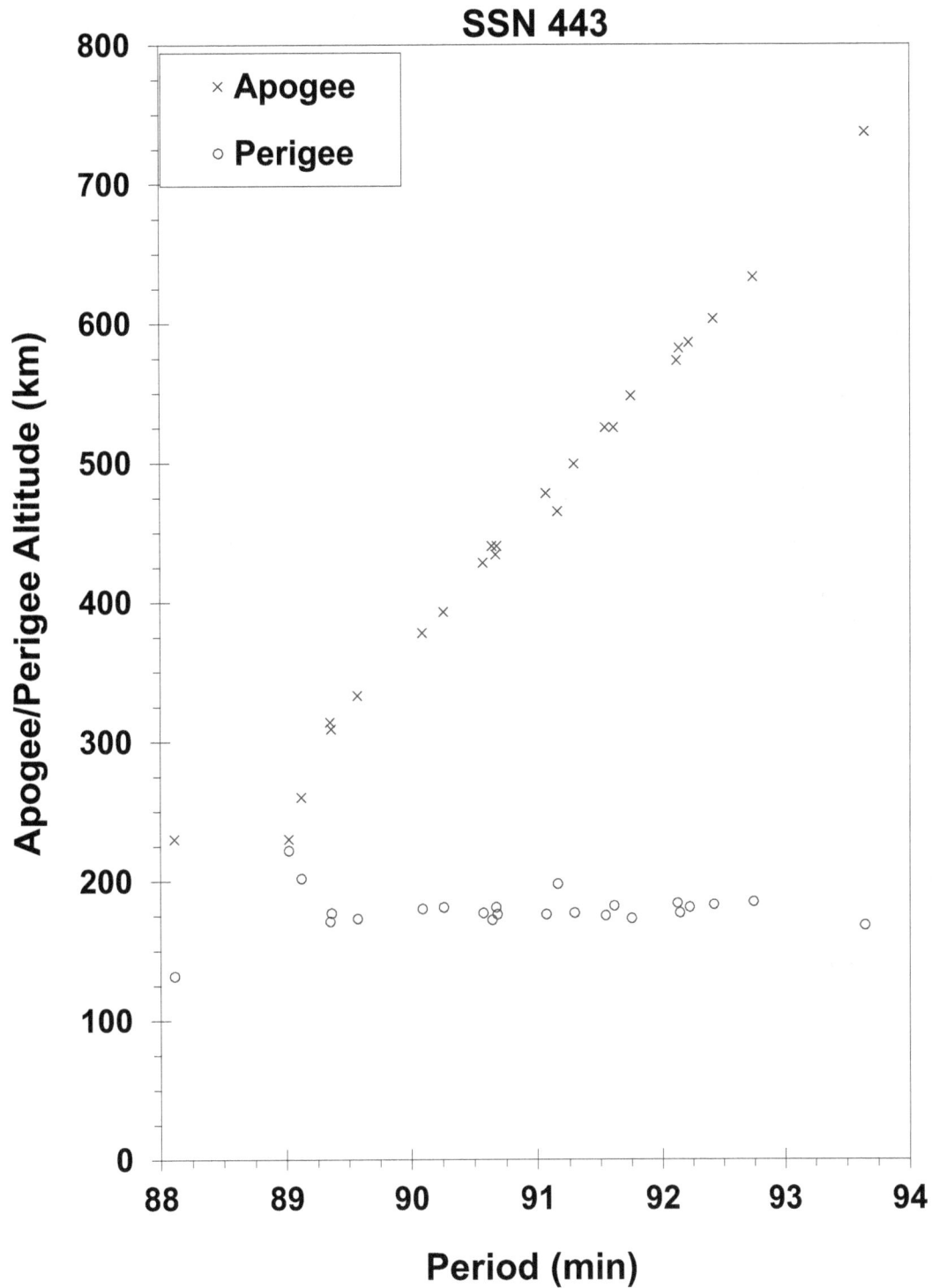

SSN 443

Sputnik 29 debris cloud of 24 fragments cataloged by mid-December 1962 as reconstructed from the US SSN database.

53

SATELLITE DATA

TYPE:	Centaur Stage
OWNER:	US
LAUNCH DATE:	27.79 Nov 1963
DRY MASS (KG):	4600
MAIN BODY:	Cylinder; 3 m diameter by 9 m length
MAJOR APPENDAGES:	None
ATTITUDE CONTROL:	Unknown at time of the event
ENERGY SOURCES:	Unknown

EVENT DATA

DATE:	27 Nov 1963	LOCATION:	Unknown
TIME:	Unknown	ASSESSED CAUSE:	Propulsion
ALTITUDE:	Unknown		

POST-EVENT ELEMENTS

EPOCH:	63336.85832214	MEAN ANOMALY:	213.1623
RIGHT ASCENSION:	135.1828	MEAN MOTION:	13.34437775
INCLINATION:	30.3440	MEAN MOTION DOT/2:	.00003262
ECCENTRICITY:	.0869282	MEAN MOTION DOT DOT/6:	.0
ARG. OF PERIGEE:	151.8246	BSTAR:	.0

DEBRIS CLOUD DATA

MAXIMUM ΔP:	0.9 min
MAXIMUM ΔI:	0.4 deg

COMMENTS

First Centaur stage to reach Earth orbit. No payload was carried. After orbital insertion, residual liquid hydrogen vaporized, resulting in an increase in tank pressurization. Venting via an aft tube then induced a pin-wheel tumble that reached 48 rpm a little more than 1 hour after launch. At the beginning of the third orbit insulation blankets around the Centaur stage were thrown off. Subsequent Centaur missions were not subject to this phenomenon that was caused by the unique configuration of Atlas Centaur 2. First six fragments were cataloged within 1 week of launch. Centaur stage retains large radar cross-section, while all debris are substantially smaller.

REFERENCE DOCUMENT

Supplementary Information on AC-2 Post-Injection Flight Events, W.S. Hicks, Memorandum BXN63-521, 27 December 1963.

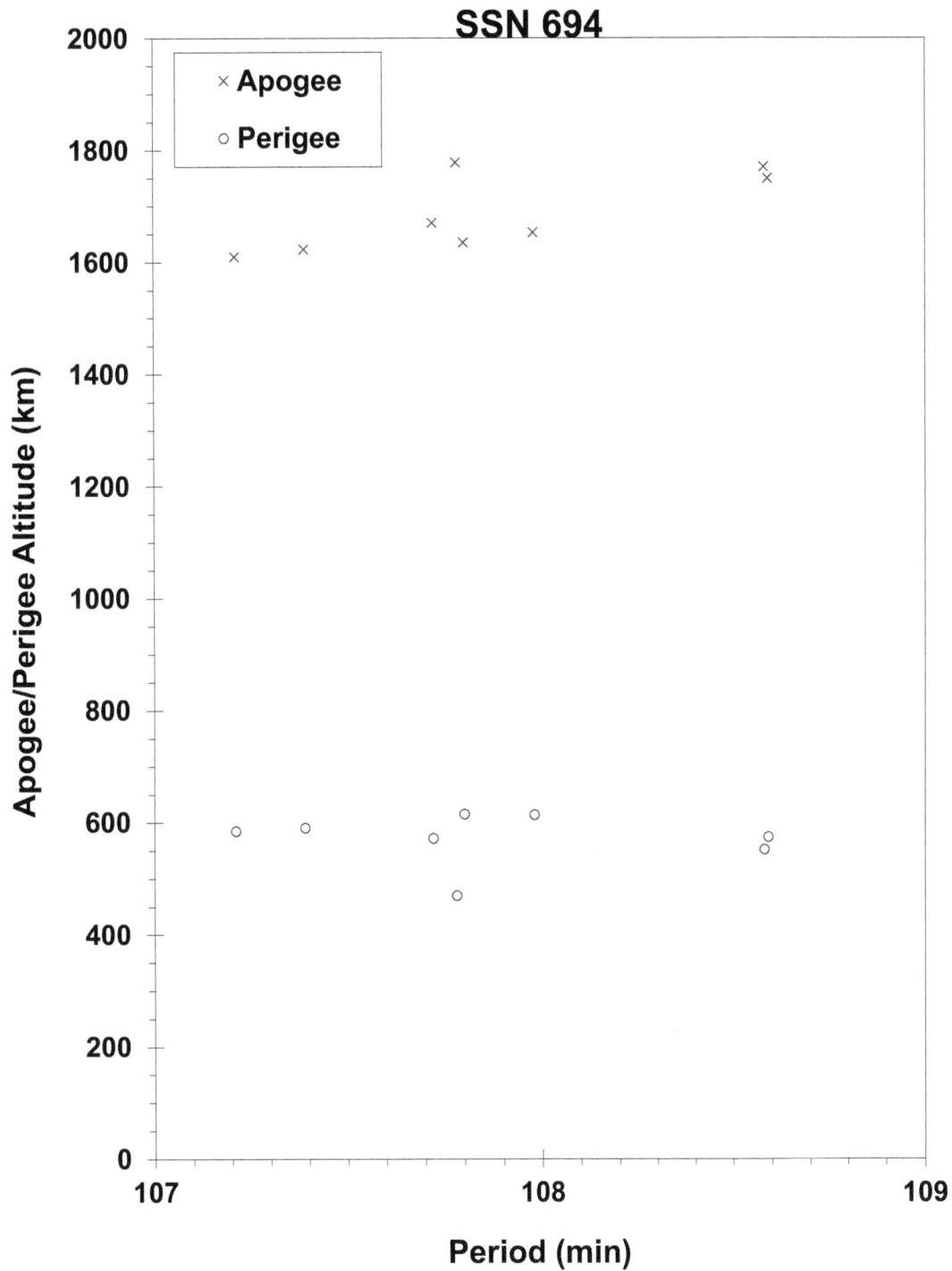

Atlas Centaur 2 debris cloud of 8 fragments 5 months after the event as reconstructed from the US SSN database.

SATELLITE DATA

TYPE:	Payload
OWNER:	CIS
LAUNCH DATE:	28.45 Oct 1964
DRY MASS (KG):	4750
MAIN BODY:	Sphere-cylinder; 2.4 m diameter by 4.3 m length
MAJOR APPENDAGES:	None
ATTITUDE CONTROL:	Active, 3-axis
ENERGY SOURCES:	On-board propellants, 10 kg TNT explosive charge

EVENT DATA

DATE:	5 Nov 1964	LOCATION:	Unknown
TIME:	Unknown	ASSESSED CAUSE:	Deliberate
ALTITUDE:	~200 km		

PRE-EVENT ELEMENTS

EPOCH:	64303.72916435	MEAN ANOMALY:	46.7488
RIGHT ASCENSION:	198.5952	MEAN MOTION:	16.23335350
INCLINATION:	51.2318	MEAN MOTION DOT/2:	.00269057
ECCENTRICITY:	.0034483	MEAN MOTION DOT DOT/6:	.0
ARG. OF PERIGEE:	312.9624	BSTAR:	.0

DEBRIS CLOUD DATA

MAXIMUM ΔP:	Unknown
MAXIMUM ΔI:	Unknown

COMMENTS

Spacecraft was destroyed after a malfunction prevented reentry and landing in the Soviet Union. Event occurred on the anticipated day of recovery. All debris were cataloged without elements. A probable fragment from this event reentered on 12 November 1964, landing in Malawi. See cited reference below.

REFERENCE DOCUMENTS

The Examination of a Sample of Space Debris, P.H.H. Bishop and K.F. Rogers, Technical Report 65165, Royal Aircraft Establishment, Farnborough Hants, August 1965.

History of Soviet/Russian Satellite Fragmentations-A Joint U.S.-Russian Investigation, N. L. Johnson, et al., Kaman Sciences Corporation, October 1995.

Insufficient data to construct a Gabbard diagram.

SATELLITE DATA

TYPE:	Payload
OWNER:	CIS
LAUNCH DATE:	22.32 Feb 1965
DRY MASS (KG):	5500
MAIN BODY:	Sphere-cylinder; 2.4 m diameter by 6 m length
MAJOR APPENDAGES:	None
ATTITUDE CONTROL:	Active, 3-axis
ENERGY SOURCES:	On-board propellants, 10 kg explosive charge

EVENT DATA

DATE:	22 Feb 1965	LOCATION:	64N, 80E (asc)
TIME:	0957 GMT	ASSESSED CAUSE:	Deliberate
ALTITUDE:	380 km		

POST-EVENT ELEMENTS

EPOCH:	65056.64509999	MEAN ANOMALY:	293.2095
RIGHT ASCENSION:	288.1532	MEAN MOTION:	15.92461677
INCLINATION:	64.7411	MEAN MOTION DOT/2:	.01501524
ECCENTRICITY:	.0182240	MEAN MOTION DOT DOT/6:	.0048063
ARG. OF PERIGEE:	68.7266	BSTAR:	.0

DEBRIS CLOUD DATA

MAXIMUM ΔP:	4.4 min
MAXIMUM ΔI:	0.9 deg

COMMENTS

Cosmos 57 was an uncrewed precursor for the crewed Voskhod 2 mission that took place in March 1965. Spacecraft fragmented a little more than 2 hours after launch when operational ground instructions were misinterpreted by the on-board command system and the self-destruct system was activated. No elements available for Cosmos 57, but the rocket body elements are provided above. The Royal Aircraft Establishment published the following parameters for Cosmos 57 for 22.4 February: 165 km by 427 km, 64.74-degree inclination, 64-degree argument of perigee. A total of 35 debris were cataloged without elements. Event may have occurred a little later than the time calculated above.

REFERENCE DOCUMENTS

The 1093 Breakup, D.J. Watson, BMEWS-ADC Systems Engineering Memorandum BSM-1000-16, 16 June 1965.

"To Save Man: A Conversation with the General Designer of Life-Support and Rescue Systems, Hero of Socialist Labor G.I. Severin", Pravda, Moscow, 26 June 1989, p. 4.

"Pages From a Diary: He Soared Freely Above the Earth", Sovetskaya Rossiya, Moscow, 17 March 1990, p. 6.

History of Soviet/Russian Satellite Fragmentations-A Joint U.S.-Russian Investigation, N. L. Johnson, et al., Kaman Sciences Corporation, October 1995.

"The Kamanin Diaries 1964-1966", B. Hendrickx, Journal of the Interplanetary Society, Vol. 51, 1998, pp. 421-422.

"Fiftieth Anniversary of First Intentional Fragmentation Event in Space", The Orbital Debris Quarterly News, NASA JSC, January 2015. Available online at https://www.orbitaldebris.jsc.nasa.gov/quarterly-news/pdfs/odqnv19i1.pdf.

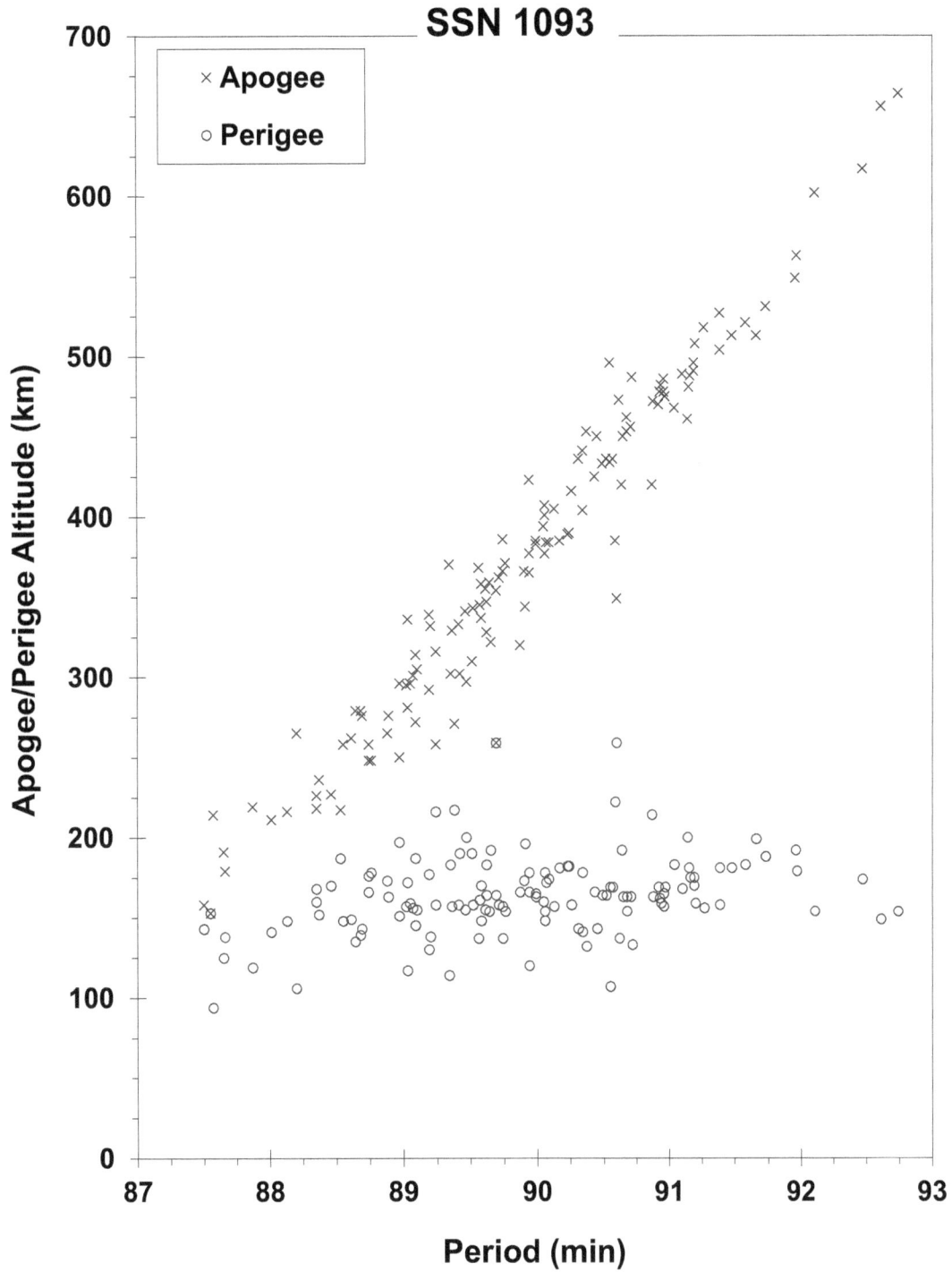

SSN 1093

Cosmos 57 debris cloud of 133 fragments cataloged within 1 month of the event as reconstructed from the US SSN database.

SATELLITE DATA

TYPE:	Cosmos Second Stage
OWNER:	CIS
LAUNCH DATE:	15.46 Mar 1965
DRY MASS (KG):	1600
MAIN BODY:	Cylinder; 2.4 m diameter by 5 m length
MAJOR APPENDAGES:	None
ATTITUDE CONTROL:	None at time of the event.
ENERGY SOURCES:	Unknown

EVENT DATA

DATE:	15 Mar 1965	LOCATION:	51S, 162E (dsc)
TIME:	1714 GMT	ASSESSED CAUSE:	Unknown
ALTITUDE:	1640 km		

POST-EVENT ELEMENTS

EPOCH:	65074.89183830	MEAN ANOMALY:	265.7165
RIGHT ASCENSION:	357.3218	MEAN MOTION:	13.57884745
INCLINATION:	56.0538	MEAN MOTION DOT/2:	.00231832
ECCENTRICITY:	.1056119	MEAN MOTION DOT DOT/6:	.0
ARG. OF PERIGEE:	106.1560	BSTAR:	.0

DEBRIS CLOUD DATA

MAXIMUM ΔP:	10.3 min
MAXIMUM ΔI:	0.4 deg

COMMENTS

This is the first confirmed case of the fragmentation of the Cosmos 3 (SL-8 or C-1) second stage. This was the third mission to deploy three payloads and was a repeat of the Cosmos 54-56 mission 3 weeks earlier. The event occurred a little more than 6 hours after the successful deployment of the three payloads. Elements above are the first developed for the rocket body and are about 4 hours after the event. Official debris cataloging did not begin for 6 weeks.

REFERENCE DOCUMENTS

"Fragmentations of Asteroids and Artificial Satellites in Orbit", W. Wiesel, Icarus, Vol. 34, 1978, pp. 99-116.

History of Soviet/Russian Satellite Fragmentations-A Joint U.S.-Russian Investigation, N. L. Johnson, *et al.,* Kaman Sciences Corporation, October 1995.

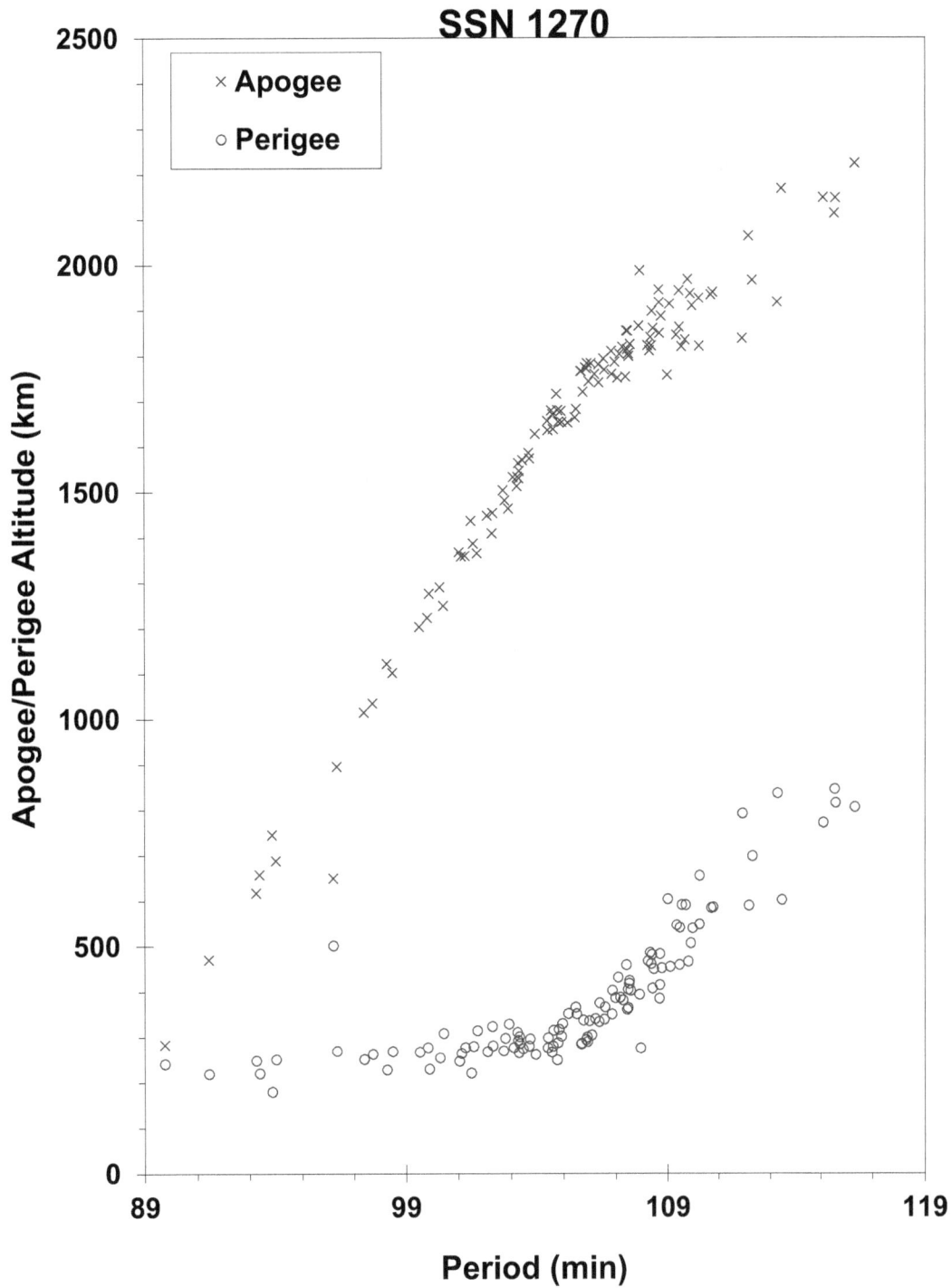

SSN 1270

Legend:
× Apogee
○ Perigee

Y-axis: Apogee/Perigee Altitude (km), ranging 0 to 2500
X-axis: Period (min), ranging 89 to 119

Cosmos 61-63 R/B debris cloud of 113 fragments 8 months after the event as reconstructed from the US SSN database.

SATELLITE DATA

TYPE:	Payload
OWNER:	CIS
LAUNCH DATE:	4.23 Nov 1965
DRY MASS (KG):	400
MAIN BODY:	Ellipsoid; 1.2 m diameter by 1.8 m length
MAJOR APPENDAGES:	Unknown
ATTITUDE CONTROL:	Unknown
ENERGY SOURCES:	Unknown

EVENT DATA

DATE:	15 Jan 1966	LOCATION:	Unknown
TIME:	Unknown	ASSESSED CAUSE:	Unknown
ALTITUDE:	Unknown		

PRE-EVENT ELEMENTS

EPOCH:	66009.5	MEAN ANOMALY:	Unknown
RIGHT ASCENSION:	Unknown	MEAN MOTION:	16.09757275
INCLINATION:	48.39	MEAN MOTION DOT/2:	Unknown
ECCENTRICITY:	0.009282	MEAN MOTION DOT DOT/6:	Unknown
ARG. OF PERIGEE:	77	BSTAR:	Unknown

DEBRIS CLOUD DATA

MAXIMUM ΔP:	Unknown
MAXIMUM ΔI:	Unknown

COMMENTS

Pre-event elements were taken from RAE Table of Earth Satellites. Cosmos 95 was placed into a low Earth orbit on 4 November 1965. Within 2 weeks nearly two dozen debris had been detected and were later cataloged. However, the nature of the debris, *i.e.* breakup versus operational, was not determined. The last of these debris decayed naturally by 6 January 1966. Russian records indicate that a breakup may have occurred on 15 January 1966, 3 days before the 400 kg spacecraft itself reentered. No other information on this event has been discovered, and no debris remains in orbit.

REFERENCE DOCUMENT

History of Soviet/Russian Satellite Fragmentations-A Joint U.S.-Russian Investigation, N. L. Johnson, *et al.,* Kaman Sciences Corporation, October 1995.

Insufficient data to construct a Gabbard diagram.

SATELLITE DATA

TYPE:	Titan 3C-4 Transtage
OWNER:	US
LAUNCH DATE:	15.72 Oct 1965
DRY MASS (KG):	2500
MAIN BODY:	Cylinder; 3 m diameter by 6 m length
MAJOR APPENDAGES:	None
ATTITUDE CONTROL:	Active, 3-axis
ENERGY SOURCES:	On-board propellants

EVENT DATA

DATE:	15 Oct 1965	LOCATION:	22S, 108E (asc)
TIME:	1820 GMT	ASSESSED CAUSE:	Propulsion
ALTITUDE:	740 km		

POST-EVENT ELEMENTS

EPOCH:	65361.23126396	MEAN ANOMALY:	237.1066
RIGHT ASCENSION:	21.5316	MEAN MOTION:	14.54928550
INCLINATION:	32.1697	MEAN MOTION DOT/2:	.00000268
ECCENTRICITY:	.0072678	MEAN MOTION DOT DOT/6:	.071801
ARG. OF PERIGEE:	123.6068	BSTAR:	.0

DEBRIS CLOUD DATA

MAXIMUM ΔP:	4.1 min
MAXIMUM ΔI:	1.4 deg

COMMENTS

This was the second test of the Titan 3C-4 Transtage with AJ10-138 engine using hypergolic propellants. Event occurred one-half revolution after launch following second ignition that may have been accompanied with vehicle tumbling. LCS 2 payload was to have been deployed at 735 km circular while OV2-1 was to have been released later in an orbit of 735 km by about 7400 km. Transtage also malfunctioned on next mission in December 1965. The main remnant of the rocket body was recently identified as Satellite No. 1822 (1965-082DM). Previous editions of this book had identified the main rocket body remnant as Satellite No. 1640 (1965-082B).

REFERENCE DOCUMENT

TRW Space Log, Winter 1965-66, Vol. 5, No. 4, T.L. Branigan, ed., TRW Systems, Redondo Beach, 1966, pp. 15-17.

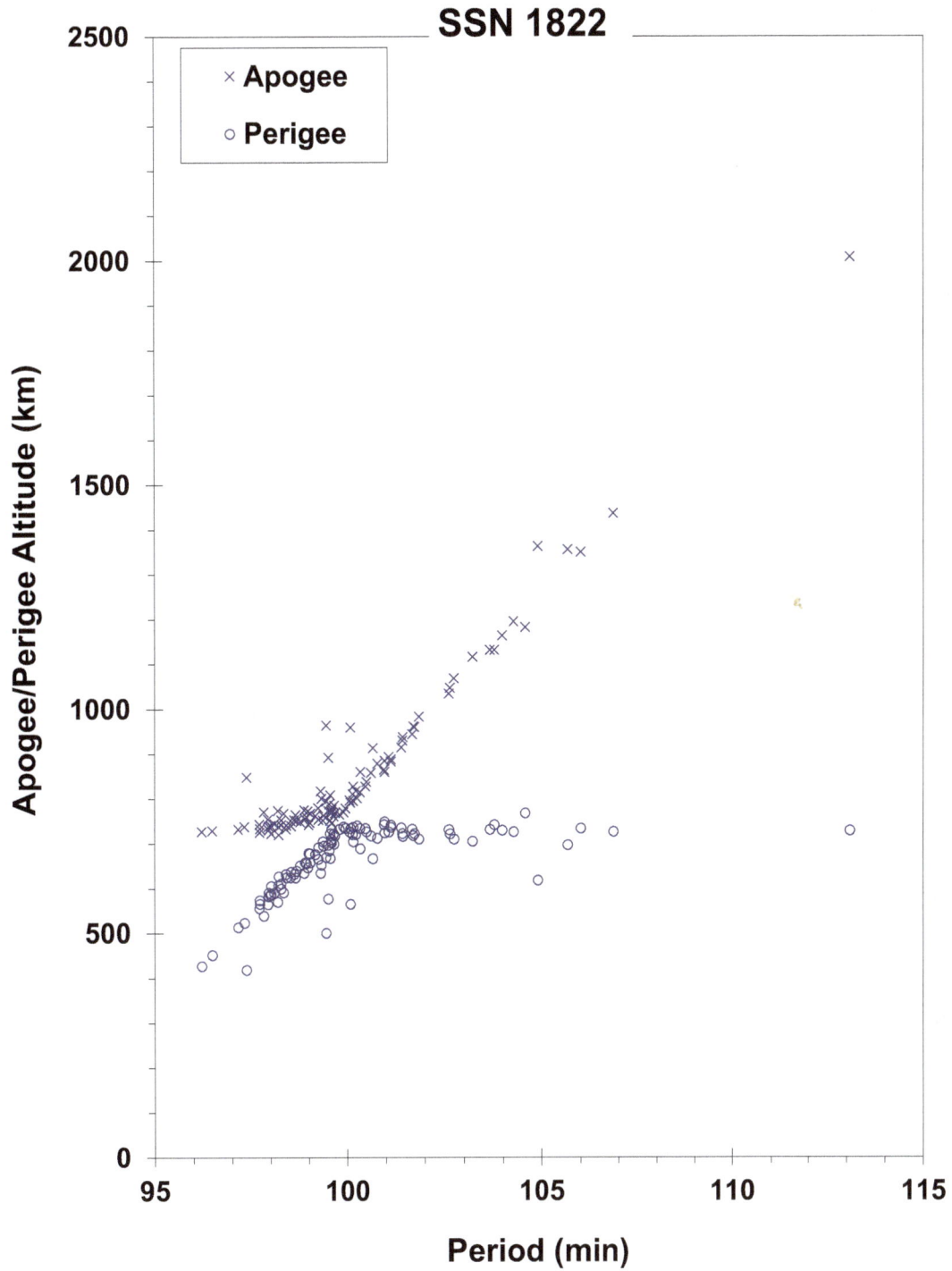

SSN 1822

Apogee/Perigee Altitude (km)

Period (min)

× Apogee
○ Perigee

OV2-1/LCS 2 R/B debris cloud of 103 cataloged fragments 6 weeks after the event as reconstructed from the US SSN database.

SATELLITE DATA

TYPE:	Titan IIIC Transtage Rocket Body
OWNER:	US
LAUNCH DATE:	21 Dec 1965
DRY MASS (KG):	2,555
MAIN BODY:	3.048 m diameter x ~5 m long right circular cylinder
MAJOR APPENDAGES:	none
ATTITUDE CONTROL:	3-axis
ENERGY SOURCES:	on-board propellants; pressure vessels; chemical batteries

EVENT DATA

DATE:	21 Dec 1965	LOCATION:	unknown
TIME:	unknown	ASSESSED CAUSE:	Propulsion
ALTITUDE:	unknown		

POST-EVENT ELEMENTS

EPOCH:	69013.84962483	MEAN ANOMALY:	2.6587
RIGHT ASCENSION:	206.0825	MEAN MOTION:	4.07936348
INCLINATION:	26.4899	MEAN MOTION DOT/2:	0.00014521
ECCENTRICITY:	0.6007336	MEAN MOTION DOT DOT/6:	0.00000
ARG. OF PERIGEE:	346.5872	BSTAR:	0.0097405

DEBRIS CLOUD DATA

MAXIMUM ΔP:	unknown
MAXIMUM ΔI:	unknown

COMMENTS

The cause of this event is assessed to be propulsion related. The 13th ed. of the History of On-Orbit Satellite Fragmentations described this event as a provisional fragmentation.

REFERENCE DOCUMENT

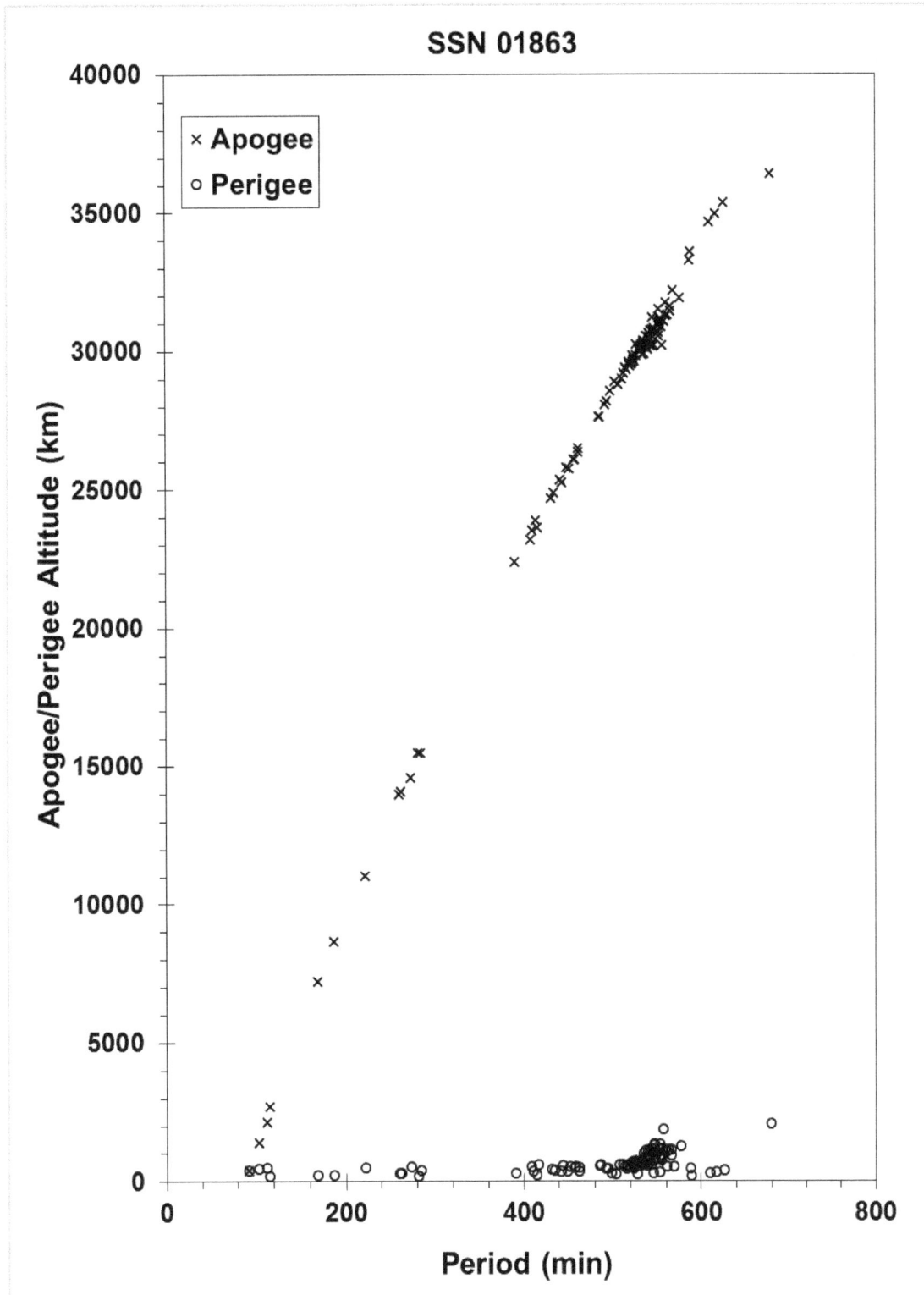

SSN 01863

Titan Transtage R/B cloud; some data cataloged up to 49 years after event.

SATELLITE DATA

TYPE:	Payload
OWNER:	US
LAUNCH DATE:	15.85 Feb 1966
DRY MASS (KG):	4
MAIN BODY:	Sphere; 0.3 m diameter
MAJOR APPENDAGES:	None
ATTITUDE CONTROL:	None
ENERGY SOURCES:	Unknown

EVENT DATA

DATE:	15 Feb 1966	LOCATION:	Unknown
TIME:	Unknown	ASSESSED CAUSE:	Unknown
ALTITUDE:	~200 km		

POST-EVENT ELEMENTS

EPOCH:	66047.01671304	MEAN ANOMALY:	234.6777
RIGHT ASCENSION:	148.6481	MEAN MOTION:	16.20030654
INCLINATION:	96.5380	MEAN MOTION DOT/2:	.01298049
ECCENTRICITY:	.0108362	MEAN MOTION DOT DOT/6:	.0053719
ARG. OF PERIGEE:	126.3670	BSTAR:	.0

DEBRIS CLOUD DATA

MAXIMUM ΔP:	Unknown
MAXIMUM ΔI:	0.6 deg

COMMENTS

OPS 3031 was an inflated sphere also known as Bluebell 2. It was deployed from satellite 2012, which was an Agena D stage carrying a separate payload. Elements above are for satellite 2012. Debris cataloging began 19 February after many debris had already decayed. Consequently, ΔP cannot be calculated. OPS 3031 and all debris decayed within 1 week of launch.

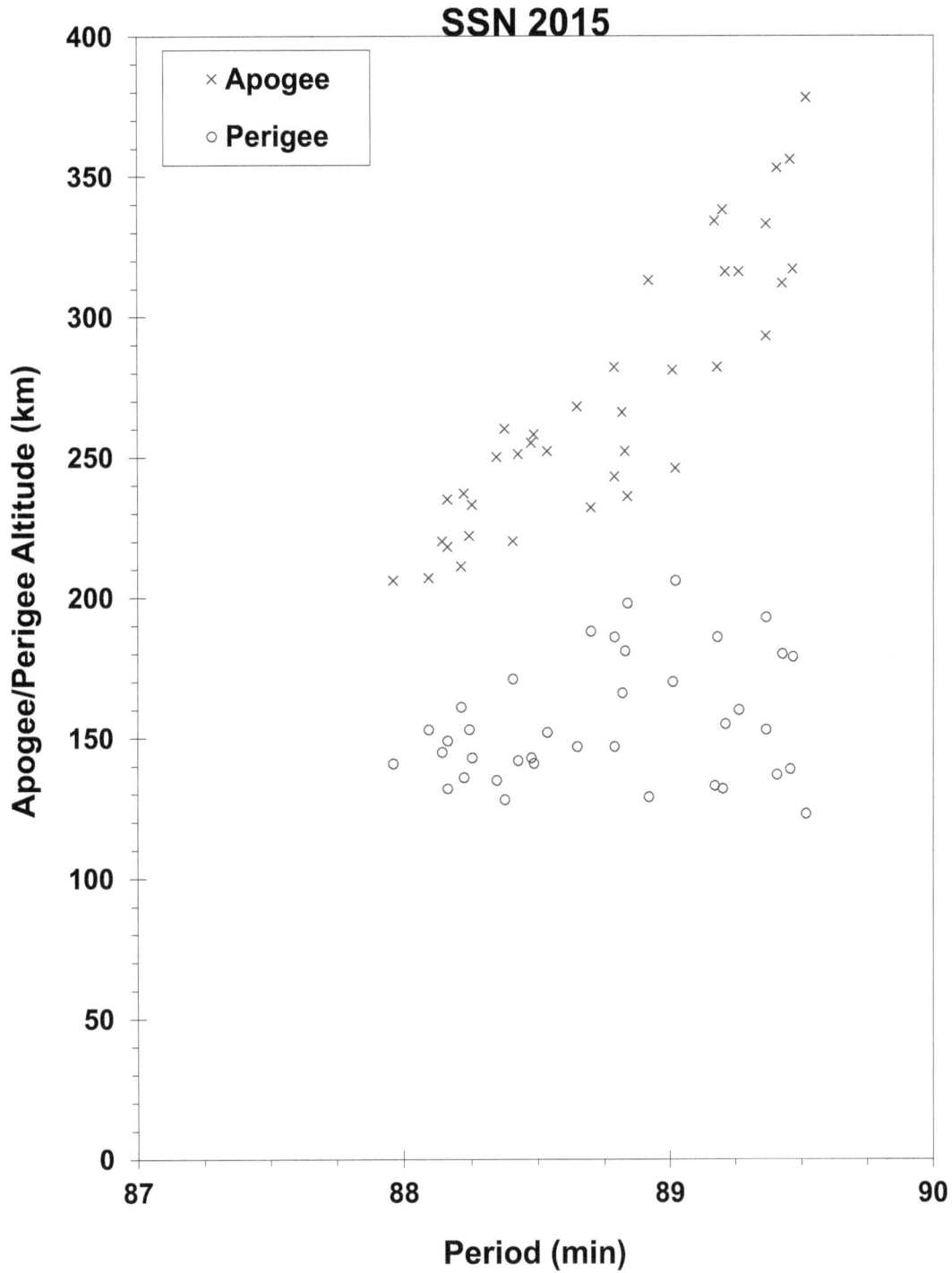

SSN 2015

OPS 3031 debris cloud of 38 fragments as initially cataloged
by the US SSN during February 1966.

SATELLITE DATA

TYPE:	Atlas Core Stage
OWNER:	US
LAUNCH DATE:	1.63 Jun 1966
DRY MASS (KG):	3400
MAIN BODY:	Cylinder; 3 m diameter by 20 m length
MAJOR APPENDAGES:	None
ATTITUDE CONTROL:	None at time of the event.
ENERGY SOURCES:	Unknown

EVENT DATA

DATE:	Mid-Jun 1966	LOCATION:	Unknown
TIME:	Unknown	ASSESSED CAUSE:	Unknown
ALTITUDE	~250 km		

PRE-EVENT ELEMENTS

EPOCH:	66164.96883397	MEAN ANOMALY:	224.9775
RIGHT ASCENSION:	223.9064	MEAN MOTION:	16.05545399
INCLINATION:	28.7968	MEAN MOTION DOT/2:	.00654808
ECCENTRICITY:	.0025152	MEAN MOTION DOT DOT/6:	.0010778
ARG. OF PERIGEE:	135.2510	BSTAR:	.0

DEBRIS CLOUD DATA

MAXIMUM ΔP:	5.5 min
MAXIMUM ΔI:	1.5 deg

COMMENTS

This stage successfully deployed the Augmented Target Docking Adapter (ATDA) for the Gemini 9 mission. The elements above are the last available for the rocket body. Debris cataloging began on 21 June. Debris decay dates ranged from 21 June to 4 July with the rocket body officially decaying on 22 June. A review of NASA archives for this mission revealed no documented anomaly with the Atlas booster. Discussions in 1989 with General Dynamics personnel involved in the mission (Mr. Phil Genser of General Dynamics, San Diego) also failed to uncover any knowledge of the event. Pressure relief valves should have relieved pressurization increases, particularly in the oxygen tank. Possible failure of the oxygen relief valve could not be ruled out.

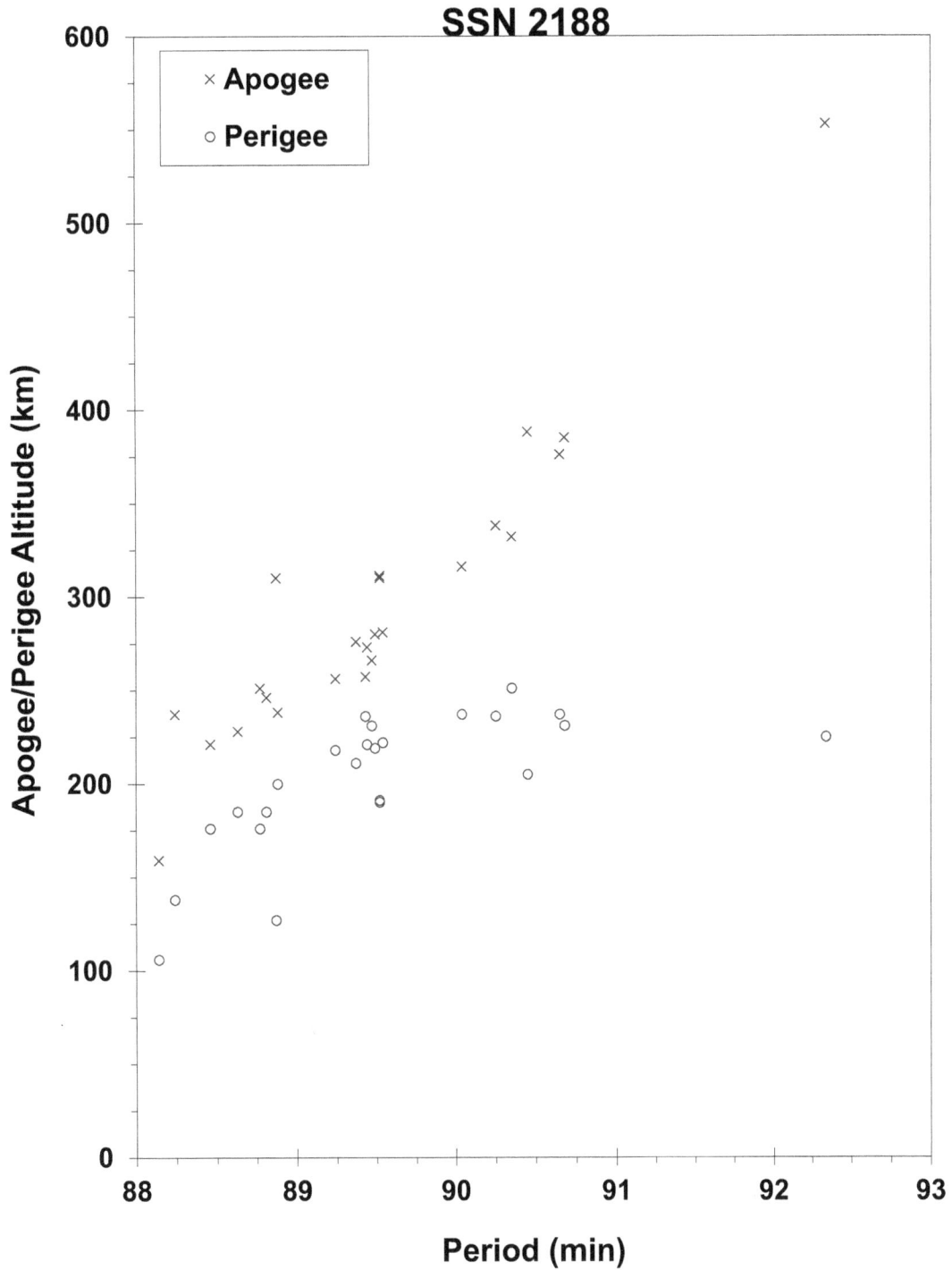

Gemini 9 ATDA R/B debris cloud of 24 fragments cataloged between 21 and 24 June as reconstructed from the US SSN database.

SATELLITE DATA

TYPE:	Payload
OWNER:	US
LAUNCH DATE:	24.01 Jun 1966
DRY MASS (KG):	55
MAIN BODY:	Sphere; 30 m diameter
MAJOR APPENDAGES:	None
ATTITUDE CONTROL:	None
ENERGY SOURCES:	None

EVENT DATA (1)

DATE:	12 Jul 1975	LOCATION:	67N, 135E (dsc)
TIME:	2248 GMT	ASSESSED CAUSE:	Unknown
ALTITUDE:	5145 km		

PRE-EVENT ELEMENTS (1)

EPOCH:	75192.78059719	MEAN ANOMALY:	67.9594
RIGHT ASCENSION:	238.7429	MEAN MOTION:	7.99684492
INCLINATION:	85.2811	MEAN MOTION DOT/2:	.00001217
ECCENTRICITY:	.0931904	MEAN MOTION DOT DOT/6:	.0
ARG. OF PERIGEE:	281.8264	BSTAR:	.77087

EVENT DATA (2)

DATE:	20 Jan 1976	LOCATION:	Unknown
TIME:	Unknown	ASSESSED CAUSE:	Unknown
ALTITUDE:	Unknown		

PRE-EVENT ELEMENTS (2)

EPOCH:	76019.86486339	MEAN ANOMALY:	305.5539
RIGHT ASCENSION:	209.8639	MEAN MOTION:	8.00368182
INCLINATION:	85.0720	MEAN MOTION DOT/2:	.0
ECCENTRICITY:	.1179567	MEAN MOTION DOT DOT/6:	.0
ARG. OF PERIGEE:	66.4633	BSTAR:	.0

DEBRIS CLOUD DATA

MAXIMUM ΔP:	0.1 min*
MAXIMUM ΔI:	0.7 deg*

*Based on 1st event data

COMMENTS

PAGEOS (Passive Geodetic Earth-Orbiting Satellite) was an inflated balloon made of thin Mylar with an aluminum coating. The first fragmentation event occurred 9 years after launch and resulted in 11 new cataloged objects. The second event was detected by D.G. King-Hele of the RAE, and NAVSPASUR confirmed 44 additional fragments. By August 1976, no additional debris had been cataloged but 19 objects were being tracked in orbits with mean motions near 8 and eccentricities between 0.16 and 0.34. Due to the character of PAGEOS and its subsequent debris, natural perturbations had little effect on orbital period but strongly increased eccentricity by simultaneously lowering perigee and raising apogee. About 10 September 1976 one of the 19 unofficial objects is believed to have broken up into perhaps more than 250 new pieces, none of which were cataloged prior to reentry. Eighteen objects were later cataloged during 7-8 October 1976. On the first anniversary of the second fragmentation (20 Jan 1977), 45 fragments were cataloged without elements and immediately decayed administratively. Additional fragmentations are suspected to have taken place in June 1978, September 1984, and December 1985. Historically, radar tracking of PAGEOS debris has been extremely difficult and cross-tagging frequent. Cause for the second and subsequent events may be material deterioration under environmental stress. A suspected PAGEOS fragment, SSN 5994, which was cataloged as a Westford Needles object, fragmented on 8 September 1995 and again on 14 September 1995 with 12 associated objects.

REFERENCE DOCUMENT

Spacetrack System Data Related to Some Non-Routine Events Through May 1981, J.R. Gabbard, Technical Memorandum 81-6, DCS/Plans, Hdqtrs NORAD/ADCOM, Colorado Springs, 30 June 1981.

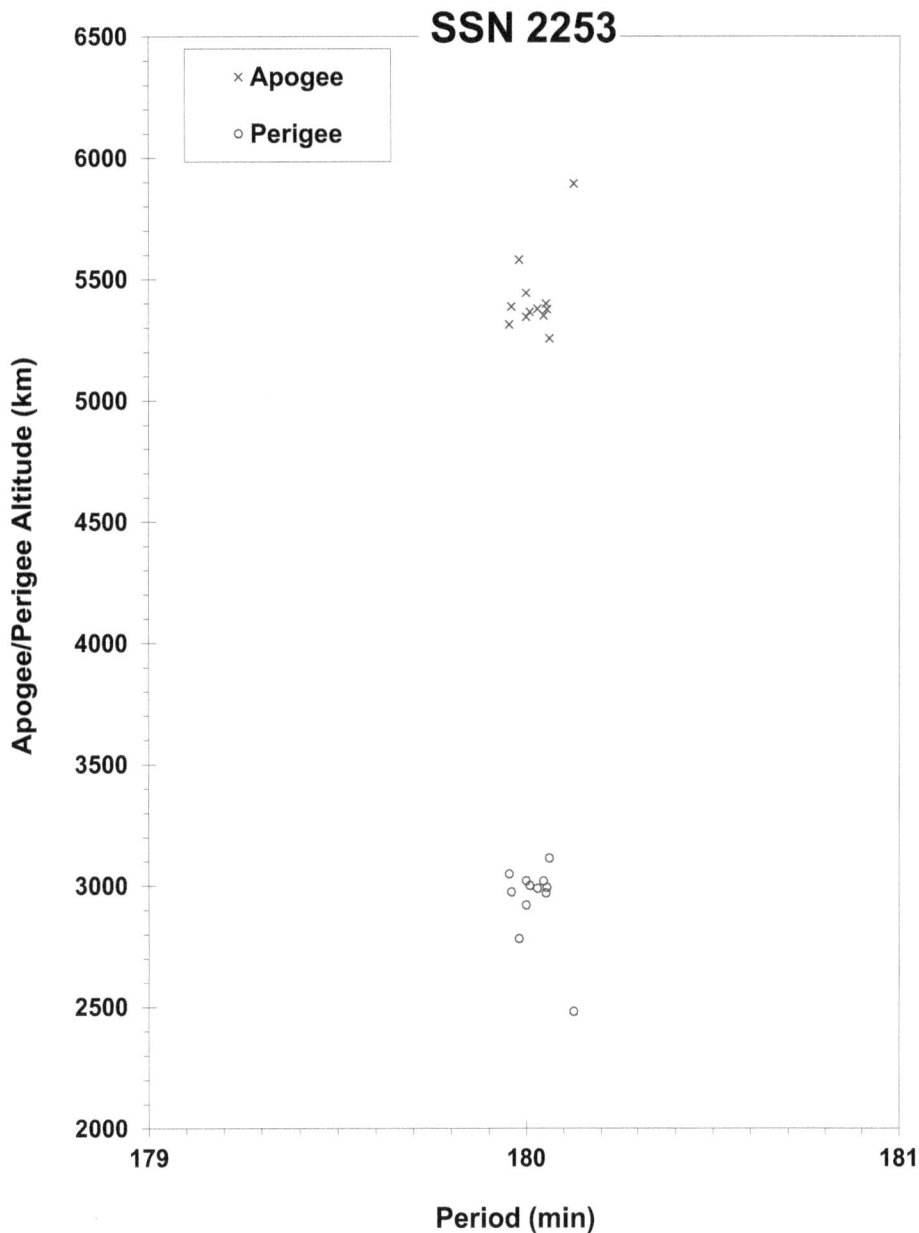

PAGEOS debris cloud of 12 fragments 5 weeks after the first event as reconstructed from the US SSN database.

SATELLITE DATA

TYPE:	Saturn SIVB Stage
OWNER:	US
LAUNCH DATE:	5.62 Jul 1966
DRY MASS (KG):	26,600
MAIN BODY:	Cylinder; 6.6 m diameter by 28.3 m length
MAJOR APPENDAGES:	None
ATTITUDE CONTROL:	Active, 3-axis
ENERGY SOURCES:	Attitude control and pressurization systems

EVENT DATA

DATE:	5 Jul 1966	LOCATION:	20N, 277E (dsc)
TIME:	2111 GMT	ASSESSED CAUSE:	Deliberate
ALTITUDE:	205 km		

PRE-EVENT ELEMENTS

EPOCH:	66186.73481847	MEAN ANOMALY:	353.9219
RIGHT ASCENSION:	5.5870	MEAN MOTION:	16.27379993
INCLINATION:	31.9810	MEAN MOTION DOT/2:	.03796193
ECCENTRICITY:	.0022272	MEAN MOTION DOT DOT/6:	.17429
ARG. OF PERIGEE:	6.1632	BSTAR:	.0

DEBRIS CLOUD DATA

MAXIMUM ΔP:	3.5 min
MAXIMUM ΔI:	1.4 deg

COMMENTS

This was the second flight of the SIVB stage. After orbital insertion, the vehicle was intentionally subjected to dynamic integrity tests, including high gravity loadings during attitude control maneuvers and high-pressure tests. The vehicle finally broke up after exceeding structural design limits with a propellant tank bulkhead differential pressure in excess of 23.7 N/cm^2. The fragmentation occurred early on the fifth revolution. Elements for the first fragments were not cataloged until 8 July.

REFERENCE DOCUMENT

Saturn AS-203 Evaluation Bulletin, No. 2, R-AERO-F-142-66, J.P. Lindberg, NASA Marshall Space Flight Center, Alabama, 21 July 1966.

AS-203 debris cloud of 25 fragments using orbits developed within 1 week of the event as reconstructed from the US SSN database.

SATELLITE DATA

TYPE:	Unknown
OWNER:	CIS
LAUNCH DATE:	17.94 Sep 1966
DRY MASS (KG):	Unknown
MAIN BODY:	Cone-cylinder; 1.5 m diameter by 6 m length
MAJOR APPENDAGES:	None
ATTITUDE CONTROL:	Unknown
ENERGY SOURCES:	Explosive device

EVENT DATA

DATE:	17 Sep 1966	LOCATION:	Unknown
TIME:	Unknown	ASSESSED CAUSE:	Deliberate
ALTITUDE:	~300 km		

POST-EVENT ELEMENTS

EPOCH:	66261.0	MEAN ANOMALY:	283
RIGHT ASCENSION:	338	MEAN MOTION:	14.879
INCLINATION:	49.63	MEAN MOTION DOT/2:	.0
ECCENTRICITY:	.063	MEAN MOTION DOT DOT/6:	.0
ARG. OF PERIGEE:	83	BSTAR:	.0

DEBRIS CLOUD DATA

MAXIMUM ΔP:	Unknown
MAXIMUM ΔI:	Unknown

COMMENTS

This was the first of two missions of this type flown in 1966 and not acknowledged by the USSR. The identity of the parent orbit is uncertain. Satellite 2437 was the first cataloged fragment. The above elements are taken or derived from the RAE Table of Earth Satellites. The debris distribution is consistent with a fragmentation near 300 km. Failure of the payload led to immediate activation of the self-destruct system.

REFERENCE DOCUMENT

History of Soviet/Russian Satellite Fragmentations-A Joint U.S.-Russian Investigation, N. L. Johnson, *et al.,* Kaman Sciences Corporation, October 1995.

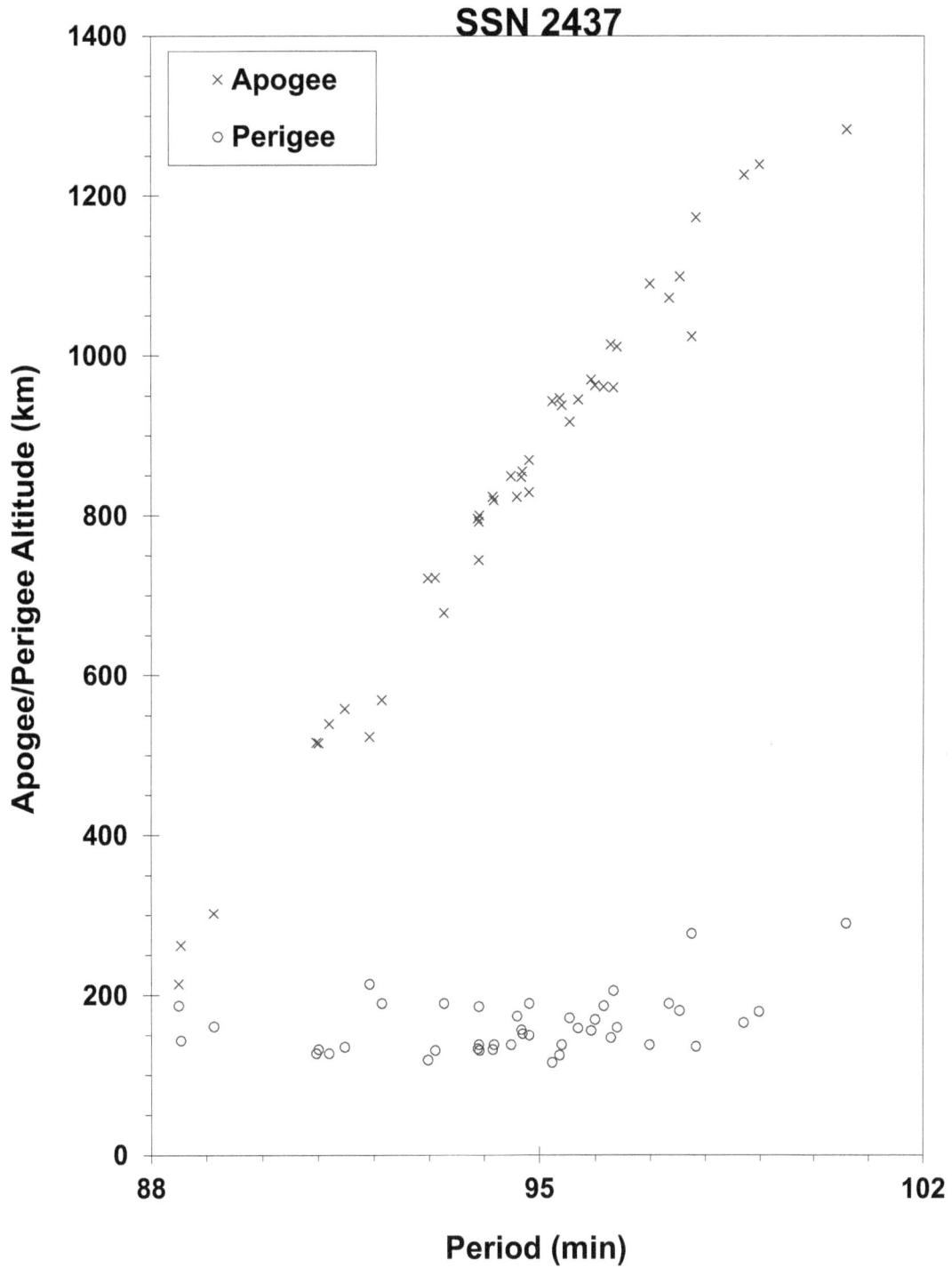

SSN 2437

Cosmos U-1 debris cloud of 43 fragments cataloged by 5 October 1966 as reconstructed from the US SSN database.

SATELLITE DATA

TYPE:	Unknown
OWNER:	CIS
LAUNCH DATE:	2.03 Nov 1966
DRY MASS (KG):	Unknown
MAIN BODY:	Cone-cylinder; 1.5 m diameter by 6 m length
MAJOR APPENDAGES:	None
ATTITUDE CONTROL:	Unknown
ENERGY SOURCES:	Explosive device

EVENT DATA

DATE:	2 Nov 1966	LOCATION:	Unknown
TIME:	Unknown	ASSESSED CAUSE:	Deliberate
ALTITUDE:	~225 km		

POST-EVENT ELEMENTS

EPOCH:	66309.99121234	MEAN ANOMALY:	265.7893
RIGHT ASCENSION:	35.2944	MEAN MOTION:	15.17033022
INCLINATION:	49.5617	MEAN MOTION DOT/2:	.01866914
ECCENTRICITY:	.05339049	MEAN MOTION DOT DOT/6:	.0043309
ARG. OF PERIGEE:	100.3324	BSTAR:	.0

DEBRIS CLOUD DATA

MAXIMUM ΔP:	Unknown
MAXIMUM ΔI:	Unknown

COMMENTS

This was the second mission of this type flown in 1966 and not acknowledged by the USSR. No elements were cataloged until 3 days after the launch. The identity of the parent orbit is uncertain. Satellite 2536 was the first object cataloged and was near the center of the debris cloud. The debris distribution is consistent with a fragmentation near 225 km. Failure of the payload led to immediate activation of the self-destruct system.

REFERENCE DOCUMENT

History of Soviet/Russian Satellite Fragmentations-A Joint U.S.-Russian Investigation, N. L. Johnson, *et al.,* Kaman Sciences Corporation, October 1995.

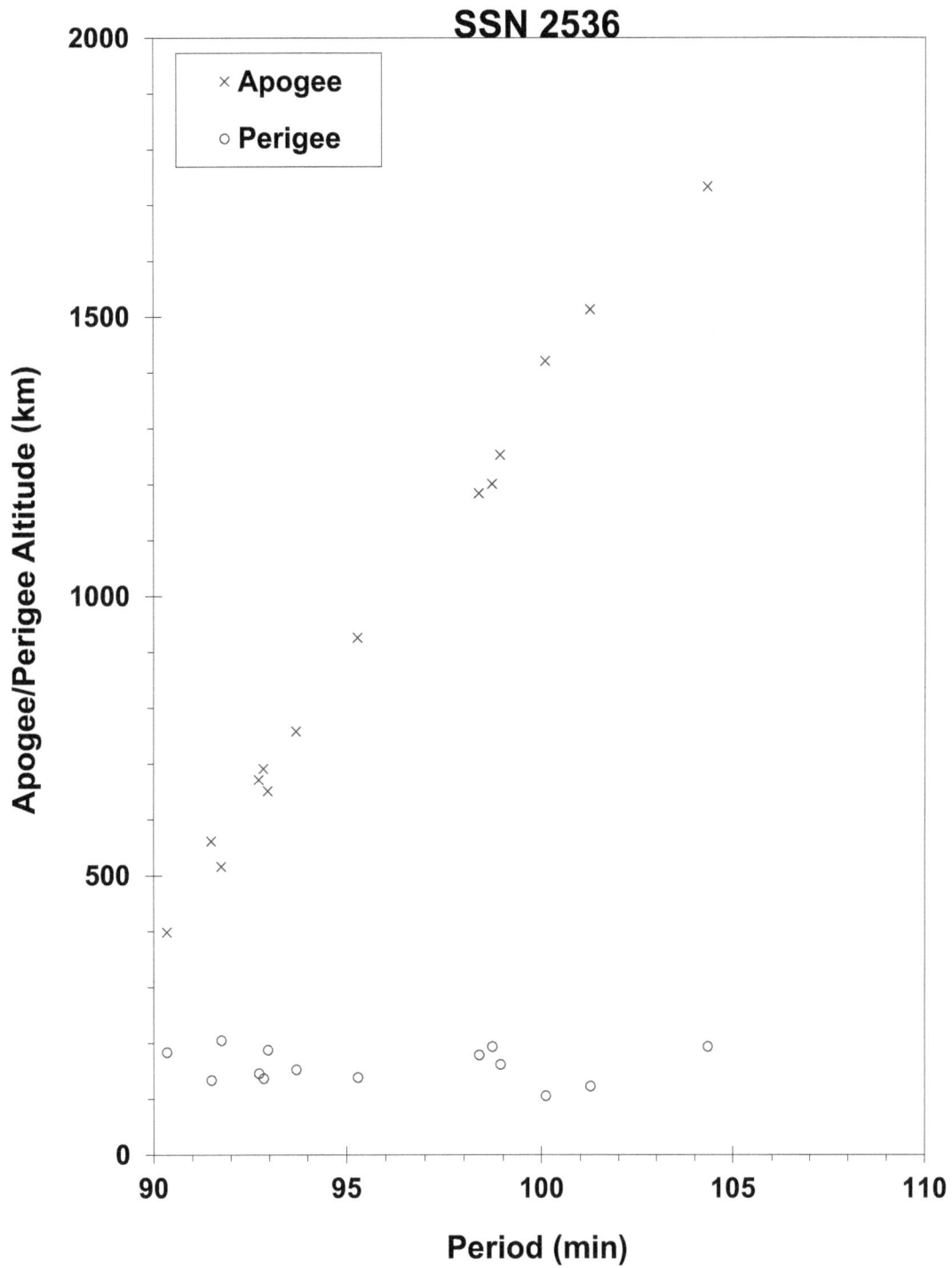

SSN 2536

Cosmos U-2 debris cloud composed of 14 different orbits as developed by the US SSN within 1 week of the event.

SATELLITE DATA

TYPE:	Payload
OWNER:	CIS
LAUNCH DATE:	23.96 Nov 1967
DRY MASS (KG):	750
MAIN BODY:	2.035 m diameter x 2 m long right circular cylinder
MAJOR APPENDAGES:	gravity gradient stabilization boom
ATTITUDE CONTROL:	gravity gradient (passive)
ENERGY SOURCES:	pressurized payload compartment; chemical batteries

EVENT DATA

DATE:	30 Aug 2009	LOCATION:	33.4S, 18.0E (asc)
TIME:	2150 GMT	ASSESSED CAUSE:	Unknown
ALTITUDE:	728.2. km		

PRE-EVENT ELEMENTS

EPOCH:	09241.21736132	MEAN ANOMALY:	353.9164
RIGHT ASCENSION:	137.1436	MEAN MOTION:	14.53942052
INCLINATION:	74.0172	MEAN MOTION DOT/2:	0. 00000028
ECCENTRICITY:	0. 0005259	MEAN MOTION DOT DOT/6:	0.0
ARG. OF PERIGEE:	6.2065	BSTAR:	0.000013364

DEBRIS CLOUD DATA

MAXIMUM ΔP:	5.9 min
MAXIMUM ΔI:	0.4 deg

COMMENTS

Cosmos 192 was a prototype or early spacecraft in the Tsyklon low altitude navigation satellite system. The spacecraft utilized the versatile KAUR-1 bus (*Kosmicheskiy Apparat Universalnogo Ryada*-1, (Космический Аппарат Универсального Ряда), which can be translated as Spacecraft Bus from the Standardized Line-1). This bus consists of a central pressurized cylinder for the payload, an outer cylinder mounting solar cells (the Solar Battery), and a gravity gradient boom for passive stabilization. The cause of this event, occurring almost 42 years after launch, is unknown.

REFERENCE DOCUMENT

"Old Spacecraft Suffers Minor Fragmentation," The Orbital Debris Quarterly News, NASA JSC, October 2009. Available online at https://orbitaldebris.jsc.nasa.gov/quarterly-news/pdfs/ODQNv13i4.pdf.

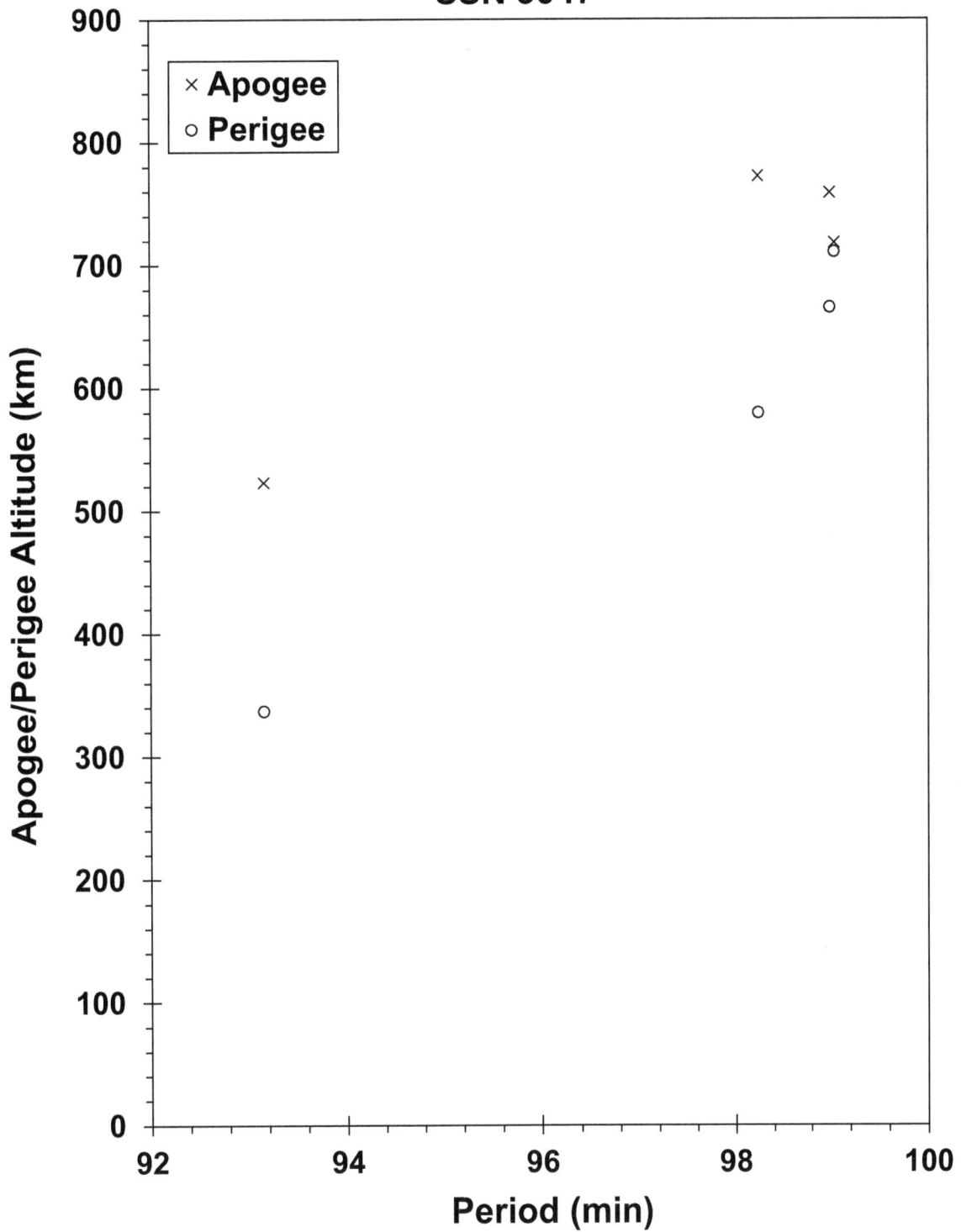

SSN 3047

Four cataloged fragments three weeks after the event.

SATELLITE DATA

TYPE:	Payload
OWNER:	CIS
LAUNCH DATE:	16.50 Jan 1968
DRY MASS (KG):	5500
MAIN BODY:	Sphere-cylinder; 2.4 m diameter by 6.0 m length
MAJOR APPENDAGES:	None
ATTITUDE CONTROL:	Active, 3-axis
ENERGY SOURCES:	On-board propellants, 10 kg TNT explosive charge

EVENT DATA

DATE:	24 Jan 1968	LOCATION:	Unknown
TIME:	Unknown	ASSESSED CAUSE:	Deliberate
ALTITUDE:	Unknown		

PRE-EVENT ELEMENTS

EPOCH:	68024. 25242706	MEAN ANOMALY:	305.4920
RIGHT ASCENSION:	247.4278	MEAN MOTION:	15.98596524
INCLINATION:	65.6289	MEAN MOTION DOT/2:	0.00196964
ECCENTRICITY:	0.0118074	MEAN MOTION DOT DOT/6:	0
ARG. OF PERIGEE:	55.7254	BSTAR:	0

DEBRIS CLOUD DATA

MAXIMUM ΔP:	Unknown
MAXIMUM ΔI:	Unknown

COMMENTS

Spacecraft was destroyed after a malfunction prevented reentry and landing in the Soviet Union.

REFERENCE DOCUMENT

History of Soviet/Russian Satellite Fragmentations-A Joint U.S.-Russian Investigation, N. L. Johnson, *et al.*, Kaman Sciences Corporation, October 1995.

Insufficient data to construct a Gabbard diagram.

SATELLITE DATA

TYPE:	Saturn SIVB Stage
OWNER:	US
LAUNCH DATE:	4.50 Apr 1968
DRY MASS (KG):	30,000
MAIN BODY:	Cylinder; 6.6 m diameter by 30 m length (?)
MAJOR APPENDAGES:	None
ATTITUDE CONTROL:	None at time of the event.
ENERGY SOURCES:	On-board propellants

EVENT DATA

DATE:	13 Apr 1968	LOCATION:	32N, 245E (asc)
TIME:	1054 GMT	ASSESSED CAUSE:	Propulsion
ALTITUDE:	330 km		

PRE-EVENT ELEMENTS

EPOCH:	68103.56521409	MEAN ANOMALY:	151.0074
RIGHT ASCENSION:	177.3270	MEAN MOTION:	15.97292993
INCLINATION:	32.5869	MEAN MOTION DOT/2:	.00302835
ECCENTRICITY:	.0120930	MEAN MOTION DOT DOT/6:	.0
ARG. OF PERIGEE:	208.3921	BSTAR:	.0

DEBRIS CLOUD DATA

MAXIMUM ΔP:	0.7 min
MAXIMUM ΔI:	0.1 deg

COMMENTS

This Saturn SIVB Stage was fitted with an 11,800 kg mock Lunar Module (LM). The SIVB stage was programmed for a second firing to place the Apollo 6 vehicle into a more eccentric orbit, but the restart did not occur. The Apollo 6 payload was separated, leaving the SIVB stage and the LM in a low Earth orbit. Vaporization and venting of residual liquid oxygen induced a tumble to the SIVB stage that reached 30 rpm by 13 April. On this date the axial loads on the LM attach strap fittings and support struts were exceeded, resulting in separation of the LM from the SIVB along with numerous debris. Five fragments were cataloged without elements.

REFERENCE DOCUMENT

Apollo 6 Mission Anomaly Report No. 6, Unexpected Structural Indications During Launch Phase (Review Copy), MSC-PT-R-68-22, prepared by Apollo 6 Mission Evaluation Team, Marshall Space Flight Center, Alabama, and Manned Spacecraft Center, Texas, 1968.

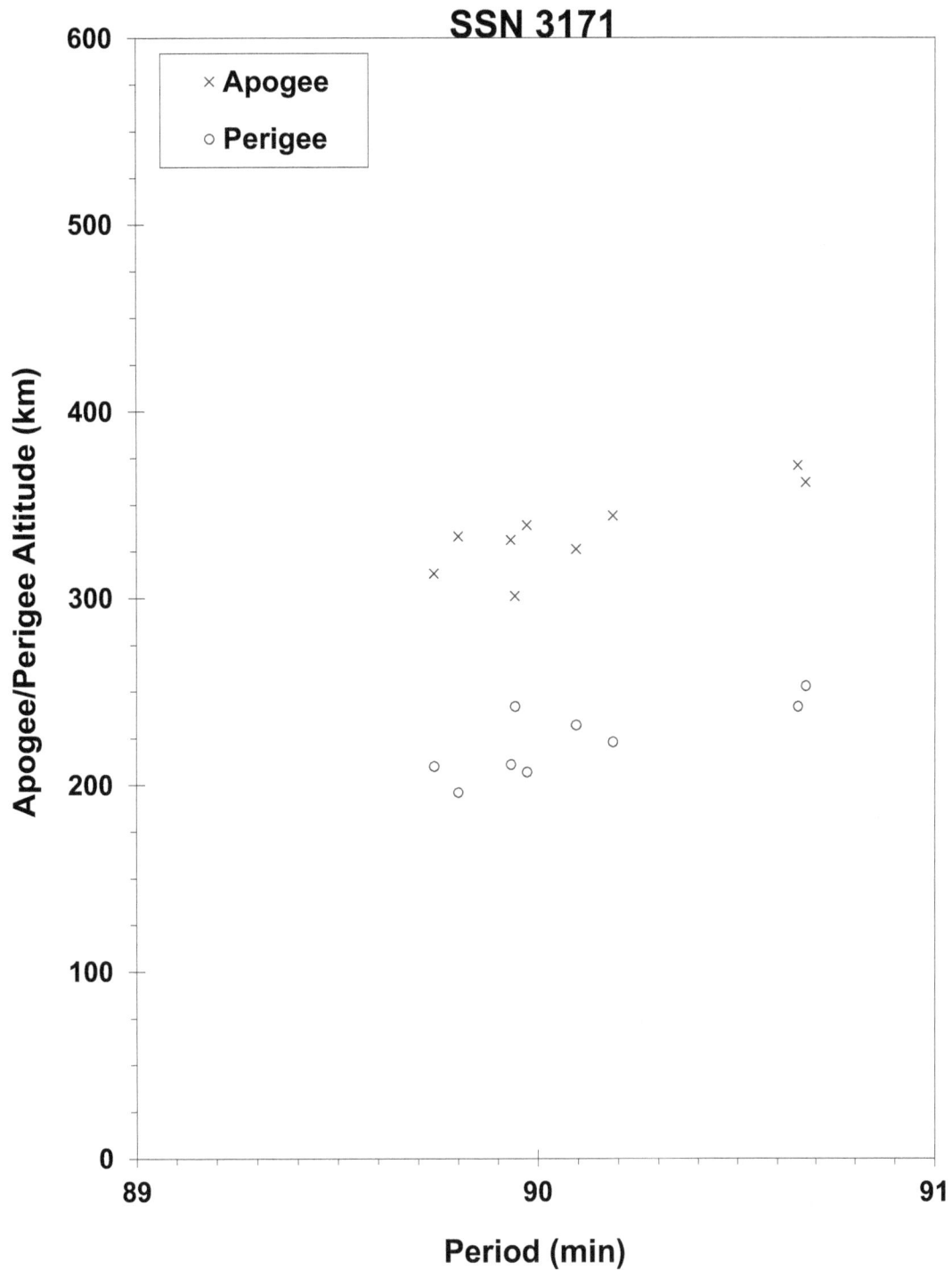

Apollo 6 R/B debris cloud of 9 fragments 4 days after the event as reconstructed from the US SSN database.

SATELLITE DATA

TYPE:	Titan 3C Transtage
OWNER:	US
LAUNCH DATE:	26.32 Sep 1968
DRY MASS (KG):	2500
MAIN BODY:	Cylinder; 3 m diameter by 6 m length
MAJOR APPENDAGES:	None
ATTITUDE CONTROL:	Active, 3-axis
ENERGY SOURCES:	On-board propellants

EVENT DATA

DATE:	21 Feb 1992	LOCATION:	Unknown (~ 197E)
TIME:	0931 GMT	ASSESSED CAUSE:	Propulsion
ALTITUDE:	~ 35600		

PRE-EVENT ELEMENTS

EPOCH:	92043.23217642	MEAN ANOMALY:	284.5600
RIGHT ASCENSION:	21.8025	MEAN MOTION:	1.01459126
INCLINATION:	11.9035	MEAN MOTION DOT/2:	.00000174
ECCENTRICITY:	.0084771	MEAN MOTION DOT DOT/6:	.0
ARG. OF PERIGEE:	76.2786	BSTAR:	.0

DEBRIS CLOUD DATA

MAXIMUM ΔP:	Unknown
MAXIMUM ΔI:	Unknown

COMMENTS

This was the second major fragmentation of a Titan 3C Transtage (the first was 1965-082DM). This transtage released ERS-28 (also known as OV5-2) in highly eccentric transfer orbit, then released LES-6 and ERS-21 (also known as OV5-4) in synchronous orbit, before slightly decelerating and releasing OV2-5 into a slightly lower orbit. This rocket body successfully completed its mission and remained on-orbit for 281 months before fragmenting. Mr. Bob Brock, operating the Maui GEODSS sensor, observed this transtage as it fragmented, liberating a reported 20 objects.

REFERENCE DOCUMENTS

TRW Space Log, Winter 1968-69 edition, Vol. 8, No. 4, H. T. Seaborn, ed., TRW Systems Group, Redondo Beach, pp. 32-35.

"Debris in Geosynchronous Orbits," A.F. Pensa, *et al.,* Space Forum, special issue, 1st International Workshop on Space Debris, Moscow, October 1995.

Insufficient data to construct a Gabbard diagram.

SATELLITE DATA

TYPE:	Payload
OWNER:	CIS
LAUNCH DATE:	19.18 Oct 1968
DRY MASS (KG):	1400
MAIN BODY:	Irregular; 1.8 m by 4.2 m
MAJOR APPENDAGES:	Unknown
ATTITUDE CONTROL:	Active, 3-axis
ENERGY SOURCES:	On-board propellants

EVENT DATA

DATE:	1 Nov 1968	LOCATION:	55N, 104E (dsc)
TIME:	0412 GMT	ASSESSED CAUSE:	Deliberate
ALTITUDE:	540 km		

PRE-EVENT ELEMENTS

EPOCH:	68304.83833772	MEAN ANOMALY:	61.1261
RIGHT ASCENSION:	82.2502	MEAN MOTION:	15.19330723
INCLINATION:	62.2495	MEAN MOTION DOT/2:	.00016932
ECCENTRICITY:	.0050333	MEAN MOTION DOT DOT/6:	.0
ARG. OF PERIGEE:	298.4670	BSTAR:	.0

DEBRIS CLOUD DATA

MAXIMUM ΔP:	Unknown
MAXIMUM ΔI:	0.1 deg

COMMENTS

Cosmos 248 was the target of rendezvous for the Cosmos 249 and Cosmos 252 tests. Calculations suggest the few fragments detected from Cosmos 248 were released within 10 minutes of the Cosmos 252 event that took place in the vicinity of Cosmos 248. The four observed fragments were not cataloged until 4-6 weeks after the event, preventing an accurate assessment of the event due to drag effects. It is possible that the Cosmos 248 event occurred immediately after the rendezvous and was a direct result of interaction with Cosmos 252 debris.

REFERENCE DOCUMENTS

"Artificial Satellite Break-Ups (Part 2): Soviet Anti-Satellite Program", N.L. Johnson, <u>Journal of the British Interplanetary Society</u>, August 1983, pp. 357-362.

<u>History of Soviet/Russian Satellite Fragmentations-A Joint U.S.-Russian Investigation</u>, N. L. Johnson, *et al.*, Kaman Sciences Corporation, October 1995.

Insufficient data to construct a Gabbard diagram.

SATELLITE DATA

TYPE:	Payload
OWNER:	CIS
LAUNCH DATE:	20.17 Oct 1968
DRY MASS (KG):	1400
MAIN BODY:	Irregular; 1.8 m by 4.2 m
MAJOR APPENDAGES:	None
ATTITUDE CONTROL:	Active, 3-axis
ENERGY SOURCES:	On-board propellants, explosive charge

EVENT DATA

DATE:	20 Oct 1968	LOCATION:	57S, 181E (asc)
TIME:	1427 GMT	ASSESSED CAUSE:	Deliberate
ALTITUDE:	1995 km		

POST-EVENT ELEMENTS

EPOCH:	68294.85197372	MEAN ANOMALY:	295.3555
RIGHT ASCENSION:	118.4255	MEAN MOTION:	12.83515528
INCLINATION:	62.3313	MEAN MOTION DOT/2:	.0
ECCENTRICITY:	.1088260	MEAN MOTION DOT DOT/6:	.0
ARG. OF PERIGEE:	76.6147	BSTAR:	.0

DEBRIS CLOUD DATA

MAXIMUM ΔP:	3.9 min
MAXIMUM ΔI:	0.4 deg

COMMENTS

Cosmos 249 was the first of a class of maneuverable spacecraft flown to rendezvous within 4 hours with another Cosmos satellite. In 9 of 20 such missions, orbital debris clouds were created by the active spacecraft, and in one case a passive (target) spacecraft also spawned a few fragments. Fragmentations occurred either in the vicinity of the passive satellite or a few hours after the rendezvous. In the case of Cosmos 249, the spacecraft was launched on a two-revolution rendezvous with Cosmos 248. After a close approach, Cosmos 249 continued on before its warhead was intentionally fired. The elements above are the first available for the final orbit. Some debris from Cosmos 249 and Cosmos 252 have been cross-tagged.

REFERENCE DOCUMENTS

"Artificial Satellite Break-Ups (Part 2): Soviet Anti-Satellite Program", N.L. Johnson, Journal of the British Interplanetary Society, August 1983, pp. 357-362.

History of Soviet/Russian Satellite Fragmentations-A Joint U.S.-Russian Investigation, N. L. Johnson, et al., Kaman Sciences Corporation, October 1995.

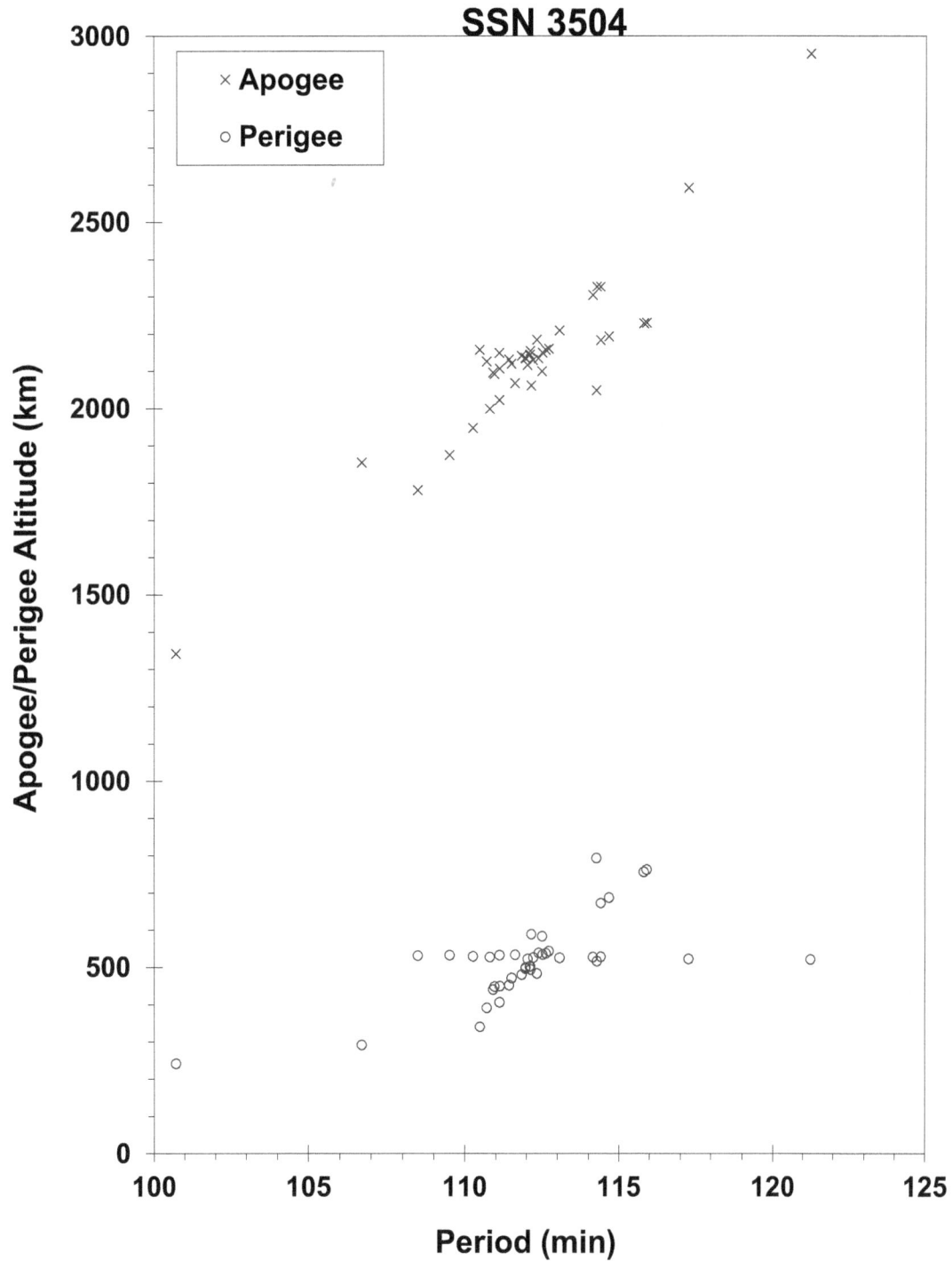

SSN 3504

Cosmos 249 cataloged debris cloud of 43 fragments 4 months after the event as reconstructed from the US SSN database. Cross-tagging with Cosmos 252 debris is evident.

SATELLITE DATA

TYPE:	Payload
OWNER:	CIS
LAUNCH DATE:	1.02 Nov 1968
DRY MASS (KG):	1400
MAIN BODY:	Irregular; 1.8 m by 4.2 m
MAJOR APPENDAGES:	None
ATTITUDE CONTROL:	Active, 3-axis
ENERGY SOURCES:	On-board propellants, explosive charge

EVENT DATA

DATE:	1 Nov 1968	LOCATION:	58N, 34E (asc)
TIME:	0402 GMT	ASSESSED CAUSE:	Deliberate
ALTITUDE:	535 km		

POST-EVENT ELEMENTS

EPOCH:	68306.70122094	MEAN ANOMALY:	297.5777
RIGHT ASCENSION:	76.5565	MEAN MOTION:	12.81276799
INCLINATION:	62.3351	MEAN MOTION DOT/2:	.00811969
ECCENTRICITY:	.1040368	MEAN MOTION DOT DOT/6:	.0
ARG. OF PERIGEE:	73.6953	BSTAR:	.0

DEBRIS CLOUD DATA

MAXIMUM ΔP:	8.7 min
MAXIMUM ΔI:	0.5 deg

COMMENTS

Cosmos 252 was launched on a two-revolution rendezvous with Cosmos 248. The fragmentation occurred in the vicinity of Cosmos 248. Cosmos 252 was part of the test series begun with Cosmos 249. Elements above are for the orbit of the spacecraft after final maneuver, which took place immediately before fragmentation. Some debris from Cosmos 249 and Cosmos 252 have been cross-tagged.

REFERENCE DOCUMENTS

"Artificial Satellite Break-Ups (Part 2): Soviet Anti-Satellite Program", N.L. Johnson, Journal of the British Interplanetary Society, August 1983, pp. 357-362.

History of Soviet/Russian Satellite Fragmentations-A Joint U.S.-Russian Investigation, N. L. Johnson, et al., Kaman Sciences Corporation, October 1995.

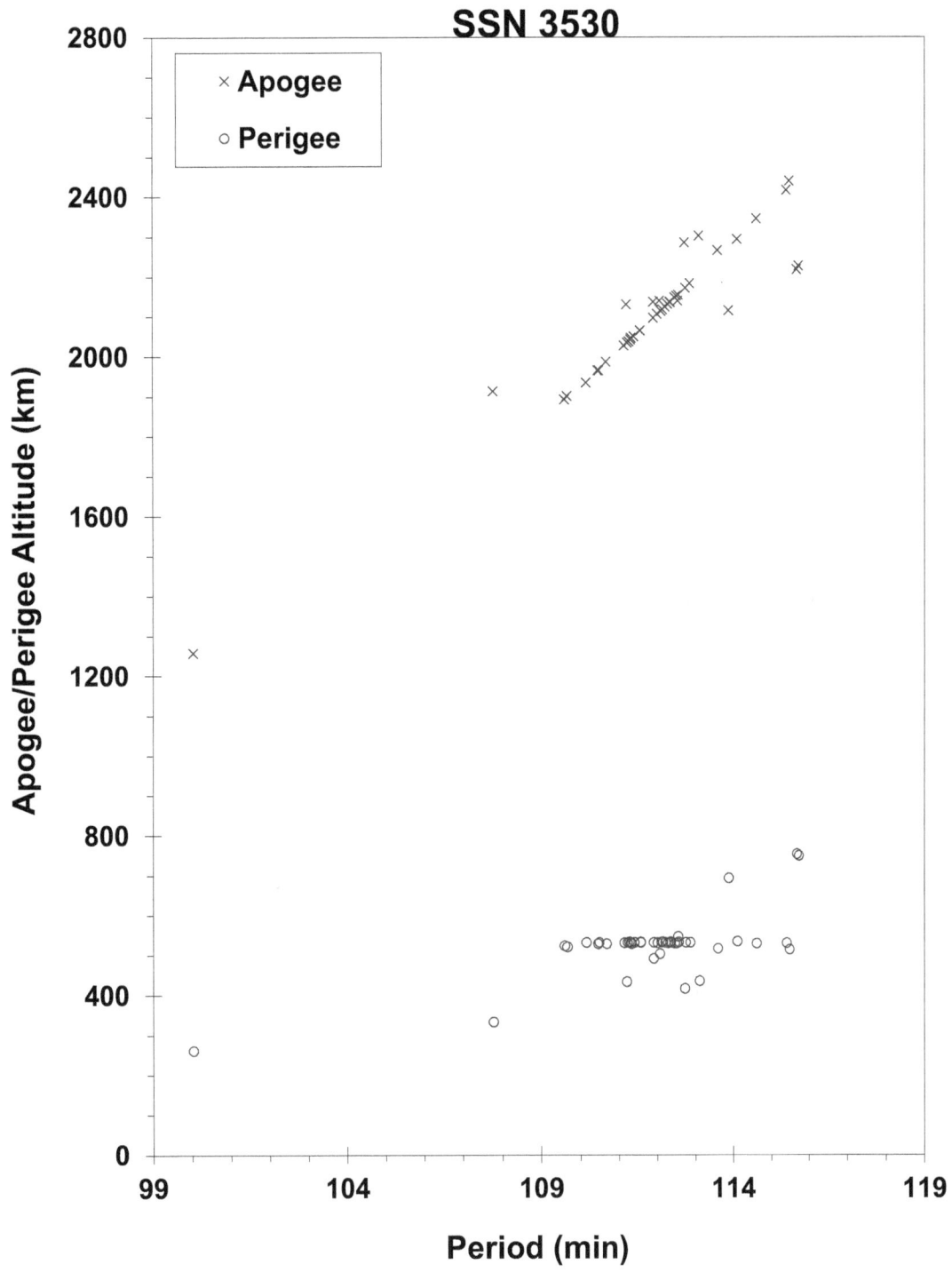

SSN 3530

× Apogee

○ Perigee

Apogee/Perigee Altitude (km)

Period (min)

Cosmos 252 cataloged debris cloud of 43 fragments 4 months after the event as reconstructed from the US SSN database. Cross-tagging with the Cosmos 249 cloud is evident.

SATELLITE DATA

TYPE:	Delta 1 (model N) second stage Rocket Body
OWNER:	US
LAUNCH DATE:	15.72 Dec 1968
DRY MASS (KG):	765
MAIN BODY:	cylinder-nozzle; 1.4 m diameter x 4.8 m long right circular cylinder
MAJOR APPENDAGES:	none
ATTITUDE CONTROL:	none at time of event
ENERGY SOURCES:	on-board propellants; pressure vessels; chemical batteries

EVENT DATA

DATE:	15 Nov 1973	LOCATION:	Unknown
TIME:	Unknown	ASSESSED CAUSE:	Propulsion
ALTITUDE:	Unknown		

PRE-EVENT ELEMENTS

EPOCH:	73318.68417963	MEAN ANOMALY:	337.0891
RIGHT ASCENSION:	26.4043	MEAN MOTION:	12.56868935
INCLINATION:	101.6374	MEAN MOTION DOT/2:	0.00000260
ECCENTRICITY:	0.0031347	MEAN MOTION DOT DOT/6:	7.694E-8
ARG. OF PERIGEE:	23.1544	BSTAR:	0.00000

DEBRIS CLOUD DATA

MAXIMUM ΔP:	5.7 min
MAXIMUM ΔI:	0.8 deg

COMMENTS

The cause of this event, occurring approximately 4.9 years after launch, is assessed to be propulsion-related. This event is recognized, as of this edition, as the first breakup of a Delta 1 rocket body, predating the previously acknowledged first Delta event, the NOAA 3 R/B [*q.v.*], by approximately 43 days.

REFERENCE DOCUMENT

"First Delta Rocket Body Fragmentation Discovered," The Orbital Debris Quarterly News, NASA JSC, August 2017. Available online at https://www.orbitaldebris.jsc.nasa.gov/quarterly-news/pdfs/odqnv21i3.pdf.

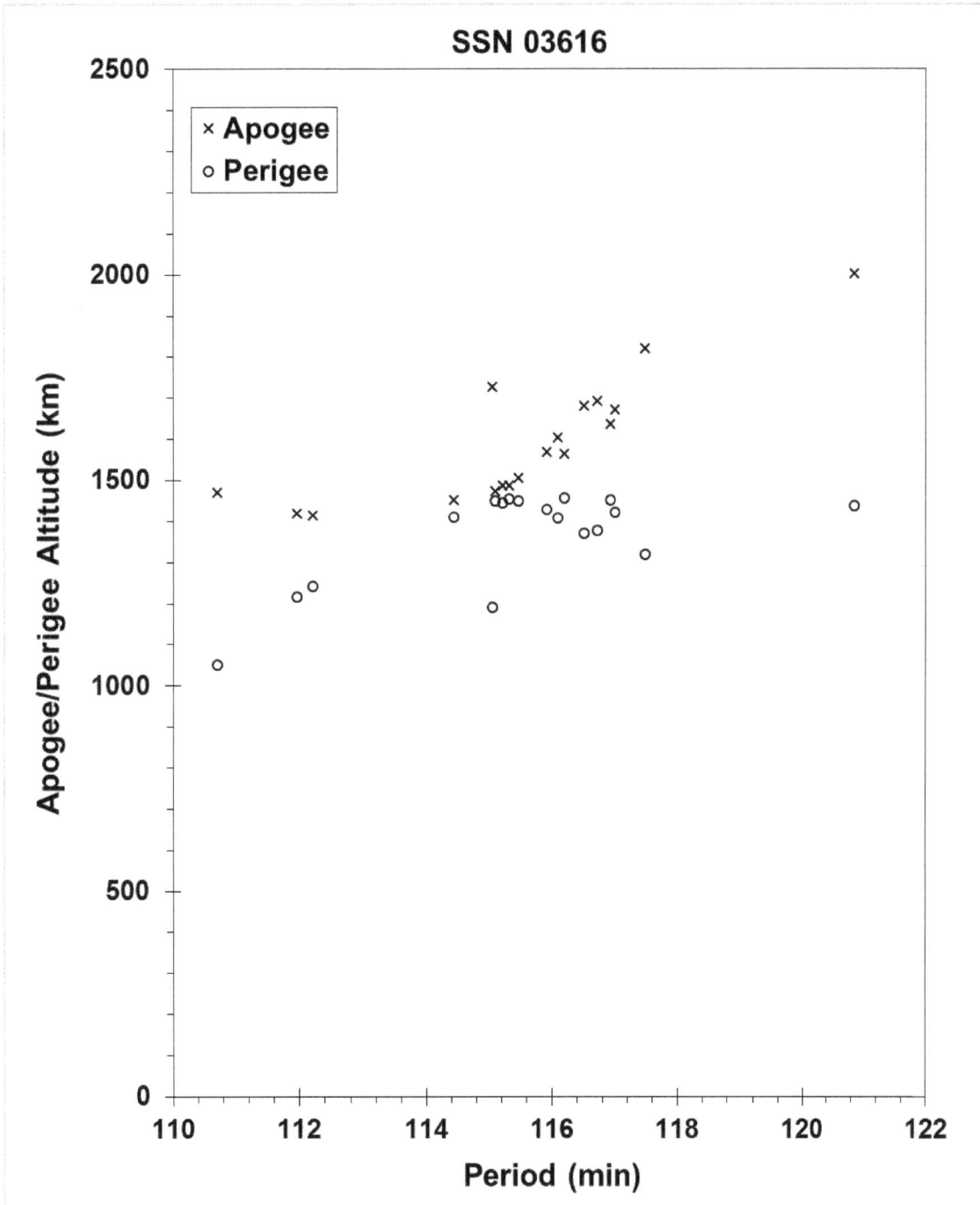

SSN 03616

ESSA 8 R/B debris cloud of 17 fragments as reconstructed from the US SSN database.
These debris entered the public catalog approximately 43.4 years after the event.

SATELLITE DATA

TYPE:	Transtage Rocket Body
OWNER:	US
LAUNCH DATE:	9 Feb 1969
DRY MASS (KG):	1,874
MAIN BODY:	3.048 m diameter x ~5 m long right circular cylinder
MAJOR APPENDAGES:	none
ATTITUDE CONTROL:	inactive 3-axis
ENERGY SOURCES:	on-board propellants; pressure vessels; chemical batteries

EVENT DATA

DATE:	28 Feb 2018	LOCATION:	5.26S, 129.97E (asc)
TIME:	2101 GMT	ASSESSED CAUSE:	Propulsion
ALTITUDE:	37247.3 km		

PRE-EVENT ELEMENTS

EPOCH:	18059.72033524	MEAN ANOMALY:	119.8108
RIGHT ASCENSION:	300.9448	MEAN MOTION:	0.97608930
INCLINATION:	6.2739	MEAN MOTION DOT/2:	-0.00000274
ECCENTRICITY:	0.0164003	MEAN MOTION DOT DOT/6:	0.00000
ARG. OF PERIGEE:	128.1168	BSTAR:	0.00000

DEBRIS CLOUD DATA

MAXIMUM ΔP:	134 min
MAXIMUM ΔI:	0.9 deg

COMMENTS

The cause of this event, occurring approximately 49 years after launch, is likely due to on-board stored energy sources. This is the fourth breakup of a Titan Transtage rocket body, and the second Transtage fragmentation near geosynchronous orbit. Due to difficulties in cataloging highly elliptical and deep-space orbit many more fragments may be resident on-orbit than those cataloged.

REFERENCE DOCUMENTS

"Fragmentation of Titan Transtage in GEO," The Orbital Debris Quarterly News, NASA JSC, May 2018.
Available online at https://www.orbitaldebris.jsc.nasa.gov/quarterly-news/pdfs/odqnv22i2.pdf.

Cowardin, H., P. Anz-Meador, and J. Reyes, "Characterizing GEO Titan IIIC Transtage Fragmentations using Ground-based and Telescopic Measurements." In Proceedings of the AMOS 2017 Technical Conference. Available online at https://amostech.com/TechnicalPapers/2017/Orbital-Debris/Cowardin.pdf

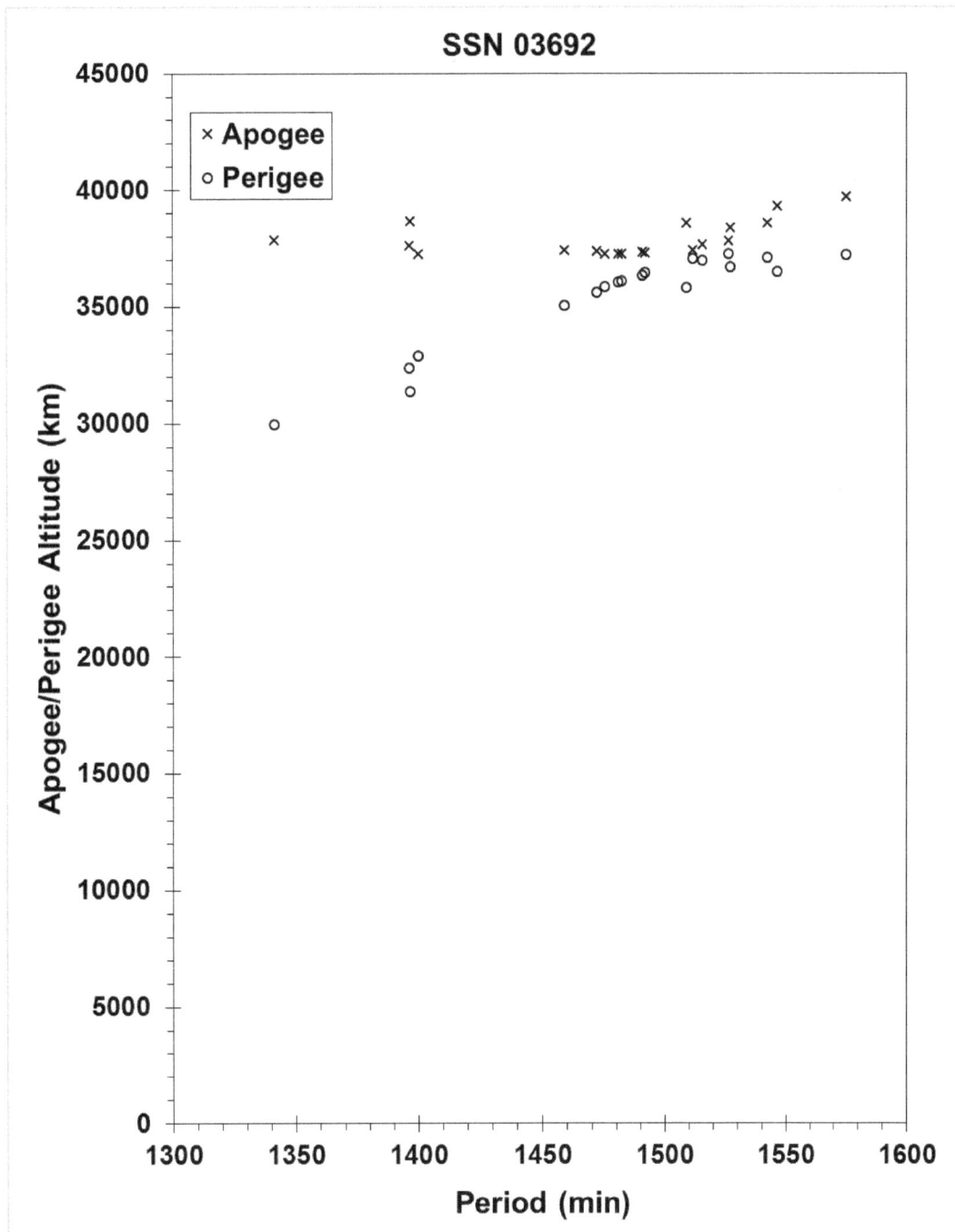

SSN 03692

Transtage R/B debris cloud of parent body and 18 fragments 4 months after the event as reconstructed from the US SSN database.

SATELLITE DATA

TYPE:	Vostok Second Stage
OWNER:	CIS
LAUNCH DATE:	26.52 Mar 1969
DRY MASS (KG):	1440
MAIN BODY:	Cylinder; 2.6 m diameter by 3.8 m length
MAJOR APPENDAGES:	None
ATTITUDE CONTROL:	None at time of the event.
ENERGY SOURCES:	Unknown

EVENT DATA

DATE:	28 Mar 1969	LOCATION:	59N, 91E (dsc)
TIME:	1845 GMT	ASSESSED CAUSE:	Unknown
ALTITUDE:	555 km		

PRE-EVENT ELEMENTS

EPOCH:	69087.21308063	MEAN ANOMALY:	175.1148
RIGHT ASCENSION:	33.3926	MEAN MOTION:	14.71400174
INCLINATION:	81.1687	MEAN MOTION DOT/2:	.0
ECCENTRICITY:	.0276787	MEAN MOTION DOT DOT/6:	.0
ARG. OF PERIGEE:	184.7318	BSTAR:	.0

DEBRIS CLOUD DATA

MAXIMUM ΔP:	2.4 min
MAXIMUM ΔI:	0.5 deg

COMMENTS

The vehicle successfully deployed the Meteor 1-1 payload into the desired orbit. An object believed to be the rocket body was found on 27 March in an orbit (1) of 565 km by 755 km, similar to earlier missions of the Vostok second stage. Early on 28 March an object was found in an orbit (2) of 460 km by 850 km with elements as indicated above. Analysis indicates that a transition from orbit (1) to orbit (2) was possible during the latter part of 27 March. Debris analysis clearly indicates that the orbit of the parent satellite had to be similar to orbit (2). Radar cross-section data supports the belief that the post-event object in the center of the debris cloud is the rocket body. No object was found in orbit (1) after the event.

REFERENCE DOCUMENT

History of the Soviet/Russian Satellite Fragmentations-A Joint U.S.-Russian Investigation, N. L. Johnson, *et al.,* Kaman Sciences Corporation, October 1995.

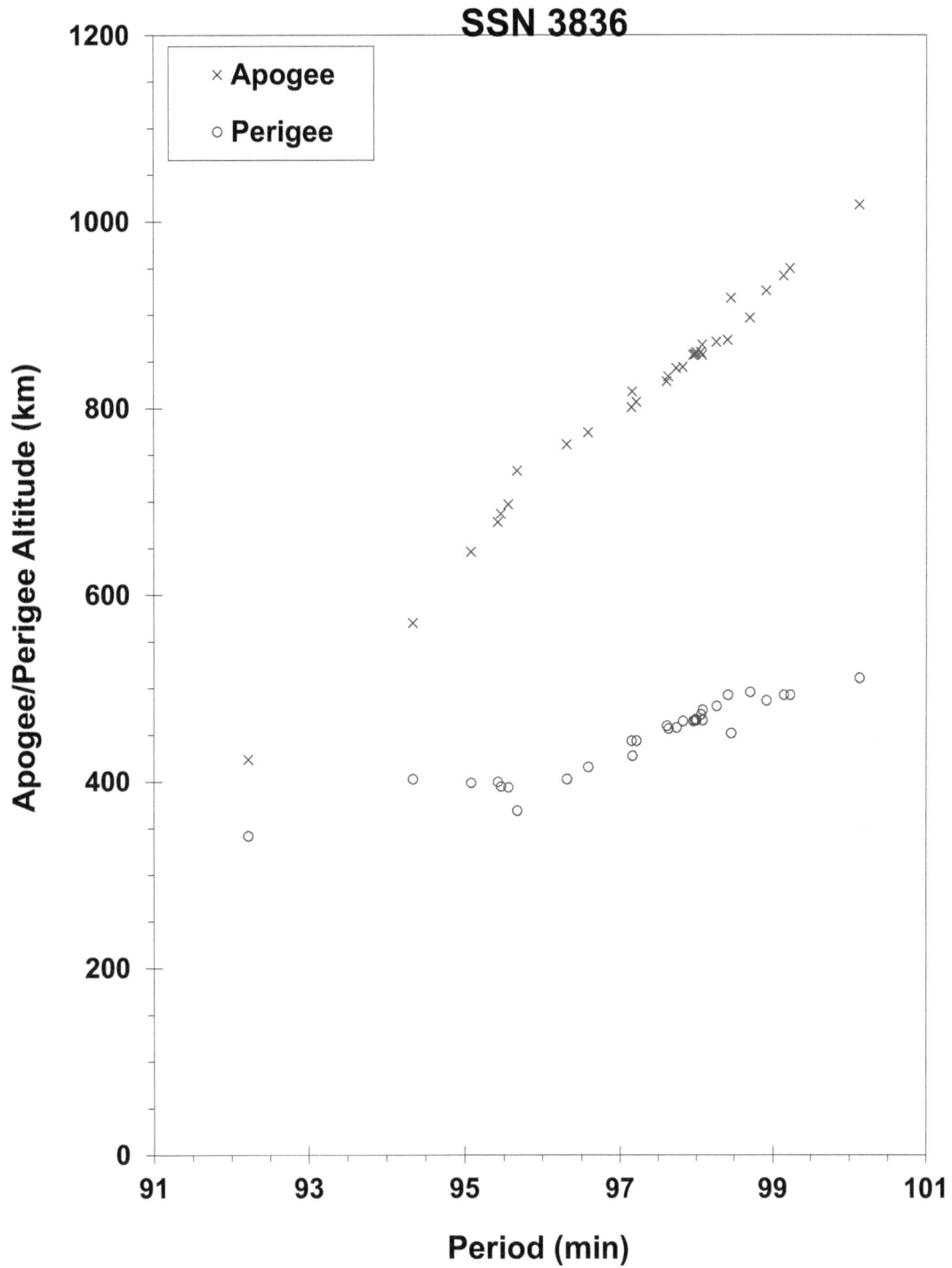

Meteor 1-1 R/B debris cloud of 31 fragments 2 months after the event as reconstructed from the US SSN database.

SATELLITE DATA

TYPE:	TE 364-4 (STAR 37E motor)
OWNER:	US
LAUNCH DATE:	26.09 Jul 1969
DRY MASS (KG):	1100 (70 without solid propellants)
MAIN BODY:	Sphere-nozzle; 1.0 m by 1.8 m
MAJOR APPENDAGES:	None
ATTITUDE CONTROL:	Active, 3-axis
ENERGY SOURCES:	On-board propellants

EVENT DATA

DATE:	26 Jul 1969	LOCATION:	0N, 333E (dsc)
TIME:	0228 GMT	ASSESSED CAUSE:	Propulsion
ALTITUDE:	270 km		

POST-EVENT ELEMENTS

EPOCH:	69208.17261261	MEAN ANOMALY:	166.4542
RIGHT ASCENSION:	130.0186	MEAN MOTION:	9.78100102
INCLINATION:	30.3692	MEAN MOTION DOT/2:	.00000270
ECCENTRICITY:	.2800849	MEAN MOTION DOT DOT/6:	.0
ARG. OF PERIGEE:	187.9970	BSTAR:	.0

DEBRIS CLOUD DATA

MAXIMUM ΔP:	Unknown
MAXIMUM ΔI:	1.2 deg

COMMENTS

This solid-propellant upper stage failed soon after ignition, following a normal launch. The cause of the failure is assessed to be a possible rupture of the motor casing or nozzle. See similar failures of two PAM-D upper stages in 1984. Elements above are first developed for the rocket body about 1 day after the event. Rocket body may later have been cross-tagged with satellite 4053. Validity of debris identification and cataloging after 1969 is suspect.

REFERENCE DOCUMENT

TRW Space Log, Winter 1969-70 edition, Vol. 9, No. 4, W.A. Donop, ed., TRW Systems Group, Redondo Beach, pp. 34-36.

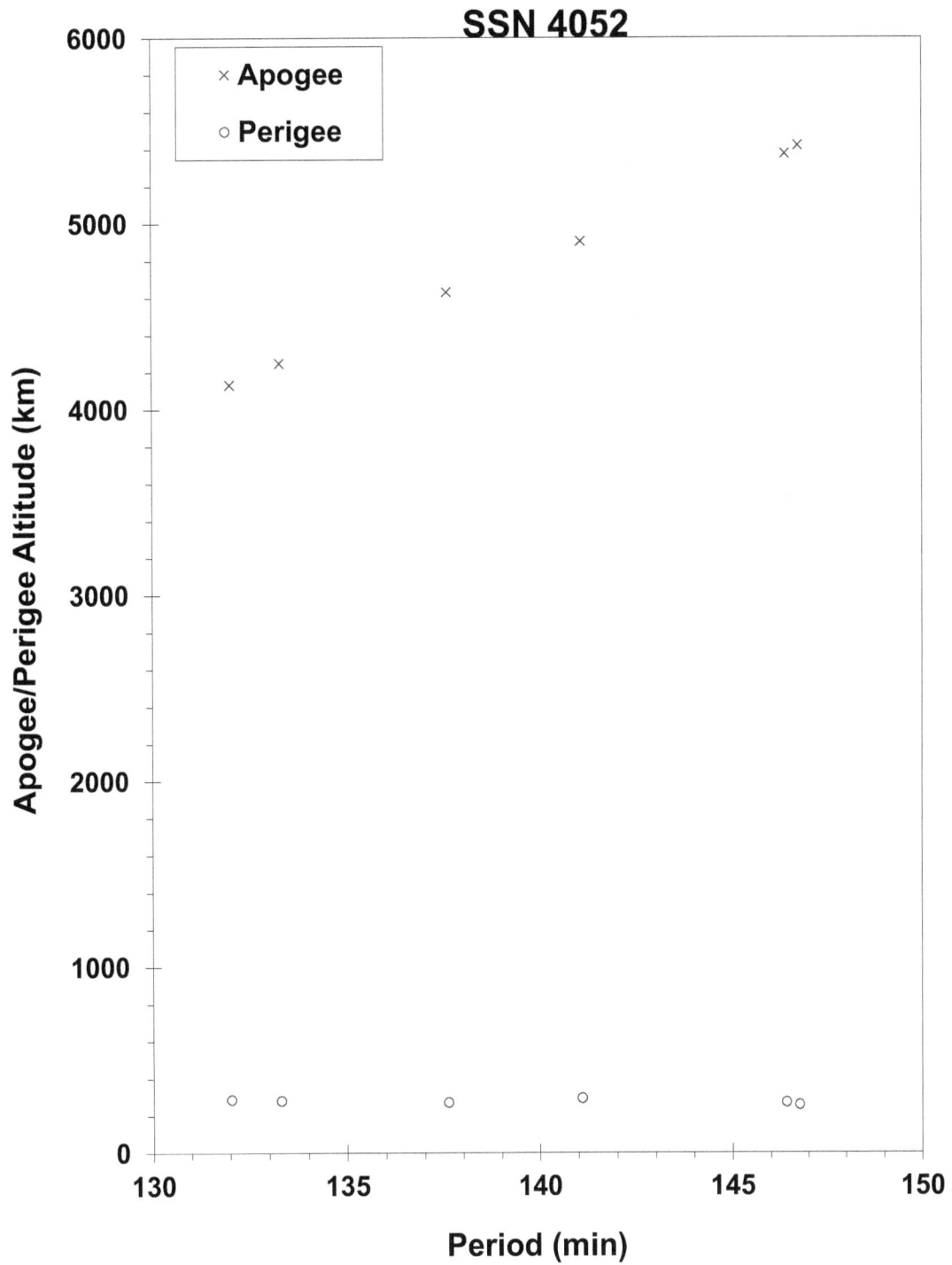

Intelsat 3 F-5 R/B debris cloud of 6 fragments 10 days after the event as reconstructed from the US SSN database.

SATELLITE DATA

TYPE:	Agena D Stage
OWNER:	US
LAUNCH DATE:	30.57 Sep 1969
DRY MASS (KG):	600
MAIN BODY:	Cylinder; 1.5 m diameter by 7.1 m length
MAJOR APPENDAGES:	None
ATTITUDE CONTROL:	None at time of the event.
ENERGY SOURCES:	Unknown

EVENT DATA

DATE:	4 Oct 1969	LOCATION:	54N, 178E (dsc)
TIME:	1553 GMT	ASSESSED CAUSE:	Unknown
ALTITUDE:	920 km		

POST-EVENT ELEMENTS

EPOCH:	69295.54249482	MEAN ANOMALY:	274.0514
RIGHT ASCENSION:	243.5157	MEAN MOTION:	13.68701087
INCLINATION:	69.9611	MEAN MOTION DOT/2:	.00000064
ECCENTRICITY:	.0117819	MEAN MOTION DOT DOT/6:	.0
ARG. OF PERIGEE:	87.4011	BSTAR:	.0

DEBRIS CLOUD DATA

MAXIMUM ΔP:	3.1 min
MAXIMUM ΔI:	1.0 deg

COMMENTS

This was the first of two Agena D stages to fragment in a span of only 12 months. The vehicle delivered ten payloads to an orbit of about 905 km by 940 km. Four days later, before the rocket body had been cataloged, a large fragmentation occurred. What appeared to be the largest piece of the rocket body was found in the orbit described by the elements above almost 3 weeks after the event. See 1967-53 as a reference to an earlier mission of this type. Both missions were sponsored by DOD and public information is limited.

REFERENCE DOCUMENT

"Fragmentations of Asteroids and Artificial Satellites in Orbit", W. Wiesel, Icarus, Vol. 34, 1978, pp. 99-116.

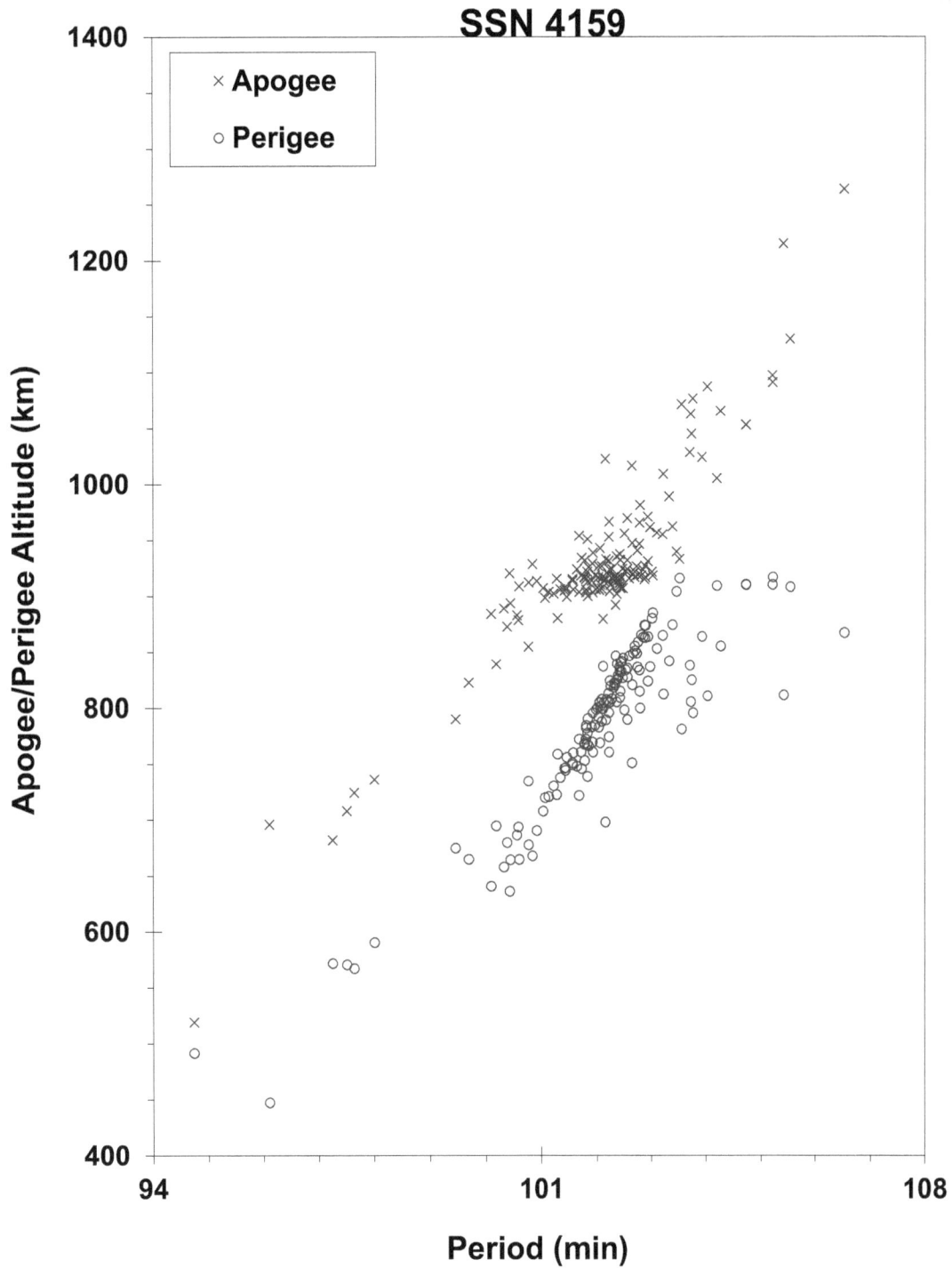

SSN 4159

OPS 7613 R/B debris cloud (excluding 10 payloads) of 152 fragments 8 months after the event. The largest fragment was found in an eccentric orbit with an orbital period of more than 105 min and is presumed to be the rocket body remnant.

SATELLITE DATA

TYPE:	Agena D Stage
OWNER:	US
LAUNCH DATE:	8.35 Apr 1970
DRY MASS (KG):	600
MAIN BODY:	Cylinder; 1.5 m diameter by 7.1 m length
MAJOR APPENDAGES:	None
ATTITUDE CONTROL:	None at time of the event.
ENERGY SOURCES:	Unknown

EVENT DATA

DATE:	17 Oct 1970	LOCATION:	50S, 142E (asc)
TIME:	0317 GMT	ASSESSED CAUSE:	Unknown
ALTITUDE:	1075 km		

PRE-EVENT ELEMENTS

EPOCH:	70289.33183878	MEAN ANOMALY:	141.3434
RIGHT ASCENSION:	203.5235	MEAN MOTION:	13.49254887
INCLINATION:	99.8780	MEAN MOTION DOT/2:	.0
ECCENTRICITY:	.0016616	MEAN MOTION DOT DOT/6:	.0
ARG. OF PERIGEE:	218.6463	BSTAR:	.0

DEBRIS CLOUD DATA

MAXIMUM ΔP:	14.2 min
MAXIMUM ΔI:	0.8 deg

COMMENTS

This was the second Agena D stage to fragment in a span of only 12 months. The event occurred 6 months after the successful deployment of the Nimbus 4 payload. Twice in 1985, again in 1986, once in 1991, and twice in 1995 Nimbus 4 R/B debris spawned a few additional fragments, accounting for an additional 16 new debris objects between the 6 sub-events.

REFERENCE DOCUMENTS

"Fragmentations of Asteroids and Artificial Satellites in Orbit", W. Wiesel, <u>Icarus</u>, Vol. 34, 1978, pp. 99-116.

"Analysis of the Nimbus 4 Rocket Body Breakup and Subsequent Debris Anomalies", N.L. Johnson, Kaman Sciences Corporation, February 1992.

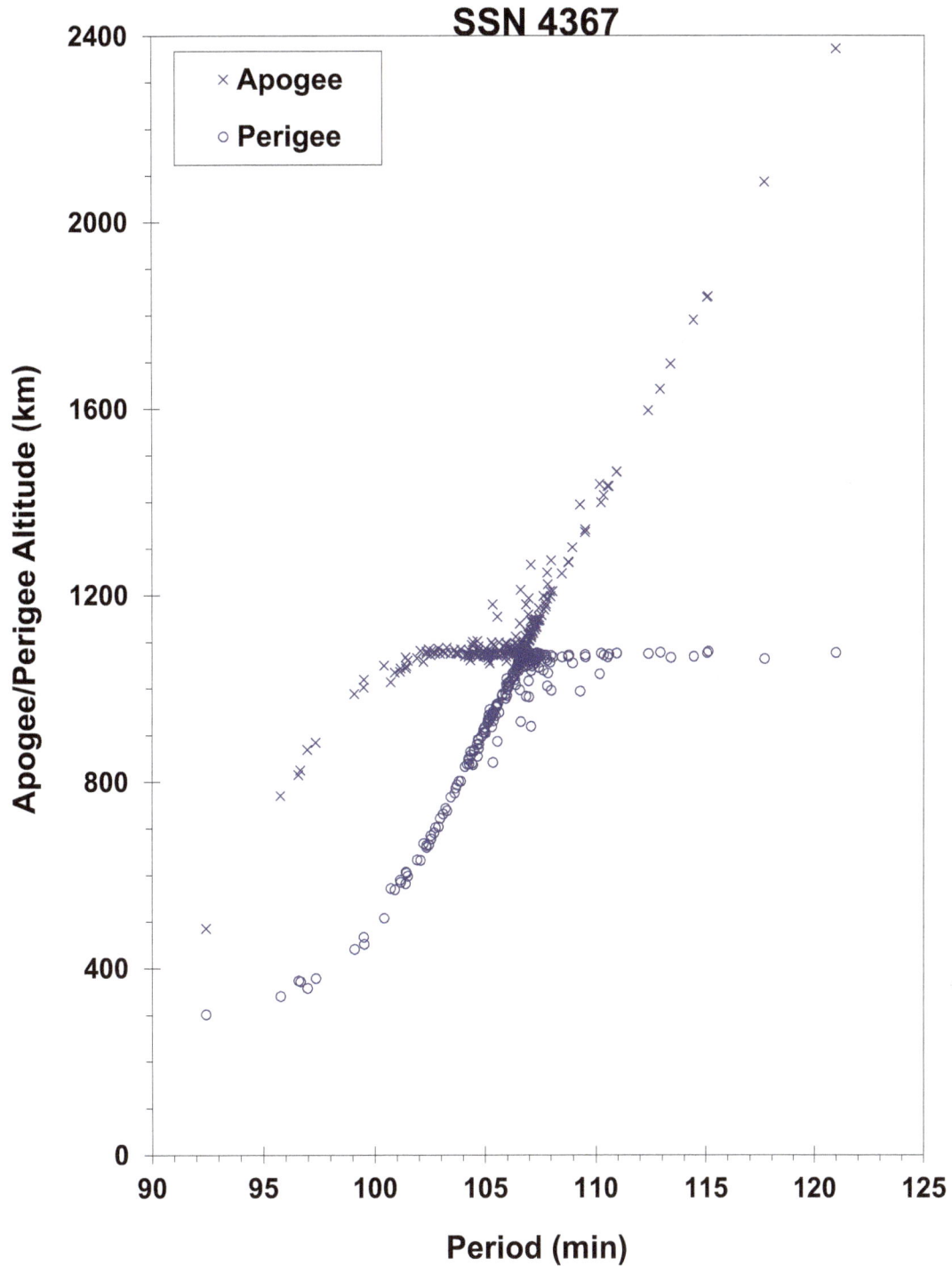

Nimbus 4 R/B debris cloud of 246 fragments 8 months after the event as reconstructed from the US SSN database. Some lower period fragments already exhibit the effects of natural decay.

SATELLITE DATA

TYPE:	Payload
OWNER:	CIS
LAUNCH DATE:	23.18 Oct 1970
DRY MASS (KG):	1400
MAIN BODY:	Irregular; 1.8 m by 4.2 m
MAJOR APPENDAGES:	None
ATTITUDE CONTROL:	Active, 3-axis
ENERGY SOURCES:	On-board propellants, explosive charge

EVENT DATA

DATE:	23 Oct 1970	LOCATION:	22S, 217E (asc)
TIME:	1513 GMT	ASSESSED CAUSE:	Deliberate
ALTITUDE:	1195 km		

PRE-EVENT ELEMENTS

EPOCH:	70296.40542099	MEAN ANOMALY:	309.5623
RIGHT ASCENSION:	129.1049	MEAN MOTION:	12.82808179
INCLINATION:	62.9380	MEAN MOTION DOT/2:	.00019973
ECCENTRICITY:	.1039489	MEAN MOTION DOT DOT/6:	.0
ARG. OF PERIGEE:	60.4933	BSTAR:	.0

DEBRIS CLOUD DATA

MAXIMUM ΔP:	Unknown
MAXIMUM ΔI:	Unknown

COMMENTS

Cosmos 374 was launched on a two-revolution rendezvous with Cosmos 373. After a close approach, Cosmos 374 continued on before its warhead was intentionally fired. Cosmos 374 was part of test series begun with Cosmos 249. Considerable cross-cataloging of Cosmos 374 and Cosmos 375 debris occurred; therefore, ΔP and ΔI are not calculated.

REFERENCE DOCUMENTS

"Artificial Satellite Break-Ups (Part 2): Soviet Anti-Satellite Program", N.L. Johnson, Journal of the British Interplanetary Society, August 1983, pp. 357-362.

History of Soviet/Russian Satellite Fragmentations-A Joint U.S.-Russian Investigation, N. L. Johnson, et al., Kaman Sciences Corporation, October 1995.

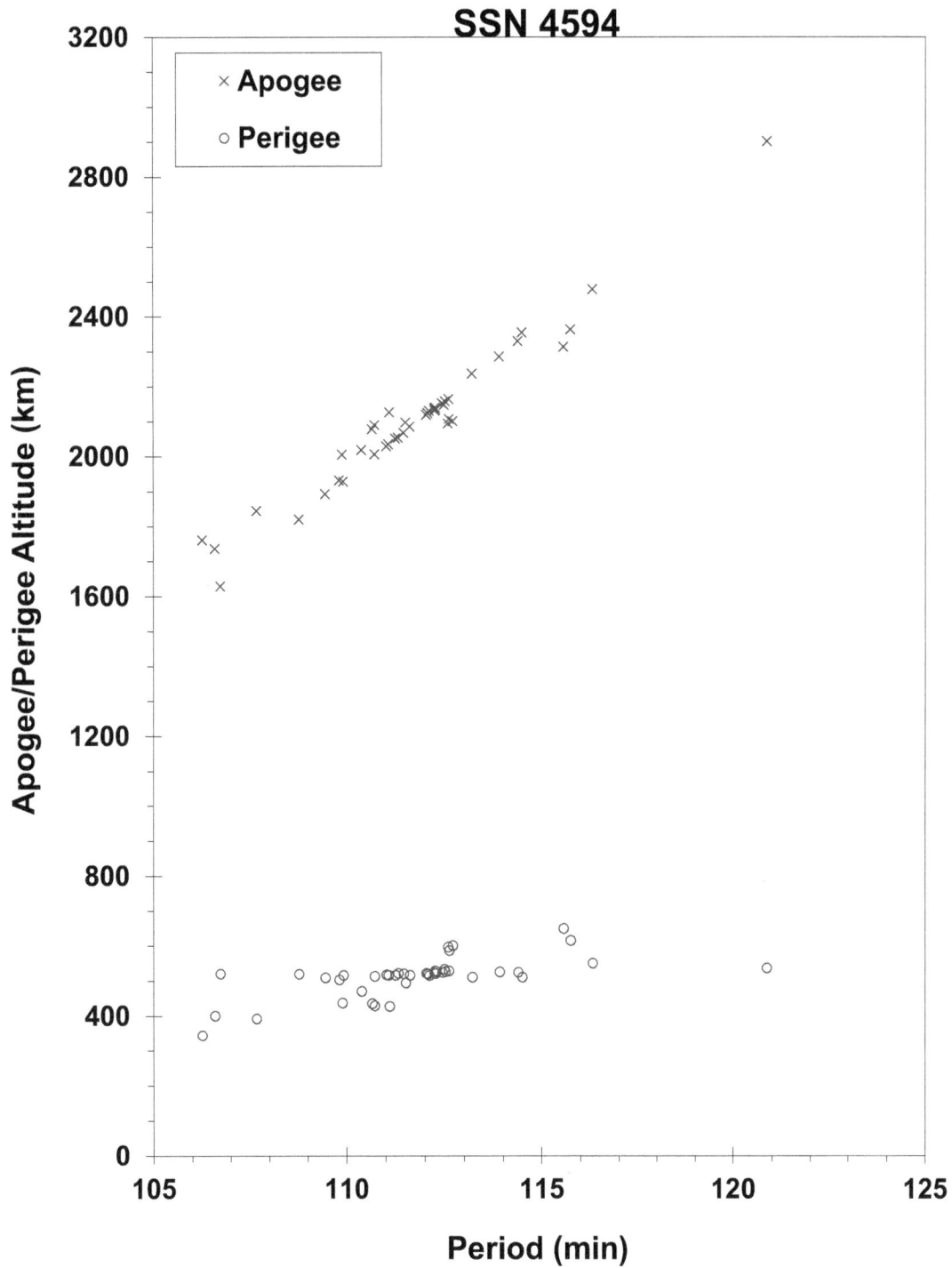

SSN 4594

Cosmos 374 official debris cloud of 43 fragments 5 months after the event as reconstructed from the US SSN database. All fragments were cataloged after the Cosmos 375 fragmentation, and some contamination exists.

SATELLITE DATA

TYPE:	Payload
OWNER:	CIS
LAUNCH DATE:	30.09 Oct 1970
DRY MASS (KG):	1400
MAIN BODY:	Irregular; 1.8 m by 4.2 m
MAJOR APPENDAGES:	None
ATTITUDE CONTROL:	Active, 3-axis
ENERGY SOURCES:	On-board propellants, explosive charge

EVENT DATA

DATE:	30 Oct 1970	LOCATION:	54N, 23E (asc)
TIME:	0600 GMT	ASSESSED CAUSE:	Deliberate
ALTITUDE:	535 km		

POST-EVENT ELEMENTS

EPOCH:	70306.81102869	MEAN ANOMALY:	313.3102
RIGHT ASCENSION:	96.4080	MEAN MOTION:	12.87482205
INCLINATION:	62.8057	MEAN MOTION DOT/2:	.00009999
ECCENTRICITY:	.1022289	MEAN MOTION DOT DOT/6:	.0
ARG. OF PERIGEE:	56.0864	BSTAR:	.0

DEBRIS CLOUD DATA

MAXIMUM ΔP:	Unknown
MAXIMUM ΔI:	Unknown

COMMENTS

Cosmos 375 was launched on a two-revolution rendezvous with Cosmos 373. The fragmentation occurred in the vicinity of Cosmos 373. Cosmos 375 was part of test series begun with Cosmos 249. Elements above are first reliable ones for orbit after final maneuver that took place immediately before fragmentation. Considerable cross-cataloging of Cosmos 374 and Cosmos 375 debris occurred; therefore, ΔP and ΔI are not calculated.

REFERENCE DOCUMENTS

"Artificial Satellite Break-Ups (Part 2): Soviet Anti-Satellite Program", N.L. Johnson, Journal of the British Interplanetary Society, August 1983, pp. 357-362.

History of Soviet/Russian Satellite Fragmentations-A Joint U.S.-Russian Investigation, N. L. Johnson, et al., Kaman Sciences Corporation, October 1995.

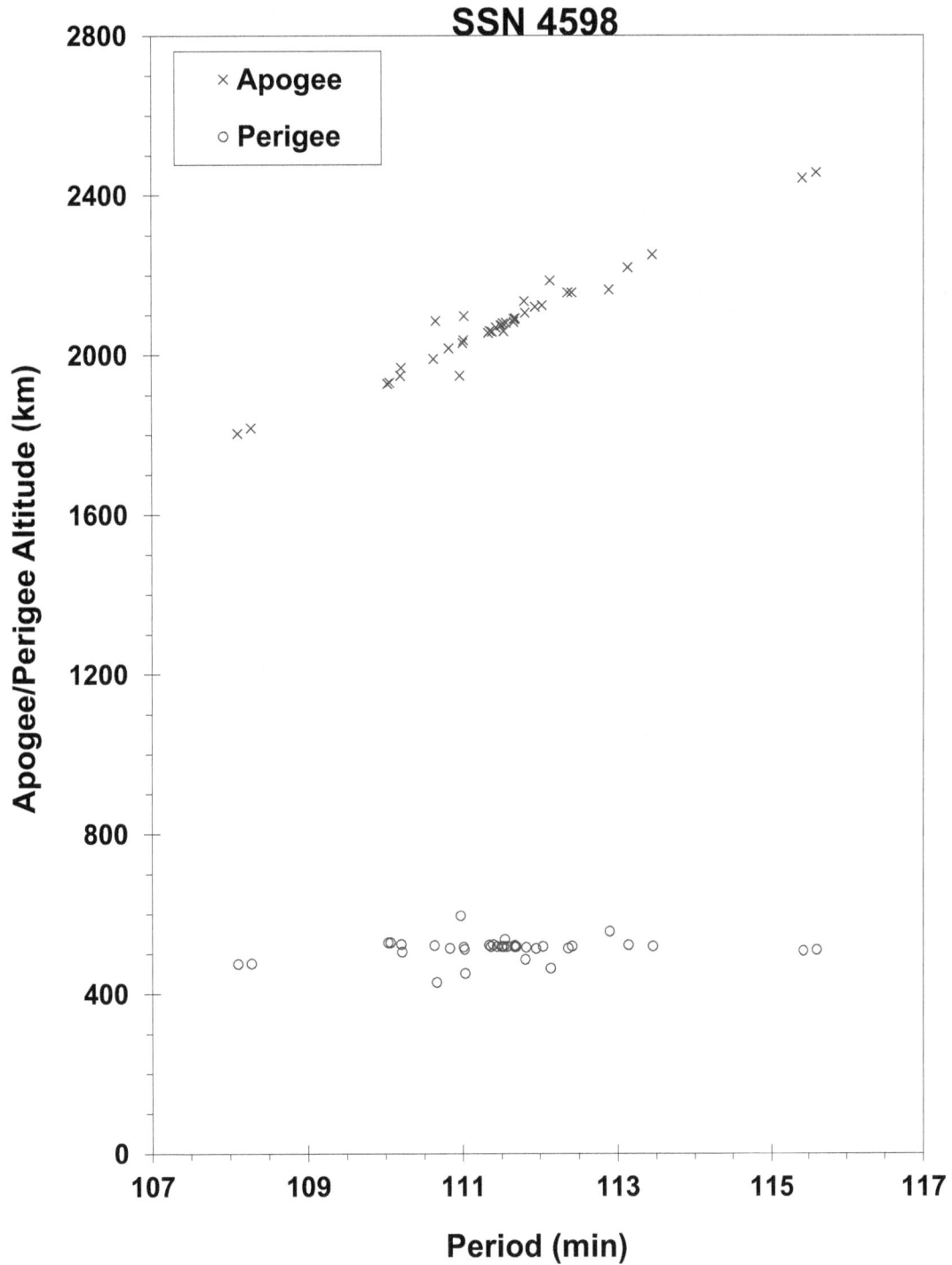

SSN 4598

Cosmos 375 debris cloud of 38 fragments about 4 months after the event as reconstructed from the US SSN database. Some contamination exists with Cosmos 374 debris.

SATELLITE DATA

TYPE:	Payload
OWNER:	CIS
LAUNCH DATE:	25.47 Feb 1971
DRY MASS (KG):	1400
MAIN BODY:	Irregular; 1.8 m by 4.2 m
MAJOR APPENDAGES:	None
ATTITUDE CONTROL:	Active, 3-axis
ENERGY SOURCES:	On-board propellants, explosive charge

EVENT DATA

DATE:	25 Feb 1971	LOCATION:	54N, 21E (asc)
TIME:	1431 GMT	ASSESSED CAUSE:	Deliberate
ALTITUDE:	585 km		

POST-EVENT ELEMENTS

EPOCH:	71057.77590281	MEAN ANOMALY:	318.5528
RIGHT ASCENSION:	352.8670	MEAN MOTION:	12.68709606
INCLINATION:	65.7618	MEAN MOTION DOT/2:	.00013192
ECCENTRICITY:	.1046189	MEAN MOTION DOT DOT/6:	.0
ARG. OF PERIGEE:	50.3064	BSTAR:	.0

DEBRIS CLOUD DATA

MAXIMUM ΔP:	2.8 min
MAXIMUM ΔI:	1.2 deg

COMMENTS

Cosmos 397 was launched on a two-revolution rendezvous with Cosmos 394. The fragmentation occurred in the vicinity of Cosmos 394. Cosmos 397 was part of the test series begun with Cosmos 249. Elements above are first available for orbit after final maneuver that took place immediately before fragmentation.

REFERENCE DOCUMENTS

"Artificial Satellite Break-Ups (Part 2): Soviet Anti-Satellite Program", N.L. Johnson, Journal of the British Interplanetary Society, August 1983, pp. 357-362.

History of Soviet/Russian Satellite Fragmentations-A Joint U.S.-Russian Investigation, N. L. Johnson, *et al.,* Kaman Sciences Corporation, October 1995.

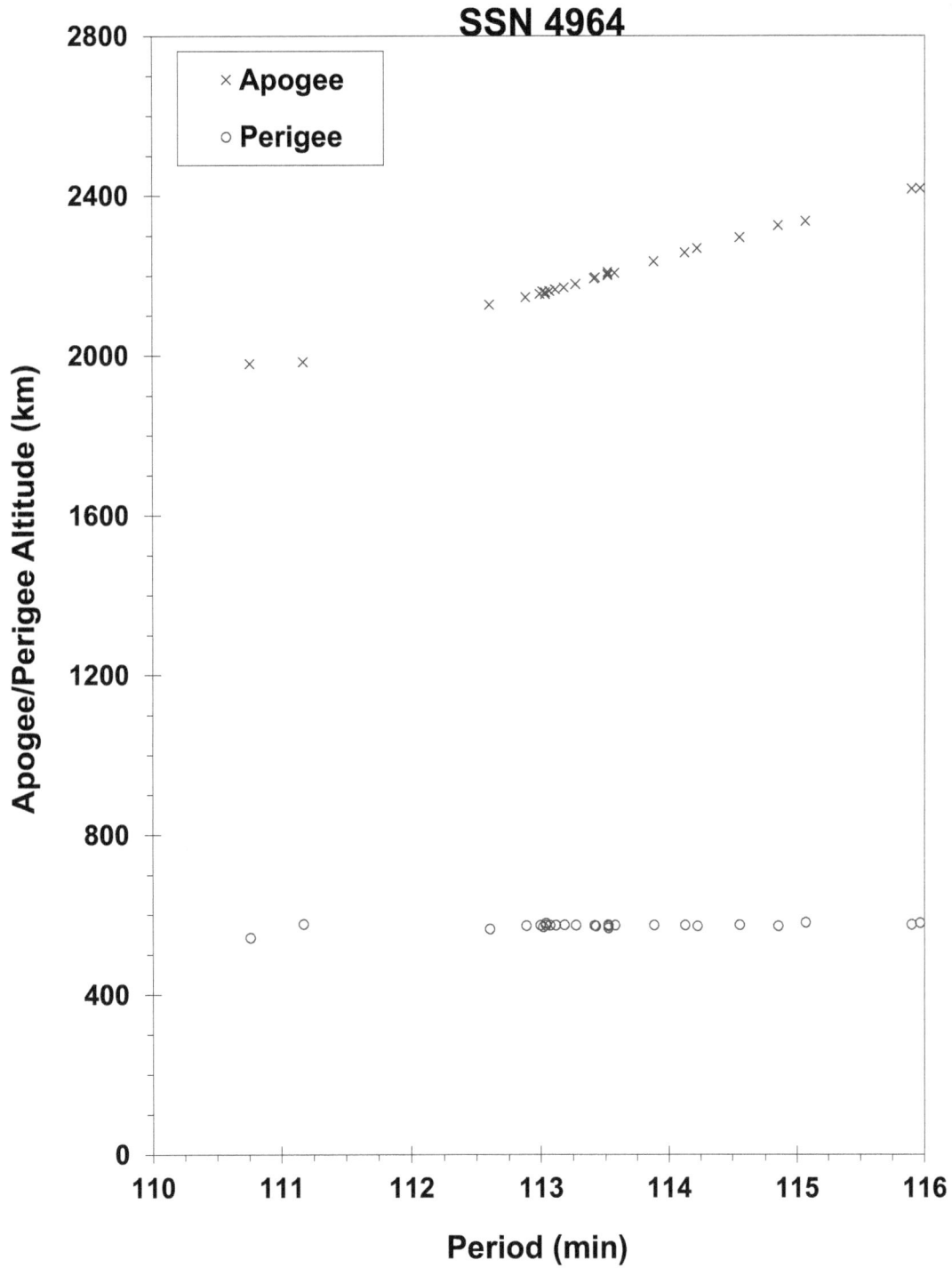

SSN 4964

Cosmos 397 cataloged debris cloud of 26 fragments about 7 weeks after the event as reconstructed from the US SSN database.

SATELLITE DATA

TYPE:	Payload
OWNER:	CIS
LAUNCH DATE:	3.55 Dec 1971
DRY MASS (KG):	1400
MAIN BODY:	Irregular; 1.8 m by 4.2 m
MAJOR APPENDAGES:	None
ATTITUDE CONTROL:	Active, 3-axis
ENERGY SOURCES:	On-board propellants, explosive charge

EVENT DATA

DATE:	3 Dec 1971	LOCATION:	51N, 7E (asc)
TIME:	1651 GMT	ASSESSED CAUSE:	Deliberate
ALTITUDE:	230 km		

POST-EVENT ELEMENTS

EPOCH:	71339.01001769	MEAN ANOMALY:	316.0762
RIGHT ASCENSION:	294.0999	MEAN MOTION:	13.65823046
INCLINATION:	65.7483	MEAN MOTION DOT/2:	.00001349
ECCENTRICITY:	.1062360	MEAN MOTION DOT DOT/6:	.0
ARG. OF PERIGEE:	53.3215	BSTAR:	.0

DEBRIS CLOUD DATA

MAXIMUM ΔP:	3.6 min
MAXIMUM ΔI:	0.7 deg

COMMENTS

Cosmos 462 was launched on a two-revolution rendezvous with Cosmos 459. The fragmentation occurred in the vicinity of Cosmos 459. Cosmos 462 was part of test series begun with Cosmos 249. Elements above are first available for orbit after final maneuver that took place immediately before fragmentation.

REFERENCE DOCUMENTS

"Artificial Satellite Break-Ups (Part 2): Soviet Anti-Satellite Program", N.L. Johnson, Journal of the British Interplanetary Society, August 1983, pp. 357-362.

History of Soviet/Russian Satellite Fragmentations-A Joint U.S.-Russian Investigation, N. L. Johnson, *et al.,* Kaman Sciences Corporation, October 1995.

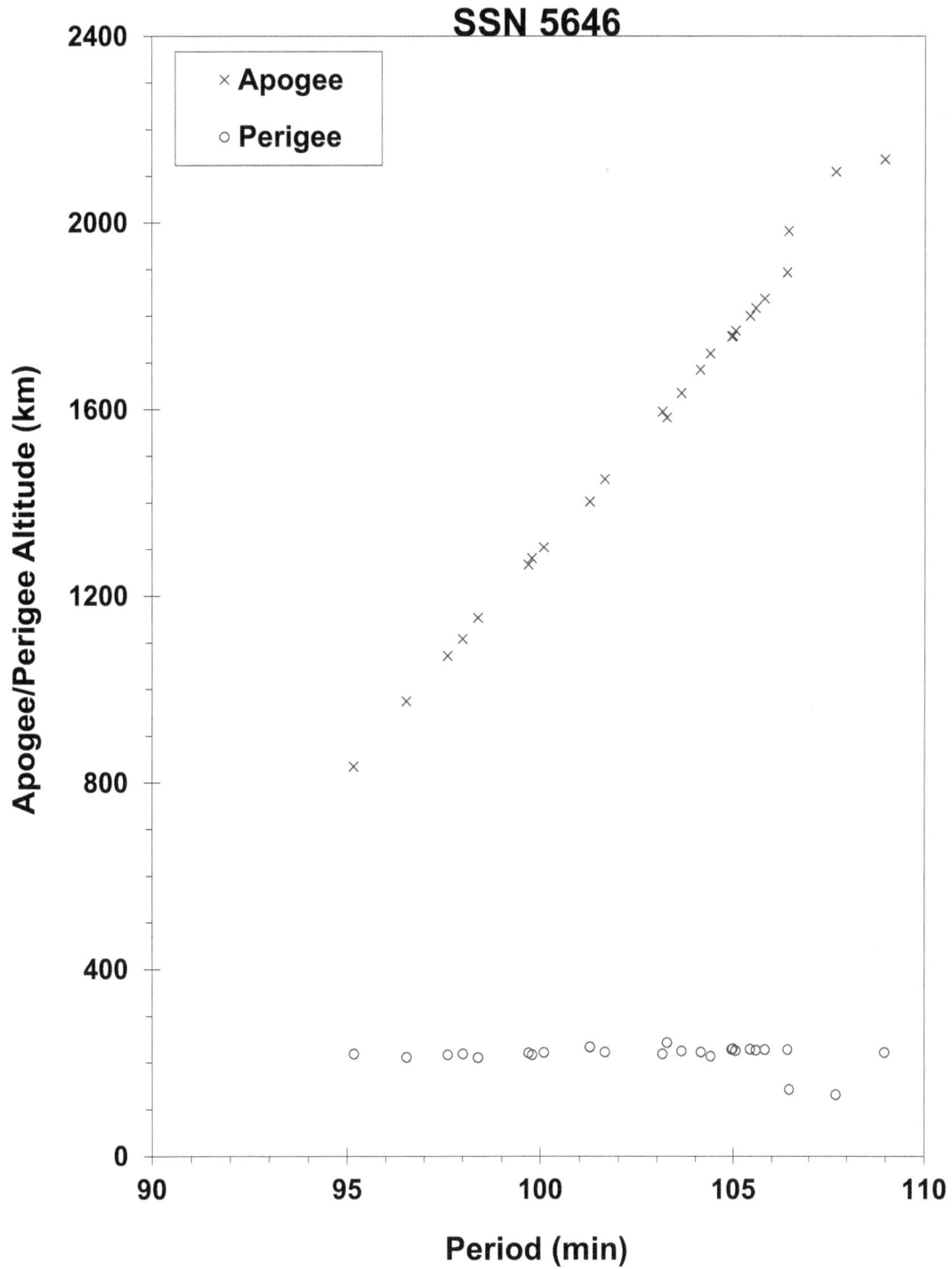

Cosmos 462 debris cloud of 25 cataloged fragments within 1 week of the event as reconstructed from the US SSN database.

LANDSAT 1 R/B — **1972-058B** — **6127**

SATELLITE DATA

TYPE: Delta Second Stage (900)
OWNER: US
LAUNCH DATE: 23.75 Jul 1972
DRY MASS (KG): 800
MAIN BODY: Cylinder-nozzle; 1.4 m diameter by 6.3 m length
MAJOR APPENDAGES: None
ATTITUDE CONTROL: None at time of the event.
ENERGY SOURCES: On-board propellants, range safety device

EVENT DATA

DATE: 22 May 1975
TIME: 1827 GMT
ALTITUDE: 730 km
LOCATION: 34S, 46E (asc)
ASSESSED CAUSE: Propulsion

PRE-EVENT ELEMENTS

EPOCH: 75142.56642671
RIGHT ASCENSION: 196.3353
INCLINATION: 98.3439
ECCENTRICITY: .0193108
ARG. OF PERIGEE: 38.1650
MEAN ANOMALY: 323.2981
MEAN MOTION: 14.36209995
MEAN MOTION DOT/2: .00000060
MEAN MOTION DOT DOT/6: .0
BSTAR: .000027579

DEBRIS CLOUD DATA

MAXIMUM ΔP: 9.3 min
MAXIMUM ΔI: 1.0 deg

COMMENTS

This was the second Delta Second Stage to experience a severe fragmentation. The event occurred 34 months after the successful deployment of the Landsat 1 payload. Cause of the explosion is assessed to be related to the nearly 150 kg of residual propellants and characteristics of the sun-synchronous orbit.

REFERENCE DOCUMENTS

Dynamics of Satellite Disintegration, R. Dasenbrock, B. Kaufman, and W. Heard, NRL Report 7954, Naval Research Laboratory, Washington, 30 January 1976.

"Fragmentations of Asteroids and Artificial Satellites in Orbit", W. Wiesel, Icarus, Vol. 34, 1978, pp. 99-116.

Explosion of Satellite 10704 and other Delta Second Stage Rockets, J.R. Gabbard, Technical Memorandum 81-5, DCS Plans, Hdqtrs NORAD/ADCOM, Colorado Springs, May 1981.

Investigation of Delta Second Stage On-Orbit Explosions, C.S. Gumpel, Report MDC-H0047, McDonnell Douglas Astronautics Company - West, Huntington Beach, April 1982.

A Later Look at Delta Second Stage On-Orbit Explosions, J.R. Gabbard, Technical Report CS85-BMDSC-00-24, Teledyne Brown Engineering, Colorado Springs, March 1985.

114

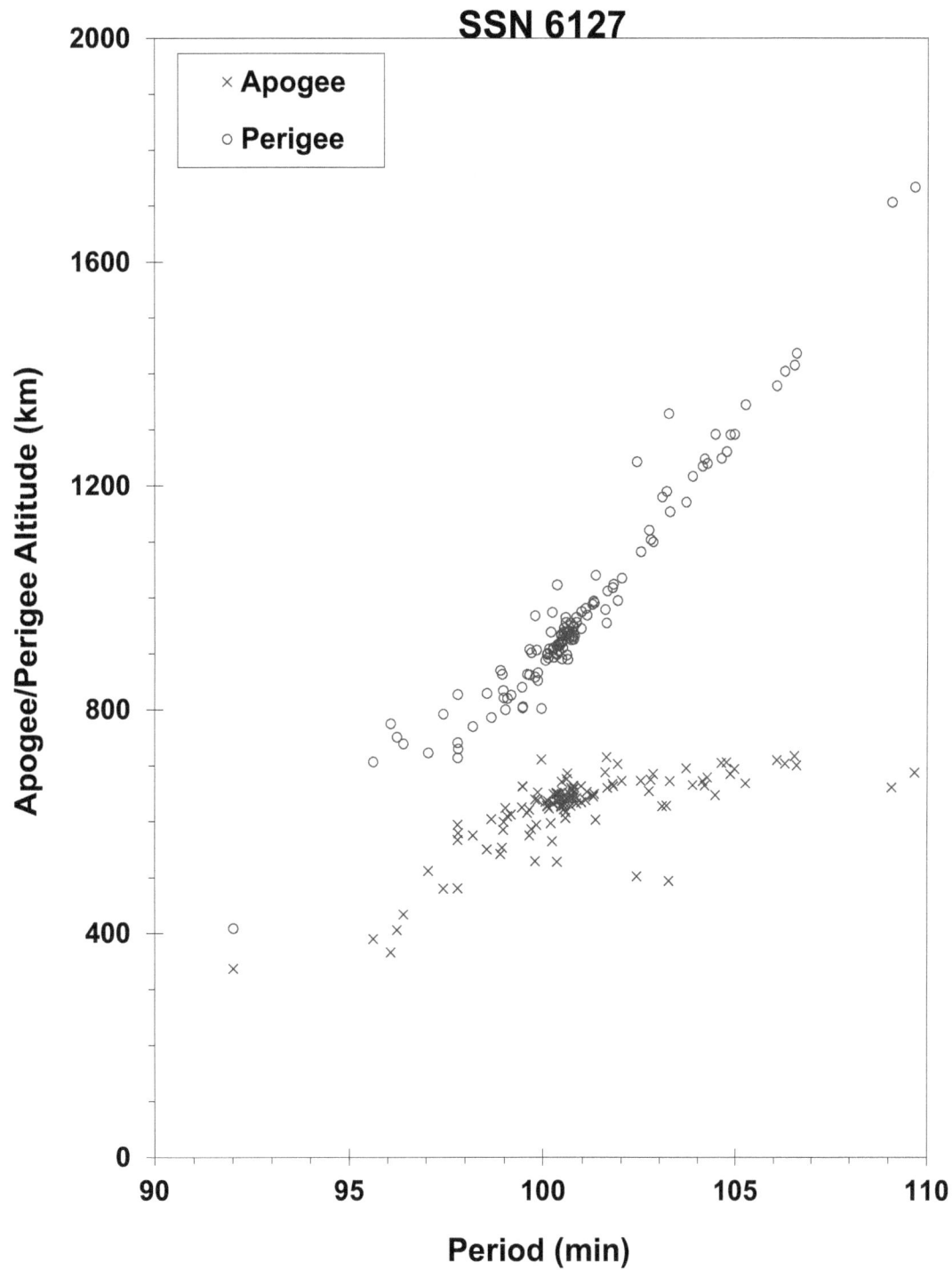

SSN 6127

Landsat 1 R/B debris cloud of 133 fragments 4 months after the event as reconstructed from the US SSN database.

SATELLITE DATA

TYPE:	Proton Third Stage
OWNER:	CIS
LAUNCH DATE:	3.38 Apr 1973
DRY MASS (KG):	4000
MAIN BODY:	Cylinder; 4.0 m diameter by 12.0 m length
MAJOR APPENDAGES:	None
ATTITUDE CONTROL:	None at time of the event.
ENERGY SOURCES:	On-board propellants

EVENT DATA

DATE:	3 Apr 1973	LOCATION:	45N, 290E (dsc)
TIME:	2236 GMT	ASSESSED CAUSE:	Propulsion
ALTITUDE:	225 km		

PRE-EVENT ELEMENTS

EPOCH:	73093.61404736	MEAN ANOMALY:	357.9254
RIGHT ASCENSION:	334.5652	MEAN MOTION:	16.20127597
INCLINATION:	51.4798	MEAN MOTION DOT/2:	.00508885
ECCENTRICITY:	.0037670	MEAN MOTION DOT DOT/6:	.0
ARG. OF PERIGEE:	2.1878	BSTAR:	.0

DEBRIS CLOUD DATA

MAXIMUM ΔP:	3.8 min
MAXIMUM ΔI:	0.5 deg

COMMENTS

This is the only known fragmentation of the Proton third stage. The event occurred less than 14 hours after reaching orbit. The event was apparently unrelated to the later payload malfunction. NAVSPASUR counted at least 95 objects shortly after the event, but most reentered before being officially cataloged. Information uncovered by Mr. Nicholas Johnson during an information exchange with Russian officials in the Spring of 1993 revealed that residual propellants resulted in an over-pressurization of the rocket body, causing this fragmentation. After this event, the Russians reported that the Proton third stage has been vented to avoid future events of this nature.

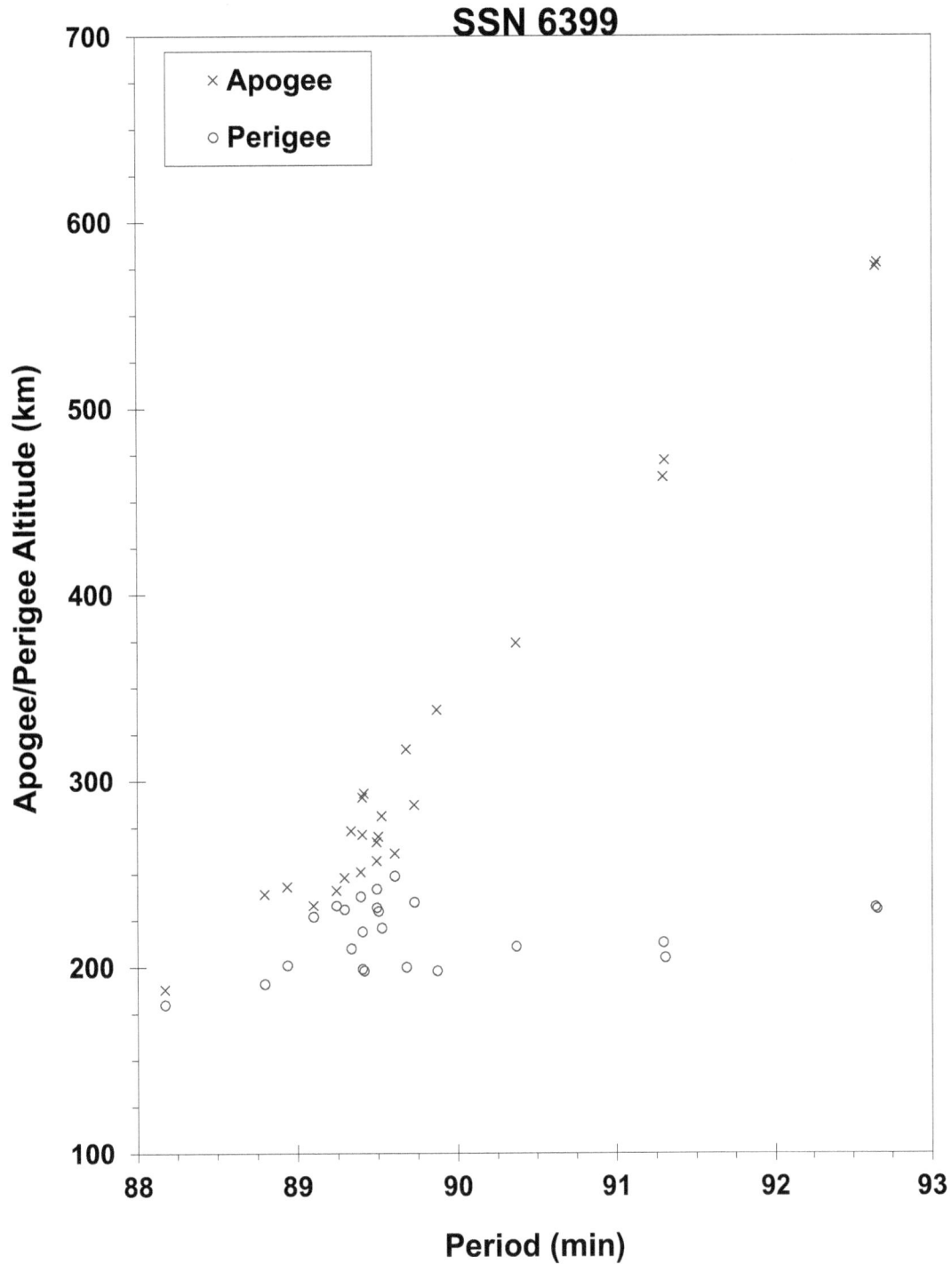

SSN 6399

Salyut 2 R/B debris cloud of 25 fragments as reconstructed from the US SSN database. Most elements were developed within 2 days of the event.

117

SATELLITE DATA

TYPE:	Payload
OWNER:	CIS
LAUNCH DATE:	19.38 Apr 1973
DRY MASS (KG):	6300
MAIN BODY:	Sphere-cylinder; 2.8 m diameter by 6.5 m length
MAJOR APPENDAGES:	None
ATTITUDE CONTROL:	Active, 3-axis
ENERGY SOURCES:	On-board propellants, explosive charge

EVENT DATA

DATE:	6 May 1973	LOCATION:	71S, 215E (asc)
TIME:	0724 GMT	ASSESSED CAUSE:	Deliberate
ALTITUDE:	310 km		

PRE-EVENT ELEMENTS

EPOCH:	73125.63953480	MEAN ANOMALY:	337.7411
RIGHT ASCENSION:	305.5573	MEAN MOTION:	16.05578988
INCLINATION:	72.8514	MEAN MOTION DOT/2:	.00433078
ECCENTRICITY:	.0137599	MEAN MOTION DOT DOT/6:	.00010923
ARG. OF PERIGEE:	22.9846	BSTAR:	.0

DEBRIS CLOUD DATA

MAXIMUM ΔP:	6.0 min
MAXIMUM ΔI:	1.3 deg

COMMENTS

Spacecraft was destroyed after a malfunction prevented controlled reentry and landing in the Soviet Union. A total of 88 fragments were cataloged without elements.

REFERENCE DOCUMENT

History of Soviet/Russian Satellite Fragmentations-A Joint U.S.-Russian Investigation, N. L. Johnson, *et al.*, Kaman Sciences Corporation, October 1995.

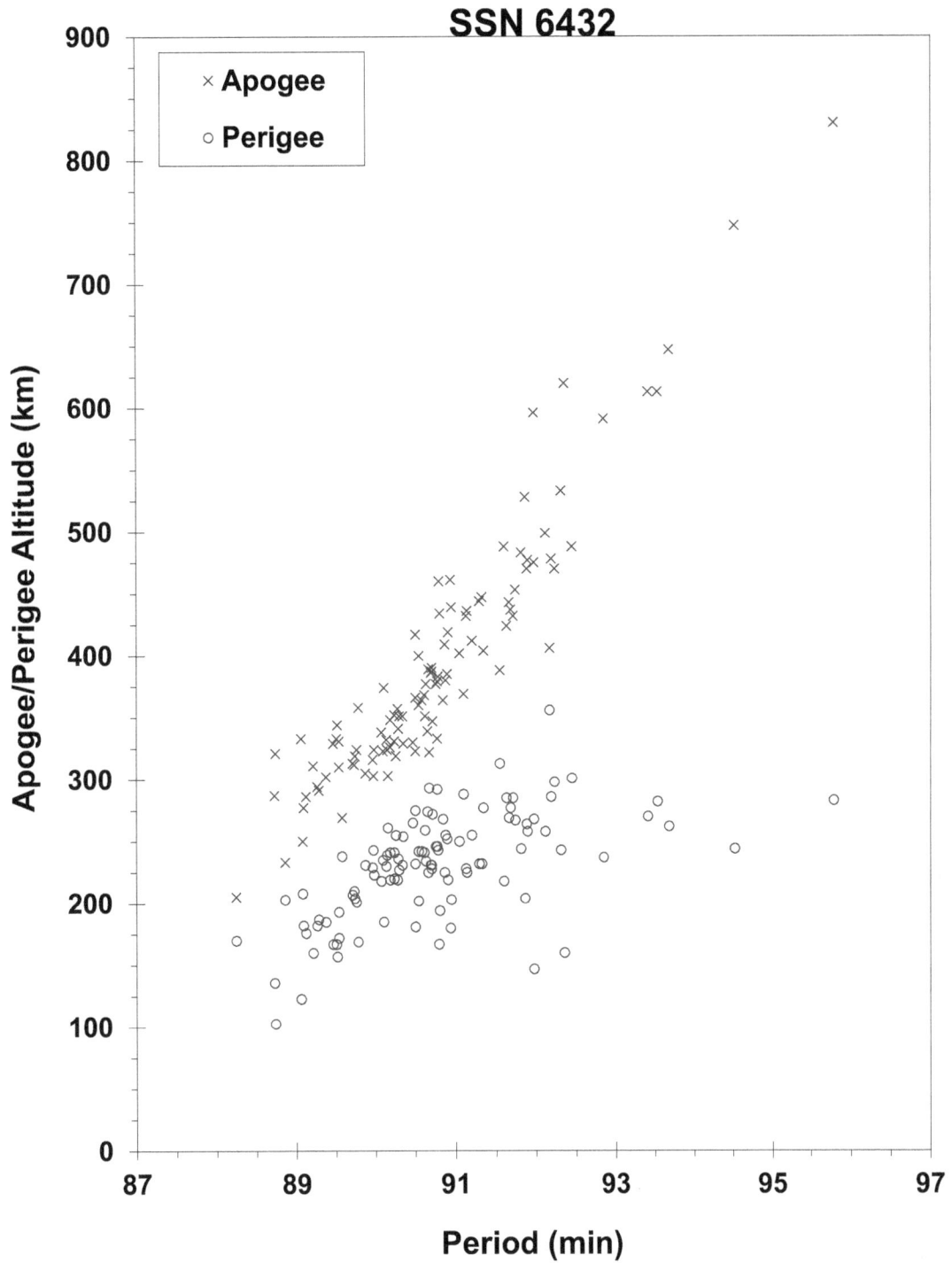

SSN 6432

Cosmos 554 debris cloud of 107 fragments using initial elements as developed over several weeks. Some decay effects are present. Source is the US SSN database.

SATELLITE DATA

TYPE:	Delta Second Stage (300)
OWNER:	US
LAUNCH DATE:	6.71 Nov 1973
DRY MASS (KG):	840
MAIN BODY:	Cylinder-nozzle; 1.4 m diameter by 6.3 m length
MAJOR APPENDAGES:	None
ATTITUDE CONTROL:	None at time of the event.
ENERGY SOURCES:	On-board propellants, range safety device

EVENT DATA

DATE:	28 Dec 1973	LOCATION:	37S, 181E (asc)
TIME:	0904 GMT	ASSESSED CAUSE:	Propulsion
ALTITUDE:	1515 km		

PRE-EVENT ELEMENTS

EPOCH:	73359.56303028	MEAN ANOMALY:	202.2816
RIGHT ASCENSION:	41.7242	MEAN MOTION:	12.40088347
INCLINATION:	102.0500	MEAN MOTION DOT/2:	.00000577
ECCENTRICITY:	.0005689	MEAN MOTION DOT DOT/6:	.000000056523
ARG. OF PERIGEE:	157.8450	BSTAR:	.0

DEBRIS CLOUD DATA

MAXIMUM ΔP:	10.4 min
MAXIMUM ΔI:	1.4 deg

COMMENTS

This was the first of seven Delta Second Stages to experience severe fragmentations between 1973 and 1981. Six of the seven stages were left in mid-morning, sun-synchronous orbits with residual propellants. Fragmentations occurred from 2-35 months after launch. The seventh stage exploded within hours of launch on a geosynchronous mission. The assessed cause in all cases is a propellant-induced explosion. Depletion burns to remove residual propellants were initiated in 1981, and no vented Delta Second Stages have fragmented since. In the case of the NOAA 3 R/B, fragmentation took place nearly 2 months after successful deployment of the NOAA 3 payload. Approximately 130 kg of propellants were left on board.

REFERENCE DOCUMENTS

Dynamics of Satellite Disintegration, R. Dasenbrock, B. Kaufman, and W. Heard, NRL Report 7954, Naval Research Laboratory, Washington, 30 January 1976.

Explosion of Satellite 10704 and other Delta Second Stage Rockets, J.R. Gabbard, Technical Memorandum 81-5, DCS Plans, Hdqtrs NORAD/ADCOM, Colorado Springs, May 1981.

Investigation of Delta Second Stage On-Orbit Explosions, C.S. Gumpel, Report MDC-H0047, McDonnell Douglas Astronautics Company - West, Huntington Beach, April 1982.

A Later Look at Delta Second Stage On-Orbit Explosions, J.R. Gabbard, Technical Report CS85-BMDSC-00-24, Teledyne Brown Engineering, Colorado Springs, March 1985.

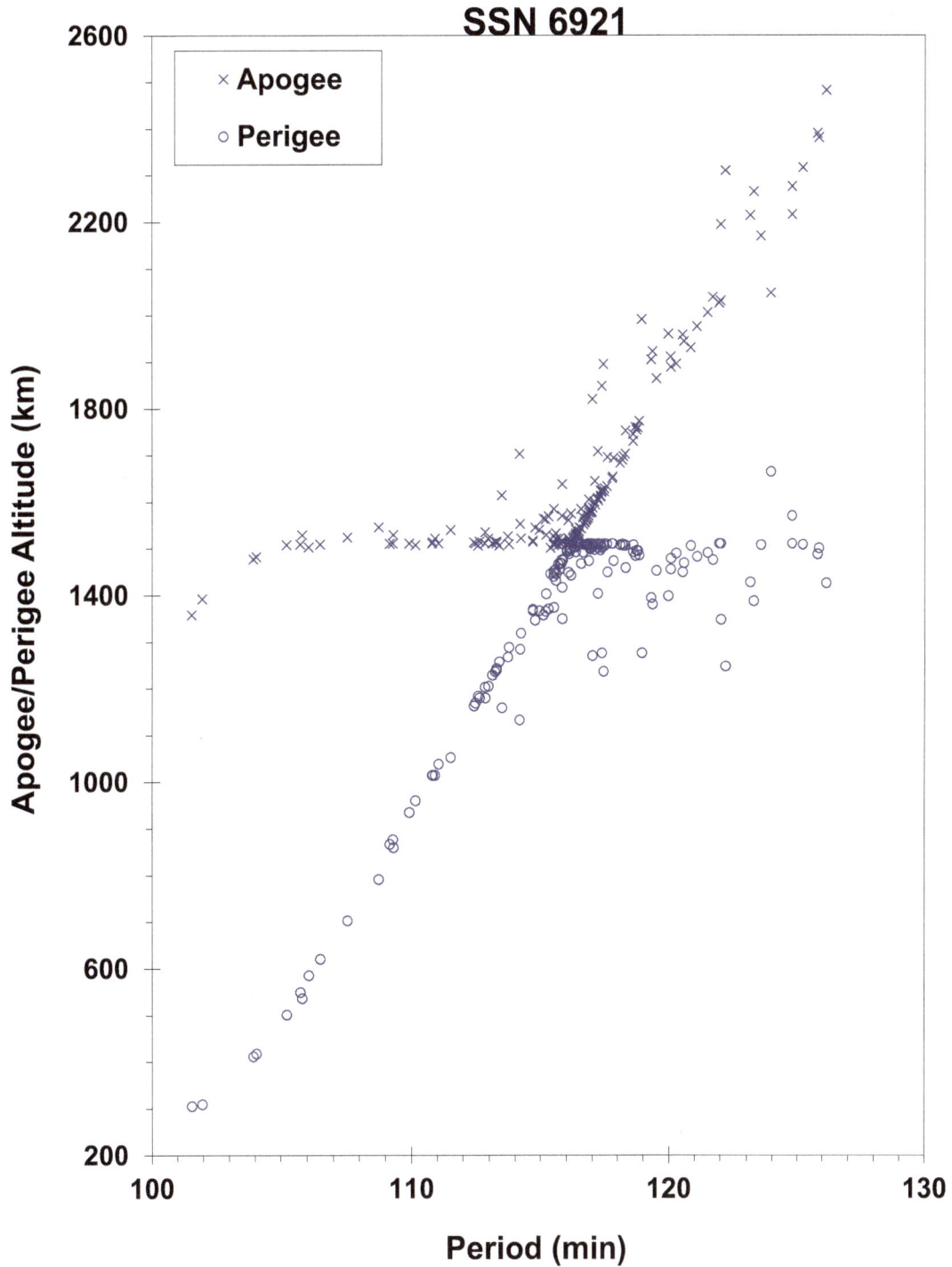

SSN 6921

NOAA 3 R/B debris cloud of 160 fragments 4 months after the event as reconstructed from the US SSN database.

SATELLITE DATA

TYPE:	Rocket Body
OWNER:	US
LAUNCH DATE:	16.33 Mar 1974
DRY MASS (KG):	50
MAIN BODY:	Box; 1 m long by 1 m wide by 1 m high
MAJOR APPENDAGES:	None
ATTITUDE CONTROL:	None at time of event
ENERGY SOURCES:	None at time of event

EVENT DATA

DATE:	17 Jan 2005	LOCATION:	80.6S, 306.8E
TIME:	0214 GMT	ASSESSED CAUSE:	Accidental Collision
ALTITUDE:	885 km		

PRE-EVENT ELEMENTS

EPOCH:	05016.54972523	MEAN ANOMALY:	256.1717
RIGHT ASCENSION:	350.2846	MEAN MOTION:	14.24162249
INCLINATION:	99.0928	MEAN MOTION DOT/2:	0.00000028
ECCENTRICITY:	.0066248	MEAN MOTION DOT DOT/6:	0.0
ARG. OF PERIGEE:	104.6813	BSTAR:	0.000031607

DEBRIS CLOUD DATA

MAXIMUM ΔP:	Unknown
MAXIMUM ΔI:	Unknown

COMMENTS

The THOR 2A upper stage collided with a piece of fragmentation debris (1999-57CV, SSN# 26207) from the March 2000 explosion of the third stage of a Chinese CZ-4 launch vehicle (1999-57C, SSN# 25942). This was the third historical collision of cataloged objects; the collision occurred over Antarctica as both objects were near the southernmost point in their respective retrograde orbits. The relative velocity of the collision was just under 6 km/s. 1999-57CV is believed to be relatively small, with a radar cross-section of 600 cm^2. The collision produced less than 10 cataloged debris. Because the upper stage had been in orbit for over 30 years, it is believed there was no on-board propellant remaining at the time of the event.

REFERENCE DOCUMENT

"Accidental Collisions of Cataloged Satellites Identified," The Orbital Debris Quarterly News, NASA JSC, April 2005. Available online at https://orbitaldebris.jsc.nasa.gov/quarterly-news/pdfs/ODQNv9i2.pdf.

Insufficient data to construct a Gabbard diagram.

SATELLITE DATA

TYPE:	Delta Second Stage (2310)
OWNER:	US
LAUNCH DATE:	15.72 Nov 1974
DRY MASS (KG):	840
MAIN BODY:	Cylinder-nozzle; 1.4 m diameter by 5.8 m length
MAJOR APPENDAGES:	Mini-skirt; 2.4m by 0.3 m
ATTITUDE CONTROL:	None at time of the event.
ENERGY SOURCES:	On-board propellants, range safety device

EVENT DATA

DATE:	20 Aug 1975	LOCATION:	52S, 278E (dsc)
TIME:	1307 GMT	ASSESSED CAUSE:	Propulsion
ALTITUDE:	1465 km		

PRE-EVENT ELEMENTS

EPOCH:	75231.53619619	MEAN ANOMALY:	309.0001
RIGHT ASCENSION:	277.2201	MEAN MOTION:	12.52826370
INCLINATION:	101.6940	MEAN MOTION DOT/2:	.00000083
ECCENTRICITY:	.0009694	MEAN MOTION DOT DOT/6:	.0
ARG. OF PERIGEE:	51.1891	BSTAR:	.0

DEBRIS CLOUD DATA

MAXIMUM ΔP:	15.7 min
MAXIMUM ΔI:	1.8 deg

COMMENTS

This was the third Delta Second Stage to experience a severe fragmentation. The event occurred 9 months after the successful deployment of the NOAA 4 payload. Cause of the explosion is assessed to be related to the estimated more than 200 kg of residual propellants and characteristics of the sun-synchronous orbit. A fragment from this event (satellite number 8138) may have generated six or more additional pieces in September 1981.

REFERENCE DOCUMENTS

Explosion of Satellite 10704 and other Delta Second Stage Rockets, J.R. Gabbard, Technical Memorandum 81-5, DCS Plans, Hdqtrs NORAD/ADCOM, Colorado Springs, May 1981.

Investigation of Delta Second Stage On-Orbit Explosions, C.S. Gumpel, Report MDC-H0047, McDonnell Douglas Astronautics Company - West, Huntington Beach, April 1982.

A Later Look at Delta Second Stage On-Orbit Explosions, J.R. Gabbard, Technical Report CS85-BMDSC-00-24, Teledyne Brown Engineering, Colorado Springs, March 1985.

SSN 7532

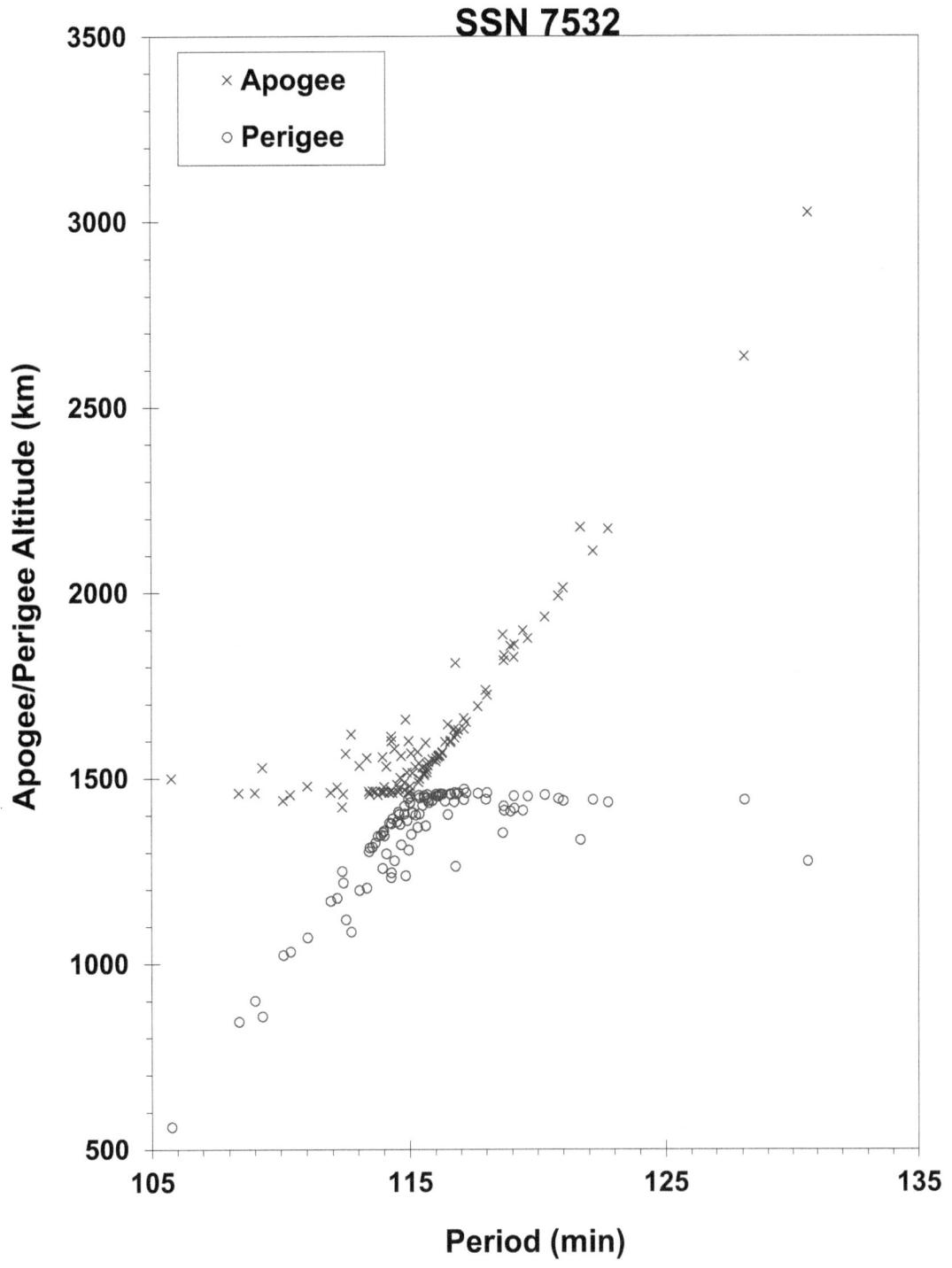

NOAA 4 R/B debris cloud of 101 fragments 6 months after the event as reconstructed from the US SSN database.

SATELLITE DATA

TYPE:	Payload
OWNER:	CIS
LAUNCH DATE:	24.46 Dec 1974
DRY MASS (KG):	3000
MAIN BODY:	Cylinder; 1.3 m diameter by 17 m length
MAJOR APPENDAGES:	Solar panels
ATTITUDE CONTROL:	Active, 3-axis
ENERGY SOURCES:	On-board propellants, explosive charge (?)

EVENT DATA (1)

DATE:	17 Apr 1975	LOCATION:	01N, 278E (dsc)
TIME:	2148 GMT	ASSESSED CAUSE:	Unknown
ALTITUDE:	440 km		

PRE-EVENT ELEMENTS (1)

EPOCH:	75107.81173798	MEAN ANOMALY:	71.8460
RIGHT ASCENSION:	271.0743	MEAN MOTION:	15.44155646
INCLINATION:	65.0355	MEAN MOTION DOT/2:	.00007106
ECCENTRICITY:	.0014224	MEAN MOTION DOT DOT/6:	.0
ARG. OF PERIGEE:	288.1084	BSTAR:	.0

EVENT DATA (2)

DATE:	2 Aug 1975	LOCATION:	02S, 258E (dsc)
TIME:	1623 GMT	ASSESSED CAUSE:	Unknown
ALTITUDE:	435 km		

PRE-EVENT ELEMENTS (2)

EPOCH:	75214.45597981	MEAN ANOMALY:	68.4232
RIGHT ASCENSION:	274.3453	MEAN MOTION:	15.46205523
INCLINATION:	65.0458	MEAN MOTION DOT/2:	.00001715
ECCENTRICITY:	.0020980	MEAN MOTION DOT DOT/6:	.0
ARG. OF PERIGEE:	291.4623	BSTAR:	.0

DEBRIS CLOUD DATA

MAXIMUM ΔP:	3.5 min*
MAXIMUM ΔI:	0.9 deg*

*Based on NRL analysis

COMMENTS

Cosmos 699 was the first of a new type spacecraft. Many members of this class have experienced breakups. Beginning in 1988 old spacecraft have been commanded to lower perigee at end of life, resulting in an accelerated natural decay with fewer fragmentations. For several spacecraft, two distinct events have been detected and observational data suggest that the spacecraft remain essentially intact after each event. In most cases, breakups occur after spacecraft has ceased orbit maintenance and entered natural decay. Debris are sometimes highly unidirectional. In the case of Cosmos 699, the spacecraft had been in a regime of natural decay for 1 month at the time of the event.

REFERENCE DOCUMENTS

An Analysis of the Breakup of Satellite 1974-103A (Cosmos 699), W. B. Heard, NRL Report 7991, Naval Research Laboratory, Washington, 23 April 1976.

"Artificial Satellite Break-Ups (Part 1): Soviet Ocean Surveillance Satellites", N. L. Johnson, Journal of the British Interplanetary Society, February 1983, pp. 51-58.

History of the Soviet/Russian Satellite Fragmentations-A Joint U.S.-Russian Investigation, N. L. Johnson, et al., Kaman Sciences Corporation, October 1995.

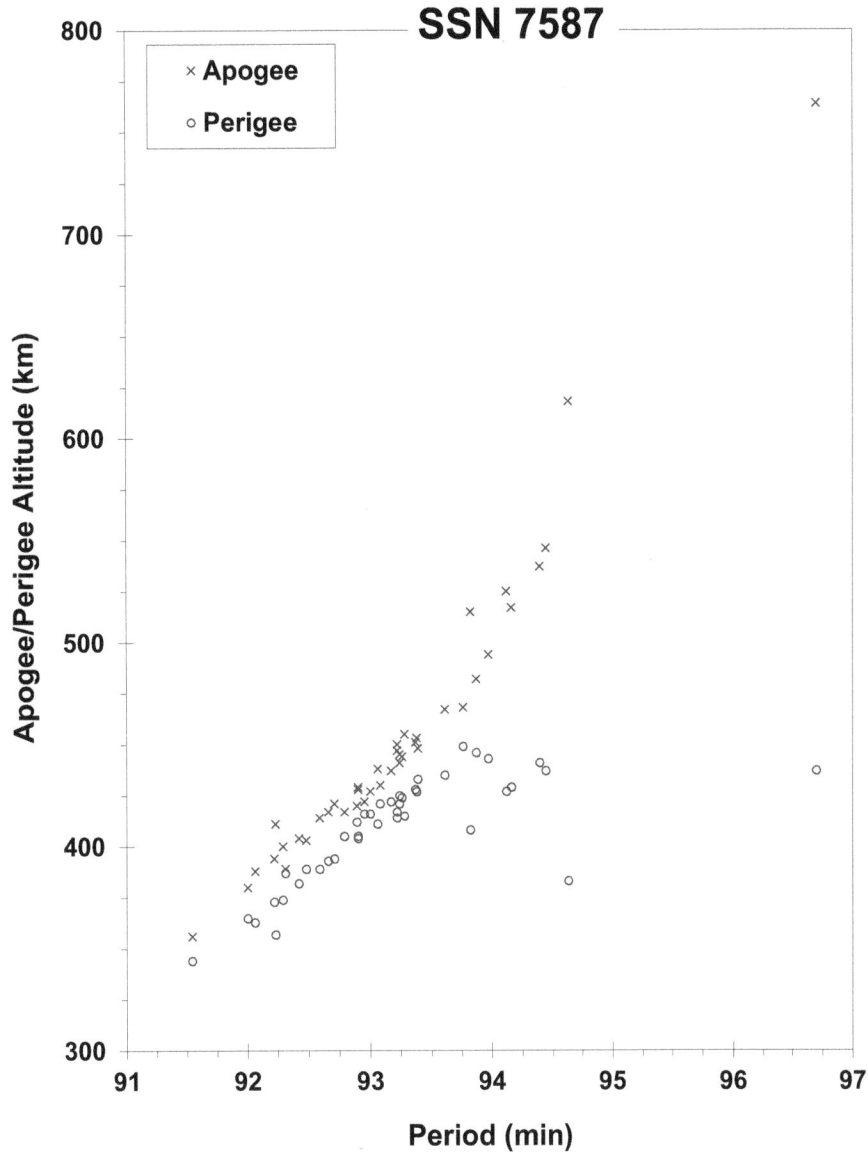

Cosmos 699 debris cloud of 41 fragments after the first breakup event as reconstructed from radar observations following the first breakup event.

SATELLITE DATA

TYPE:	Delta Second Stage (2910)
OWNER:	US
LAUNCH DATE:	22.75 Jan 1975
DRY MASS (KG):	840
MAIN BODY:	Cylinder-nozzle; 1.4 m diameter by 5.8 m length
MAJOR APPENDAGES:	Mini-skirt; 2.4 m by 0.2 m
ATTITUDE CONTROL:	None at time of the event.
ENERGY SOURCES:	On-board propellants, range safety device

EVENT DATA (1)

DATE:	9 Feb 1976	LOCATION:	Unknown
TIME:	Unknown	ASSESSED CAUSE:	Propulsion
ALTITUDE:	Unknown		

PRE-EVENT ELEMENTS (1)

EPOCH:	76040.08509016	MEAN ANOMALY:	189.3492
RIGHT ASCENSION:	60.2329	MEAN MOTION:	14.19373945
INCLINATION:	97.7751	MEAN MOTION DOT/2:	.0
ECCENTRICITY:	.0120730	MEAN MOTION DOT DOT/6:	.0
ARG. OF PERIGEE:	170.9843	BSTAR:	.0

EVENT DATA (2)

DATE:	19 Jun 1976	LOCATION:	7N, 344E (dsc)
TIME:	0659 GMT	ASSESSED CAUSE:	Propulsion
ALTITUDE:	750 km		

PRE-EVENT ELEMENTS (2)

EPOCH:	76170.97576375	MEAN ANOMALY:	217.2433
RIGHT ASCENSION:	175.3897	MEAN MOTION:	14.19574919
INCLINATION:	97.7497	MEAN MOTION DOT/2:	.0
ECCENTRICITY:	.0115288	MEAN MOTION DOT DOT/6:	.0
ARG. OF PERIGEE:	143.6594	BSTAR:	.0

DEBRIS CLOUD DATA

MAXIMUM ΔP:	5.6 min
MAXIMUM ΔI:	2.3 deg

COMMENTS

This was the fourth Delta Second Stage to experience a severe fragmentation. The first event occurred almost 13 months after the successful deployment of the Landsat 2 payload. Only 14 fragments were cataloged after the first event and all possessed orbital period changes of less than 0.6 min. Four months later a much larger fragmentation occurred. The cause of the second event is assessed to be related to the estimated 150 kg of residual propellants on board and characteristics of the sun-synchronous orbit.

REFERENCE DOCUMENTS

Explosion of Satellite 10704 and other Delta Second Stage Rockets, J.R. Gabbard, Technical Memorandum 81-5, DCS Plans, Hdqtrs NORAD/ADCOM, Colorado Springs, May 1981.

Investigation of Delta Second Stage On-Orbit Explosions, C.S. Gumpel, Report MDC-H0047, McDonnell Douglas Astronautics Company - West, Huntington Beach, April 1982.

A Later Look at Delta Second Stage On-Orbit Explosions, J.R. Gabbard, Technical Report CS85-BMDSC-00-24, Teledyne Brown Engineering, Colorado Springs, March 1985.

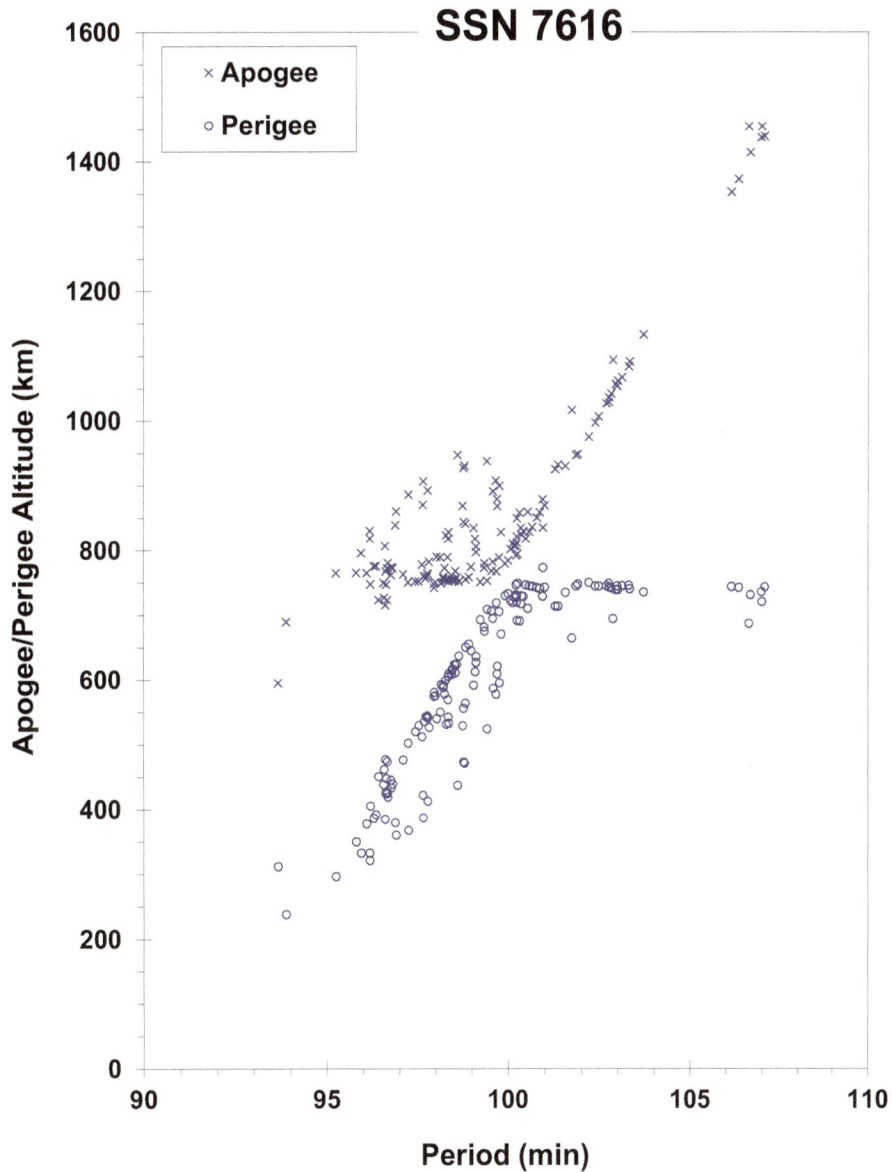

Landsat 2 R/B debris cloud of 147 fragments about 6 weeks after the second event as reconstructed from the US SSN database.

SATELLITE DATA

TYPE:	Delta Second Stage (2910)
OWNER:	US
LAUNCH DATE:	12.34 Jun 1975
DRY MASS (KG):	840
MAIN BODY:	Cylinder-nozzle; 1.4 m diameter by 5.8 m length
MAJOR APPENDAGES:	Mini-skirt; 2.4 m by 0.3 m
ATTITUDE CONTROL:	None at time of the event.
ENERGY SOURCES:	On-board propellants, range safety device

EVENT DATA

DATE:	1 May 1991	LOCATION:	66N, 322E (asc)
TIME:	0856 GMT	ASSESSED CAUSE:	Propulsion
ALTITUDE:	1090 km		

PRE-EVENT ELEMENTS

EPOCH:	91112.56709963	MEAN ANOMALY:	211.7525
RIGHT ASCENSION:	329.2109	MEAN MOTION:	13.43007146
INCLINATION:	99.5801	MEAN MOTION DOT/2:	.00000050
ECCENTRICITY:	.0006217	MEAN MOTION DOT DOT/6:	0.0
ARG. OF PERIGEE:	148.3989	BSTAR:	.0055458

DEBRIS CLOUD DATA

MAXIMUM ΔP:	27.4 min*
MAXIMUM ΔI:	2.4 min*

*Based on uncataloged debris data

COMMENTS

This was the eighth Delta Second Stage to experience a severe fragmentation. The event occurred nearly 191 months after the successful deployment of the Nimbus 6 payload. Cause of the explosion is assessed to be related to the estimated 245 kg of residual propellants on board and characteristics of the sun-synchronous orbit.

REFERENCE DOCUMENTS

The Fragmentation of the Nimbus 6 Rocket Body, D. J. Nauer and N. L. Johnson, Technical Report CS91-TR-JSC-017, Teledyne Brown Engineering, Colorado Springs, Colorado, November 1991.

Nimbus 6 Delta Upper Stage Rocket Body Breakup Report, E. L. Jenkins and H. V. Reynolds, Naval Space Surveillance Center, Dahlgren, Virginia, 1991.

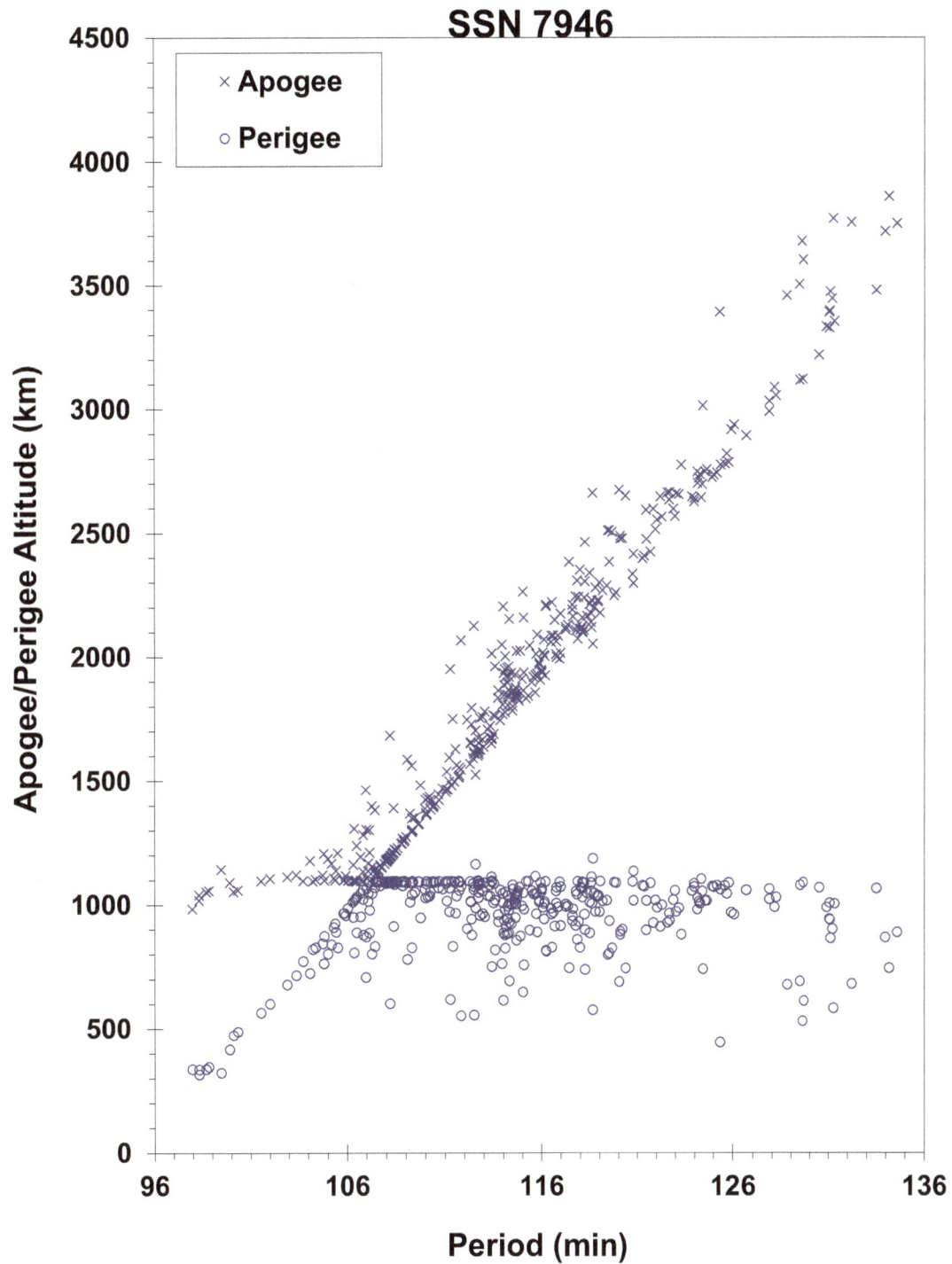

SSN 7946

Nimbus 6 R/B debris cloud of 386 identified fragments within 1 week after the event as reconstructed from Naval Space Surveillance System database. This diagram is taken from the first cited reference.

SATELLITE DATA

TYPE:	Payload
OWNER:	CIS
LAUNCH DATE:	5.62 Sep 1975
DRY MASS (KG):	5700
MAIN BODY:	Cone-cylinder; 2.7 m diameter by 6.3 m length
MAJOR APPENDAGES:	Solar panels
ATTITUDE CONTROL:	Active, 3-axis
ENERGY SOURCES:	On-board propellants, explosive charge

EVENT DATA

DATE:	6 Sep 1975	LOCATION:	32N, 293E (asc)
TIME:	1906 GMT	ASSESSED CAUSE:	Deliberate
ALTITUDE:	185 km		

PRE-EVENT ELEMENTS

EPOCH:	75249.72782895	MEAN ANOMALY:	294.2107
RIGHT ASCENSION:	189.2795	MEAN MOTION:	16.09422927
INCLINATION:	67.1445	MEAN MOTION DOT/2:	.00430774
ECCENTRICITY:	.0113994	MEAN MOTION DOT DOT/6:	.0
ARG. OF PERIGEE:	67.1020	BSTAR:	.0

DEBRIS CLOUD DATA

MAXIMUM ΔP:	Unknown
MAXIMUM ΔI:	Unknown

COMMENTS

Spacecraft was apparently destroyed after a malfunction prevented controlled reentry and landing in the Soviet Union. Most debris reentered before being officially cataloged. All but three official fragments were cataloged without elements.

REFERENCE DOCUMENT

History of Soviet/Russian Satellite Fragmentations-A Joint U.S.-Russian Investigation, N. L. Johnson, *et al.,* Kaman Sciences Corporation, October 1995.

Insufficient data to construct a Gabbard diagram.

SATELLITE DATA

TYPE:	Payload
OWNER:	CIS
LAUNCH DATE:	29.46 Oct 1975
DRY MASS (KG):	3000
MAIN BODY:	Cylinder; 1.3 m diameter by 17 m length
MAJOR APPENDAGES:	Solar panels
ATTITUDE CONTROL:	Active, 3-axis
ENERGY SOURCES:	On-board propellants, explosive charge (?)

EVENT DATA

DATE:	25 Jan 1976	LOCATION:	53N, 7E (asc)
TIME:	1400 GMT	ASSESSED CAUSE:	Unknown
ALTITUDE:	440 km		

PRE-EVENT ELEMENTS

EPOCH:	76025.37753295	MEAN ANOMALY:	88.9272
RIGHT ASCENSION:	303.6319	MEAN MOTION:	15.43461781
INCLINATION:	65.0177	MEAN MOTION DOT/2:	.00000373
ECCENTRICITY:	.0009065	MEAN MOTION DOT DOT/6:	.0
ARG. OF PERIGEE:	271.0782	BSTAR:	.0

DEBRIS CLOUD DATA

MAXIMUM ΔP:	1.6 min
MAXIMUM ΔI:	0.4 deg

COMMENTS

Cosmos 777 was the second spacecraft of the Cosmos 699-type to experience a fragmentation. It is the only one to breakup before terminating its precise orbit maintenance pattern and entering a regime of natural decay. A second event may have occurred about 90 minutes after the event cited above.

REFERENCE DOCUMENTS

"Artificial Satellite Break-Ups (Part 1): Soviet Ocean Surveillance Satellites", N. L. Johnson, Journal of the British Interplanetary Society, February 1983, pp. 51-58.

History of Soviet/Russian Satellite Fragmentations-A Joint U.S.-Russian Investigation, N. L. Johnson, et al., Kaman Sciences Corporation, October 1995.

SSN 8416

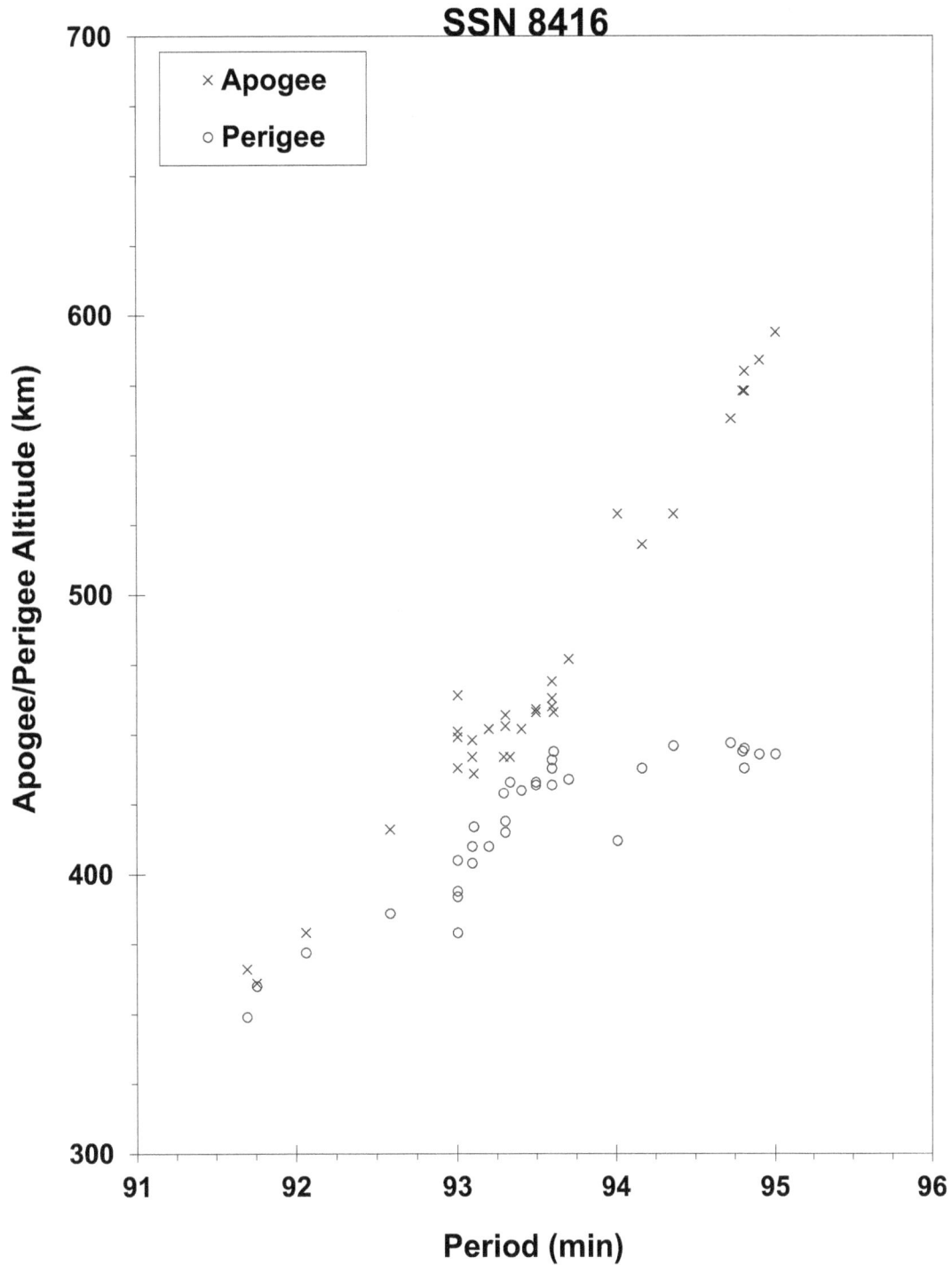

Cosmos 777 debris cloud of 35 fragments about 10 days after the event as reconstructed from the US SSN database. Some drag effects are already evident.

SATELLITE DATA

TYPE:	Transtage Rocket Body
OWNER:	US
LAUNCH DATE:	15 Mar. 1976
DRY MASS (KG):	1,874
MAIN BODY:	3.048 m diameter x ~5 m long right circular cylinder
MAJOR APPENDAGES:	none
ATTITUDE CONTROL:	inactive 3-axis
ENERGY SOURCES:	on-board propellants; pressure vessels; chemical batteries

EVENT DATA

DATE:	07 May 2019	LOCATION:	14.26N, 346.89E (asc)
TIME:	1902 GMT	ASSESSED CAUSE:	Propulsion
ALTITUDE:	35898.4 km		

PRE-EVENT ELEMENTS

EPOCH:	19127.12410370	MEAN ANOMALY:	163.4933
RIGHT ASCENSION:	77.1522	MEAN MOTION:	0.98264106
INCLINATION:	16.2502	MEAN MOTION DOT/2:	-0.00000220
ECCENTRICITY:	0.0142404	MEAN MOTION DOT DOT/6:	0.00000
ARG. OF PERIGEE:	20.3519	BSTAR:	0.00000

DEBRIS CLOUD DATA

MAXIMUM ΔP:	Unknown
MAXIMUM ΔI:	Unknown

COMMENTS

The cause of this event, occurring approximately 43 years after launch, is likely due to on-board stored energy sources. This is the fifth breakup of a Titan Transtage rocket body, and the third Transtage fragmentation near geosynchronous orbit. Due to difficulties in cataloging highly elliptical and deep-space orbit many more fragments may be resident on-orbit; to date, no additional debris have entered the public catalog.

REFERENCE DOCUMENTS

"Three Recent Rocket Body Breakups," The Orbital Debris Quarterly News, NASA JSC, November 2019. Available online at https://www.orbitaldebris.jsc.nasa.gov/quarterly-news/pdfs/odqnv23i4.pdf.

Cowardin, H., P. Anz-Meador, and J. Reyes, "Characterizing GEO Titan IIIC Transtage Fragmentations using Ground-based and Telescopic Measurements."   In Proceedings of the AMOS 2017 Technical Conference. Available online at https://amostech.com/TechnicalPapers/2017/Orbital-Debris/Cowardin.pdf

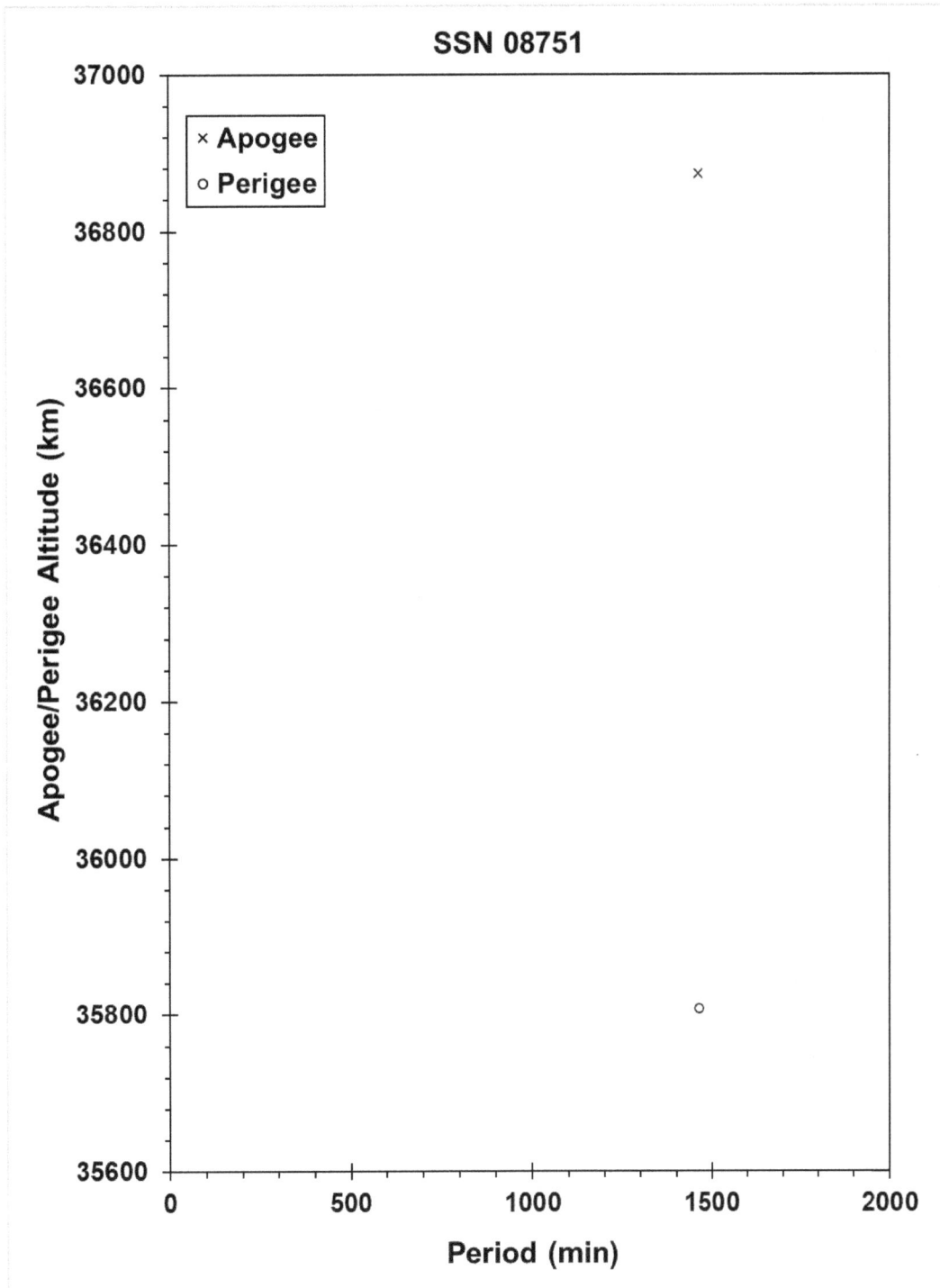

SSN 08751

Transstage R/B debris "cloud" of parent body as reconstructed from the US SSN database.

SATELLITE DATA

TYPE:	Payload
OWNER:	CIS
LAUNCH DATE:	2.44 Jul 1976
DRY MASS (KG):	3000
MAIN BODY:	Cylinder; 1.3 m diameter by 17 m length
MAJOR APPENDAGES:	Solar panels
ATTITUDE CONTROL:	Active, 3-axis
ENERGY SOURCES:	On-board propellants, explosive charge (?)

EVENT DATA

DATE:	17 May 1977	LOCATION:	9S, 284E (dsc)
TIME:	1018 GMT	ASSESSED CAUSE:	Unknown
ALTITUDE:	430 km		

PRE-EVENT ELEMENTS

EPOCH:	77136.94211102	MEAN ANOMALY:	73.5502
RIGHT ASCENSION:	131.3837	MEAN MOTION:	15.45822335
INCLINATION:	65.0556	MEAN MOTION DOT/2:	.00007521
ECCENTRICITY:	.0021270	MEAN MOTION DOT DOT/6:	.0
ARG. OF PERIGEE:	286.3253	BSTAR:	.0

DEBRIS CLOUD DATA

MAXIMUM ΔP:	8.0 min*
MAXIMUM ΔI:	1.1 deg*

*Based on uncataloged debris data

COMMENTS

Cosmos 838 was the third spacecraft of the Cosmos 699-type to experience a fragmentation. Spacecraft had been in a regime of natural decay for 6 months prior to the event. Many debris reentered before being officially cataloged.

REFERENCE DOCUMENTS

"Artificial Satellite Break-Ups (Part 1): Soviet Ocean Surveillance Satellites", N. L. Johnson, <u>Journal of the British Interplanetary Society</u>, February 1983, pp. 51-58.

<u>History of Soviet/Russian Satellite Fragmentations-A Joint U.S.-Russian Investigation</u>, N. L. Johnson, *et al.,* Kaman Sciences Corporation, October 1995.

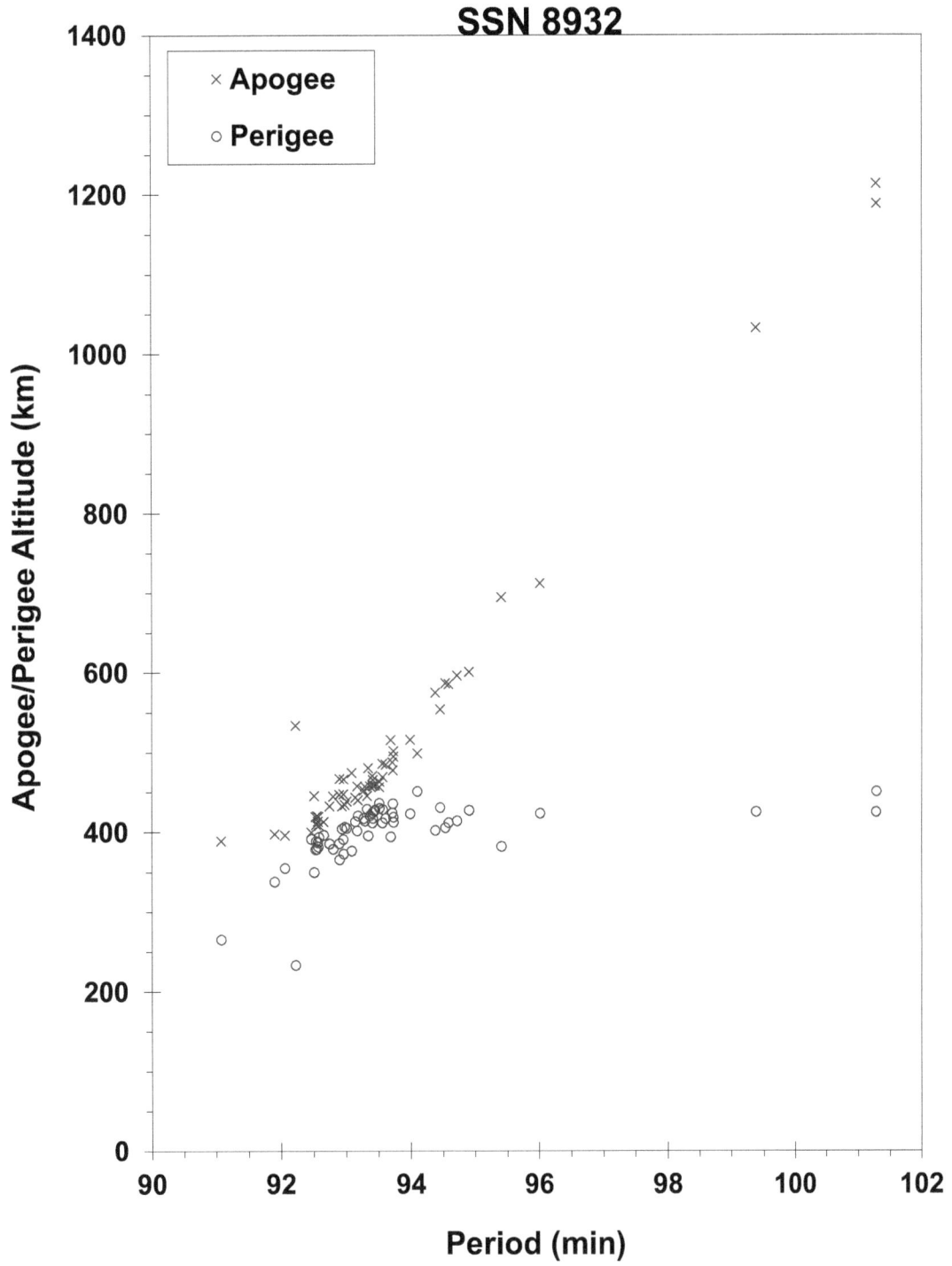

Cosmos 838 debris cloud of 59 fragments about 1 week after the event as reconstructed from the US SSN database.

SATELLITE DATA

TYPE:	Payload
OWNER:	CIS
LAUNCH DATE:	8.88 Jul 1976
DRY MASS (KG):	650
MAIN BODY:	Polyhedron; 1.4 m by 1.4 m
MAJOR APPENDAGES:	Solar panels, gravity-gradient boom (?)
ATTITUDE CONTROL:	Gravity gradient (?)
ENERGY SOURCES:	Battery

EVENT DATA

DATE:	29 Sep 1977	LOCATION:	33S, 162E (dsc)
TIME:	0717 GMT	ASSESSED CAUSE:	Battery
ALTITUDE:	1910 km		

PRE-EVENT ELEMENTS

EPOCH:	77270.46732078	MEAN ANOMALY:	7.6996
RIGHT ASCENSION:	85.9347	MEAN MOTION:	12.32137908
INCLINATION:	65.8538	MEAN MOTION DOT/2:	.00000367
ECCENTRICITY:	.0706585	MEAN MOTION DOT DOT/6:	.0
ARG. OF PERIGEE:	351.1444	BSTAR:	.0

DEBRIS CLOUD DATA

MAXIMUM ΔP:	2.7 min
MAXIMUM ΔI:	0.3 deg

COMMENTS

Cosmos 839 was the first of three satellites of the same class to experience unexplained fragmentations. These satellites are used in conjunction with the Cosmos 249-type spacecraft, which are deliberately fragmented; but the cause of the Cosmos 839-type events appears to be unrelated since they occur more than 1 year after tests with Cosmos 249-type spacecraft. In the case of Cosmos 839, 14 months elapsed between its test with a Cosmos 249-type spacecraft and its fragmentation. Russian officials have determined that battery malfunctions were the causes of these events.

REFERENCE DOCUMENTS

"Artificial Satellite Break-Ups (Part 2): Soviet Anti-Satellite Program", N.L. Johnson, <u>Journal of the British Interplanetary Society</u>, August 1983, pp. 357-362.

<u>History of Soviet/Russian Satellite Fragmentations-A Joint U.S.-Russian Investigation</u>, N. L. Johnson, *et al.,* Kaman Sciences Corporation, October 1995.

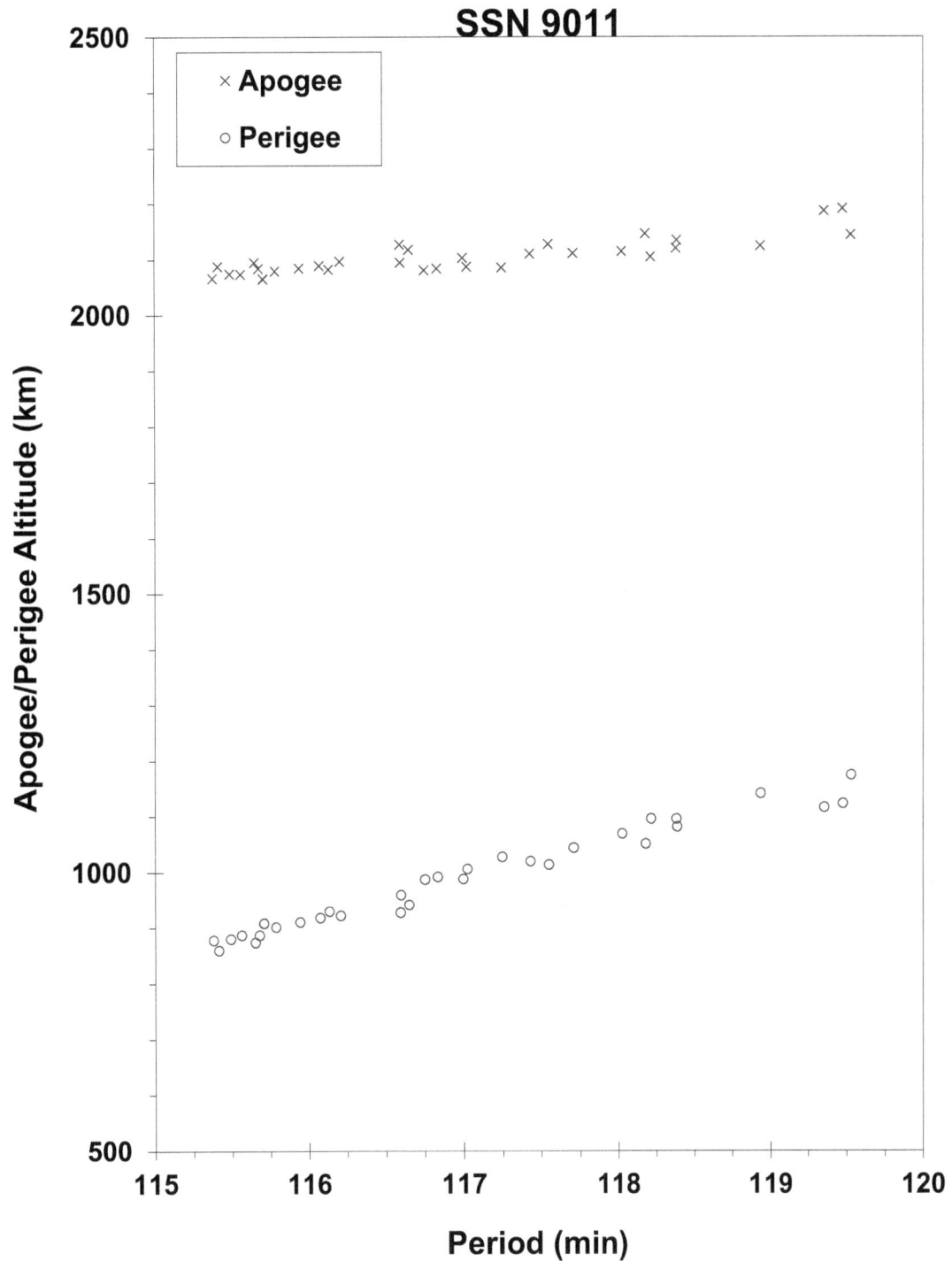

SSN 9011

Cosmos 839 debris cloud of 33 fragments about 5 weeks after the event as reconstructed from the US SSN database.

SATELLITE DATA

TYPE:	Payload
OWNER:	CIS
LAUNCH DATE:	22.66 Jul 1976
DRY MASS (KG):	5700
MAIN BODY:	Cone-cylinder; 2.7 m diameter by 6.3 m length
MAJOR APPENDAGES:	Solar panels
ATTITUDE CONTROL:	Active, 3-axis
ENERGY SOURCES:	On-board propellants, explosive charge (?)

EVENT DATA

DATE:	25 Jul 1976	LOCATION:	49N, 100E (dsc)
TIME:	1718 GMT	ASSESSED CAUSE:	Deliberate
ALTITUDE:	210 km		

PRE-EVENT ELEMENTS

EPOCH:	76207.45032150	MEAN ANOMALY:	291.2246
RIGHT ASCENSION:	152.6930	MEAN MOTION:	16.04433196
INCLINATION:	67.1467	MEAN MOTION DOT/2:	.00313532
ECCENTRICITY:	.0136374	MEAN MOTION DOT DOT/6:	.0
ARG. OF PERIGEE:	70.3553	BSTAR:	.0

DEBRIS CLOUD DATA

MAXIMUM ΔP:	Unknown
MAXIMUM ΔI:	Unknown

COMMENTS

Spacecraft was apparently destroyed after a malfunction prevented controlled reentry and landing in the Soviet Union. No elements were cataloged on any of the official debris. Most fragments reentered rapidly.

REFERENCE DOCUMENT

History of Soviet/Russian Satellite Fragmentations-A Joint U.S.-Russian Investigation, N. L. Johnson, *et al.,* Kaman Sciences Corporation, October 1995.

Insufficient data to construct a Gabbard diagram.

SATELLITE DATA

TYPE:	Delta Second Stage (2310)
OWNER:	US
LAUNCH DATE:	29.71 Jul 1976
DRY MASS (KG):	840
MAIN BODY:	Cylinder-nozzle; 1.4 m diameter by 5.8 m length
MAJOR APPENDAGES:	Mini-skirt; 2.4 m by 0.3 m
ATTITUDE CONTROL:	None at time of the event.
ENERGY SOURCES:	On-board propellants, range safety device

EVENT DATA

DATE:	24 Dec 1977	LOCATION:	40S, 146E (asc)
TIME:	1133 GMT	ASSESSED CAUSE:	Propulsion
ALTITUDE:	1510 km		

PRE-EVENT ELEMENTS

EPOCH:	77354.53228225	MEAN ANOMALY:	330.8663
RIGHT ASCENSION:	38.5560	MEAN MOTION:	12.38394892
INCLINATION:	102.0192	MEAN MOTION DOT/2:	.0
ECCENTRICITY:	.0010085	MEAN MOTION DOT DOT/6:	.0
ARG. OF PERIGEE:	29.2920	BSTAR:	.0

DEBRIS CLOUD DATA

MAXIMUM ΔP:	12.5 min
MAXIMUM ΔI:	3.0 deg

COMMENTS

This was the sixth Delta Second Stage to experience a severe fragmentation. The event occurred 17 months after the successful deployment of the NOAA 5 payload. Cause of the explosion is assessed to be related to the estimated 250 kg of residual propellants on board and characteristics of the sun-synchronous orbit.

REFERENCE DOCUMENTS

Explosion of Satellite 10704 and other Delta Second Stage Rockets, J.R. Gabbard, Technical Memorandum 81-5, DCS Plans, Hdqtrs NORAD/ADCOM, Colorado Springs, May 1981.

Investigation of Delta Second Stage On-Orbit Explosions, C.S. Gumpel, Report MDC-H0047, McDonnell Douglas Astronautics Company - West, Huntington Beach, April 1982.

A Later Look at Delta Second Stage On-Orbit Explosions, J.R. Gabbard, Technical Report CS85-BMDSC-00-24, Teledyne Brown Engineering, Colorado Springs, March 1985.

SSN 9063

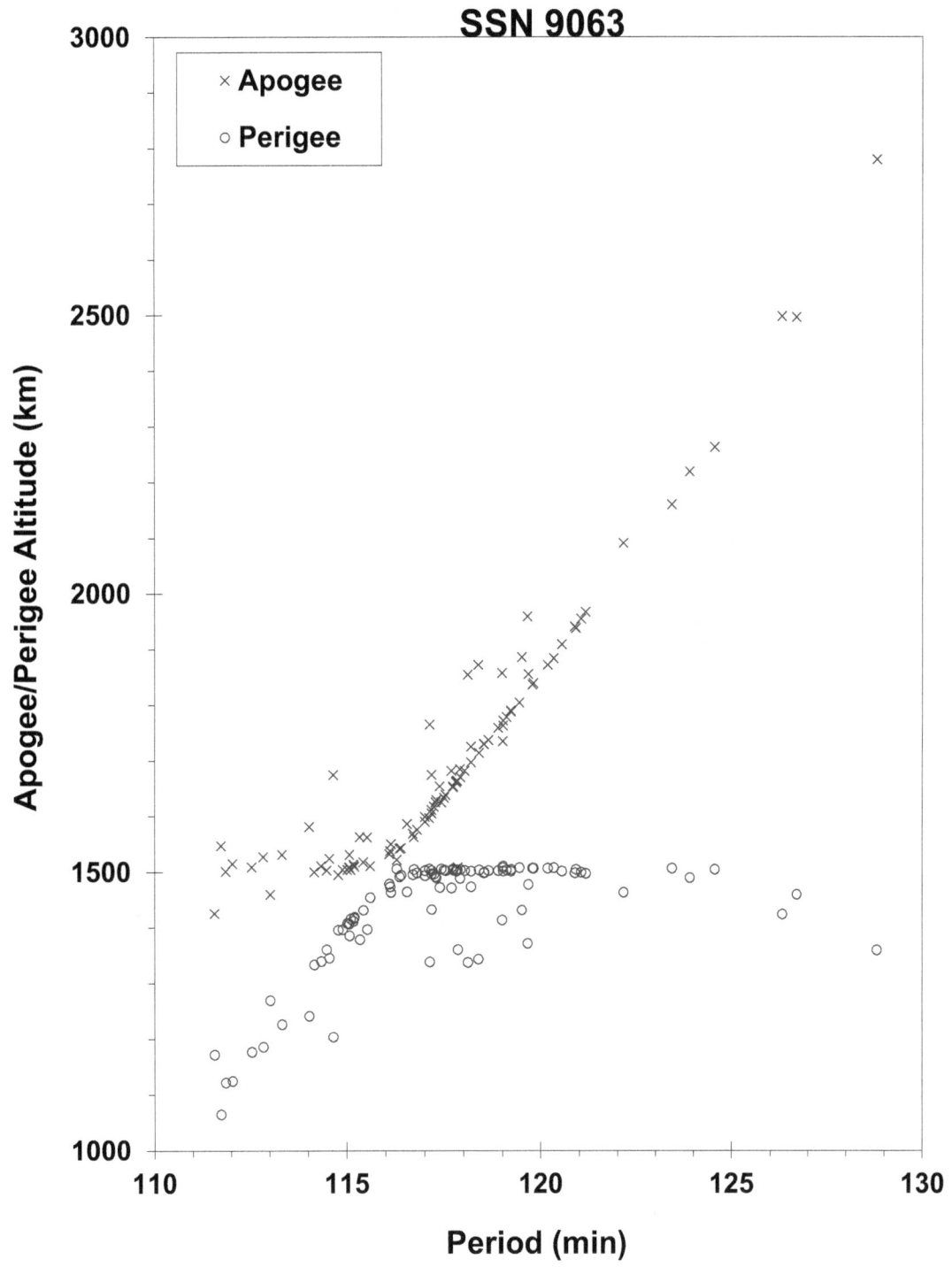

NOAA 5 R/B debris cloud of 98 fragments about 4 months after the event as reconstructed from the US SSN database.

SATELLITE DATA

TYPE:	Payload
OWNER:	CIS
LAUNCH DATE:	22.38 Oct 1976
DRY MASS (KG):	1250
MAIN BODY:	Cylinder; 1.7 m diameter by 2 m length
MAJOR APPENDAGES:	Solar panels
ATTITUDE CONTROL:	Active, 3-axis
ENERGY SOURCES:	On-board propellants, explosive device

EVENT DATA

DATE:	15 Mar 1977	LOCATION:	39N, 114E (asc)
TIME:	1256 GMT	ASSESSED CAUSE:	Deliberate
ALTITUDE:	5375 km		

PRE-EVENT ELEMENTS

EPOCH:	77066.03986408	MEAN ANOMALY:	4.4196
RIGHT ASCENSION:	98.8078	MEAN MOTION:	2.00311741
INCLINATION:	63.1553	MEAN MOTION DOT/2:	.0
ECCENTRICITY:	.7312859	MEAN MOTION DOT DOT/6:	.0
ARG. OF PERIGEE:	318.6653	BSTAR:	.0

DEBRIS CLOUD DATA

MAXIMUM ΔP:	5.7 min
MAXIMUM ΔI:	0.4 deg

COMMENTS

Cosmos 862 was the first of a new class of operational satellites in highly elliptical, semi-synchronous orbits that experienced a total of 18 fragmentations during the period 1977-1986. Due to the nature of these orbits, which result in high altitudes over the Northern Hemisphere where most surveillance sensors are located, debris detection and tracking is extremely difficult. Only the largest fragments can be seen. Cosmos 862 spacecraft were equipped with self-destruct packages in the event that spacecraft control was lost; this was the cause of breakups until the explosives were removed after Cosmos 1481.

REFERENCE DOCUMENT

History of Soviet/Russian Satellite Fragmentations-A Joint U.S.-Russian Investigation, N. L. Johnson, et al., Kaman Sciences Corporation, October 1995.

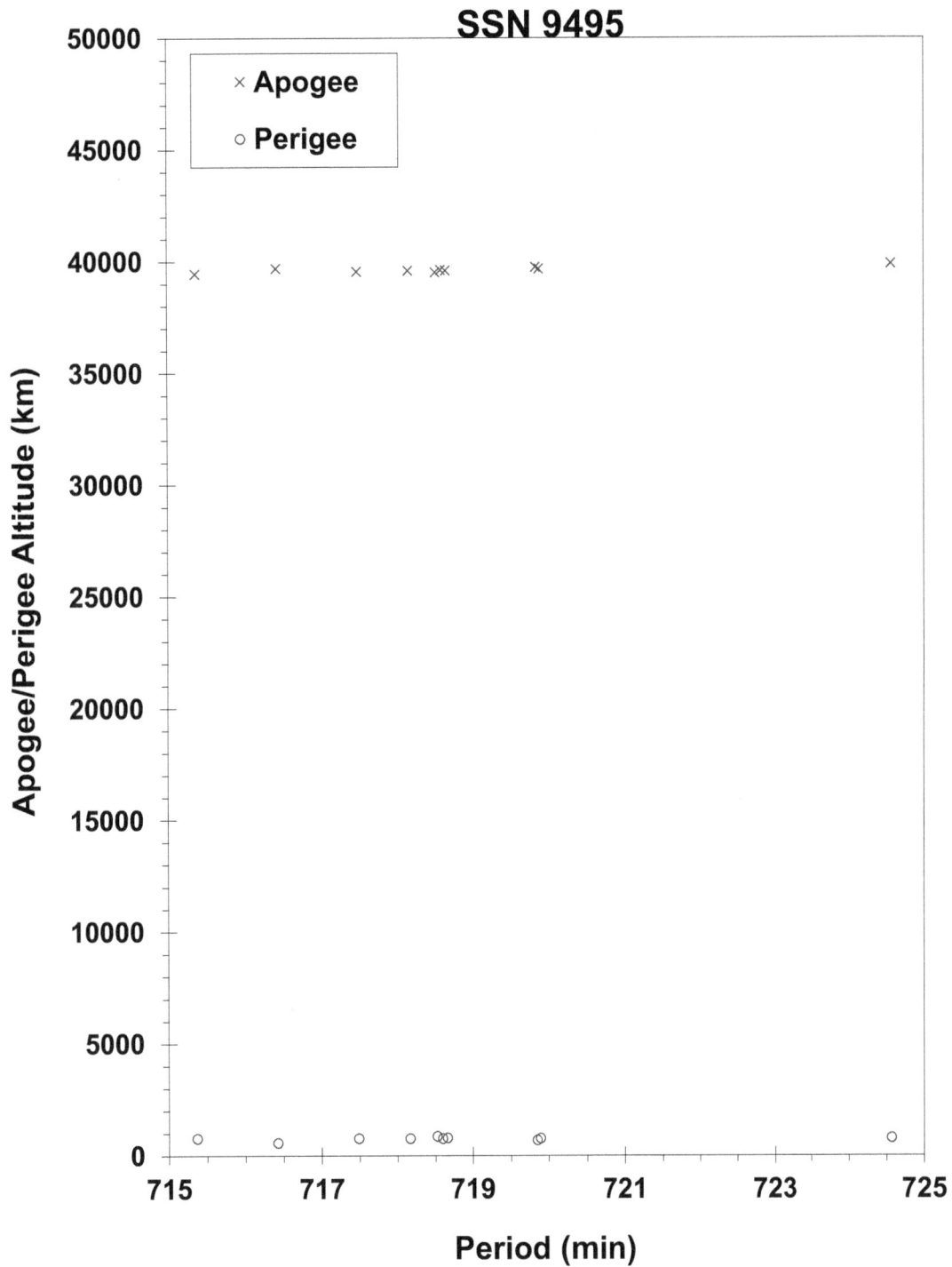

SSN 9495

Cosmos 862 debris cloud of 10 cataloged fragments 2 weeks after the event as reconstructed from the US SSN database.

SATELLITE DATA

TYPE:	Payload
OWNER:	CIS
LAUNCH DATE:	9.84 Dec 1976
DRY MASS (KG):	650
MAIN BODY:	Polyhedron; 1.4 m by 1.4 m
MAJOR APPENDAGES:	Solar panels, gravity-gradient boom (?)
ATTITUDE CONTROL:	Gravity gradient (?)
ENERGY SOURCES:	Battery

EVENT DATA

DATE:	27 Nov 1978	LOCATION:	65S, 306E (dsc)
TIME:	1703 GMT	ASSESSED CAUSE:	Battery
ALTITUDE:	560 km		

PRE-EVENT ELEMENTS

EPOCH:	78331.59395829	MEAN ANOMALY:	55.5772
RIGHT ASCENSION:	11.0317	MEAN MOTION:	14.93841919
INCLINATION:	65.8440	MEAN MOTION DOT/2:	.00000004
ECCENTRICITY:	.0050108	MEAN MOTION DOT DOT/6:	.0
ARG. OF PERIGEE:	304.0553	BSTAR:	.0

DEBRIS CLOUD DATA

MAXIMUM ΔP:	1.3 min*
MAXIMUM ΔI:	0.0 deg*

*Based on uncataloged debris data

COMMENTS

Cosmos 880 was the second spacecraft of the Cosmos 839-type to experience a fragmentation. Although these satellites are used in conjunction with the Cosmos 249-type spacecraft that are deliberately fragmented, the cause of the Cosmos 839-type events appears to be unrelated. In the case of Cosmos 880, 23 months elapsed since its test with a Cosmos 249-type spacecraft. Russian officials have determined that battery malfunctions were the cause of these events.

REFERENCE DOCUMENTS

"Artificial Satellite Break-Ups (Part 2): Soviet Anti-Satellite Program", N.L. Johnson, Journal of the British Interplanetary Society, August 1983, pp. 357-362.

History of Soviet/Russian Satellite Fragmentations-A Joint U.S.-Russian Investigation, N. L. Johnson, *et al.,* Kaman Sciences Corporation, October 1995.

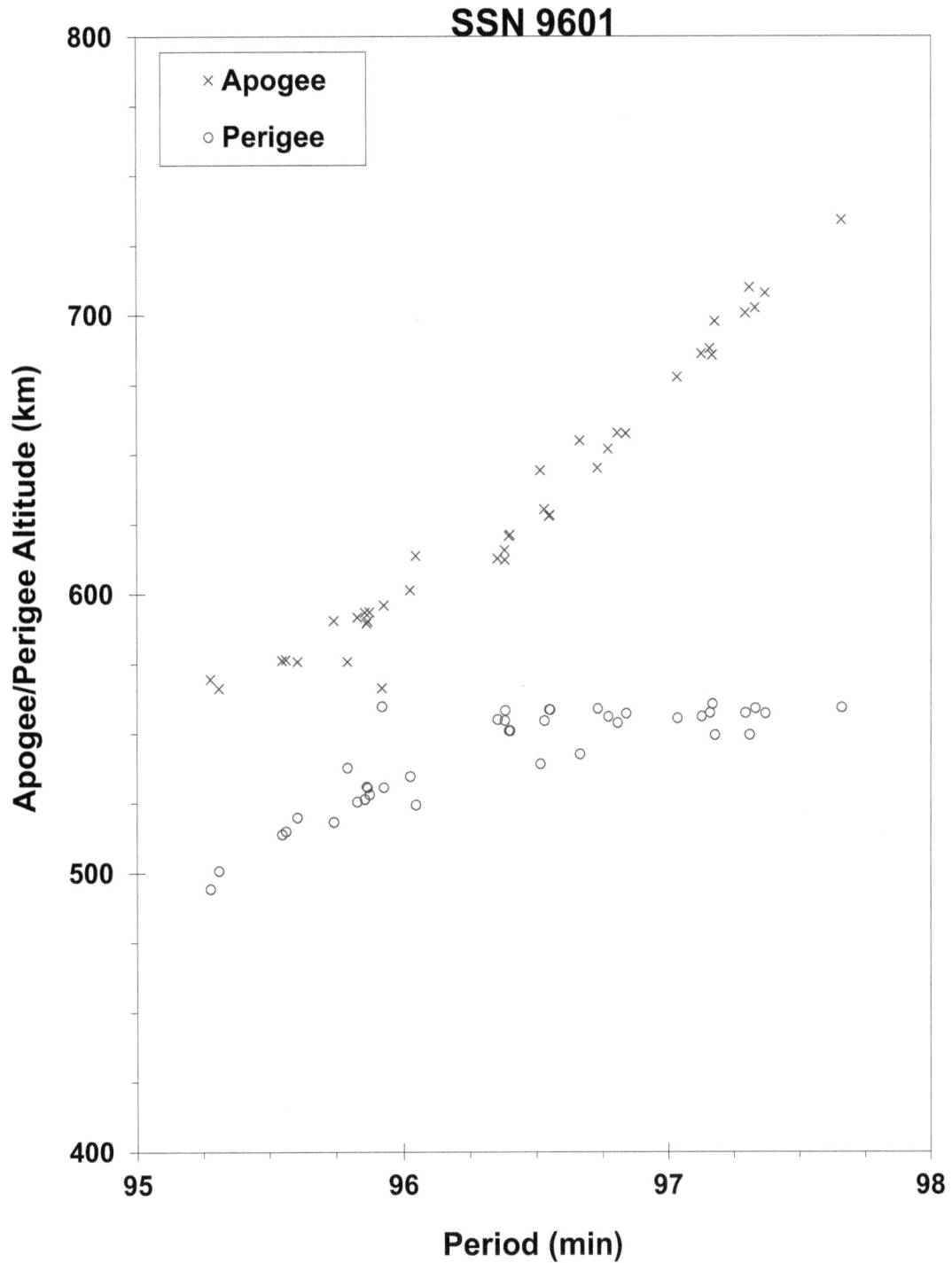

Cosmos 880 debris cloud of 40 fragments 2 days after the event as reconstructed from the US SSN database.

SATELLITE DATA

TYPE:	Payload
OWNER:	CIS
LAUNCH DATE:	17.40 Dec 1976
DRY MASS (KG):	6300
MAIN BODY:	Sphere-cylinder; 2.4 m diameter by 6.5 m length
MAJOR APPENDAGES:	None
ATTITUDE CONTROL:	Active, 3-axis
ENERGY SOURCES:	On-board propellants, explosive charge

EVENT DATA

DATE:	29 Dec 1976	LOCATION:	Unknown
TIME:	Unknown	ASSESSED CAUSE:	Deliberate
ALTITUDE:	Unknown		

PRE-EVENT ELEMENTS

EPOCH:	76362.45360574	MEAN ANOMALY:	302.3648
RIGHT ASCENSION:	227.6719	MEAN MOTION:	16.11011505
INCLINATION:	65.0214	MEAN MOTION DOT/2:	0.00147448
ECCENTRICITY:	0.0113306	MEAN MOTION DOT DOT/6:	0
ARG. OF PERIGEE:	58.8529	BSTAR:	0

DEBRIS CLOUD DATA

MAXIMUM ΔP:	Unknown
MAXIMUM ΔI:	Unknown

COMMENTS

Spacecraft was destroyed after a malfunction prevented reentry and landing in the Soviet Union. Event identified by Russian officials during investigation cited below.

REFERENCE DOCUMENT

History of Soviet/Russian Satellite Fragmentations-A Joint U.S.-Russian Investigation, N. L. Johnson, *et al.,* Kaman Sciences Corporation, October 1995.

Insufficient data to construct a Gabbard diagram.

SATELLITE DATA

TYPE:	Payload
OWNER:	CIS
LAUNCH DATE:	27.53 Dec 1976
DRY MASS (KG):	1400
MAIN BODY:	Irregular; 1.8 m by 4.2 m
MAJOR APPENDAGES:	None
ATTITUDE CONTROL:	Active, 3-axis
ENERGY SOURCES:	On-board propellants, explosive charge

EVENT DATA

DATE:	27 Dec 1976	LOCATION:	65S, 210E (asc)
TIME:	1840 GMT	ASSESSED CAUSE:	Deliberate
ALTITUDE:	2090 km		

POST-EVENT ELEMENTS

EPOCH:	76362.79720829	MEAN ANOMALY:	313.0540
RIGHT ASCENSION:	306.5669	MEAN MOTION:	12.54457816
INCLINATION:	65.8434	MEAN MOTION DOT/2:	.00004000
ECCENTRICITY:	.1087102	MEAN MOTION DOT DOT/6:	.0
ARG. OF PERIGEE:	57.0236	BSTAR:	.0

DEBRIS CLOUD DATA

MAXIMUM ΔP:	4.3 min
MAXIMUM ΔI:	0.2 deg

COMMENTS

Cosmos 886 was launched on a two-revolution rendezvous with Cosmos 880. After a close approach, Cosmos 886 continued on before its warhead was intentionally fired. Cosmos 886 was part of test series begun with Cosmos 249. The elements above are the first available after the final maneuver of Cosmos 886 but represent the revolution immediately after the event.

REFERENCE DOCUMENTS

"Artificial Satellite Break-Ups (Part 2): Soviet Anti-Satellite Program", N.L. Johnson, Journal of the British Interplanetary Society, August 1983, pp. 357-362.

History of Soviet/Russian Satellite Fragmentations-A Joint U.S.-Russian Investigation, N. L. Johnson, *et al.,* Kaman Sciences Corporation, October 1995.

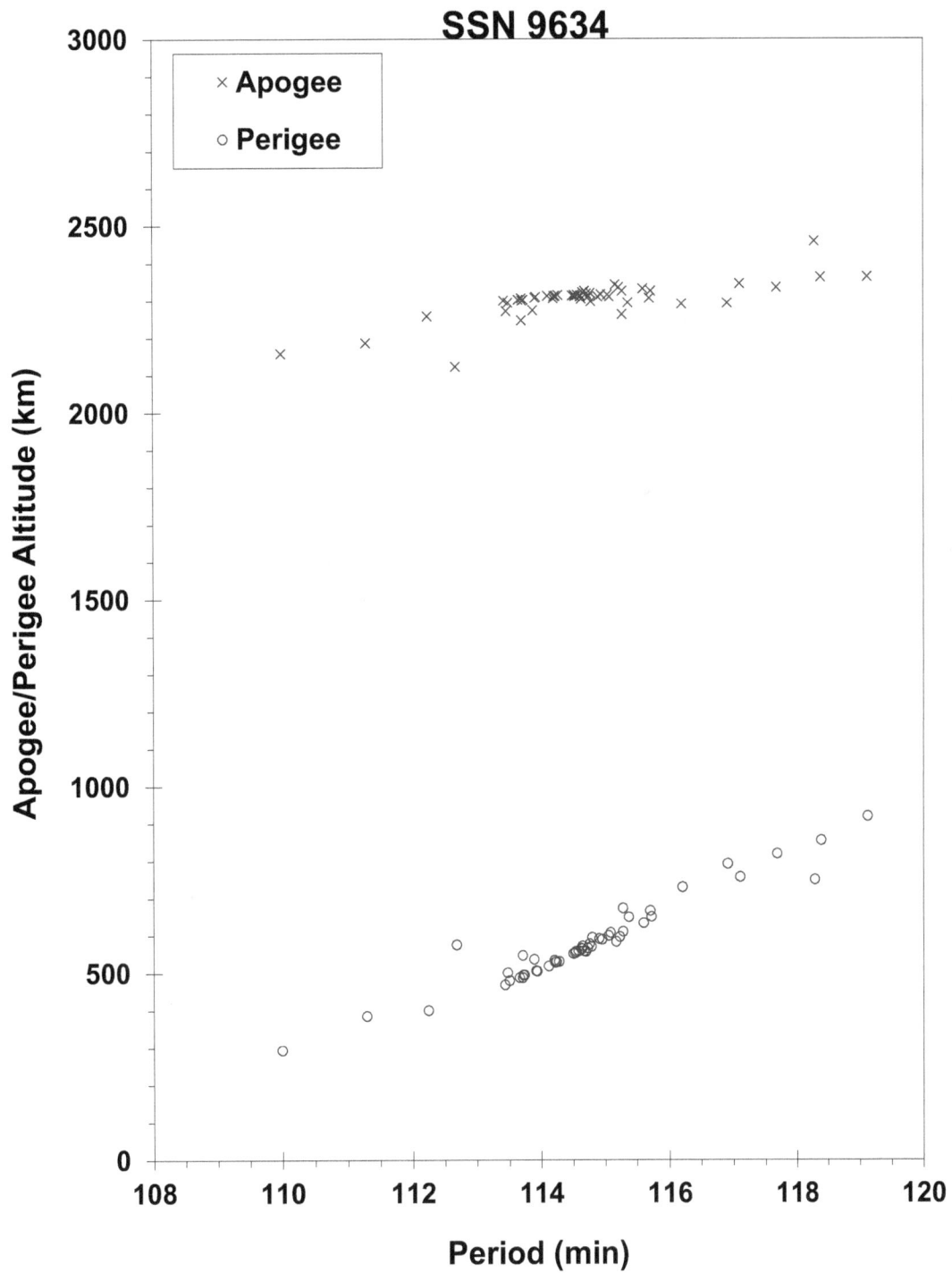

SSN 9634

Cosmos 886 debris cloud of 53 fragments 5 months after the event as reconstructed from the US SSN database.

153

COSMOS 903   1977-027A   9911

SATELLITE DATA

TYPE:	Payload
OWNER:	CIS
LAUNCH DATE:	11.07 Apr 1977
DRY MASS (KG):	1250
MAIN BODY:	Cylinder; 1.7 m diameter by 2 m length
MAJOR APPENDAGES:	Solar panels
ATTITUDE CONTROL:	Active, 3-axis
ENERGY SOURCES:	On-board propellants, explosive charge

EVENT DATA

DATE:	8 Jun 1978	LOCATION:	Unknown
TIME:	Unknown	ASSESSED CAUSE:	Deliberate
ALTITUDE:	Unknown		

PRE-EVENT ELEMENTS

EPOCH:	78156.86414074	MEAN ANOMALY:	5.0496
RIGHT ASCENSION:	115.5660	MEAN MOTION:	2.00599850
INCLINATION:	63.1514	MEAN MOTION DOT/2:	.0
ECCENTRICITY:	.7100107	MEAN MOTION DOT DOT/6:	.0
ARG. OF PERIGEE:	319.7397	BSTAR:	.0

DEBRIS CLOUD DATA

MAXIMUM ΔP:	2.6 min*
MAXIMUM ΔI:	0.5 deg*

*See Comments

COMMENTS

Cosmos 903 was another spacecraft of the Cosmos 862-type to experience a fragmentation. One new fragment was cataloged within a week of the event. The ΔP and ΔI values above are based on the lower period (717.5 min) orbit of Cosmos 903 after the event.

REFERENCE DOCUMENT

History of Soviet/Russian Satellite Fragmentations-A Joint U.S.-Russian Investigation, N. L. Johnson, et al., Kaman Sciences Corporation, October 1995.

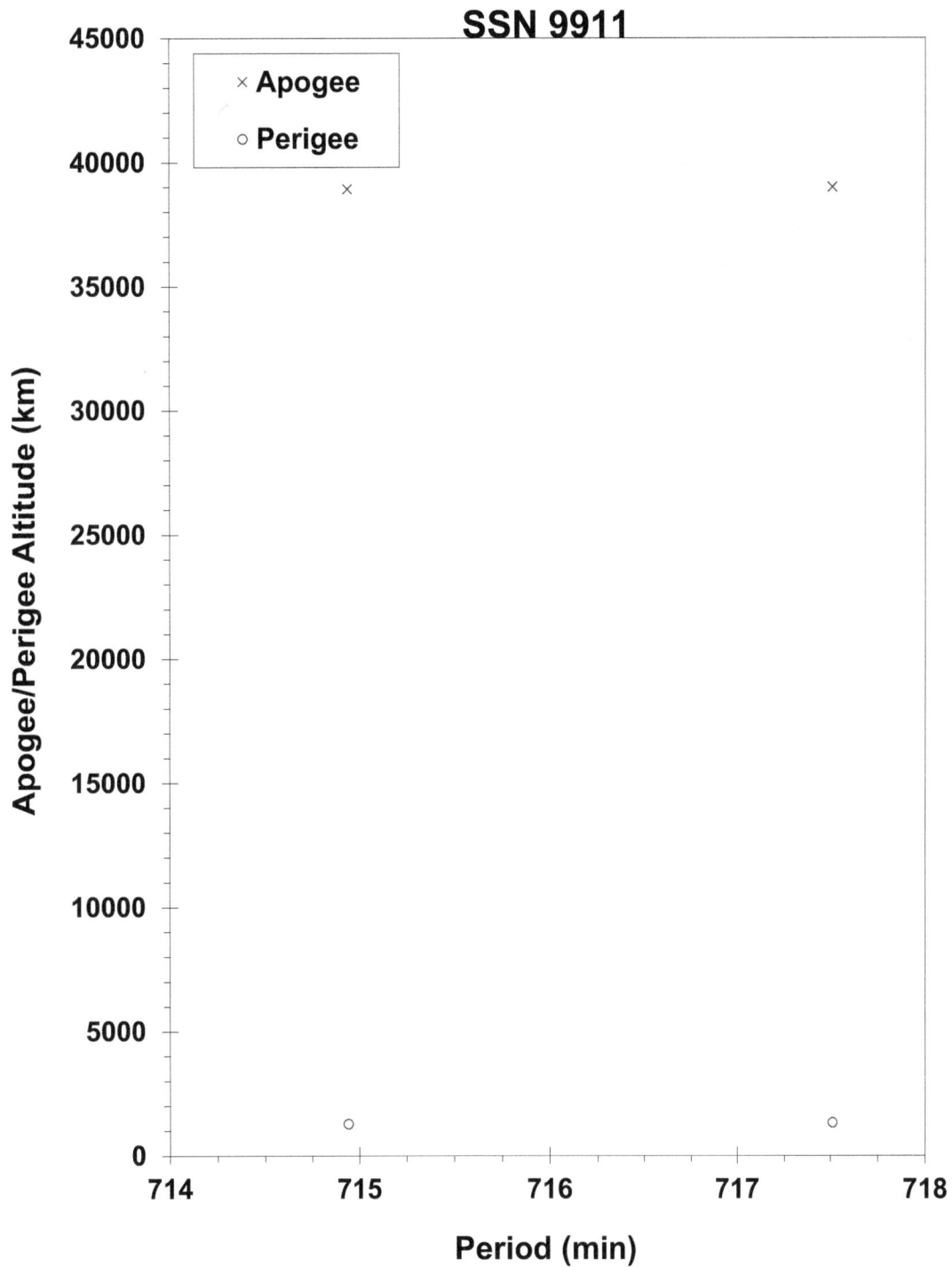

SSN 9911

Cosmos 903 and a single piece of debris 3 weeks after the event as reconstructed from the US SSN database.

SATELLITE DATA

TYPE:	Payload
OWNER:	CIS
LAUNCH DATE:	16.08 Jun 1977
DRY MASS (KG):	1250
MAIN BODY:	Irregular; 1.7 m by 2 m
MAJOR APPENDAGES:	Solar panels
ATTITUDE CONTROL:	Active, 3-axis
ENERGY SOURCES:	On-board propellants, explosive charge

EVENT DATA

DATE:	30 March 1979	LOCATION:	63S, 0E (dsc)
TIME:	1545 GMT	ASSESSED CAUSE:	Deliberate
ALTITUDE:	3280 km		

PRE-EVENT ELEMENTS

EPOCH:	79089.17562851	MEAN ANOMALY:	5.2297
RIGHT ASCENSION:	156.1576	MEAN MOTION:	2.00553521
INCLINATION:	62.9498	MEAN MOTION DOT/2:	.0
ECCENTRICITY:	.6980052	MEAN MOTION DOT DOT/6:	.0
ARG. OF PERIGEE:	322.3289	BSTAR:	.0

DEBRIS CLOUD DATA

MAXIMUM ΔP:	22.6 min*
MAXIMUM ΔI:	0.6 deg*

*Based on uncataloged debris data

COMMENTS

Cosmos 917 was another spacecraft of the Cosmos 862-type to experience a fragmentation.

REFERENCE DOCUMENT

History of Soviet/Russian Satellite Fragmentations-A Joint U.S.-Russian Investigation, N. L. Johnson, *et al.,* Kaman Sciences Corporation, October 1995.

SSN 10059

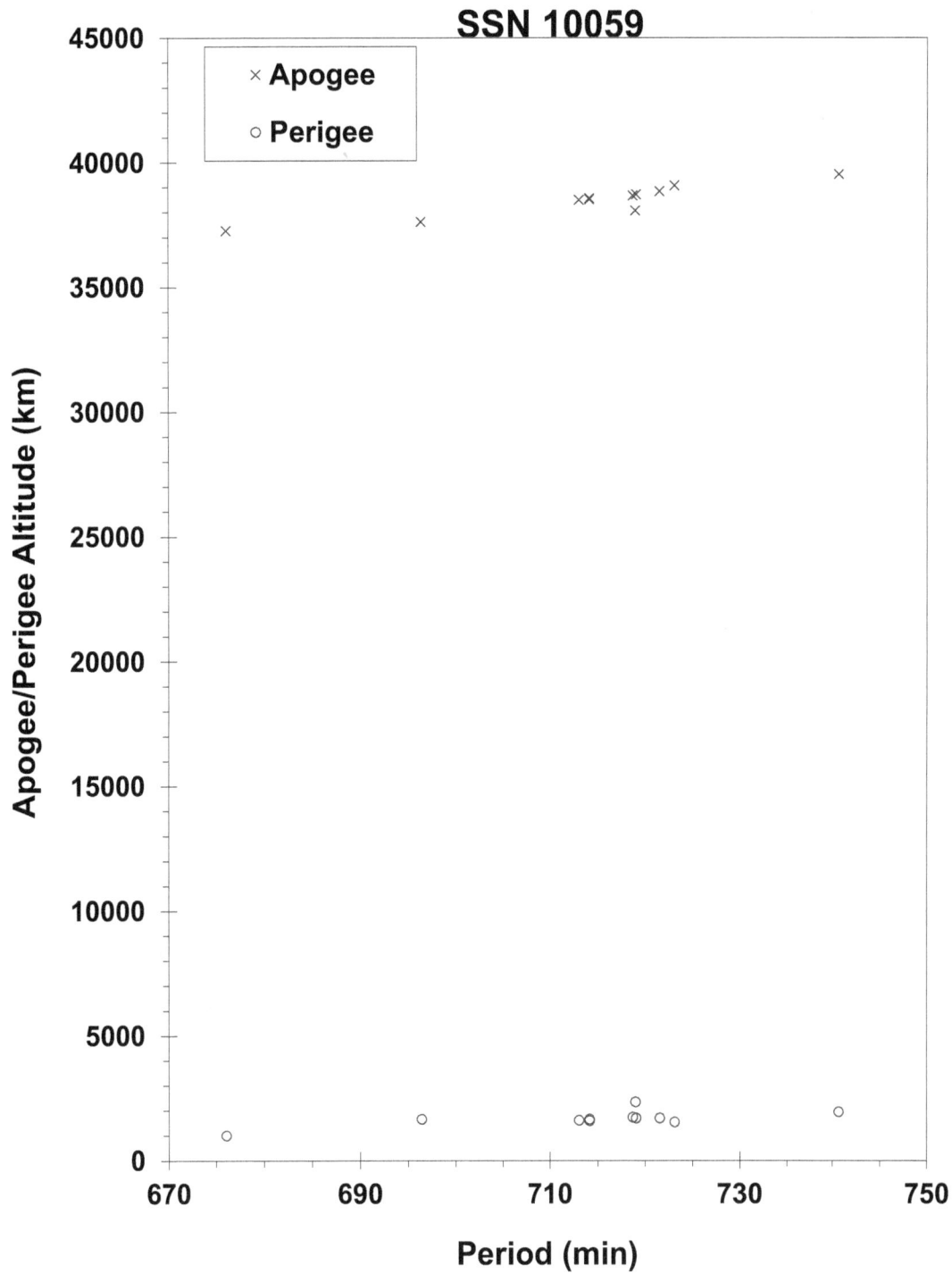

Cosmos 917 debris cloud of 12 fragments about 3 weeks after the event as reconstructed from the US SSN database.

HIMAWARI 1 R/B   1977-065B   10144

SATELLITE DATA

TYPE:	Delta Second Stage (2914)
OWNER:	US
LAUNCH DATE:	14.44 Jul 1977
DRY MASS (KG):	900
MAIN BODY:	Cylinder-nozzle; 1.4 m diameter by 5.8 m length
MAJOR APPENDAGES:	Mini-skirt; 2.4 m by 0.3 m
ATTITUDE CONTROL:	None at time of the event.
ENERGY SOURCES:	On-board propellants, range safety device

EVENT DATA

DATE:	14 Jul 1977	LOCATION:	14N, 249E (dsc)
TIME:	1612 GMT	ASSESSED CAUSE:	Propulsion
ALTITUDE:	1450 km		

POST-EVENT ELEMENTS

EPOCH:	77197.57445278	MEAN ANOMALY:	303.2693
RIGHT ASCENSION:	262.0317	MEAN MOTION:	12.95114397
INCLINATION:	29.0493	MEAN MOTION DOT/2:	.00007335
ECCENTRICITY:	.0973469	MEAN MOTION DOT DOT/6:	.0
ARG. OF PERIGEE:	66.7255	BSTAR:	.0

DEBRIS CLOUD DATA

MAXIMUM ΔP:	9.7 min*
MAXIMUM ΔI:	3.0 deg*

*Based on uncataloged debris data

COMMENTS

This was the fifth Delta Second Stage to experience a severe fragmentation. It is also the only one that was not in a sun-synchronous orbit, which had performed a depletion burn, and which fragmented on the day of launch. This rocket body did perform its mission successfully, carrying the third stage and the payload into a low Earth orbit. The energy for the breakup is assessed to have been the 40 kg of propellants (mainly oxidizer) remaining after the depletion burn. The elements above are the first available after the depletion burn although also after the event.

REFERENCE DOCUMENTS

Explosion of Satellite 10704 and other Delta Second Stage Rockets, J.R. Gabbard, Technical Memorandum 81-5, DCS Plans, Hdqtrs NORAD/ADCOM, Colorado Springs, May 1981.

Investigation of Delta Second Stage On-Orbit Explosions, C.S. Gumpel, Report MDC-H0047, McDonnell Douglas Astronautics Company - West, Huntington Beach, April 1982.

A Later Look at Delta Second Stage On-Orbit Explosions, J.R. Gabbard, Technical Report CS85-BMDSC-00-24, Teledyne Brown Engineering, Colorado Springs, March 1985.

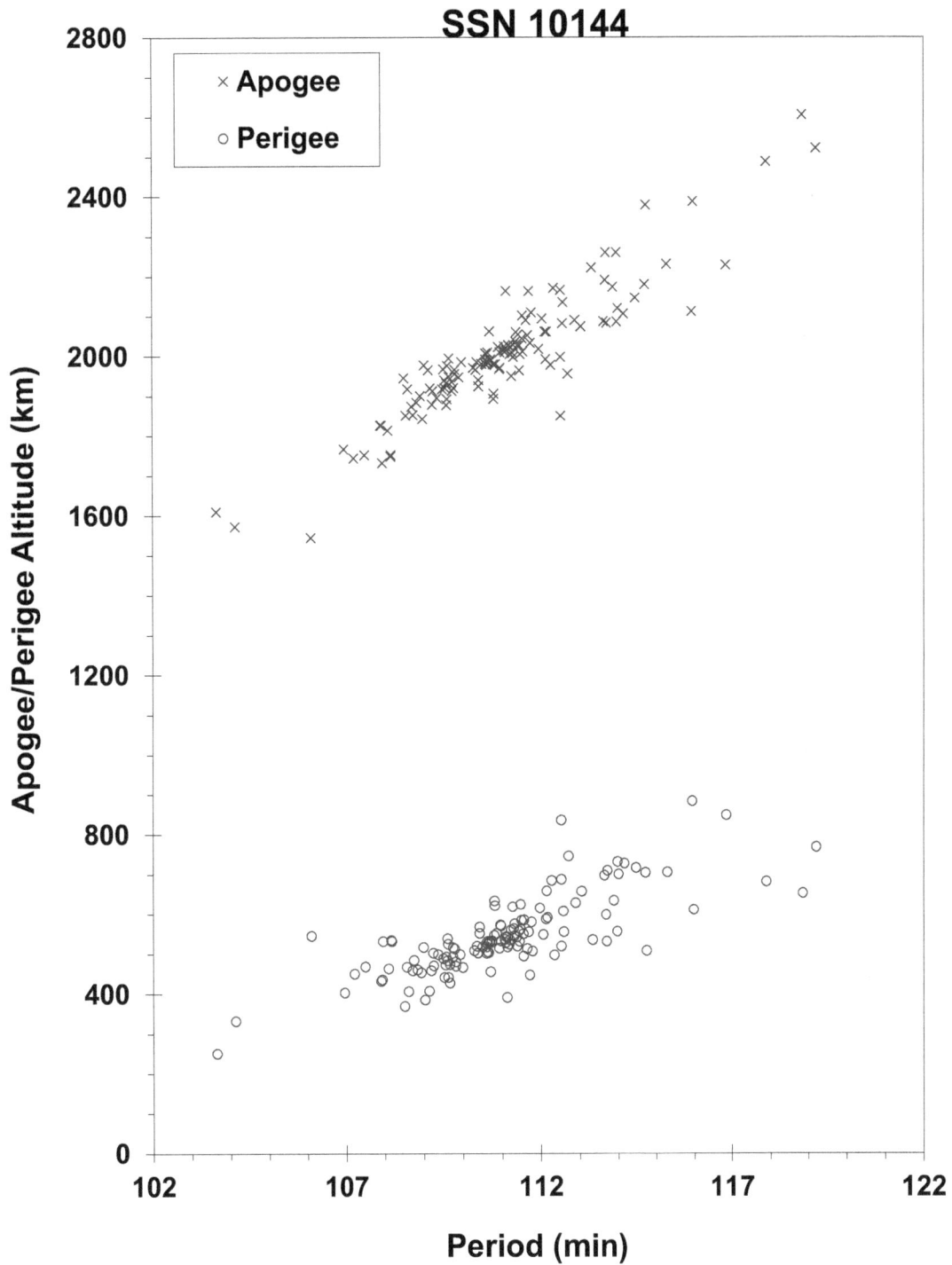

SSN 10144

Himawari 1 R/B debris cloud of 132 fragments 5 months after the event as reconstructed from the US SSN database.

SATELLITE DATA

TYPE:	Payload
OWNER:	CIS
LAUNCH DATE:	20.20 Jul 1977
DRY MASS (KG):	1250
MAIN BODY:	Cylinder; 1.7 m diameter by 2 m length
MAJOR APPENDAGES:	Solar panels
ATTITUDE CONTROL:	Active, 3-axis
ENERGY SOURCES:	On-board propellants, explosive charge

EVENT DATA

DATE:	24 Oct 1977	LOCATION:	Unknown
TIME:	Unknown	ASSESSED CAUSE:	Deliberate
ALTITUDE:	Unknown		

PRE-EVENT ELEMENTS

EPOCH:	77289.02131186	MEAN ANOMALY:	4.2624
RIGHT ASCENSION:	305.6648	MEAN MOTION:	2.00651833
INCLINATION:	62.9440	MEAN MOTION DOT/2:	.0
ECCENTRICITY:	.7341055	MEAN MOTION DOT DOT/6:	.0
ARG. OF PERIGEE:	318.8771	BSTAR:	.0

DEBRIS CLOUD DATA

MAXIMUM ΔP:	5.3 min*
MAXIMUM ΔI:	0.7 deg*

*Based on uncataloged debris data

COMMENTS

Cosmos 931 was another spacecraft of the Cosmos 862-type to experience a fragmentation. Debris were not officially cataloged until 4 years after the event.

REFERENCE DOCUMENT

History of Soviet/Russian Satellite Fragmentations-A Joint U.S.-Russian Investigation, N. L. Johnson, *et al.*, Kaman Sciences Corporation, October 1995.

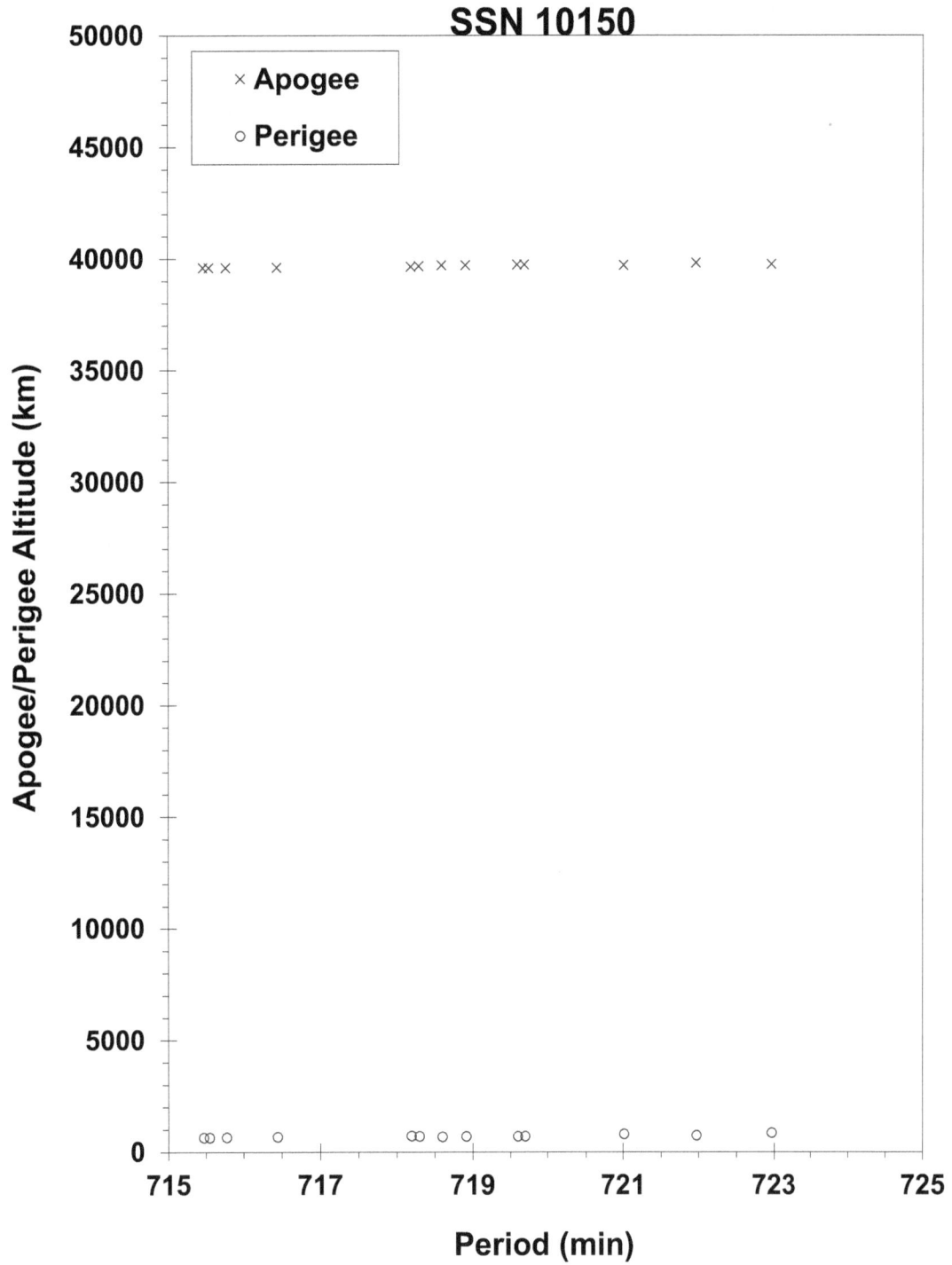

SSN 10150

× Apogee

○ Perigee

Apogee/Perigee Altitude (km)

Period (min)

Cosmos 931 debris cloud of 13 fragments 2 weeks after the event as reconstructed from the US SSN database.

SATELLITE DATA

TYPE:	Payload
OWNER:	CIS
LAUNCH DATE:	20.73 Sept 1977
DRY MASS (KG):	1970
MAIN BODY:	Cylinder; 2 m by 4 m
MAJOR APPENDAGES:	Plate + 2 solar panels
ATTITUDE CONTROL:	Active, 3-axis
ENERGY SOURCES:	On-board propellants, Battery

EVENT DATA

DATE:	23 Jun 1978	LOCATION:	0.0N, 98.7E
TIME:	Unknown	ASSESSED CAUSE:	Battery
ALTITUDE:	35790 km		

PRE-EVENT ELEMENTS

EPOCH:	88166.03647595	MEAN ANOMALY:	78.3897
RIGHT ASCENSION:	78.3897	MEAN MOTION:	1.00252588
INCLINATION:	0.1137	MEAN MOTION DOT/2:	.0
ECCENTRICITY	.0001436	MEAN MOTION DOT DOT/6:	.0
ARG. OF PERIGEE:	325.2771	BSTAR:	.0

DEBRIS CLOUD DATA

MAXIMUM ΔP:	Unknown
MAXIMUM ΔI:	Unknown

COMMENTS

This event was revealed by Russian officials in an orbital debris meeting in February 1992 in Moscow. This is the first known geostationary orbit fragmentation and was not detected by the Space Surveillance Network (SSN). Russian photographs originally linked to the breakup were later determined to have been misidentified.

REFERENCE DOCUMENTS

History of Soviet/Russian Satellite Fragmentations-A Joint U.S.-Russian Investigation, N. L. Johnson, *et al.,* Kaman Sciences Corporation, October 1995.

"Breakup in Review-Two GEO Breakups," Orbital Debris Monitor, April 1992, p 35-36.

Insufficient data to construct a Gabbard diagram.

SATELLITE DATA

TYPE:	Payload
OWNER:	CIS
LAUNCH DATE:	21.44 Dec 1977
DRY MASS (KG):	1400
MAIN BODY:	Cylinder; 2 m diameter by 4 m length
MAJOR APPENDAGES:	None
ATTITUDE CONTROL:	Active, 3-axis
ENERGY SOURCES:	On-board propellants, explosive charge

EVENT DATA

DATE:	21 Dec 1977	LOCATION:	38S, 274E (asc)
TIME:	1710 GMT	ASSESSED CAUSE:	Deliberate
ALTITUDE:	1135 km		

PRE-EVENT ELEMENTS

EPOCH:	77355.65049149	MEAN ANOMALY:	245.5638
RIGHT ASCENSION:	282.1792	MEAN MOTION:	13.58084598
INCLINATION:	65.8467	MEAN MOTION DOT/2:	.00023007
ECCENTRICITY:	.0129854	MEAN MOTION DOT DOT/6:	.0
ARG. OF PERIGEE:	116.3098	BSTAR:	.0

DEBRIS CLOUD DATA

MAXIMUM ΔP:	4.7 min
MAXIMUM ΔI:	1.1 deg

COMMENTS

Cosmos 970 was launched on a two-revolution rendezvous with Cosmos 967. After a close approach, Cosmos 970 continued on before its warhead was intentionally fired. Cosmos 970 was part of test series begun with Cosmos 249.

REFERENCE DOCUMENTS

"Artificial Satellite Break-Ups (Part 2): Soviet Anti-Satellite Program", N. L. Johnson, Journal of the British Interplanetary Society, August 1983, p. 357-362.

History of Soviet/Russian Satellite Fragmentations-A Joint U.S.-Russian Investigation, N. L. Johnson, *et al.,* Kaman Sciences Corporation, October 1995.

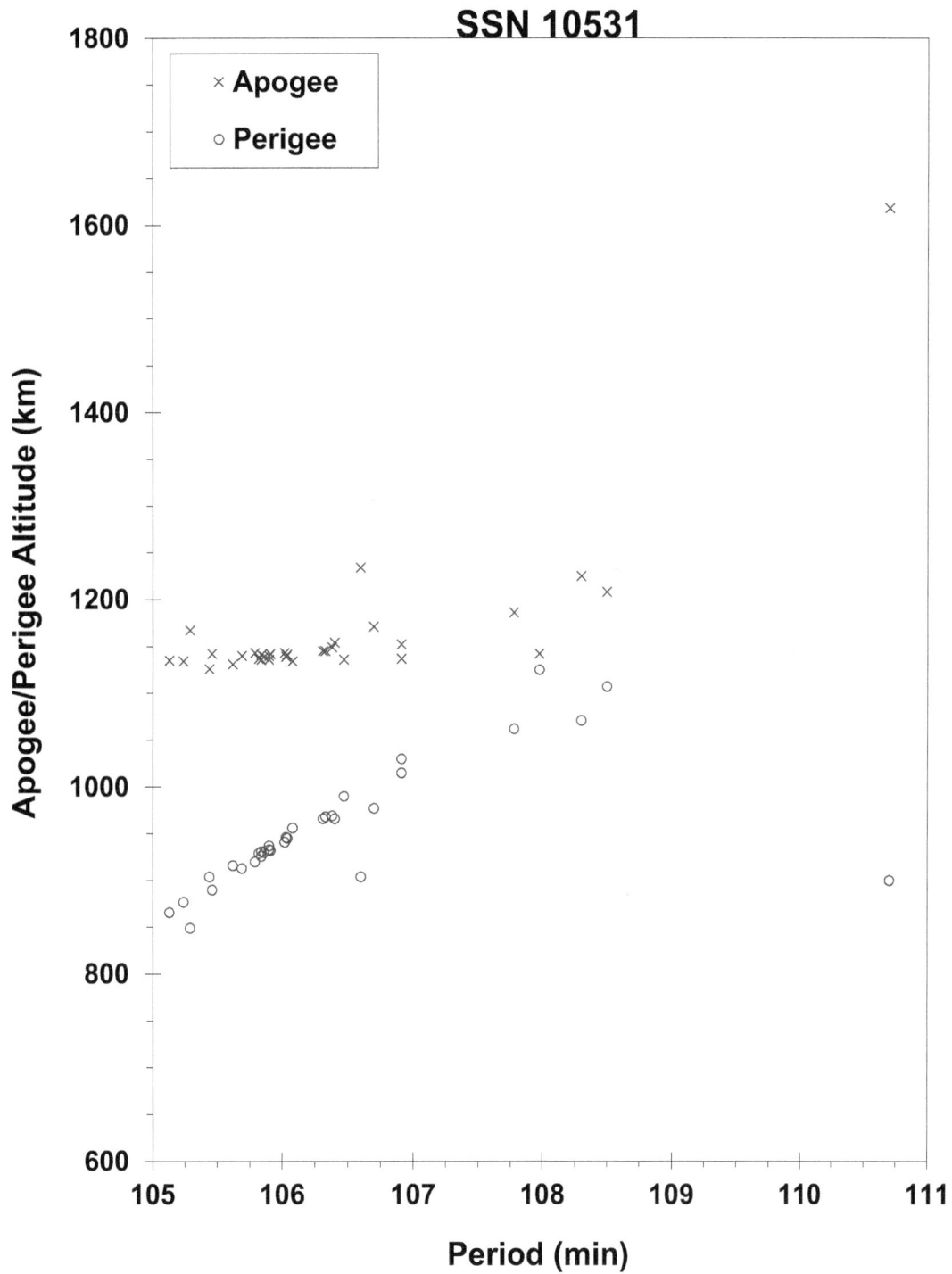

SSN 10531

Cosmos 970 debris cloud of 34 fragments about 5 months after the event as reconstructed from the US SSN database.

SATELLITE DATA

TYPE:	Delta Second Stage (2910)
OWNER:	US
LAUNCH DATE:	5.75 Mar 1978
DRY MASS (KG):	900
MAIN BODY:	Cylinder-nozzle; 1.4 m diameter by 5.8 m length
MAJOR APPENDAGES:	Mini-skirt; 2.4 m by 0.3 m
ATTITUDE CONTROL:	None at time of the event.
ENERGY SOURCES:	On-board propellants, range safety device

EVENT DATA

DATE:	27 Jan 1981	LOCATION:	80S, 301E (asc)
TIME:	0432 GMT	ASSESSED CAUSE:	Propulsion
ALTITUDE:	910 km		

PRE-EVENT ELEMENTS

EPOCH:	81026.99107090	MEAN ANOMALY:	147.0549
RIGHT ASCENSION:	68.7927	MEAN MOTION:	13.96108433
INCLINATION:	98.8485	MEAN MOTION DOT/2:	.00000434
ECCENTRICITY:	.0006255	MEAN MOTION DOT DOT/6:	.0
ARG. OF PERIGEE:	212.9842	BSTAR:	.00032708

DEBRIS CLOUD DATA

MAXIMUM ΔP:	9.1 min
MAXIMUM ΔI:	0.5 deg

COMMENTS

This was the seventh Delta Second Stage to experience a severe fragmentation. The event occurred nearly 35 months after the successful deployment of the Landsat 3 payload. Cause of the explosion is assessed to be related to the estimated 100 kg of residual propellants on board and characteristics of the sun-synchronous orbit.

REFERENCE DOCUMENTS

Explosion of Satellite 10704 and other Delta Second Stage Rockets, J.R. Gabbard, Technical Memorandum 81-5, DCS Plans, Hdqtrs NORAD/ADCOM, Colorado Springs, May 1981.

Analysis of PARCS Recorded Data on the Breakup of Satellite 10704 on 27 January 1981, S.F. Hoffman and P.P. Shinkunas, Technical Report MSB82-ADC-0138, Teledyne Brown Engineering, Huntsville, February 1982.

Investigation of Delta Second Stage On-Orbit Explosions, C.S. Gumpel, Report MDC-H0047, McDonnell Douglas Astronautics Company - West, Huntington Beach, April 1982.

A Later Look at Delta Second Stage On-Orbit Explosions, J.R. Gabbard, Technical Report CS85-BMDSC-00-24, Teledyne Brown Engineering, Colorado Springs, March 1985.

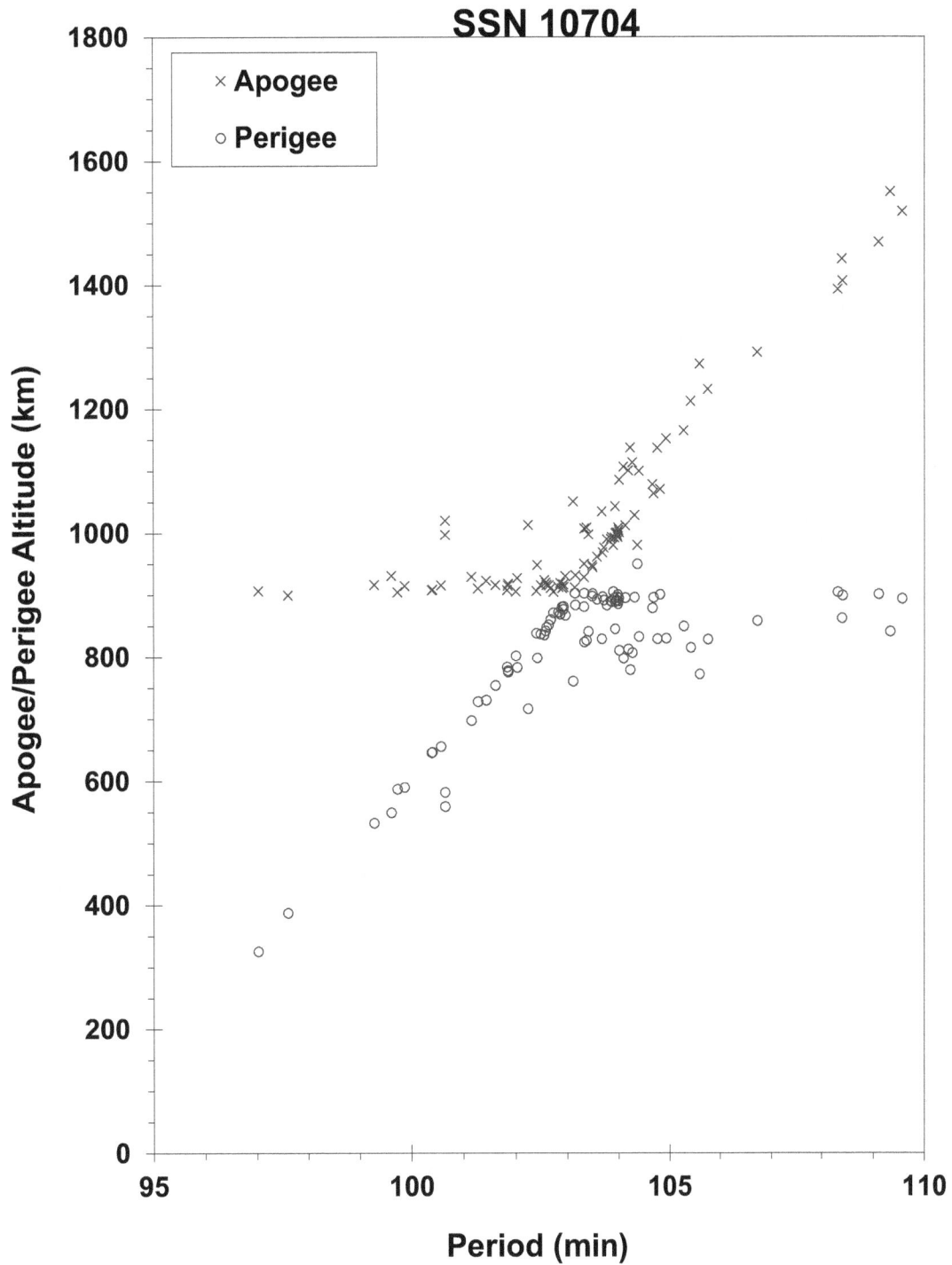

SSN 10704

Landsat 3 R/B debris cloud of 90 identified fragments 4 days after the event as reconstructed from the US SSN database.

COSMOS 1030   **1978-083A**   **11015**

SATELLITE DATA

TYPE:	Payload
OWNER:	CIS
LAUNCH DATE:	6.13 Sep 1978
DRY MASS (KG):	1250
MAIN BODY:	Cylinder; 1.7 m diameter by 2 m length
MAJOR APPENDAGES:	Solar panels (?)
ATTITUDE CONTROL:	Active, 3-axis
ENERGY SOURCES:	On-board propellants, explosive charge

EVENT DATA

DATE:	10 Oct 1978	LOCATION:	Unknown
TIME:	Unknown	ASSESSED CAUSE:	Deliberate
ALTITUDE:	Unknown		

PRE-EVENT ELEMENTS

EPOCH:	78277.19859350	MEAN ANOMALY:	4.9827
RIGHT ASCENSION:	336.7676	MEAN MOTION:	2.00213289
INCLINATION:	62.8388	MEAN MOTION DOT/2:	.0
ECCENTRICITY:	.7350882	MEAN MOTION DOT DOT/6:	.0
ARG. OF PERIGEE:	318.4262	BSTAR:	.0

DEBRIS CLOUD DATA

MAXIMUM ΔP:	Unknown
MAXIMUM ΔI:	Unknown

COMMENTS

Cosmos 1030 was another spacecraft of the Cosmos 862-type to experience a fragmentation. After entering a Molniya-type transfer orbit on 6 September, Cosmos 1030 maneuvered about 14 September to enter an operational orbit. Elements on the first identifiable fragment did not appear until a year after the event. Official cataloging of debris did not begin until 3 years after the event.

REFERENCE DOCUMENT

History of Soviet/Russian Satellite Fragmentations-A Joint U.S.-Russian Investigation, N. L. Johnson, *et al.,* Kaman Sciences Corporation, October 1995.

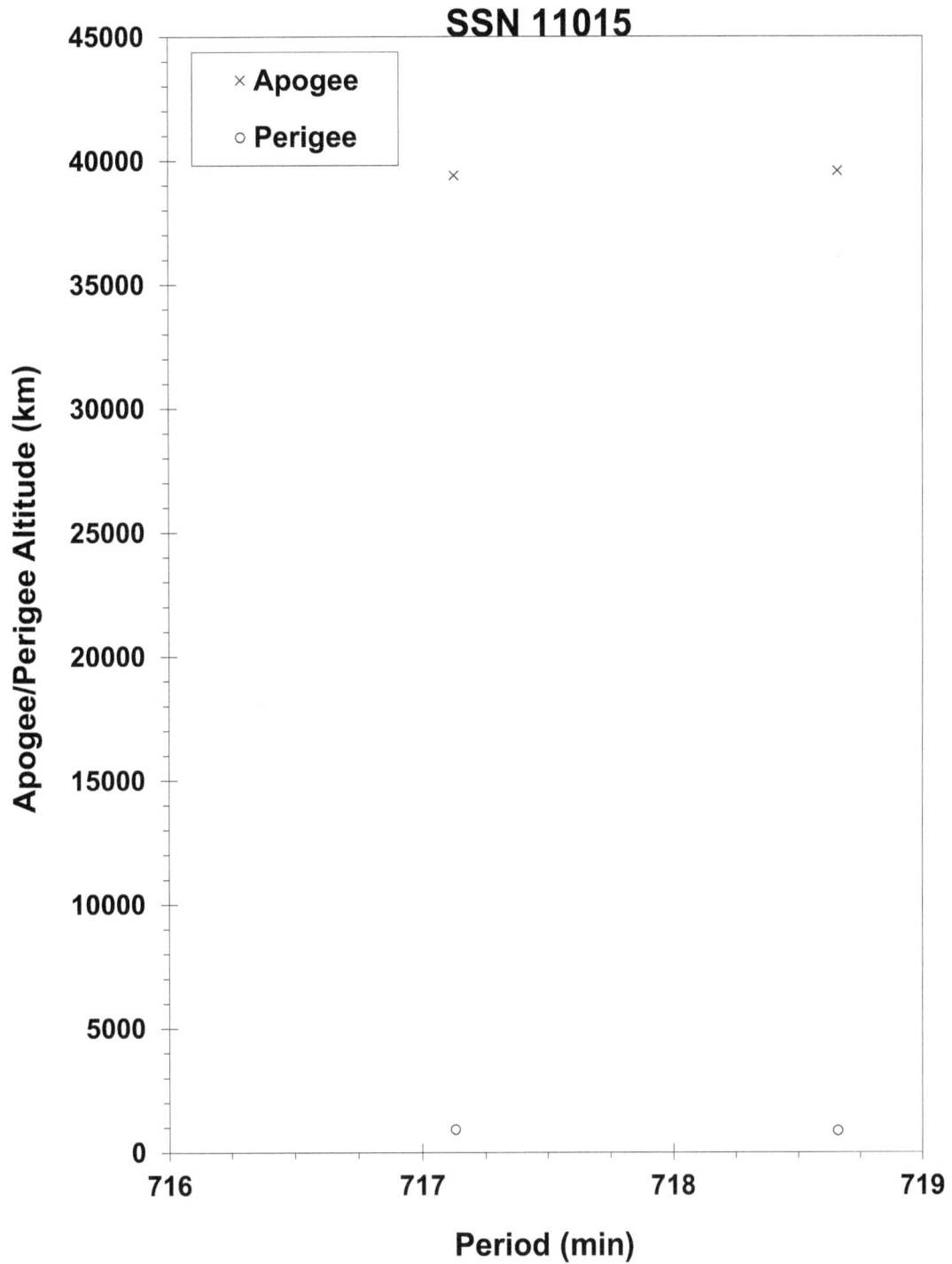

SSN 11015

Cosmos 1030 and a single debris fragment 1 year after the event as reconstructed from the US SSN database.

SATELLITE DATA

TYPE:	Delta Second Stage (2910)
OWNER:	US
LAUNCH DATE:	24.34 Oct 1978
DRY MASS (KG):	900
MAIN BODY:	Cylinder-nozzle; 2.4 m diameter by 8 m length
MAJOR APPENDAGES:	None
ATTITUDE CONTROL:	None at time of the event.
ENERGY SOURCES:	On-board propellants, range safety device

EVENT DATA

DATE:	26 Dec 1981	LOCATION:	Unknown
TIME:	Unknown	ASSESSED CAUSE:	Propulsion
ALTITUDE:	Unknown		

PRE-EVENT ELEMENTS

EPOCH:	81360.19972720	MEAN ANOMALY:	311.8261
RIGHT ASCENSION:	277.7553	MEAN MOTION:	13.85390161
INCLINATION:	99.3003	MEAN MOTION DOT/2:	.000000425
ECCENTRICITY:	.0010821	MEAN MOTION DOT DOT/6:	.0
ARG. OF PERIGEE:	48.3801	BSTAR:	.00004426123

DEBRIS CLOUD DATA

MAXIMUM ΔP:	Unknown
MAXIMUM ΔI:	0.6 deg*

*Based on uncataloged debris data

COMMENTS

Nimbus 7 R/B is designated Cameo in US Space Command Satellite Catalog in reference to scientific piggy-back payload attached to the Delta second stage. This satellite experienced an anomalous event prior to and after the event cited above (See Section 3). Most fragments decayed very rapidly, preventing an accurate assessment of the event and its resulting debris cloud. No new objects were cataloged as a result of this event. The event apparently occurred prior to 0700 GMT.

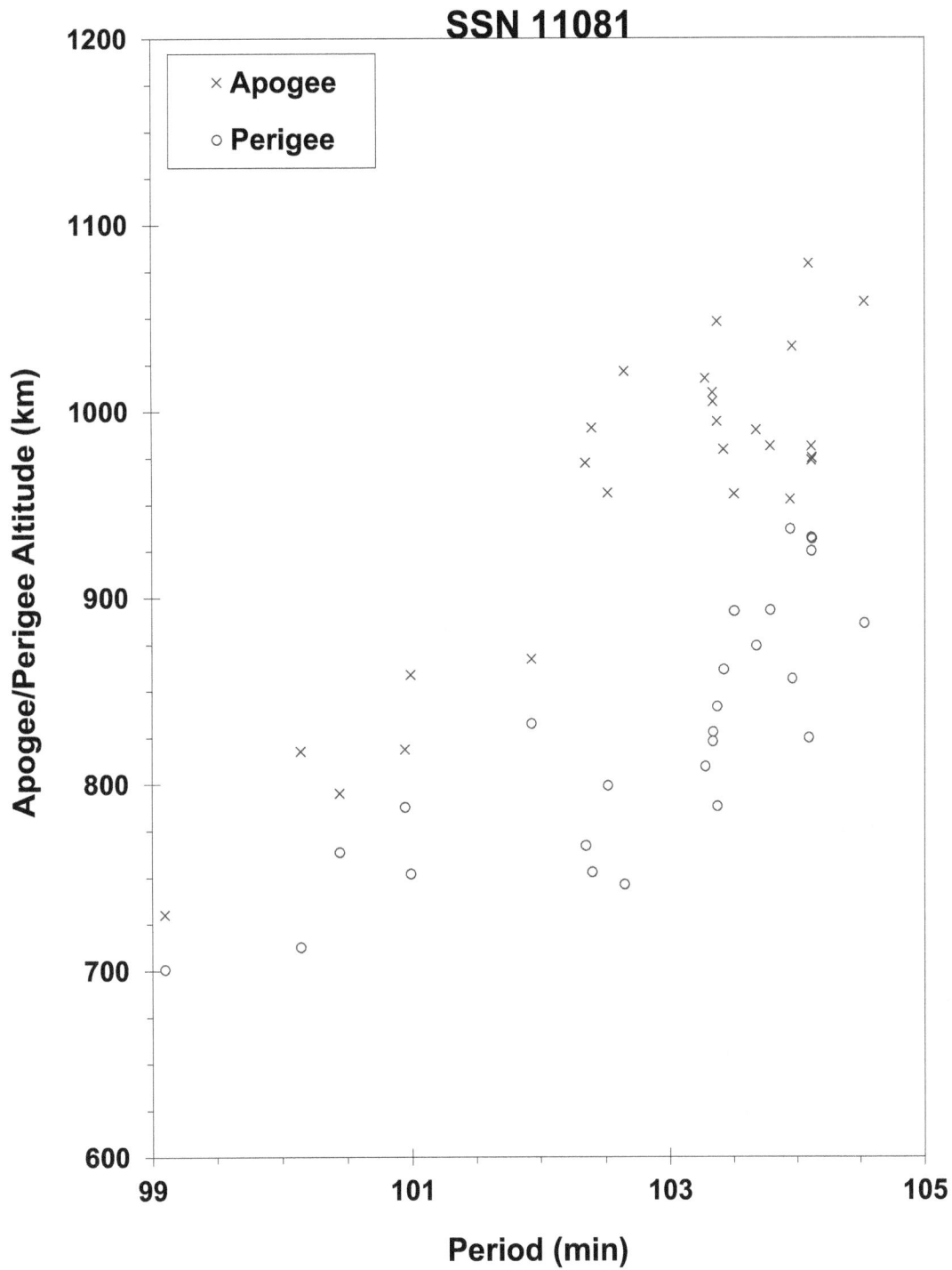

SSN 11081

The Nimbus 7 R/B debris cloud remnant of 27 fragments a few days after the event as reconstructed from the US SSN database. Most fragments have already experienced considerable drag effects.

COSMOS 1045 R/B   **1978-100D**   **11087**

SATELLITE DATA

TYPE:	Tsyklon Third Stage
OWNER:	CIS
LAUNCH DATE:	26.29 Oct 1978
DRY MASS (KG):	1360
MAIN BODY:	Cone-cylinder; 2.1 m diameter by 3.3 m length
MAJOR APPENDAGES:	None
ATTITUDE CONTROL:	None at time of the event.
ENERGY SOURCES:	Unknown

EVENT DATA

DATE:	9 May 1988	LOCATION:	29S, 126E (dsc)
TIME:	1218 GMT	ASSESSED CAUSE:	Propulsion
ALTITUDE:	1705 km		

PRE-EVENT ELEMENTS

EPOCH:	88121.02005933	MEAN ANOMALY:	279.0818
RIGHT ASCENSION:	359.3059	MEAN MOTION:	11.97080974
INCLINATION:	82.5543	MEAN MOTION DOT/2:	.000000208
ECCENTRICITY:	.0011463	MEAN MOTION DOT DOT/6:	.0
ARG. OF PERIGEE:	81.1553	BSTAR:	.0

DEBRIS CLOUD DATA

MAXIMUM ΔP:	7.8 min
MAXIMUM ΔI:	0.9 deg

COMMENTS

This flight, which successfully carried three separate payloads, was the fifth orbital mission of the Tsyklon third stage. Propellants used were N_2O_4 and UDMH. Nearly 10 years elapsed from launch to breakup. A second Tsyklon third stage breakup after a similar length of time; see 1987-068B.

REFERENCE DOCUMENT

History of Soviet/Russian Satellite Fragmentations-A Joint U.S.-Russian Investigation, N. L. Johnson, *et al.,* Kaman Sciences Corporation, October 1995.

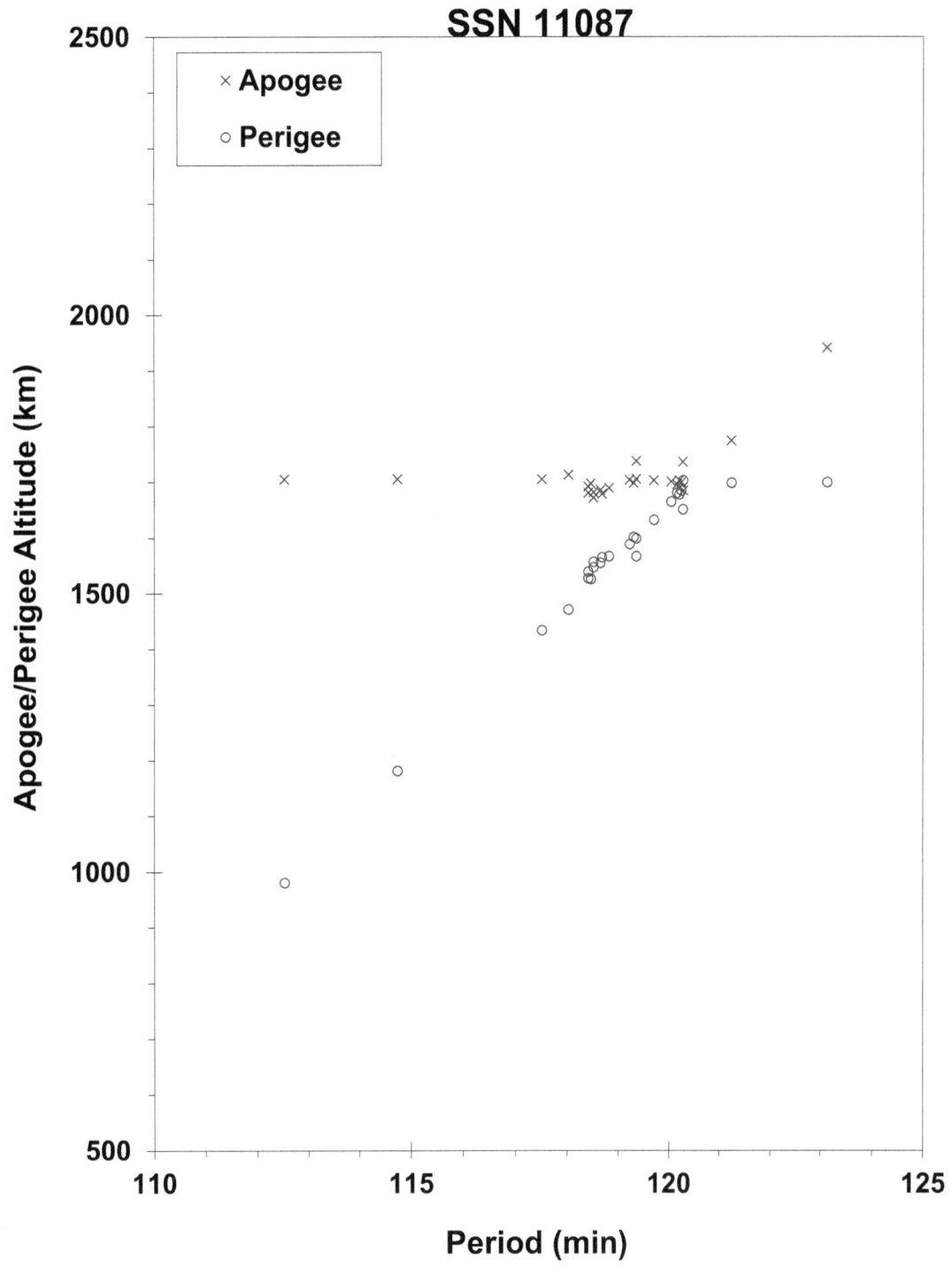

SSN 11087

Cosmos 1045 R/B debris cloud of 25 fragments as determined 1 week after the event.

SATELLITE DATA

TYPE:	Payload
OWNER:	US
LAUNCH DATE:	24.35 Feb 1979
DRY MASS (KG):	850
MAIN BODY:	Cylinder; 2.1 m diameter by 1.3 m length
MAJOR APPENDAGES:	1 solar panel
ATTITUDE CONTROL:	Spin-stabilized
ENERGY SOURCES:	None

EVENT DATA

DATE:	13 Sep 1985	LOCATION:	35N, 234E (asc)
TIME:	2043 GMT	ASSESSED CAUSE:	Deliberate
ALTITUDE:	525 km		

PRE-EVENT ELEMENTS

EPOCH:	85256.72413718	MEAN ANOMALY:	260.9644
RIGHT ASCENSION:	182.5017	MEAN MOTION:	15.11755304
INCLINATION:	97.6346	MEAN MOTION DOT/2:	.00000616
ECCENTRICITY:	.0022038	MEAN MOTION DOT DOT/6:	.0
ARG. OF PERIGEE:	99.4081	BSTAR:	.000037918

DEBRIS CLOUD DATA

MAXIMUM ΔP:	12.7 min
MAXIMUM ΔI:	1.4 deg

COMMENTS

P-78 was impacted by a sub-orbital object at high velocity as part of a planned test.

REFERENCE DOCUMENT

Postmortem of a Hypervelocity Impact: Summary, R. L. Kling, Technical Report CS86-LKD-001, Teledyne Brown Engineering, Colorado Springs, September 1986.

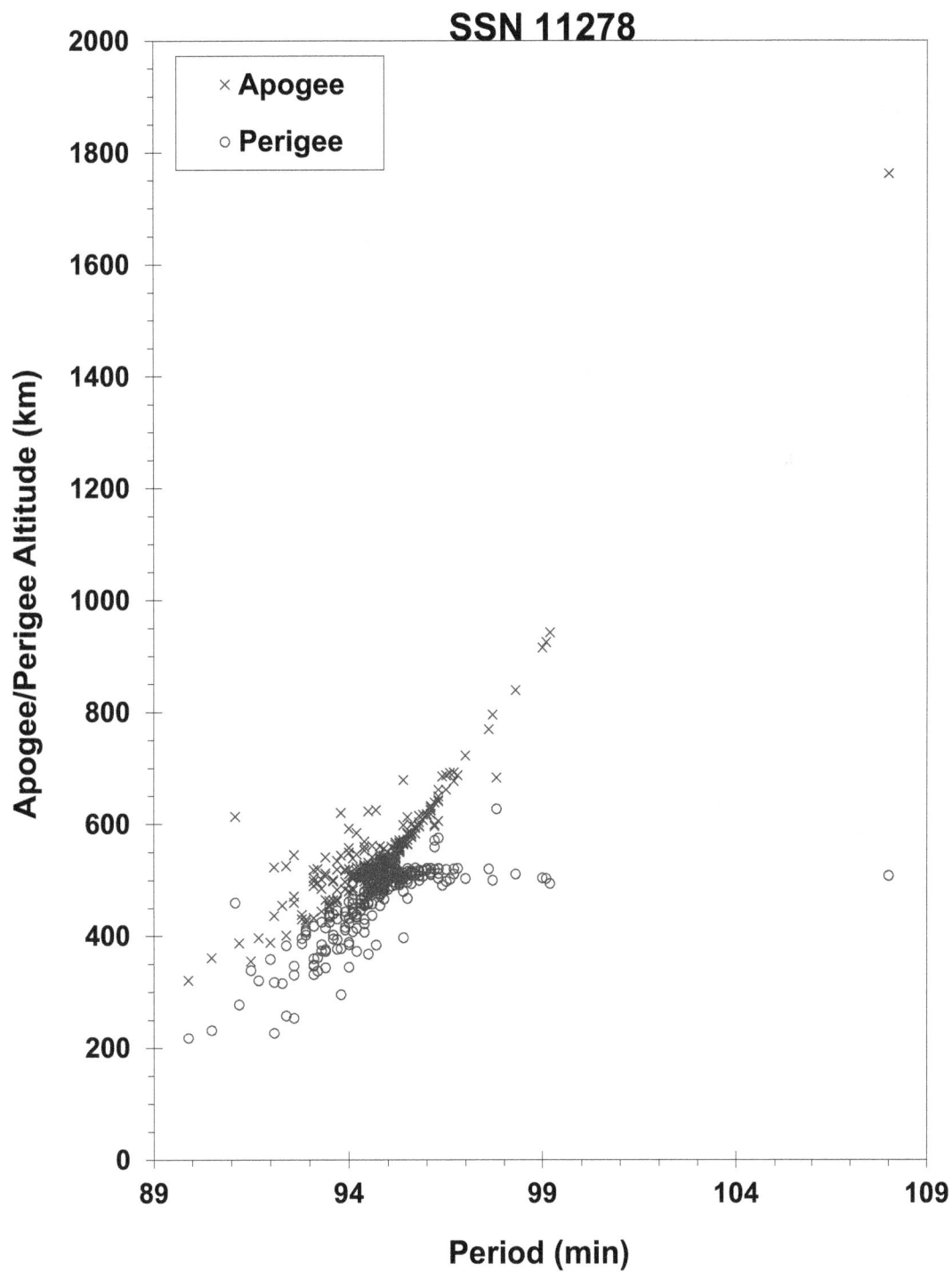

P-78 debris cloud of 267 fragments seen 11 hours after the event by the
US SSN PARCS radar.

SATELLITE DATA

TYPE:	Payload
OWNER:	CIS
LAUNCH DATE:	18.50 Apr 1979
DRY MASS (KG):	3000
MAIN BODY:	Cylinder; 1.3 m diameter by 17 m length
MAJOR APPENDAGES:	Solar panels
ATTITUDE CONTROL:	Active, 3-axis
ENERGY SOURCES:	On-board propellants, explosive charge (?)

EVENT DATA

DATE:	17 Sep 1979	LOCATION:	53S, 336E (dsc)
TIME:	1039 GMT	ASSESSED CAUSE:	Unknown
ALTITUDE:	385 km		

PRE-EVENT ELEMENTS

EPOCH:	79260.33615661	MEAN ANOMALY:	61.9566
RIGHT ASCENSION:	271.8638	MEAN MOTION:	15.58096051
INCLINATION:	65.0398	MEAN MOTION DOT/2:	.00102640
ECCENTRICITY:	.0016936	MEAN MOTION DOT DOT/6:	.0
ARG. OF PERIGEE:	297.9871	BSTAR:	.0013492

DEBRIS CLOUD DATA

MAXIMUM ΔP:	7.1 min*
MAXIMUM ΔI:	0.3 deg*

*Based on uncataloged debris data

COMMENTS

Cosmos 1094 was the fourth spacecraft of the Cosmos 699-type to experience a fragmentation. Spacecraft had been in a regime of natural decay for 4 months prior to the event. All new debris decayed before being officially cataloged.

REFERENCE DOCUMENTS

"Artificial Satellite Break-Ups (Part 1): Soviet Ocean Surveillance Satellites", N. L. Johnson, <u>Journal of the British Interplanetary Society</u>, February 1983, pp. 51-58.

<u>History of Soviet/Russian Satellite Fragmentations-A Joint U.S-Russian Investigation</u>, N. L. Johnson, *et al.,* Kaman Sciences Corporation, October 1995.

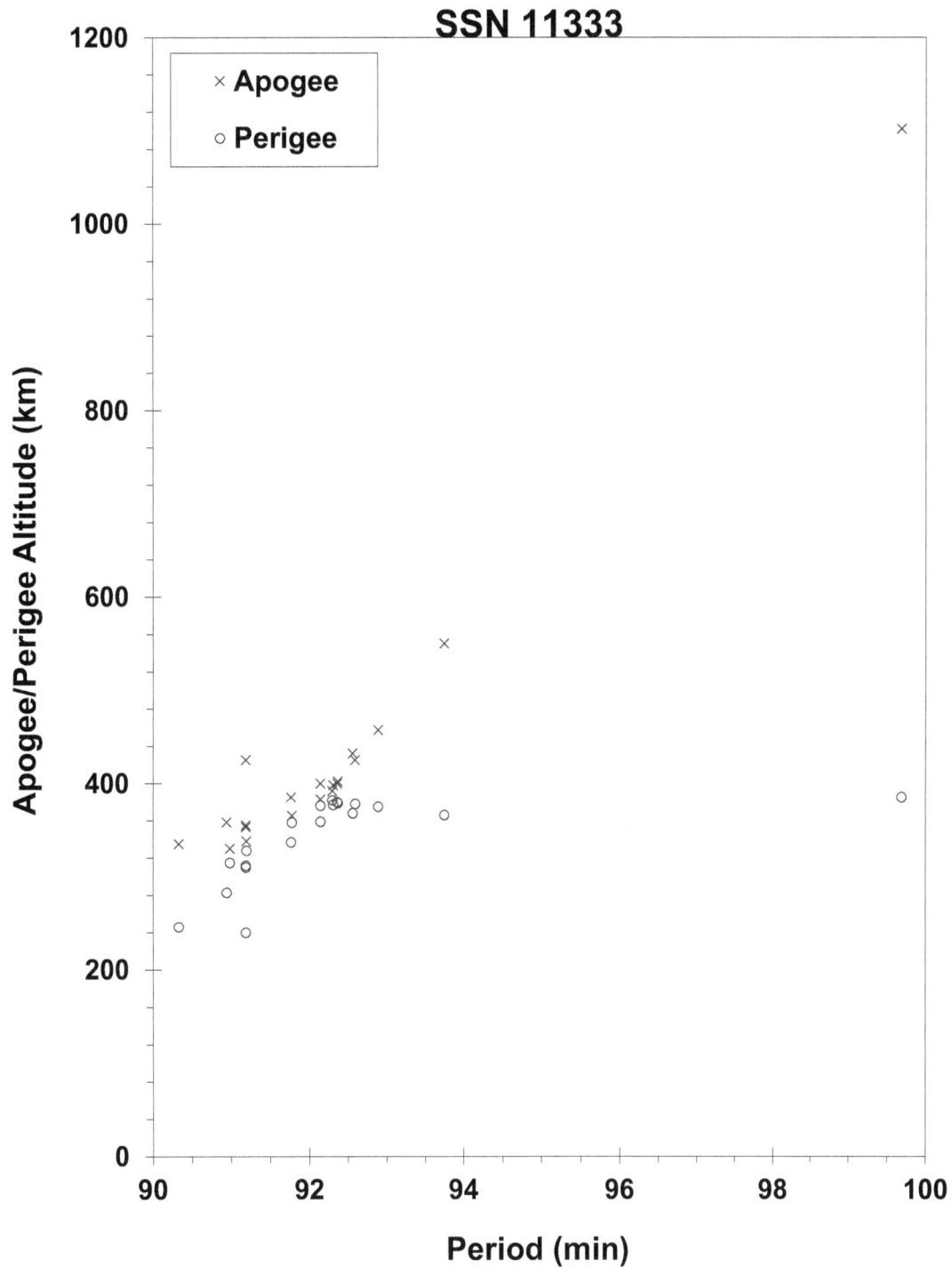

SSN 11333

Cosmos 1094 debris cloud of 20 fragments within 1 week of the event as reconstructed from the US SSN database.

SATELLITE DATA

TYPE:	Payload
OWNER:	CIS
LAUNCH DATE:	27.76 Jun 1979
DRY MASS (KG):	1250
MAIN BODY:	Cylinder; 1.7 m diameter by 2 m length
MAJOR APPENDAGES:	Solar panels
ATTITUDE CONTROL:	Active, 3-axis
ENERGY SOURCES:	On-board propellants, explosive charge

EVENT DATA

DATE:	Mid-Feb 1980	LOCATION:	Unknown
TIME:	Unknown	ASSESSED CAUSE:	Deliberate
ALTITUDE:	Unknown		

PRE-EVENT ELEMENTS

EPOCH:	80048.26161234	MEAN ANOMALY:	5.0375
RIGHT ASCENSION:	104.4713	MEAN MOTION:	2.00453352
INCLINATION:	63.3495	MEAN MOTION DOT/2:	.0
ECCENTRICITY:	.7238911	MEAN MOTION DOT DOT/6:	.0
ARG. OF PERIGEE:	318.4445	BSTAR:	.0

DEBRIS CLOUD DATA

MAXIMUM ΔP:	3.5 min*
MAXIMUM ΔI:	0.2 deg*

*Based on uncataloged debris data

COMMENTS

Cosmos 1109 was another spacecraft of the Cosmos 862-type to experience a fragmentation. Cosmos 1109 maneuvered into an operational orbit about 19 July. The payload was "lost" after 17 February 1980 and three pieces of debris were soon found that could be traced back to that period.

REFERENCE DOCUMENT

History of Soviet/Russian Satellite Fragmentations-A Joint U.S.-Russian Investigation, N. L. Johnson, *et al.,* Kaman Sciences Corporation, October 1995.

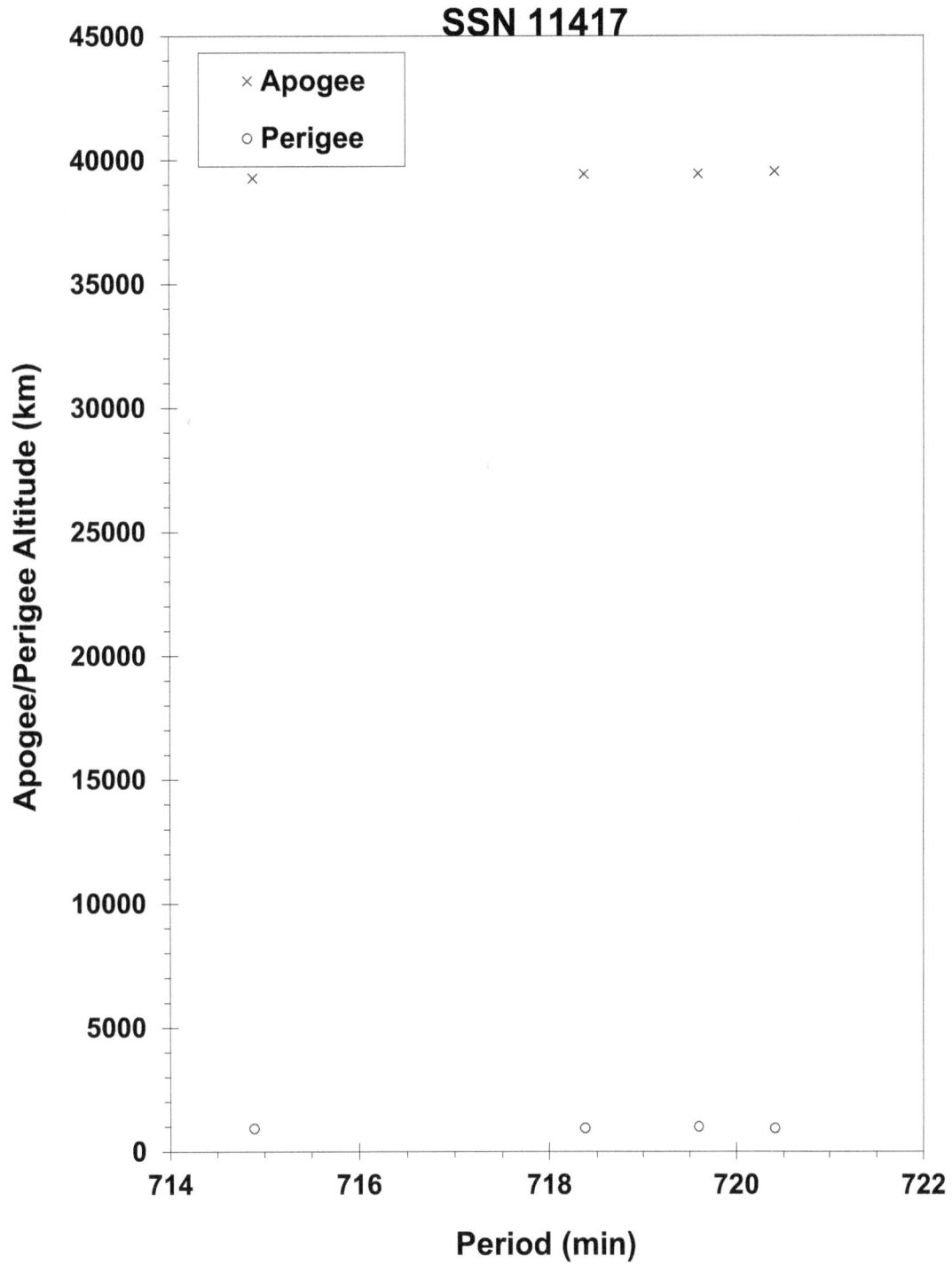

SSN 11417

Cosmos 1109 and three fragments in February 1980 as reconstructed
from the US SSN database.

SATELLITE DATA

TYPE:	Payload
OWNER:	CIS
LAUNCH DATE:	28.01 Aug 1979
DRY MASS (KG):	1250
MAIN BODY:	Cylinder; 1.7 m diameter by 2 m length
MAJOR APPENDAGES:	Solar panels
ATTITUDE CONTROL:	Active, 3-axis
ENERGY SOURCES:	On-board propellants, explosive charge

EVENT DATA

DATE:	9 Sep 1979	LOCATION:	52N, 304E (asc)
TIME:	0230 GMT	ASSESSED CAUSE:	Deliberate
ALTITUDE:	8375 km		

PRE-EVENT ELEMENTS

EPOCH:	79249.09448656	MEAN ANOMALY:	3.7678
RIGHT ASCENSION:	288.1742	MEAN MOTION:	2.00548359
INCLINATION:	63.0212	MEAN MOTION DOT/2:	.0
ECCENTRICITY:	.7383335	MEAN MOTION DOT DOT/6:	.0
ARG. OF PERIGEE:	318.3799	BSTAR:	.0

DEBRIS CLOUD DATA

MAXIMUM ΔP:	4.0 min*
MAXIMUM ΔI:	0.1 deg*

*Based on uncataloged debris data

COMMENTS

Cosmos 1124 was another spacecraft of the Cosmos 862-type to experience a fragmentation. After insertion into a Molniya-type transfer orbit on 28 August, Cosmos 1124's ascending node was allowed to drift until 3 September when a maneuver placed the spacecraft into an operational, semi-synchronous orbit. The fragmentation occurred 6 days later. The spacecraft never maneuvered again and soon drifted off station.

REFERENCE DOCUMENT

History of Soviet/Russian Satellite Fragmentations-A Joint U.S.-Russian Investigation, N. L. Johnson, *et al.,* Kaman Sciences Corporation, October 1995.

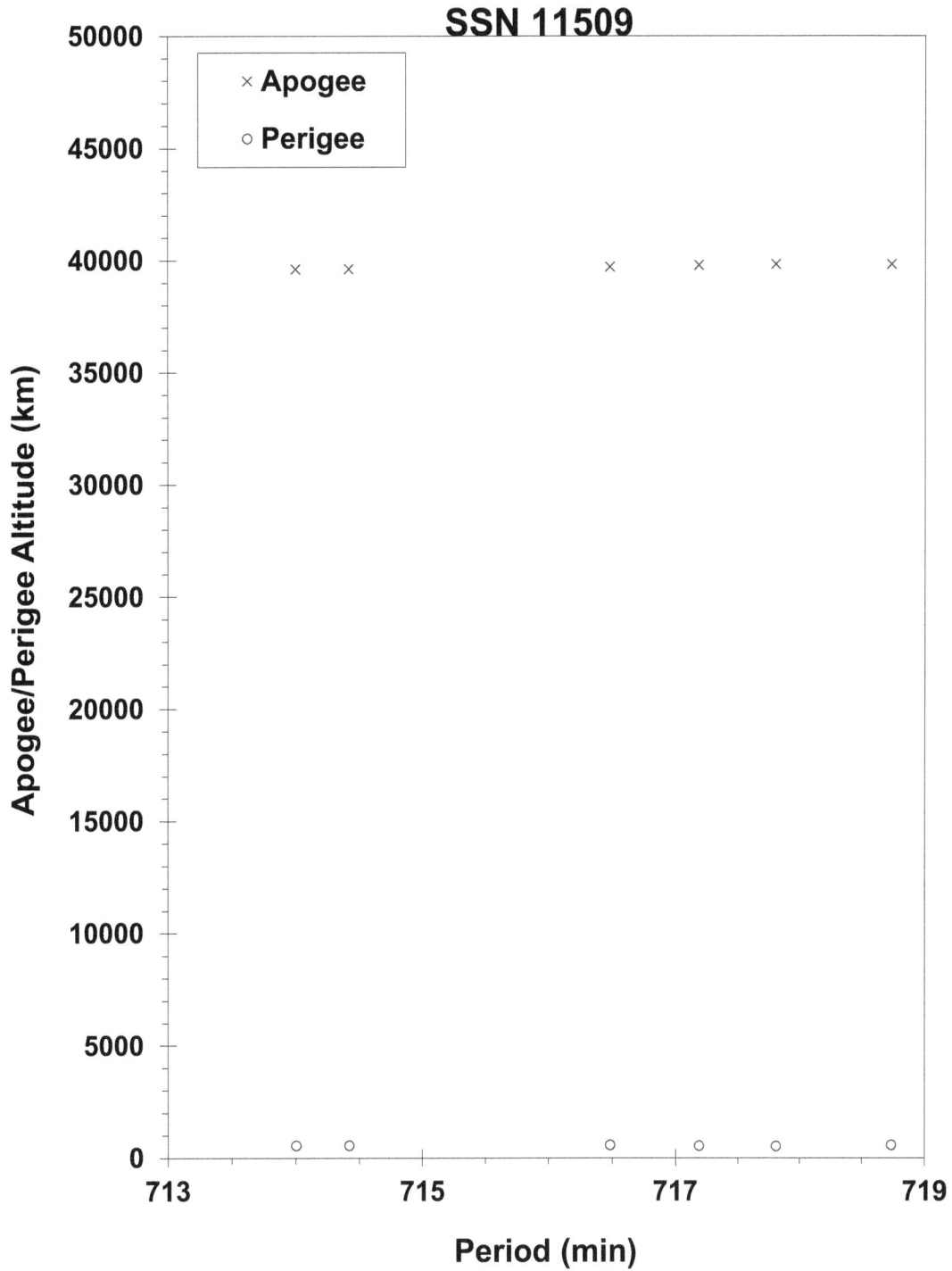

Cosmos 1124 debris cloud of 6 fragments about 1 week after the event as reconstructed from the US SSN database.

SATELLITE DATA

TYPE:	Payload
OWNER:	CIS
LAUNCH DATE:	3.72 Oct 1979
DRY MASS (KG):	1970
MAIN BODY:	Cylinder; 2 m by 4 m
MAJOR APPENDAGES:	Plate + 2 solar panels
ATTITUDE CONTROL:	Active, 3-axis
ENERGY SOURCES:	On-board propellants, Battery

EVENT DATA

DATE:	23 Apr 1981	LOCATION:	Unknown
TIME:	Unknown	ASSESSED CAUSE:	Battery?
ALTITUDE:	Unknown		

EVENT-CONTEMPORARY ELEMENTS

EPOCH:	81113.14159148	MEAN ANOMALY:	80.5434
RIGHT ASCENSION:	94.9928	MEAN MOTION:	1.00255486
INCLINATION:	0.7416	MEAN MOTION DOT/2:	.00000180
ECCENTRICITY	.00000180	MEAN MOTION DOT DOT/6:	.0
ARG. OF PERIGEE:	184.2970	BSTAR:	.0

DEBRIS CLOUD DATA

MAXIMUM ΔP:	Unknown
MAXIMUM ΔI:	Unknown

COMMENTS

This event was alluded to by Russian officials in an orbital debris meeting in April 2005 in Darmstadt and confirmed by the 18th SpCS. The event date was provided by the 18th SpCS and the observed cessation of East-West station keeping, or maintenance of the sub-satellite longitude, is generally consistent with this date. This event is now the second known geostationary orbit fragmentation. Root cause may be similar to the acknowledged cause of Ekran 2 [*q.v.*].

REFERENCE DOCUMENTS

History of Soviet/Russian Satellite Fragmentations-A Joint U.S.-Russian Investigation, N. L. Johnson, *et al.,* Kaman Sciences Corporation, October 1995.

Agapov, V., V. Biryukov, R. Kiladze et al., "Faint GEO Objects Search and Orbital Analysis," in Proc. Of the Fourth European Conf. on Space Debris (ESA SP-587, August 2005). Accessed at https://conference.sdo.esoc.esa.int/proceedings/sdc4/paper/106/SDC4-paper106.pdf, June 2022.

Insufficient data to construct a Gabbard diagram.

SATELLITE DATA

TYPE:	Payload
OWNER:	US
LAUNCH DATE:	7.07 Dec 1979
DRY MASS (KG):	463
MAIN BODY:	Box; 1.2 m length by 1.62 m length by 1.17 m height
MAJOR APPENDAGES:	2 Solar Panels, antenna/feeds
ATTITUDE CONTROL:	Active, 3 axis
ENERGY SOURCES:	On-board solid AKM and liquid monopropellants

EVENT DATA

DATE:	~ 11 Dec 1979	LOCATION:	Unknown
TIME:	Unknown	ASSESSED CAUSE:	Propulsion (solid)
ALTITUDE:	Unknown		

PRE-EVENT ELEMENTS

EPOCH:	79345.24087748	MEAN ANOMALY:	174.6228
RIGHT ASCENSION:	276.7452	MEAN MOTION:	2.28199220
INCLINATION:	23.7341	MEAN MOTION DOT/2:	.00044295
ECCENTRICITY:	.7298759	MEAN MOTION DOT DOT/6:	.0
ARG. OF PERIGEE:	181.2679	BSTAR:	.000009999

POST-EVENT ELEMENTS

EPOCH:	82327.71145422	MEAN ANOMALY:	349.7392
RIGHT ASCENSION:	164.2803	MEAN MOTION:	1.82544069
INCLINATION:	8.1767	MEAN MOTION DOT/2:	.00000069
ECCENTRICITY:	.4789421	MEAN MOTION DOT DOT/6:	.0
ARG. OF PERIGEE:	31.9453	BSTAR:	.0

DEBRIS CLOUD DATA

MAXIMUM ΔP:	Unknown
MAXIMUM ΔI:	Unknown

COMMENTS

Detected by software.

This mission was the third launch of the RCA SATCOM series. Communications with payload was lost upon firing of Aerojet SVM-7 solid apogee kick motor. Detection and tracking of debris have always been extremely difficult in part due to low inclination and highly elliptical orbit. Debris entered the SSN catalog in 2006 and after. The magnitude of the event and the total number of pieces created are unknown.

REFERENCE DOCUMENT

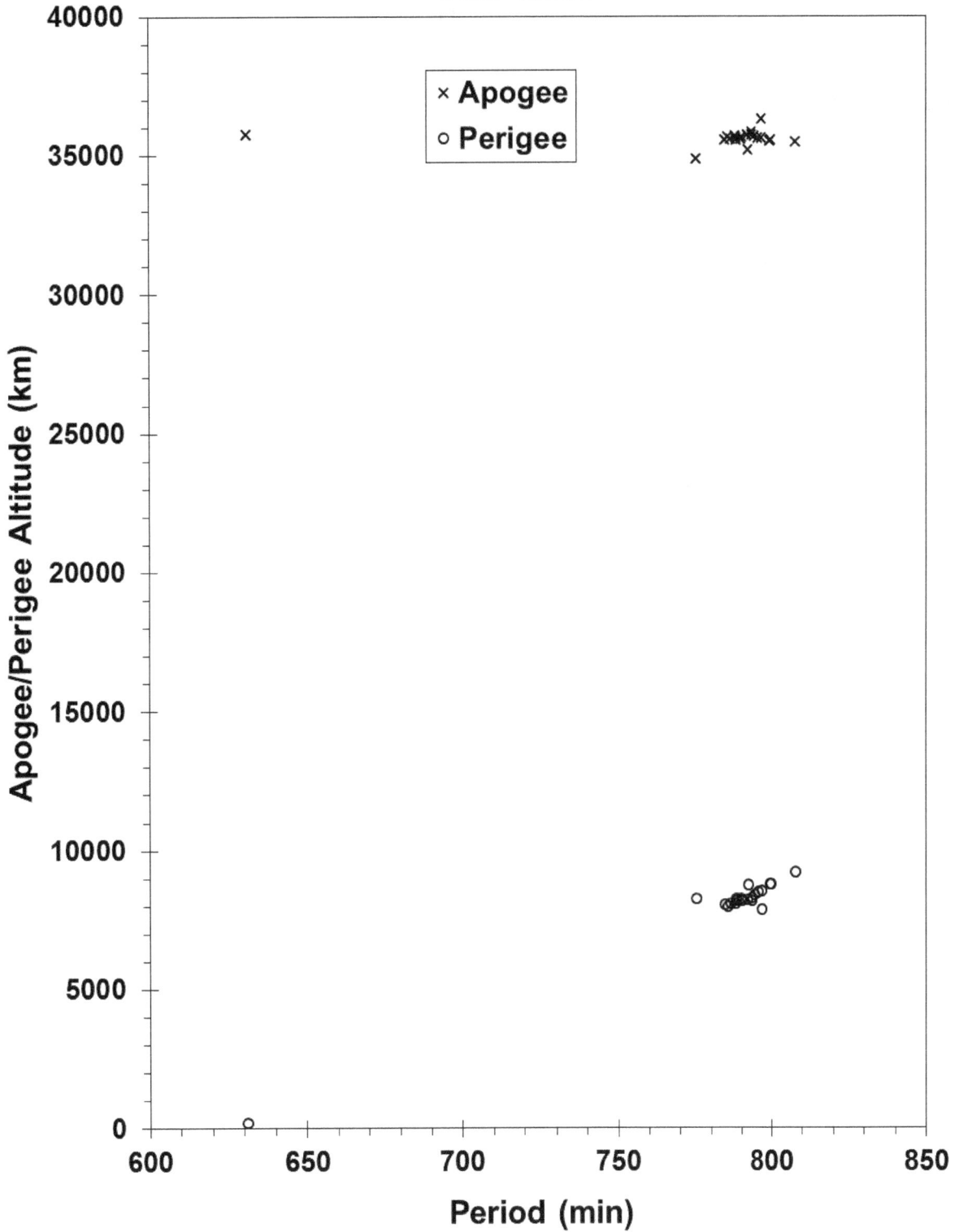

SATCOM 3 debris cloud of 22 fragments (including SATCOM 3) over 26 years after the event as reconstructed from the US SSN database.

SATELLITE DATA

TYPE:	Ariane 1 Third Stage
OWNER:	ESA
LAUNCH DATE:	24.72 Dec 1979
DRY MASS (KG):	1400
MAIN BODY:	Cylinder; 2.6 m diameter by 10.3 m length
MAJOR APPENDAGES:	None
ATTITUDE CONTROL:	None at time of the event.
ENERGY SOURCES:	On-board propellants, range safety device

EVENT DATA

DATE:	Apr 1980	LOCATION:	Unknown
TIME:	Unknown	ASSESSED CAUSE:	Propulsion
ALTITUDE:	Unknown		

PRE-EVENT ELEMENTS

EPOCH:	80088.55565320	MEAN ANOMALY:	17.6019
RIGHT ASCENSION:	101.5521	MEAN MOTION:	2.48253031
INCLINATION:	17.9092	MEAN MOTION DOT/2:	.001764977
ECCENTRICITY:	.7152375	MEAN MOTION DOT DOT/6:	.0
ARG. OF PERIGEE:	264.7858	BSTAR:	.001078542

DEBRIS CLOUD DATA

MAXIMUM ΔP:	Unknown
MAXIMUM ΔI:	Unknown

COMMENTS

This mission was the inaugural flight of the Ariane 1 launch vehicle. Payload and R/B were apparently cross-tagged until mid-January 1980. Detection and tracking of debris have always been extremely difficult in part due to low inclination and highly elliptical orbit. Debris data were first developed in the second half of April, and calculations suggest the fragmentation occurred during the first week of April. The magnitude of the event and the total number of pieces created are unknown. Many debris had high decay rates.

REFERENCE DOCUMENT

A Preliminary Analysis of the Fragmentation of the Spot 1 Ariane Third Stage, N. L. Johnson, Technical Report CS87-LKD-003, Teledyne Brown Engineering, Colorado Springs, March 1987.

SSN 11659

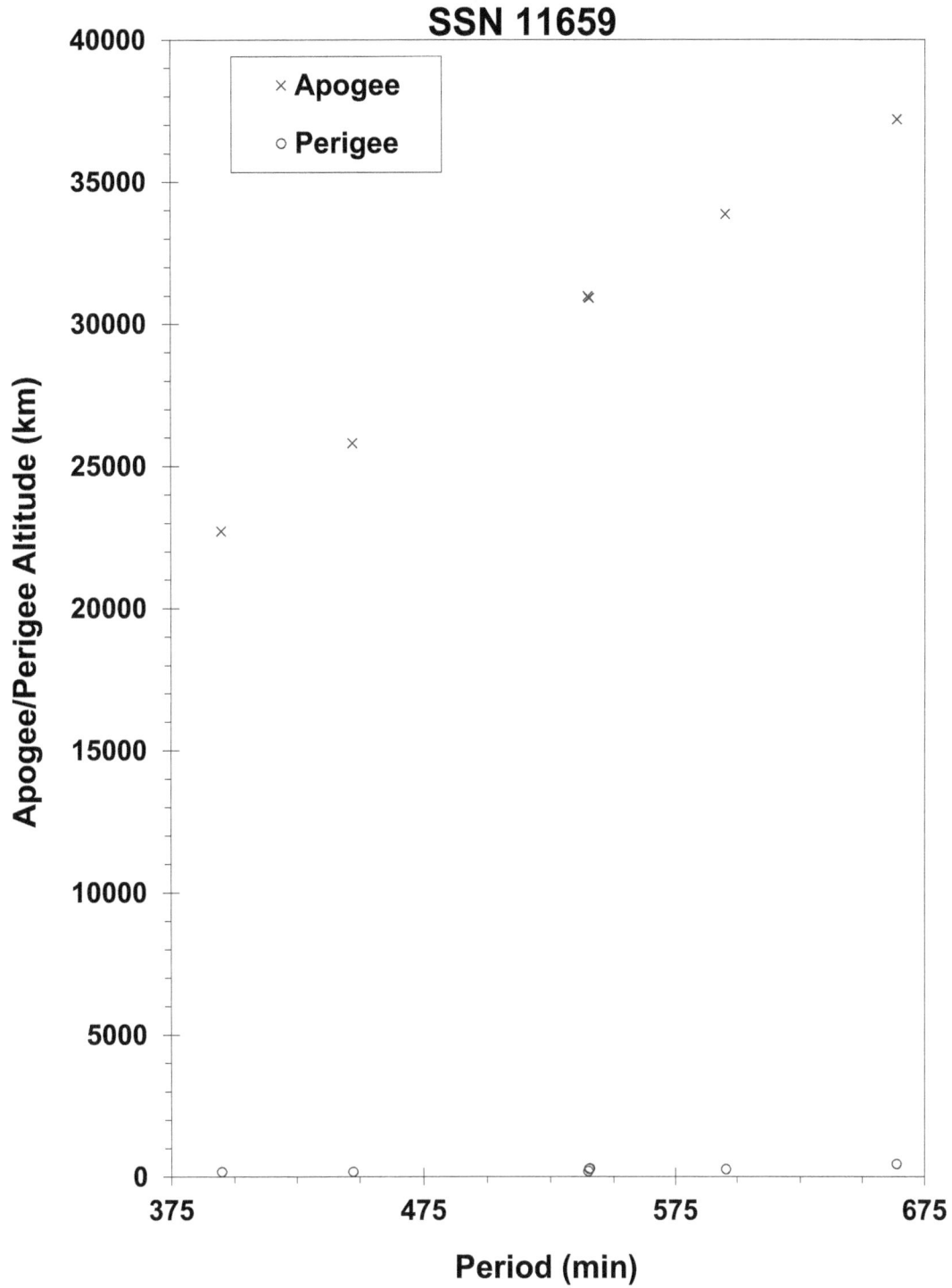

CAT R/B debris cloud of 7 fragments about 8 weeks after the event as reconstructed from the US SSN database.

COSMOS 1167   1980-021A   11729

SATELLITE DATA

TYPE:	Payload
OWNER:	CIS
LAUNCH DATE:	14.44 Mar 1980
DRY MASS (KG):	3000
MAIN BODY:	Cylinder; 1.3 m diameter by 17 m length
MAJOR APPENDAGES:	Solar panels
ATTITUDE CONTROL:	Active, 3-axis
ENERGY SOURCES:	On-board propellants, explosive charge (?)

EVENT DATA

DATE:	15 Jul 1981	LOCATION:	10N, 106E (asc)
TIME:	0921 GMT	ASSESSED CAUSE:	Unknown
ALTITUDE:	430 km		

PRE-EVENT ELEMENTS

EPOCH:	81196.19449955	MEAN ANOMALY:	110.8351
RIGHT ASCENSION:	174.9184	MEAN MOTION:	15.54665775
INCLINATION:	65.0101	MEAN MOTION DOT/2:	.00025375
ECCENTRICITY:	.0068471	MEAN MOTION DOT DOT/6:	.0
ARG. OF PERIGEE:	248.6139	BSTAR:	.00034595

DEBRIS CLOUD DATA

MAXIMUM ΔP: 1.0 min*
MAXIMUM ΔI: 0.5 deg*

*Based on uncataloged debris data

COMMENTS

Cosmos 1167 was the fifth spacecraft of the Cosmos 699-type to experience a fragmentation. The spacecraft had been in a regime of natural decay for 3 months prior to the event. Most debris reentered before being officially cataloged.

REFERENCE DOCUMENTS

The Fragmentations of USSR Satellites 11729 and 12504 (U), J. R. Gabbard and P. M. Landry, Technical Memorandum 82-S-03, DCS/Plans, Hdqtrs NORAD/ADCOM, Colorado Springs, August 1982 (Secret).

"Artificial Satellite Break-Ups (Part 1): Soviet Ocean Surveillance Satellites", N. L. Johnson, Journal of the British Interplanetary Society, February 1983, pp. 51-58.

History of Soviet/Russian Satellite Fragmentations-A Joint U.S.-Russian Investigation, N. L Johnson, et al., Kaman Sciences Corporation, October 1995.

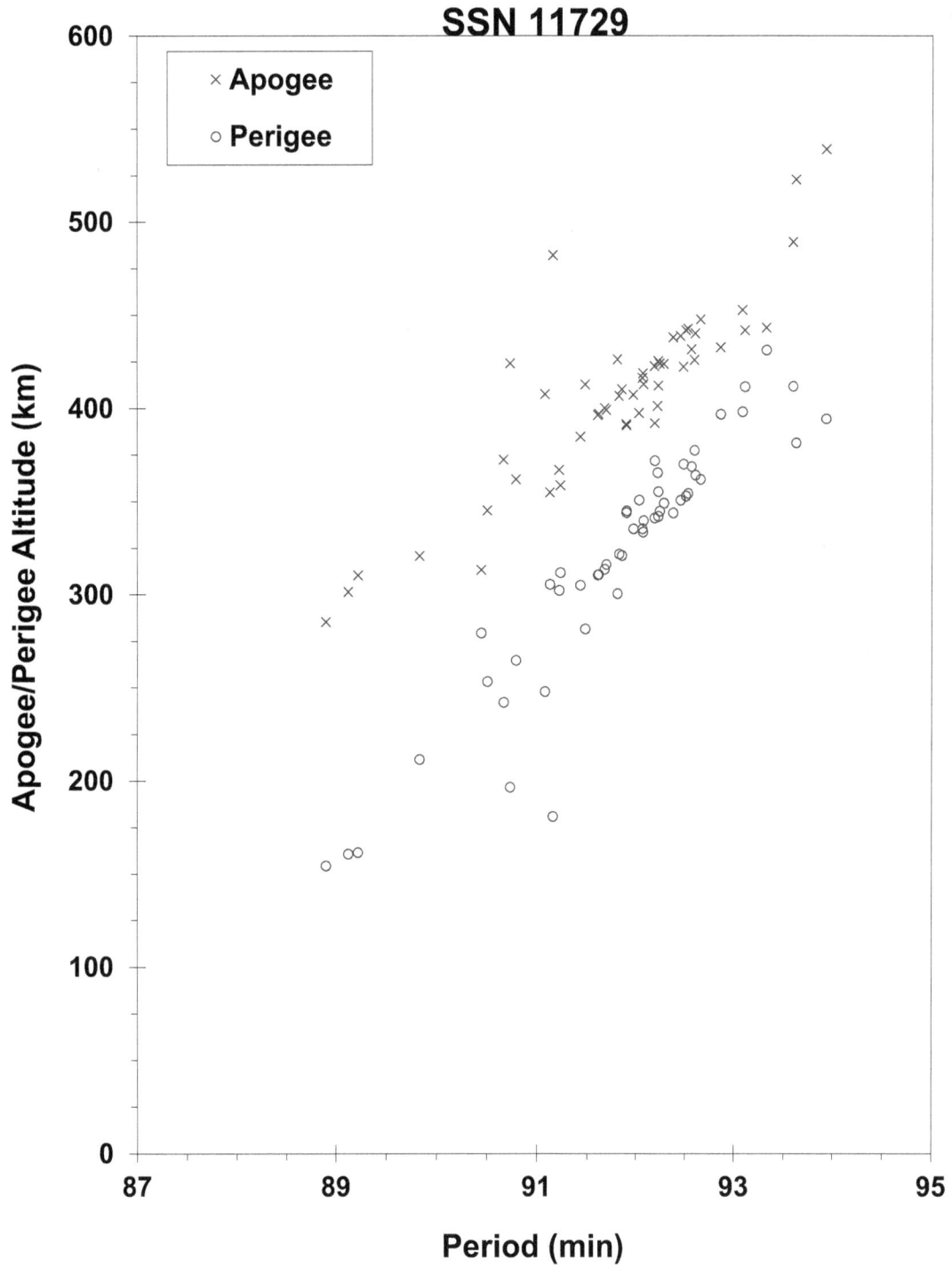

SSN 11729

× Apogee
○ Perigee

Apogee/Perigee Altitude (km)

Period (min)

Cosmos 1167 debris cloud remnant of 53 fragments about 2 weeks after the event as reconstructed from the US SSN database.

SATELLITE DATA

TYPE:	Payload
OWNER:	CIS
LAUNCH DATE:	18.04 Apr 1980
DRY MASS (KG):	1400
MAIN BODY:	Cylinder; 2 m diameter by 4 m length
MAJOR APPENDAGES:	None
ATTITUDE CONTROL:	Active, 3-axis
ENERGY SOURCES:	On-board propellants, explosive charge

EVENT DATA

DATE:	18 Apr 1980	LOCATION:	47N, 322E (asc)
TIME:	0726 GMT	ASSESSED CAUSE:	Deliberate
ALTITUDE:	1625 km		

POST-EVENT ELEMENTS

EPOCH:	80109.51771250	MEAN ANOMALY:	102.2095
RIGHT ASCENSION:	250.9679	MEAN MOTION:	13.64414319
INCLINATION:	66.1153	MEAN MOTION DOT/2:	.0
ECCENTRICITY:	.0865337	MEAN MOTION DOT DOT/6:	.0
ARG. OF PERIGEE:	248.5294	BSTAR:	.0

DEBRIS CLOUD DATA

MAXIMUM ΔP:	5.4 min
MAXIMUM ΔI:	0.6 deg

COMMENTS

Cosmos 1174 was launched on a two-revolution rendezvous with Cosmos 1171. After a close approach, Cosmos 1174 performed a final maneuver shortly before its warhead was intentionally fired. Elements above are first data available after the final maneuver but also following the fragmentation. Cosmos 1174 was part of test series begun with Cosmos 249.

REFERENCE DOCUMENTS

"Artificial Satellite Break-Ups (Part 2): Soviet Anti-Satellite Program", N.L. Johnson, Journal of the British Interplanetary Society, August 1983, pp. 357-362.

History of Soviet/Russian Satellite Fragmentations-A Joint U.S.-Russian Investigation, N. L. Johnson, et al., Kaman Sciences Corporation, October 1995.

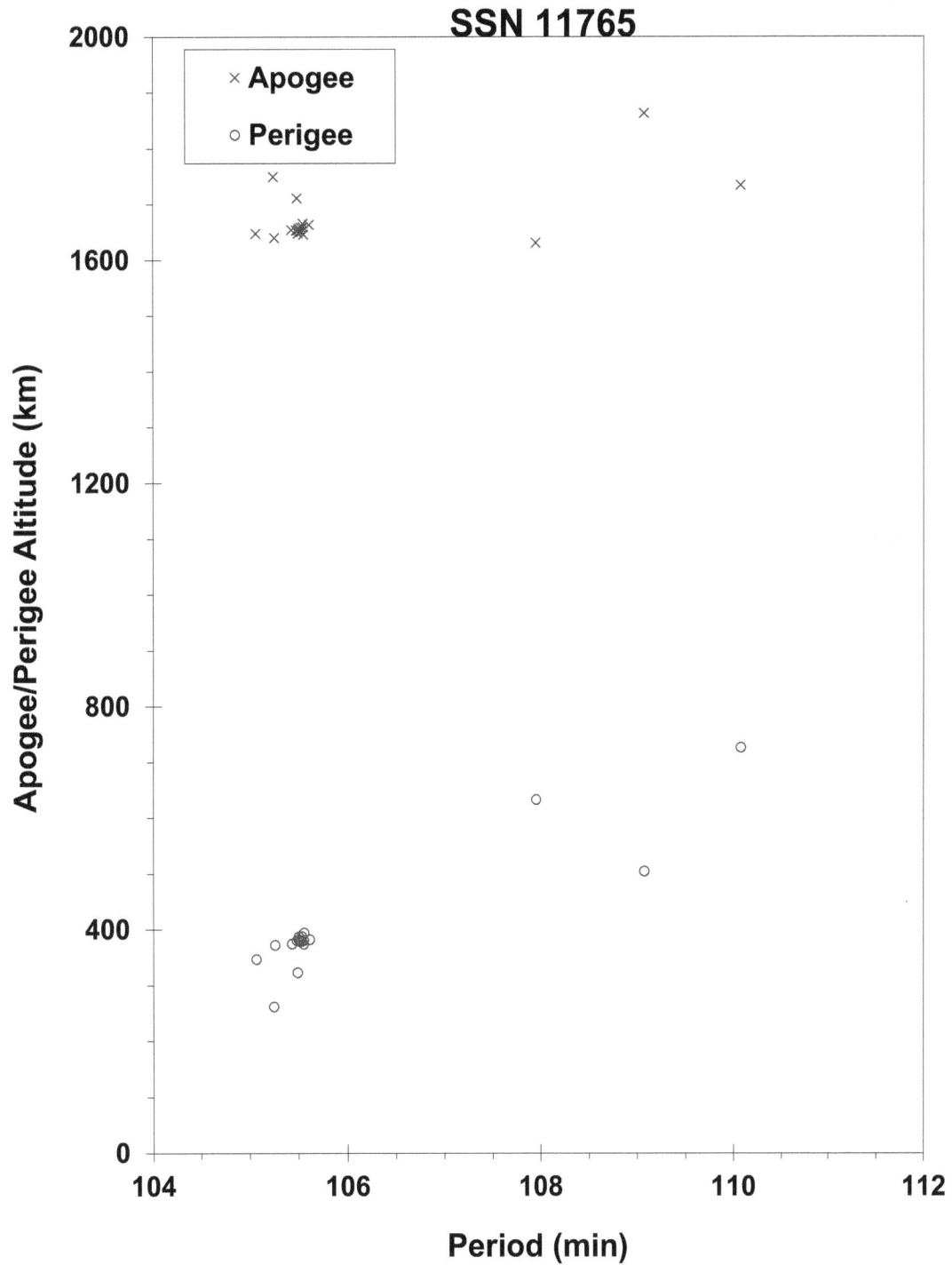

SSN 11765

Cosmos 1174 debris cloud of 18 identified fragments about 10 days after the event as reconstructed from the US SSN database.

SATELLITE DATA

TYPE:	Payload
OWNER:	CIS
LAUNCH DATE:	14.87 June 1980
DRY MASS (KG):	1250
MAIN BODY:	Cylinder; 1.7 m diameter by 2 m length
MAJOR APPENDAGES:	Solar panels
ATTITUDE CONTROL:	Active, 3-axis
ENERGY SOURCES:	On-board propellants, explosive device

EVENT DATA

DATE:	26 August 1980	LOCATION:	Unknown
TIME:	Unknown	ASSESSED CAUSE:	Deliberate
ALTITUDE:	Unknown		

PRE-EVENT ELEMENTS

EPOCH:	80238.74437670	MEAN ANOMALY:	4.7027
RIGHT ASCENSION:	155.4878	MEAN MOTION:	2.00554276
INCLINATION:	62.9033	MEAN MOTION DOT/2:	.00000217
ECCENTRICITY:	.7321456	MEAN MOTION DOT DOT/6:	.0
ARG. OF PERIGEE:	318.3182	BSTAR:	.0

DEBRIS CLOUD DATA

MAXIMUM ΔP:	42.6 min
MAXIMUM ΔI:	7.1 deg

COMMENTS

Cosmos 1188 was a member of the Cosmos 862 class of operational satellites in highly elliptical, semi-synchronous orbits that experienced a total of 18 fragmentations during the period 1977-1986. Due to the nature of these orbits, which result in high altitudes over the Northern Hemisphere where most surveillance sensors are located, debris detection and tracking is extremely difficult. Only the largest fragments can be seen. Cosmos 862 class spacecraft were equipped with self-destruct packages in the event that spacecraft control was lost; this was the cause of breakups until the explosives were removed after Cosmos 1481.

REFERENCE DOCUMENT

History of Soviet/Russian Satellite Fragmentations-A Joint U.S.-Russian Investigation, N. L. Johnson,, *et al.,* Kaman Sciences Corporation, October 1995.

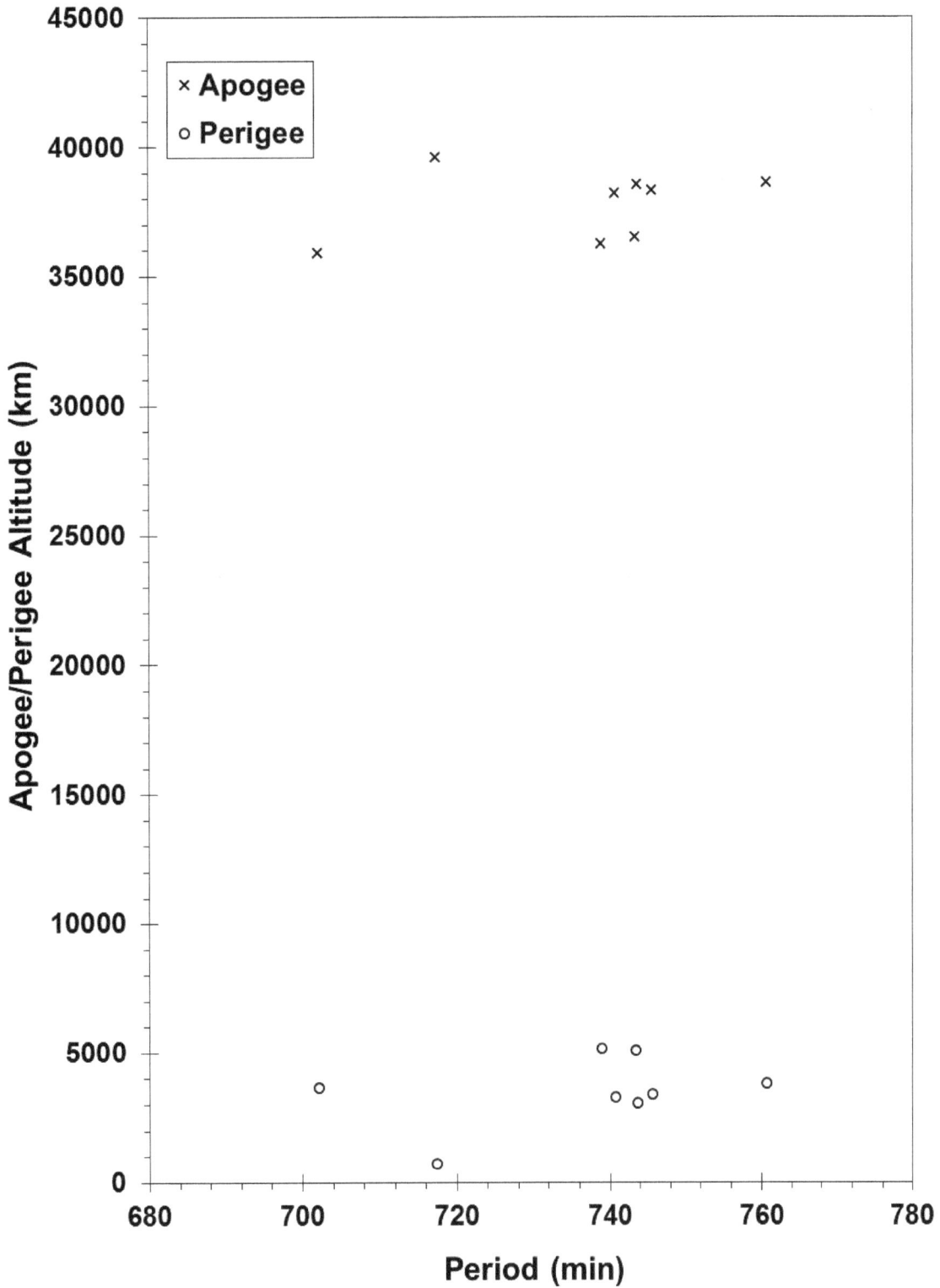

SSN 11844

Cosmos 1188 debris cloud; fragments cataloged up to 23 years after the event as reconstructed from the US SSN database.

COSMOS 1191   1980-057A   11871

SATELLITE DATA

TYPE:	Payload
OWNER:	CIS
LAUNCH DATE:	2.04 Jul 1980
DRY MASS (KG):	1250
MAIN BODY:	Cylinder; 1.7 m diameter by 2 m length
MAJOR APPENDAGES:	Solar panels
ATTITUDE CONTROL:	Active, 3-axis
ENERGY SOURCES:	On-board propellants, explosive charge

EVENT DATA

DATE:	14 May 1981	LOCATION:	Unknown
TIME:	Unknown	ASSESSED CAUSE:	Deliberate
ALTITUDE:	Unknown		

PRE-EVENT ELEMENTS

EPOCH:	81133.07322634	MEAN ANOMALY:	5.1166
RIGHT ASCENSION:	198.5704	MEAN MOTION:	2.00555560
INCLINATION:	62.6448	MEAN MOTION DOT/2:	.00001257
ECCENTRICITY:	.7180863	MEAN MOTION DOT DOT/6:	.0
ARG. OF PERIGEE:	319.4330	BSTAR:	.0

DEBRIS CLOUD DATA

MAXIMUM ΔP:	6.0 min*
MAXIMUM ΔI:	0.1 deg*

*Based on uncataloged debris data

COMMENTS

Cosmos 1191 was another spacecraft of the Cosmos 862-type to experience a fragmentation. The first debris elements were developed for 25 May.

REFERENCE DOCUMENT

History of Soviet/Russian Satellite Fragmentations-A Joint U.S.-Russian Investigation, N. L. Johnson, *et al.,* Kaman Sciences Corporation, October 1995.

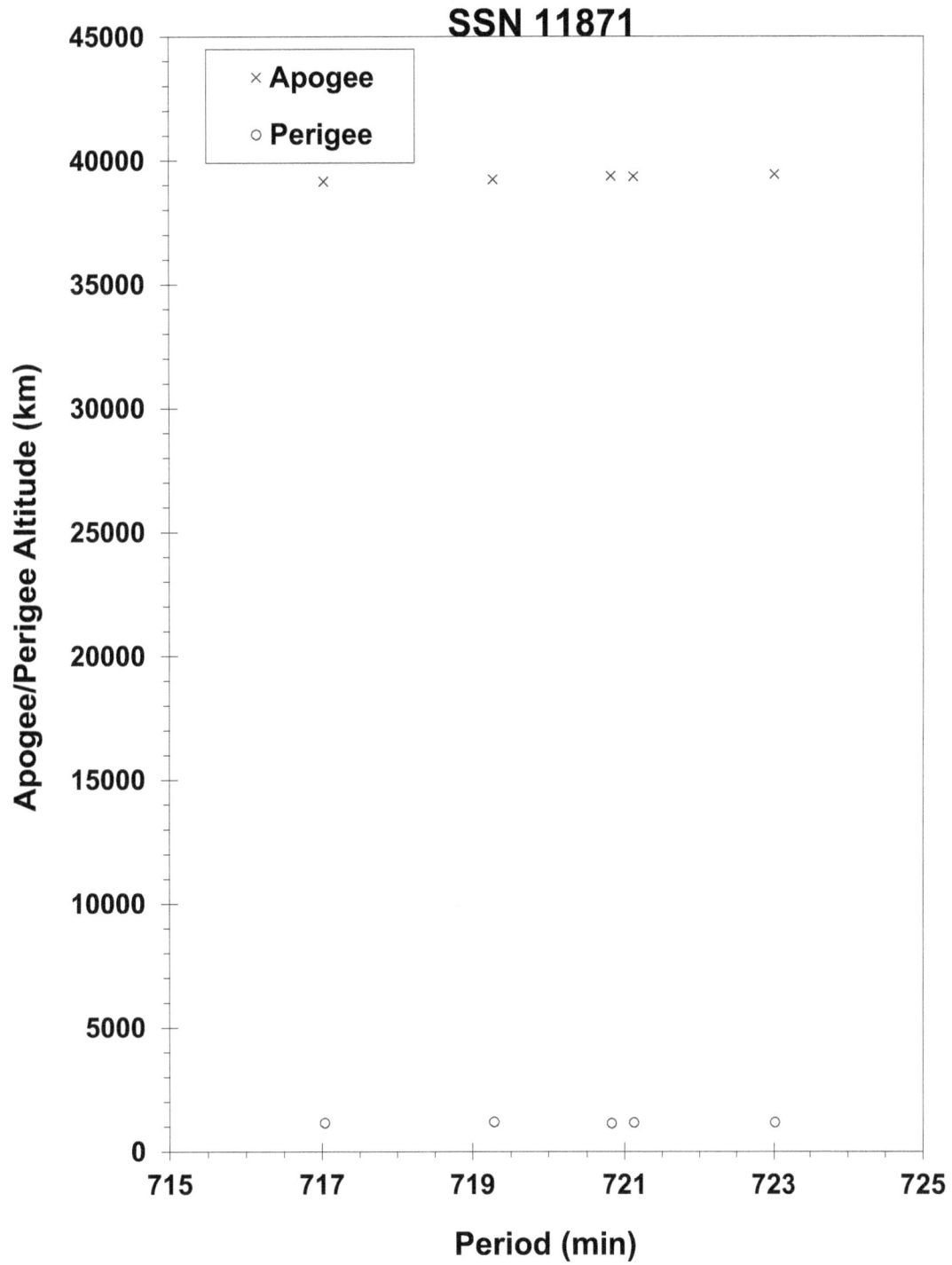

SSN 11871

Cosmos 1191 debris cloud of 5 identified fragments 1 month after the event as reconstructed from the US SSN database.

SATELLITE DATA

TYPE:	Payload
OWNER:	CIS
LAUNCH DATE:	24.46 Oct 1980
DRY MASS (KG):	1250
MAIN BODY:	Cylinder; 1.7 m diameter by 2 m length
MAJOR APPENDAGES:	Solar panels
ATTITUDE CONTROL:	Active, 3-axis
ENERGY SOURCES:	On-board propellants, explosive charge

EVENT DATA

DATE:	12 Feb 1983	LOCATION:	Unknown
TIME:	Unknown	ASSESSED CAUSE:	Deliberate
ALTITUDE:	Unknown		

PRE-EVENT ELEMENTS

EPOCH:	83042.34036514	MEAN ANOMALY:	6.0502
RIGHT ASCENSION:	36.1600	MEAN MOTION:	2.00587025
INCLINATION:	65.2478	MEAN MOTION DOT/2:	0.00001154
ECCENTRICITY:	0.7021051	MEAN MOTION DOT DOT/6:	0
ARG. OF PERIGEE:	314.5975	BSTAR:	0

DEBRIS CLOUD DATA

MAXIMUM ΔP:	Unknown
MAXIMUM ΔI:	Unknown

COMMENTS

Cosmos 1217 was another member of the Cosmos 862-type to experience a fragmentation.

REFERENCE DOCUMENT

History of Soviet/Russian Satellite Fragmentation-A Joint U.S.-Russian Investigation, N. L. Johnson, *et al.*, Kaman Sciences Corporation, October 1995.

Insufficient data to construct a Gabbard diagram.

SATELLITE DATA

TYPE:	Payload
OWNER:	CIS
LAUNCH DATE:	4.63 Nov 1980
DRY MASS (KG):	3000
MAIN BODY:	Cylinder; 1.3 m diameter by 17 m length
MAJOR APPENDAGES:	Solar panels
ATTITUDE CONTROL:	Active, 3-axis
ENERGY SOURCES:	On-board propellants, explosive charge (?)

EVENT DATA (1)

DATE:	20 Jun 1982	LOCATION:	10S, 332E (dsc)
TIME:	1818 GMT	ASSESSED CAUSE:	Unknown
ALTITUDE:	875 km		

PRE-EVENT ELEMENTS (1)

EPOCH:	82171.72558670	MEAN ANOMALY:	0.2166
RIGHT ASCENSION:	330.3811	MEAN MOTION:	14.49658466
INCLINATION:	65.0033	MEAN MOTION DOT/2:	.00000066
ECCENTRICITY:	.0219432	MEAN MOTION DOT DOT/6:	.0
ARG. OF PERIGEE:	357.8883	BSTAR:	.000025640

EVENT DATA (2)

DATE:	25 Aug 1982	LOCATION:	65S, 238E (dsc)
TIME:	1231 GMT	ASSESSED CAUSE:	Unknown
ALTITUDE:	665 km		

PRE-EVENT ELEMENTS (2)

EPOCH:	82230.91714195	MEAN ANOMALY:	22.7965
RIGHT ASCENSION:	159.4489	MEAN MOTION:	14.49745561
INCLINATION:	65.0025	MEAN MOTION DOT/2:	.0
ECCENTRICITY:	.0225583	MEAN MOTION DOT DOT/6:	.0
ARG. OF PERIGEE:	336.3217	BSTAR:	.0

DEBRIS CLOUD DATA

MAXIMUM ΔP:	3.4 min*
MAXIMUM ΔI:	1.8 deg*

*Based on uncataloged debris data

COMMENTS

Cosmos 1220 was the seventh spacecraft of the Cosmos 699-type to experience a fragmentation. The spacecraft had been in a natural decay regime for more than 14 months at the time of the first event. A total of 47 fragments had been officially cataloged by the time of the second event that occurred 2 months later. See similar dual events happening in the summer of 1982 with Cosmos 1306 and Cosmos 1260.

REFERENCE DOCUMENTS

Analysis of PARCS Recorded Data on the Breakup of Satellite 12054, J.W. Rider, Technical Report MSB83-ADC-0162, Teledyne Brown Engineering, Huntsville, January 1983.

Analysis of Cosmos 1220 and Cosmos 1306 Fragments (U), D. Fennessy, Report AH-23, FTD/OLAI, Cheyenne Mountain, Colorado, 12 January 1983 (Secret).

"Artificial Satellite Break-Ups (Part 1): Soviet Ocean Surveillance Satellites", N. L. Johnson, Journal of the British Interplanetary Society, February 1983, pp. 51-58.

History of Soviet/Russian Satellite Fragmentations-A Joint U.S.-Russian Investigation, N. L. Johnson, et al., Kaman Sciences Corporation, October 1995.

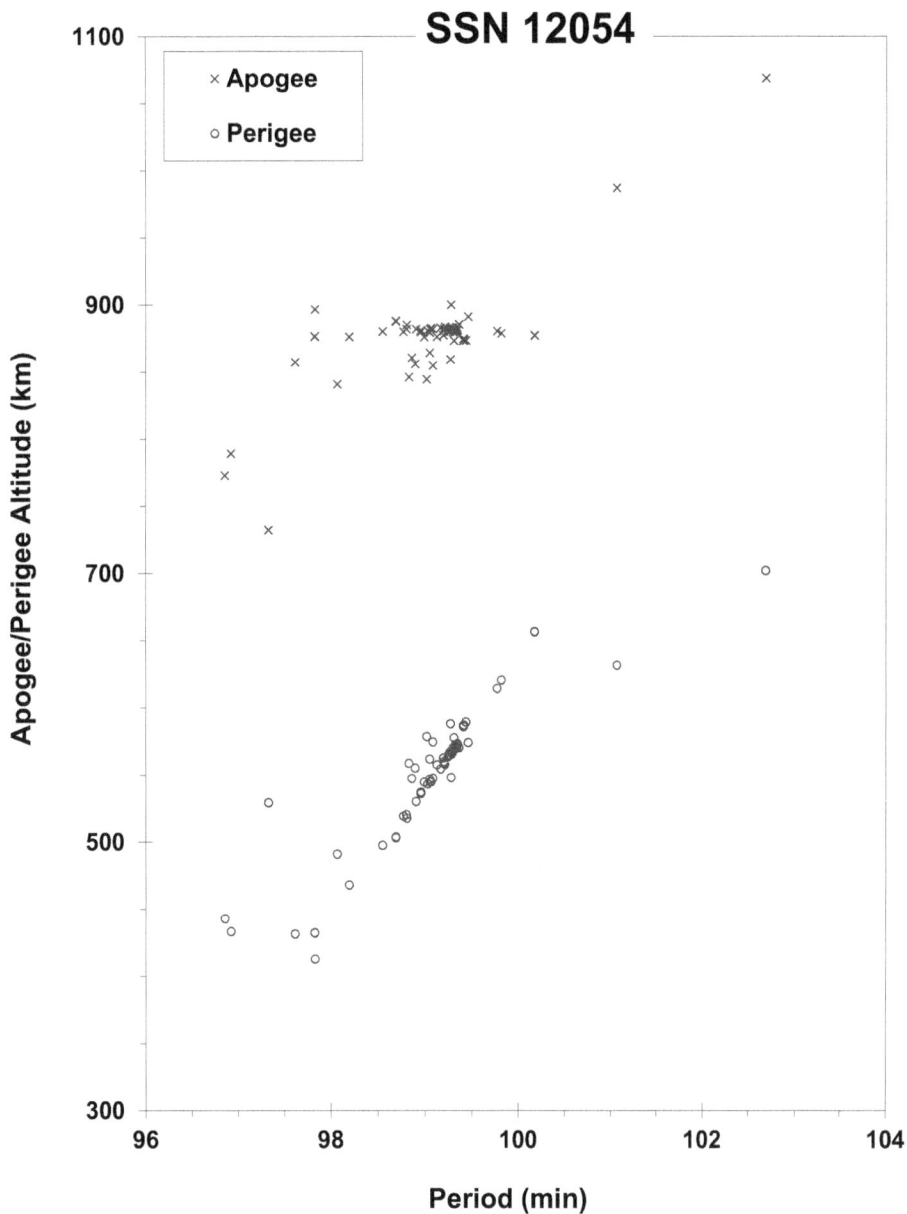

Cosmos 1220 debris cloud of 72 fragments about 1 week after the first event as reconstructed from the US SSN database.

COSMOS 1247   1981-016A   12303

SATELLITE DATA

TYPE:	Payload
OWNER:	CIS
LAUNCH DATE:	19.41 Feb 1981
DRY MASS (KG):	1250
MAIN BODY:	Cylinder; 1.7 m diameter by 2 m length
MAJOR APPENDAGES:	Solar panels
ATTITUDE CONTROL:	Active, 3-axis
ENERGY SOURCES:	On-board propellants, explosive charge

EVENT DATA

DATE:	20 Oct 1981	LOCATION:	Unknown
TIME:	Unknown	ASSESSED CAUSE:	Deliberate
ALTITUDE:	Unknown		

PRE-EVENT ELEMENTS

EPOCH:	81293.17083627	MEAN ANOMALY:	5.0298
RIGHT ASCENSION:	214.2278	MEAN MOTION:	2.00570861
INCLINATION:	62.9685	MEAN MOTION DOT/2:	.0
ECCENTRICITY:	.7233048	MEAN MOTION DOT DOT/6:	.0
ARG. OF PERIGEE:	318.2473	BSTAR:	.0

DEBRIS CLOUD DATA

MAXIMUM ΔP: 2.7 min*
MAXIMUM ΔI: 0.4 deg*

*See comments below

COMMENTS

Cosmos 1247 was another spacecraft of the Cosmos 862-type to experience a fragmentation. Cosmos 1247 appears to have completed the first burn of a 2-phase maneuver sequence on the event date, followed by debris generation. The ΔP and ΔI values above are based on the post-maneuver, 711-minute orbit of 12303 rather than the pre-maneuver, 718-minute orbit cited above.

REFERENCE DOCUMENT

History of Soviet/Russian Satellite Fragmentations-A Joint U.S.-Russian Investigation, N. L Johnson, et al., Kaman Sciences Corporation, October 1995.

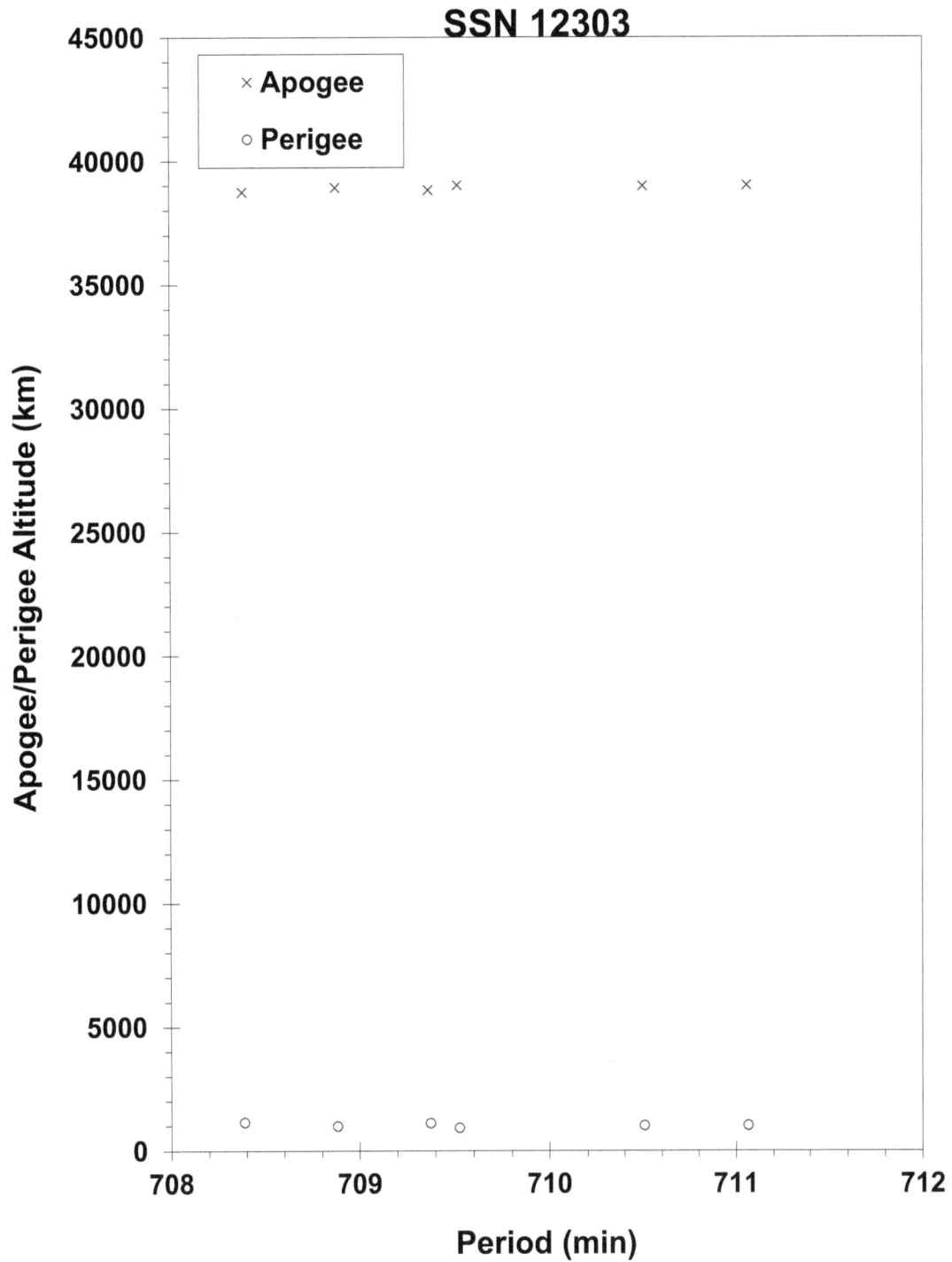

Cosmos 1247 debris cloud of 6 fragments about 6 weeks after the event as reconstructed from the US SSN database.

COSMOS 1260   **1981-028A**   **12364**

SATELLITE DATA

TYPE:	Payload
OWNER:	CIS
LAUNCH DATE:	20.99+ Mar 1981
DRY MASS (KG):	3000
MAIN BODY:	Cylinder; 1.3 m diameter by 17 m length
MAJOR APPENDAGES:	Solar panels
ATTITUDE CONTROL:	Active, 3-axis
ENERGY SOURCES:	On-board propellants, explosive charge (?)

EVENT DATA (1)

DATE:	8 May 1982	LOCATION:	40N, 62E (asc)
TIME:	0444 GMT	ASSESSED CAUSE:	Unknown
ALTITUDE:	555 km		

PRE-EVENT ELEMENTS (1)

EPOCH:	82127.98788154	MEAN ANOMALY:	28.1726
RIGHT ASCENSION:	337.2406	MEAN MOTION:	14.88799005
INCLINATION:	65.0246	MEAN MOTION DOT/2:	.00003980
ECCENTRICITY:	.0214690	MEAN MOTION DOT DOT/6:	.0
ARG. OF PERIGEE:	330.7493	BSTAR:	.00028791

EVENT DATA (2)

DATE:	10 Aug 1982	LOCATION:	51N, 238E (dsc)
TIME:	2335 GMT	ASSESSED CAUSE:	Deliberate
ALTITUDE:	750 km		

PRE-EVENT ELEMENTS (2)

EPOCH:	82222.89259484	MEAN ANOMALY:	62.7628
RIGHT ASCENSION:	45.7388	MEAN MOTION:	14.89366232
INCLINATION:	65.0248	MEAN MOTION DOT/2:	.00004369
ECCENTRICITY:	.0219155	MEAN MOTION DOT DOT/6:	.0
ARG. OF PERIGEE:	295.0884	BSTAR:	.00030390

DEBRIS CLOUD DATA

MAXIMUM ΔP:	5.2 min
MAXIMUM ΔI:	1.0 deg

COMMENTS

Cosmos 1260 was the sixth spacecraft of the Cosmos 699-type to experience a fragmentation. The spacecraft had been in a regime of natural decay for 8 months before the first event. After the event, the main remnant became satellite 13183, which then fragmented 3 months later. A total of 40 new fragments were officially cataloged prior to the second event. See also Cosmos 1220 and Cosmos 1306 for similar dual fragmentations of Cosmos 699-type spacecraft during this period.

REFERENCE DOCUMENTS

"Artificial Satellite Break-Ups (Part 1): Soviet Ocean Surveillance Satellites", N. L. Johnson, <u>Journal of the British Interplanetary Society</u>, February 1983, pp. 51-58.

<u>History of Soviet/Russian Satellite Fragmentations-A Joint U.S.-Russian Investigation</u>, N. L. Johnson, *et al.,* Kaman Sciences Corporation, October 1995.

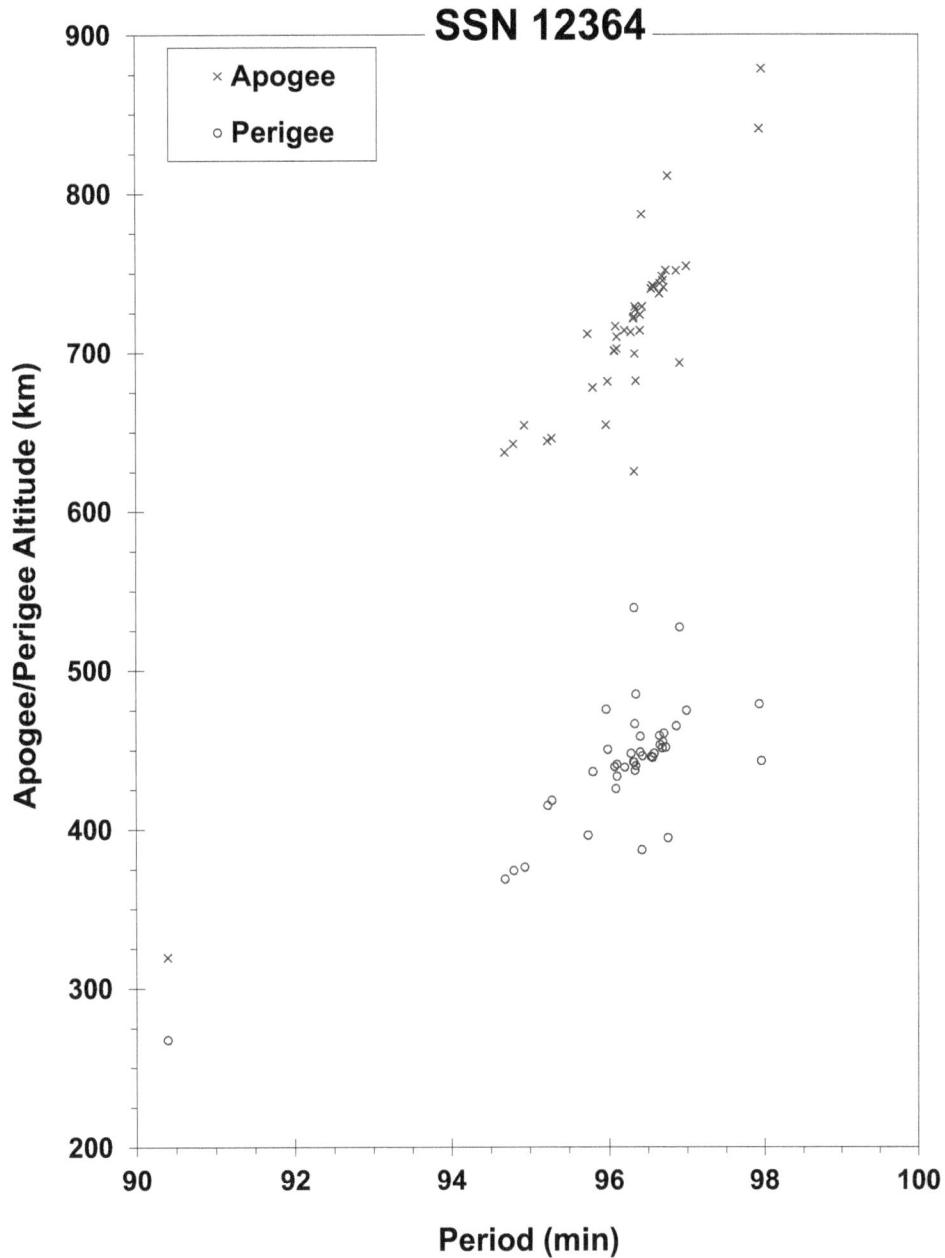

Cosmos 1260 debris cloud of 43 fragments 3 weeks after the first event from the US SSN database.

SATELLITE DATA

TYPE:	Payload
OWNER:	CIS
LAUNCH DATE:	31.40 Mar 1981
DRY MASS (KG):	1250
MAIN BODY:	Cylinder; 1.7 m diameter by 2 m length
MAJOR APPENDAGES:	Solar panels
ATTITUDE CONTROL:	Active, 3-axis
ENERGY SOURCES:	On-board propellants, explosive charge

EVENT DATA

DATE:	Apr-May 1981	LOCATION:	Unknown
TIME:	Unknown	ASSESSED CAUSE:	Deliberate
ALTITUDE:	Unknown		

PRE-EVENT ELEMENTS

EPOCH:	81095.90157023	MEAN ANOMALY:	4.6715
RIGHT ASCENSION:	282.6240	MEAN MOTION:	2.00494188
INCLINATION:	63.0386	MEAN MOTION DOT/2:	.0
ECCENTRICITY:	.7369210	MEAN MOTION DOT DOT/6:	.0
ARG. OF PERIGEE:	316.4347	BSTAR:	.0

DEBRIS CLOUD DATA

MAXIMUM ΔP:	2.3 min*
MAXIMUM ΔI:	0.3 deg*

*Based on uncataloged debris data

COMMENTS

Cosmos 1261 was another spacecraft of the Cosmos 862-type to experience a fragmentation. The spacecraft attempted to maneuver from its transfer orbit to an operational orbit 3 days after launch. The maneuver appears to have been unsuccessful, and the spacecraft never became ground track-stabilized. Some debris appeared immediately after the maneuver, while additional debris were discovered in mid-May. More than one event may have occurred. The element set above is the first available after the unsuccessful maneuver.

REFERENCE DOCUMENT

History of Soviet/Russian Satellite Fragmentations-A Joint U.S.-Russian Investigation, N. L. Johnson, et al., Kaman Sciences Corporation, October 1995.

SSN 12376

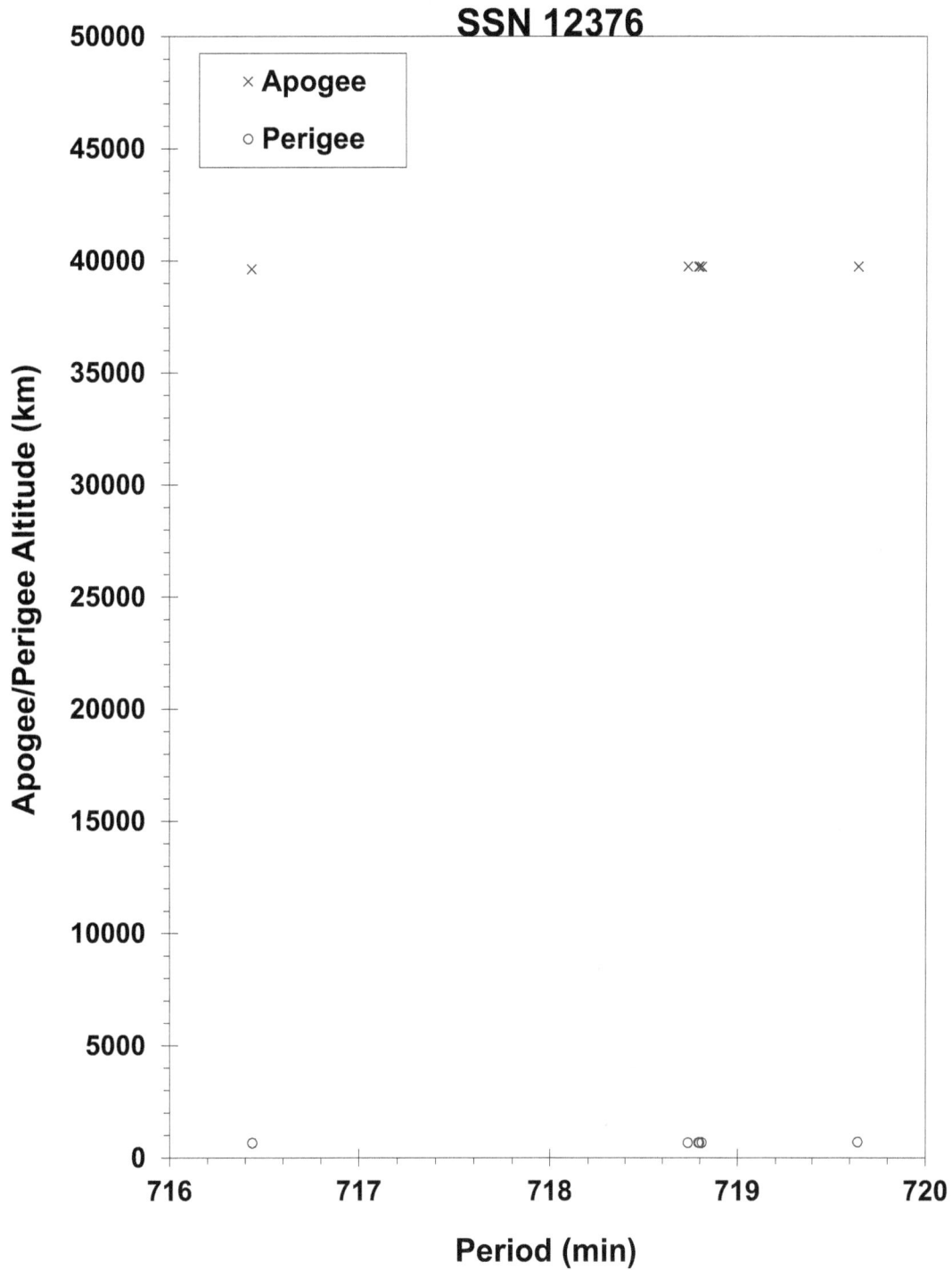

Cosmos 1261 debris cloud of 6 fragments about 8 weeks after (initial) event as reconstructed from the US SSN database.

SATELLITE DATA

TYPE:	Payload
OWNER:	CIS
LAUNCH DATE:	4.66 Jun 1981
DRY MASS (KG):	800
MAIN BODY:	Cylinder; 2.4 m diameter by 2 m length
MAJOR APPENDAGES:	Gravity-gradient boom
ATTITUDE CONTROL:	Gravity gradient
ENERGY SOURCES:	Unknown

EVENT DATA

DATE:	24 Jul 1981	LOCATION:	68N, 197E (asc)
TIME:	2351 GMT	ASSESSED CAUSE:	Battery
ALTITUDE:	980 km		

PRE-EVENT ELEMENTS

EPOCH:	81205.39693092	MEAN ANOMALY:	221.3567
RIGHT ASCENSION:	119.8245	MEAN MOTION:	13.73455672
INCLINATION:	82.9633	MEAN MOTION DOT/2:	.000000580
ECCENTRICITY:	.0036415	MEAN MOTION DOT DOT/6:	.0
ARG. OF PERIGEE:	139.0334	BSTAR:	.00004538900

DEBRIS CLOUD DATA

MAXIMUM ΔP:	4.9 min
MAXIMUM ΔI:	0.4 deg

COMMENTS

Cosmos 1275 is the only member of its class, Parus, to explosively fragment; the Cosmos 1934 event was a collision with tracked, mission-related debris. The satellite was only 50 days old at the time of the event. During the February 1992 Space Debris Conference in Moscow, Russian analysts discussed independent studies about the probable cause of the breakup. Later, the official Russian assessment asserted that a battery malfunction was the likely culprit.

REFERENCE DOCUMENTS

The Fragmentations of USSR Satellites 11729 and 12504 (U), J.R. Gabbard and P.M. Landry, Technical Memorandum 82-S-03, DCS/Plans, Hdqtrs NORAD/ADCOM, Colorado Springs, August 1982 (Secret).

Determining the Cause of a Satellite Breakup: A Case Study of the Kosmos 1275 Breakup, D.S. McKnight, IAA-87-573, 38th Congress of the International Astronautical Federation, Brighton, England, October 1987.

History of Soviet/Russian Satellite Fragmentations-A Joint U.S.-Russian Investigation, N. L. Johnson, et al., Kaman Sciences Corporation, October 1995.

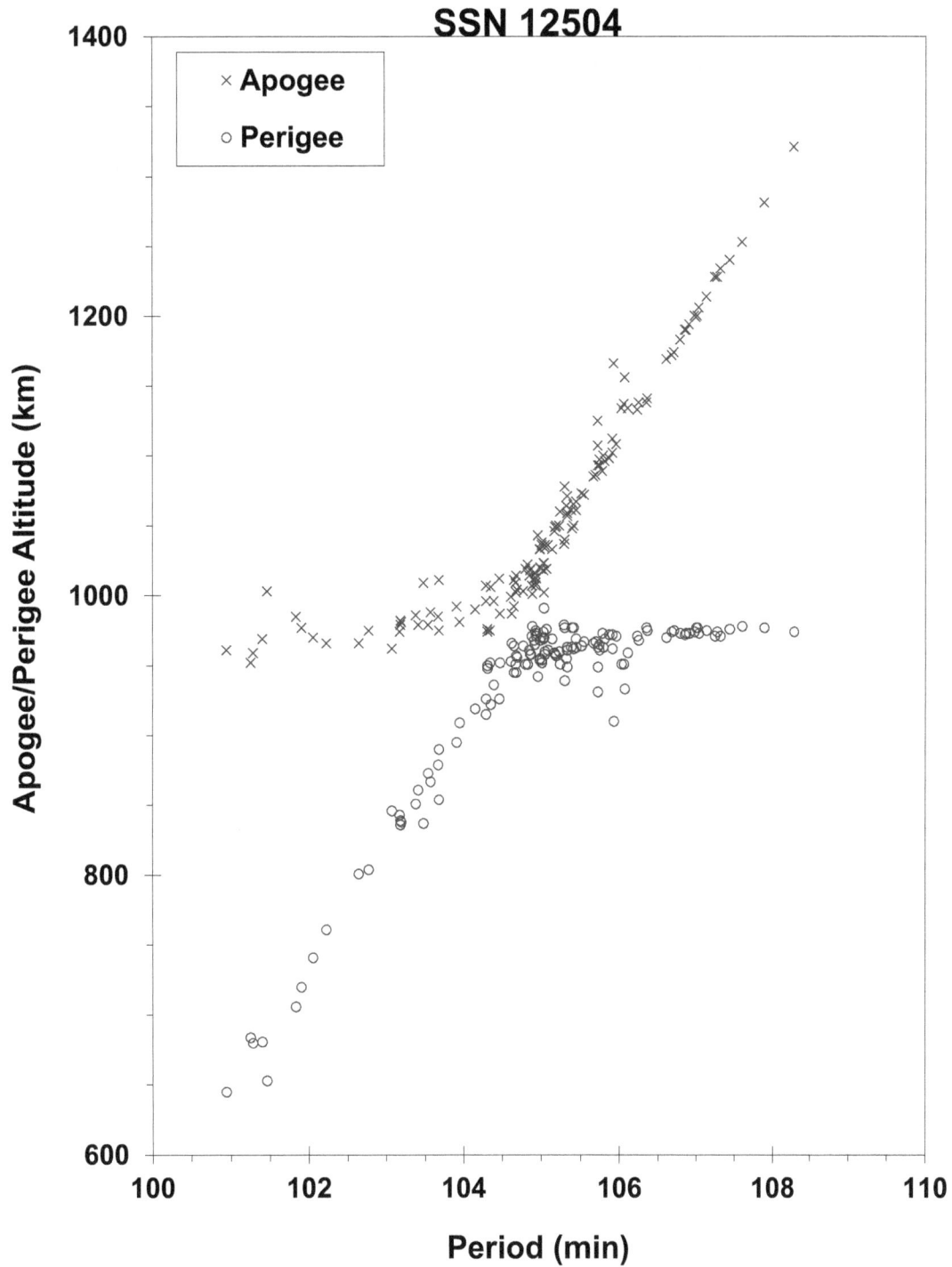

SSN 12504

Cosmos 1275 debris cloud of 136 identified fragments 1 week after the event as reconstructed from the US SSN database.

SATELLITE DATA

TYPE:	Payload
OWNER:	CIS
LAUNCH DATE:	19.81 Jun 1981
DRY MASS (KG):	1250
MAIN BODY:	Cylinder; 1.7 m diameter by 2 m length
MAJOR APPENDAGES:	Solar panels
ATTITUDE CONTROL:	Active, 3-axis
ENERGY SOURCES:	On-board propellants, explosive charge

EVENT DATA

DATE:	Early Dec 1986	LOCATION:	Unknown
TIME:	Unknown	ASSESSED CAUSE:	Deliberate
ALTITUDE:	Unknown		

PRE-EVENT ELEMENTS

EPOCH:	86334.22199701	MEAN ANOMALY:	12.7886
RIGHT ASCENSION:	288.0814	MEAN MOTION:	2.00618298
INCLINATION:	67.1073	MEAN MOTION DOT/2:	.0
ECCENTRICITY:	.6594262	MEAN MOTION DOT DOT/6:	.0
ARG. OF PERIGEE:	291.9890	BSTAR:	.0

DEBRIS CLOUD DATA

MAXIMUM	ΔP:	0.1 min
MAXIMUM	ΔI:	0.0 deg

COMMENTS

Cosmos 1278 was another spacecraft of the Cosmos 862-type to experience a fragmentation. Spacecraft had apparently been inactive since early 1984. Additional fragments may exist, but surveillance for small objects in this high eccentricity orbit is difficult.

REFERENCE DOCUMENT

History of Soviet/Russian Satellite Fragmentations-A Joint U.S.-Russian Investigation, N. L. Johnson, *et al.,* Kaman Sciences Corporation, October 1995.

SSN 12547

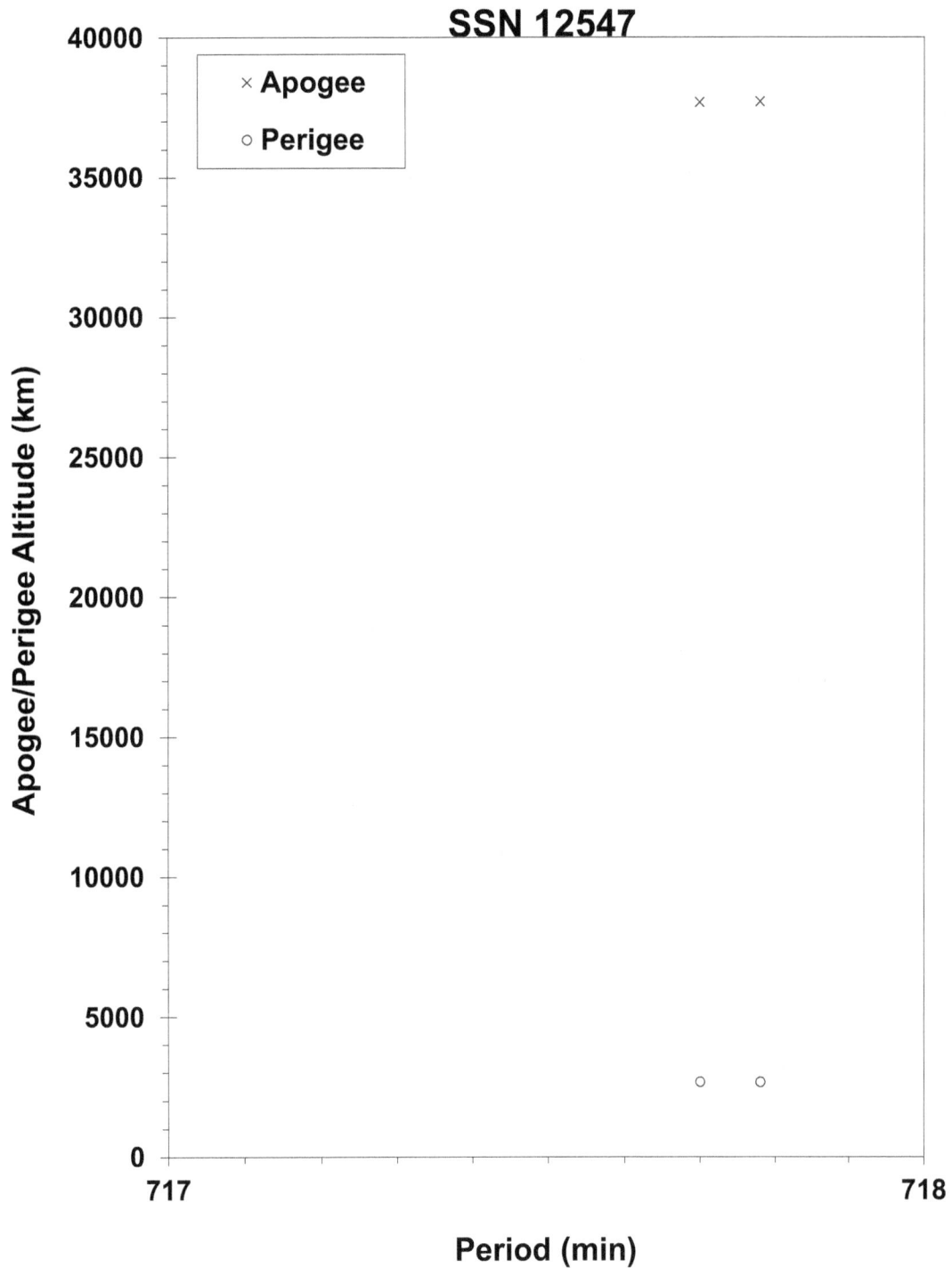

Cosmos 1278 and additional fragment in mid-December 1986. Elements from the US SSN as published by the NASA Goddard Space Flight Center.

SATELLITE DATA

TYPE:	Payload
OWNER:	CIS
LAUNCH DATE:	4.01 Aug 1981
DRY MASS (KG):	1250
MAIN BODY:	Cylinder; 1.7 m diameter by 2 m length
MAJOR APPENDAGES:	Solar panels
ATTITUDE CONTROL:	Active, 3-axis
ENERGY SOURCES:	On-board propellants, explosive charge

EVENT DATA

DATE:	21 Nov 1981	LOCATION:	Unknown
TIME:	Unknown	ASSESSED CAUSE:	Deliberate
ALTITUDE:	Unknown		

PRE-EVENT ELEMENTS

EPOCH:	81324.16708257	MEAN ANOMALY:	4.8196
RIGHT ASCENSION:	249.5852	MEAN MOTION:	1.98014597
INCLINATION:	63.1086	MEAN MOTION DOT/2:	.00000781
ECCENTRICITY:	.7350717	MEAN MOTION DOT DOT/6:	.0
ARG. OF PERIGEE:	317.0022	BSTAR:	.0

DEBRIS CLOUD DATA

MAXIMUM ΔP:	8.6 min*
MAXIMUM ΔI:	0.2 deg*

*Based on uncataloged debris data

COMMENTS

Cosmos 1285 was another spacecraft of the Cosmos 862-type to experience a fragmentation. Spacecraft was placed in a temporary transfer orbit on the day of launch by its launch vehicle but never maneuvered to an operational orbit, suggesting an early fatal spacecraft malfunction. Event occurred 3.5 months after the launch.

REFERENCE DOCUMENT

History of Soviet/Russian Satellite Fragmentations-A Joint U.S.-Russian Investigation, N. L. Johnson, *et al.,* Kaman Sciences Corporation, October 1995.

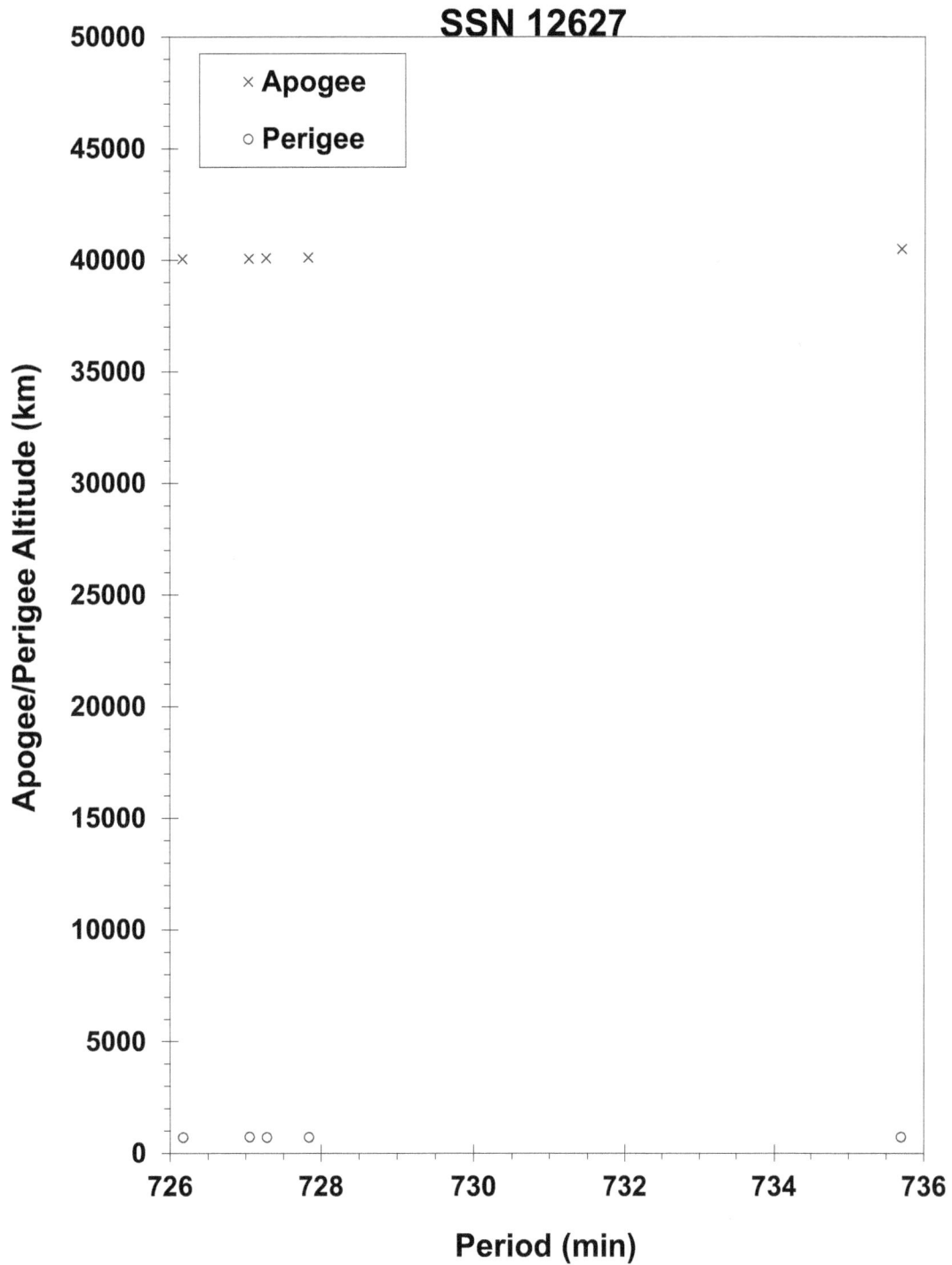

SSN 12627

Cosmos 1285 debris cloud of 5 fragments less than 1 week after the event as reconstructed from the US SSN database.

SATELLITE DATA

TYPE:	Payload
OWNER:	CIS
LAUNCH DATE:	4.35 Aug 1981
DRY MASS (KG):	3000
MAIN BODY:	Cylinder; 1.3 m diameter by 17 m length
MAJOR APPENDAGES:	Solar panels
ATTITUDE CONTROL:	Active, 3-axis
ENERGY SOURCES:	On-board propellants, explosive charge (?)

EVENT DATA

DATE:	29 Sep 1982	LOCATION:	51N, 80E (asc)
TIME:	0520 GMT	ASSESSED CAUSE:	Unknown
ALTITUDE:	325 km		

PRE-EVENT ELEMENTS

EPOCH:	82272.21193719	MEAN ANOMALY:	92.4681
RIGHT ASCENSION:	132.9736	MEAN MOTION:	15.86141247
INCLINATION:	65.0071	MEAN MOTION DOT/2:	.00400345
ECCENTRICITY:	.0017215	MEAN MOTION DOT DOT/6:	.0
ARG. OF PERIGEE:	267.4145	BSTAR:	.0015199

DEBRIS CLOUD DATA

MAXIMUM ΔP:	0.9 min*
MAXIMUM ΔI:	0.2 deg*

*Based on uncataloged debris data

COMMENTS

Cosmos 1286 was the ninth spacecraft of the Cosmos 699-type to experience a fragmentation. The spacecraft had been in a regime of natural decay for more than 6 months at the time of the event. The low altitude and high drag conditions made determination of the precise breakup time uncertain. The breakup or a precursor event may have occurred earlier on 29 September 1982. Most fragments decayed before being officially cataloged.

REFERENCE DOCUMENTS

"Artificial Satellite Break-Ups (Part 1): Soviet Ocean Surveillance Satellites", N. L. Johnson, Journal of the British Interplanetary Society, February 1983, pp. 51-58.

History of Soviet/Russian Satellite Fragmentation-A Joint U.S.-Russian Investigation, N. L. Johnson, et al., Kaman Sciences Corporation, October 1995.

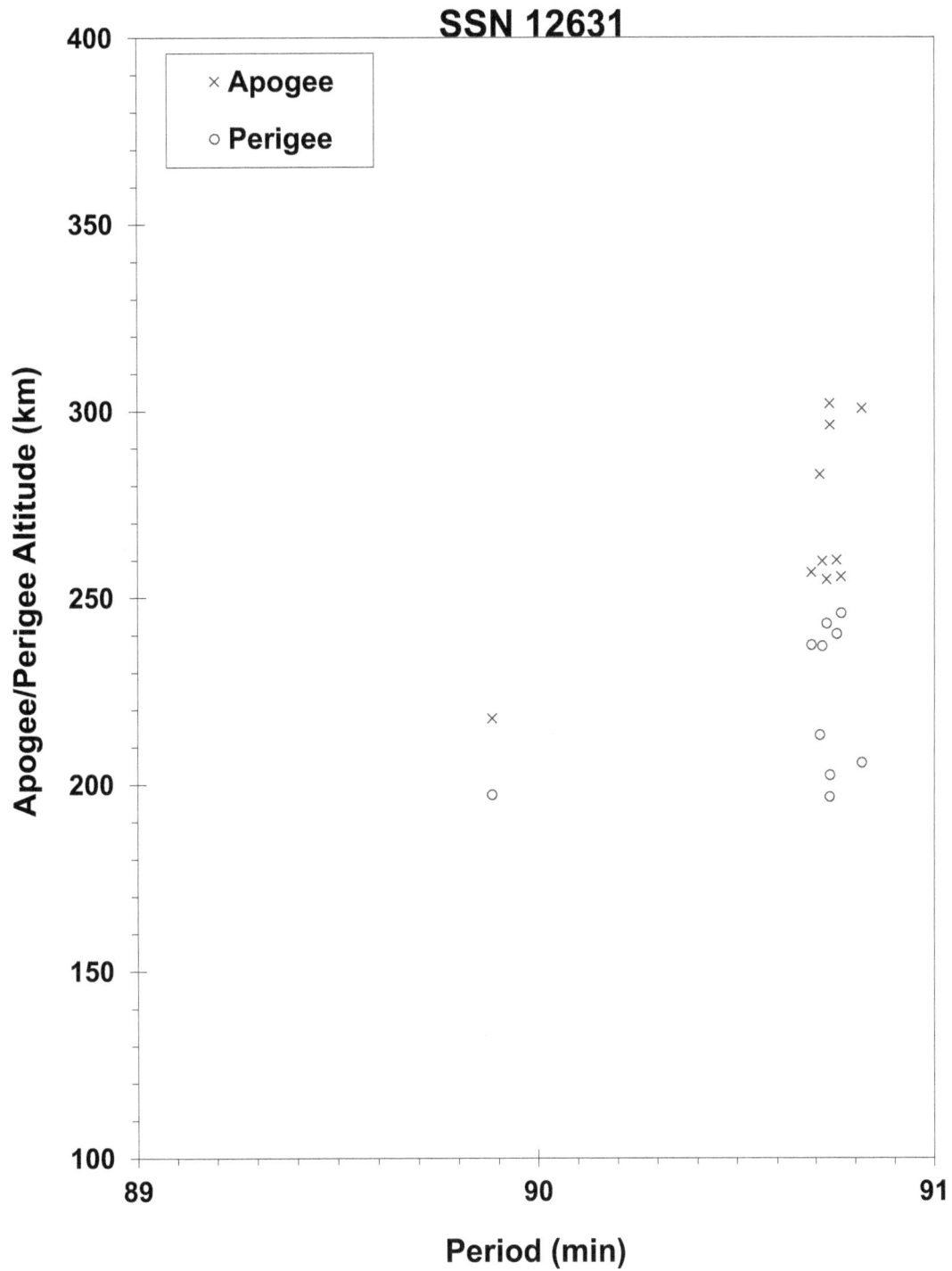

SSN 12631

Cosmos 1286 debris cloud of 10 fragments 1 day after the event as reconstructed from the US SSN database.

SATELLITE DATA

TYPE:	Molniya Final Stage
OWNER:	CIS
LAUNCH DATE:	11.36 Sep 1981
DRY MASS (KG):	1100
MAIN BODY:	Cylinder; 2.7 m diameter by 3 m length
MAJOR APPENDAGES:	None
ATTITUDE CONTROL:	Active, 3-axis
ENERGY SOURCES:	On-board propellants

EVENT DATA

DATE:	11 Sep 1981	LOCATION:	Unknown
TIME:	Unknown	ASSESSED CAUSE:	Propulsion
ALTITUDE:	Unknown		

POST-EVENT ELEMENTS

EPOCH:	81258.60717998	MEAN ANOMALY:	26.9249
RIGHT ASCENSION:	68.6245	MEAN MOTION:	5.48678032
INCLINATION:	62.8166	MEAN MOTION DOT/2:	.0
ECCENTRICITY:	.4855644	MEAN MOTION DOT DOT/6:	.0
ARG. OF PERIGEE:	286.6972	BSTAR:	.0

DEBRIS CLOUD DATA

MAXIMUM ΔP:	Unknown
MAXIMUM ΔI:	Unknown

COMMENTS

Cosmos 1305 R/B malfunctioned about 1 hour after launch during a maneuver from a LEO parking orbit to a Molniya-type orbit. The maneuver was initiated at approximately 0937 GMT near 58S, 245E (asc) at an altitude of 600 km. Apogee was raised to less than 14,000 km. Debris tracking after the event was limited, preventing an accurate assessment of magnitude of the event. First debris officially cataloged in June 1983. Debris generation is assumed to have occurred during or immediately after the unsuccessful maneuver. The element set above is for the rocket body after burn termination.

REFERENCE DOCUMENT

History of Soviet/Russian Satellite Fragmentations-A Joint U.S.-Russian Investigation, N. L. Johnson, et al., Kaman Sciences Corporation, October 1995.

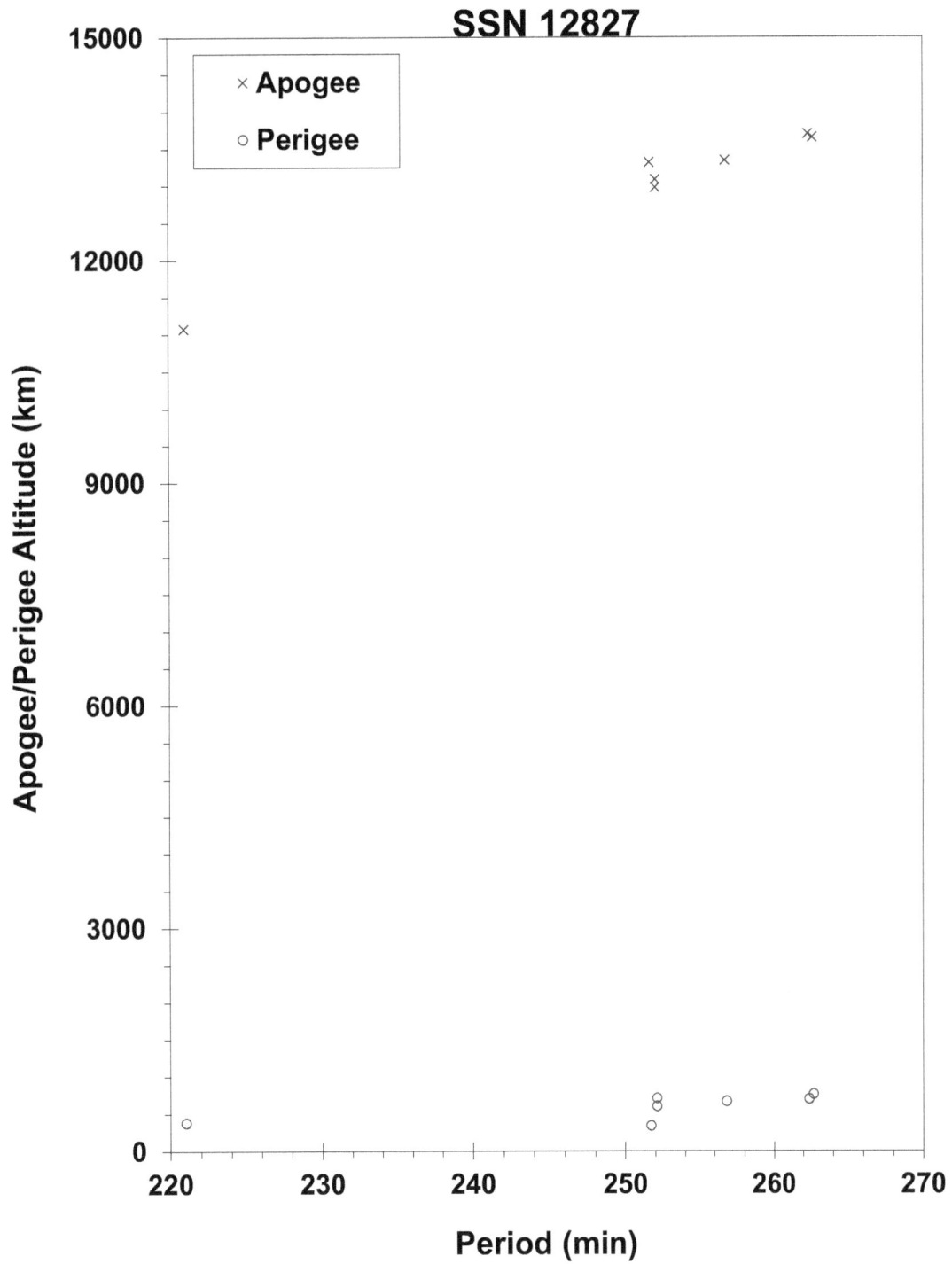

Cosmos 1305 R/B debris cloud of 7 fragments about 2 years after the event as reconstructed from the US SSN database.

SATELLITE DATA

TYPE:	Payload
OWNER:	CIS
LAUNCH DATE:	14.85 Sep 1981
DRY MASS (KG):	3000
MAIN BODY:	Cylinder; 1.3 m diameter by 17 m length
MAJOR APPENDAGES:	Solar panels
ATTITUDE CONTROL:	Active, 3-axis
ENERGY SOURCES:	On-board propellants, explosive charge (?)

EVENT DATA (1)

DATE:	12 Jul 1982	LOCATION:	65S, 40E (asc)
TIME:	2325 GMT	ASSESSED CAUSE:	Unknown
ALTITUDE:	380 km		

PRE-EVENT ELEMENTS (1)

EPOCH:	82193.22052182	MEAN ANOMALY:	72.7640
RIGHT ASCENSION:	43.8843	MEAN MOTION:	15.58171668
INCLINATION:	64.9399	MEAN MOTION DOT/2:	.00042116
ECCENTRICITY:	.0019953	MEAN MOTION DOT DOT/6:	.0
ARG. OF PERIGEE:	287.2390	BSTAR:	.00055055

EVENT DATA (2)

DATE:	18 Sep 1982	LOCATION:	32N, 293E (asc)
TIME:	1702 GMT	ASSESSED CAUSE:	Unknown
ALTITUDE:	370 km		

PRE-EVENT ELEMENTS (2)

EPOCH:	82260.17037940	MEAN ANOMALY:	44.8033
RIGHT ASCENSION:	173.7764	MEAN MOTION:	15.65882738
INCLINATION:	64.9408	MEAN MOTION DOT/2:	.00076164
ECCENTRICITY:	.0002181	MEAN MOTION DOT DOT/6:	.0
ARG. OF PERIGEE:	315.2578	BSTAR:	.00073994

DEBRIS CLOUD DATA

MAXIMUM ΔP:	2.1 min*
MAXIMUM ΔI:	0.2 deg

*Based on uncataloged debris data

COMMENTS

Cosmos 1306 was the eighth spacecraft of the Cosmos 699-type to experience a fragmentation. The first event occurred 5 months after the spacecraft had entered a regime of natural decay. After the event, the main remnant was tagged as satellite 13369, while a piece of debris tagged as 12828 decayed on 16 July 1982. Only 5 new fragments were officially cataloged prior to the second event when satellite 13369 experienced a fragmentation. Three long-lived fragments cataloged with 1981-89 (13393, 13404, and 14837) were actually part of the breakup of 1980-89, another Cosmos 699-type satellite. Most Cosmos 1306 debris reentered quickly, and elements were developed for only a few fragments.

REFERENCE DOCUMENTS

Analysis of Cosmos 1220 and Cosmos 1306 Fragments (U), D. Fennessy, Report AH-23, FTD/OLAI, Cheyenne Mountain, Colorado, 12 January 1983 (Secret)

"Artificial Satellite Break-Ups (Part 1): Soviet Ocean Surveillance Satellites", N. L. Johnson, Journal of the British Interplanetary Society, February 1983, pp. 51-58.

History of Soviet/Russian Satellite Fragmentations-A Joint U.S.-Russian Investigation, N. L. Johnson, et al., Kaman Sciences Corporation, October 1995.

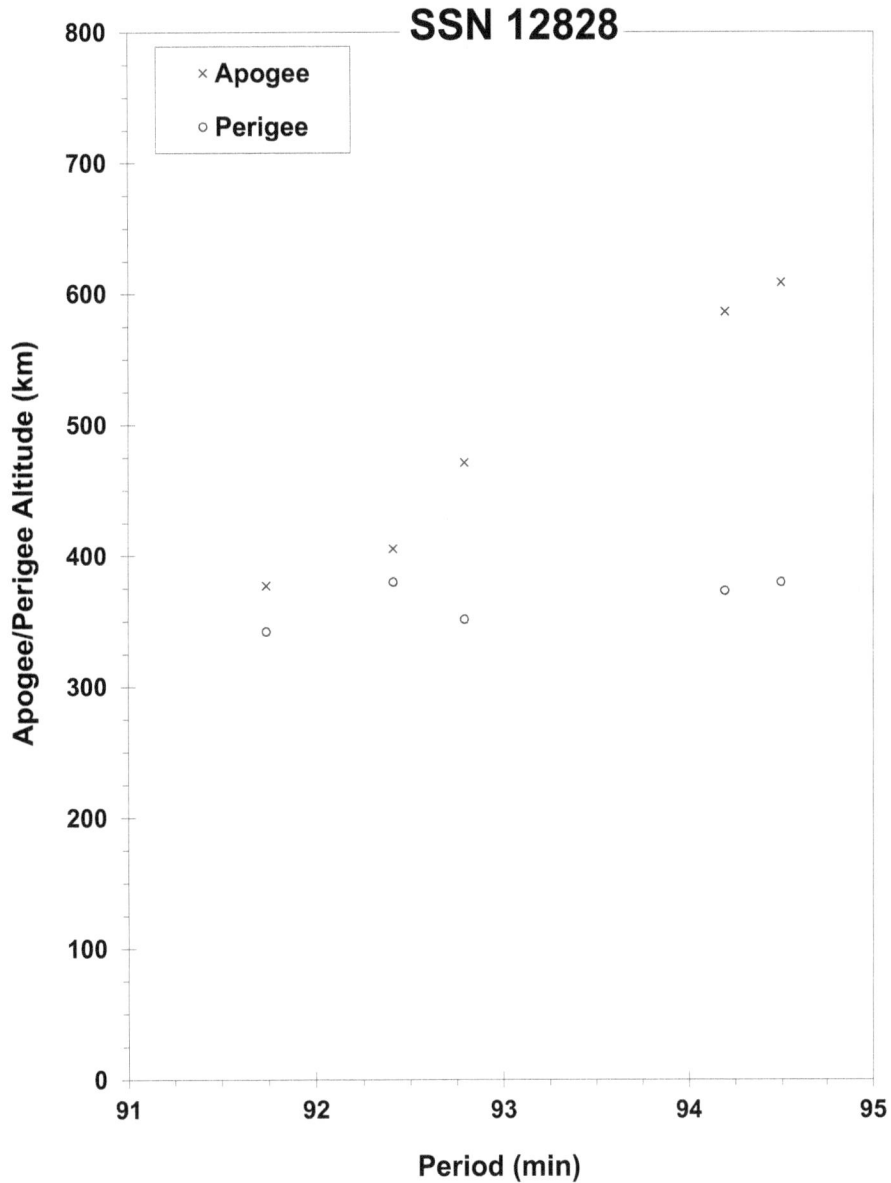

Cosmos 1306 debris cloud of 5 identified fragments 1 day after the event as reconstructed from the US SSN database.

SATELLITE DATA

TYPE:	Payload
OWNER:	CIS
LAUNCH DATE:	31.95 Oct 1981
DRY MASS (KG):	1250
MAIN BODY:	Cylinder; 1.7 m diameter by 2 m length
MAJOR APPENDAGES:	Solar panels
ATTITUDE CONTROL:	Active, 3-axis
ENERGY SOURCES:	On-board propellants, explosive

EVENT DATA

DATE:	25-28 Jan 1984	LOCATION:	Unknown
TIME:	Unknown	ASSESSED CAUSE:	Deliberate
ALTITUDE:	Unknown		

PRE-EVENT ELEMENTS

EPOCH:	84024.46309667	MEAN ANOMALY:	4.4900
RIGHT ASCENSION:	219.5352	MEAN MOTION:	2.00535027
INCLINATION:	62.8286	MEAN MOTION DOT/2:	.0
ECCENTRICITY:	.7103977	MEAN MOTION DOT DOT/6:	.0
ARG. OF PERIGEE:	324.1891	BSTAR:	.0

DEBRIS CLOUD DATA

MAXIMUM ΔP:	1.8 min
MAXIMUM ΔI:	0.3 deg

COMMENTS

Cosmos 1317 was another spacecraft of the Cosmos 862-type to experience a fragmentation

REFERENCE DOCUMENT

History of Soviet/Russian Satellite Fragmentations-A Joint U.S.-Russian Investigation, N. L. Johnson, *et al.,* Kaman Sciences Corporation, October 1995.

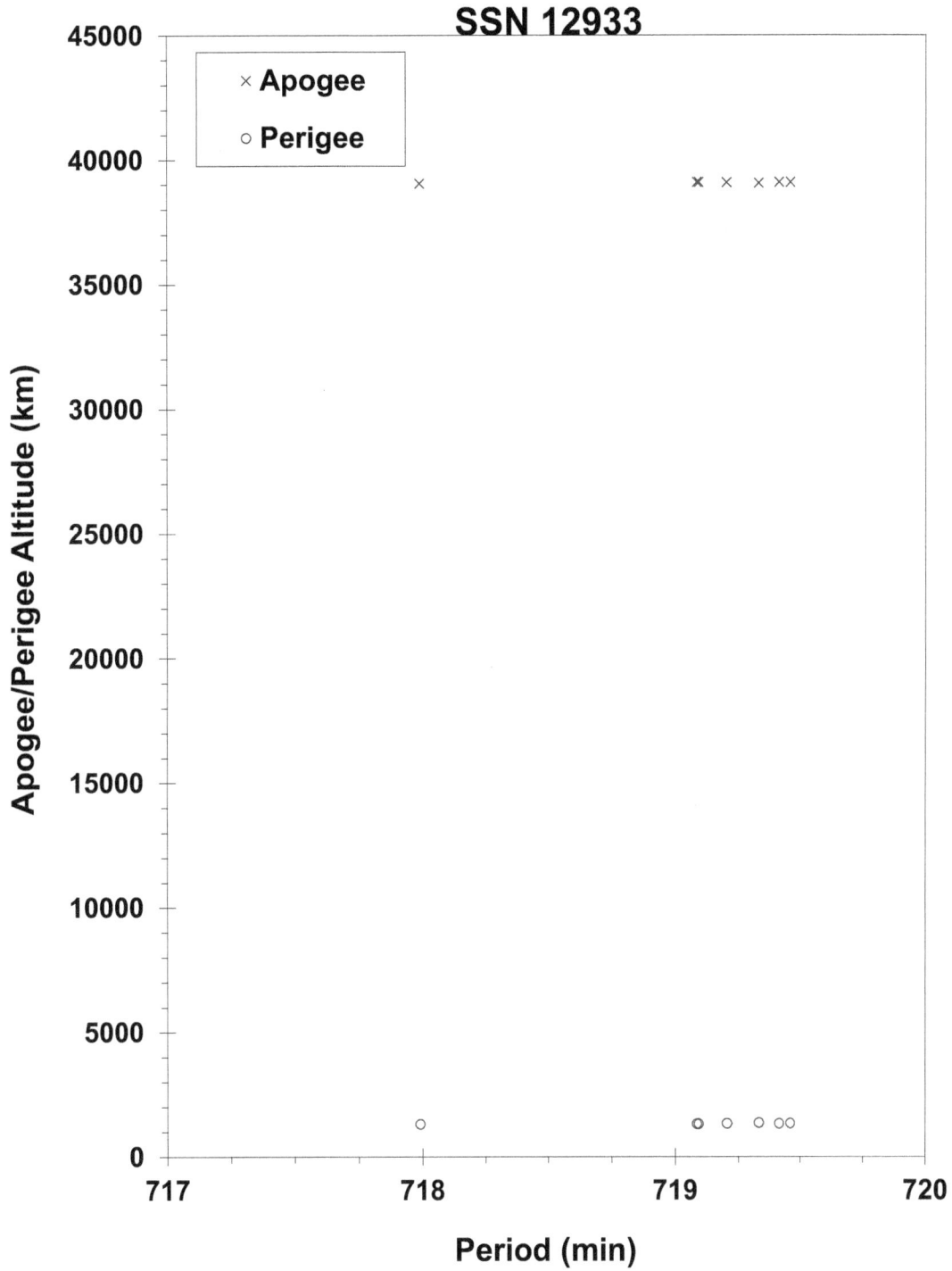

SSN 12933

Cosmos 1317 debris cloud of 7 fragments about 2 weeks after the event as reconstructed from the US SSN database.

SATELLITE DATA

TYPE:	Payload
OWNER:	CIS
LAUNCH DATE:	7.57 April 1982
DRY MASS (KG):	1250
MAIN BODY:	Cylinder; 1.7 m diameter by 2 m length
MAJOR APPENDAGES:	Solar panels
ATTITUDE CONTROL:	Active, 3-axis
ENERGY SOURCES:	On-board propellants, explosive device

EVENT DATA

DATE:	2 September 1984	LOCATION:	39.5N, 121.3E (asc)
TIME:	0957 GMT	ASSESSED CAUSE:	Deliberate
ALTITUDE:	5020 km		

PRE-EVENT ELEMENTS

EPOCH:	84246.40302854	MEAN ANOMALY:	3.8405
RIGHT ASCENSION:	227.2447	MEAN MOTION:	2.00449045
INCLINATION:	62.8163	MEAN MOTION DOT/2:	.00000970
ECCENTRICITY:	.7154927	MEAN MOTION DOT DOT/6:	.0
ARG. OF PERIGEE:	326.7879	BSTAR:	.0

DEBRIS CLOUD DATA

MAXIMUM ΔP:	25.5 min
MAXIMUM ΔI:	7.3 deg

COMMENTS

Cosmos 1348 was a member of the Cosmos 862 class of operational satellites in highly elliptical, semi-synchronous orbits that experienced a total of 18 fragmentations during the period 1977-1986. Due to the nature of these orbits, which result in high altitudes over the Northern Hemisphere where most surveillance sensors are located, debris detection and tracking is extremely difficult. Only the largest fragments can be seen. Cosmos 862 class spacecraft were equipped with self-destruct packages in the event that spacecraft control was lost; this was the cause of breakups until the explosives were removed after Cosmos 1481.

Podvig reported an estimate end of life for Cosmos 1348 as 22 July 1984. If correct, this could imply a breakup mechanism different that other members of the Cosmos 862 class. An examination of the mean motion history of Cosmos 1348 indicated, however, that it was under active control until the event date of 2 September 1984, implying that the cause attributed to the fragmentation of other class members was similarly responsible for this spacecraft's fragmentation.

REFERENCE DOCUMENT

History of Soviet/Russian Satellite Fragmentations-A Joint U.S.-Russian Investigation, N. L. Johnson et al., Kaman Sciences Corporation, October 1995.

Podvig, P., "History and the Current Status of the Russian Early-Warning System." Science and Global Security 10 (2002): 21-60.

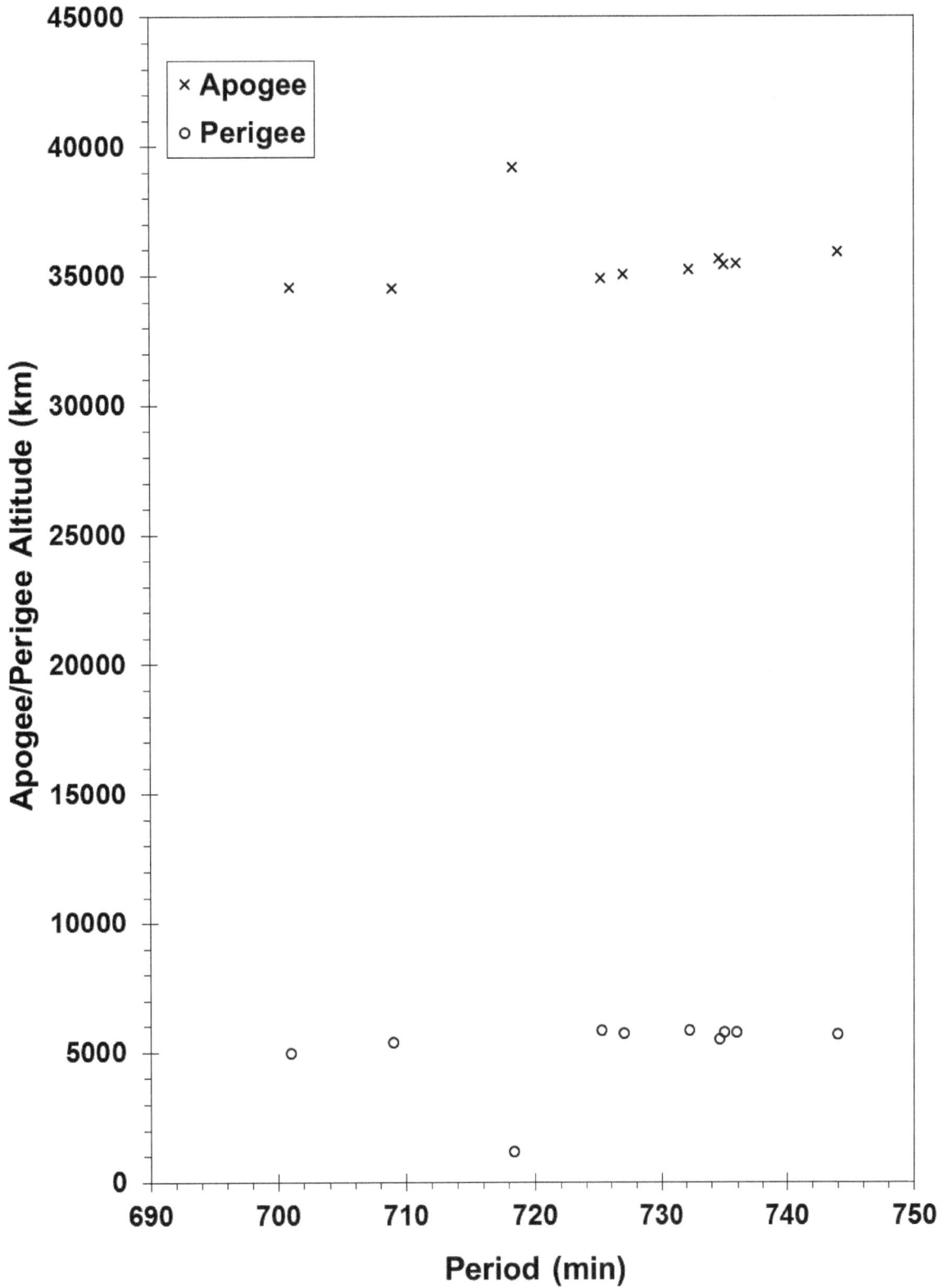

SSN 13124

Cosmos 1348 debris cloud; fragments cataloged over 23 years after the event as reconstructed from the US SSN database.

SATELLITE DATA

TYPE:	Payload
OWNER:	CIS
LAUNCH DATE:	29.41 Apr 1982
DRY MASS (KG):	3000
MAIN BODY:	Cylinder; 1.3 m diameter by 17 m length
MAJOR APPENDAGES:	Solar panels
ATTITUDE CONTROL:	Active, 3-axis
ENERGY SOURCES:	On-board propellants, explosive charge (?)

EVENT DATA (1)

DATE:	8 Aug 1983	LOCATION:	32S, 310E (asc)
TIME:	2331 GMT	ASSESSED CAUSE:	Unknown
ALTITUDE:	365 km		

PRE-EVENT ELEMENTS (1)

EPOCH:	83220.21851552	MEAN ANOMALY:	66.8795
RIGHT ASCENSION:	279.4096	MEAN MOTION:	15.63233551
INCLINATION:	65.0504	MEAN MOTION DOT/2:	.00048258
ECCENTRICITY:	.0024043	MEAN MOTION DOT DOT/6:	.0
ARG. OF PERIGEE:	292.8515	BSTAR:	.00051620

EVENT DATA (2)

DATE:	1 Feb 1984	LOCATION:	4S, 200E (asc)
TIME:	0322 GMT	ASSESSED CAUSE:	Unknown
ALTITUDE:	320 km		

PRE-EVENT ELEMENTS (2)

EPOCH:	84031.38369465	MEAN ANOMALY:	81.7159
RIGHT ASCENSION:	25.3553	MEAN MOTION:	15.84652631
INCLINATION:	65.0404	MEAN MOTION DOT/2:	.00119378
ECCENTRICITY:	.0017572	MEAN MOTION DOT DOT/6:	.0
ARG. OF PERIGEE:	278.1110	BSTAR:	.00050318

EVENT DATA (3)

DATE:	20 Feb 1984	LOCATION:	Unknown
TIME:	Before 0340 GMT	ASSESSED CAUSE:	Unknown
ALTITUDE:	Unknown		

PRE-EVENT ELEMENTS (3)

EPOCH:	84050.69015256	MEAN ANOMALY:	105.8772
RIGHT ASCENSION:	316.3115	MEAN MOTION:	15.97914042
INCLINATION:	65.0338	MEAN MOTION DOT/2:	.00430956
ECCENTRICITY:	.0014134	MEAN MOTION DOT DOT/6:	.000083799
ARG. OF PERIGEE:	254.0517	BSTAR:	.00093344

DEBRIS CLOUD DATA

MAXIMUM ΔP:	36.8 min*
MAXIMUM ΔI:	2.3 deg*

*Based on uncataloged debris data (Event 1)

COMMENTS

Cosmos 1355 was the tenth spacecraft of the Cosmos 699-type to experience a fragmentation. The spacecraft had been in a regime of natural decay for 6 months prior to the first event. Twenty-one fragments were cataloged following the first event, and the main body became satellite 14275. This object spawned at least seven more fragments on 1 February. The parent was then retagged to the original 13150 satellite number. The third event resulted in the development of 13 new fragment element sets, but none were cataloged, and the low altitude prevented an estimate of a precise breakup location.

REFERENCE DOCUMENTS

"Artificial Satellite Break-Ups (Part 1): Soviet Ocean Surveillance Satellites", N. L. Johnson, Journal of the British Interplanetary Society, February 1983, pp. 51-58.

Analysis of the Fragmentation of Kosmos 1355, N. L. Johnson, Technical Report CS84-SPACECMD-28, Teledyne Brown Engineering, Colorado Springs, January 1985.

History of Soviet/Russian Satellite Fragmentations-A Joint U.S.-Russian Investigation, N. L. Johnson, et al., Kaman Sciences Corporation, October 1995.

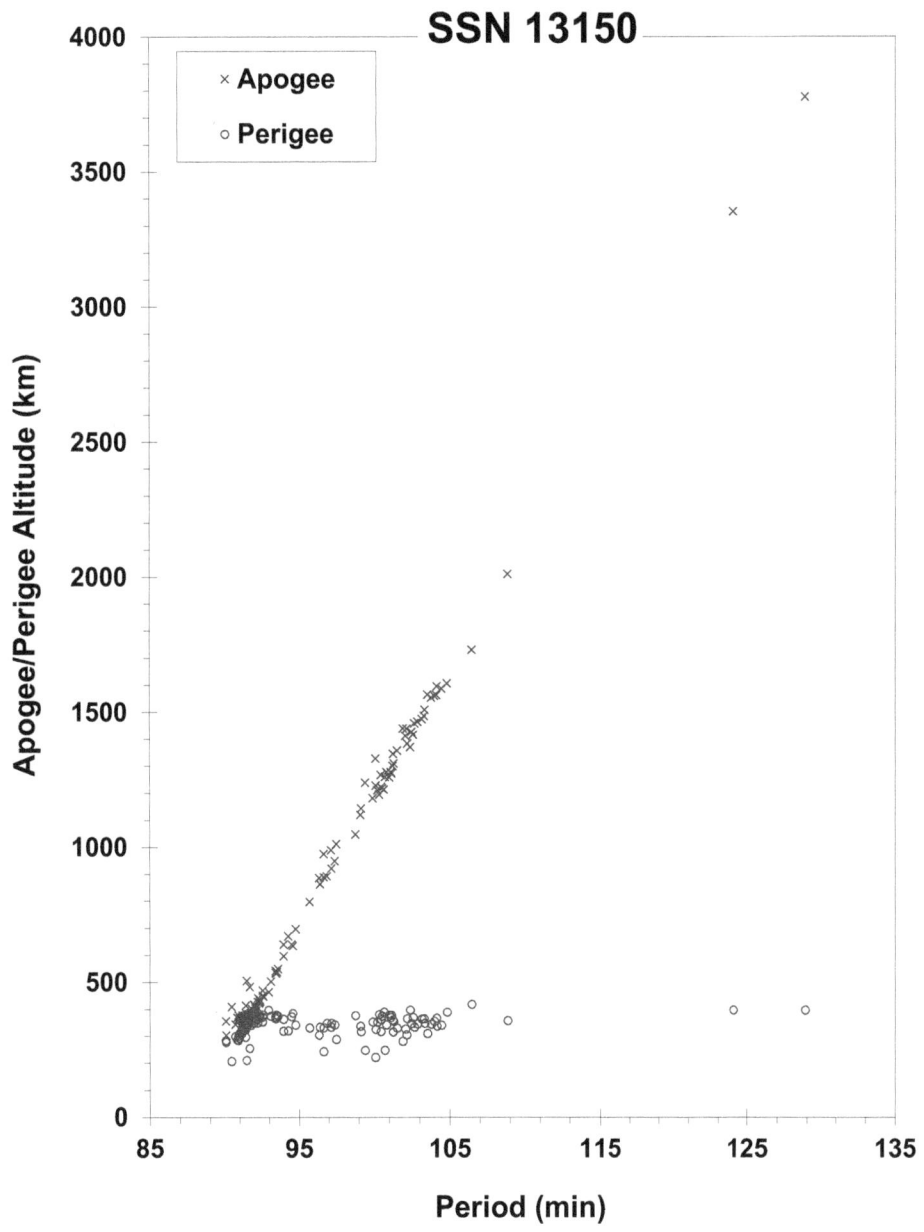

Cosmos 1355 debris cloud of 150 fragments about 7 hours after the first event in August 1983 as seen by the US SSN PARCS radar. Figure from the cited reference.

SATELLITE DATA

TYPE:	Payload
OWNER:	CIS
LAUNCH DATE:	6.72 Jun 1982
DRY MASS (KG):	650
MAIN BODY:	Polyhedron; 1.4 m by 1.4 m
MAJOR APPENDAGES:	Solar panels, gravity-gradient boom (?)
ATTITUDE CONTROL:	Gravity gradient (?)
ENERGY SOURCES:	Battery

EVENT DATA

DATE:	21 Oct 1985	LOCATION:	66N, 351E (asc)
TIME:	0353 GMT	ASSESSED CAUSE:	Battery
ALTITUDE:	995 km		

PRE-EVENT ELEMENTS

EPOCH:	85293.85195210	MEAN ANOMALY:	333.5602
RIGHT ASCENSION:	350.2805	MEAN MOTION:	13.71079597
INCLINATION:	65.8390	MEAN MOTION DOT/2:	.00000158
ECCENTRICITY:	.0005355	MEAN MOTION DOT DOT/6:	.0
ARG. OF PERIGEE:	26.5667	BSTAR:	.00023894

DEBRIS CLOUD DATA

MAXIMUM ΔP:	2.3 min*
MAXIMUM ΔI:	0.1 deg*

*Based on uncataloged debris data

COMMENTS

Cosmos 1375 was the third spacecraft of the Cosmos 839-type to experience a fragmentation. Although these satellites are used in conjunction with the Cosmos 249-type spacecraft that are deliberately fragmented, the cause of Cosmos 839-type events appears to be unrelated. In the case of Cosmos 1375, 40 months elapsed since its test with a Cosmos 249-type spacecraft.

REFERENCE DOCUMENTS

"Artificial Satellite Break-Ups (Part 2): Soviet Anti-Satellite Program", N.L. Johnson, Journal of the British Interplanetary Society, August 1983, pp. 357-362.

Analysis of the Kosmos 1375 Fragmentation, J. M. Koskella and R. L. Kling, Technical Report CS86-USASDC-0006, Teledyne Brown Engineering, Colorado Springs, March 1986.

History of Soviet/Russian Satellite Fragmentations-A Joint U.S.-Russian Investigation, N. L. Johnson, et al., Kaman Sciences Corporation, October 1995.

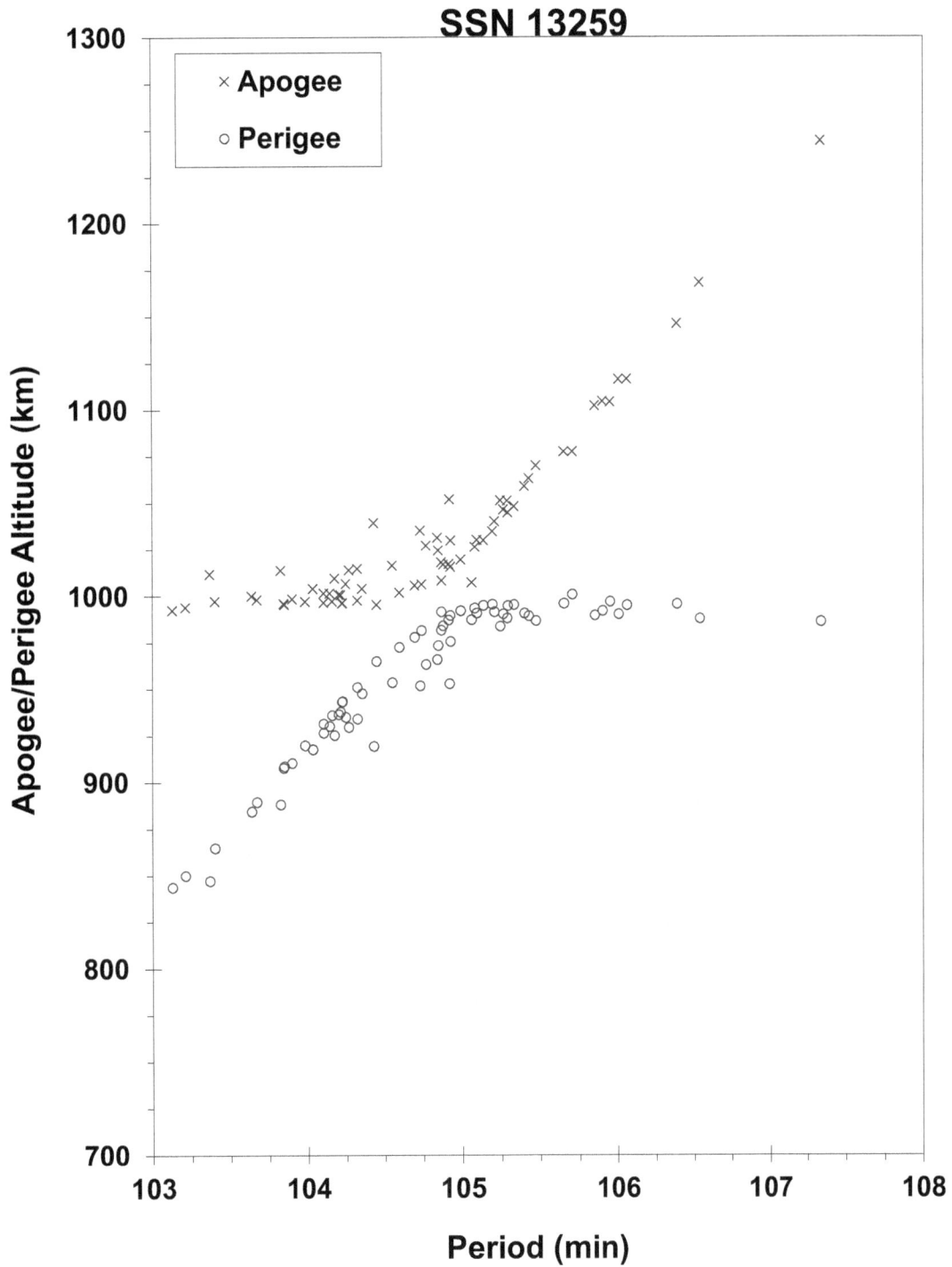

Cosmos 1375 debris cloud of 68 fragments seen a few hours after the event
by the US SSN PARCS radar.

COSMOS 1405   1982-088A   13508

SATELLITE DATA

TYPE:	Payload
OWNER:	CIS
LAUNCH DATE:	4.74 Sep 1982
DRY MASS (KG):	3000
MAIN BODY:	Cylinder; 1.3 m diameter by 17 m length
MAJOR APPENDAGES:	Solar panels
ATTITUDE CONTROL:	Active, 3-axis
ENERGY SOURCES:	On-board propellants, explosive charge (?)

EVENT DATA

DATE:	20 Dec 1983	LOCATION:	25S, 45E (dsc)
TIME:	1215 GMT	ASSESSED CAUSE:	Unknown
ALTITUDE:	330 km		

PRE-EVENT ELEMENTS

EPOCH:	83354.22079767	MEAN ANOMALY:	42.0375
RIGHT ASCENSION:	126.1259	MEAN MOTION:	15.81899265
INCLINATION:	65.0055	MEAN MOTION DOT/2:	.00186341
ECCENTRICITY:	.0020774	MEAN MOTION DOT DOT/6:	.0
ARG. OF PERIGEE:	318.0927	BSTAR:	.00088277

DEBRIS CLOUD DATA

MAXIMUM ΔP:	7.3 min*
MAXIMUM ΔI:	2.0 deg*

*Based on uncataloged debris data

COMMENTS

Cosmos 1405 was the eleventh spacecraft of the Cosmos 699-type to experience a fragmentation. Spacecraft had been in natural decay for 12 months prior to the event. Most debris reentered before being officially cataloged.

REFERENCE DOCUMENTS

"Artificial Satellite Break-Ups (Part 1): Soviet Ocean Surveillance Satellites", N. L. Johnson, Journal of the British Interplanetary Society, February 1983, pp. 51-58.

Separation of Objects from Cosmos 1405, F.T. Lipp, NAVSPASUR Technical Note 1-84, Naval Space Surveillance System, Dahlgren, 2 April 1984.

Analysis of the Fragmentation of Kosmos 1405, N.L. Johnson, Technical Report CS84-SPACECMD-10, Teledyne Brown Engineering, Colorado Springs, September 1984.

History of Soviet/Russian Satellite Fragmentations-A Joint U.S.-Russian Investigation, N. L. Johnson, et al., Kaman Sciences Corporation, October 1995.

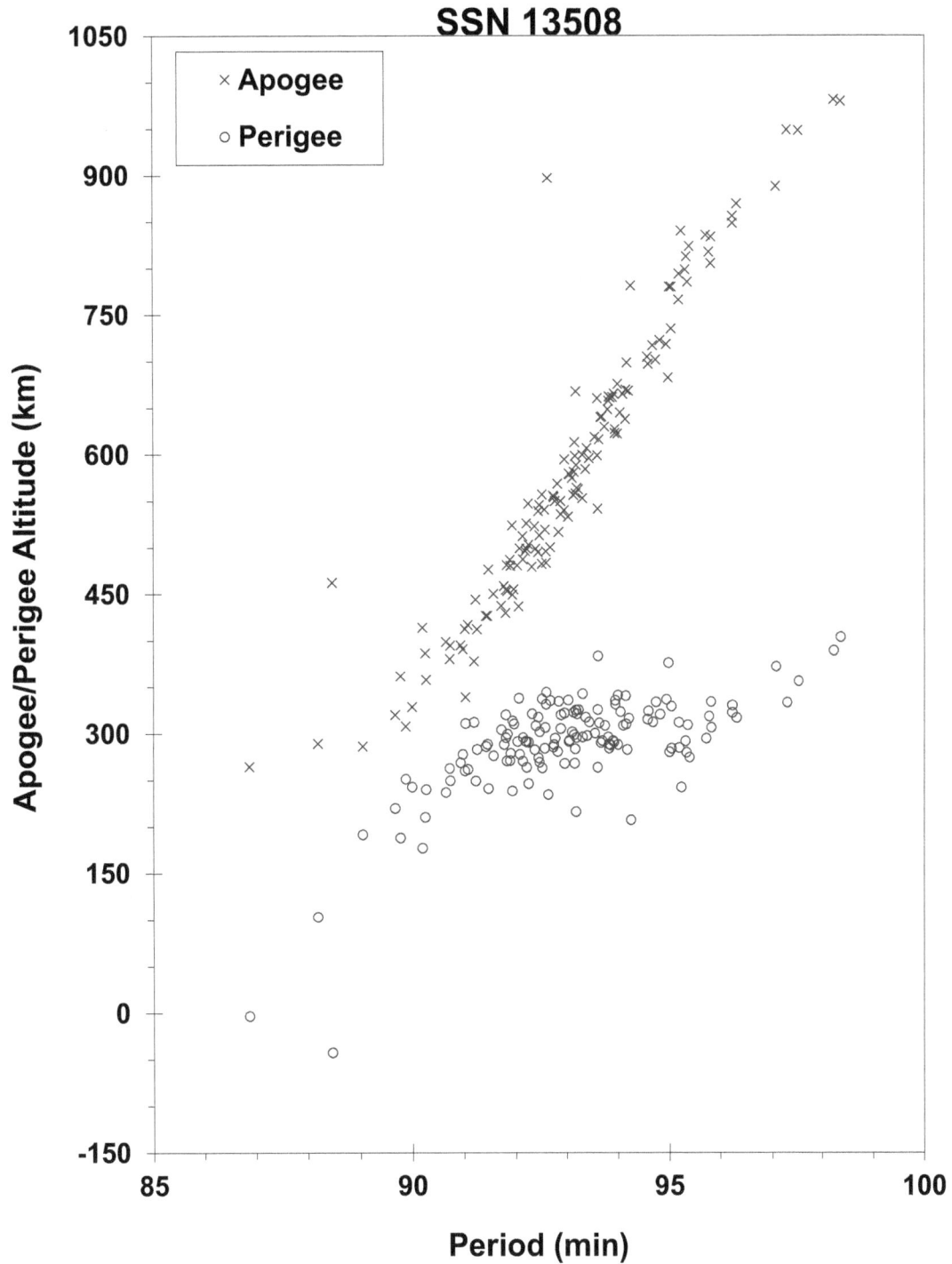

Cosmos 1405 debris cloud of 143 fragments 1 hour after the event as seen by the US SSN PARCS radar.

227

COSMOS 1408   **1982-092A**   **13552**

SATELLITE DATA

TYPE:	Payload
OWNER:	CIS
LAUNCH DATE:	16.20 Sep 1982
DRY MASS (KG):	1750
MAIN BODY:	Cylinder; 1.4 m diameter by 3 m length
MAJOR APPENDAGES:	Solar panels, antennas, gravity gradient boom
ATTITUDE CONTROL:	Inactive at time of event; 3 axis stabilization & gravity gradient
ENERGY SOURCES:	batteries, attitude control (?)

EVENT DATA

DATE:	15 Nov. 2021	LOCATION:	70.25 N, 48.39 E (asc)
TIME:	0248 GMT	ASSESSED CAUSE:	Intentional Collision
ALTITUDE:	479.5 km		

PRE-EVENT ELEMENTS

EPOCH:	21317.79514133	MEAN ANOMALY:	248.1629
RIGHT ASCENSION:	124.9329	MEAN MOTION:	15.29386133
INCLINATION:	82.5640	MEAN MOTION DOT/2:	.00002627
ECCENTRICITY:	.0018496	MEAN MOTION DOT DOT/6:	.0
ARG. OF PERIGEE:	112.1573	BSTAR:	.0000905

DEBRIS CLOUD DATA

MAXIMUM ΔP:	9.7 min
MAXIMUM ΔI:	1.5 deg

COMMENTS

Cosmos 1408 was a member of the Soviet Union's *Tselina-D*-class electronic intelligence spacecraft. The Russian Federation conducted a direct-ascent Anti-Satellite (ASAT) test in the early hours of 15 Nov. 2021 which catastrophically fragmented the long-derelict spacecraft. Debris cataloging continues at the time of this edition.

REFERENCE DOCUMENTS

"The Intentional Destruction of Cosmos 1408," The Orbital Debris Quarterly News, NASA JSC, March 2022. Available online at https://orbitaldebris.jsc.nasa.gov/quarterly-news/pdfs/odqnv26i1.pdf.

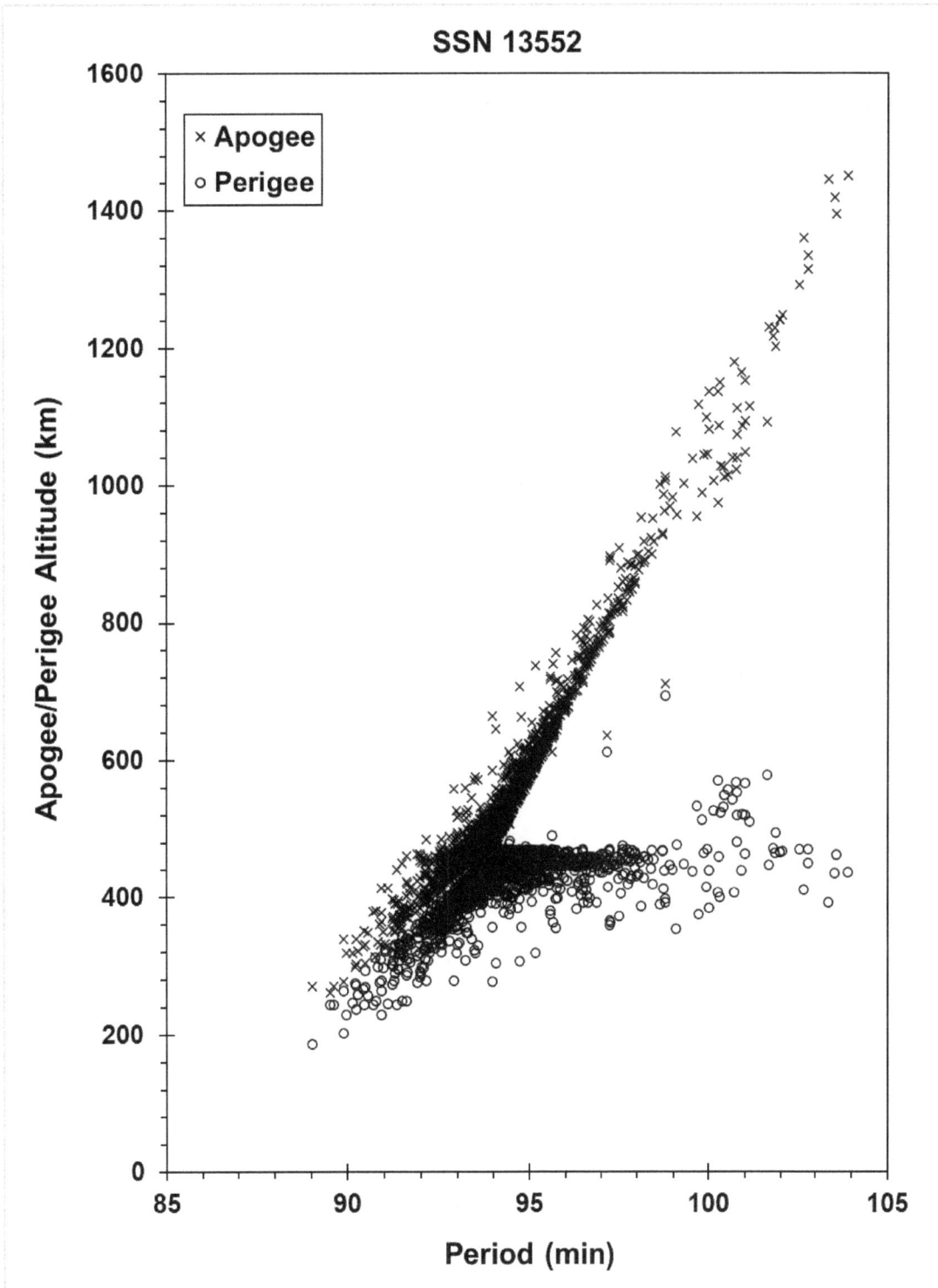

SSN 13552

Cosmos 1408 debris cloud of the parent body and 1762 fragments cataloged up to over 200 days after the event.

SATELLITE DATA

TYPE:	Payload
OWNER:	CIS
LAUNCH DATE:	16.77 Sept 1982
DRY MASS (KG):	1970
MAIN BODY:	Cylinder; 2 m by 4 m
MAJOR APPENDAGES:	Plate + 2 solar panels
ATTITUDE CONTROL:	Active, 3-axis
ENERGY SOURCES:	On-board propellants, Battery

EVENT DATA

DATE:	23 Dec 1983	LOCATION:	Unknown
TIME:	Unknown	ASSESSED CAUSE:	Battery?
ALTITUDE:	Unknown		

PRE-EVENT ELEMENTS

EPOCH:	83354.48292712	MEAN ANOMALY:	306.8102
RIGHT ASCENSION:	86.8977	MEAN MOTION:	1.00252516
INCLINATION:	0.7423	MEAN MOTION DOT/2:	.00000180
ECCENTRICITY	.0000873	MEAN MOTION DOT DOT/6:	.0
ARG. OF PERIGEE:	326.3126	BSTAR:	.0

DEBRIS CLOUD DATA

MAXIMUM ΔP:	Unknown
MAXIMUM ΔI:	Unknown

COMMENTS

This event is now the third known geostationary orbit fragmentation. The event date was provided by the 18th SpCS and the observed cessation of East-West station keeping, or maintenance of the sub-satellite longitude, is generally consistent with this date. Root cause may be similar to the acknowledged cause of Ekran 2 and/or the Ekran 4 event [*q.v.*].

REFERENCE DOCUMENTS

History of Soviet/Russian Satellite Fragmentations-A Joint U.S.-Russian Investigation, N. L. Johnson, *et al.,* Kaman Sciences Corporation, October 1995.

Agapov, V., V. Biryukov, R. Kiladze et al., "Faint GEO Objects Search and Orbital Analysis," in Proc. Of the Fourth European Conf. on Space Debris (ESA SP-587, August 2005). Accessed at https://conference.sdo.esoc.esa.int/proceedings/sdc4/paper/106/SDC4-paper106.pdf, June 2022.

Insufficient data to construct a Gabbard diagram.

SATELLITE DATA

TYPE:	Molniya Final Stage
OWNER:	CIS
LAUNCH DATE:	8.58 Dec 1982
DRY MASS (KG):	1100
MAIN BODY:	Cylinder; 2.7 m diameter by 3 m length
MAJOR APPENDAGES:	None
ATTITUDE CONTROL:	Active, 3-axis
ENERGY SOURCES:	On-board propellants

EVENT DATA

DATE:	8 Dec 1982	LOCATION:	62S, 302E (asc)
TIME:	1448 GMT	ASSESSED CAUSE:	Propulsion
ALTITUDE:	400 km		

PRE-EVENT ELEMENTS

EPOCH:	82342.56790507	MEAN ANOMALY:	305.2204
RIGHT ASCENSION:	316.3789	MEAN MOTION:	15.79849844
INCLINATION:	62.9496	MEAN MOTION DOT/2:	.0
ECCENTRICITY:	.0143321	MEAN MOTION DOT DOT/6:	.0
ARG. OF PERIGEE:	56.2493	BSTAR:	.0

DEBRIS CLOUD DATA

MAXIMUM ΔP:	4.9 min
MAXIMUM ΔI:	0.2 deg

COMMENTS

Fragmentation occurred at the time the Molniya final stage was fired to move the payload from a parking orbit to a Molniya-type transfer orbit. Pre-event elements are taken from satellite 13686 for first revolution parking orbit. A second fragmentation may have occurred on 9 December 1982.

REFERENCE DOCUMENT

History of Soviet/Russian Satellite Fragmentations-A Joint U.S.-Russian Investigation, N. L. Johnson, *et al.,* Kaman Sciences Corporation, October 1995.

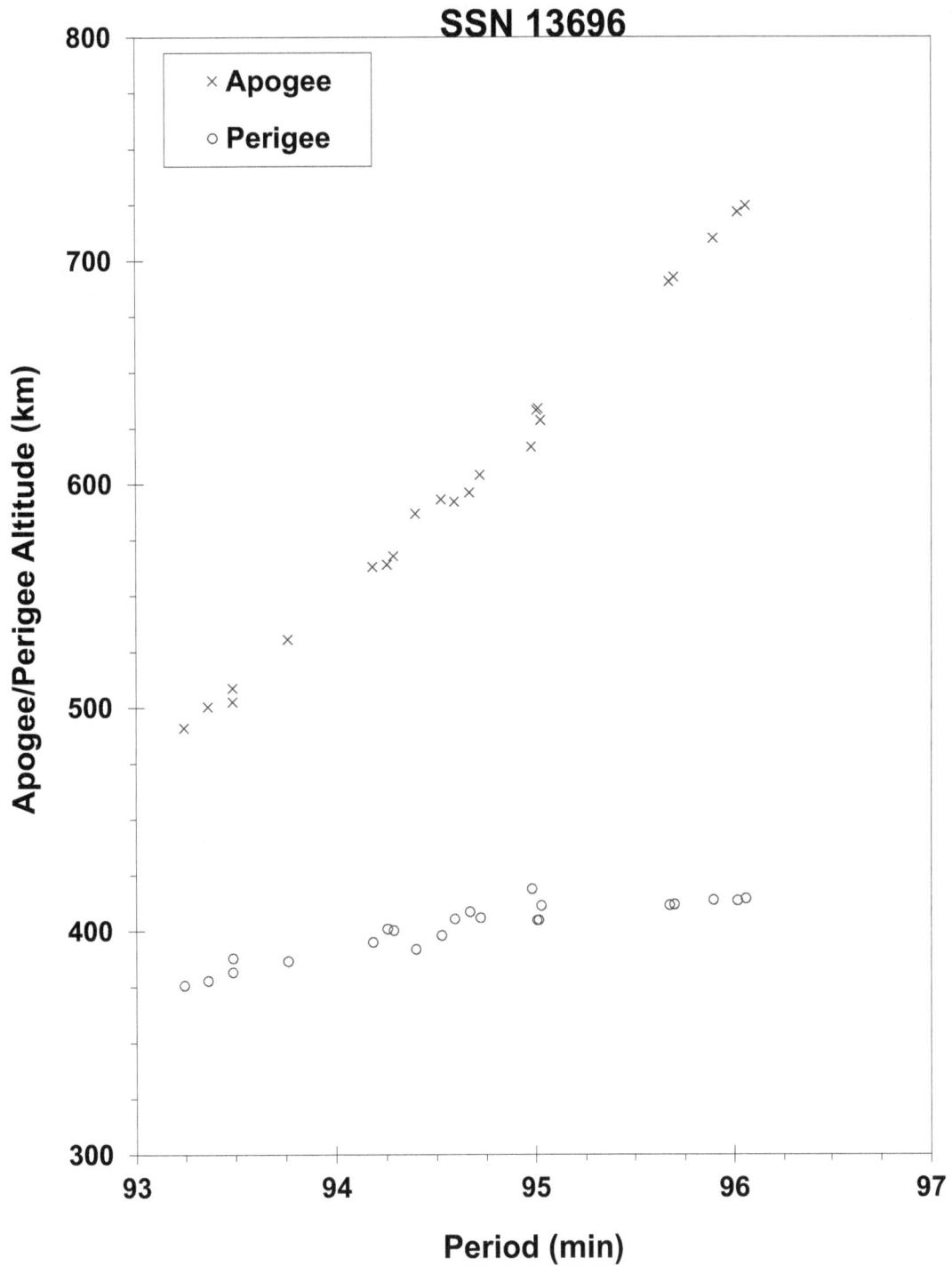

SSN 13696

Cosmos 1423 R/B debris cloud of 22 fragments soon after the event(s) as reconstructed from the US SSN database.

ASTRON ULLAGE MOTOR   1983-020B   13902

SATELLITE DATA

TYPE:	Mission Related Debris
OWNER:	CIS
LAUNCH DATE:	23.53 Mar 1983
DRY MASS (KG):	55
MAIN BODY:	Ellipsoid; 0.6 m diameter by 1.0 m length
MAJOR APPENDAGES:	None
ATTITUDE CONTROL:	None
ENERGY SOURCES:	On-board propellants

EVENT DATA

DATE:	3 Sep 1984	LOCATION:	12S, 352E (dsc)
TIME:	2023 GMT	ASSESSED CAUSE:	Propulsion
ALTITUDE:	400 km		

PRE-EVENT ELEMENTS

EPOCH:	84247.05150886	MEAN ANOMALY:	106.3279
RIGHT ASCENSION:	94.4099	MEAN MOTION:	14.50264973
INCLINATION:	51.5306	MEAN MOTION DOT/2:	.00079313
ECCENTRICITY:	.0710960	MEAN MOTION DOT DOT/6:	.0000075234
ARG. OF PERIGEE:	246.1573	BSTAR:	.00035531

DEBRIS CLOUD DATA

MAXIMUM ΔP:	2.4 min*
MAXIMUM ΔI:	0.3 deg*

*Based on uncataloged debris data

COMMENTS

Parent satellite was one of two small engine units that are routinely released after the first burn of the Proton fourth stage. The nature of these objects was identified by Dr. Boris V. Cherniatiev, Deputy Constructor for the Energiya NPO, in October 1992. The cause of this fragmentation is assumed to be related to the residual hypergolic propellants on board and failure of the membrane separating the fuel and oxidizer. Element sets on 16 fragments were developed. None were officially cataloged. This was the first in a series of fragmentations of this object type.

REFERENCE DOCUMENTS

The Fragmentation of Proton Debris, D. J. Nauer, TBE Technical Report CS93-LKD-004, Teledyne Brown Engineering, Colorado Springs, 31 December 1992.

"Identification and Resolution of an Orbital Debris Problem with the Proton Launch Vehicle," B.V. Cherniatiev, et al., Proceedings of the First European Conference on Space Debris, April 1993.

History of Soviet/Russian Satellite Fragmentations-A Joint U.S.-Russian Investigation, N. L. Johnson, et al., Kaman Sciences Corporation, October 1995.

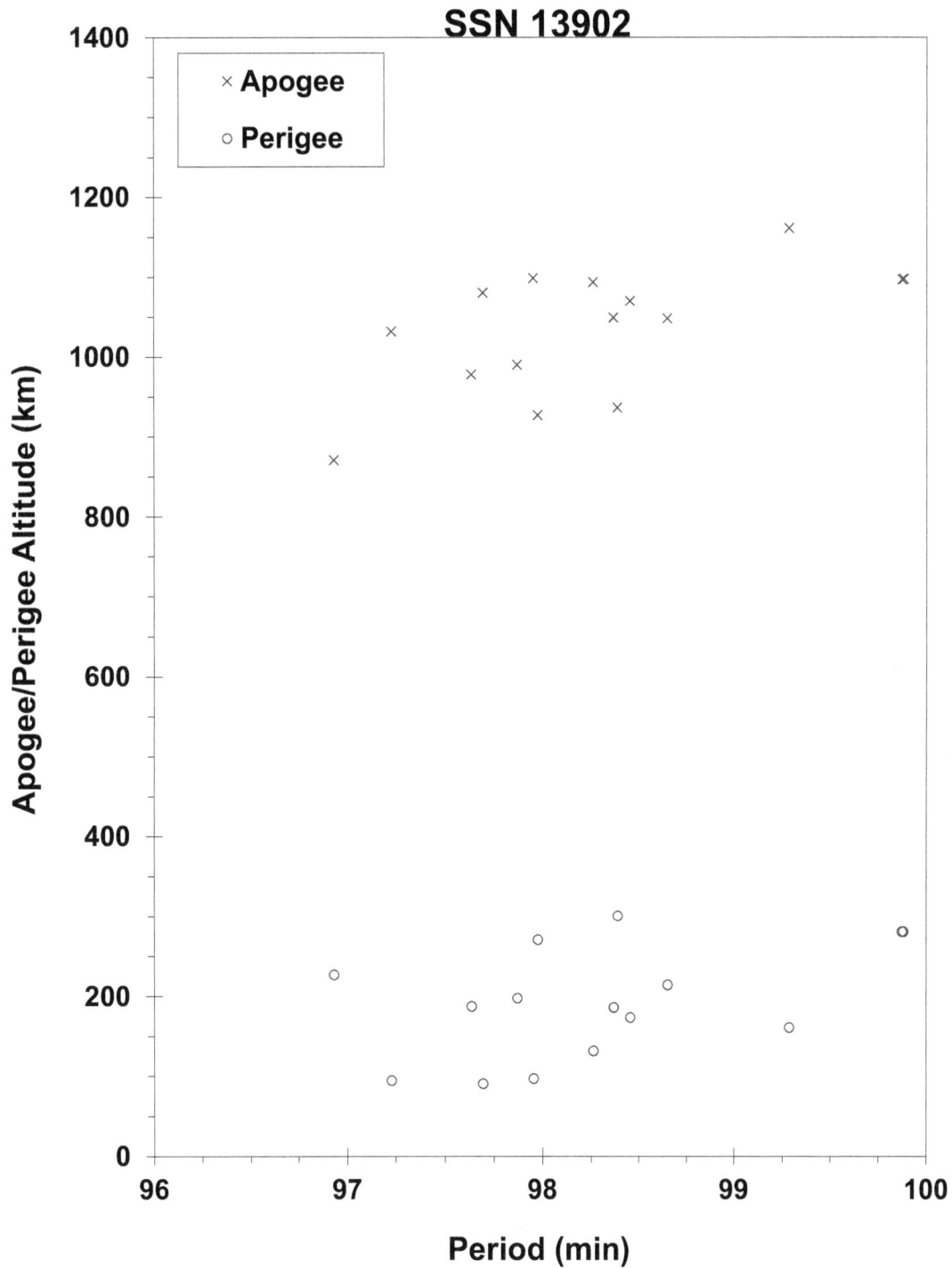

SSN 13902

Astron ullage motor debris cloud of 16 fragments as determined within a few days of the event. Elements from the US SSN database.

SATELLITE DATA

TYPE:	Payload
OWNER:	US
LAUNCH DATE:	28.66 Mar 1983
DRY MASS (KG):	1000
MAIN BODY:	Cylinder-box; 1.9 m diameter by 7.5 m length
MAJOR APPENDAGES:	1 solar panel
ATTITUDE CONTROL:	Active, 3-axis
ENERGY SOURCES:	On-board propellants

EVENT DATA

DATE:	30 Dec 1985	LOCATION:	68S, 300E (dsc)
TIME:	1005 GMT	ASSESSED CAUSE:	Battery
ALTITUDE:	825 km		

PRE-EVENT ELEMENTS

EPOCH:	85348.40460348	MEAN ANOMALY:	83.2801
RIGHT ASCENSION:	16.9717	MEAN MOTION:	14.22481975
INCLINATION:	98.6488	MEAN MOTION DOT/2:	.00000037
ECCENTRICITY:	.0015724	MEAN MOTION DOT DOT/6:	.0
ARG. OF PERIGEE:	276.6589	BSTAR:	.000025130

DEBRIS CLOUD DATA

MAXIMUM ΔP:	4.7 min
MAXIMUM ΔI:	0.1 deg

COMMENTS

A malfunction on NOAA 8 caused a battery to overcharge, resulting in a minor explosion of the battery. The spacecraft was operational at the time of the event. Six new fragments were detected and cataloged. All decayed by February 1989, leaving the parent still in orbit.

REFERENCE DOCUMENT

"NOAA Turns Off Satellite Following Malfunction", Aviation Week and Space Technology, 13 January 1986, p. 21.

SSN 13923

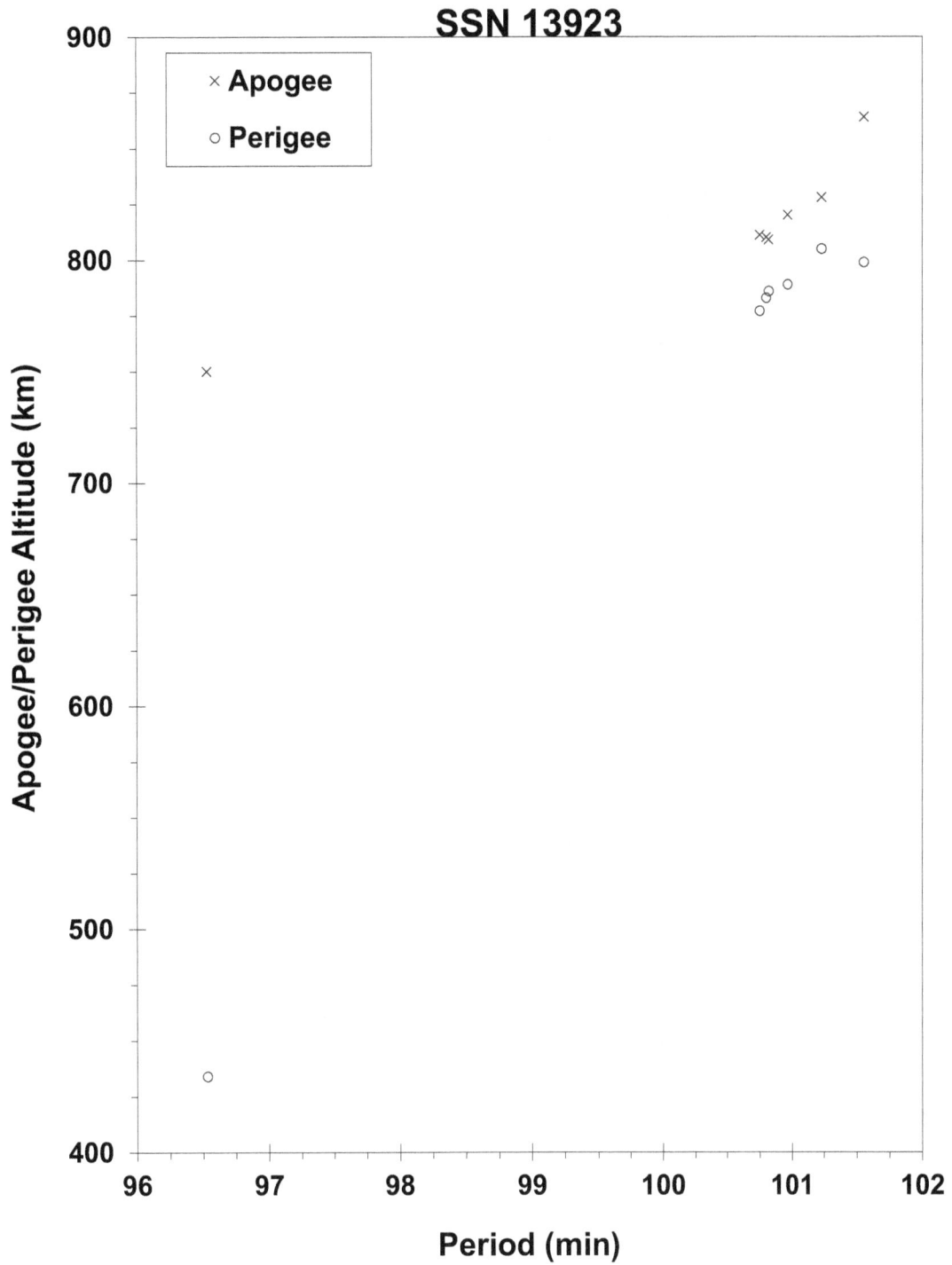

NOAA 8 debris cloud of 6 fragments plus the parent satellite 1 day after the event as reconstructed from the Naval Space Surveillance System database.

SATELLITE DATA

TYPE:	Payload
OWNER:	CIS
LAUNCH DATE:	25.81 Apr 1983
DRY MASS (KG):	1250
MAIN BODY:	Cylinder; 1.7 m diameter by 2 m length
MAJOR APPENDAGES:	Solar panels
ATTITUDE CONTROL:	Active, 3-axis
ENERGY SOURCES:	On-board propellants, explosive charge

EVENT DATA

DATE:	13 Aug 1983	LOCATION:	Unknown
TIME:	Unknown	ASSESSED CAUSE:	Deliberate
ALTITUDE:	Unknown		

PRE-EVENT ELEMENTS

EPOCH:	83225.00107283	MEAN ANOMALY:	4.5332
RIGHT ASCENSION:	79.8630	MEAN MOTION:	2.00589678
INCLINATION:	63.3076	MEAN MOTION DOT/2:	.0
ECCENTRICITY:	.7324437	MEAN MOTION DOT DOT/6:	.0
ARG. OF PERIGEE:	320.0041	BSTAR:	.0068163

DEBRIS CLOUD DATA

MAXIMUM	ΔP:	4.8 min*
MAXIMUM	ΔI:	0.4 deg*

*Based on uncataloged debris data

COMMENTS

Cosmos 1456 was another spacecraft of the Cosmos 862-type to experience a fragmentation. The spacecraft may have been active at the time of the event, having last made a station-keeping maneuver on 22 June 1983. The next station-keeping maneuver should have occurred in the second half of August or early September 1983. The spacecraft began drifting off station immediately after the event and never recovered.

REFERENCE DOCUMENT

History of Soviet/Russian Satellite Fragmentations-A Joint U.S.-Russian Investigation, N. L. Johnson, et al., Kaman Sciences Corporation, October 1995.

SSN 14034

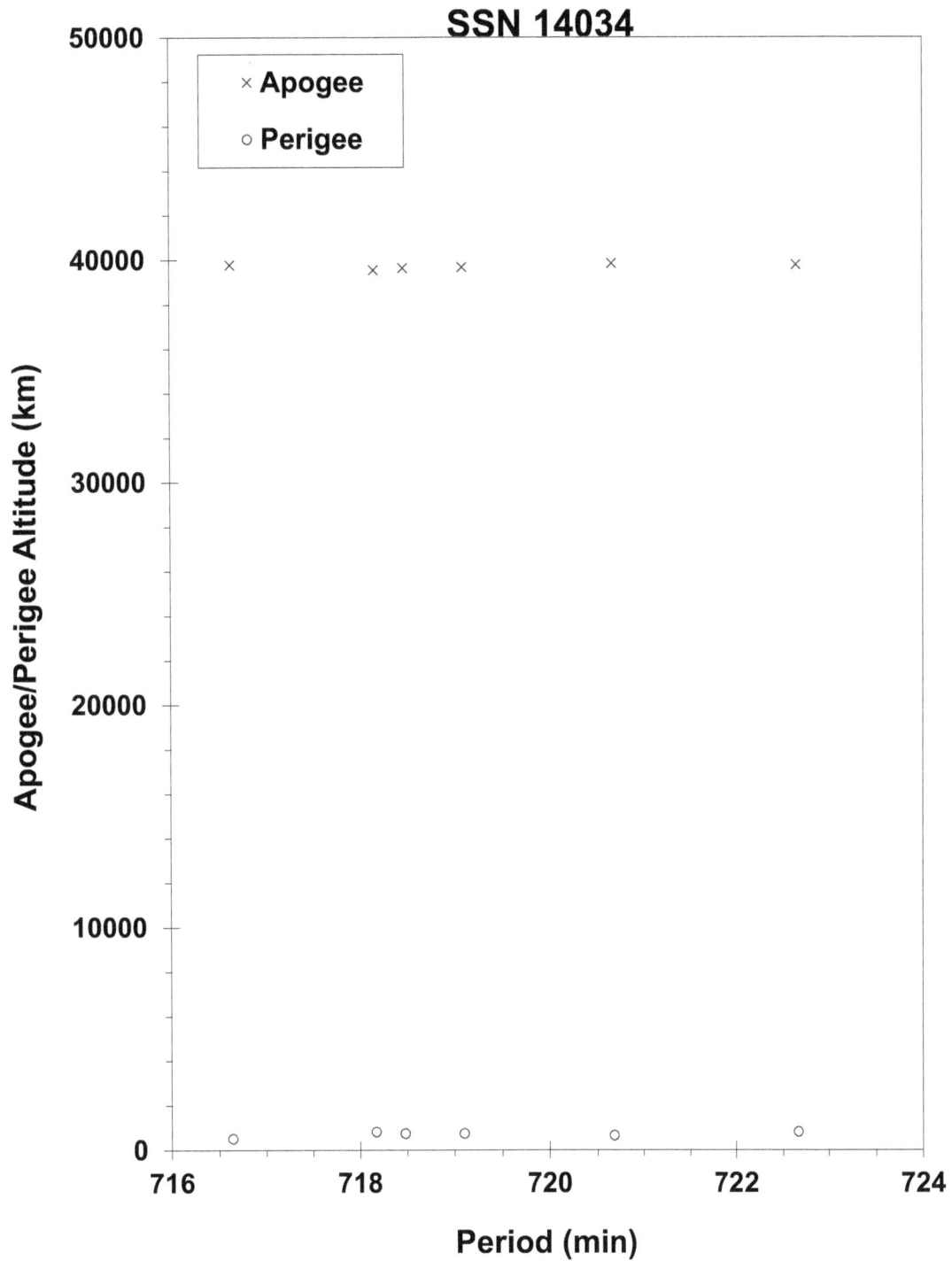

Cosmos 1456 debris cloud of 6 fragments less than 3 weeks after the event as reconstructed from the US SSN database.

SATELLITE DATA

TYPE:	Payload
OWNER:	CIS
LAUNCH DATE:	7.44 May 1983
DRY MASS (KG):	3000
MAIN BODY:	Cylinder; 1.3 m diameter by 17 m length
MAJOR APPENDAGES:	Solar panels
ATTITUDE CONTROL:	Active, 3-axis
ENERGY SOURCES:	On-board propellants, explosive charge (?)

EVENT DATA (1)

DATE:	11 Mar 1985	LOCATION:	4S, 196E (asc)
TIME:	0940 GMT	ASSESSED CAUSE:	Unknown
ALTITUDE:	750 km		

PRE-EVENT ELEMENTS (1)

EPOCH:	85068.60956125	MEAN ANOMALY:	101.2285
RIGHT ASCENSION:	157.6403	MEAN MOTION:	14.49322542
INCLINATION:	65.0244	MEAN MOTION DOT/2:	.00000357
ECCENTRICITY:	.0224980	MEAN MOTION DOT DOT/6:	.0
ARG. OF PERIGEE:	256.3703	BSTAR:	.000080310

EVENT DATA (2)

DATE:	13 May 1985	LOCATION:	10N, 82E (asc)
TIME:	0133 GMT	ASSESSED CAUSE:	Unknown
ALTITUDE:	845 km		

PRE-EVENT ELEMENTS (2)

EPOCH:	85125.54047130	MEAN ANOMALY:	121.1528
RIGHT ASCENSION:	353.4544	MEAN MOTION:	14.49239036
INCLINATION:	65.0248	MEAN MOTION DOT/2:	.0
ECCENTRICITY:	.0222492	MEAN MOTION DOT DOT/6:	.0
ARG. OF PERIGEE:	236.8082	BSTAR:	.0

DEBRIS CLOUD DATA

MAXIMUM ΔP:	5.9 min*
MAXIMUM ΔI:	1.0 deg*

*Based on uncataloged debris data

COMMENTS

Cosmos 1461 was the twelfth spacecraft of the Cosmos 699-type to experience a fragmentation. Cosmos 1461 entered a natural decay regime more than 13 months prior to first event. After the first event as many as 20 fragments were detected but only six new objects were cataloged. The second event occurred 2 months later and produced considerably more debris. These events followed the pattern set by Cosmos 1220 and Cosmos 1260.

REFERENCE DOCUMENTS

"Artificial Satellite Break-Ups (Part 1): Soviet Ocean Surveillance Satellites", N. L. Johnson, Journal of the British Interplanetary Society, February 1983, pp. 51-58.

Analysis of the Fragmentation of Kosmos 1461, G.T. DeVere and N.L. Johnson, Technical Report CS85-BMDSC-0056, Teledyne Brown Engineering, Colorado Springs, September 1985.

History of Soviet/Russian Satellite Fragmentations-A Joint U.S.-Russian Investigation, N. L. Johnson, et al., Kaman Sciences Corporation, October 1995.

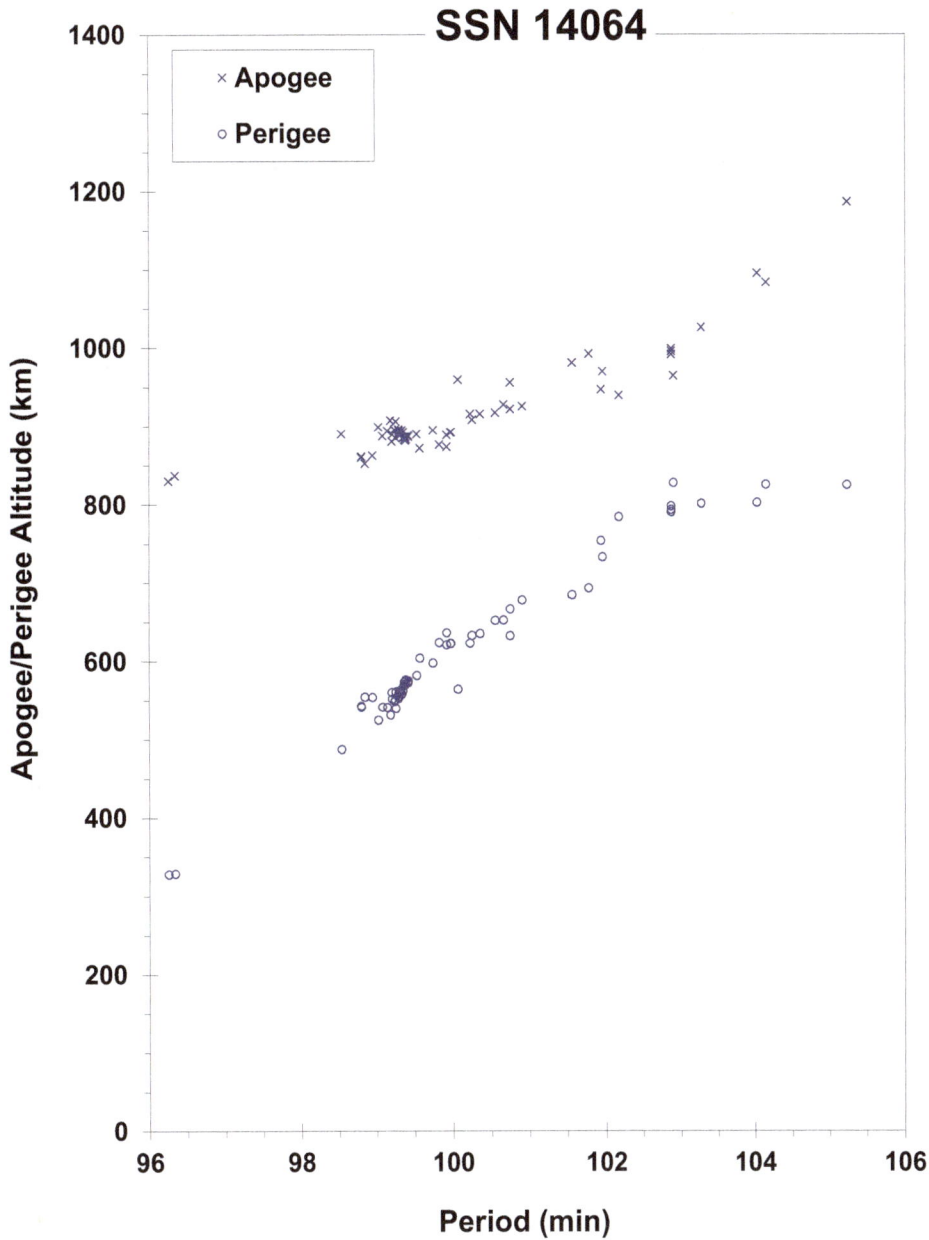

Cosmos 1461 debris cloud remnant of 65 fragments 4 days after the second event as reconstructed from the US SSN database.

SATELLITE DATA

TYPE:	Payload
OWNER:	CIS
LAUNCH DATE:	8.80 Jul 1983
DRY MASS (KG):	1250
MAIN BODY:	Cylinder; 1.7 m diameter by 2 m length
MAJOR APPENDAGES:	Solar panels
ATTITUDE CONTROL:	Active, 3-axis
ENERGY SOURCES:	On-board propellants, explosive charge

EVENT DATA

DATE:	9 Jul 1983	LOCATION:	Unknown
TIME:	Unknown	ASSESSED CAUSE:	Deliberate
ALTITUDE:	Unknown		

PRE-EVENT ELEMENTS

EPOCH:	83189.85702098	MEAN ANOMALY:	4.6462
RIGHT ASCENSION:	166.3194	MEAN MOTION:	2.03523282
INCLINATION:	62.9394	MEAN MOTION DOT/2:	.00000702
ECCENTRICITY:	.7337681	MEAN MOTION DOT DOT/6:	.0
ARG. OF PERIGEE:	317.9301	BSTAR:	.0

DEBRIS CLOUD DATA

MAXIMUM ΔP:	8.7 min*
MAXIMUM ΔI:	0.8 deg*

*Based on uncataloged debris data

COMMENTS

Cosmos 1481 was the twelfth spacecraft of the Cosmos 862-type to experience a fragmentation. The event apparently occurred within a day of launch. An expected orbital maneuver by Cosmos 1481 to move from its transfer orbit to an operational orbit about 3 days after launch was never performed.

REFERENCE DOCUMENT

History of Soviet/Russian Satellite Fragmentations-A Joint U.S.-Russian Investigation, N. L. Johnson, *et al.*, Kaman Sciences Corporation, October 1995.

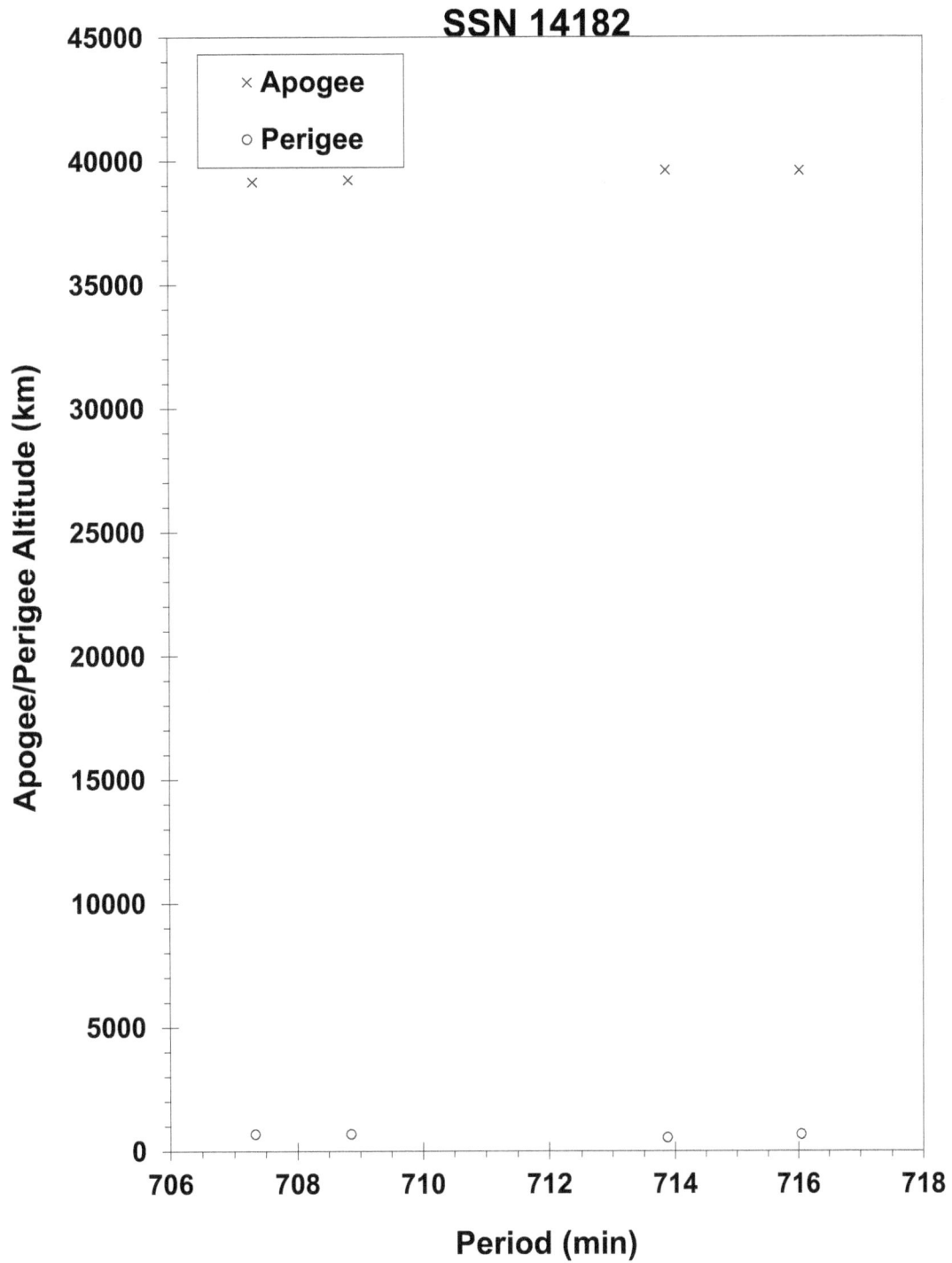

SSN 14182

Cosmos 1481 debris cloud of 4 objects 1 month after the event as reconstructed from the US SSN database.

243

SATELLITE DATA

TYPE:	Payload
OWNER:	CIS
LAUNCH DATE:	24.25 Jul 1983
DRY MASS (KG):	1800
MAIN BODY:	Cylinder; 1.5 m diameter by 5.0 m length
MAJOR APPENDAGES:	Solar panels, antenna
ATTITUDE CONTROL:	Gravity gradient; momentum wheels
ENERGY SOURCES:	Electrical system (?); pressurized vessels

EVENT DATA

DATE:	18 Oct 1993	LOCATION:	7S, 111E (asc)
TIME:	1204 GMT	ASSESSED CAUSE:	Unknown
ALTITUDE:	605 km		

PRE-EVENT ELEMENTS

EPOCH:	93289.76777232	MEAN ANOMALY:	40.8047
RIGHT ASCENSION:	316.3082	MEAN MOTION:	14.98254133
INCLINATION:	97.5219	MEAN MOTION DOT/2:	.00001299
ECCENTRICITY:	.0033451	MEAN MOTION DOT DOT/6:	.0
ARG. OF PERIGEE:	319.0655	BSTAR:	.00011294

DEBRIS CLOUD DATA

MAXIMUM ΔP: 14.0 min *
MAXIMUM ΔI: 2.5 deg *

* Based on uncataloged debris data

COMMENTS

Cosmos 1484 was the third of four Resurs-0 prototypes flown in sun-synchronous orbits, and the only one to fragment. This is the first sun-synchronous Russian satellite to ever fragment. The NAVSPOC generated 79 analyst satellites on this event.

REFERENCE DOCUMENTS

The Fragmentation of Cosmos 1484, D. J. Nauer, Technical Report CS94-LKD-003, Teledyne Brown Engineering, Colorado Springs, 17 November 1993.

The Soviet Year in Space, 1990, N. L. Johnson, Teledyne Brown Engineering, 1991.

History of Soviet/Russian Satellite Fragmentations-A Joint U.S.-Russian Investigation, N. L. Johnson, et al., Kaman Sciences Corporation, October 1995.

SSN 14207

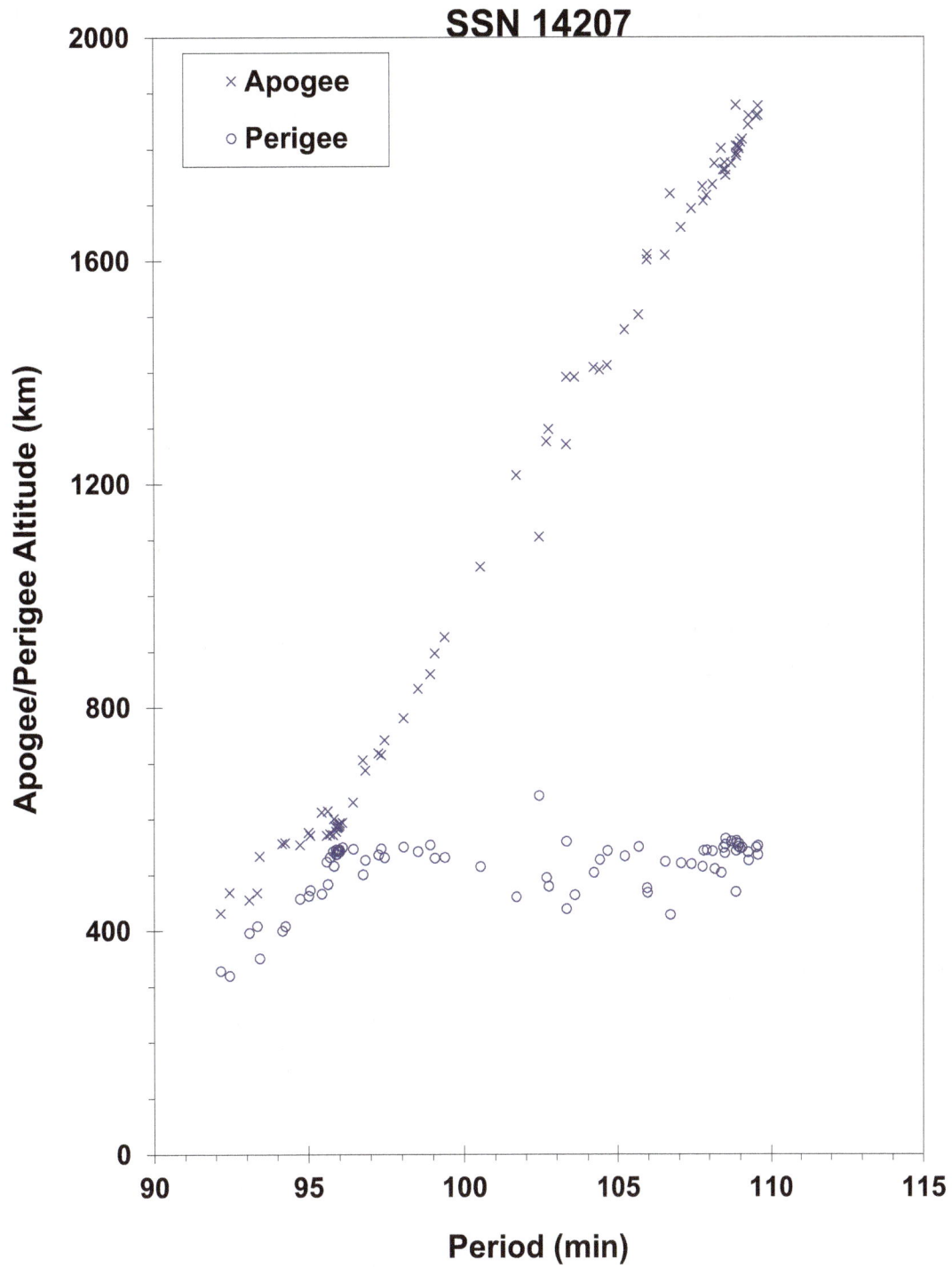

Cosmos 1484 debris cloud of 79 fragments assembled by the NAVSPOC.

SATELLITE DATA

TYPE:	Mission Related Debris
OWNER:	CIS
LAUNCH DATE:	29.04 Dec 1983
DRY MASS (KG):	55
MAIN BODY:	Ellipsoid; 0.6 m diameter by 1 m length
MAJOR APPENDAGES:	None
ATTITUDE CONTROL:	None
ENERGY SOURCES:	On-board propellants

EVENT DATA

DATE:	4 Feb 1991	LOCATION:	28N, 106E (dsc)
TIME:	0312 GMT	ASSESSED CAUSE:	Propulsion
ALTITUDE:	18550 km		

PRE-EVENT ELEMENTS

EPOCH:	91032.22560633	MEAN ANOMALY:	10.4843
RIGHT ASCENSION:	133.4557	MEAN MOTION:	4.30882556
INCLINATION:	51.9464	MEAN MOTION DOT/2:	.00004140
ECCENTRICITY:	.5787304	MEAN MOTION DOT DOT/6:	.0
ARG. OF PERIGEE:	315.5487	BSTAR:	.0018354

DEBRIS CLOUD DATA

MAXIMUM ΔP:	Unknown
MAXIMUM ΔI:	Unknown

COMMENTS

Parent satellite was one of two small engine units that are routinely released after the first burn of the Proton fourth stage. The nature of these objects was identified by Dr. Boris V. Chernlatiev, Deputy Constructor for the Energiya NPO, in October 1992. The cause of this fragmentation is assumed to be related to the residual hypergolic propellants on board and failure of the membrane separating the fuel and oxidizer. NAVSPASUR observed at least 12 fragments on the day of the event and approximately three dozen on 7 February. An element set was initially developed on only one new fragment. This was the third in a series of fragmentations of this object type.

REFERENCE DOCUMENTS

The Fragmentation of Proton Debris, D. J. Nauer, TBE Technical Report CS93-LKD-004, Teledyne Brown Engineering, Colorado Springs, 31 December 1992.

History of Soviet/Russian Satellite Fragmentations-A Joint U.S.-Russian Investigation, N. L. Johnson, et al., Kaman Sciences Corporation, October 1995.

"Identification and Resolution of an Orbital Debris Problem with the Proton Launch Vehicle," B. V. Cherniatiev, et al., Proceedings of the First European Conference on Space Debris, April 1993.

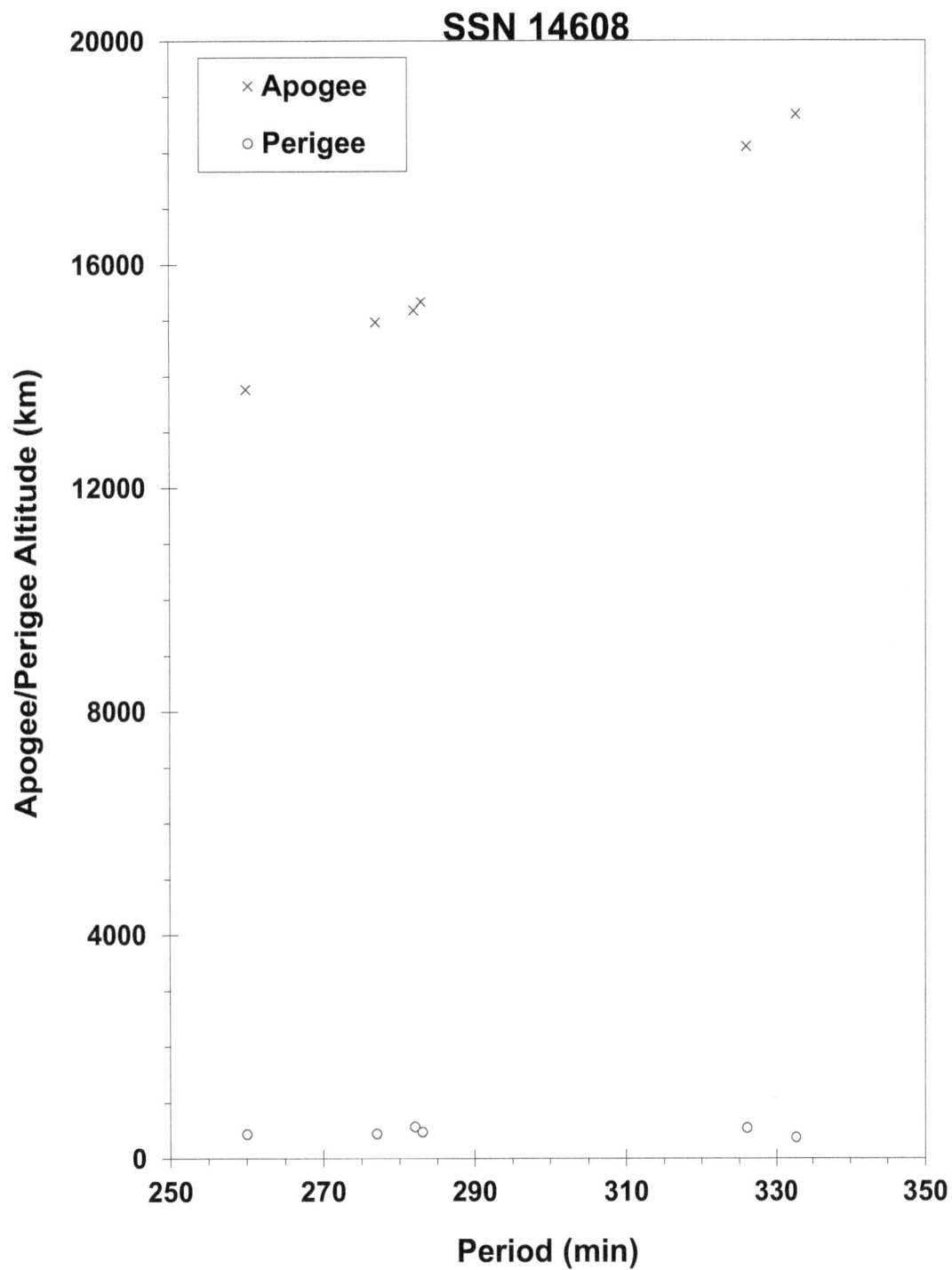

SSN 14608

Apogee/Perigee Altitude (km) vs Period (min)

Gabbard diagram of the five cataloged and single analyst satellite generated on the Cosmos 1519-21 debris cloud.

SATELLITE DATA

TYPE:	PAM-D Upper Stage (STAR 48 motor)
OWNER:	US
LAUNCH DATE:	3.54 Feb 1984
DRY MASS (KG):	2200
MAIN BODY:	Sphere-nozzle; 1.2 m by 2.1 m
MAJOR APPENDAGES:	None
ATTITUDE CONTROL:	Spin-stabilized
ENERGY SOURCES:	On-board propellants

EVENT DATA

DATE:	6 Feb 1984	LOCATION:	0N, 120E (asc)
TIME:	1600 GMT	ASSESSED CAUSE:	Propulsion
ALTITUDE:	280 km		

PRE-EVENT ELEMENTS

EPOCH:	84037.35377144	MEAN ANOMALY:	82.4657
RIGHT ASCENSION:	138.8370	MEAN MOTION:	15.97451864
INCLINATION:	28.4669	MEAN MOTION DOT/2:	.00197501
ECCENTRICITY:	.0006481	MEAN MOTION DOT DOT/6:	.0
ARG. OF PERIGEE:	277.3659	BSTAR:	.00040999

DEBRIS CLOUD DATA

MAXIMUM ΔP:	9.4 min*
MAXIMUM ΔI:	0.3 deg*

*Based on uncataloged debris data

COMMENTS

Palapa B2 and its PAM-D upper stage were deployed from the Space Shuttle Challenger at 1513 GMT, 6 February 1984. Ignition of the upper stage occurred on schedule at 1600 GMT, but the nozzle fragmented within 10 seconds. Without the nozzle the burn could not be sustained, and a natural shutdown quickly followed. The PAM-D then separated from Palapa B2. The above elements are for the Shuttle prior to deployment. The Shuttle made a small posigrade evasive maneuver after deployment and before ignition of the PAM-D. See also Westar 6 R/B fragmentation.

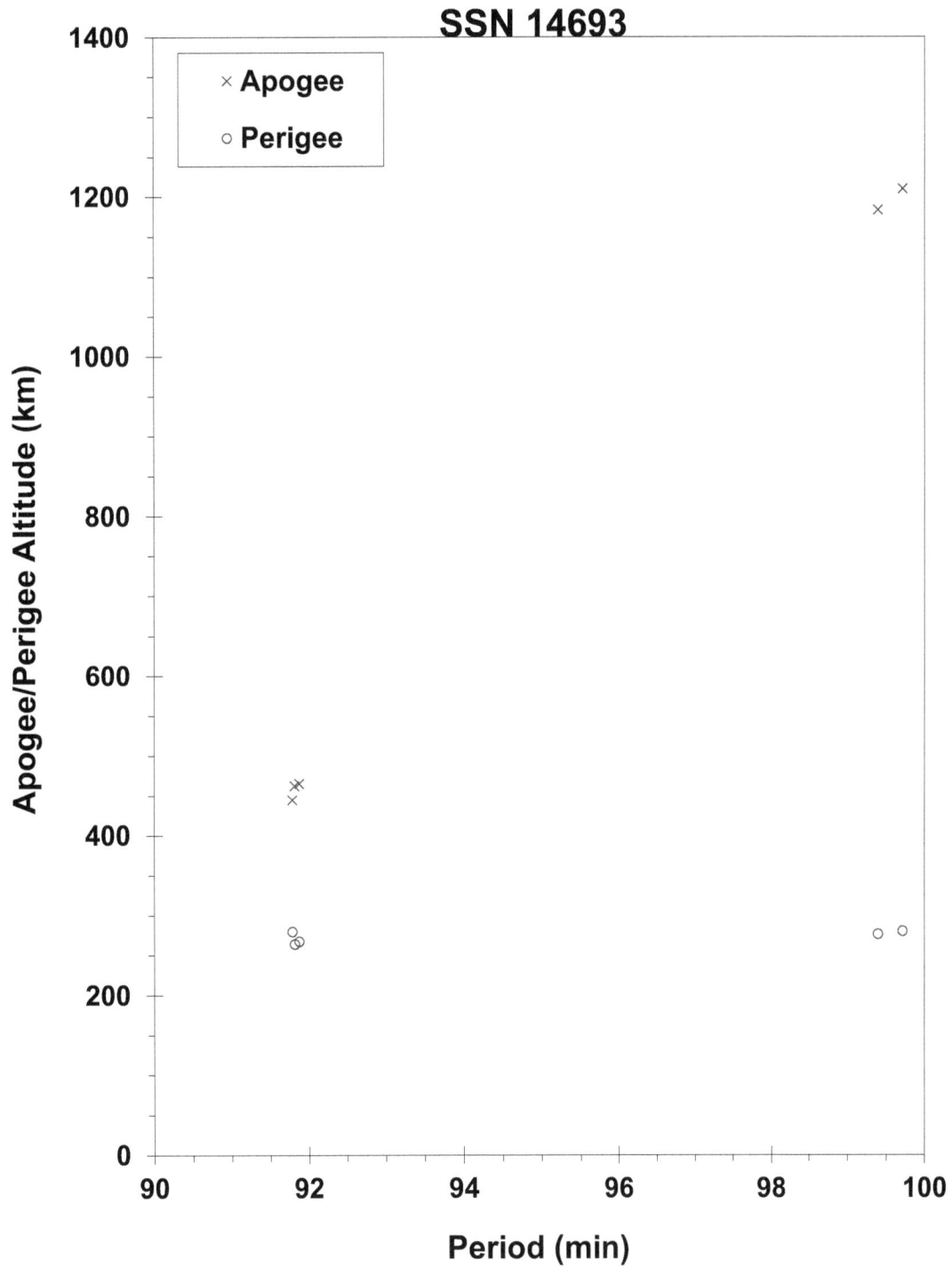

SSN 14693

Palapa B2 R/B debris cloud of 5 fragments about 3 days after the event as reconstructed from the US SSN database. The Palapa B2 R/B is the object with the second highest orbital period.

SATELLITE DATA

TYPE:	PAM-D Upper Stage (STAR 48 motor)
OWNER:	US
LAUNCH DATE:	3.54 Feb 1984
DRY MASS (KG):	2200
MAIN BODY:	Sphere-nozzle; 1.2 m by 2.1 m
MAJOR APPENDAGES:	None
ATTITUDE CONTROL:	Spin-stabilized
ENERGY SOURCES:	On-board propellants

EVENT DATA

DATE:	3 Feb 1984	LOCATION:	0N, 56E (asc)
TIME:	2145 GMT	ASSESSED CAUSE:	Propulsion
ALTITUDE:	305 km		

PRE-EVENT ELEMENTS

EPOCH:	84034.84362284	MEAN ANOMALY:	48.7355
RIGHT ASCENSION:	157.5848	MEAN MOTION:	15.88299499
INCLINATION:	28.4660	MEAN MOTION DOT/2:	.00000250
ECCENTRICITY:	.0006644	MEAN MOTION DOT DOT/6:	.0
ARG. OF PERIGEE:	311.2683	BSTAR:	.0

DEBRIS CLOUD DATA

MAXIMUM ΔP:	9.7 min
MAXIMUM ΔI:	0.8 deg

COMMENTS

Westar 6 and its PAM-D upper stage were deployed from the Space Shuttle Challenger at 2100 GMT, 3 February 1984. Ignition of the upper stage occurred on schedule at 2145 GMT, but the nozzle fragmented within 10 seconds. Without the nozzle the burn could not be sustained, and a natural shutdown quickly followed. The PAM-D then separated from Westar 6. See also Palapa B2 R/B fragmentation.

REFERENCE DOCUMENT

Westar Failure, Technical Memorandum from N.L. Johnson, Teledyne Brown Engineering, to Preston Landry, NORAD/ADCOM/XPYS, Colorado Springs, 7 February 1984.

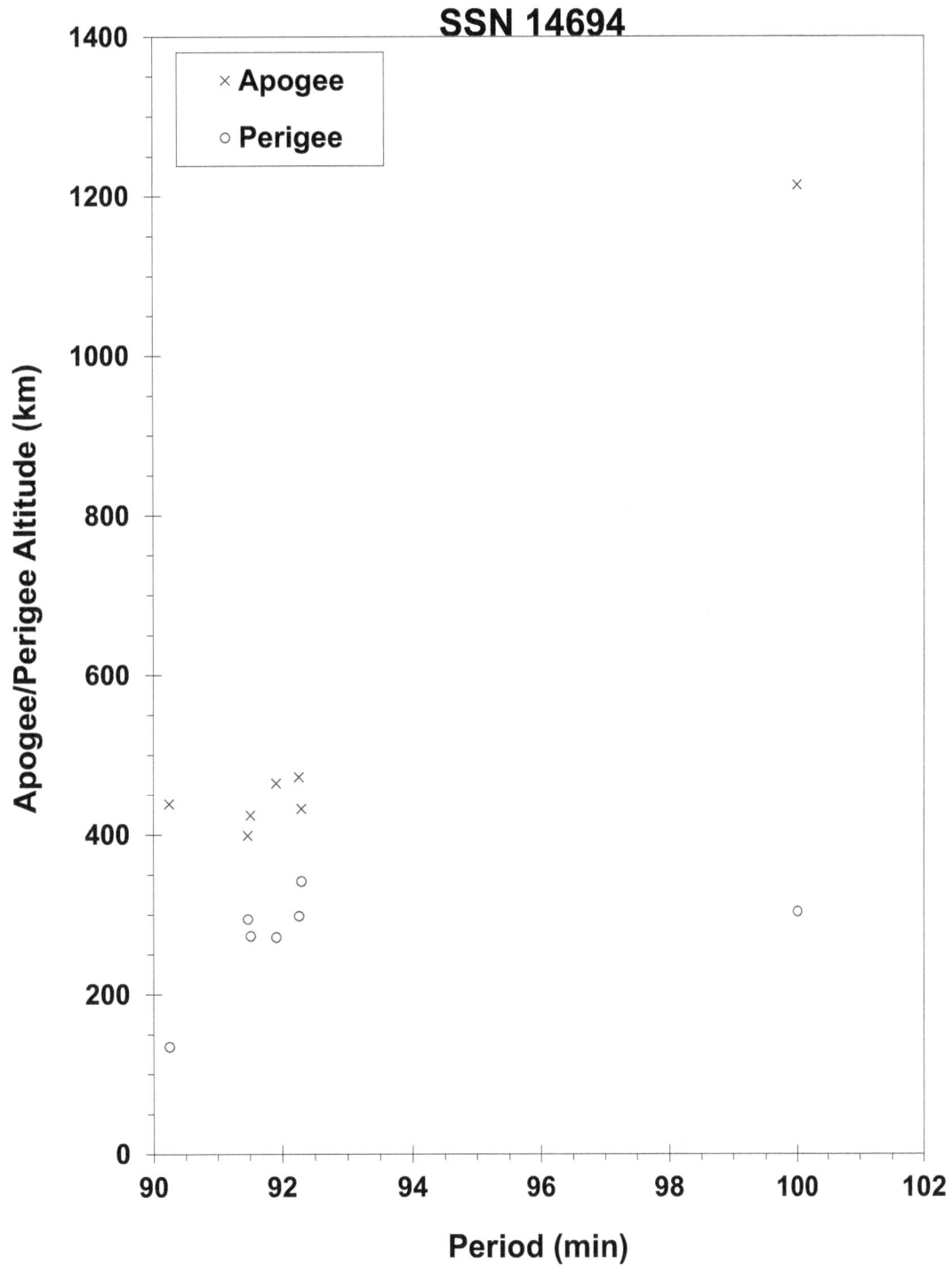

SSN 14694

Westar 6 R/B debris cloud of 7 fragments less than 2 days after the event as reconstructed from the US SSN database. The Westar 6 R/B is the object in the high, 100-min orbit.

SATELLITE DATA

TYPE:	Payload
OWNER:	CIS
LAUNCH DATE:	7.95 Aug 1984
DRY MASS (KG):	3000
MAIN BODY:	Cylinder; 1.3 m diameter by 17 m length
MAJOR APPENDAGES:	Solar panels
ATTITUDE CONTROL:	Active, 3-axis
ENERGY SOURCES:	On-board propellants, explosive charge (?)

EVENT DATA

DATE:	23 Feb 1986	LOCATION:	29N, 187E (asc)
TIME:	1850 GMT	ASSESSED CAUSE:	Unknown
ALTITUDE:	430 km		

PRE-EVENT ELEMENTS

EPOCH:	86048.57631415	MEAN ANOMALY:	72.5463
RIGHT ASCENSION:	268.3025	MEAN MOTION:	15.47795866
INCLINATION:	65.0271	MEAN MOTION DOT/2:	.00005888
ECCENTRICITY:	.0022403	MEAN MOTION DOT DOT/6:	.0
ARG. OF PERIGEE:	287.3230	BSTAR:	.00011680

DEBRIS CLOUD DATA

MAXIMUM ΔP:	2.0 min
MAXIMUM ΔI:	0.4 deg

COMMENTS

Cosmos 1588 was the thirteenth spacecraft of the Cosmos 699-type to experience a fragmentation. Spacecraft had been in natural decay for 7 months prior to the event.

REFERENCE DOCUMENTS

"Artificial Satellite Break-Ups (Part 1): Soviet Ocean Surveillance Satellites", N. L. Johnson, <u>Journal of the British Interplanetary Society</u>, February 1983, pp. 51-58.

<u>History of Soviet/Russian Satellite Fragmentations-A Joint U.S.-Russian Investigation</u>, N. L. Johnson, *et al.,* Kaman Sciences Corporation, October 1995.

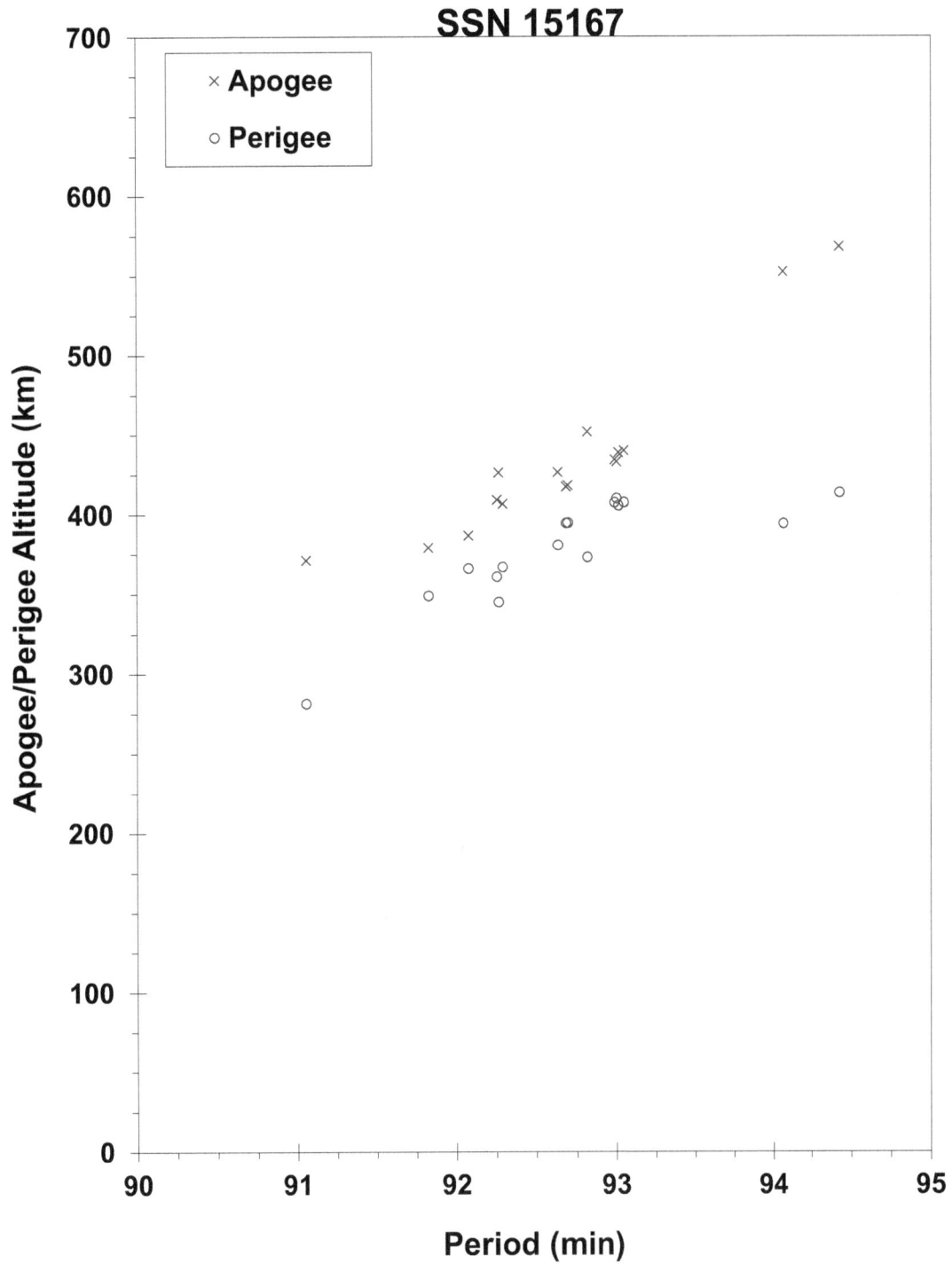

SSN 15167

Apogee/Perigee Altitude (km) vs Period (min)

× Apogee
○ Perigee

Cosmos 1588 cataloged debris cloud of 16 fragments 3 weeks after the event as reconstructed from the US SSN database.

COSMOS 1603 ULLAGE MOTOR   1984-106F   15338

SATELLITE DATA

TYPE:	Mission Related Debris
OWNER:	CIS
LAUNCH DATE:	28.58 Sep 1984
DRY MASS (KG):	55
MAIN BODY:	Ellipsoid; 0.6 m diameter by 1 m length
MAJOR APPENDAGES:	None
ATTITUDE CONTROL:	None
ENERGY SOURCES:	On-board propellants

EVENT DATA

DATE:	5 Sep 1992	LOCATION:	46.1S, 351.8E
TIME:	1451 GMT	ASSESSED CAUSE:	Propulsion
ALTITUDE:	835 km		

PRE-EVENT ELEMENTS

EPOCH:	92249.36121283	MEAN ANOMALY:	6.5555
RIGHT ASCENSION:	353.4197	MEAN MOTION:	14.15474339
INCLINATION:	66.5712	MEAN MOTION DOT/2:	-.00009086
ECCENTRICITY:	.0007128	MEAN MOTION DOT DOT/6:	.0
ARG. OF PERIGEE:	353.5641	BSTAR:	-.004641

DEBRIS CLOUD DATA

MAXIMUM ΔP:	2.9 min
MAXIMUM ΔI:	0.5 deg

COMMENTS

Parent satellite was one of two small engine units that are routinely released after the first burn of the Proton fourth stage. The nature of these objects was identified by Dr. Boris V. Chernlatiev, Deputy Constructor for the Energiya NPO, in October 1992. The cause of this fragmentation is assumed to be related to the residual hypergolic propellants on board and failure of the membrane separating the fuel and oxidizer. NAVSPASUR has observed 62 objects associated with this breakup. This was the sixth in a series of fragmentations of this object type.

REFERENCE DOCUMENTS

Soviet Space Programs 1980-1985, Science and Technology Series, Volume 66, Nicholas L. Johnson, American Astronautical Society, Univelt, Inc., 1987.

The Fragmentation of Proton Debris, D. J. Nauer, TBE Technical Report CS93-LKD-004, Teledyne Brown Engineering, Colorado Springs, 31 December 1992.

History of Soviet/Russian Satellite Fragmentations-A Joint U.S.-Russian Investigation, N. L. Johnson, et al., Kaman Sciences Corporation, October 1995.

"Identification and Resolution of an Orbital Debris Problem with the Proton Launch Vehicle," B. V. Cherniatiev, et al., Proceedings of the First European Conference on Space Debris, April 1993.

SSN 15338

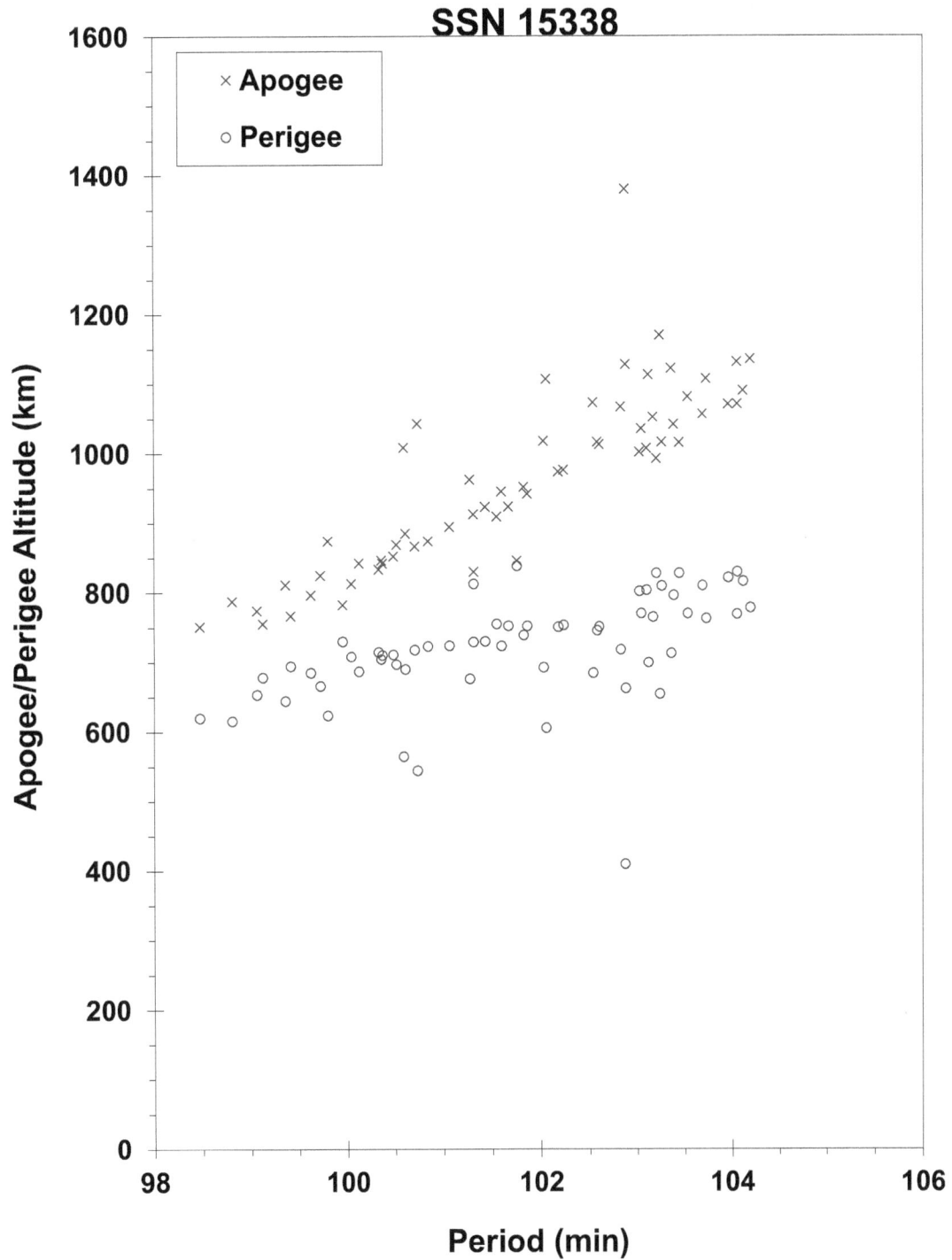

Naval Space Surveillance System tracked 62 objects within the Cosmos 1603 debris cloud, with 22 appearing in the Satellite Catalog.

255

SPACENET 2/MARECS B2 R/B   **1984-114C**   **15388**

SATELLITE DATA

TYPE:	Ariane 3 Final Stage
OWNER:	France
LAUNCH DATE:	10.05 Nov 1984
DRY MASS (KG):	~1100
MAIN BODY:	Unknown
MAJOR APPENDAGES:	None
ATTITUDE CONTROL:	None
ENERGY SOURCES:	On-board propellants

EVENT DATA

DATE:	20 Nov 1984	LOCATION:	Unknown
TIME:	1425Z	ASSESSED CAUSE:	Propulsion
ALTITUDE:	Unknown		

PRE-EVENT ELEMENTS

EPOCH:	84325.41617	MEAN ANOMALY:	146.5463
RIGHT ASCENSION:	236.1289	MEAN MOTION:	2.26087292
INCLINATION:	7.0293	MEAN MOTION DOT/2:	.00001128
ECCENTRICITY:	.7265710	MEAN MOTION DOT DOT/6:	.0
ARG. OF PERIGEE:	187.8823	BSTAR:	.0010954

DEBRIS CLOUD DATA

MAXIMUM ΔP:	Unknown
MAXIMUM ΔI:	Unknown

COMMENTS

This Ariane R/B fragmentation occurred to 10 days after launch but not detected until 2003. This is the first Ariane Rocket Body fragmentation that is attributed to France. Previous Ariane Stages were attributed to ESA.

Insufficient data to construct a Gabbard diagram.

SATELLITE DATA

TYPE:	Payload
OWNER:	CIS
LAUNCH DATE:	18.90 Apr 1985
DRY MASS (KG):	3000
MAIN BODY:	Cylinder; 1.3 m diameter by 17 m length
MAJOR APPENDAGES:	Solar panels
ATTITUDE CONTROL:	Active, 3-axis
ENERGY SOURCES:	On-board propellants, explosive charge (?)

EVENT DATA

DATE:	20 Nov 1987	LOCATION:	65N, 300E (dsc)
TIME:	0131 GMT	ASSESSED CAUSE:	Unknown
ALTITUDE:	410 km		

PRE-EVENT ELEMENTS

EPOCH:	87323.98216942	MEAN ANOMALY:	105.3951
RIGHT ASCENSION:	286.0367	MEAN MOTION:	15.56048984
INCLINATION:	65.0306	MEAN MOTION DOT/2:	.00039428
ECCENTRICITY:	.0018658	MEAN MOTION DOT DOT/6:	.0
ARG. OF PERIGEE:	254.4728	BSTAR:	.00055895

DEBRIS CLOUD DATA

MAXIMUM ΔP:	5.5 min*
MAXIMUM ΔI:	0.2 deg*

*Based on cataloged and uncataloged debris data

COMMENTS

Cosmos 1646 was the sixteenth spacecraft of the Cosmos 699-type to experience a fragmentation. Spacecraft had been in natural decay for nearly 20 months prior to the event. Many debris reentered before being officially cataloged.

REFERENCE DOCUMENTS

"Artificial Satellite Break-Ups (Part 1): Soviet Ocean Surveillance Satellites", N. L. Johnson, Journal of the British Interplanetary Society, February 1983, pp. 51-58.

History of Soviet/Russian Satellite Fragmentations-A Joint U.S.-Russian Investigation, N. L. Johnson, et al., Kaman Sciences Corporation, October 1995.

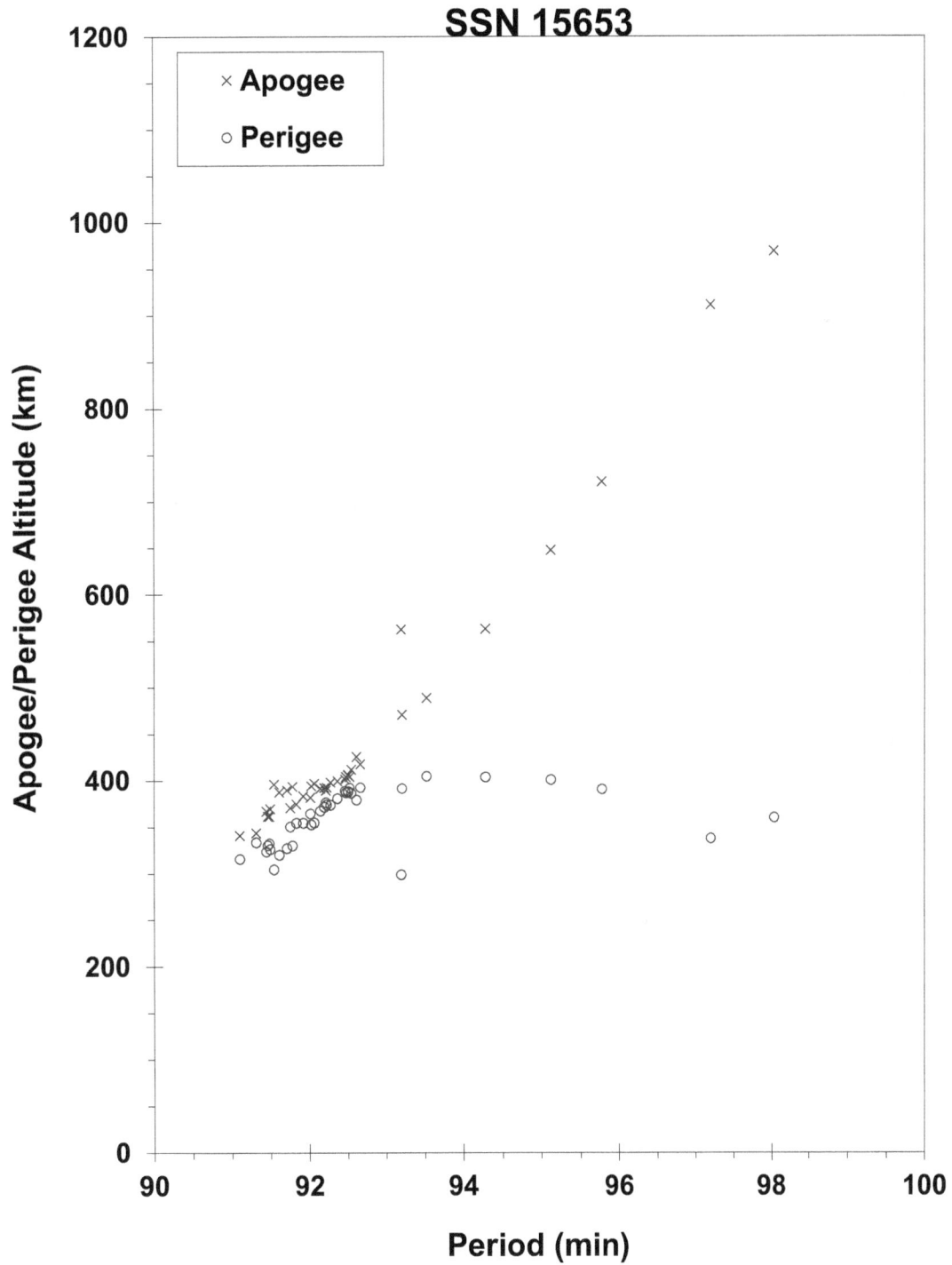

Cosmos 1646 debris cloud remnant of 38 fragments about 10 days after the event as reconstructed from the US SSN database.

SATELLITE DATA

TYPE:	Mission Related Debris
OWNER:	CIS
LAUNCH DATE:	17 May 1985
DRY MASS (KG):	55
MAIN BODY:	Ellipsoid; 0.6 m diameter by 1 m length
MAJOR APPENDAGES:	None
ATTITUDE CONTROL:	None
ENERGY SOURCES:	On-board propellants

EVENT DATA

DATE:	29 Nov 1998	LOCATION:	38.3N, 172.6E
TIME:	0925 GMT	ASSESSED CAUSE:	Propulsion
ALTITUDE:	16420 km		

POST-EVENT ELEMENTS

EPOCH:	98332.38151447	MEAN ANOMALY:	98.9018
RIGHT ASCENSION:	344.4719	MEAN MOTION:	4.35077855212150
INCLINATION:	52.0277	MEAN MOTION DOT/2:	.00009109
ECCENTRICITY:	.5772516	MEAN MOTION DOT DOT/6:	00000-0
ARG. OF PERIGEE:	209.7130	BSTAR:	.030939

DEBRIS CLOUD DATA

MAXIMUM ΔP:	91.18 min
MAXIMUM ΔI:	.76 deg

COMMENTS

This is the 18th event of the Proton Block DM SOZ Ullage Motor class identified to date; it is the seventh associated with a GLONASS mission. This mission was conducted before the engineering defect was identified and passivation measures implemented. In this orbit, debris may be long-lived but hard to track. A total of 60 debris objects were detected.

REFERENCE DOCUMENT

"1998 Ends with Eighth Satellite Breakup", The Orbital Debris Quarterly News, NASA JSC, January 1999. Available online at https://orbitaldebris.jsc.nasa.gov/quarterly-news/pdfs/ODQNv4i1.pdf.

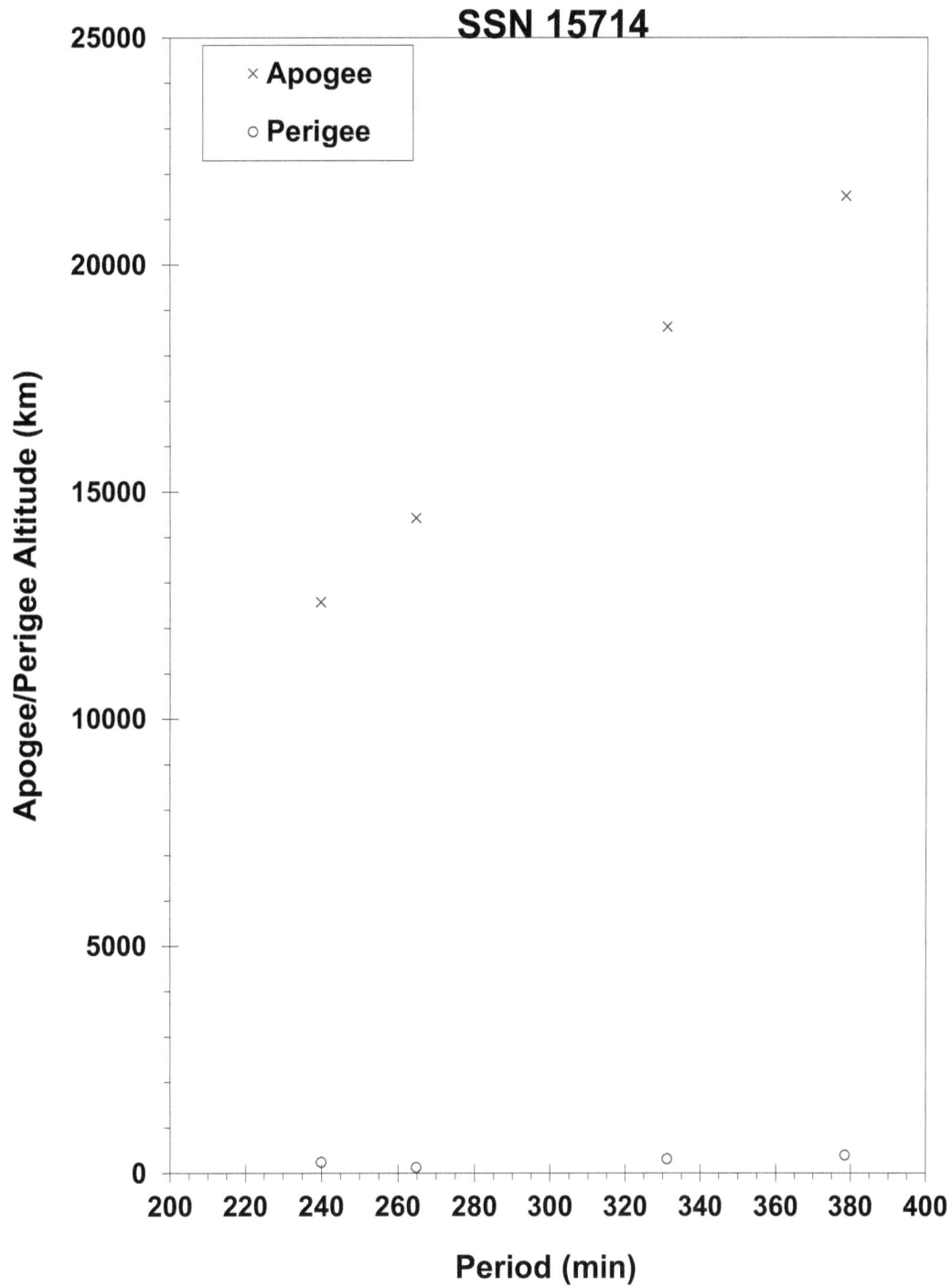

SSN 15714

Cosmos 1650-1652 ullage motor debris cloud of 4 fragments within 1 day of the event as reconstructed from the US SSN database.

SATELLITE DATA

TYPE:	Payload
OWNER:	CIS
LAUNCH DATE:	23.53 May 1985
DRY MASS (KG):	5700
MAIN BODY:	Cone-cylinder; 2.7 m diameter by 6.3 m length
MAJOR APPENDAGES:	Solar panels
ATTITUDE CONTROL:	Active, 3-axis
ENERGY SOURCES:	On-board propellants, explosive charge

EVENT DATA

DATE:	21 Jun 1985	LOCATION:	8N, 292E (asc)
TIME:	1047 GMT	ASSESSED CAUSE:	Deliberate
ALTITUDE:	200 km		

PRE-EVENT ELEMENTS

EPOCH:	85172.01363851	MEAN ANOMALY:	313.0734
RIGHT ASCENSION:	1.2391	MEAN MOTION:	16.11890623
INCLINATION:	64.8566	MEAN MOTION DOT/2:	.00311214
ECCENTRICITY:	.0086971	MEAN MOTION DOT DOT/6:	.000034493
ARG. OF PERIGEE:	47.8764	BSTAR:	.00015520

DEBRIS CLOUD DATA

MAXIMUM ΔP:	22.1 min*
MAXIMUM ΔI:	1.5 deg*

*Based on uncataloged debris data

COMMENTS

Spacecraft was destroyed after a malfunction prevented controlled reentry and landing in the Soviet Union. Most debris reentered before being officially cataloged.

REFERENCE DOCUMENTS

Analysis of the Fragmentation of Kosmos 1654, G.T. DeVere, Technical Report CS86-BMDSC-0003, Teledyne Brown Engineering, Colorado Springs, October 1985.

History of Soviet/Russian Satellite Fragmentations-A Joint U.S.-Russian Investigation, N. L. Johnson, *et al.,* Kaman Sciences Corporation, October 1995.

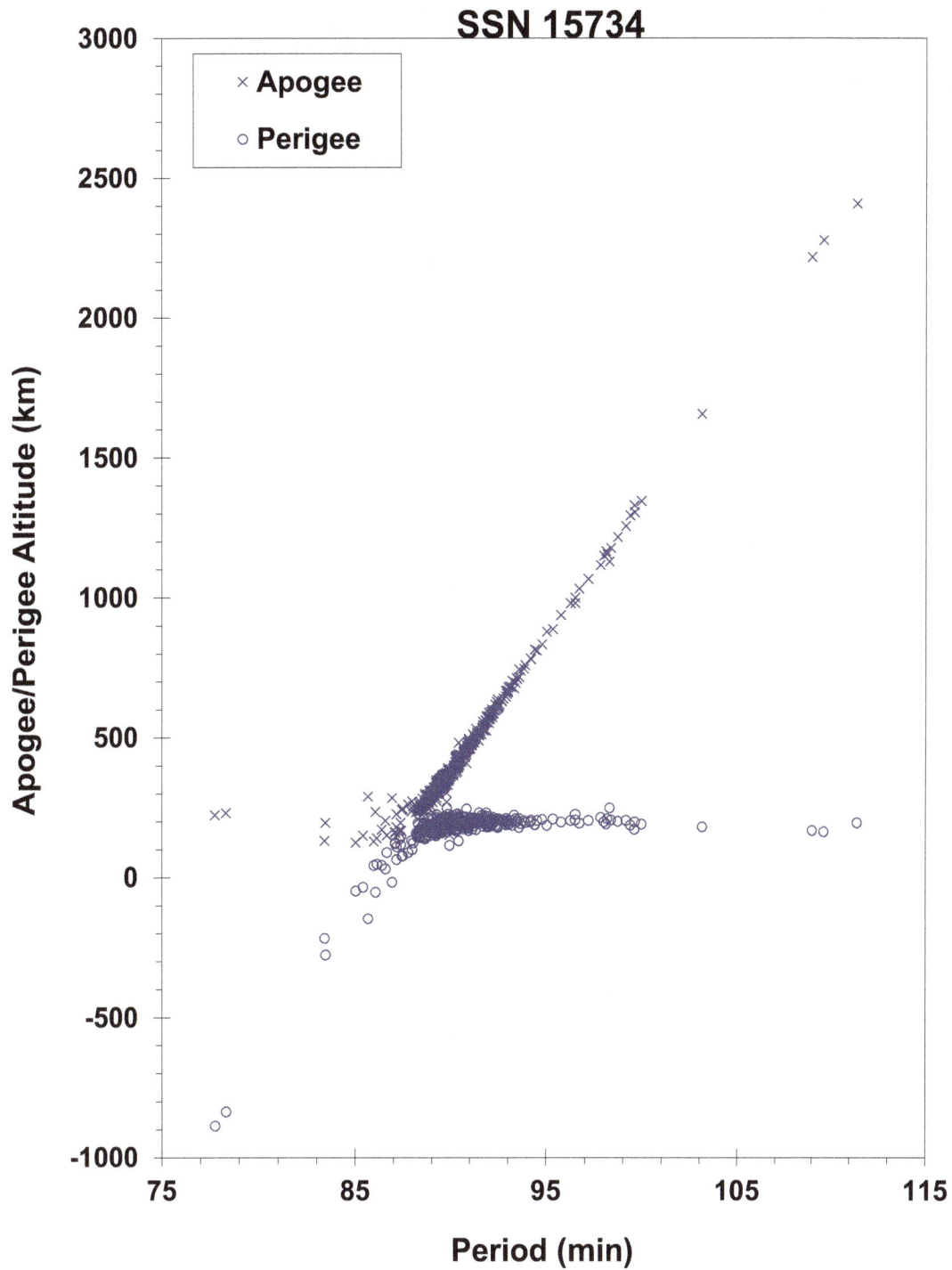

SSN 15734

Cosmos 1654 debris cloud remnant of 543 fragments seen 9 hours after the event
by the US SSN PARCS radar.

SATELLITE DATA

TYPE:	Mission Related Debris
OWNER:	CIS
LAUNCH DATE:	30.62 May 1985
DRY MASS (KG):	55
MAIN BODY:	Ellipsoid; 0.6 m diameter by 1 m length
MAJOR APPENDAGES:	None
ATTITUDE CONTROL:	None
ENERGY SOURCES:	On-board propellants

EVENT DATA

DATE:	5 Jan 1988	LOCATION:	66N, 151E (asc)
TIME:	0147 GMT	ASSESSED CAUSE:	Propulsion
ALTITUDE:	860 km		

PRE-EVENT ELEMENTS

EPOCH:	88002.58690356	MEAN ANOMALY:	91.9605
RIGHT ASCENSION:	205.7335	MEAN MOTION:	14.17143400
INCLINATION:	66.5867	MEAN MOTION DOT/2:	.00000144
ECCENTRICITY:	.0034143	MEAN MOTION DOT DOT/6:	.0
ARG. OF PERIGEE:	267.7562	BSTAR:	.000088961

DEBRIS CLOUD DATA

MAXIMUM ΔP:	3.3 min
MAXIMUM ΔI:	0.0 deg

COMMENTS

Parent satellite was one of two small engine units that are routinely released after the first burn of the Proton fourth stage. The nature of these objects was identified by Dr. Boris V. Chernlatiev, Deputy Constructor for the Energiya NPO, in October 1992. The cause of this fragmentation is assumed to be related to the residual hypergolic propellants on board and failure of the membrane separating the fuel and oxidizer. NAVSPASUR observed two additional, uncataloged fragments associated with this event. This was the second in a series of fragmentations of this object type.

REFERENCE DOCUMENTS

The Fragmentation of Proton Debris, D. J. Nauer, TBE Technical Report CS93-LKD-004, Teledyne Brown Engineering, Colorado Springs, 31 December 1992.

History of Soviet/Russian Satellite Fragmentations-A Joint U.S.-Russian Investigation, N. L. Johnson, *et al.,* Kaman Sciences Corporation, October 1995.

"Identification and Resolution of an Orbital Debris Problem with the Proton Launch Vehicle," B. V. Cherniatiev, *et al.,* Proceedings of the First European Conference on Space Debris, April 1993.

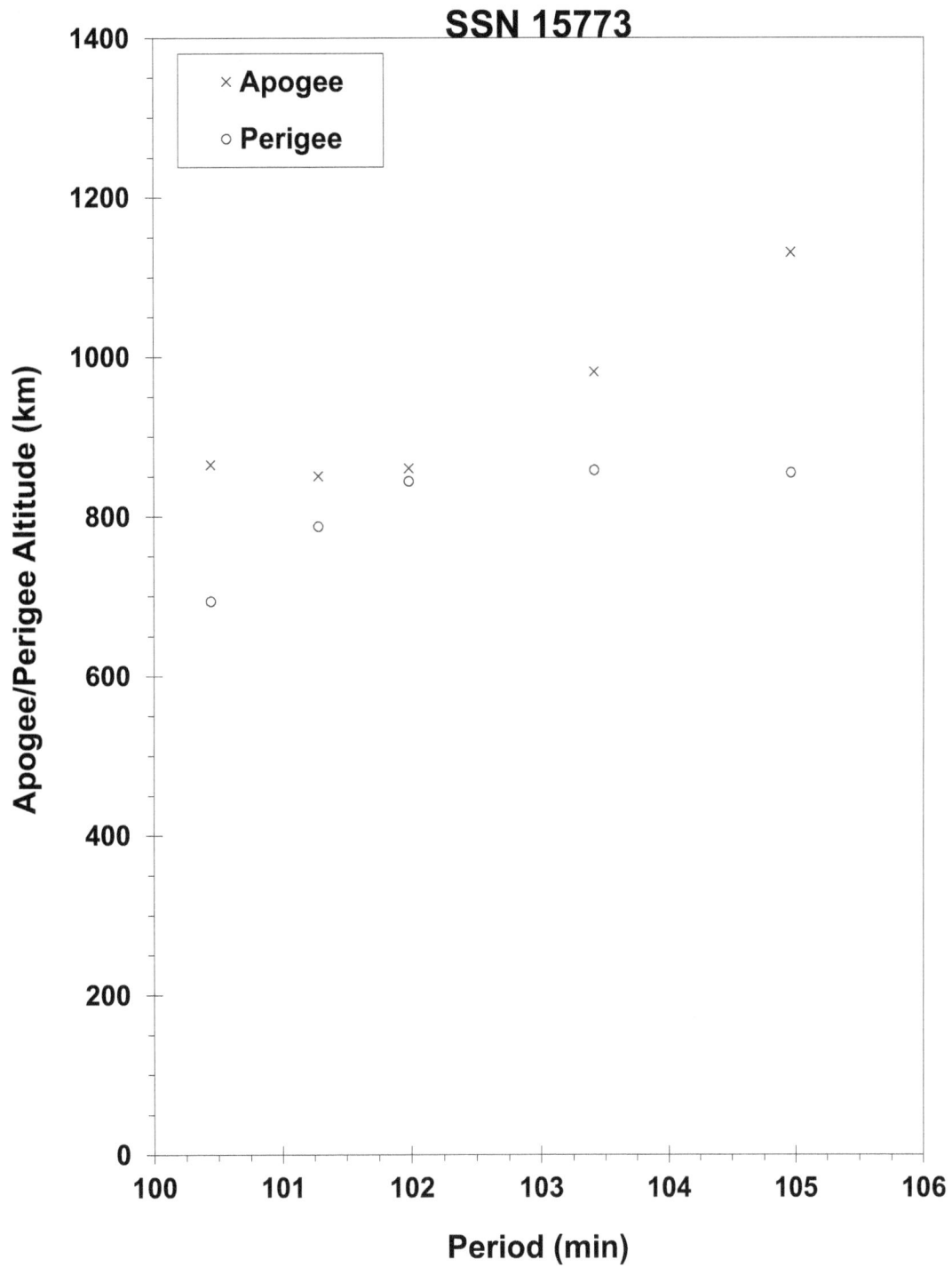

Fragments from Cosmos 1656 debris as determined 2 weeks after the event. Elements from the US SSN database as published by NASA Goddard Space Flight Center.

COSMOS 1682   **1985-082A**   **16054**

SATELLITE DATA

TYPE:	Payload
OWNER:	CIS
LAUNCH DATE:	19.07 Sep 1985
DRY MASS (KG):	3000
MAIN BODY:	Cylinder; 1.3 m diameter by 17 m length
MAJOR APPENDAGES:	Solar panels
ATTITUDE CONTROL:	Active, 3-axis
ENERGY SOURCES:	On-board propellants, explosive charge (?)

EVENT DATA

DATE:	18 Dec 1986	LOCATION:	22S, 292 E (asc)
TIME:	2017 GMT	ASSESSED CAUSE:	Unknown
ALTITUDE:	415 km		

PRE-EVENT ELEMENTS

EPOCH:	86351. 87879723	MEAN ANOMALY:	315.5258
RIGHT ASCENSION:	337.4852	MEAN MOTION:	15.45249396
INCLINATION:	65.0089	MEAN MOTION DOT/2:	.00011076
ECCENTRICITY:	.0068048	MEAN MOTION DOT DOT/6:	.0
ARG. OF PERIGEE:	45.1423	BSTAR:	.00021714

DEBRIS CLOUD DATA

MAXIMUM ΔP:	2.3 min*
MAXIMUM ΔI:	0.7 deg*

*Based on uncataloged debris data

COMMENTS

Cosmos 1682 was the 14th spacecraft of the Cosmos 699-type to experience a fragmentation. Spacecraft had been in natural decay for 2 months prior to the event. Many debris reentered before being officially cataloged.

REFERENCE DOCUMENTS

"Artificial Satellite Break-Ups (Part 1): Soviet Ocean Surveillance Satellites", N. L. Johnson, Journal of the British Interplanetary Society, February 1983, p. 51-58.

History of Soviet/Russian Satellite Fragmentations-A Joint U.S.-Russian Investigation, N. L. Johnson, et al., Kaman Sciences Corporation, October 1995.

266

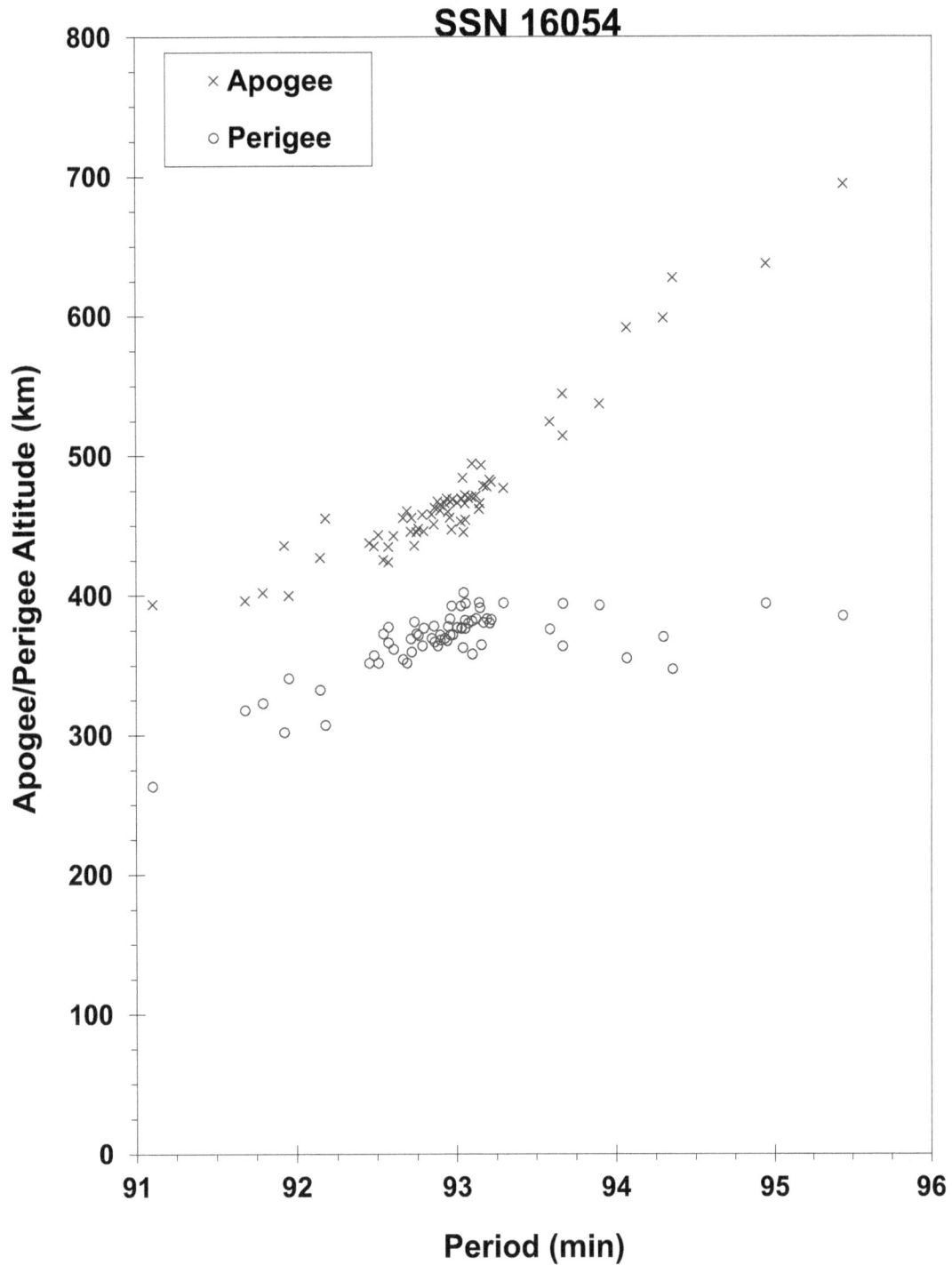

SSN 16054

Cosmos 1682 debris cloud remnant of 66 fragments about 1 week after the event as reconstructed from the US SSN database.

SATELLITE DATA

TYPE:	Payload
OWNER:	CIS
LAUNCH DATE:	9.90 Oct 1985
DRY MASS (KG):	220
MAIN BODY:	Cylinder; 1.0 m diameter by 1.5 m length
MAJOR APPENDAGES:	Gravity gradient boom
ATTITUDE CONTROL:	Gravity gradient
ENERGY SOURCES:	Battery

EVENT DATA

DATE:	22 Nov 1985	LOCATION:	31N, 326E (dsc)
TIME:	0840 GMT	ASSESSED CAUSE:	Battery
ALTITUDE:	1415 km		

PRE-EVENT ELEMENTS

EPOCH:	85320.62059878	MEAN ANOMALY:	91.0897
RIGHT ASCENSION:	345.1807	MEAN MOTION:	12.62038878
INCLINATION:	82.6124	MEAN MOTION DOT/2:	.00000022
ECCENTRICITY:	.0002812	MEAN MOTION DOT DOT/6:	.0
ARG. OF PERIGEE:	268.9870	BSTAR:	.000099999

DEBRIS CLOUD DATA

MAXIMUM ΔP:	1.0 min
MAXIMUM ΔI:	0.1 deg

COMMENTS

Cosmos 1691 was one of six independent payloads on this launch, which was only the second in this program. Cosmos 1691 was the last payload deployed and may be referred to as Cosmos 1695 in the former Soviet Union. One fragment was administratively decayed in February 1989. No other payloads in this program have fragmented. This event is assessed to be the second known NiH_2 battery failure as indicated by Dr. K. M. Suitnshev during the early 1992 Space Debris Conference in Moscow. See also reference below.

REFERENCE DOCUMENT

History of Soviet/Russian Satellite Fragmentations-A Joint U.S.-Russian Investigation, N. L. Johnson, *et al.*, Kaman Sciences Corporation, October 1995.

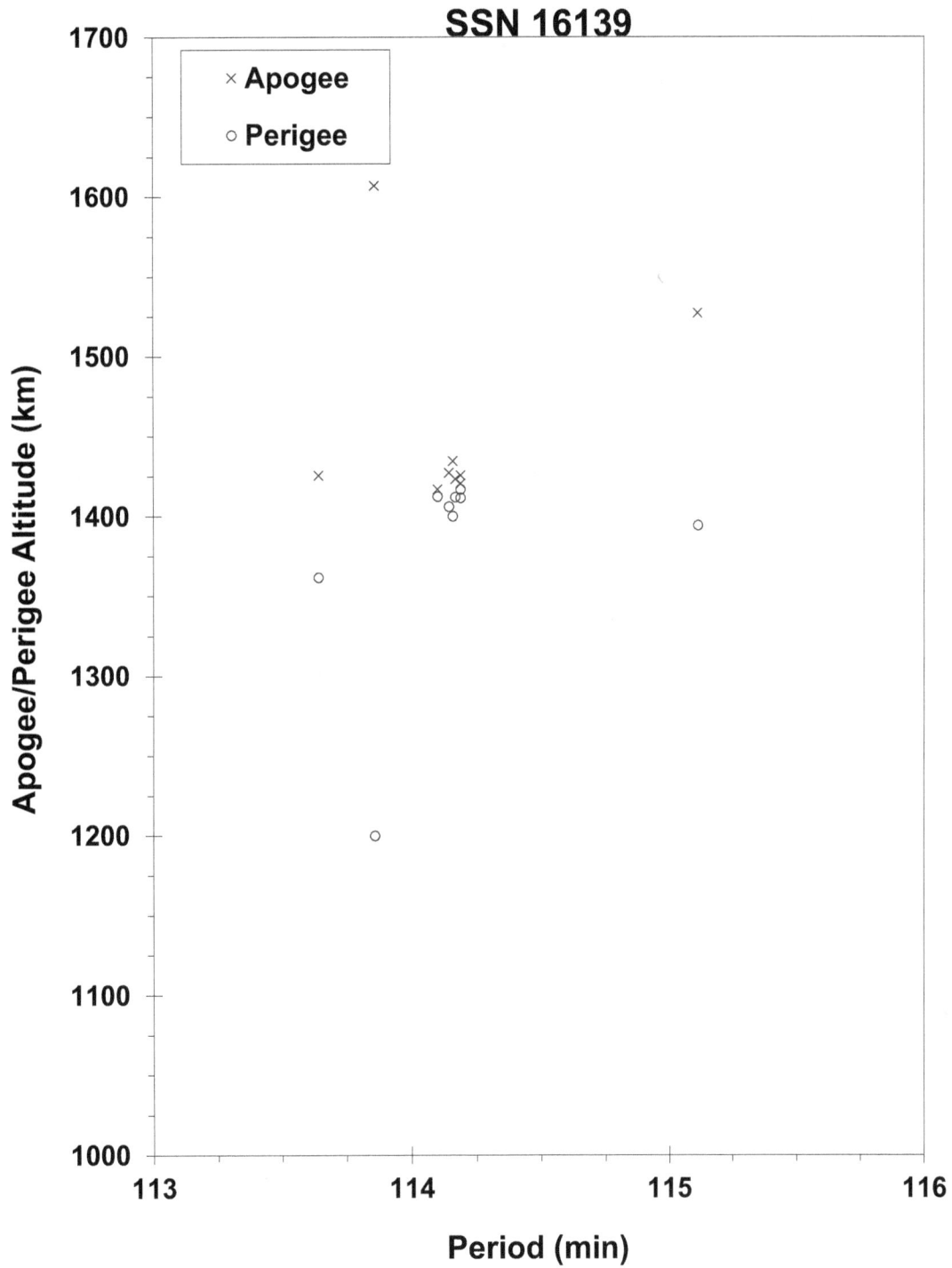

SSN 16139

Cosmos 1691 debris cloud of 9 fragments 2 days after the event as reconstructed from Naval Space Surveillance System database.

COSMOS 1703 R/B   **1985-108B**   **16263**

SATELLITE DATA

TYPE:	Rocket Body
OWNER:	CIS
LAUNCH DATE:	22.93 Nov 1985
DRY MASS (KG):	1360
MAIN BODY:	Cone-cylinder; 2.1 m diameter by 3.3 m length
MAJOR APPENDAGES:	None
ATTITUDE CONTROL:	None
ENERGY SOURCES:	Unknown

EVENT DATA

DATE:	4 May 2006	LOCATION:	67N, 17E (dsc)
TIME:	1604 GMT	ASSESSED CAUSE:	Propulsion
ALTITUDE:	635 km		

PRE-EVENT ELEMENTS

EPOCH:	06123.63561455	MEAN ANOMALY:	329.9835
RIGHT ASCENSION:	319.0019	MEAN MOTION:	14.8137473
INCLINATION:	82.5005	MEAN MOTION DOT/2:	0.00000107
ECCENTRICITY:	.0021969	MEAN MOTION DOT DOT/6:	0.0
ARG. OF PERIGEE:	30.2640	BSTAR:	0.000010657

DEBRIS CLOUD DATA

MAXIMUM ΔP:	1.6 min
MAXIMUM ΔI:	0.2 deg

COMMENTS

This is the 5th event of the Tsyklon third stage (SL-14) identified to date.

REFERENCE DOCUMENT

"First Satellite Breakups of 2006", The Orbital Debris Quarterly News, NASA JSC, July 2006.
Available online at https://orbitaldebris.jsc.nasa.gov/quarterly-news/pdfs/ODQNv10i3.pdf.

SSN 16263

Tsyklon third stage debris cloud of 44 fragments six days after the event as reconstructed from the US SSN database.

SATELLITE DATA

TYPE:	Zenit Second Stage
OWNER:	CIS
LAUNCH DATE:	28.40 Dec 1985
DRY MASS (KG):	9000
MAIN BODY:	Cylinder; 3.9 m diameter by 12 m length
MAJOR APPENDAGES:	None
ATTITUDE CONTROL:	Active, 3-axis
ENERGY SOURCES:	On-board propellants

EVENT DATA

DATE:	28 Dec 1985	LOCATION:	Unknown
TIME:	Unknown	ASSESSED CAUSE:	Propulsion
ALTITUDE:	Unknown		

PRE-EVENT ELEMENTS

EPOCH:	85363.19328410	MEAN ANOMALY:	84.6199
RIGHT ASCENSION:	281.3886	MEAN MOTION:	14.77971051
INCLINATION:	71.0178	MEAN MOTION DOT/2:	0.00065991
ECCENTRICITY:	0.0306365	MEAN MOTION DOT DOT/6:	0
ARG. OF PERIGEE:	271.9949	BSTAR:	0.0041108

DEBRIS CLOUD DATA

MAXIMUM ΔP:	Unknown
MAXIMUM ΔI:	Unknown

COMMENTS

The Zenit second stage low thrust engine used to perform final orbit insertion exploded. Four pieces of debris cataloged with this mission are probably not associated with the breakup.

REFERENCE DOCUMENT

History of Soviet/Russian Satellite Fragmentations-A Joint U.S.-Russian Investigation, N. L. Johnson, *et al.,* Kaman Sciences Corporation, October 1995.

Insufficient data to construct a Gabbard diagram.

SATELLITE DATA

TYPE:	Mission Related Debris
OWNER:	CIS
LAUNCH DATE:	24.91 Dec 1985
DRY MASS (KG):	55
MAIN BODY:	Ellipsoid; 0.6 m diameter by 1 m length
MAJOR APPENDAGES:	None
ATTITUDE CONTROL:	None
ENERGY SOURCES:	On-board propellants

EVENT DATA

DATE:	29 Dec 1991	LOCATION:	25.3N, 331.9E
TIME:	0903 GMT	ASSESSED CAUSE:	Propulsion
ALTITUDE:	4730 km		

PRE-EVENT ELEMENTS

EPOCH:	91333.40579226	MEAN ANOMALY:	46.8976
RIGHT ASCENSION:	48.0333	MEAN MOTION:	4.23089679
INCLINATION:	65.2547	MEAN MOTION DOT/2:	.00000167
ECCENTRICITY:	.5645362	MEAN MOTION DOT DOT/6:	.0
ARG. OF PERIGEE:	245.7447	BSTAR:	.0012603

DEBRIS CLOUD DATA

MAXIMUM ΔP:	5.7 min *
MAXIMUM ΔI:	0.8 deg *

                                   * based upon uncataloged debris data

COMMENTS

Parent satellite was one of two small engine units that are routinely released after the first burn of the Proton fourth stage. The nature of these objects was identified by Dr. Boris V. Chernlatiev, Deputy Constructor for the Energiya NPO, in October 1992. The cause of this fragmentation is assumed to be related to the residual hypergolic propellants on board and failure of the membrane separating the fuel and oxidizer. There were 26 objects associated with this event on 30 December per a telecon with NAVSPASUR (Edna Jenkins). Only two analyst satellites were generated, and insufficient data was available for a Gabbard diagram. This was the fourth in a series of fragmentations of this object type.

REFERENCE DOCUMENTS

The Fragmentation of Proton Debris, D. J. Nauer, TBE Technical Report CS93-LKD-004, Teledyne Brown Engineering, Colorado Springs, 31 December 1992.

History of Soviet/Russian Satellite Fragmentations-A Joint U.S.-Russian Investigation, N. L. Johnson, et al., Kaman Sciences Corporation, October 1995.

"Identification and Resolution of an Orbital Debris Problem with the Proton Launch Vehicle," B. V. Cherniatiev, et al., Proceedings of the First European Conference on Space Debris, April 1993.

Insufficient data to construct a Gabbard diagram.

SATELLITE DATA

TYPE:	Ariane 1 Third Stage
OWNER:	France
LAUNCH DATE:	22.07 Feb 1986
DRY MASS (KG):	1400
MAIN BODY:	Cylinder; 2.6 m diameter by 10.3 m length
MAJOR APPENDAGES:	None
ATTITUDE CONTROL:	None at time of the event.
ENERGY SOURCES:	On-board propellants, range safety package

EVENT DATA

DATE:	13 Nov 1986	LOCATION:	7N, 42E (asc)
TIME:	1940 GMT	ASSESSED CAUSE:	Propulsion
ALTITUDE:	805 km		

PRE-EVENT ELEMENTS

EPOCH:	86305.08337689	MEAN ANOMALY:	300.1947
RIGHT ASCENSION:	18.0087	MEAN MOTION:	14.22163662
INCLINATION:	98.6973	MEAN MOTION DOT/2:	.00000203
ECCENTRICITY:	.0021203	MEAN MOTION DOT DOT/6:	.0
ARG. OF PERIGEE:	60.1312	BSTAR:	.000099999

DEBRIS CLOUD DATA

MAXIMUM ΔP:	6.2 min
MAXIMUM ΔI:	1.2 deg

COMMENTS

Event occurred approximately 9 months after the rocket body had successfully deployed the SPOT 1 and Viking payloads. First use of Ariane launch vehicle for low Earth orbit. May be related to other Ariane fragmentations.

REFERENCE DOCUMENTS

A Preliminary Analysis of the Fragmentation of the Spot 1 Ariane Third Stage, N. L. Johnson, Technical Report CS87-LKD-003, Teledyne Brown Engineering, Colorado Springs, March 1987.

Orbital Debris from Upper Stage Breakup, J.P. Loftus, Jr., ed., Vol. 121, Progress in Astronautics and Aeronautics, AIAA, 1989.

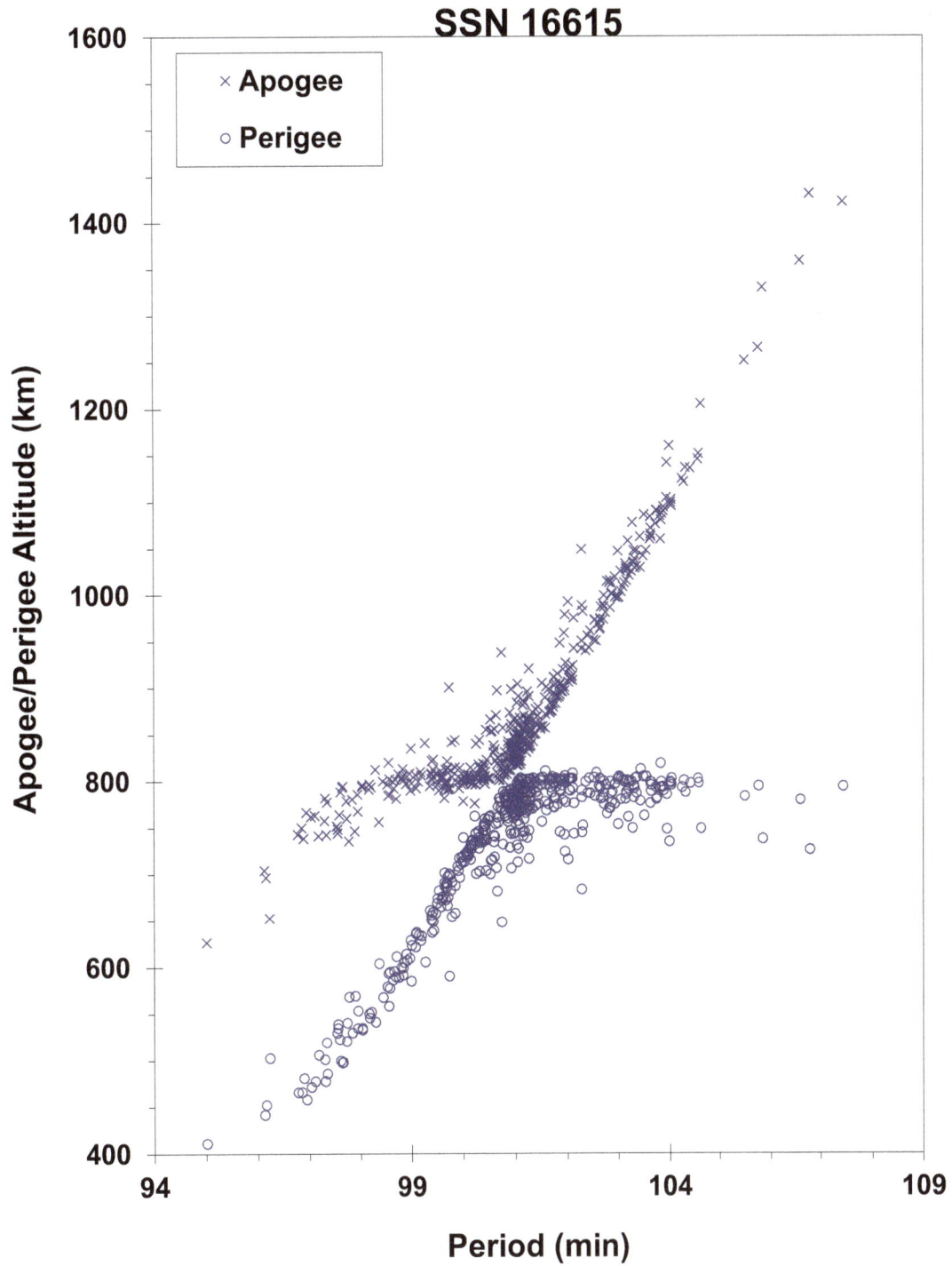

SSN 16615

Spot 1 R/B debris cloud of 463 fragments 3 months after the event as reconstructed from the US SSN database.

SATELLITE DATA

TYPE:	Payload
OWNER:	CIS
LAUNCH DATE:	4.21 Aug 1986
DRY MASS (KG):	3000
MAIN BODY:	Cylinder; 1.3 m diameter by 17 m length
MAJOR APPENDAGES:	Solar panels
ATTITUDE CONTROL:	Active, 3-axis
ENERGY SOURCES:	On-board propellants, explosive charge (?)

EVENT DATA

DATE:	21 Sep 1987	LOCATION:	60S, 174E (dsc)
TIME:	1205 GMT	ASSESSED CAUSE:	Unknown
ALTITUDE:	320 km		

PRE-EVENT ELEMENTS

EPOCH:	87263.81808697	MEAN ANOMALY:	70.4851
RIGHT ASCENSION:	122.5376	MEAN MOTION:	15.63167584
INCLINATION:	65.0147	MEAN MOTION DOT/2:	.00078200
ECCENTRICITY:	.0099296	MEAN MOTION DOT DOT/6:	.0
ARG. OF PERIGEE:	288.4915	BSTAR:	.00065556

DEBRIS CLOUD DATA

MAXIMUM ΔP:	1.9 min*
MAXIMUM ΔI:	0.0 deg*

*Based on uncataloged debris data

COMMENTS

Cosmos 1769 was the fifteenth spacecraft of the Cosmos 699-type to experience a fragmentation. Spacecraft was regularly maneuvered until 17 Sep 1987 when the vehicle began to decay naturally. Most debris reentered before being officially cataloged.

REFERENCE DOCUMENTS

"Artificial Satellite Break-Ups (Part 1): Soviet Ocean Surveillance Satellites", N. L. Johnson, Journal of the British Interplanetary Society, February 1983, pp. 51-58.

History of Soviet/Russian Satellite Fragmentations-A Joint U.S.-Russian Investigation, N. L. Johnson, et al., Kaman Sciences Corporation, October 1995.

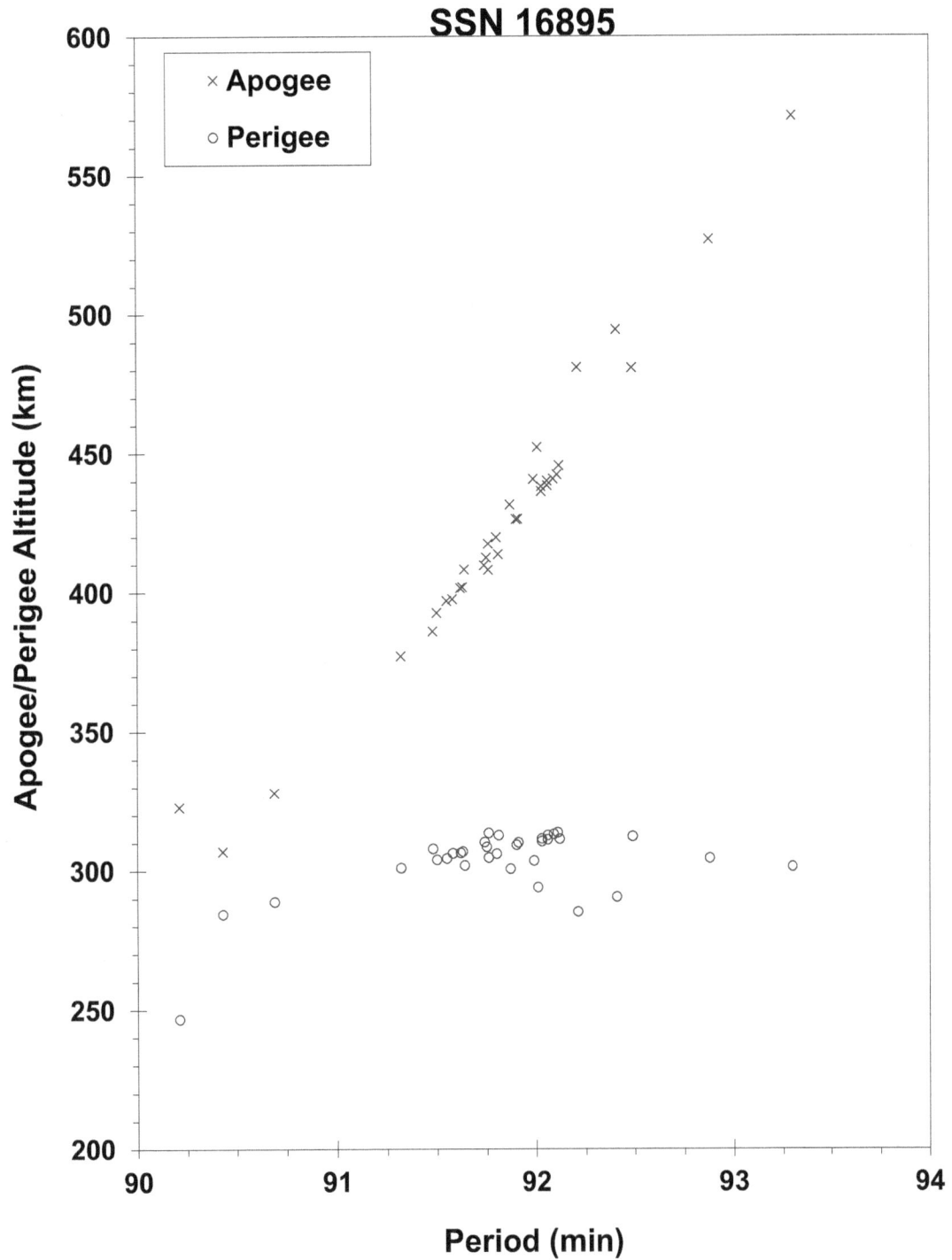

SSN 16895

Cosmos 1769 debris cloud remnant of 34 fragments 3 days after the event as reconstructed from Naval Space Surveillance System database.

SATELLITE DATA

TYPE:	Payload
OWNER:	US
LAUNCH DATE:	5.63 Sep 1986
DRY MASS (KG):	930
MAIN BODY:	Cylinder-cone; 1.2 m diameter by 4.6 m length
MAJOR APPENDAGES:	None
ATTITUDE CONTROL:	Active, 3-axis
ENERGY SOURCES:	On-board propellants, explosive charge (?)

EVENT DATA

DATE:	5 Sep 1986	LOCATION:	15N, 166E (asc)
TIME:	1752 GMT	ASSESSED CAUSE:	Deliberate
ALTITUDE:	220 km		

POST-EVENT ELEMENTS

EPOCH:	86250.63774662	MEAN ANOMALY:	335.3264
RIGHT ASCENSION:	28.1524	MEAN MOTION:	15.28976390
INCLINATION:	39.0665	MEAN MOTION DOT/2:	.01159823
ECCENTRICITY:	.0390567	MEAN MOTION DOT DOT/6:	.0000050922
ARG. OF PERIGEE:	26.7075	BSTAR:	.0028192

DEBRIS CLOUD DATA

MAXIMUM ΔP:	424.1 min*
MAXIMUM ΔI:	4.4 deg*

*Based on uncataloged debris data

COMMENTS

USA 19 deliberately collided with USA 19 R/B at high relative velocity. Both satellites were thrusting at the time of impact. Element set above is post-event and is best estimate of orbit at time of the event. Most debris reentered before being officially cataloged.

REFERENCE DOCUMENTS

The Collision of Satellites 16937 and 16938: A Preliminary Report, N. L. Johnson, Technical Report CS87-LKD-002, Teledyne Brown Engineering, Colorado Springs, 3 December 1986.

The Collision of Satellites 16937 and 16938: Debris Characterization, R. L. Kling, Technical Report CS87-LKD-005, Teledyne Brown Engineering, Colorado Springs, 15 May 1987.

Hazard Analysis of the Breakup of Satellites 16937 and 16938, Technical Report JSC 22471(U), NASA Lyndon B. Johnson Space Center, Houston, 27 February 1987.

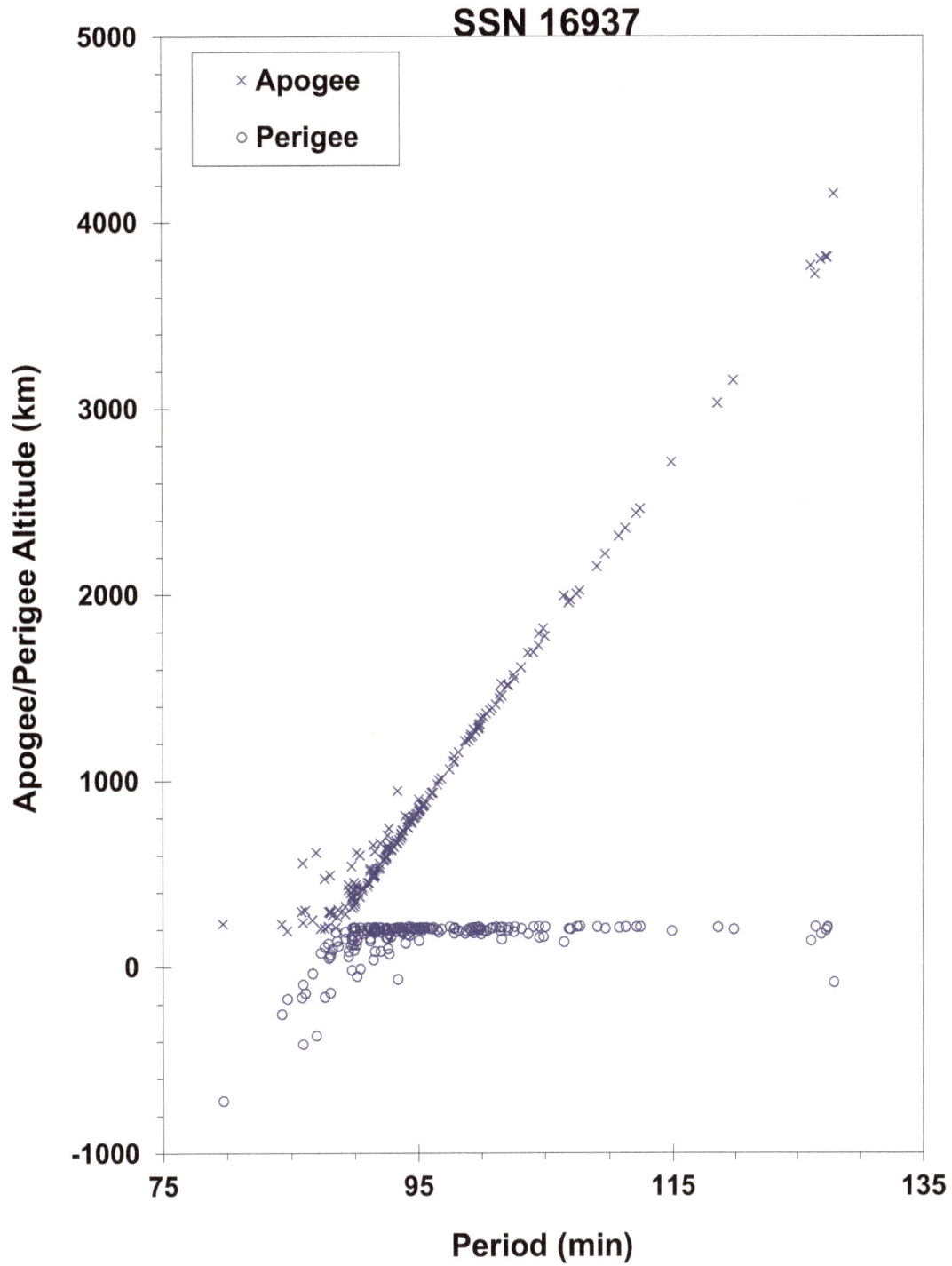

SSN 16937

USA 19 debris cloud remnant of 187 fragments 1 day after the event as seen by the US SSN radar FPS-85 at Eglin AFB, Florida.

SATELLITE DATA

TYPE:	Delta Second Stage (3920) with auxiliary payload
OWNER:	US
LAUNCH DATE:	5.63 Sep 1986
DRY MASS (KG):	1455
MAIN BODY:	Cylinder-nozzle; 1.4 m diameter by 4.8 m length
MAJOR APPENDAGES:	Mini-skirt; 2.4 m by 0.3 m
ATTITUDE CONTROL:	Active
ENERGY SOURCES:	On-board propellants

EVENT DATA

DATE:	5 Sep 1986	LOCATION:	15N, 166E (asc)
TIME:	1752 GMT	ASSESSED CAUSE:	Deliberate
ALTITUDE:	220 km		

POST-EVENT ELEMENTS

EPOCH:	86249.96053354	MEAN ANOMALY:	307.9381
RIGHT ASCENSION:	10.4654	MEAN MOTION:	15.50608380
INCLINATION:	22.7830	MEAN MOTION DOT/2:	.00138611
ECCENTRICITY:	.0288474	MEAN MOTION DOT DOT/6:	.0
ARG. OF PERIGEE:	54.7772	BSTAR:	.00033298

DEBRIS CLOUD DATA

MAXIMUM ΔP:	53.6 min*
MAXIMUM ΔI:	2.5 deg*

*Based on uncataloged debris data

COMMENTS

USA 19 R/B was deliberately struck by USA 19 at high relative velocity. Both satellites were thrusting at the time of impact. Element set above is post-event and is best estimate of orbit at time of the event. Most debris reentered before being officially cataloged.

REFERENCE DOCUMENTS

The Collision of Satellites 16937 and 16938: A Preliminary Report, N. L. Johnson, Technical Report CS87-LKD-002, Teledyne Brown Engineering, Colorado Springs, 3 December 1986.

The Collision of Satellites 16937 and 16938: Debris Characterization, R. L. Kling, Technical Report CS87-LKD-005, Teledyne Brown Engineering, Colorado Springs, 15 May 1987.

Hazard Analysis of the Breakup of Satellites 16937 and 16938, Technical Report JSC 22471(U), NASA Lyndon B. Johnson Space Center, Houston, 27 February 1987.

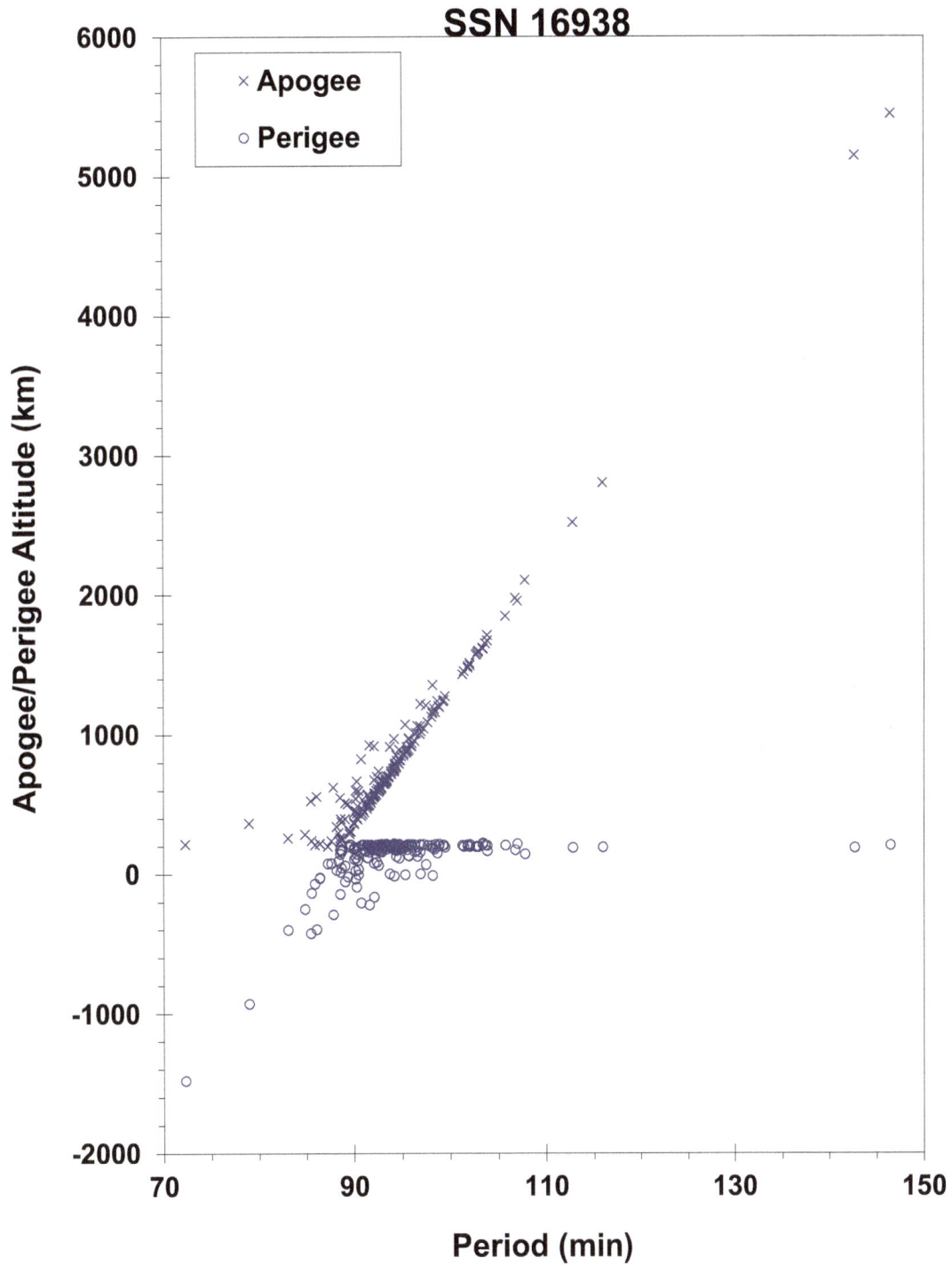

SSN 16938

USA 19 R/B debris cloud of 190 fragments 1 day after the event as seen
by the US SSN FPS-85 radar at Eglin AFB, Florida.

SATELLITE DATA

TYPE:	Payload
OWNER:	CIS
LAUNCH DATE:	15.47 Jan 1987
DRY MASS (KG):	6300
MAIN BODY:	Sphere-cylinder; 2.4 m diameter by 6.5 m length
MAJOR APPENDAGES:	None
ATTITUDE CONTROL:	Active, 3-axis
ENERGY SOURCES:	On-board propellants, explosive charge

EVENT DATA

DATE:	29 Jan 1987	LOCATION:	73N, 122E (asc)
TIME:	0555 GMT	ASSESSED CAUSE:	Deliberate
ALTITUDE:	390 km		

PRE-EVENT ELEMENTS

EPOCH:	87028.91020168	MEAN ANOMALY:	178.1696
RIGHT ASCENSION:	256.7724	MEAN MOTION:	15.60427146
INCLINATION:	72.8163	MEAN MOTION DOT/2:	.00008569
ECCENTRICITY:	.0043147	MEAN MOTION DOT DOT/6:	.0
ARG. OF PERIGEE:	182.0100	BSTAR:	.000099999

DEBRIS CLOUD DATA

MAXIMUM ΔP:	9.1 min*
MAXIMUM ΔI:	0.1 deg*

*Based on PARCS observations

COMMENTS

Spacecraft apparently destroyed after a malfunction prevented controlled reentry and landing in the Soviet Union. A total of 846 separate fragments were observed during one pass over a U.S. Space Surveillance Network radar (PARCS) 2 days after the event.

REFERENCE DOCUMENTS

The Fragmentation of Kosmos 1813, R. L. Kling and J. S. Dowdy, Technical Report CS87-LKD-004, Teledyne Brown Engineering, Colorado Springs, 8 May 1987.

History of Soviet/Russian Satellite Fragmentations-A Joint U.S.-Russian Investigation, N. L. Johnson, et al., Kaman Sciences Corporation, October 1995.

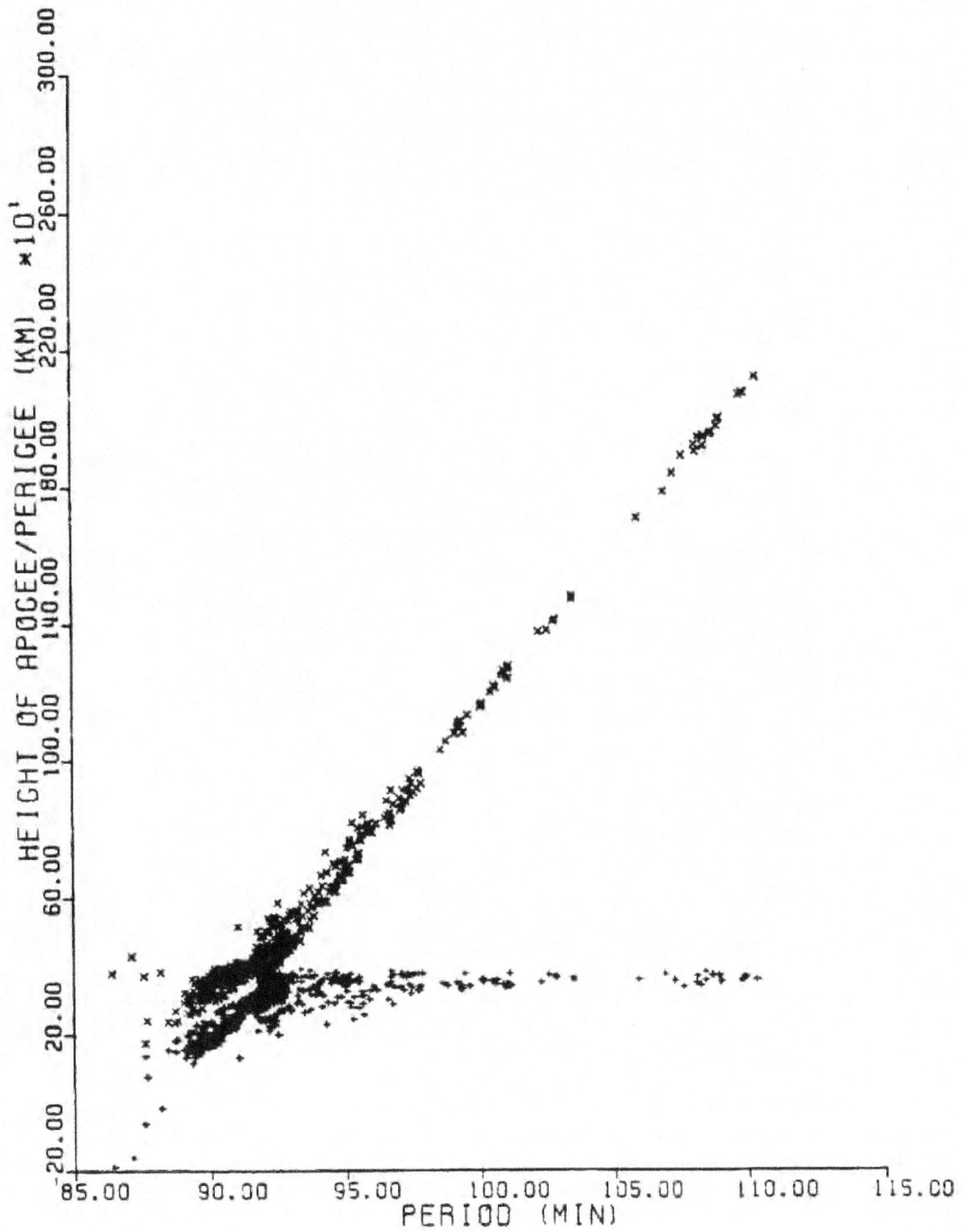

Cosmos 1813 debris cloud as reconstructed from PARCS radar observations taken about 10 hours after the breakup. A total of 846 fragments were identified with Cosmos 1813. This diagram is taken from the cited reference document.

COSMOS 1823   **1987-020A**   **17535**

SATELLITE DATA

TYPE:	Payload
OWNER:	CIS
LAUNCH DATE:	20.20 Feb 1987
DRY MASS (KG):	1500
MAIN BODY:	Cylinder; 2.4 m diameter by 4 m length
MAJOR APPENDAGES:	Gravity-gradient boom; 10 small solar panels
ATTITUDE CONTROL:	Gravity gradient
ENERGY SOURCES:	Battery

EVENT DATA

DATE:	17 Dec 1987	LOCATION:	15S, 18E (dsc)
TIME:	1739 GMT	ASSESSED CAUSE:	Battery
ALTITUDE:	1485 km		

PRE-EVENT ELEMENTS

EPOCH:	87351.61079422	MEAN ANOMALY:	147.6712
RIGHT ASCENSION:	184.5746	MEAN MOTION:	12.40947361
INCLINATION:	73.6064	MEAN MOTION DOT/2:	.0
ECCENTRICITY:	.0028819	MEAN MOTION DOT DOT/6:	.0
ARG. OF PERIGEE:	212.2988	BSTAR:	.0

DEBRIS CLOUD DATA

MAXIMUM ΔP:	4.9 min
MAXIMUM ΔI:	1.4 deg

COMMENTS

Cosmos 1823 has been acknowledged by the Soviet Union as a geodetic spacecraft, the eighth in a series that debuted in 1981. The spacecraft is known to have been operating 3 months before the event. USSR acknowledged mission termination as of 19 December 1987. Unusually strong radial velocity components are evident in cloud analyses over a period of many months. This event has been confirmed to be the third known failure of the NiH_2 battery as reported by Dr. K. M. Suitashev at the February 1992 Space Debris Conference held in Moscow.

REFERENCE DOCUMENT

History of Soviet/Russian Satellite Fragmentations-A Joint U.S.-Russian Investigation, N. L. Johnson, *et al.,* Kaman Sciences Corporation, October 1995.

SSN 17535

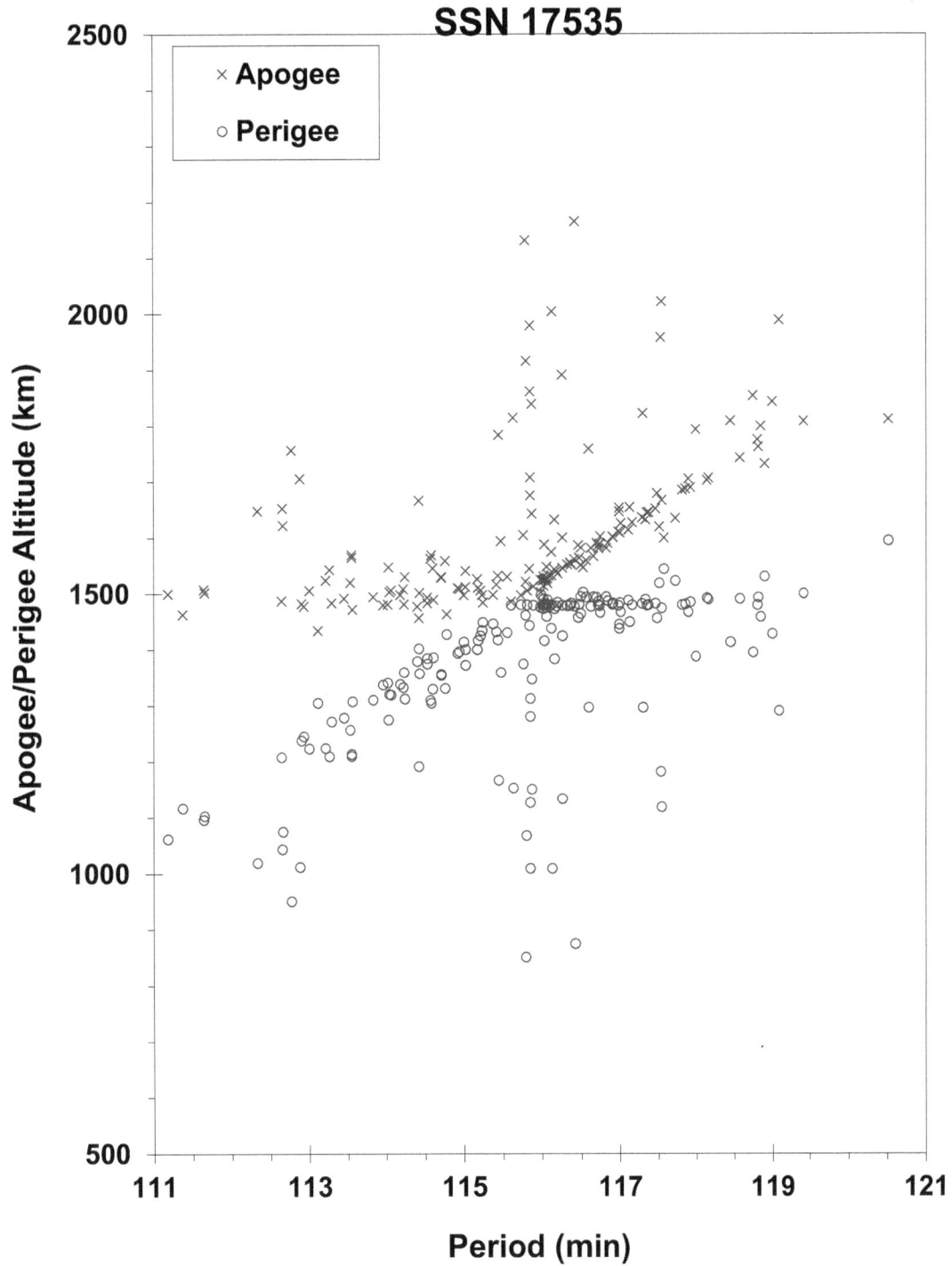

Cosmos 1823 debris cloud of 165 fragments 2 weeks after the event as reconstructed from Naval Space Surveillance System database.

SATELLITE DATA

TYPE:	Payload
OWNER:	CIS
LAUNCH DATE:	9.67 Jul 1987
DRY MASS (KG):	5700
MAIN BODY:	Cone-cylinder; 2.7 m diameter by 6.3 m length
MAJOR APPENDAGES:	Solar panels
ATTITUDE CONTROL:	Active, 3-axis
ENERGY SOURCES:	On-board propellants, explosive charge

EVENT DATA

DATE:	26 Jul 1987	LOCATION:	57S, 239E (asc)
TIME:	1539 GMT	ASSESSED CAUSE:	Deliberate
ALTITUDE:	245 km		

PRE-EVENT ELEMENTS

EPOCH:	87207.60199851	MEAN ANOMALY:	300.9577
RIGHT ASCENSION:	98.7735	MEAN MOTION:	16.25421506
INCLINATION:	67.1494	MEAN MOTION DOT/2:	.01099941
ECCENTRICITY:	.0073576	MEAN MOTION DOT DOT/6:	.000028662
ARG. OF PERIGEE:	61.7654	BSTAR:	.00016423

DEBRIS CLOUD DATA

MAXIMUM ΔP:	17.3 min
MAXIMUM ΔI:	0.5 deg

COMMENTS

Spacecraft was destroyed after a malfunction prevented controlled reentry and landing in the Soviet Union. Hundreds of fragments were detected but most reentered before being officially cataloged.

REFERENCE DOCUMENT

History of Soviet/Russian Satellite Fragmentations-A Joint U.S.-Russian Investigation, N. L. Johnson, *et al.,* Kaman Sciences Corporation, October 1995.

SSN 18184

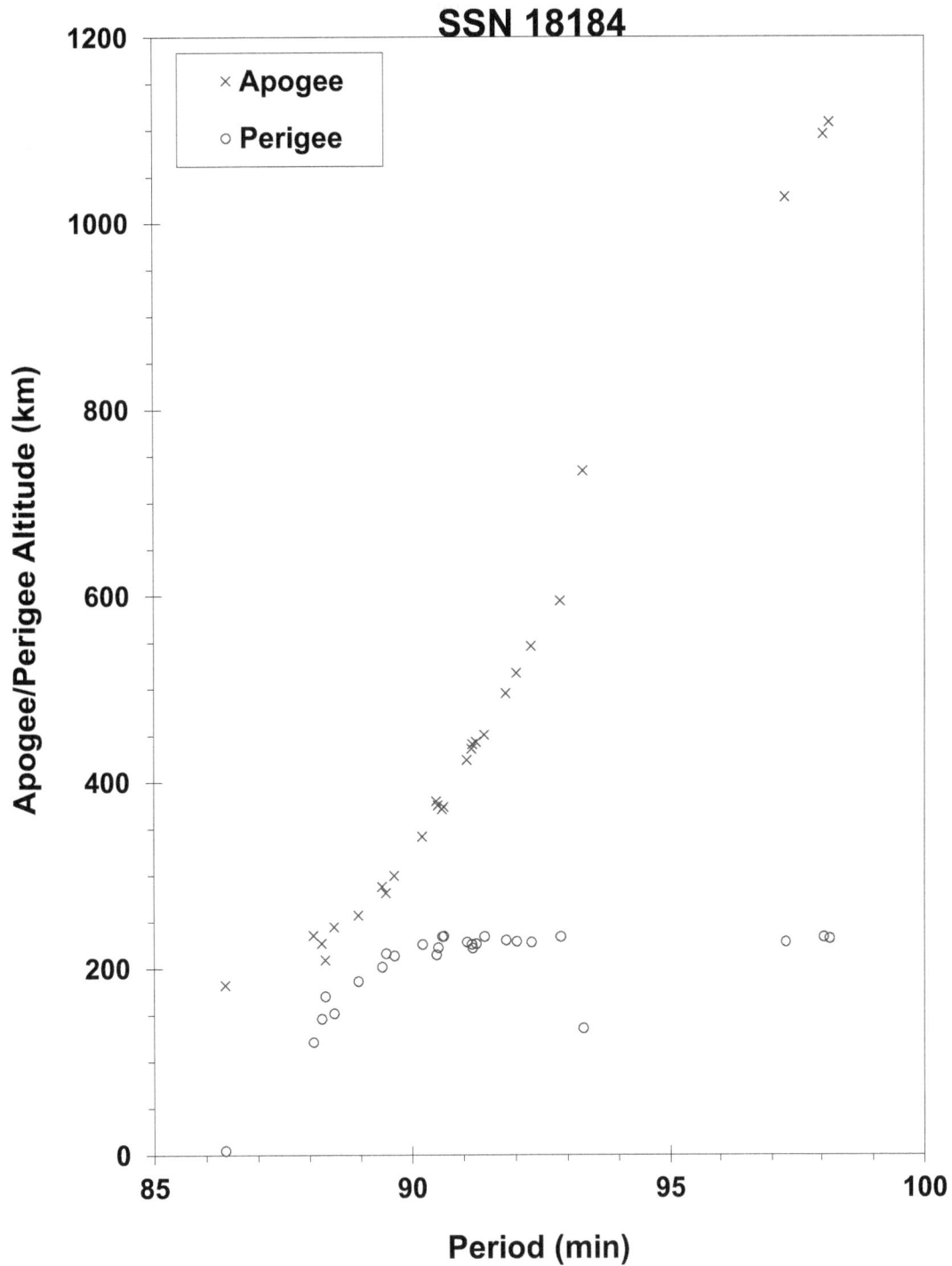

Cosmos 1866 debris cloud of 27 fragments one to 2 days after the event as reconstructed from the US SSN database. Two fragments with orbital periods greater than 103 minutes were cataloged in mid-August 1987.

SATELLITE DATA

TYPE:	Payload
OWNER:	CIS
LAUNCH DATE:	16.18 Jul 1987
DRY MASS (KG):	1900
MAIN BODY:	Cylinder; 0.8-1.4 m diameter by 3 m length
MAJOR APPENDAGES:	Solar arrays, radar, and other payload systems
ATTITUDE CONTROL:	Gravity gradient
ENERGY SOURCES:	Battery, pressurized vessels

EVENT DATA

DATE:	27 Nov 1997	LOCATION:	Unknown
TIME:	0006-0040 GMT?	ASSESSED CAUSE:	Unknown
ALTITUDE:	~630 km		

PRE-EVENT ELEMENTS

EPOCH:	97329.88487815	MEAN ANOMALY:	245.1014
RIGHT ASCENSION:	97.7878	MEAN MOTION:	14.83337853
INCLINATION:	82.5131	MEAN MOTION DOT/2:	0.00000439
ECCENTRICITY:	0.0021357	MEAN MOTION DOT DOT/6:	0
ARG. OF PERIGEE:	115.2417	BSTAR:	0.000050420

DEBRIS CLOUD DATA

MAXIMUM ΔP:	Unknown
MAXIMUM ΔI:	Unknown

COMMENTS

Cosmos 1869 suffered a failure of its radar antenna to deploy immediately after launch. The spacecraft carried other optical sensors, but the vehicle appears to have become non-operational by 1988. At least 20 debris were detected. Virtually all the debris associated with the breakup event exhibited very large area-to-mass ratios, resulting in exceptionally rapid orbital decay.

REFERENCE DOCUMENT

"Recent Satellite Fragmentation Investigations," N. Johnson, The Orbital Debris Quarterly News, NASA JSC, January 1998, p. 3. Available online at https://orbitaldebris.jsc.nasa.gov/quarterly-news/pdfs/ODQNv3i1.pdf.

Insufficient data to construct a Gabbard diagram.

SATELLITE DATA

TYPE:	Tsyklon Third Stage
OWNER:	CIS
LAUNCH DATE:	18.10 Aug 1987
DRY MASS (KG):	1360
MAIN BODY:	Cylinder; 2.1 m diameter by 2.4 m length
MAJOR APPENDAGES:	None
ATTITUDE CONTROL:	None at time of the event.
ENERGY SOURCES:	Unknown

EVENT DATA

DATE:	15 Feb 1998	LOCATION:	67.8 N, 125.6 E (asc.)
TIME:	2224 GMT	ASSESSED CAUSE:	Propulsion
ALTITUDE:	945 km		

PRE-EVENT ELEMENTS

EPOCH:	98044.02783074	MEAN ANOMALY:	25.0628
RIGHT ASCENSION:	230.9724	MEAN MOTION:	13.84031596
INCLINATION:	82.5526	MEAN MOTION DOT/2:	0.00000025
ECCENTRICITY:	0.0011144	MEAN MOTION DOT DOT/6:	0
ARG. OF PERIGEE:	334.9992	BSTAR:	0.0000096468

DEBRIS CLOUD DATA

MAXIMUM ΔP:	8.2 min
MAXIMUM ΔI:	0.6 deg

COMMENTS

This is the second time a Ukrainian Tsyklon third stage has experienced a significant breakup. The previous incident in 1988 involved the Cosmos 1045 rocket body at a higher altitude. In both cases, the vehicle was approximately 10 years old. The debris from the current breakup were ejected with a wide range of velocities, from about 15 m/s to more than 250 m/s. Some debris were thrown to altitudes below 500 km, and some exhibited high area-to-mass ratios. Naval Space Command ran COMBO to determine if a tracked object was in vicinity of Meteor 2-16 R/B at the time of the event, and the results were negative.

REFERENCE DOCUMENT

"Three Upper Stage Breakups in One Week Top February Debris Activity," The Orbital Debris Quarterly News, NASA JSC, April 1998, p. 1. Available online at https://orbitaldebris.jsc.nasa.gov/quarterly-news/pdfs/ODQNv3i2.pdf.

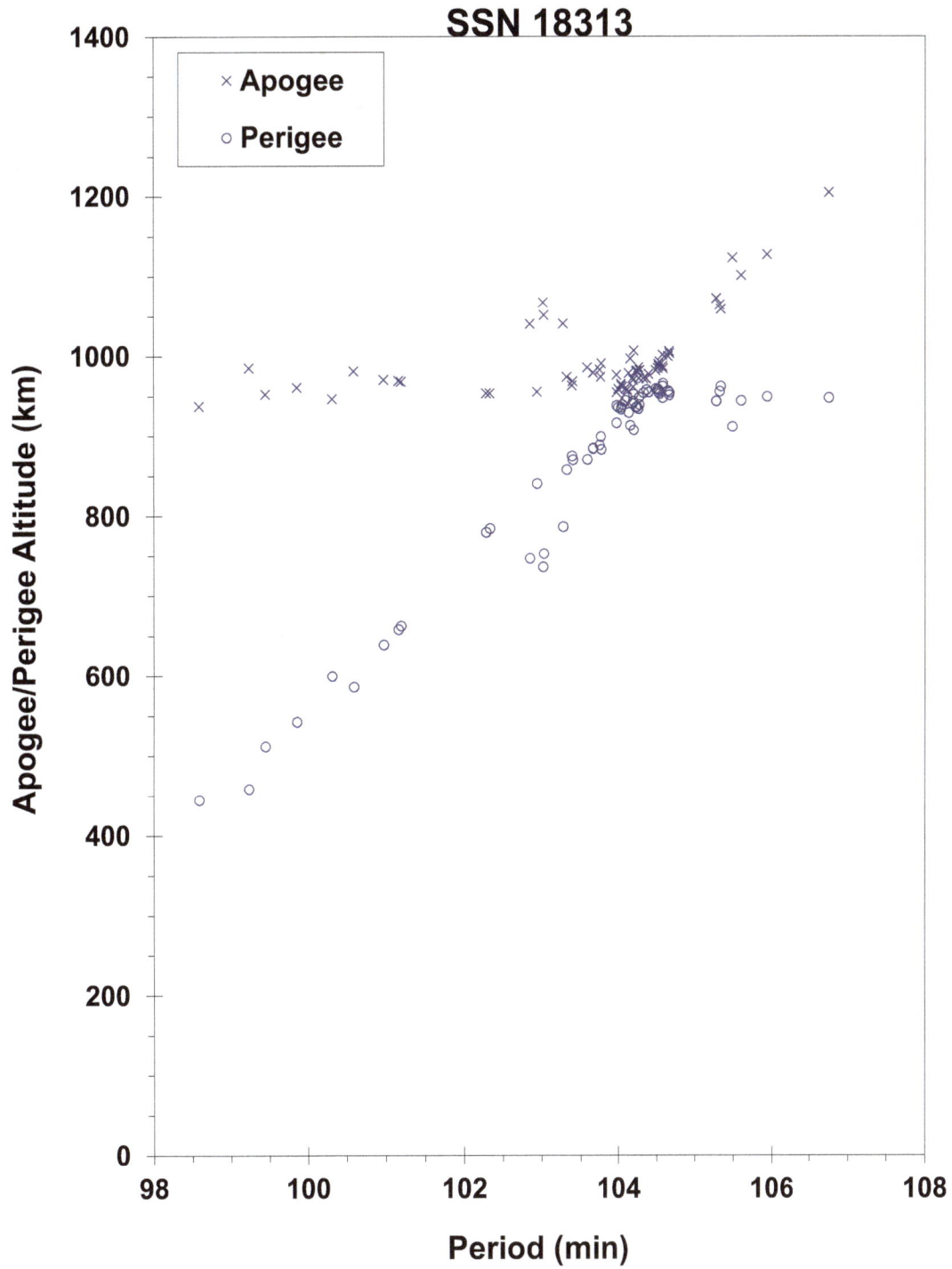

Meteor 2-16 R/B debris cloud of 67 fragments 1 week after the event
as reconstructed from the US SSN database.

SATELLITE DATA

TYPE:	Ariane 3 Third Stage
OWNER:	France
LAUNCH DATE:	16.03 Sep 1987
DRY MASS (KG):	1200
MAIN BODY:	Cylinder; 2.6 m diameter by 9.9 m length
MAJOR APPENDAGES:	None
ATTITUDE CONTROL:	None
ENERGY SOURCES:	On-board propellants

EVENT DATA

DATE:	16-19 Sep 1987	LOCATION:	Unknown
TIME:	Unknown	ASSESSED CAUSE:	Propulsion
ALTITUDE:	Unknown		

POST-EVENT ELEMENTS

EPOCH:	87264.18031994	MEAN ANOMALY:	170.9704
RIGHT ASCENSION:	176.7680	MEAN MOTION:	2.22860839
INCLINATION:	6.8720	MEAN MOTION DOT/2:	.00014489
ECCENTRICITY:	.7324768	MEAN MOTION DOT DOT/6:	.0
ARG. OF PERIGEE:	182.0665	BSTAR:	.0038829

DEBRIS CLOUD DATA

MAXIMUM ΔP:	29.1 min*
MAXIMUM ΔI:	0.9 deg*

*Based on uncataloged debris data

COMMENTS

Above elements are initial published values for the rocket body but are after the event.

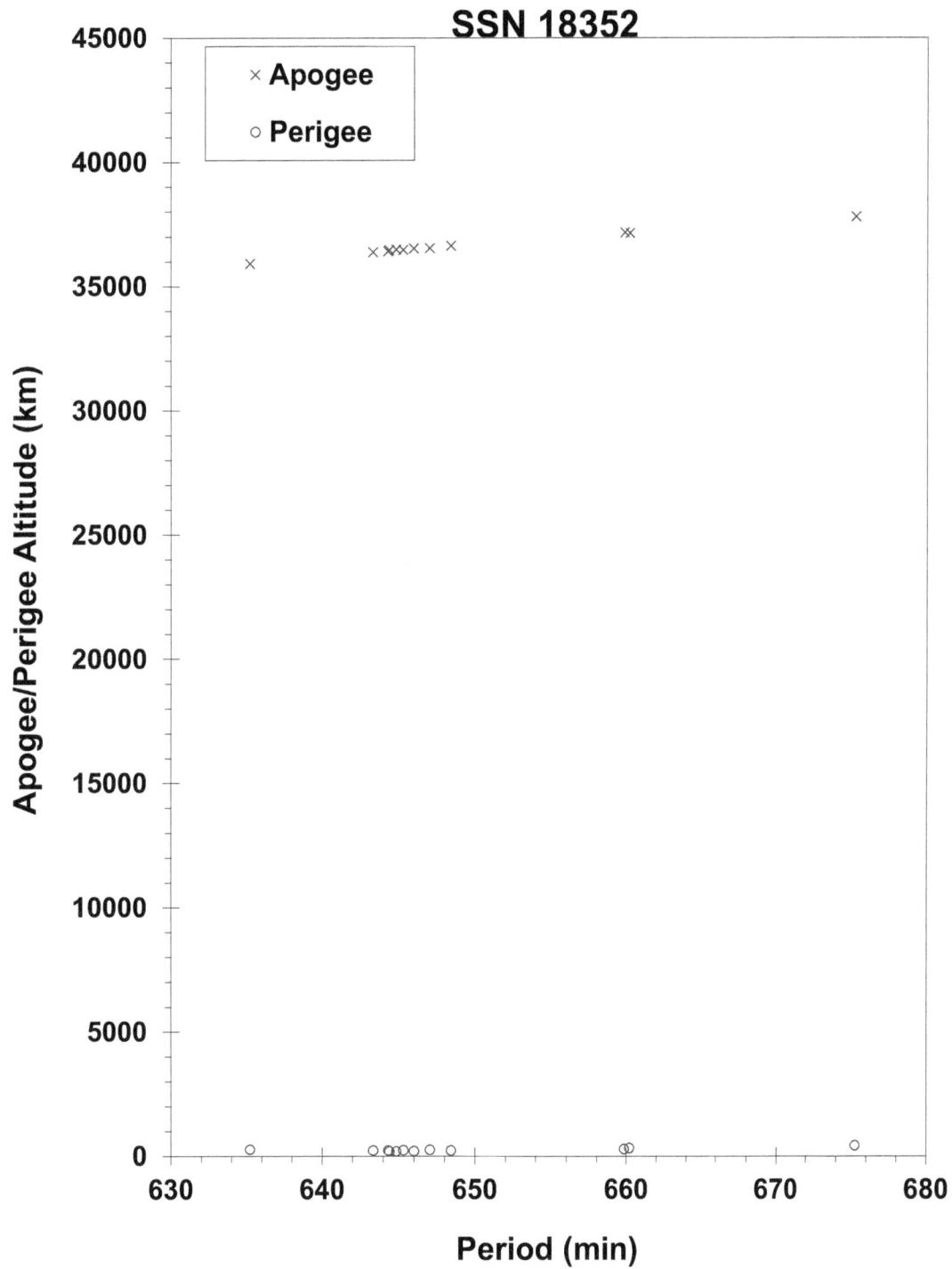

SSN 18352

AUSSAT K3/ECS 4 R/B debris cloud of 12 fragments about 4 days after launch as reconstructed from the US SSN database.

SATELLITE DATA

TYPE:	Mission Related Debris
OWNER:	CIS
LAUNCH DATE:	16.12 Sep 1987
DRY MASS (KG):	55
MAIN BODY:	Ellipsoid; 0.6 m diameter by 1 m length
MAJOR APPENDAGES:	None
ATTITUDE CONTROL:	None
ENERGY SOURCES:	On-board propellants

EVENT DATA

DATE:	~ 01 December 1996	LOCATION:	Unknown
TIME:	Unknown	ASSESSED CAUSE:	Propulsion
ALTITUDE:	Unknown		

PRE-EVENT ELEMENTS

EPOCH:	96335.26612005	MEAN ANOMALY:	175.6198
RIGHT ASCENSION:	300.4954	MEAN MOTION:	4.24439384
INCLINATION:	64.9068	MEAN MOTION DOT/2:	0.00015773
ECCENTRICITY:	0.5826382	MEAN MOTION DOT DOT/6:	0
ARG. OF PERIGEE:	181.3565	BSTAR:	0

DEBRIS CLOUD DATA

MAXIMUM ΔP:	234.1 min
MAXIMUM ΔI:	2.6 deg

COMMENTS

This is the 14th event of this class identified to date.

REFERENCE DOCUMENTS

"Identification and Resolution of an Orbital Debris Problem with the Proton Launch Vehicle", Cherniatiev, Chernyavskiy, Johnson, and McKnight, First European Conference on Space Debris, 5-7 April 1993.

"The Fragmentation of Proton Debris", Nauer, Teledyne Brown Engineering Technical Report CS93LKD-004, 31 December 1992.

History of Soviet/Russian Satellite Fragmentations-A Joint U.S.-Russian Investigation, N. L. Johnson, et al., Kaman Sciences Corporation, October 1995.

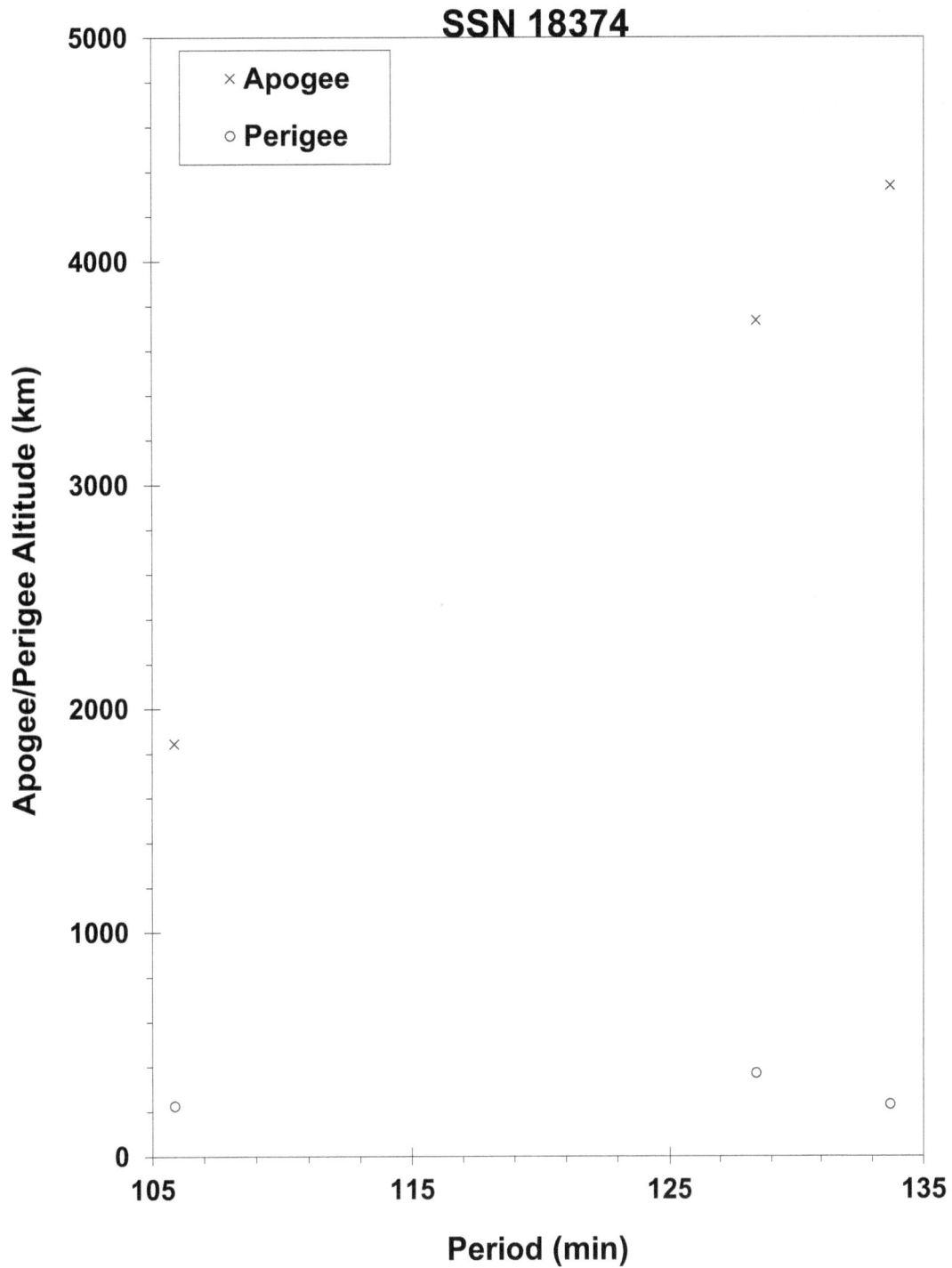

SSN 18374

Apogee/Perigee Altitude (km) vs Period (min)

Cosmos 1883-1885 auxiliary motor debris cloud of 3 fragments 11 to 14 months after the event as reconstructed from the US SSN database.

COSMOS 1883-1885 ULLAGE MOTOR   1987-079H   18375

SATELLITE DATA

TYPE:	Mission Related Debris
OWNER:	CIS
LAUNCH DATE:	16.12 Sep 1987
DRY MASS (KG):	~55
MAIN BODY:	Ellipsoid; 0.6 m diameter by 1.0 m length
MAJOR APPENDAGES:	None
ATTITUDE CONTROL:	None
ENERGY SOURCES:	On-board propellants

EVENT DATA

DATE:	23 Apr 2003	LOCATION:	Unknown
TIME:	~1800Z	ASSESSED CAUSE:	Propulsion
ALTITUDE:	Unknown		

PRE-EVENT ELEMENTS

EPOCH:	03113.46108488	MEAN ANOMALY:	332.8061
RIGHT ASCENSION:	156.9474	MEAN MOTION:	4.27871903
INCLINATION:	65.2438	MEAN MOTION DOT/2:	.00000068
ECCENTRICITY:	.5548829	MEAN MOTION DOT DOT/6:	.0
ARG. OF PERIGEE:	85.3049	BSTAR:	.00025672

DEBRIS CLOUD DATA

MAXIMUM ΔP:	26.0 min*
MAXIMUM ΔI:	1.19 deg*

*Based on uncataloged debris data

COMMENTS

This event marks the 27th known breakup of a Proton Blok DM SOZ ullage motor since 1984. This ullage motor was launched before implementation of breakup preventive measures. 31 debris objects were cataloged from this breakup.

REFERENCE DOCUMENT

"Satellite Fragmentations in 2003", The Orbital Debris Quarterly News, NASA JSC, January 2004. Available online at https://orbitaldebris.jsc.nasa.gov/quarterly-news/pdfs/ODQNv8i1.pdf.

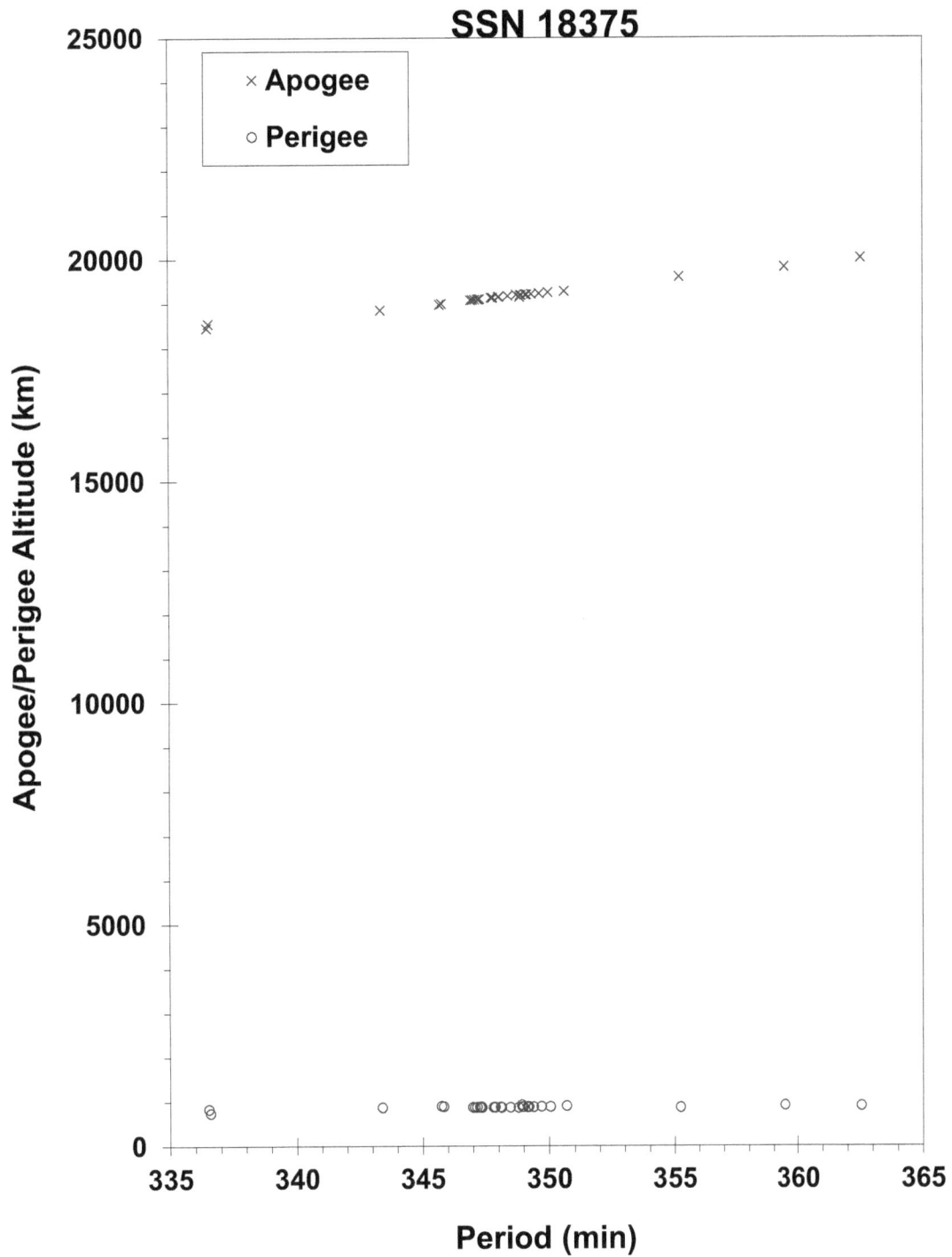

Cosmos 1883-85 auxiliary motor debris cloud of 31 fragments 2 days after the event as reconstructed from the US SSN database.

SATELLITE DATA

TYPE:	Payload
OWNER:	CIS
LAUNCH DATE:	26.48 Dec 1987
DRY MASS (KG):	6300
MAIN BODY:	Sphere-Cylinder; 2.4 m diameter by 6.5 m length
MAJOR APPENDAGES:	Solar panels
ATTITUDE CONTROL:	Active, 3-axis
ENERGY SOURCES:	On-board propellants, explosive charge

EVENT DATA

DATE:	31 Jan 1988	LOCATION:	11S, 138E (dsc)
TIME:	1109 GMT	ASSESSED CAUSE:	Deliberate
ALTITUDE:	250 km		

PRE-EVENT ELEMENTS

EPOCH:	88030.87152193	MEAN ANOMALY:	208.0352
RIGHT ASCENSION:	254.6565	MEAN MOTION:	16.07089398
INCLINATION:	82.5872	MEAN MOTION DOT/2:	.00174892
ECCENTRICITY:	.0015551	MEAN MOTION DOT DOT/6:	.000012805
ARG. OF PERIGEE:	152.1926	BSTAR:	.00022253

DEBRIS CLOUD DATA

MAXIMUM ΔP:	5.7 min*
MAXIMUM ΔI:	1.7 deg*

*Based on cataloged and uncataloged debris data

COMMENTS

Spacecraft destroyed after a malfunction prevented controlled reentry and landing in the Soviet Union. Elements for 83 objects remaining in orbit about 10 days after the event were developed. Other debris reentered before being officially cataloged.

REFERENCE DOCUMENTS

The Soviet Year in Space 1988, N. L. Johnson, Teledyne Brown Engineering, 1989, p. 27.

History of Soviet/Russian Satellite Fragmentations-A Joint U.S.-Russian Investigation, N. L. Johnson, *et al.,* Kaman Sciences Corporation, October 1995.

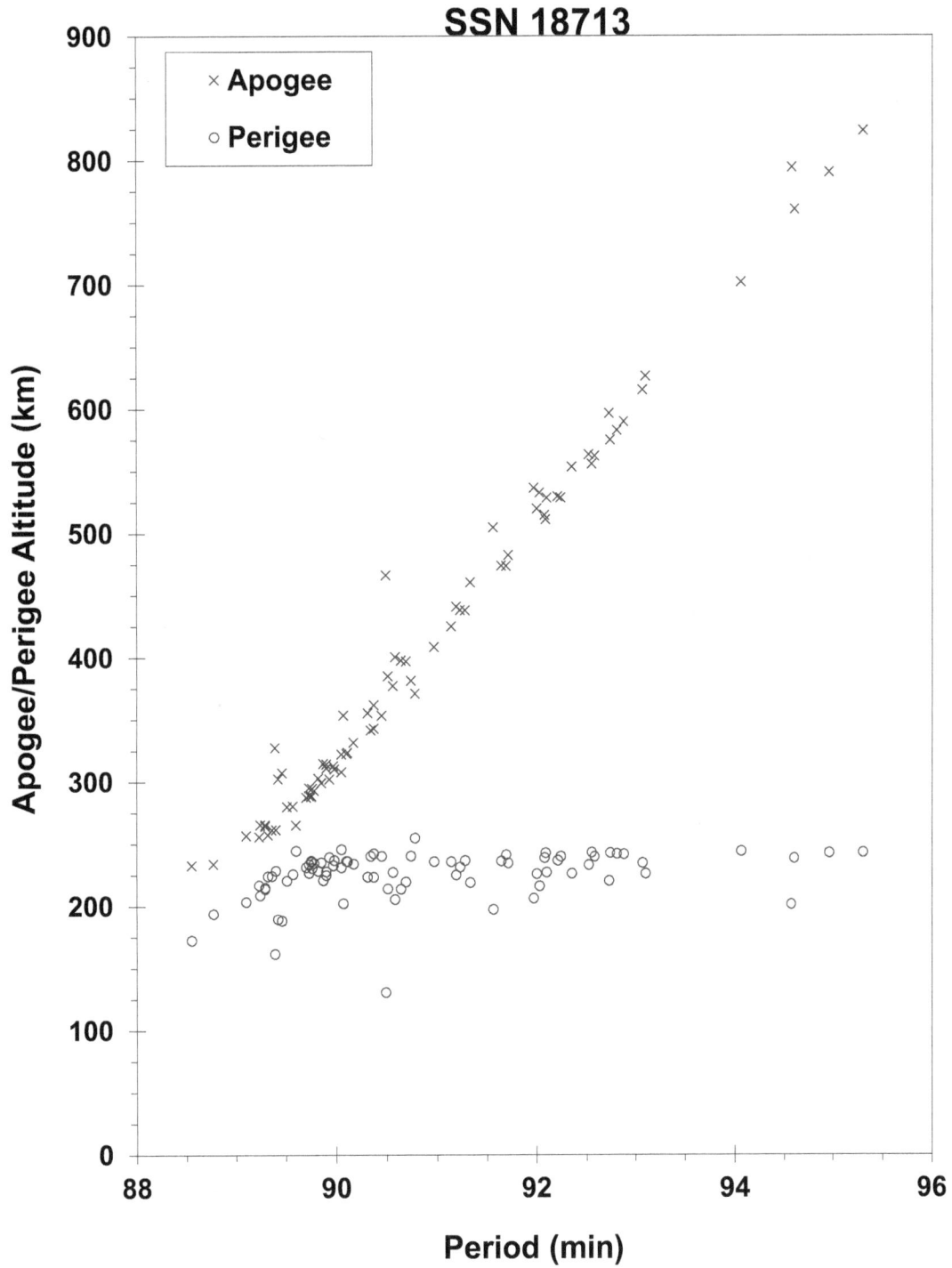

SSN 18713

Cosmos 1906 debris cloud remnant of 83 objects 10 days after the event as reconstructed from Naval Space Surveillance System database.

SATELLITE DATA

TYPE:	Mission Related Debris
OWNER:	CIS
LAUNCH DATE:	27.48 Dec 1987
DRY MASS (KG):	55
MAIN BODY:	Ellipsoid; 0.6 m diameter by 1.0 m length
MAJOR APPENDAGES:	None
ATTITUDE CONTROL:	None at time of the event.
ENERGY SOURCES:	On-board propellants

EVENT DATA

DATE:	22 May 1997	LOCATION:	Unknown
TIME:	Unknown	ASSESSED CAUSE:	Propulsion
ALTITUDE:	Unknown		

PRE-EVENT ELEMENTS

EPOCH:	97141.34020043	MEAN ANOMALY:	1.8603
RIGHT ASCENSION:	253.0389	MEAN MOTION:	3.58845480
INCLINATION:	46.6273	MEAN MOTION DOT/2:	-0.00000117
ECCENTRICITY:	0.6287941	MEAN MOTION DOT DOT/6:	0
ARG. OF PERIGEE:	349.7051	BSTAR:	0

DEBRIS CLOUD DATA

MAXIMUM ΔP:	Unknown
MAXIMUM ΔI:	Unknown

COMMENTS

This is the 15th event of this class identified to date. At least 72 debris were detected.

REFERENCE DOCUMENTS

"Three Satellite Breakups During May-June," The Orbital Debris Quarterly News, NASA JSC, July 1997, p. 2. Available online at https://orbitaldebris.jsc.nasa.gov/quarterly-news/pdfs/ODQNv2i3.pdf.

"Identification and Resolution of an Orbital Debris Problem with Proton Launch Vehicle", Cherniatiev, Chernyavskiy, Johnson, and McKnight, First European Conference on Space Debris, 5-7 April 1993.

"The Fragmentation of Proton Debris", Nauer, Teledyne Brown Engineering Technical Report CS93-LKD-004, 31 Dec 1992.

History of Soviet/Russian Satellite Fragmentations-A Joint U.S.-Russian Investigation, N. L. Johnson, et al., Kaman Sciences Corporation, October 1995.

Insufficient data to construct a Gabbard diagram.

METEOR 2-17   **1988-005A**   **18820**

SATELLITE DATA

TYPE:	Payload
OWNER:	CIS
LAUNCH DATE:	30.46 January 1988
DRY MASS (KG):	2750
MAIN BODY:	Cylinder
MAJOR APPENDAGES:	Solar panels
ATTITUDE CONTROL:	none at time of event
ENERGY SOURCES:	unknown

EVENT DATA

DATE:	21 June 2005	LOCATION:	36.5N, 74.22E (asc)
TIME:	09:14:54.5 GMT	ASSESSED CAUSE:	Unknown
ALTITUDE:	954 km		

PRE-EVENT ELEMENTS

EPOCH:	05171.27320873	MEAN ANOMALY:	320.4591
RIGHT ASCENSION:	309.0605	MEAN MOTION:	13.85140738
INCLINATION:	82.5393	MEAN MOTION DOT/2:	.00000047
ECCENTRICITY:	.0018107	MEAN MOTION DOT DOT/6:	.0
ARG. OF PERIGEE:	39.7894	BSTAR:	.000027897

DEBRIS CLOUD DATA

MAXIMUM ΔP:	1.5 min
MAXIMUM ΔI:	0.03 deg

COMMENTS

This meteorological spacecraft suffered both an anomalous event as well as a breakup event.

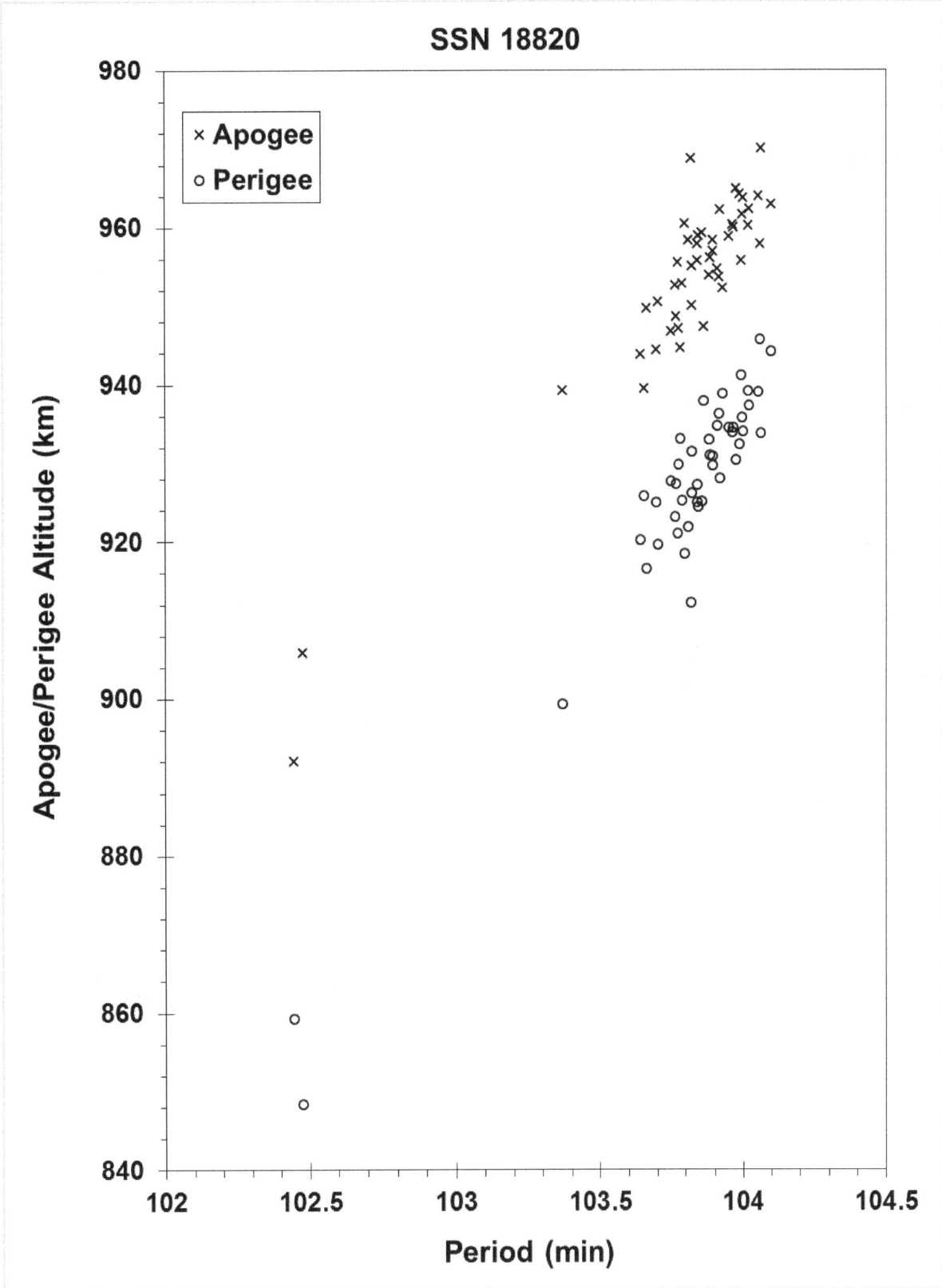

Meteor 2-17 debris cloud cataloged up to seven years after the event as reconstructed from the US SSN database.

DMSP 5D-2 F9 (USA 29)   **1988-006A**   **18822**

SATELLITE DATA

TYPE:	Payload
OWNER:	US
LAUNCH DATE:	3.25 February 1988
DRY MASS (KG):	767
MAIN BODY:	Cylinder; 1.0 m diameter by 3.6 m length
MAJOR APPENDAGES:	Solar panel and deployable instrumentation
ATTITUDE CONTROL:	none at time of event
ENERGY SOURCES:	unknown

EVENT DATA

DATE:	14-17 December 2012	LOCATION:	Unknown
TIME:	Unknown	ASSESSED CAUSE:	Unknown
ALTITUDE:	Unknown		

PRE-EVENT ELEMENTS

EPOCH:	12349.62481750	MEAN ANOMALY:	233.9639
RIGHT ASCENSION:	203.9910	MEAN MOTION:	14.26449692
INCLINATION:	98.8446	MEAN MOTION DOT/2:	.00000181
ECCENTRICITY:	.0006372	MEAN MOTION DOT DOT/6:	.0
ARG. OF PERIGEE:	126.2134	BSTAR:	.000094135

DEBRIS CLOUD DATA

MAXIMUM ΔP:	1.3 min
MAXIMUM ΔI:	0.2 deg

COMMENTS

Detected by software.

DMSP 5D-2 F9 debris objects B-F are mission-related debris typical for this spacecraft series. Pieces G-N (SSN 39041-6 inclusive, 43329) appear to be relatively high area-to-mass objects, with only two of the seven objects (39046 and 43329) remaining on orbit as of 4 July 2018. This event may be similar in root cause to the F11 (USA 73) and F13 (USA 109) events.

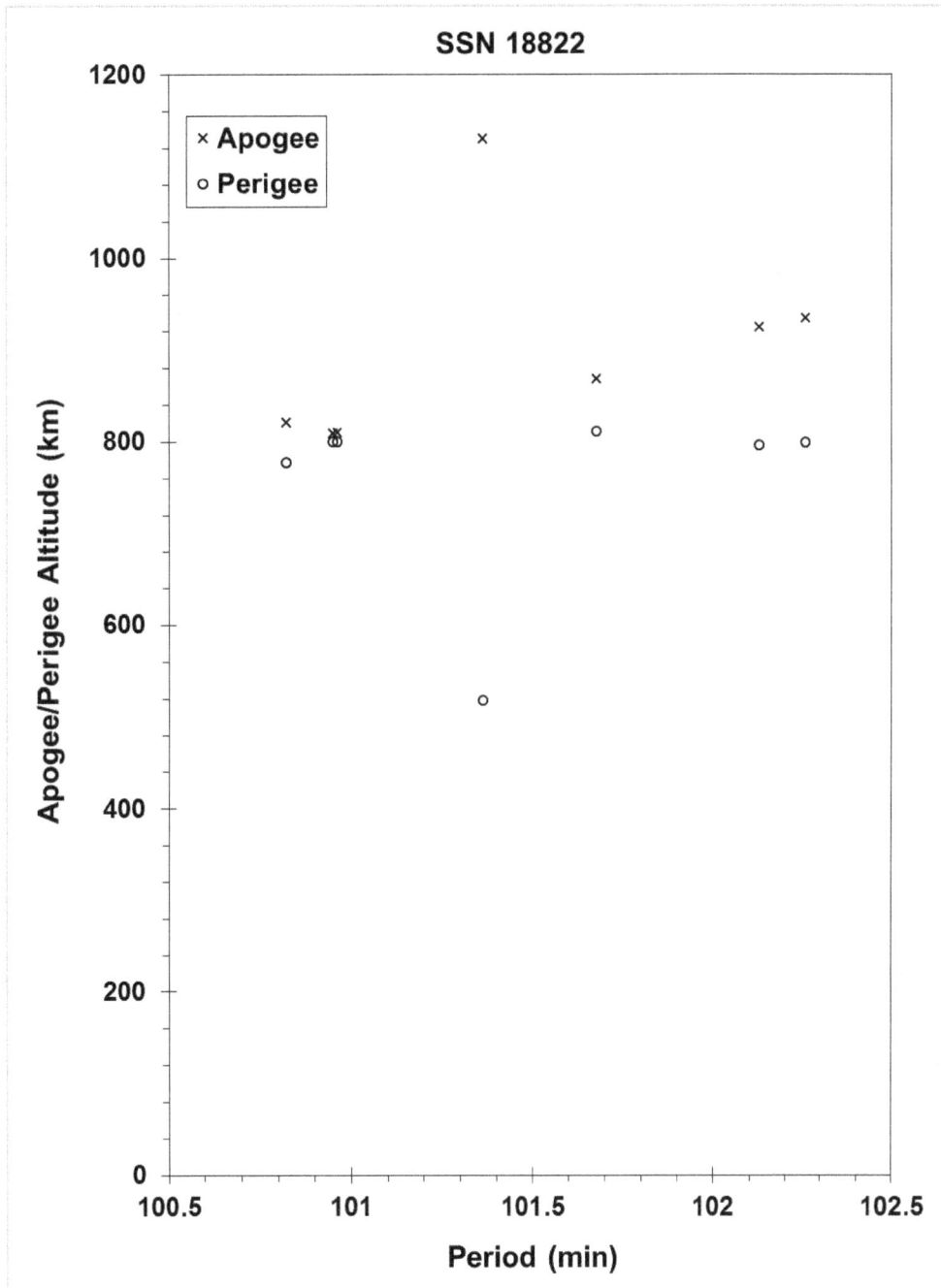

SSN 18822

USA 29 debris cloud cataloged within two weeks of the event as reconstructed from the US SSN database.

SATELLITE DATA

TYPE:	Payload
OWNER:	CIS
LAUNCH DATE:	3.15 Feb 1988
DRY MASS (KG):	5700
MAIN BODY:	Cone-cylinder; 2.7 m diameter by 6.3 m length
MAJOR APPENDAGES:	Solar panels
ATTITUDE CONTROL:	Active, 3-axis
ENERGY SOURCES:	On-board propellants, explosive charge

EVENT DATA

DATE:	27 Feb 1988	LOCATION:	62N, 98E (asc)
TIME:	0444 GMT	ASSESSED CAUSE:	Deliberate
ALTITUDE:	155 km		

PRE-EVENT ELEMENTS

EPOCH:	88058.12322153	MEAN ANOMALY:	309.0154
RIGHT ASCENSION:	264.6529	MEAN MOTION:	16.30989909
INCLINATION:	64.8359	MEAN MOTION DOT/2:	.03233928
ECCENTRICITY:	.0060041	MEAN MOTION DOT DOT/6:	.00003669
ARG. OF PERIGEE:	51.6410	BSTAR:	.00025587

DEBRIS CLOUD DATA

MAXIMUM ΔP:	4.2 min*
MAXIMUM ΔI:	1.1 deg*

*Based on uncataloged debris data

COMMENTS

Spacecraft destroyed after a malfunction prevented controlled reentry and landing in the Soviet Union. Early elements on only 6 objects available. All debris reentered before being officially cataloged.

REFERENCE DOCUMENTS

The Soviet Year in Space 1988, N. L. Johnson, Teledyne Brown Engineering, 1989, p. 31.

History of Soviet/Russian Satellite Fragmentations-A Joint U.S.-Russian Investigation, N. L. Johnson, et al., Kaman Sciences Corporation, October 1995.

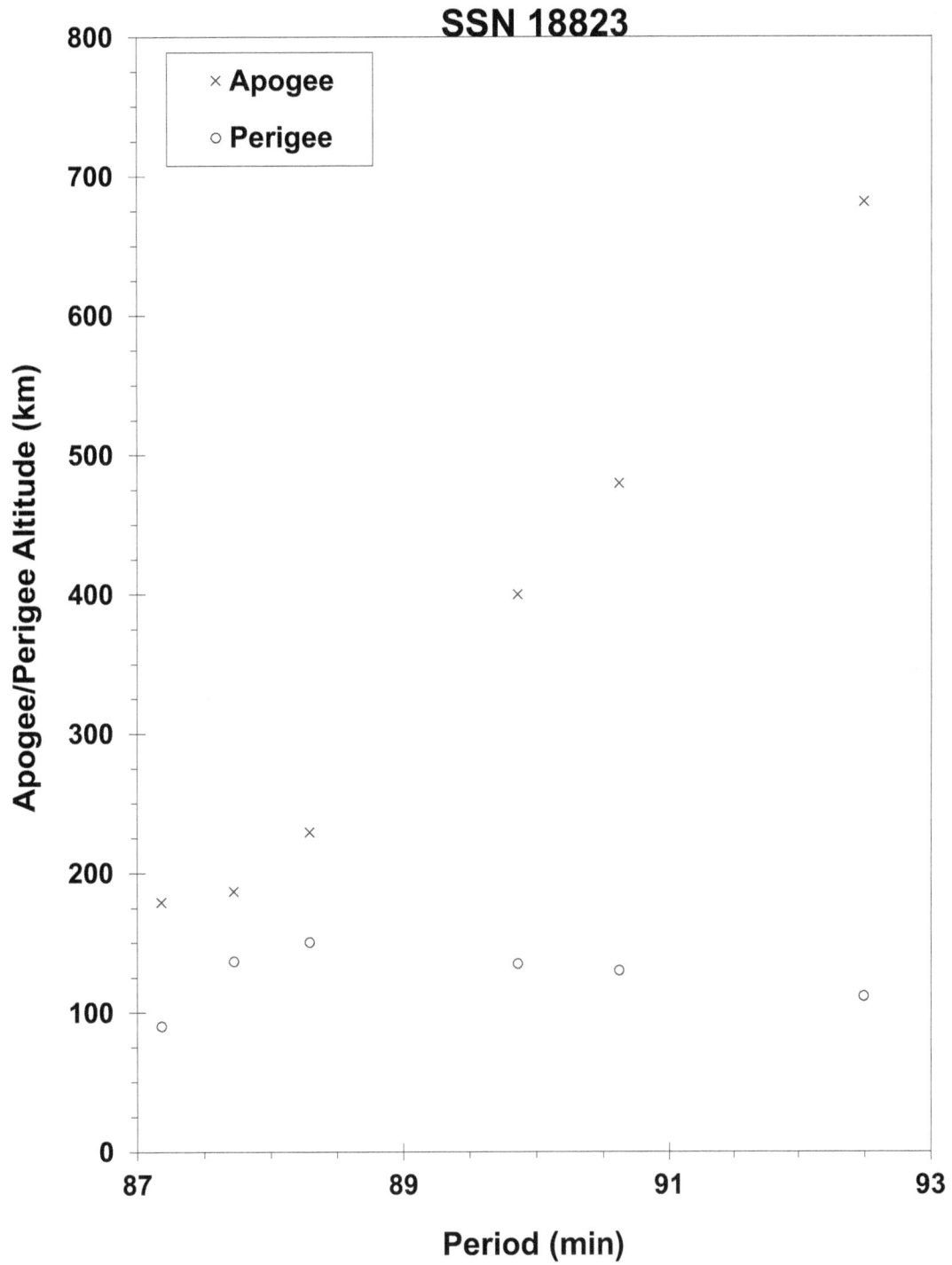

SSN 18823

Cosmos 1916 debris cloud remnant of 6 objects within 1 day of the event as reconstructed from the US SSN database.

COSMOS 1934   1988-023A   18985

SATELLITE DATA

TYPE:	Payload
OWNER:	CIS
LAUNCH DATE:	22.59 Mar 1988
DRY MASS (KG):	800
MAIN BODY:	Cylinder; 2.035 m diameter x 2 m length
MAJOR APPENDAGES:	Several short booms
ATTITUDE CONTROL:	Gravity-gradient (passive)
ENERGY SOURCES:	Batteries

EVENT DATA

DATE:	23 Dec 1991	LOCATION:	Unknown
TIME:	Unknown	ASSESSED CAUSE:	Accidental Collision
ALTITUDE:	Unknown		

PRE-EVENT ELEMENTS

EPOCH:	91356.93360267	MEAN ANOMALY:	291.3330
RIGHT ASCENSION:	126.2142	MEAN MOTION:	13.75709229
INCLINATION:	82.9564	MEAN MOTION DOT/2:	0.00000135
ECCENTRICITY:	.0041502	MEAN MOTION DOT DOT/6:	0.0
ARG. OF PERIGEE:	69.2265	BSTAR:	0.00012752

DEBRIS CLOUD DATA

MAXIMUM ΔP:	Unknown
MAXIMUM ΔI:	Unknown

COMMENTS

The collision occurred with a piece of launch debris from Cosmos 926. The debris piece (Sat. No. 13475) was evidently shattered into smaller, non-trackable debris by the impact and the public satellite catalog now notes that it "COLLIDED WITH SATELLITE" in lieu of the standard orbital elements. There were several very close conjunctions during the day in question, the exact time of the collision is unknown. Two pieces of debris were cataloged from the event long before the event was recognized as a collision.

REFERENCE DOCUMENT

"Accidental Collisions of Cataloged Satellites Identified," The Orbital Debris Quarterly News, NASA JSC, April 2005. Available online at https://orbitaldebris.jsc.nasa.gov/quarterly-news/pdfs/ODQNv9i2.pdf.

Insufficient data to construct a Gabbard diagram.

INTELSAT 513 R/B   **1988-040B**   **19122**

SATELLITE DATA

TYPE:	Ariane 2 Third Stage
OWNER:	France
LAUNCH DATE:	17.99 May 1988
DRY MASS (KG):	~1480
MAIN BODY:	Cylinder; 2.6 m diameter by 11.7 m length
MAJOR APPENDAGES:	None
ATTITUDE CONTROL:	None
ENERGY SOURCES:	On-board propellants

EVENT DATA

DATE:	9 Jul 2002	LOCATION:	4.3 N, 5.7 E
TIME:	1930Z	ASSESSED CAUSE:	Propulsion
ALTITUDE:	21,500 km		

PRE-EVENT ELEMENTS

EPOCH:	02190.22071506	MEAN ANOMALY:	172.0370
RIGHT ASCENSION:	187.4675	MEAN MOTION:	2.28211164
INCLINATION:	7.0311	MEAN MOTION DOT/2:	.00000024
ECCENTRICITY:	.7162572	MEAN MOTION DOT DOT/6:	.0
ARG. OF PERIGEE:	181.6723	BSTAR:	.0

DEBRIS CLOUD DATA

MAXIMUM ΔP:	6.60 min*
MAXIMUM ΔI:	0.33 deg*

* Based on uncataloged debris data

COMMENTS

This is the second breakup of an Ariane 2 third stage officially recognized and the 11th overall breakup of an Ariane upper stage. This stage was launched prior to the implementation of passivation measures. The age of the stage at the time of the breakup was 14 years. Six pieces of debris were initially seen by the SSN, while four were cataloged.

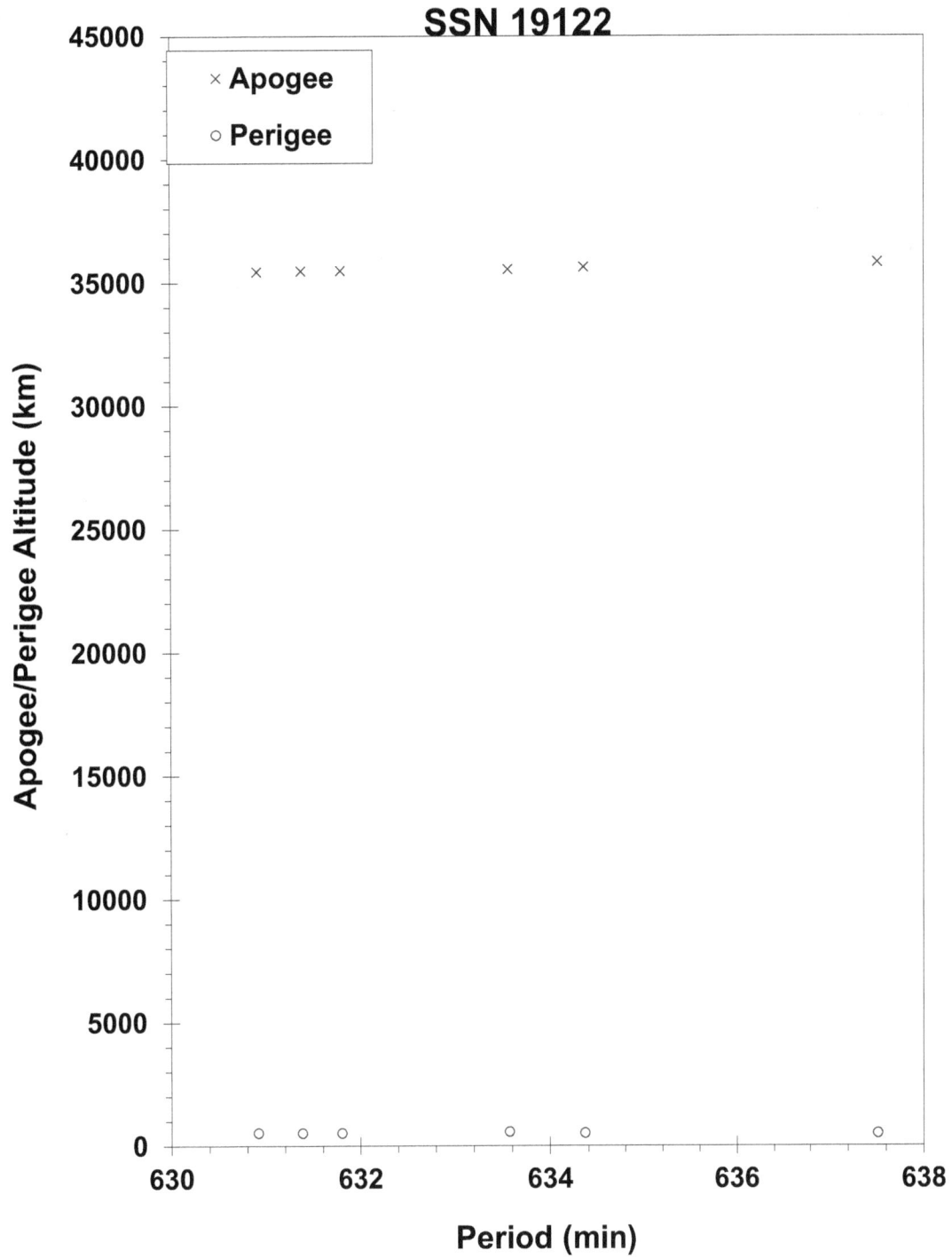

SSN 19122

Intelsat 513 R/B debris cloud of 6 fragments 2 weeks after the breakup as reconstructed from the US SSN database.

COSMOS 1970-1972 ULLAGE MOTOR   1988-085F   19535

SATELLITE DATA

TYPE:	Mission Related Debris
OWNER:	CIS
LAUNCH DATE:	16.08 Sep 1988
DRY MASS (KG):	~55
MAIN BODY:	Ellipsoid; 0.6 m diameter by 1.0 m length
MAJOR APPENDAGES:	None
ATTITUDE CONTROL:	None
ENERGY SOURCES:	On-board propellants

EVENT DATA

DATE:	04 Aug 2003	LOCATION:	Unknown
TIME:	~0725Z	ASSESSED CAUSE:	Propulsion
ALTITUDE:	Unknown		

PRE-EVENT ELEMENTS

EPOCH:	03214.47923598	MEAN ANOMALY:	334.9286
RIGHT ASCENSION:	239.4643	MEAN MOTION:	4.29128214
INCLINATION:	65.3341	MEAN MOTION DOT/2:	.00007107
ECCENTRICITY:	.5561230	MEAN MOTION DOT DOT/6:	.0
ARG. OF PERIGEE:	85.1870	BSTAR:	.071402

DEBRIS CLOUD DATA

MAXIMUM ΔP:	18.8 min*
MAXIMUM ΔI:	2.79 deg*

* Based on uncataloged debris data

COMMENTS

This event marks the 28th known breakup of a Proton Block DM SOZ ullage motor since 1984. This ullage motor was launched before implementation of breakup preventive measures. Approximately 175 objects were initially seen by the SSN 1 week after the event. 76 debris objects were cataloged.

REFERENCE DOCUMENT

"Satellite Fragmentations in 2003", The Orbital Debris Quarterly News, NASA JSC, January 2004. Available online at https://orbitaldebris.jsc.nasa.gov/quarterly-news/pdfs/ODQNv8i1.pdf.

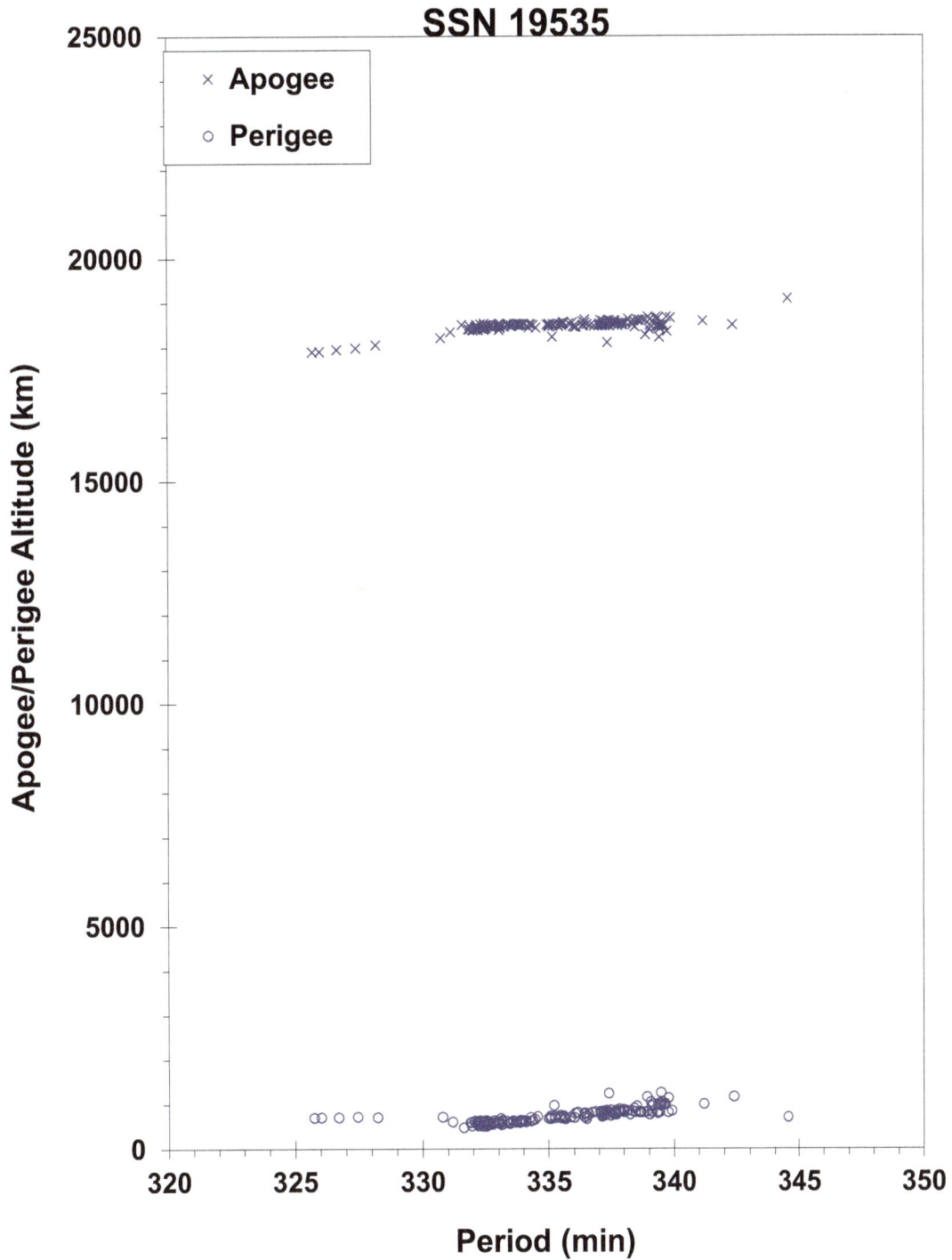

SSN 19535

Cosmos 1970-72 auxiliary motor debris cloud of 175 fragments 1 week after the event as reconstructed from the US SSN database.

SATELLITE DATA

TYPE:	Mission Related Debris
OWNER:	CIS
LAUNCH DATE:	16 Sep 1988
DRY MASS (KG):	55
MAIN BODY:	Ellipsoid; 0.6 m diameter by 1.0 m length
MAJOR APPENDAGES:	None
ATTITUDE CONTROL:	None
ENERGY SOURCES:	On-board propellants

EVENT DATA

DATE:	9 Mar 1999	LOCATION:	Unknown
TIME:	Unknown	ASSESSED CAUSE:	Propulsion
ALTITUDE:	Unknown		

POST-EVENT ELEMENTS

EPOCH:	99067.36656961	MEAN ANOMALY:	189.8576
RIGHT ASCENSION:	108.7309	MEAN MOTION:	4.28860956162171
INCLINATION:	64.6425	MEAN MOTION DOT/2:	.00000813
ECCENTRICITY:	.5827119	MEAN MOTION DOT DOT/6:	00000-0
ARG. OF PERIGEE:	176.8483	BSTAR:	.0022335

DEBRIS CLOUD DATA

MAXIMUM ΔP:	Unknown
MAXIMUM ΔI:	Unknown

COMMENTS

This is the 19th event of the Proton Block DM SOZ Ullage Motor class identified to date; it is the eighth associated with a GLONASS mission. This mission was conducted before the engineering defect was identified and passivation measures implemented. In this highly-eccentric orbit, debris may be long-lived but hard to track. 17 debris objects were detected.

REFERENCE DOCUMENT

"Abandoned Proton Ullage Motors Continue to Create Debris," The Orbital Debris Quarterly News, NASA JSC, April 1999. Available online at https://orbitaldebris.jsc.nasa.gov/quarterly-news/pdfs/ODQNv4i2.pdf.

Insufficient data to construct a Gabbard diagram.

SATELLITE DATA

TYPE:	Ariane 4 H-10 Third Stage
OWNER:	France
LAUNCH DATE:	11.02 Dec 1988
DRY MASS (KG):	1760
MAIN BODY:	Cylinder; 2.6 m diameter by 11.4 m length
MAJOR APPENDAGES:	None
ATTITUDE CONTROL:	None at time of the event.
ENERGY SOURCES:	On-board propellants

EVENT DATA

DATE:	17 Feb 1998	LOCATION:	6.9 N, 157.2 E (dsc)
TIME:	1235 GMT	ASSESSED CAUSE:	Propulsion
ALTITUDE:	19630 km		

PRE-EVENT ELEMENTS

EPOCH:	98047.29326560	MEAN ANOMALY:	25.3394
RIGHT ASCENSION:	23.7998	MEAN MOTION:	2.25942020
INCLINATION:	7.3381	MEAN MOTION DOT/2:	0.00000046
ECCENTRICITY:	0.7222736	MEAN MOTION DOT DOT/6:	0
ARG. OF PERIGEE:	248.1711	BSTAR:	0.00057969

DEBRIS CLOUD DATA

MAXIMUM ΔP:	Unknown
MAXIMUM ΔI:	Unknown

COMMENTS

This mission was the second for the Ariane 4 series and occurred prior to implementation of passivation measures. Using observations from the Eglin radar, specialists at Millstone radar found four new pieces from the upper stage. Naval Space Command personnel generated the first two debris element sets and calculated the approximate breakup time noted above.

REFERENCE DOCUMENT

"Three Upper Stage Breakups in One Week Top February Debris Activity," The Orbital Debris Quarterly News, NASA JSC, April 1998, p. 1. Available online at https://orbitaldebris.jsc.nasa.gov/quarterly-news/pdfs/ODQNv3i2.pdf.

Insufficient data to construct a Gabbard diagram.

COSMOS 1987-1989 ULLAGE MOTOR   1989-001G   19755

SATELLITE DATA

TYPE:	Mission Related Debris
OWNER:	CIS
LAUNCH DATE:	10 Jan 1989
DRY MASS (KG):	55
MAIN BODY:	Ellipsoid; 0.6 m diameter by 1 m length
MAJOR APPENDAGES:	None
ATTITUDE CONTROL:	None
ENERGY SOURCES:	On-board propellants

EVENT DATA

DATE:	3 Aug 1998	LOCATION:	Unknown
TIME:	Unknown	ASSESSED CAUSE:	Propulsion
ALTITUDE:	Unknown		

POST-EVENT ELEMENTS

EPOCH:	98211.80543118	MEAN ANOMALY:	172.2753
RIGHT ASCENSION:	16.7694	MEAN MOTION:	4.24137167
INCLINATION:	64.9243	MEAN MOTION DOT/2:	.00000287
ECCENTRICITY:	.5776927	MEAN MOTION DOT DOT/6:	00000-0
ARG. OF PERIGEE:	182.6029	BSTAR:	.0041366

DEBRIS CLOUD DATA

MAXIMUM ΔP:	162.64 min
MAXIMUM ΔI:	3.78 deg

COMMENTS

This is the 17th event of the Proton Block DM SOZ Ullage Motor class identified to date; it is the sixth associated with a GLONASS mission. This mission was conducted before the engineering defect was identified and passivation measures implemented. In this orbit, debris may be long-lived but hard to track. More than 110 debris objects were detected.

REFERENCE DOCUMENT

"Solitary Breakup and Anomalous Events in Third Quarter are Familiar," <u>The Orbital Debris Quarterly News</u>, NASA JSC, October 1998. Available online at: https://orbitaldebris.jsc.nasa.gov/quarterly-news/pdfs/ODQNv3i4.pdf.

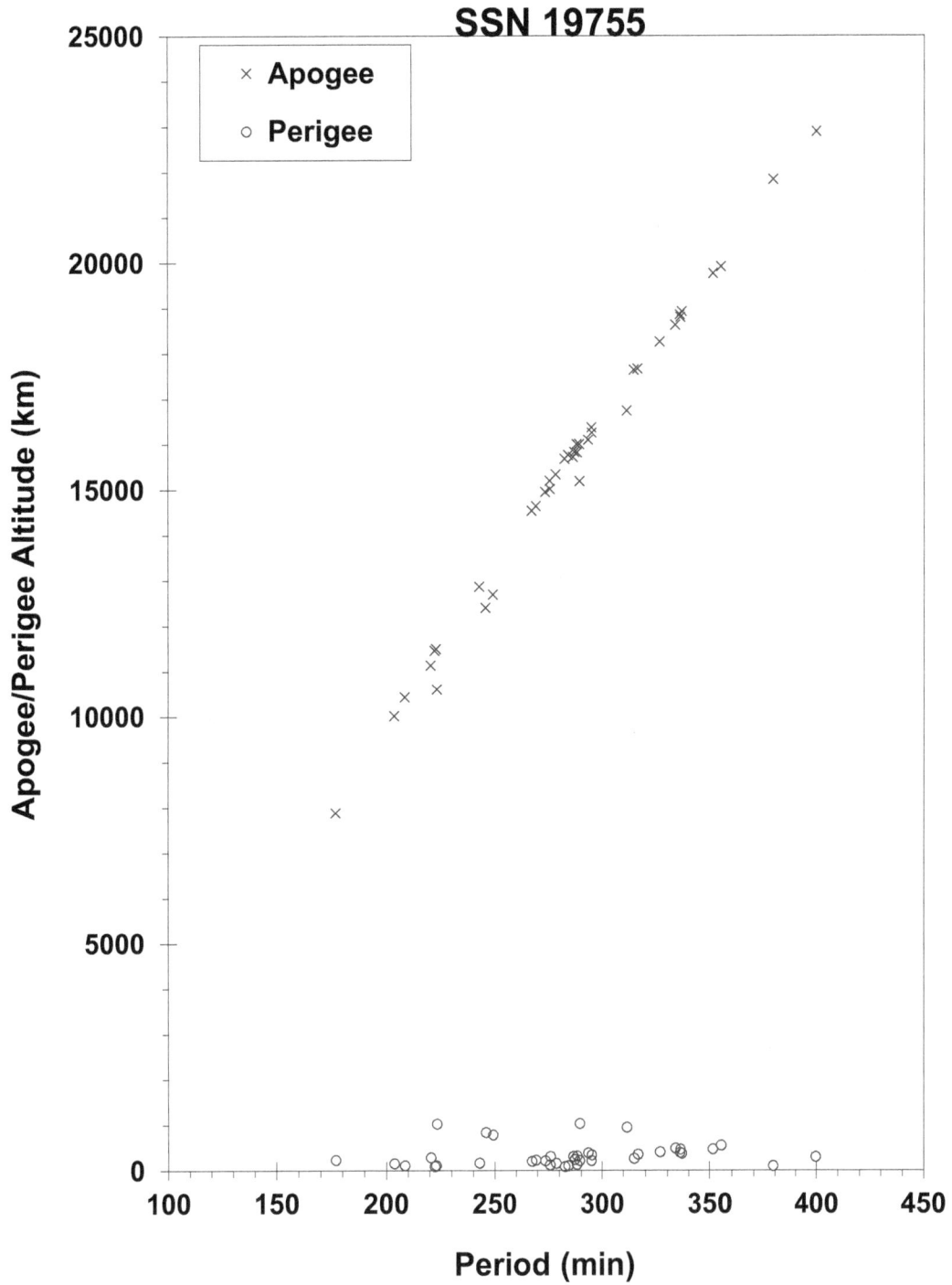

SSN 19755

Cosmos 1987-1989 ullage motor debris cloud of 39 fragments 7 days after the event as reconstructed from the US SSN database.

COSMOS 1987-1989 ULLAGE MOTOR   **1989-001H**   **19856**

SATELLITE DATA

TYPE:	Mission Related Debris
OWNER:	CIS
LAUNCH DATE:	10 Jan 1989
DRY MASS (KG):	~55
MAIN BODY:	Ellipsoid; 0.6 m diameter by 1.0 m length
MAJOR APPENDAGES:	None
ATTITUDE CONTROL:	None
ENERGY SOURCES:	On-board propellants

EVENT DATA

DATE:	13 Nov 2003	LOCATION:	Unknown
TIME:	Unknown	ASSESSED CAUSE:	Propulsion
ALTITUDE:	Unknown		

PRE-EVENT ELEMENTS

EPOCH:	03317.76135862	MEAN ANOMALY:	339.1502
RIGHT ASCENSION:	52.9695	MEAN MOTION:	4.24824637
INCLINATION:	65.4357	MEAN MOTION DOT/2:	.00000161
ECCENTRICITY:	.5599025	MEAN MOTION DOT DOT/6:	.0
ARG. OF PERIGEE:	72.44443	BSTAR:	.0017638

DEBRIS CLOUD DATA

MAXIMUM ΔP:	Unknown
MAXIMUM ΔI:	Unknown

COMMENTS

This event marks the 29th known breakup of a Proton Block DM SOZ ullage motor since 1984, although the event went undetected for over 3 months. This ullage motor was launched before implementation of breakup preventive measures. No debris objects were cataloged from this breakup.

Insufficient data to construct a Gabbard diagram.

SATELLITE DATA

TYPE:	Mission Related Debris
OWNER:	CIS
LAUNCH DATE:	26.39 Jan 1989
DRY MASS (KG):	55
MAIN BODY:	Ellipsoid; 0.6 m diameter by 1 m diameter
MAJOR APPENDAGES:	None
ATTITUDE CONTROL:	None
ENERGY SOURCES:	On-board propellants

EVENT DATA

DATE:	17-18 Dec 1992	LOCATION:	Unknown
TIME:	Unknown	ASSESSED CAUSE:	Propulsion
ALTITUDE:	Unknown		

PRE-EVENT ELEMENTS

EPOCH:	92351.90838995	MEAN ANOMALY:	1.4295
RIGHT ASCENSION:	266.2338	MEAN MOTION:	4.60309514
INCLINATION:	46.7001	MEAN MOTION DOT/2:	.00060784
ECCENTRICITY:	.5692927	MEAN MOTION DOT DOT/6:	.0000093219
ARG. OF PERIGEE:	353.9854	BSTAR:	.0015056

DEBRIS CLOUD DATA

MAXIMUM ΔP:	Unknown
MAXIMUM ΔI:	Unknown

COMMENTS

Parent satellite was one of two small engine units that are routinely released after the first burn of the Proton fourth stage. The nature of these objects was identified by Dr. Boris V. Chernlatiev, Deputy Constructor for the Energiya NPO, in October 1992. The cause of this fragmentation is assumed to be related to the residual hypergolic propellants on board and failure of the membrane separating the fuel and oxidizer. NAVSPASUR observed between 30-40 objects that were associated with this breakup. Only 4 element sets were generated, insufficient for a Gabbard Diagram or BLAST point. This was the seventh in a series of fragmentations of this object type and was the second located in a geosynchronous transfer orbit.

REFERENCE DOCUMENTS

The Fragmentation of Proton Debris, D. J. Nauer, TBE Technical Report CS93-LKD-004, Teledyne Brown Engineering, Colorado Springs, 31 December 1992.

Analysis of Fragmentations From December 1992 - February 1993, TBE Technical Report CS93-LKD-010, Teledyne Brown Engineering, Colorado Springs, 30 March 1993.

History of Soviet/Russian Satellite Fragmentations-A Joint U.S.-Russian Investigation, N. L. Johnson, et al., Kaman Sciences Corporation, October 1995.

"Identification and Resolution of an Orbital Debris Problem with the Proton Launch Vehicle," B. V. Cherniatiev, et al., Proceedings of the First European Conference on Space Debris, April 1993.

Insufficient data to construct a Gabbard diagram.

SATELLITE DATA

TYPE:	Ariane 2 third stage with VEB
OWNER:	France
LAUNCH DATE:	27.06 Jan 1989
DRY MASS (KG):	~1480 kg
MAIN BODY:	2.6 m diameter by 11.7 m length
MAJOR APPENDAGES:	None
ATTITUDE CONTROL:	None
ENERGY SOURCES:	On-board propellants?

EVENT DATA

DATE:	~1 Jan 2001	LOCATION:	Unknown
TIME:	Unknown	ASSESSED CAUSE:	Propulsion
ALTITUDE:	Unknown		

PRE-EVENT ELEMENTS

EPOCH:	00366.06151127	MEAN ANOMALY:	45.8970
RIGHT ASCENSION:	73.3900	MEAN MOTION:	2.26500973
INCLINATION:	8.3781	MEAN MOTION DOT/2:	.00000580
ECCENTRICITY:	.7188412	MEAN MOTION DOT DOT/6:	.0000000
ARG. OF PERIGEE:	225.8250	BSTAR:	.0040973

DEBRIS CLOUD DATA

MAXIMUM ΔP:	Unknown*
MAXIMUM ΔI:	Unknown*

* Not calculated due to provisional nature
of orbital data at event time.

COMMENTS

This is the first breakup of an Ariane 2 third stage officially recognized. One Ariane 3 third stage (same as Ariane 2) is known to have broken-up within a few days of launch in 1987. Both vehicles were launched before passivation measures were incorporated with Ariane third stages. Ariane third stage passivation was introduced in January 1990 and has been employed on all Ariane missions since October 1993. The age of the Ariane 2 third stage at the time of the breakup was nearly 12 years.

SSN 19773

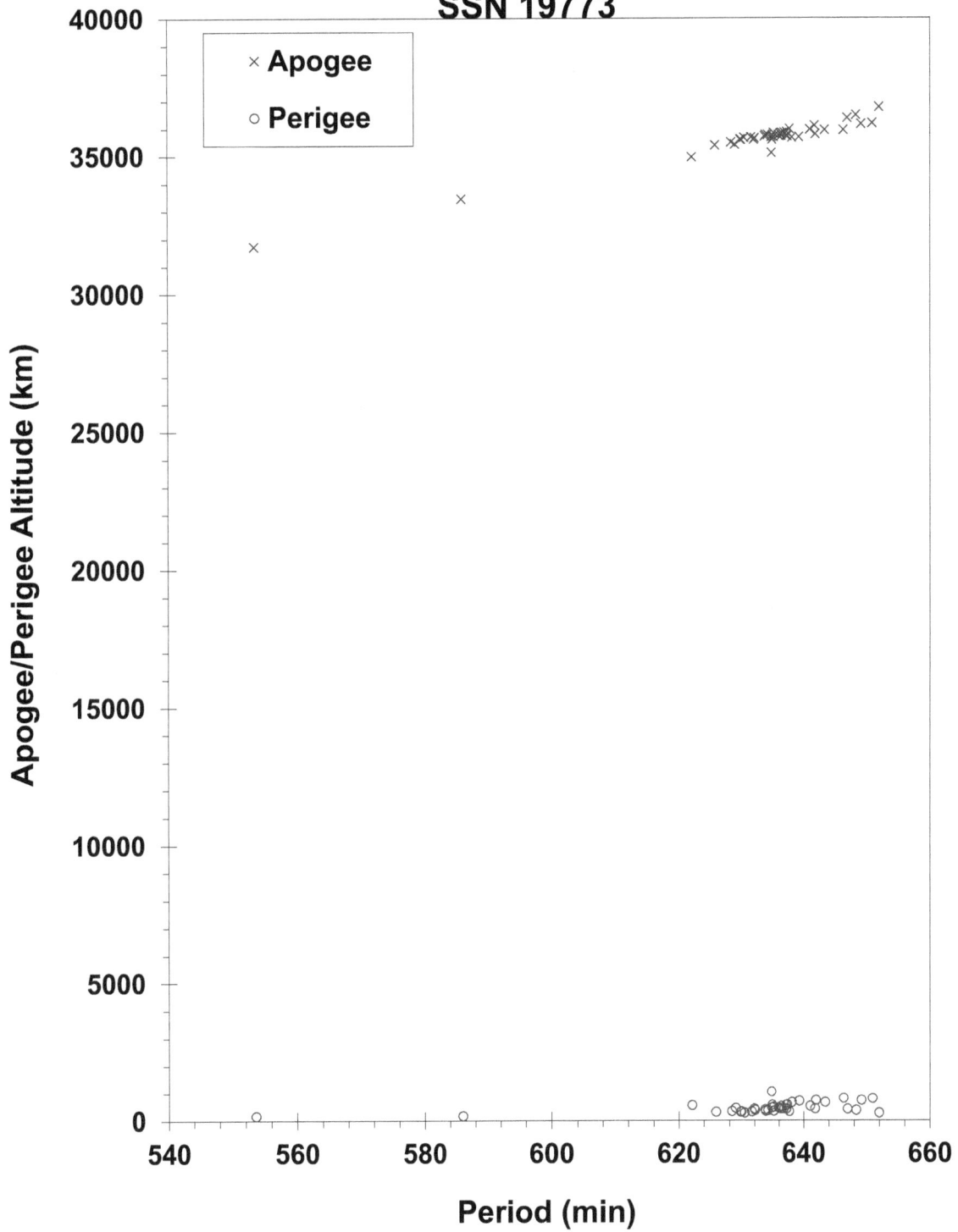

INTELSAT V F-15 R/B (Ariane 2) debris cloud as constructed using SSN 8XXXX series elements sets (10 January 2001 and before).

COSMOS 2022-2024 ULLAGE MOTOR   1989-039G   20081

SATELLITE DATA

TYPE:	Mission Related Debris
OWNER:	CIS
LAUNCH DATE:	31.36 May 1989
DRY MASS (KG):	55
MAIN BODY:	Ellipsoid; 0.6 m diameter by 1 m length
MAJOR APPENDAGES:	None
ATTITUDE CONTROL:	None
ENERGY SOURCES:	On-board propellant

EVENT DATA

DATE:	10 Jun 2006	LOCATION:	65S, 100E (dsc)
TIME:	1320 GMT	ASSESSED CAUSE:	Propulsion
ALTITUDE:	17,375 km		

PRE-EVENT ELEMENTS

EPOCH:	06161.39815228	MEAN ANOMALY:	338.3349
RIGHT ASCENSION:	289.1150	MEAN MOTION:	4.32576815
INCLINATION:	65.0603	MEAN MOTION DOT/2:	-0.00002708
ECCENTRICITY:	.5578964	MEAN MOTION DOT DOT/6:	0.0
ARG. OF PERIGEE:	74.2422	BSTAR:	-0.018697

DEBRIS CLOUD DATA

MAXIMUM ΔP:	12.9 min
MAXIMUM ΔI:	2.1 deg

COMMENTS

This event marks the 34th known breakup of a Proton Block DM SOZ ullage motor since 1984. There were two more fragmentation events for this object during July 2006; the second event was on July 3; the third event was July 27. These events resulted in the most debris cataloged of any SOZ ullage motor in history with over 100 pieces cataloged. The majority of debris (>75) were created during the second event on 3 July 2006.

REFERENCE DOCUMENTS

History of Soviet/Russian Satellite Fragmentations-A Joint U.S.-Russian Investigation, N. L. Johnson, *et al.*, Kaman Sciences Corporation, October 1995.

"First Satellite Breakups of 2006", The Orbital Debris Quarterly News, NASA JSC, July 2006.
Available online at https://orbitaldebris.jsc.nasa.gov/quarterly-news/pdfs/ODQNv10i3.pdf.

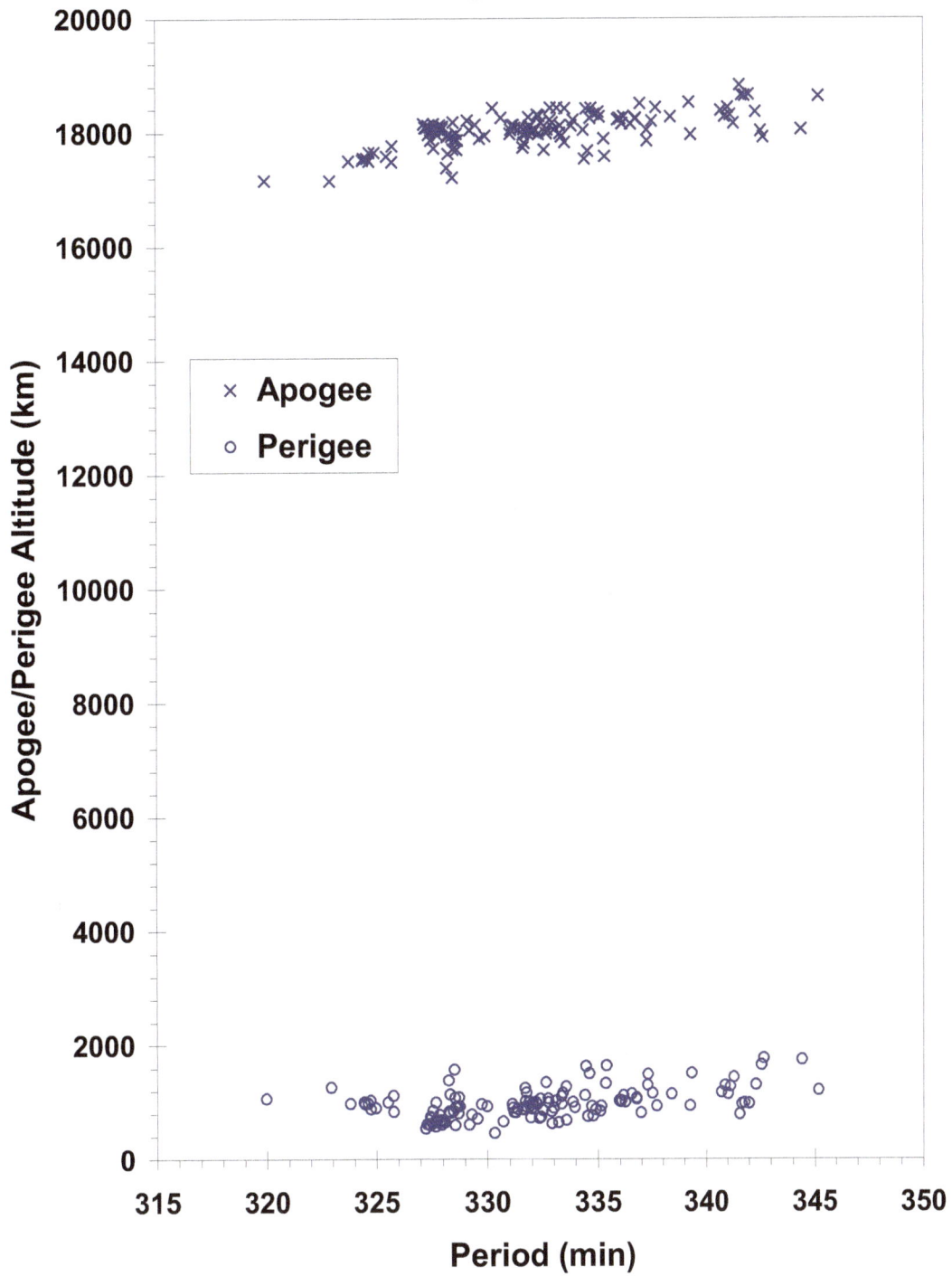

SSN 20081

SOZ motor debris cloud around September 2006 with 131 objects, reflecting debris from all three fragmentation events as reconstructed from the US SSN database.

GORIZONT 18 ULLAGE MOTOR   **1989-052F**   **20116**

SATELLITE DATA

TYPE:	Mission Related Debris
OWNER:	CIS
LAUNCH DATE:	5.95 Jul 1989
DRY MASS (KG):	55
MAIN BODY:	Ellipsoid; 0.6 m diameter by 1 m length
MAJOR APPENDAGES:	None
ATTITUDE CONTROL:	None
ENERGY SOURCES:	On-board propellants

EVENT DATA

DATE:	12 Jan 1993	LOCATION:	Unknown
TIME:	Unknown	ASSESSED CAUSE:	Propulsion
ALTITUDE:	Unknown		

PRE-EVENT ELEMENTS

EPOCH:	93004.96424013	MEAN ANOMALY:	353.7659
RIGHT ASCENSION:	215.2912	MEAN MOTION:	2.68234049
INCLINATION:	46.7556	MEAN MOTION DOT/2:	.00007021
ECCENTRICITY:	.6967354	MEAN MOTION DOT DOT/6:	.0
ARG. OF PERIGEE:	45.1358	BSTAR:	.0017532

DEBRIS CLOUD DATA

MAXIMUM ΔP:	Unknown
MAXIMUM ΔI:	Unknown

COMMENTS

Parent satellite was one of two small engine units that are routinely released after the first burn of the Proton fourth stage. The nature of these objects was identified by Dr. Boris V. Chernlatiev, Deputy Constructor for the Energiya NPO, in October 1992. The cause of this fragmentation is assumed to be related to the residual hypergolic propellants on board and failure of the membrane separating the fuel and oxidizer. NAVSPASUR observed 18 objects that were associated with this breakup. Only five element sets were generated and were of insufficient quality for a credible Gabbard Diagram or BLAST point. This was the eighth in a series of fragmentations of this object type and was the third located in a geosynchronous transfer orbit.

REFERENCE DOCUMENTS

The Fragmentation of Proton Debris, D. J. Nauer, TBE Technical Report CS93-LKD-004, Teledyne Brown Engineering, Colorado Springs, 31 December 1992.

Analysis of Fragmentations From December 1992 - February 1993, TBE Technical Report CS93-LKD-010, Teledyne Brown Engineering, Colorado Springs, 30 March 1993.

History of Soviet/Russian Satellite Fragmentations-A Joint U.S.-Russian Investigation, N. L. Johnson, et al., Kaman Sciences Corporation, October 1995.

"Identification and Resolution of an Orbital Debris Problem with the Proton Launch Vehicle," B. V. Cherniatiev, et al., Proceedings of the First European Conference on Space Debris, April 1993.

Insufficient data to construct a Gabbard diagram.

SATELLITE DATA

TYPE:	Payload
OWNER:	CIS
LAUNCH DATE:	12.63 Jul 1989
DRY MASS (KG):	5700
MAIN BODY:	Cone-cylinder; 2.7 m diameter by 6.3 m length
MAJOR APPENDAGES:	Solar panels
ATTITUDE CONTROL:	Active, 3-axis
ENERGY SOURCES:	On-board propellants, explosive charge

EVENT DATA

DATE:	28 Jul 1989	LOCATION:	35-65N, 95-140E (asc)
TIME:	0410-0420 GMT	ASSESSED CAUSE:	Deliberate
ALTITUDE:	150 km		

PRE-EVENT ELEMENTS

EPOCH:	89208.98384568	MEAN ANOMALY:	302.7810
RIGHT ASCENSION:	89.7470	MEAN MOTION:	16.33519268
INCLINATION:	67.1441	MEAN MOTION DOT/2:	.03079561
ECCENTRICITY:	.0048139	MEAN MOTION DOT DOT/6:	.000029506
ARG. OF PERIGEE:	57.9032	BSTAR:	.00023479

DEBRIS CLOUD DATA

MAXIMUM ΔP:	7.1 min*
MAXIMUM ΔI:	1.3 deg*

*Based on uncataloged debris data

COMMENTS

Spacecraft was destroyed after a malfunction prevented controlled reentry and landing in the Soviet Union. Early element sets on only 20 objects available. Rapid decay of objects made calculation of breakup time and location difficult.

REFERENCE DOCUMENTS

The Fragmentation of Kosmos 2030, N. L. Johnson, Technical Report CS89-TR-JSC-002, Teledyne Brown Engineering, Colorado Springs, Colorado, September 1989.

History of Soviet/Russian Satellite Fragmentations-A Joint U.S.-Russian Investigation, N. L. Johnson, et al., Kaman Sciences Corporation, October 1995.

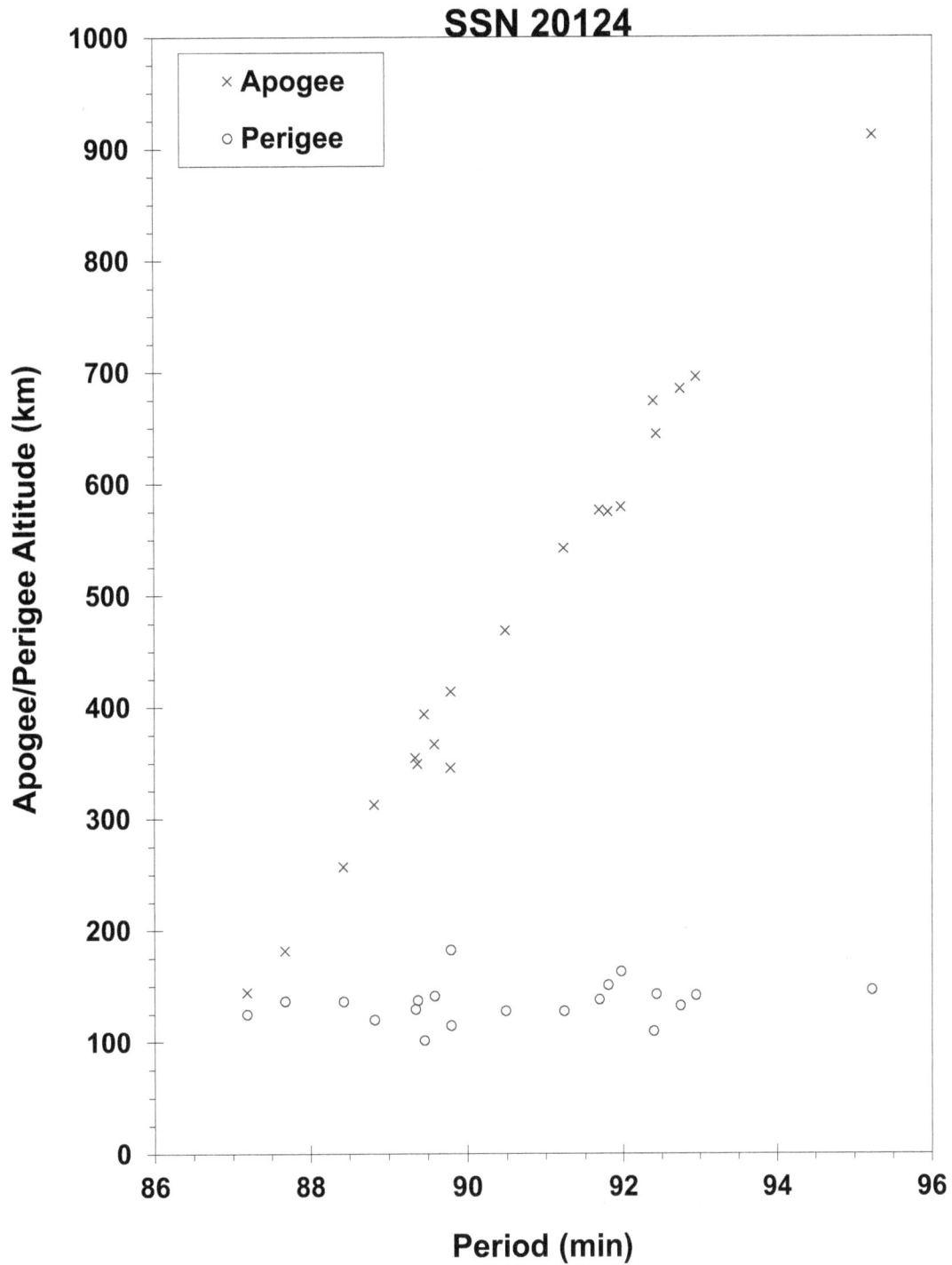

Cosmos 2030 debris cloud remnant of 20 objects two to 3 days after the event as reconstructed from the US SSN database. This diagram is taken from the cited reference.

SATELLITE DATA

TYPE:	Payload
OWNER:	CIS
LAUNCH DATE:	18.51 Jul 1989
DRY MASS (KG):	6000
MAIN BODY:	Cylinder; 2.4 m diameter by 7 m length
MAJOR APPENDAGES:	Solar panels
ATTITUDE CONTROL:	Active, 3-axis
ENERGY SOURCES:	On-board propellants, explosive charge

EVENT DATA

DATE:	31 Aug 1989	LOCATION:	43N, 111E (dsc)
TIME:	1851 GMT	ASSESSED CAUSE:	Deliberate
ALTITUDE:	270 km		

PRE-EVENT ELEMENTS

EPOCH:	89243.76468690	MEAN ANOMALY:	305.4386
RIGHT ASCENSION:	242.9132	MEAN MOTION:	15.89273241
INCLINATION:	50.5464	MEAN MOTION DOT/2:	.00196451
ECCENTRICITY:	.0093577	MEAN MOTION DOT DOT/6:	.00002154
ARG. OF PERIGEE:	55.5300	BSTAR:	.00045172

DEBRIS CLOUD DATA

MAXIMUM ΔP:	7.4 min*
MAXIMUM ΔI:	0.9 deg*

*Based on uncataloged debris data

COMMENTS

Spacecraft was destroyed with a planned detonation. Cosmos 2031 was the first of a new series of spacecraft that employs end-of-mission detonation as standard operating procedure. Early elements on 43 objects available. Most debris reentered before being officially cataloged.

REFERENCE DOCUMENTS

The Fragmentation of Kosmos 2031, N. L. Johnson, Technical Report CS89-TR-JSC-003, Teledyne Brown Engineering, Colorado Springs, Colorado, September 1989.

History of Soviet/Russian Satellite Fragmentations-A Joint U.S.-Russian Investigation, N. L. Johnson, et al., Kaman Sciences Corporation, October 1995.

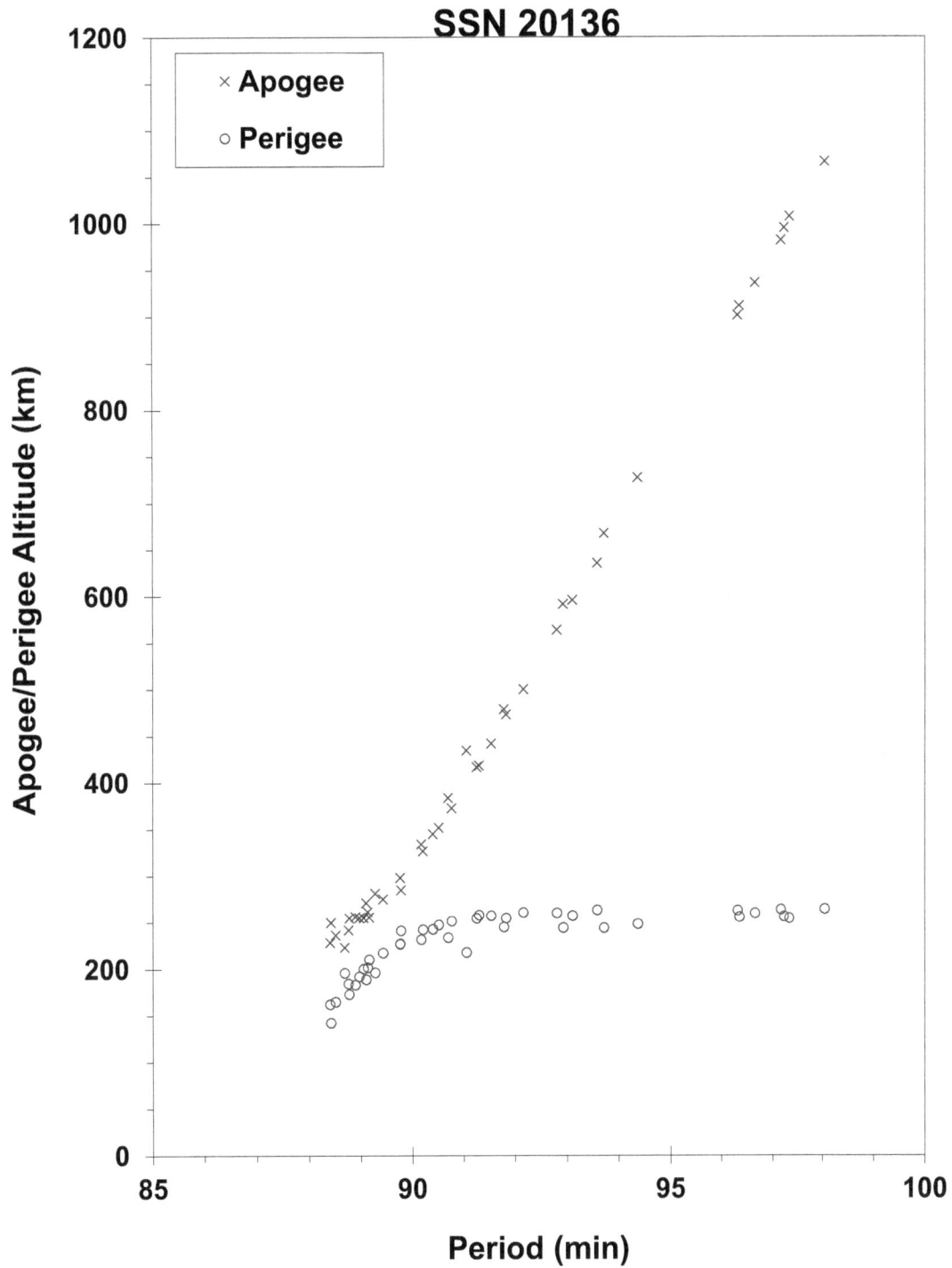

SSN 20136

Legend:
- × Apogee
- ○ Perigee

Y-axis: Apogee/Perigee Altitude (km)

X-axis: Period (min)

Cosmos 2031 debris cloud remnant of 43 objects 3 days after the event as reconstructed from Naval Space Surveillance System database. This diagram is taken from the cited reference.

SATELLITE DATA

TYPE:	Rocket Body
OWNER:	US
LAUNCH DATE:	18.61 Nov 1989
DRY MASS (KG):	920
MAIN BODY:	Cylinder; 2.4 m diameter by 6.0 m length
MAJOR APPENDAGES:	None
ATTITUDE CONTROL:	None at time of event
ENERGY SOURCES:	None at time of event

EVENT DATA

DATE:	3 Dec 2006	LOCATION:	52S, 168E (dsc)
TIME:	0337 GMT	ASSESSED CAUSE:	Unknown
ALTITUDE:	730 km		

PRE-EVENT ELEMENTS

EPOCH:	06336.48315357	MEAN ANOMALY:	65.8381
RIGHT ASCENSION:	123.6830	MEAN MOTION:	14.46527792
INCLINATION:	97.0839	MEAN MOTION DOT/2:	0.00000076
ECCENTRICITY:	.0073269	MEAN MOTION DOT DOT/6:	0.0
ARG. OF PERIGEE:	293.5127	BSTAR:	0.000029963

DEBRIS CLOUD DATA

MAXIMUM ΔP:	2.1 min
MAXIMUM ΔI:	0.3 deg

COMMENTS

The Delta 2 rocket body had been passivated and dormant for 17 years. Observation of the object showed a high tumbling rate after the event. A collision with a smaller particle has not been ruled out. For the most part, the cataloged objects from this event were short-lived, *i.e.*, less than 1 month.

REFERENCE DOCUMENT

"Significant Increase in Satellite Breakups During 2006", The Orbital Debris Quarterly News, NASA JSC, January 2007. Available online at https://orbitaldebris.jsc.nasa.gov/quarterly-news/pdfs/ODQNv11i1.pdf.

SSN 20323

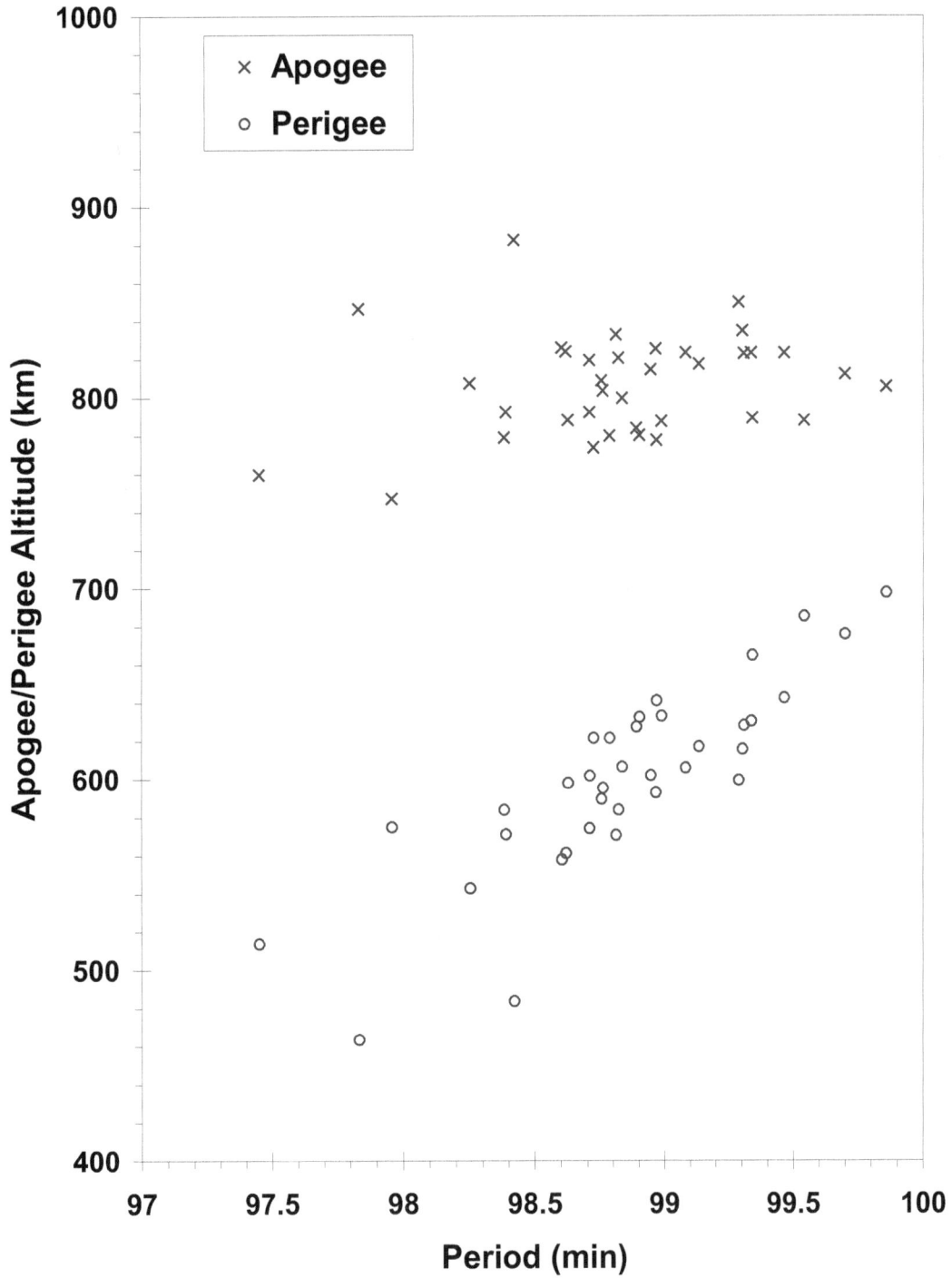

COBE R/B debris cloud with 36 fragments, 6 days after the event
as reconstructed from the US SSN database.

SATELLITE DATA

TYPE:	Tsyklon Third Stage
OWNER:	CIS
LAUNCH DATE:	27 Dec 1989
DRY MASS (KG):	1360
MAIN BODY:	Cone-cylinder; 2.1 m diameter by 3.3 m length
MAJOR APPENDAGES:	None
ATTITUDE CONTROL:	None
ENERGY SOURCES:	Unknown

EVENT DATA

DATE:	18 Apr 1999	LOCATION:	16.9S, 234.1E
TIME:	0119 GMT	ASSESSED CAUSE:	Propulsion
ALTITUDE:	485 km		

POST-EVENT ELEMENTS

EPOCH:	99107.56102679	MEAN ANOMALY:	26.3814
RIGHT ASCENSION:	275.5509	MEAN MOTION:	15.29126555517603
INCLINATION:	73.5159	MEAN MOTION DOT/2:	.00003667
ECCENTRICITY:	.0010450	MEAN MOTION DOT DOT/6:	00000-0
ARG. OF PERIGEE:	333.6852	BSTAR:	.0013164

DEBRIS CLOUD DATA

MAXIMUM ΔP:	4.18 min
MAXIMUM ΔI:	.66 deg

COMMENTS

This is the 3rd event of the Tsyklon third stage (SL-14) identified to date, and the second within 14 months. All stages have been about 10 years old at the time of breakup. The vehicle is a Ukrainian-produced stage with unknown end-of-mission passivation. Its propellants are UDMH and N204. More than 60 debris objects were detected.

REFERENCE DOCUMENT

"Third Tsyklon Upper Stage Breaks Up," The Orbital Debris Quarterly News, NASA JSC, July 1999. Available online at https://orbitaldebris.jsc.nasa.gov/quarterly-news/pdfs/ODQNv4i3.pdf

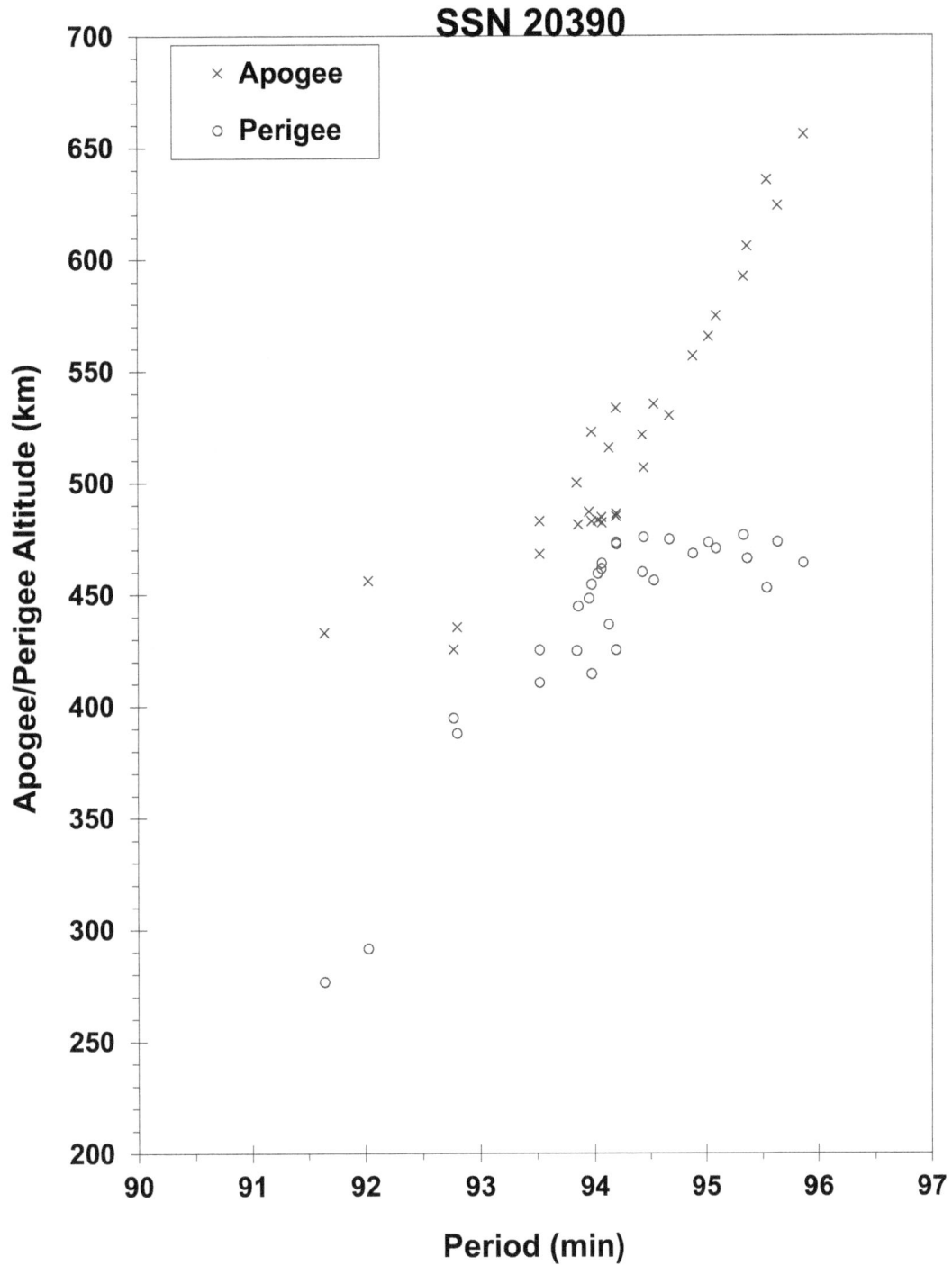

SSN 20390

Cosmos 2053 rocket body debris cloud of 31 fragments 9 days after the event as reconstructed from the US SSN database.

SATELLITE DATA

TYPE:	Mission Related Debris
OWNER:	CIS
LAUNCH DATE:	27.47 Dec 1989
DRY MASS (KG):	55
MAIN BODY:	Ellipsoid; 0.6 m diameter by 1 m length
MAJOR APPENDAGES:	None
ATTITUDE CONTROL:	None
ENERGY SOURCES:	On-board propellants

EVENT DATA

DATE:	Jul 1992 (?)	LOCATION:	Unknown
TIME:	Unknown	ASSESSED CAUSE:	Propulsion
ALTITUDE:	Unknown		

PRE-EVENT ELEMENTS

EPOCH:	92182.661921495	MEAN ANOMALY:	6.2737
RIGHT ASCENSION:	305.7529	MEAN MOTION:	2.98492104
INCLINATION:	47.1115	MEAN MOTION DOT/2:	.00001757
ECCENTRICITY:	.6700939	MEAN MOTION DOT DOT/6:	.0
ARG. OF PERIGEE:	319.3202	BSTAR:	.0014976

DEBRIS CLOUD DATA

MAXIMUM ΔP:	Unknown
MAXIMUM ΔI:	Unknown

COMMENTS

Parent satellite was one of two small engine units that are routinely released after the first burn of the Proton fourth stage. The nature of these objects was identified by Dr. Boris V. Chernlatiev, Deputy Constructor for the Energiya NPO, in October 1992. The cause of this fragmentation is assumed to be related to the residual hypergolic propellants on board and failure of the membrane separating the fuel and oxidizer. NAVSPASUR observed 18 objects that were associated with this breakup. Twelve element sets were generated but were of insufficient quality for a credible Gabbard Diagram or BLAST point. One object was cataloged on this event in early August 1992. This was the fifth in a series of fragmentations of this object type and was the first located in a geosynchronous transfer orbit.

REFERENCE DOCUMENTS

The Fragmentation of Proton Debris, D. J. Nauer, TBE Technical Report CS93-LKD-004, Teledyne Brown Engineering, Colorado Springs, 31 December 1992.

Analysis of Fragmentations From December 1992 - February 1993, TBE Technical Report CS93-LKD-010, Teledyne Brown Engineering, Colorado Springs, 30 March 1993.

History of Soviet/Russian Satellite Fragmentations-A Joint U.S.-Russian Investigation, N. L. Johnson, et al., Kaman Sciences Corporation, October 1995.

"Identification and Resolution of an Orbital Debris Problem with the Proton Launch Vehicle," B. V. Cherniatiev, et al., Proceedings of the First European Conference on Space Debris, April 1993.

Insufficient data to construct a Gabbard diagram.

COSMOS 2079-81 ULLAGE MOTOR   **1990-045F**   **20630**

SATELLITE DATA

TYPE:	Mission Related Debris
OWNER:	CIS
LAUNCH DATE:	20.0 May 1990
DRY MASS (KG):	56
MAIN BODY:	ellipsoid; 0.6 m diameter x 1 m length
MAJOR APPENDAGES:	none
ATTITUDE CONTROL:	none
ENERGY SOURCES:	on-board propellants

EVENT DATA

DATE:	18 Nov 2011	LOCATION:	60.2S, 217.8E (asc)
TIME:	0344 GMT	ASSESSED CAUSE:	Propulsion
ALTITUDE:	1292.55 km		

PRE-EVENT ELEMENTS

EPOCH:	11321.47402816	MEAN ANOMALY:	6.2919
RIGHT ASCENSION:	24.8398	MEAN MOTION:	4.33127778
INCLINATION:	65.0258	MEAN MOTION DOT/2:	0.00000578
ECCENTRICITY:	0.5722498	MEAN MOTION DOT DOT/6:	0.0
ARG. OF PERIGEE:	332.9281	BSTAR:	0.00072502

DEBRIS CLOUD DATA

MAXIMUM ΔP:	0 min
MAXIMUM ΔI:	0 deg

COMMENTS

This is the 39th known breakup of a Proton-DM SOZ ullage unit since 1984. The breakups are believed to be caused by residual propellants. Russian officials have been aware of the problem since 1992 and have made design changes, although the date of full implementation is unknown. Only one object has been officially cataloged. Due to the difficulty in observing and cataloging debris in deep-space, highly elliptical orbits, this event may have created a much larger debris ensemble.

REFERENCE DOCUMENT

History of Soviet/Russian Satellite Fragmentations: A Joint U.S.-Russian Investigation. N.L. Johnson et al. Kaman Sciences Corporation, October 2005.

"Only a Few Minor Satellite Breakups in 2011", The Orbital Debris Quarterly News, NASA JSC, January 2012. Available online at https://orbitaldebris.jsc.nasa.gov/quarterly-news/ODQNv16i1.pdf.

Insufficient data to construct a Gabbard diagram.

SATELLITE DATA

TYPE:	Mission Related Debris
OWNER:	CIS
LAUNCH DATE:	19 May 1990
DRY MASS (KG):	55
MAIN BODY:	Ellipsoid; 0.6 m diameter by 1 m length
MAJOR APPENDAGES:	None
ATTITUDE CONTROL:	None
ENERGY SOURCES:	On-board propellants

EVENT DATA

DATE:	~28 Mar 1999	LOCATION:	Unknown
TIME:	Unknown	ASSESSED CAUSE:	Propulsion
ALTITUDE:	Unknown		

POST-EVENT ELEMENTS

EPOCH:	99087.88291821	MEAN ANOMALY:	123.5812
RIGHT ASCENSION:	319.9610	MEAN MOTION:	4.24414150137202
INCLINATION:	64.8090	MEAN MOTION DOT/2:	0.00000311
ECCENTRICITY:	0.5789417	MEAN MOTION DOT DOT/6:	0.0
ARG. OF PERIGEE:	199.4305	BSTAR:	0.0040281

DEBRIS CLOUD DATA

MAXIMUM ΔP:	Unknown
MAXIMUM ΔI:	Unknown

COMMENTS

This is the 20th event of the Proton Block DM SOZ Ullage Motor class identified to date; it is the ninth associated with a GLONASS mission. This mission was conducted before the engineering defect was identified and passivation measures implemented. In this orbit, debris may be long-lived but hard to track. 76 debris objects were detected.

REFERENCE DOCUMENT

"Abandoned Proton Ullage Motors Continue to Create Debris," The Orbital Debris Quarterly News, NASA JSC, April 1999. Available online at https://orbitaldebris.jsc.nasa.gov/quarterly-news/pdfs/ODQNv4i2.pdf.

Insufficient data to construct a Gabbard diagram.

SATELLITE DATA

TYPE:	CZ-4A Final Stage
OWNER:	PRC
LAUNCH DATE:	3.04 Sep 1990
DRY MASS (KG):	1700
MAIN BODY:	Cylinder-Nozzle; 2.9 m diameter by ~5 m length
MAJOR APPENDAGES:	none
ATTITUDE CONTROL:	none
ENERGY SOURCES:	On-board propellants

EVENT DATA

DATE:	4 Oct 1990	LOCATION:	81S, 68E (asc)
TIME:	2014 GMT	ASSESSED CAUSE:	Propulsion
ALTITUDE:	895 km		

PRE-EVENT ELEMENTS

EPOCH:	90276.6451544	MEAN ANOMALY:	162.6773
RIGHT ASCENSION:	310.6975	MEAN MOTION:	14.01192890
INCLINATION:	98.9340	MEAN MOTION DOT/2:	.000003118
ECCENTRICITY:	.0010179	MEAN MOTION DOT DOT/6:	.0
ARG. OF PERIGEE:	197.4122	BSTAR:	.0002183343

DEBRIS CLOUD DATA

MAXIMUM ΔP:	5.8 min
MAXIMUM ΔI:	0.1 deg

COMMENTS

This second flight of the CZ-4 final stage successfully deployed three payloads (one weather satellite and two inflated balloons) into a sun-synchronous orbit. Propellants used were N_2O_4 and UDMH. An estimated 70-75 fragments were detected soon after the event.

REFERENCE DOCUMENTS

The Fragmentation of Fengyun 1-2 R/B, N. L. Johnson, Technical Report CS90-TR-JSC-013, Teledyne Brown Engineering, Colorado Springs, Colorado, November 1990.

"Analyzing the Cause of LM-4 (A)'s Upper Stage's Disintegration and the Countermeasures", W. X. Zhang and S. Y. Liao, 5th International Conference of Pacific Basin Societies, 6-9 Jun 1993, Shanghai.

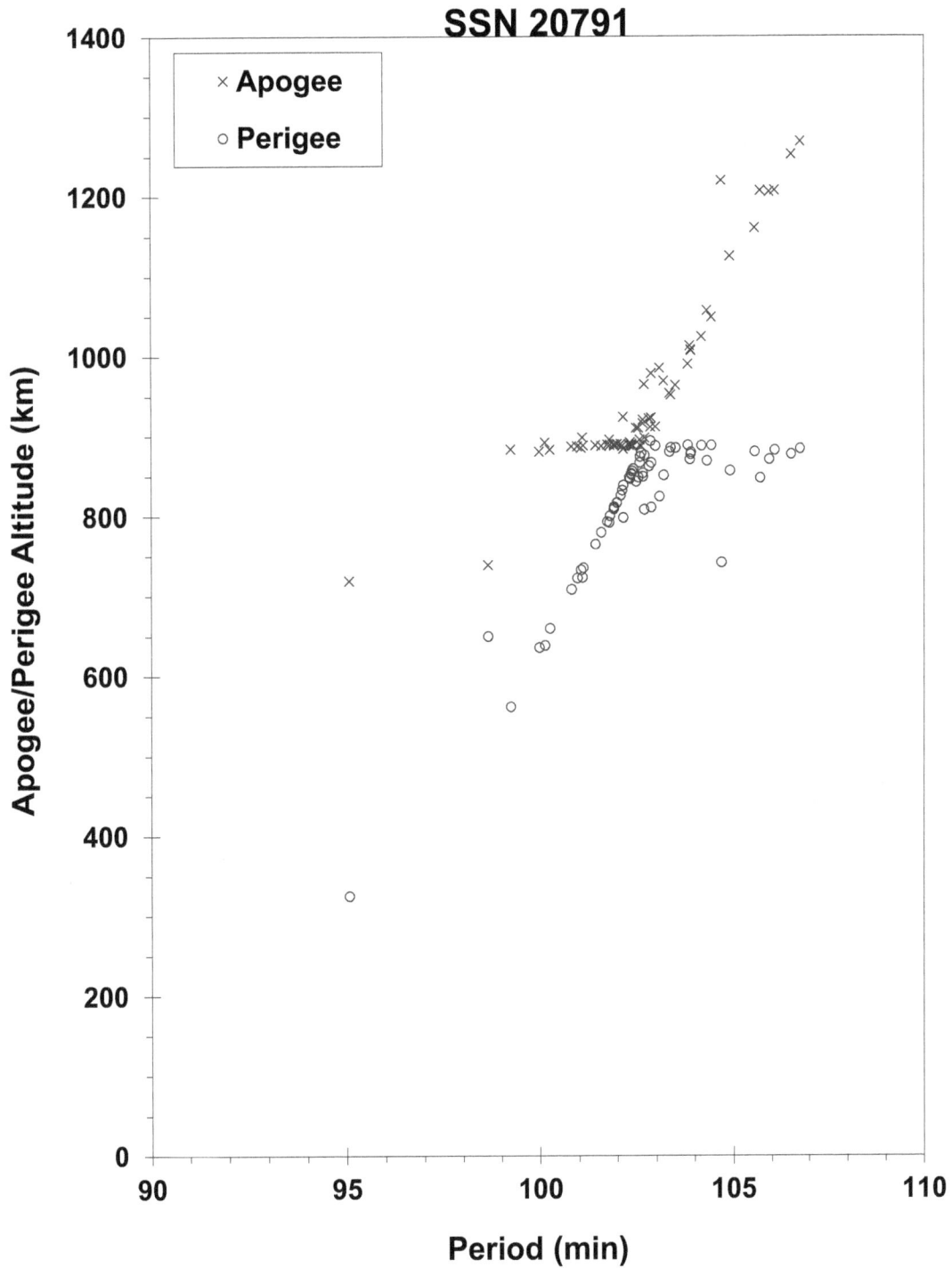

SSN 20791

Fengyun 1-2 R/B debris cloud remnant of 65 objects 5 days after the event as reconstructed from Naval Space Surveillance System database.

SATELLITE DATA

TYPE:	Payload
OWNER:	CIS
LAUNCH DATE:	1.46 Oct 1990
DRY MASS (KG):	6000
MAIN BODY:	Cylinder; 2.4 m diameter by 7 m length
MAJOR APPENDAGES:	Solar panels
ATTITUDE CONTROL:	Active, 3-axis
ENERGY SOURCES:	On-board propellants, explosive charge

EVENT DATA

DATE:	30 Nov 1990	LOCATION:	54N, 157E (dsc)
TIME:	1720 GMT	ASSESSED CAUSE:	Deliberate
ALTITUDE:	210 km		

PRE-EVENT ELEMENTS

EPOCH:	90334.45391019	MEAN ANOMALY:	205.3252
RIGHT ASCENSION:	347.9431	MEAN MOTION:	16.12811753
INCLINATION:	64.7547	MEAN MOTION DOT/2:	.00671617
ECCENTRICITY:	.0065418	MEAN MOTION DOT DOT/6:	.000035339
ARG. OF PERIGEE:	155.2258	BSTAR:	.00040815

DEBRIS CLOUD DATA

MAXIMUM ΔP:	>7.3 min*
MAXIMUM ΔI:	0.3 deg*

*Based on uncataloged debris data

COMMENTS

Spacecraft was destroyed with a planned detonation. Second fragmentation of the Cosmos 2031 subclass. Early elements on only 7 objects available. Most debris reentered before being officially cataloged.

REFERENCE DOCUMENTS

The Fragmentation of Kosmos 2101, N. L. Johnson, Technical Report CS91-TR-JSC-002, Teledyne Brown Engineering, Colorado Springs, Colorado, January 1991.

History of Soviet/Russian Satellite Fragmentations-A Joint U.S.-Russian Investigation, N. L. Johnson, *et al.,* Kaman Sciences Corporation, October 1995.

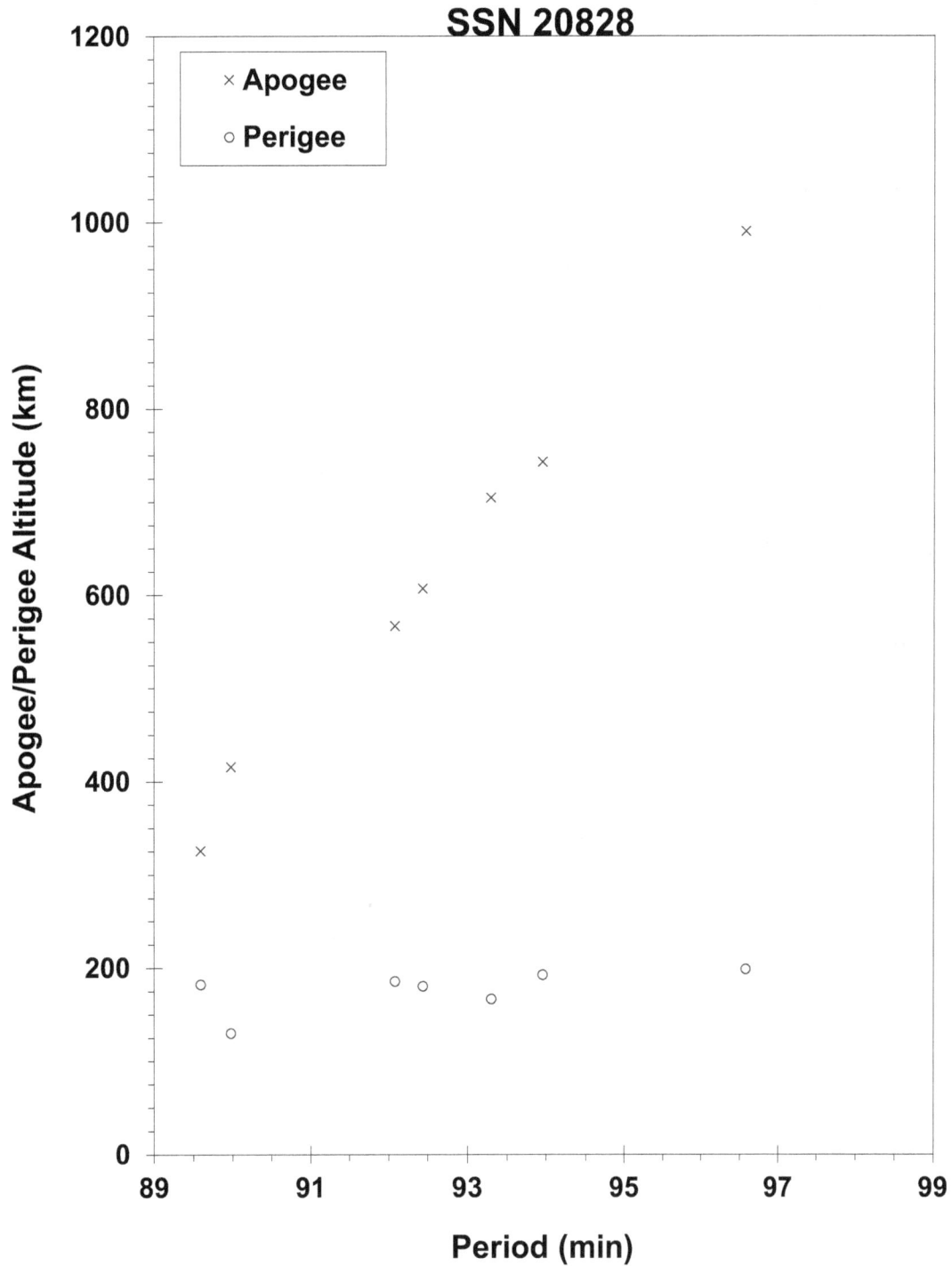

SSN 20828

Cosmos 2101 debris cloud remnant of 7 objects 3 days after the event was reconstructed from Naval Space Surveillance System database. This diagram is taken from the cited reference.

349

SATELLITE DATA

TYPE:	Mission Related Debris
OWNER:	CIS
LAUNCH DATE:	23 Nov 1990
DRY MASS (KG):	55
MAIN BODY:	Ellipsoid; 0.6 m diameter by 1 m length
MAJOR APPENDAGES:	None
ATTITUDE CONTROL:	None
ENERGY SOURCES:	On-board propellants

EVENT DATA

DATE:	Approx. 14 Dec 1995	LOCATION:	Unknown
TIME:	Unknown	ASSESSED CAUSE:	Propulsion
ALTITUDE:	Unknown		

PRE-EVENT ELEMENTS

EPOCH:	95348.79476448	MEAN ANOMALY:	300.3633
RIGHT ASCENSION:	140.3319	MEAN MOTION:	5.84898259
INCLINATION:	46.4887	MEAN MOTION DOT/2:	0.00111293
ECCENTRICITY:	0.4967539	MEAN MOTION DOT DOT/6:	0.00000006
ARG. OF PERIGEE:	117.7610	BSTAR:	0.00074791

DEBRIS CLOUD DATA

MAXIMUM ΔP:	Unknown
MAXIMUM ΔI:	Unknown

COMMENTS

Parent satellite was one of two small engine units that are routinely released after the first burn of the Proton fourth stage. The nature of these objects was identified by Dr. Boris V. Chernlatiev, Deputy Constructor for the Energiya NPO, in October 1992. The cause of this fragmentation is assumed to be related to the residual hypergolic propellants on board and failure of the membrane separating the fuel and oxidizer. NAVSPASUR observed 69 objects that were associated with this breakup.

REFERENCE DOCUMENTS

The Fragmentation of Proton Debris, D. J. Nauer, TBE Technical Report CS93-LKD-004, Teledyne Brown Engineering, Colorado Springs, 31 December 1992.

Analysis of Fragmentations From December 1992 - February 1993, TBE Technical Report CS93-LKD-010, Teledyne Brown Engineering, Colorado Springs, 30 March 1993.

History of Soviet/Russian Satellite Fragmentations-A Joint U.S.-Russian Investigation, N. L. Johnson, *et al.*, Kaman Sciences Corporation, October 1995.

"Identification and Resolution of an Orbital Debris Problem with the Proton Launch Vehicle," B. V. Cherniatiev, *et al.*, Proceedings of the First European Conference on Space Debris, April 1993.

Insufficient data to construct a Gabbard diagram.

SATELLITE DATA

TYPE:	Payload
OWNER:	US
LAUNCH DATE:	1.66 Dec 1990
DRY MASS (KG):	767
MAIN BODY:	Cylinder; 1.0 m diameter by 3.6 m length
MAJOR APPENDAGES:	1 solar panel
ATTITUDE CONTROL:	Active, 3 axis
ENERGY SOURCES:	On-board propellants

EVENT DATA

DATE:	1 Dec 1990	LOCATION:	6N, 232E (dsc)
TIME:	1610 GMT	ASSESSED CAUSE:	Propulsion
ALTITUDE:	850 km		

POST-EVENT ELEMENTS

EPOCH:	90335.71008487	MEAN ANOMALY:	0.9090
RIGHT ASCENSION:	4.0350	MEAN MOTION:	14.29892145
INCLINATION:	98.8600	MEAN MOTION DOT/2:	-0.00000049
ECCENTRICITY:	0.0080986	MEAN MOTION DOT DOT/6:	0.0
ARG. OF PERIGEE:	359.1948	BSTAR:	-0.000010171

DEBRIS CLOUD DATA

MAXIMUM ΔP:	>2.0 min*
MAXIMUM ΔI:	1.0 deg*

*Based on uncataloged debris data

COMMENTS

During the burn of USA 68's solid-fuel apogee kick motor (STAR-37S, TE-M-364-15), the 20 kg nozzle came apart, terminating thrust. At shutdown USA 68 was in an orbit of 610 km by 850 km. Immediately, a hydrazine orbit make-up system was activated, providing an additional 32.3 m/s DV. More than 40 pieces of non-Mission related debris were observed within a day of the event. The observed debris may include components of the USA 68 sun shield and AKM nozzle shield (total mass 2 kg). Most debris decayed very rapidly. The payload remained operational.

REFERENCE DOCUMENT

The Fragmentation of USA 68, N.L. Johnson, Technical Report CS91-TR-JSC-005, Teledyne Brown Engineering, Colorado Springs, Colorado, March 1991.

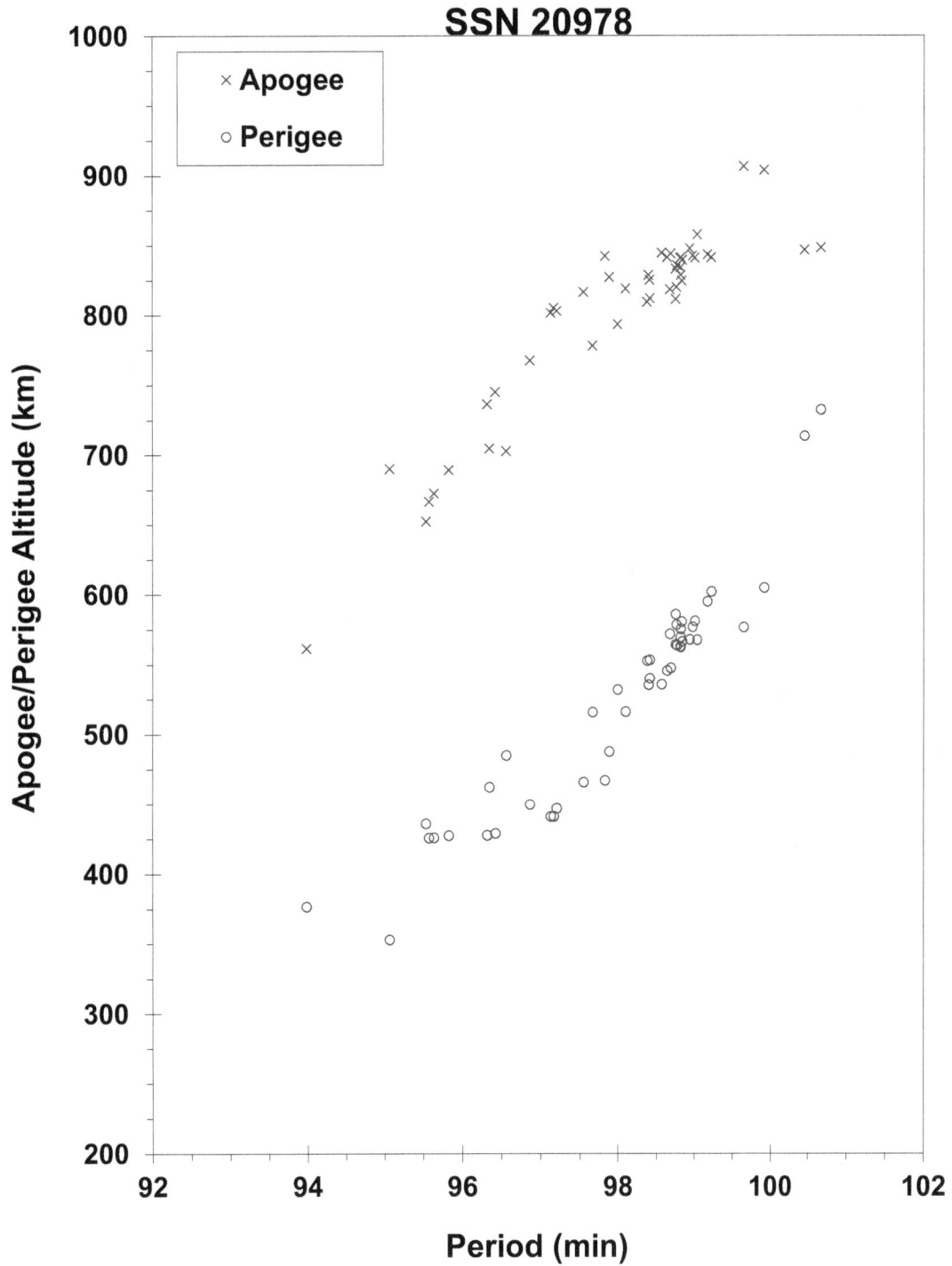

USA 68 debris cloud remnant of 48 fragments 12 days after the event as reconstructed from the US SSN database.

SATELLITE DATA

TYPE:	Mission Related Debris
OWNER:	CIS
LAUNCH DATE:	8.11 Dec 1990
DRY MASS (KG):	55
MAIN BODY:	Ellipsoid; 0.6 m diameter by 1.0 m length
MAJOR APPENDAGES:	None
ATTITUDE CONTROL:	None
ENERGY SOURCES:	On-board propellants

EVENT DATA

DATE:	21 Feb 2003	LOCATION:	34.11 S, 208.61 E
TIME:	~0300Z	ASSESSED CAUSE:	Propulsion
ALTITUDE:	~17650 km		

PRE-EVENT ELEMENTS

EPOCH:	03051.93857279	MEAN ANOMALY:	300.1330
RIGHT ASCENSION:	5.3297	MEAN MOTION:	4.24855437
INCLINATION:	65.3642	MEAN MOTION DOT/2:	-0.00000082
ECCENTRICITY:	0.5638383	MEAN MOTION DOT DOT/6:	0.0
ARG. OF PERIGEE:	126.1785	BSTAR:	-0.00014197

DEBRIS CLOUD DATA

MAXIMUM ΔP:	Unknown
MAXIMUM ΔI:	Unknown

COMMENTS

No debris was cataloged for this breakup. This is the 26th event of the Proton Block DM SOZ ullage motor since 1984. This ullage motor was launched prior to Russian recognition of the problem and before implementation of preventive measures.

REFERENCE DOCUMENT

"Satellite Fragmentations in 2003", The Orbital Debris Quarterly News, NASA JSC, January 2004. Available online at https://orbitaldebris.jsc.nasa.gov/quarterly-news/pdfs/ODQNv8i1.pdf.

Insufficient data to construct a Gabbard diagram.

SATELLITE DATA

TYPE:	Mission Related Debris
OWNER:	CIS
LAUNCH DATE:	8.11 Dec 1990
DRY MASS (KG):	55
MAIN BODY:	Ellipsoid; 0.6 m diameter by 1.0 m length
MAJOR APPENDAGES:	None
ATTITUDE CONTROL:	None
ENERGY SOURCES:	On-board propellants

EVENT DATA

DATE:	14 Mar 1998	LOCATION:	Unknown
TIME:	Unknown	ASSESSED CAUSE:	Propulsion
ALTITUDE:	Unknown		

PRE-EVENT ELEMENTS

EPOCH:	98072.07217599	MEAN ANOMALY:	85.4178
RIGHT ASCENSION:	306.4512	MEAN MOTION:	4.23530449
INCLINATION:	65.0803	MEAN MOTION DOT/2:	0.00000895
ECCENTRICITY:	0.5724061	MEAN MOTION DOT DOT/6:	0
ARG. OF PERIGEE:	216.7168	BSTAR:	0.0025728

DEBRIS CLOUD DATA

MAXIMUM ΔP:	Unknown
MAXIMUM ΔI:	Unknown

COMMENTS

This is the 16th event of this class identified to date; it is the fifth associated with a GLONASS mission. More than 110 debris detected, but element sets developed for only a few.

REFERENCE DOCUMENTS

"Identification and Resolution of an Orbital Debris Problem with the Proton Launch Vehicle", Cherniatiev, Chernyavskiy, Johnson, and McKnight, First European Conference on Space Debris, 5-7 April 1993.

"The Fragmentation of Proton Debris", Nauer, Teledyne Brown Engineering Technical Report CS93-LKD-004, 31 December 1992.

History of Soviet/Russian Satellite Fragmentations-A Joint U.S.-Russian Investigation, N. L. Johnson, *et al.,* Kaman Sciences Corporation, October 1995.

"Three Upper Stage Breakups in One Week Top February Debris Activity," The Orbital Debris Quarterly News, NASA JSC, April 1998, p. 1-2. Available online at https://orbitaldebris.jsc.nasa.gov/quarterly-news/pdfs/ODQNv3i2.pdf.

Insufficient data to construct a Gabbard diagram.

ITALSAT 1/EUTELSAT 2 F2 R/B　　　　　**1991-003C**　　　　　**21057**

SATELLITE DATA

TYPE:	Ariane 4 H-10 Third Stage
OWNER:	France
LAUNCH DATE:	15.97 Jan 1991
DRY MASS (KG):	1760
MAIN BODY:	Cylinder; 2.6 m diameter by 11.4 m length
MAJOR APPENDAGES:	None
ATTITUDE CONTROL:	None at time of the event.
ENERGY SOURCES:	On-board propellants

EVENT DATA

DATE:	Late Apr-early May 1996	LOCATION:	Unknown
TIME:	Unknown	ASSESSED CAUSE:	Propulsion
ALTITUDE:	Unknown		

PRE-EVENT ELEMENTS

EPOCH:	96106.15481796	MEAN ANOMALY:	312.6005
RIGHT ASCENSION:	104.8696	MEAN MOTION:	2.66496263
INCLINATION:	6.7146	MEAN MOTION DOT/2:	0.00007071
ECCENTRICITY:	0.6989841	MEAN MOTION DOT DOT/6:	0
ARG. OF PERIGEE:	132.7372	BSTAR:	0.0012265

DEBRIS CLOUD DATA

MAXIMUM ΔP:	147.3 min
MAXIMUM ΔI:	1.3 deg

COMMENTS

The event was first recognized by Naval Space Command analysts in early May 1996. Element sets for as many as 20 debris were developed. Since deliberate passivation of Ariane GTO stages was not implemented until 1993, the vehicle was not purged of its residual propellants or pressurants.

REFERENCE DOCUMENT

"Newly Recognized 1996 Breakup", N. L. Johnson, <u>The Orbital Debris Quarterly News</u>, April 1997, p. 2. Available online at https://orbitaldebris.jsc.nasa.gov/quarterly-news/pdfs/ODQNv2i2.pdf.

SSN 21057

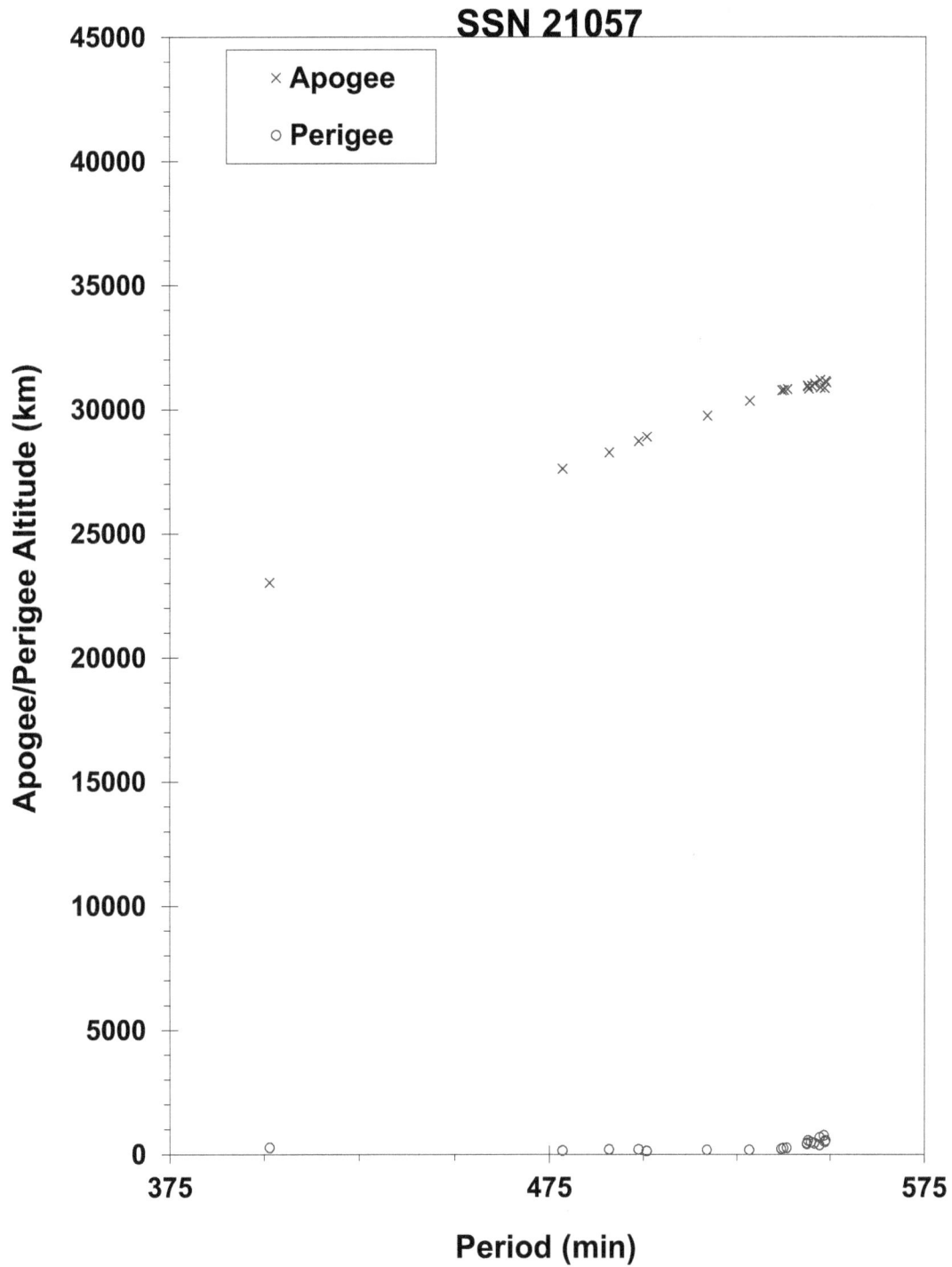

Italsat 1/Eutelsat 2 F2 R/B debris cloud of 20 fragments 1 year after
the event as reconstructed from the US SSN database.

SATELLITE DATA

TYPE:	Cosmos Second Stage
OWNER:	CIS
LAUNCH DATE:	12.12 Feb 1991
DRY MASS (KG):	1435
MAIN BODY:	Cylinder; 2.4 m diameter by 6.6 m length
MAJOR APPENDAGES:	Payload deployment mechanism
ATTITUDE CONTROL:	None at time of the event.
ENERGY SOURCES:	Unknown

EVENT DATA

DATE:	5 Mar 1991	LOCATION:	43S, 140E (asc)
TIME:	1345 GMT	ASSESSED CAUSE:	Propulsion
ALTITUDE:	1560 km		

PRE-EVENT ELEMENTS

EPOCH:	91062.94236834	MEAN ANOMALY:	112.8991
RIGHT ASCENSION:	166.0317	MEAN MOTION:	12.19552620
INCLINATION:	74.0386	MEAN MOTION DOT/2:	0.00000005
ECCENTRICITY:	0.0166507	MEAN MOTION DOT DOT/6:	0.0
ARG. OF PERIGEE:	245.0348	BSTAR:	0.000099999

DEBRIS CLOUD DATA

MAXIMUM ΔP:	4.3 min*
MAXIMUM ΔI:	0.3 deg*

*Based on uncataloged debris data

COMMENTS

This is the second known fragmentation of the Cosmos second stage and the first in more than 25 years and 370 missions. Like the earlier event (Cosmos 61-63 R/B), this rocket body successfully completed its multiple payload delivery before breakup. NAVSPASUR determined that several minor separations occurred both prior to and after the main breakup cited above (see NAVSPASUR report referenced below).

REFERENCE DOCUMENTS

Cosmos 2125-2132 Rocket Body (U), Fragmentation and Breakup Report (U), E.L. Jenkins and R.E. Farmer, Naval Space Surveillance Center, Dahlgren, Virginia, April 1991.

A Preliminary Analysis of the Fragmentations of the Kosmos 2125-2132 Rocket Body, N.L. Johnson, Technical Report CS91-TR-JSC-007, Teledyne Brown Engineering, Colorado Springs, Colorado, April 1991.

History of Soviet/Russian Satellite Fragmentations-A Joint U.S.-Russian Investigation, N. L. Johnson, et al., Kaman Sciences Corporation, October 1995.

"The Recent Fragmentations of LEO Upper Stages," G. Chernyavskiy, et al., 45th IAF Congress, 1994.

SSN 21108

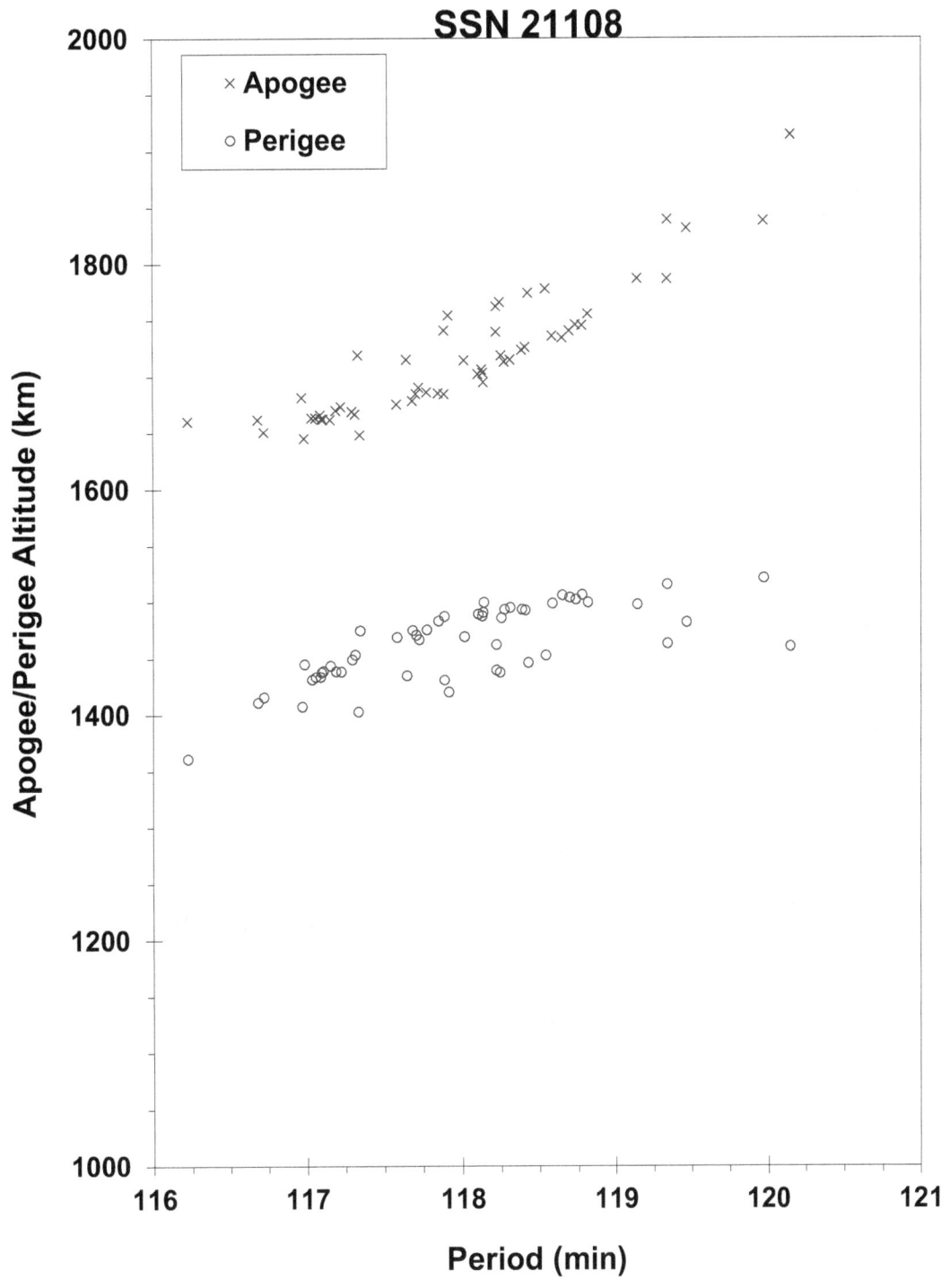

Cosmos 2125-32 R/B debris cloud of 54 objects 5 days after the major breakup event as reconstructed from Naval Space Surveillance System database. This diagram is taken from the reference cited at the top of this page.

SATELLITE DATA

TYPE:	Mission Related Debris
OWNER:	CIS
LAUNCH DATE:	12 Feb 1991
DRY MASS (KG):	55
MAIN BODY:	Ellipsoid; 0.6 m diameter by 1 m length
MAJOR APPENDAGES:	None
ATTITUDE CONTROL:	Unknown
ENERGY SOURCES:	On-board propellants

EVENT DATA

DATE:	7 May 1994	LOCATION:	10N, 112E
TIME:	0930 GMT	ASSESSED CAUSE:	Propulsion
ALTITUDE:	16195 km		

PRE-EVENT ELEMENTS

EPOCH:	94126.31580012	MEAN ANOMALY:	240.6661
RIGHT ASCENSION:	110.6447	MEAN MOTION:	3.78477656
INCLINATION:	46.6223	MEAN MOTION DOT/2:	0.000127656
ECCENTRICITY:	0.6204369	MEAN MOTION DOT DOT/6:	0.0
ARG. OF PERIGEE:	160.8637	BSTAR:	0.00086951

DEBRIS CLOUD DATA

MAXIMUM ΔP:	Unknown
MAXIMUM ΔI:	Unknown

COMMENTS

Parent satellite was one of two small engine units that are routinely released after the first burn of the Proton fourth stage. The nature of these objects was identified by Dr. Boris V. Chernlatiev, Deputy Constructor for the Energiya NPO, in October 1992. The cause of this fragmentation appears to be related to the residual hypergolic propellants on board and failure of the membrane separating the fuel and oxidizer. NAVSPASUR observed 38 objects that were associated with this breakup. Only 6 element sets were generated. This was the ninth in a series of fragmentations of this object type and was the fourth located in a geosynchronous transfer orbit. Two possible fragmentation locations were calculated by the NAVSPOC. The numbers above represent the first possible calculated location.

REFERENCE DOCUMENTS

The Fragmentation of Proton Debris, D. J. Nauer, TBE Technical Report CS93-LKD-004, Teledyne Brown Engineering, Colorado Springs, 31 December 1992.

Analysis of Fragmentations From December 1992 - February 1993, TBE Technical Report CS93-LKD-010, Teledyne Brown Engineering, Colorado Springs, 30 March 1993.

The Fragmentation of Cosmos 2133 Debris, I. W. Grissom and D. J. Nauer, TBE Technical Report CS94-LKD-016, Teledyne Brown Engineering, Colorado Springs, 30 June 1994.

History of Soviet/Russian Satellite Fragmentations-A Joint U.S.-Russian Investigation, N. L. Johnson, *et al.,* Kaman Sciences Corporation, October 1995.

"Identification and Resolution of an Orbital Debris Problem with the Proton Launch Vehicle," B. V. Cherniatiev, *et al., Proceedings of the First European Conference on Space Debris,* April 1993.

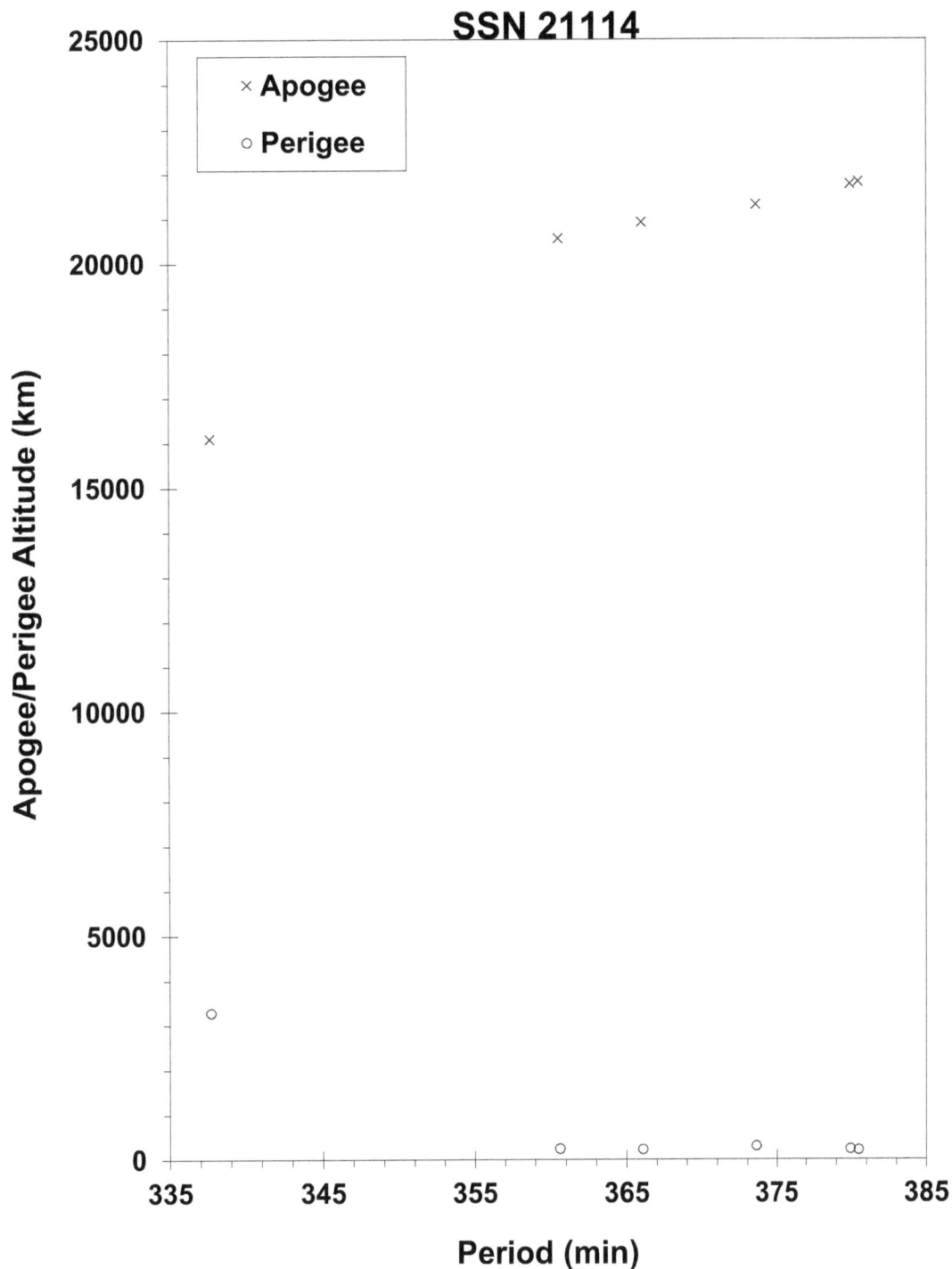

Gabbard diagram of six objects from the Cosmos 2133 debris fragmentation.

SATELLITE DATA

TYPE:	Ariane 4 H10 Third Stage
OWNER:	France
LAUNCH DATE:	2.98 Mar 1991
DRY MASS (KG):	1760
MAIN BODY:	Cylinder; 2.6 m diameter by 10 m length
MAJOR APPENDAGES:	None
ATTITUDE CONTROL:	None at time of the event.
ENERGY SOURCES:	On-board propellants

EVENT DATA

DATE:	27 Apr 1994	LOCATION:	0.5S, 79E (dsc)
TIME:	0144 GMT	ASSESSED CAUSE:	Propulsion
ALTITUDE:	270 km		

PRE-EVENT ELEMENTS

EPOCH:	94116.17965845	MEAN ANOMALY:	157.2349
RIGHT ASCENSION:	136.1778	MEAN MOTION:	2.86975555
INCLINATION:	6.5808	MEAN MOTION DOT/2:	0.00006058
ECCENTRICITY:	0.6829164	MEAN MOTION DOT DOT/6:	0.0
ARG. OF PERIGEE:	185.9406	BSTAR:	0.001267

DEBRIS CLOUD DATA

MAXIMUM ΔP:	148 min
MAXIMUM ΔI:	1.1 deg

COMMENTS

The fragmentation of this R/B occurred over 37 months after launch. Since deliberate passivation of Ariane GTO stages was not implemented until 1993, the vehicle was not purged of its residual propellants or pressurants. As many as 30 debris fragments were detected.

REFERENCE DOCUMENTS

TRW Space Log 1957-1991. TRW Space and Defense Sector, Space and Technology Group. Redondo Beach, CA. 1992.

Space News, "Ariane Rocket Flies After Electrical Glitch Delay, Volume 2, Number 8, 11-17 March 1991.

Rockets of the World. Peter Alway, Ann Arbor, MI, 1993.

The Fragmentation of the Astra 1B/MOP 2 (1) Rocket Body, I. W. Grissom and D. J. Nauer, TBE Technical Report CS94-LKD-014, Teledyne Brown Engineering, Colorado Springs, 15 May 1994.

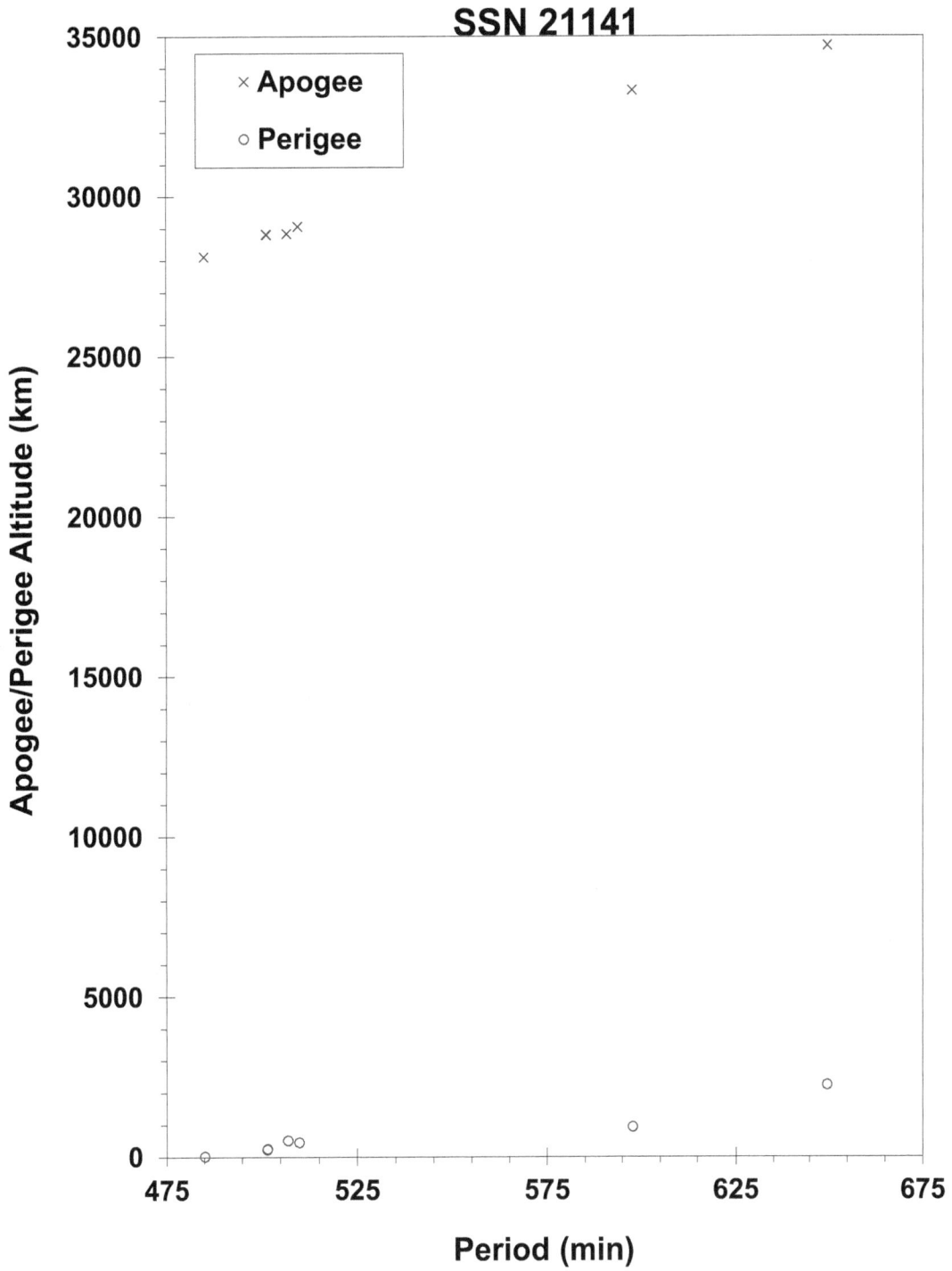

Astra 1B/MOP 2 R/B debris cloud of 7 fragments as reconstructed from the US SSN database.

COSMOS 2139-41 ULLAGE MOTOR   **1991-025F**   **21220**

SATELLITE DATA

TYPE:	Mission Related Debris
OWNER:	CIS
LAUNCH DATE:	04.0 Apr 1991
DRY MASS (KG):	55
MAIN BODY:	ellipsoid; 0.6 m diameter x 1 m length
MAJOR APPENDAGES:	none
ATTITUDE CONTROL:	none
ENERGY SOURCES:	on-board propellants

EVENT DATA

DATE:	08 Mar 2009	LOCATION:	35.5S, 21.8E (dsc)
TIME:	1000 GMT	ASSESSED CAUSE:	Propulsion
ALTITUDE:	18,544.0 km		

PRE-EVENT ELEMENTS

EPOCH:	09066.60605701	MEAN ANOMALY:	350.1859
RIGHT ASCENSION:	138.9363	MEAN MOTION:	4.33919111
INCLINATION:	64.9463	MEAN MOTION DOT/2:	0.00000158
ECCENTRICITY:	0.5689328	MEAN MOTION DOT DOT/6:	0.0
ARG. OF PERIGEE:	41.0118	BSTAR:	0.00042936

DEBRIS CLOUD DATA

MAXIMUM ΔP:	200.0 min
MAXIMUM ΔI:	19.3 deg

COMMENTS

This is the 37th known breakup of a Proton-DM SOZ ullage unit since 1984. The breakups are believed to be caused by residual propellants. Russian officials have been aware of the problem since 1992 and have made design changes, although the date of full implementation is unknown. Due to the difficulty in observing and cataloging debris in deep-space, highly elliptical orbits, a debris ensemble much larger than that cataloged may have been created by this event.

REFERENCE DOCUMENT

History of Soviet/Russian Satellite Fragmentations: A Joint U.S.-Russian Investigation. N.L. Johnson et al. Kaman Sciences Corporation, October 2005.

"Minor March Satellite Break-Up," The Orbital Debris Quarterly News, NASA JSC, April 2009. Available online at https://orbitaldebris.jsc.nasa.gov/quarterly-news/ODQNv13i2.pdf.

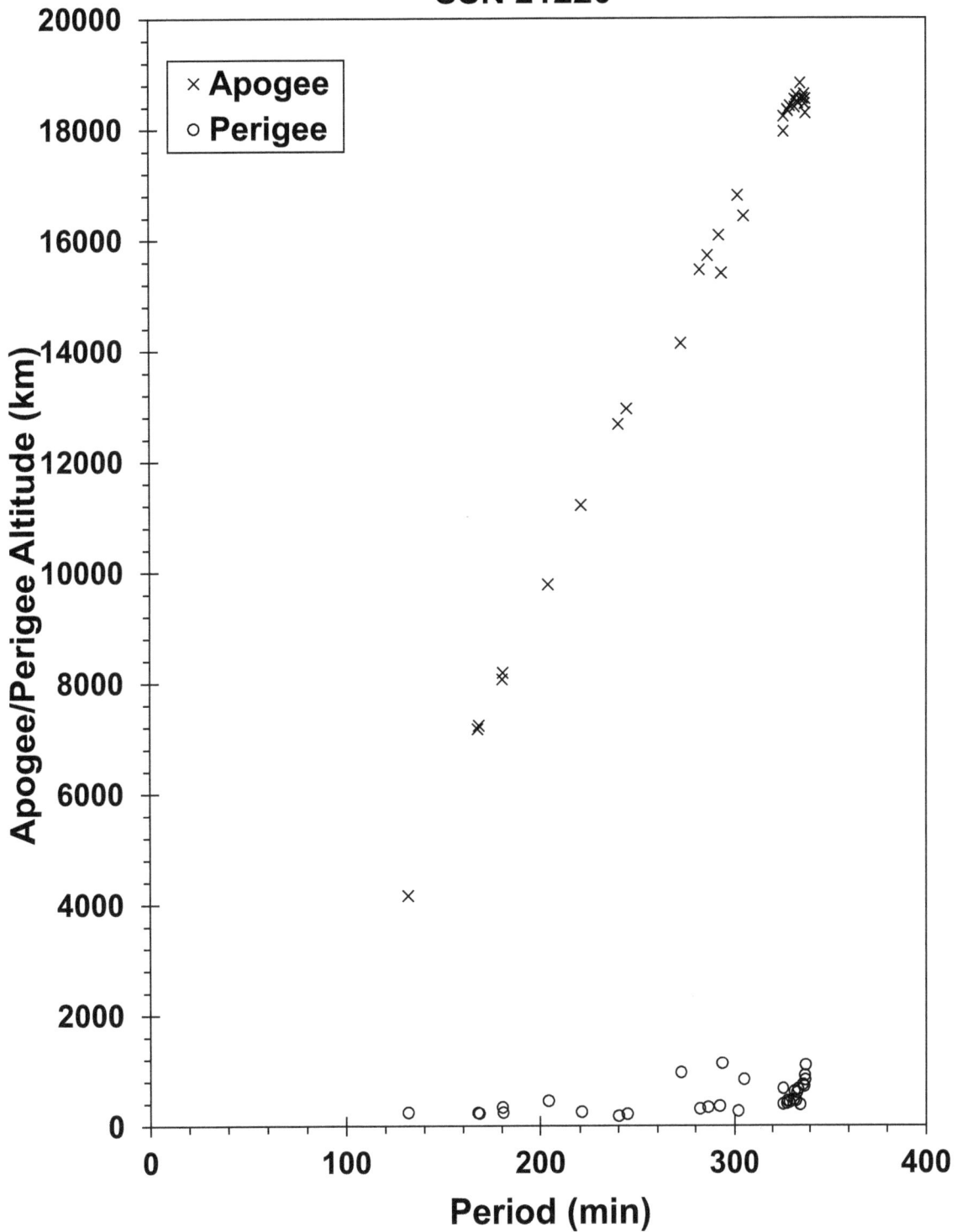

SSN 21220

The COSMOS ullage motor debris cloud, including 32 fragments cataloged up to 3 months after the event.

COSMOS 2139-2141 ULLAGE MOTOR   1991-025G   21226

SATELLITE DATA

TYPE:	Mission Related Debris
OWNER:	CIS
LAUNCH DATE:	4.45 Apr 1991
DRY MASS (KG):	~55
MAIN BODY:	Ellipsoid; 0.6 m diameter by 1.0 m length
MAJOR APPENDAGES:	None
ATTITUDE CONTROL:	None
ENERGY SOURCES:	On-board propellants

EVENT DATA

DATE:	16 Jun 2001	LOCATION:	Unknown
TIME:	~0700Z	ASSESSED CAUSE:	Propulsion
ALTITUDE:	Unknown		

PRE-EVENT ELEMENTS

EPOCH:	01165.32627059	MEAN ANOMALY:	158.6980
RIGHT ASCENSION:	118.5521	MEAN MOTION:	4.28587592
INCLINATION:	64.5545	MEAN MOTION DOT/2:	0.00004370
ECCENTRICITY:	0.5826262	MEAN MOTION DOT DOT/6:	0.0
ARG. OF PERIGEE:	187.0212	BSTAR:	0.0011075

DEBRIS CLOUD DATA

MAXIMUM ΔP:	236.7 min*
MAXIMUM ΔI:	2.13 deg*

* Based on uncataloged debris data

COMMENTS

The debris from this breakup were difficult for the US Space Surveillance Network to track. Although over 100 debris were initially tracked, none were cataloged. This is the 24th event of the Proton Block DM SOZ ullage motor since 1984. This ullage motor was launched prior to Russian recognition of the problem and before implementation of preventive measures.

REFERENCE DOCUMENT

"Two More Satellite Breakups Detected," The Orbital Debris Quarterly News, NASA JSC, July 2001. Available online at https://orbitaldebris.jsc.nasa.gov/quarterly-news/pdfs/ODQNv6i3.pdf.

SSN 21226

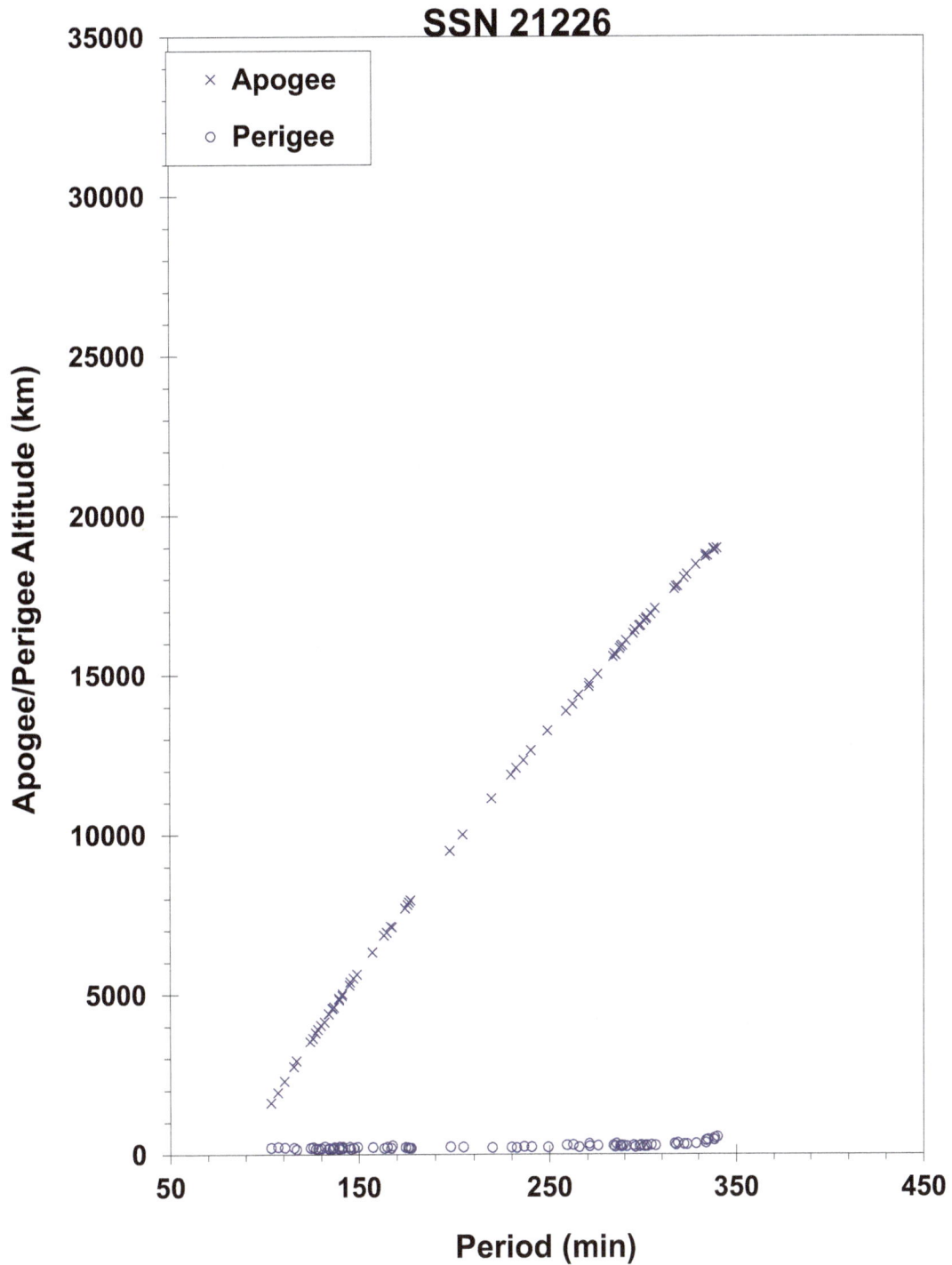

Cosmos 2139-41 auxiliary motor debris cloud of 77 fragments within 2 weeks after the event as reconstructed from the US SSN database.

369

SATELLITE DATA

TYPE:	Tsyklon Third Stage
OWNER:	CIS
LAUNCH DATE:	15 Aug. 1991
DRY MASS (KG):	1360
MAIN BODY:	Cylinder; 2.1 m diameter by 2.4 m length
MAJOR APPENDAGES:	None
ATTITUDE CONTROL:	None at time of the event.
ENERGY SOURCES:	Unknown

EVENT DATA

DATE:	12 Feb. 2020	LOCATION:	80.3 S, 262.1 E (asc.)
TIME:	1046 GMT	ASSESSED CAUSE:	Propulsion
ALTITUDE:	1225.3 km		

PRE-EVENT ELEMENTS

EPOCH:	20042.59392153	MEAN ANOMALY:	4.0309
RIGHT ASCENSION:	336.6347	MEAN MOTION:	13.17056770
INCLINATION:	82.5589	MEAN MOTION DOT/2:	-0.00000003
ECCENTRICITY:	0.0013558	MEAN MOTION DOT DOT/6:	0.0
ARG. OF PERIGEE:	169.2046	BSTAR:	-0.000038746

DEBRIS CLOUD DATA

MAXIMUM ΔP:	7.8 min
MAXIMUM ΔI:	0.5 deg

COMMENTS

REFERENCE DOCUMENT

"Three Recent Breakup Events," The Orbital Debris Quarterly News, NASA JSC, April 2020. Available online at https://orbitaldebris.jsc.nasa.gov/quarterly-news/pdfs/ODQNv24i2.pdf.

Meteor 3-5 R/B debris cloud of 111 fragments 2 weeks to over 280 days after the event as reconstructed from the US SSN database.

SATELLITE DATA

TYPE:	Tsyklon Third Stage
OWNER:	CIS
LAUNCH DATE:	28.30 Sep 1991
DRY MASS (KG):	1360
MAIN BODY:	Cone-cylinder; 2.1 m diameter by 2.4 m length
MAJOR APPENDAGES:	None
ATTITUDE CONTROL:	None
ENERGY SOURCES:	Unknown

EVENT DATA

DATE:	09 Oct 1999	LOCATION:	41.5N, 217.9E (dsc)
TIME:	1508 GMT	ASSESSED CAUSE:	Propulsion
ALTITUDE:	1460 km		

PRE-EVENT ELEMENTS

EPOCH:	99281.98318497	MEAN ANOMALY:	220.2415
RIGHT ASCENSION:	96.5043	MEAN MOTION:	12.54216420
INCLINATION:	82.5731	MEAN MOTION DOT/2:	0.00000027
ECCENTRICITY:	0.0046780	MEAN MOTION DOT DOT/6:	0.0
ARG. OF PERIGEE:	140.1600	BSTAR:	0.00010000

DEBRIS CLOUD DATA

MAXIMUM ΔP:	2.716 min
MAXIMUM ΔI:	0.79 deg

COMMENTS

This is the 4th event of this class identified to date and the second of 1999. All stages have been about 8-10 years old at the time of breakup. The vehicle is a Ukrainian-produced stage, using UDMH and N_2O_4 as propellants. To date these stages have not been passivated at end of mission and may contain up to 300 kg of residual propellants. The issue of Tsyklon orbital stage breakups was discussed with representatives of the National Space Agency of Ukraine during 11-13 October 1999 in Darmstadt, Germany. More than 100 of these stages are currently in Earth orbit. Although the exact cause of these breakups remains unknown, all four events have occurred during periods of high solar flux, *i.e.,* near solar maximum.

REFERENCE DOCUMENT

"Third Tsyklon Upper Stage Breaks Up," The Orbital Debris Quarterly News, NASA JSC, July 1999. Available online at https://orbitaldebris.jsc.nasa.gov/quarterly-news/pdfs/ODQNv4i3.pdf.

SSN 21734

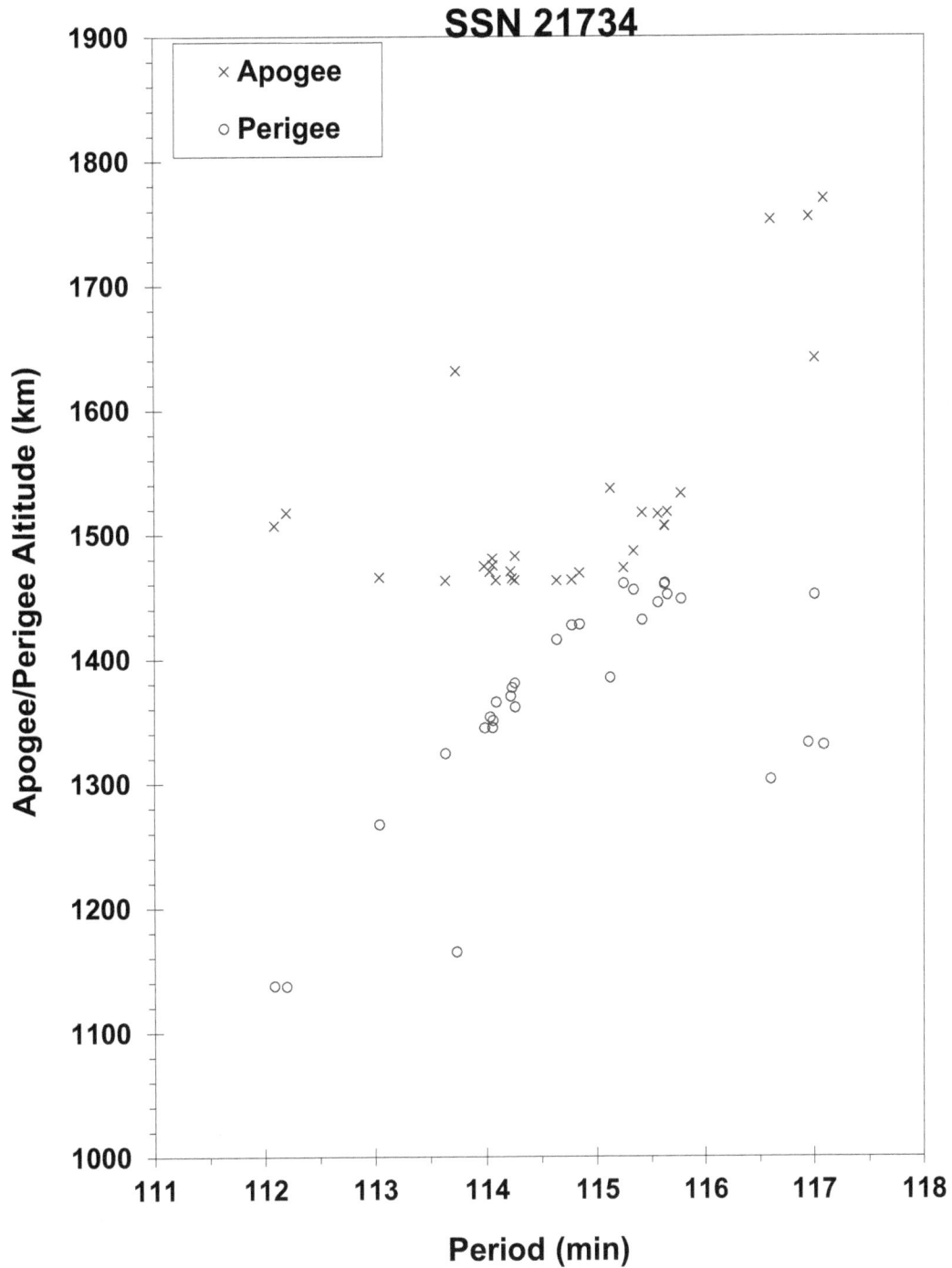

Cosmos 2157-2162 rocket body debris cloud of 31 fragments within 1 day of the event as reconstructed from the US SSN database.

SATELLITE DATA

TYPE:	Payload
OWNER:	CIS
LAUNCH DATE:	9.55 October 1991
DRY MASS (KG):	6000
MAIN BODY:	Cylinder; 2.4 m diameter by 7 m length
MAJOR APPENDAGES:	Solar panels
ATTITUDE CONTROL:	Active, 3-axis
ENERGY SOURCES:	On-board propellants, explosive charge

EVENT DATA

DATE:	6 December 1991	LOCATION:	55N, 154E (dsc)
TIME:	2021 GMT	ASSESSED CAUSE:	Deliberate
ALTITUDE:	210 km		

PRE-EVENT ELEMENTS

EPOCH:	91340.51933896	MEAN ANOMALY:	213.3470
RIGHT ASCENSION:	37.7884	MEAN MOTION:	16.18797546
INCLINATION:	64.7678	MEAN MOTION DOT/2:	0.00862876
ECCENTRICITY:	0.0054670	MEAN MOTION DOT DOT/6:	0.000035685
ARG. OF PERIGEE:	147.5032	BSTAR:	0.00035926

DEBRIS CLOUD DATA

MAXIMUM ΔP:	>9.8 min*
MAXIMUM ΔI:	0.2 deg*

*Based on uncataloged debris data

COMMENTS

Spacecraft was destroyed with a planned detonation. Third fragmentation of the Cosmos 2031 subclass. Early elements on only 8 objects available. All debris reentered before being officially cataloged.

REFERENCE DOCUMENTS

The Fragmentation of Kosmos 2163, Technical Report CS92-TR-JSC-002, Teledyne Brown Engineering, Colorado Springs, Colorado, January 1992.

History of Soviet/Russian Satellite Fragmentations-A Joint U.S.-Russian Investigation, N. L. Johnson, et al., Kaman Sciences Corporation, October 1995.

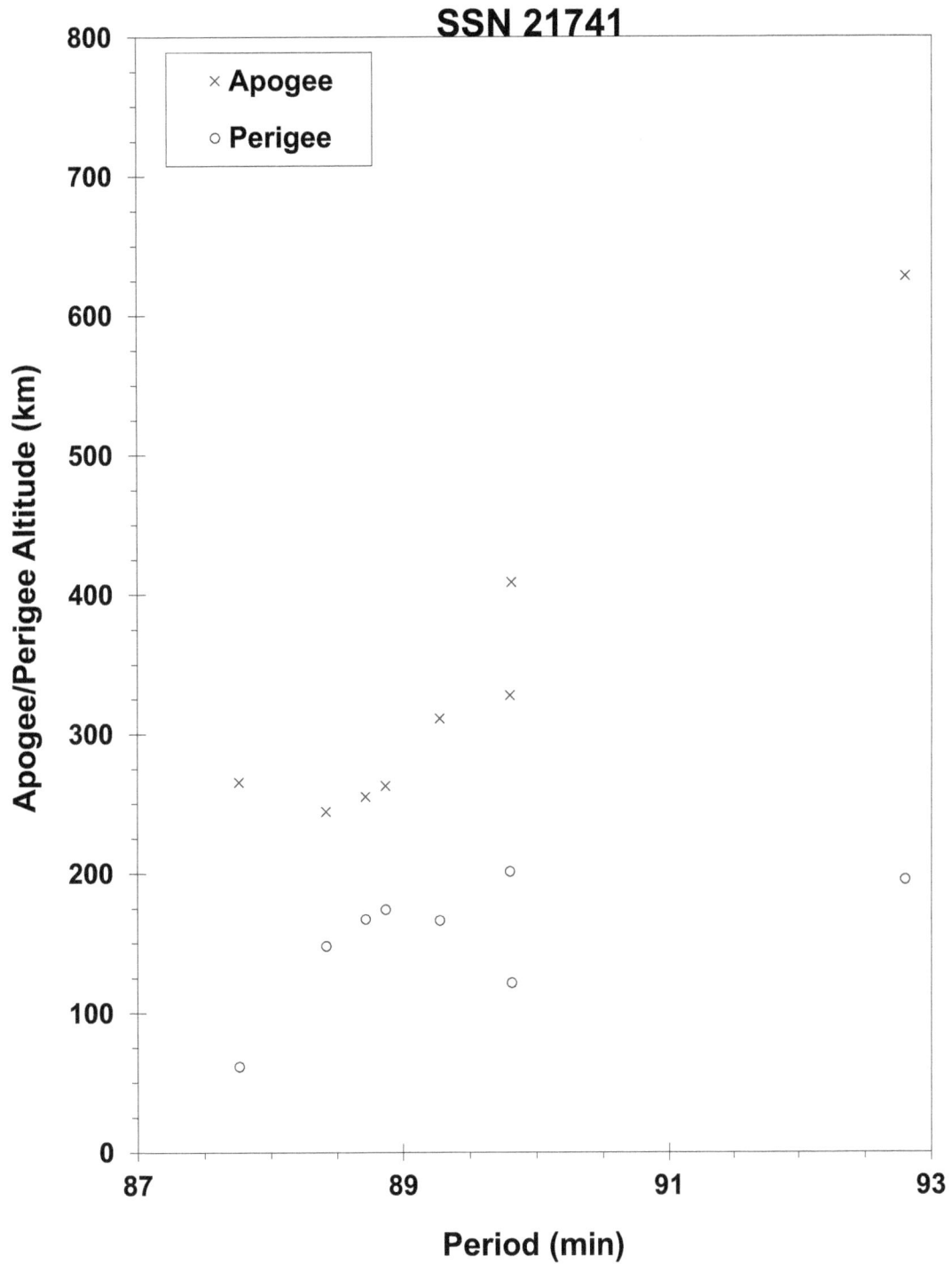

Cosmos 2163 debris cloud remnant of 8 objects 1 day after the event as reconstructed from the US SSN database. This diagram is taken from the cited reference.

INTELSAT 601 R/B   **1991-075B**   **21766**

SATELLITE DATA

TYPE:	Ariane 4 Third Stage
OWNER:	France
LAUNCH DATE:	29.96 Oct 1991
DRY MASS (KG):	~1760
MAIN BODY:	Cylinder: 2.6 m diameter by 11.4 m length
MAJOR APPENDAGES:	None
ATTITUDE CONTROL:	None
ENERGY SOURCES:	On-board propellants

EVENT DATA

DATE:	24 Dec 2001	LOCATION:	4.0 N, 344.4 E (asc)
TIME:	~2212Z	ASSESSED CAUSE:	Propulsion
ALTITUDE:	22,100 km		

PRE-EVENT ELEMENTS

EPOCH:	01358.15120659	MEAN ANOMALY:	1.1953
RIGHT ASCENSION:	264.6850	MEAN MOTION:	2.90501578
INCLINATION:	7.1968	MEAN MOTION DOT/2:	0.00042976
ECCENTRICITY:	0.6814056	MEAN MOTION DOT DOT/6:	0.0
ARG. OF PERIGEE:	351.9651	BSTAR:	0.0055981

DEBRIS CLOUD DATA

MAXIMUM ΔP:	4.45 min*
MAXIMUM ΔI:	0.08 deg*

                                   * Based on uncataloged debris data

COMMENTS

This 10-year-old Ariane 4 third stage appears to have suffered a minor fragmentation. The low inclination and high eccentricity of the orbit made debris detection and tracking difficult. Three pieces were initially detected by the SSN and ten objects cataloged a month after the event. The object was seen intact at about 2030 UTC, 24 December. Approximately 25 hours later a debris cloud of eight objects was seen by the same sensor. The perigee of the vehicle prior to breakup was sufficiently high that aerodynamic forces should not have been a factor in the event.

REFERENCE DOCUMENT

"Two Major Satellite Breakups Near End of 2001," The Orbital Debris Quarterly News, NASA JSC, January 2002. Available online at https://orbitaldebris.jsc.nasa.gov/quarterly-news/pdfs/ODQNv7i1.pdf.

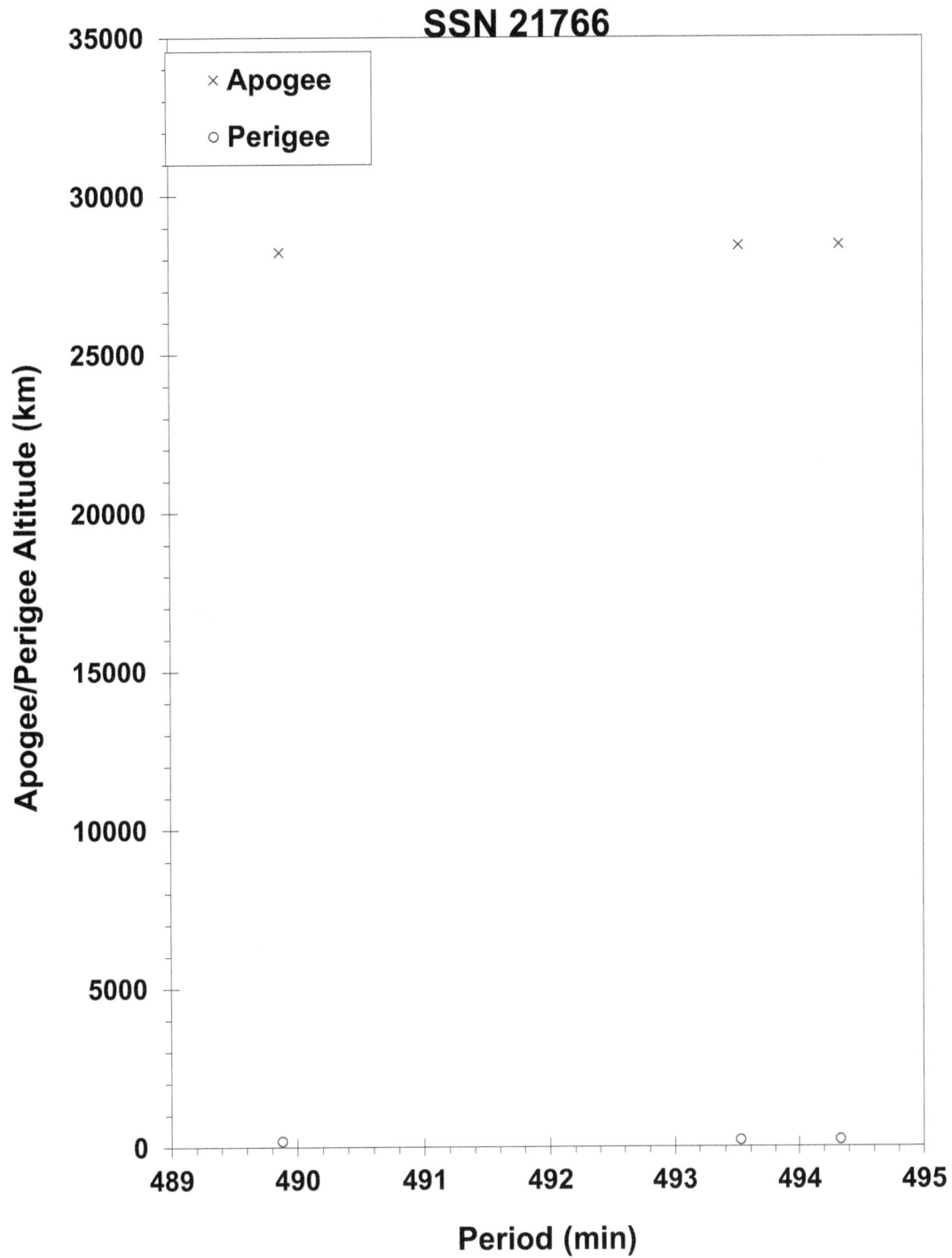

Intelsat 601 R/B debris cloud of 3 fragments 3 days after the breakup as reconstructed from the US SSN database.

SATELLITE DATA

TYPE:	Payload
OWNER:	US
LAUNCH DATE:	28.56 Nov 1991
DRY MASS (KG):	767
MAIN BODY:	Cylinder; 1.0 m diameter by 3.6 m length
MAJOR APPENDAGES:	1 solar panel
ATTITUDE CONTROL:	None at time of event
ENERGY SOURCES:	On-board propellant

EVENT DATA

DATE:	15 Apr 2004	LOCATION:	31N, 56E (asc)
TIME:	1454 GMT	ASSESSED CAUSE:	Unknown
ALTITUDE:	835 km		

POST-EVENT ELEMENTS

EPOCH:	04106.47330773	MEAN ANOMALY:	346.2855
RIGHT ASCENSION:	129.0059	MEAN MOTION:	14.15516644
INCLINATION:	98.6744	MEAN MOTION DOT/2:	0.00000244
ECCENTRICITY:	.0012890	MEAN MOTION DOT DOT/6:	0.0
ARG. OF PERIGEE:	13.8671	BSTAR:	0.0014668

DEBRIS CLOUD DATA

MAXIMUM ΔP:	2.8 min
MAXIMUM ΔI:	0.4 deg

COMMENTS

The spacecraft was non-operational at the time of the event. The electrical power generation system had been passivated by discharging the batteries and disconnecting them from the charging circuit. Virtually no nitrogen remained on board due to a leak detected early in the mission. The only energy source assessed to be on the spacecraft at the time of the event was approximately 6 kg of hydrazine.

REFERENCE DOCUMENT

"Recent Satellite Breakups," The Orbital Debris Quarterly News, NASA JSC, October 2004.
Available online at https://orbitaldebris.jsc.nasa.gov/quarterly-news/pdfs/ODQNv8i4.pdf.

SSN 21798

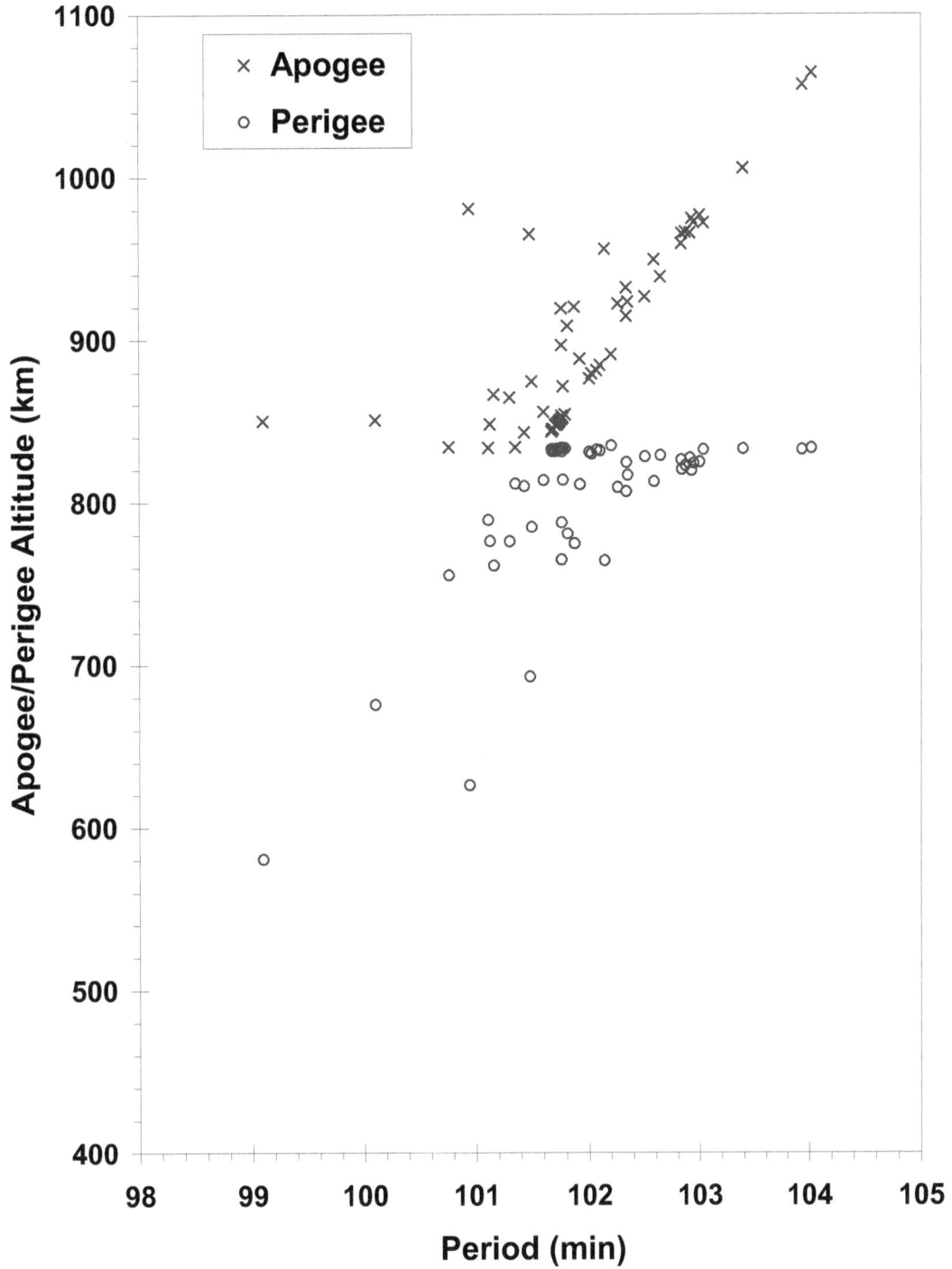

USA 73 debris cloud of 56 cataloged fragments 6 weeks after the event as reconstructed from the US SSN database.

SATELLITE DATA

TYPE:	Ariane 4 H10$^+$ Third Stage
OWNER:	France
LAUNCH DATE:	15.98 Apr 1992
DRY MASS (KG):	1800
MAIN BODY:	Cylinder; 2.6 m diameter by 10 m length
MAJOR APPENDAGES:	None
ATTITUDE CONTROL:	None at time of the event.
ENERGY SOURCES:	On-board propellants

EVENT DATA

DATE:	21 Apr 1993 (EST)	LOCATION:	Unknown
TIME:	Unknown	ASSESSED CAUSE:	Propulsion
ALTITUDE:	Unknown		

PRE-EVENT ELEMENTS

EPOCH:	93110.33659871	MEAN ANOMALY:	335.5551
RIGHT ASCENSION:	224.3451	MEAN MOTION:	2.28914093
INCLINATION:	4.03	MEAN MOTION DOT/2:	0.000024
ECCENTRICITY:	0.7248434	MEAN MOTION DOT DOT/6:	0.0
ARG. OF PERIGEE:	110.6851	BSTAR:	0.0020699

DEBRIS CLOUD DATA

MAXIMUM ΔP:	Unknown
MAXIMUM ΔI:	Unknown

COMMENTS

Although analysis indicates that 92-021C fragmented around 21 April 1993, the event was not recognized until eight pieces were cataloged during the period from 30 August 1994 to 2 September 1994. Naval Space Command reported that the eight objects were discovered during the period from July 1993 to July 1994.

Insufficient data to construct a Gabbard diagram.

SATELLITE DATA

TYPE:	Ariane 4 H10 Final Stage
OWNER:	France
LAUNCH DATE:	9.95 Jul 1992
DRY MASS (KG):	~1760
MAIN BODY:	Cylinder: 2.6 m diameter by 11.4 m length
MAJOR APPENDAGES:	None
ATTITUDE CONTROL:	None
ENERGY SOURCES:	On-board propellants

EVENT DATA

DATE:	Feb 2002	LOCATION:	Unknown
TIME:	Unknown	ASSESSED CAUSE:	Propulsion
ALTITUDE:	Unknown		

PRE-EVENT ELEMENTS

EPOCH:	02032.98792301	MEAN ANOMALY:	23.4497
RIGHT ASCENSION:	196.5922	MEAN MOTION:	3.10885568
INCLINATION:	7.0154	MEAN MOTION DOT/2:	.00036687
ECCENTRICITY:	.6663885	MEAN MOTION DOT DOT/6:	.0
ARG. OF PERIGEE:	261.6162	BSTAR:	.0

DEBRIS CLOUD DATA

MAXIMUM ΔP:	41.05 min*
MAXIMUM ΔI:	0.25 deg*

* Based on uncataloged debris data

COMMENTS

This marks the sixth known fragmentation of an Ariane 4 third stage. The last three vehicles involved in such events (1988-109C, 1991-075C, and 1992-041C) had been in orbit 9-10 years at the time of their respective breakups. All flights were conducted prior to the implementation of passivation measures for Ariane GTO missions in September 1993. No Ariane launch vehicle launched since that time is known to have experienced an on-orbit fragmentation.

REFERENCE DOCUMENT

"Second Identified Satellite Breakup of 2002", The Orbital Debris Quarterly News, NASA JSC, July 2002. Available online at https://orbitaldebris.jsc.nasa.gov/quarterly-news/pdfs/ODQNv7i3.pdf.

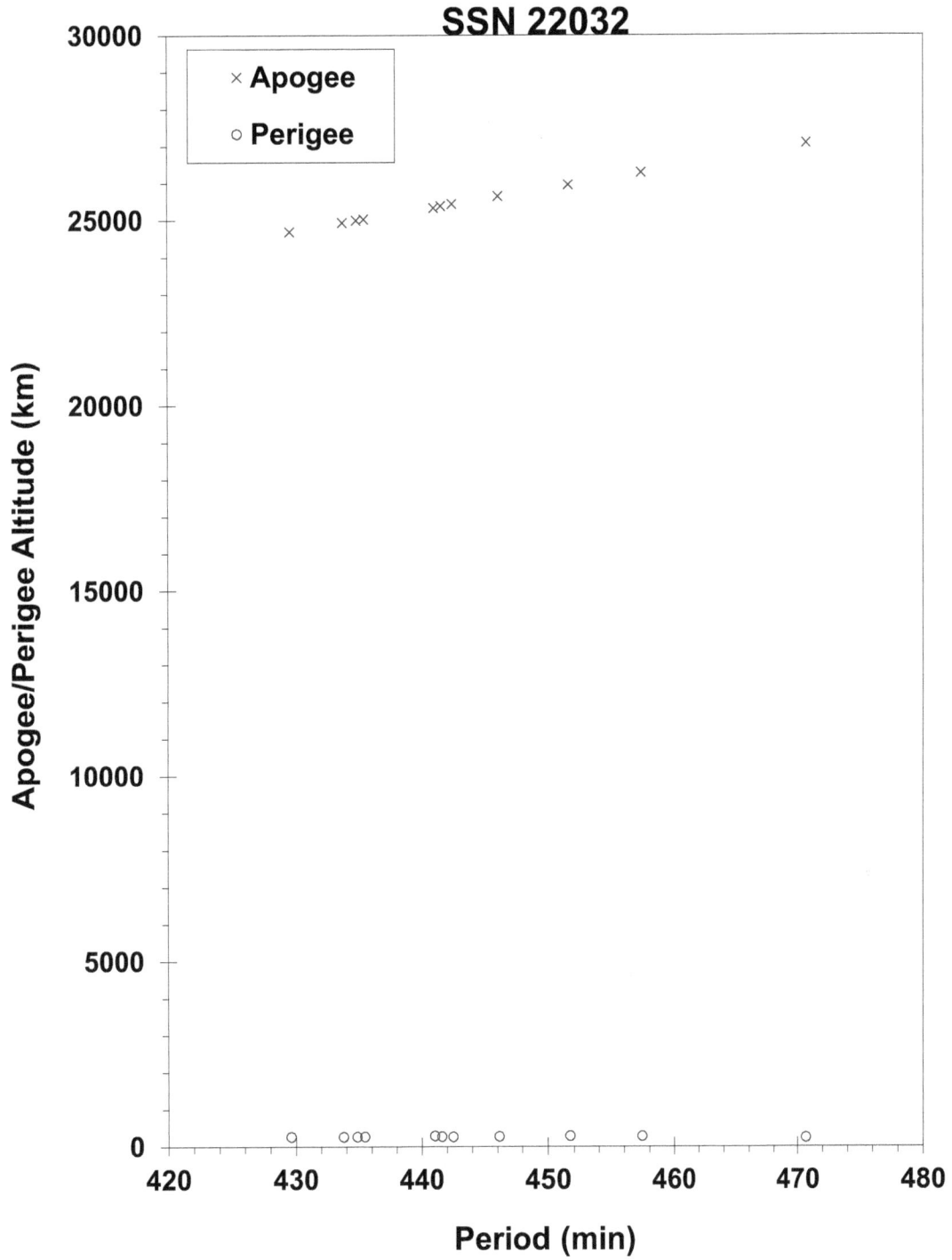

SSN 22032

Insat 2A/Eutelsat 2F4 R/B debris cloud of 11 fragments 4 months after the breakup as reconstructed from the US SSN database.

COSMOS 2204-2206 ULLAGE MOTOR   1992-047G   22066

SATELLITE DATA

TYPE:	Mission Related Debris
OWNER:	CIS
LAUNCH DATE:	30.08 Jul 1992
DRY MASS (KG):	55
MAIN BODY:	Ellipsoid; 0.6 m diameter by 1 m length
MAJOR APPENDAGES:	None
ATTITUDE CONTROL:	None
ENERGY SOURCES:	On-board propellant

EVENT DATA

DATE:	10 Jul 2004	LOCATION:	34S, 240E (asc)
TIME:	0240 GMT (est.)	ASSESSED CAUSE:	Propulsion
ALTITUDE:	18,525 km		

POST-EVENT ELEMENTS

EPOCH:	04190.17957430	MEAN ANOMALY:	263.0579
RIGHT ASCENSION:	223.5224	MEAN MOTION:	4.29145980
INCLINATION:	64.8832	MEAN MOTION DOT/2:	0.00005202
ECCENTRICITY:	0.5757033	MEAN MOTION DOT DOT/6:	0.0
ARG. OF PERIGEE:	149.3553	BSTAR:	0.0055262

DEBRIS CLOUD DATA

MAXIMUM ΔP:	37.6 min
MAXIMUM ΔI:	0.5 deg

COMMENTS

The object was in a decaying, eccentric orbit; this event marks the 30th known breakup of a Proton Block DM SOZ ullage motor since 1984. More than 100 debris were detected from Sat. No. 22066 by the US SSN. The sister SOZ motor for this launch broke up in an unrelated event ten years earlier (22067).

REFERENCE DOCUMENT

History of Soviet/Russian Satellite Fragmentations-A Joint U.S.-Russian Investigation, N. L. Johnson, et al., Kaman Sciences Corporation, October 1995.

"Recent Satellite Breakups," The Orbital Debris Quarterly News, NASA JSC, October 2004. Available online at https://orbitaldebris.jsc.nasa.gov/quarterly-news/pdfs/ODQNv8i4.pdf.

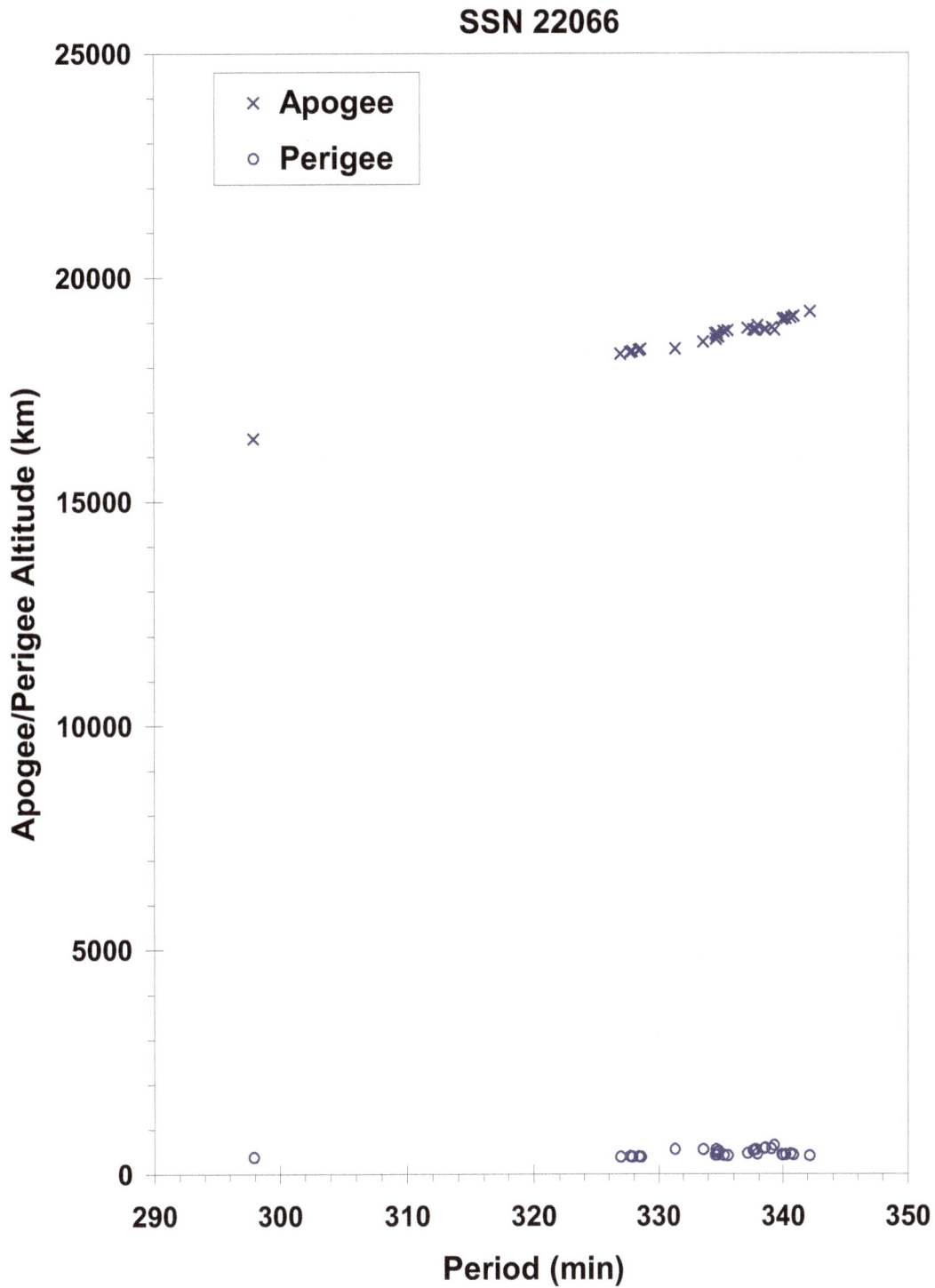

SSN 22066

SOZ motor debris cloud of 31 fragments less than 1 week after the event as reconstructed from the US SSN database.

SATELLITE DATA

TYPE:	Mission Related Debris
OWNER:	CIS
LAUNCH DATE:	30.08 Jul 1992
DRY MASS (KG):	55
MAIN BODY:	Ellipsoid; 0.6 m diameter by 1.0 m length
MAJOR APPENDAGES:	None
ATTITUDE CONTROL:	None
ENERGY SOURCES:	On-board propellants

EVENT DATA

DATE:	Prior to 0824 GMT, 8 Nov 1994	LOCATION:	Unknown
TIME:	Unknown	ASSESSED CAUSE:	Propulsion
ALTITUDE:	Unknown		

PRE-EVENT ELEMENTS

EPOCH:	94310.28602258	MEAN ANOMALY:	9.8460
RIGHT ASCENSION:	65.2049	MEAN MOTION:	4.23571466
INCLINATION:	64.8556	MEAN MOTION DOT/2:	0.00001002
ECCENTRICITY:	0.5708388	MEAN MOTION DOT DOT/6:	0.0
ARG. OF PERIGEE:	316.7786	BSTAR:	0.0033777

DEBRIS CLOUD DATA

MAXIMUM ΔP:	4.2 min
MAXIMUM ΔI:	0.9 deg

COMMENTS

Parent satellite was one of two small engine units that are routinely released after the first burn of the Proton fourth stage. The nature of these objects was identified by Dr. Boris V. Cherniatiev, Deputy Constructor for the Energiya NPO and Mr. Nicholas Johnson of Kaman Sciences, in October 1992. The cause of this fragmentation appears to be related to the residual hypergolic propellants on board and failure of the membrane separating the fuel and oxidizer. NAVSPASUR observed 32 objects that were associated with this breakup on 8 Nov 94, 36 objects on 9 Nov and 31 objects on 10 Nov. This was the tenth in a series of fragmentations of this object type.

REFERENCE DOCUMENTS

The Fragmentation of Proton Debris, D. J. Nauer, TBE Technical Report CS93-LKD-004, Teledyne Brown Engineering, Colorado Springs, 31 December 1992.

Identification and Resolution of an Orbital Debris Problem with the Proton Launch Vehicle, B. V. Cherniatiev,, et al., First European Conference on Space Debris, 5-7 April 1993.

History of Soviet/Russian Satellite Fragmentations-A Joint U.S.-Russian Investigation, N. L. Johnson, et al., Kaman Sciences Corporation, October 1995.

SSN 22067

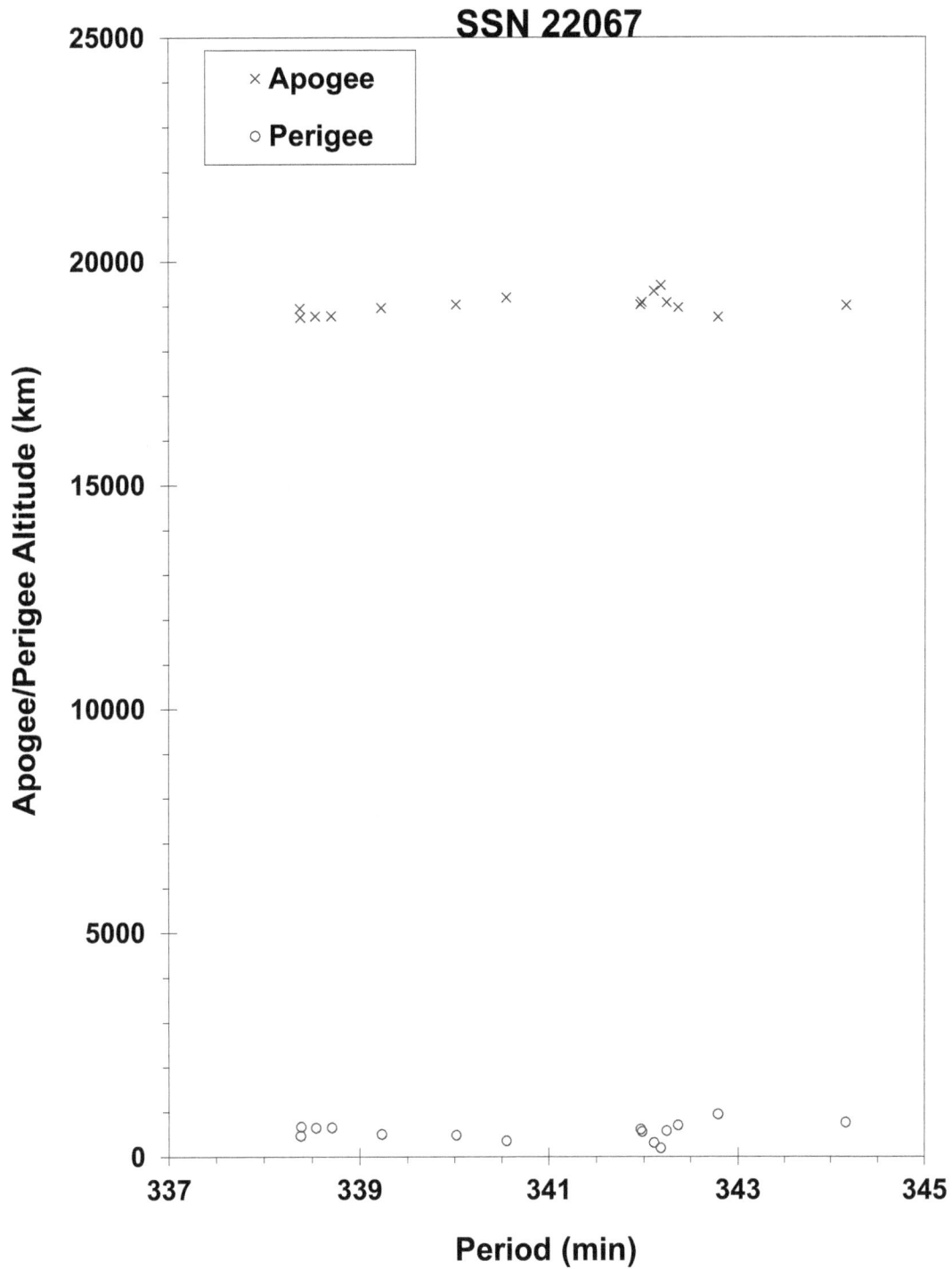

Gabbard Diagram from the Cosmos 2204-2206 debris fragmentation.

SATELLITE DATA

TYPE:	Ariane 4 Third Stage
OWNER:	France
LAUNCH DATE:	10 Aug. 1992
DRY MASS (KG):	~1720
MAIN BODY:	Cylinder: 2.6 m diameter by 11.4 m length
MAJOR APPENDAGES:	None
ATTITUDE CONTROL:	None
ENERGY SOURCES:	Status unknown

EVENT DATA

DATE:	22 July 2019	LOCATION:	66.17 S, 300.86 E (dsc)
TIME:	1902 Z	ASSESSED CAUSE:	Unknown
ALTITUDE:	1373.99 km		

PRE-EVENT ELEMENTS

EPOCH:	19203.35822740	MEAN ANOMALY:	139.2833
RIGHT ASCENSION:	110.4854	MEAN MOTION:	12.77863800
INCLINATION:	66.0677	MEAN MOTION DOT/2:	-0.00000040
ECCENTRICITY:	0.0068265	MEAN MOTION DOT DOT/6:	0.0
ARG. OF PERIGEE:	359.8522	BSTAR:	0.000089165

DEBRIS CLOUD DATA

MAXIMUM ΔP:	3.7 min
MAXIMUM ΔI:	0.3 deg

COMMENTS

It is understood that this vehicle (Ariane 42P third stage, launch V-52) performed a LEO-flight (prior to the general introduction of mitigation procedures for all orbit profiles) mitigation procedure at end of life, though it was not in a sun-synchronous orbit. Therefore, the root cause is currently Unknown.

REFERENCE DOCUMENT

"Three Recent Rocket Body Breakups," The Orbital Debris Quarterly News, NASA JSC, November 2019. Available online at https://orbitaldebris.jsc.nasa.gov/quarterly-news/pdfs/ODQNv23i4.pdf.

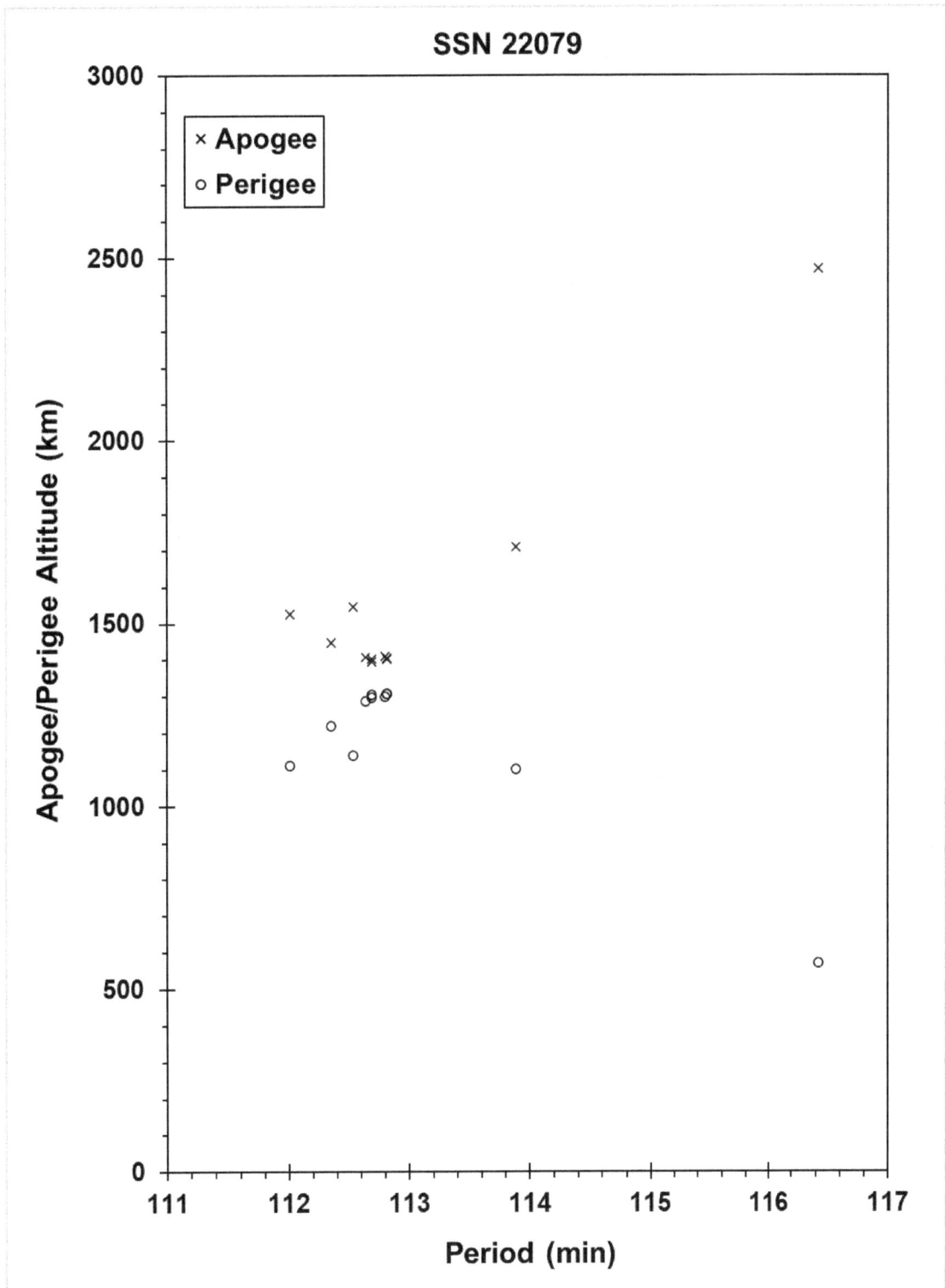

SSN 22079

TOPEX-POSEIDON R/B debris cloud of 10 fragments up to 116 days after the breakup as reconstructed from the US SSN database.

GORIZONT 27 ULLAGE MOTOR   1992-082F   22250

SATELLITE DATA

TYPE:	Mission Related Debris
OWNER:	CIS
LAUNCH DATE:	27.55 Nov 1992
DRY MASS (KG):	~55
MAIN BODY:	Ellipsoid; 0.6 m diameter by 1.0 m length
MAJOR APPENDAGES:	None
ATTITUDE CONTROL:	None
ENERGY SOURCES:	On-board propellants

EVENT DATA

DATE:	14 Jul 2001	LOCATION:	Unknown
TIME:	~1800Z	ASSESSED CAUSE:	Propulsion
ALTITUDE:	Unknown		

PRE-EVENT ELEMENTS

EPOCH:	01194.12977223	MEAN ANOMALY:	232.1640
RIGHT ASCENSION:	101.3588	MEAN MOTION:	9.96766196
INCLINATION:	46.4697	MEAN MOTION DOT/2:	0.01023324
ECCENTRICITY:	0.2850325	MEAN MOTION DOT DOT/6:	0.0000014714
ARG. OF PERIGEE:	148.6125	BSTAR:	0.00050789

DEBRIS CLOUD DATA

MAXIMUM ΔP:	Unknown
MAXIMUM ΔI:	Unknown

COMMENTS

This event marks the 25th known breakup of a Proton Blok DM SOZ ullage motor since 1984. Due to the moderate eccentricity and altitude of the orbits, the debris were difficult for the US Space Surveillance Network to track. This ullage motor was launched before implementation of breakup preventive measures. No debris were cataloged from this breakup.

REFERENCE DOCUMENT

"New Satellite Breakups Detected," The Orbital Debris Quarterly News, NASA JSC, October 2001. Available online at https://orbitaldebris.jsc.nasa.gov/quarterly-news/pdfs/ODQNv6i4.pdf.

Insufficient data to construct a Gabbard diagram.

SATELLITE DATA

TYPE:	Mission Related Debris
OWNER:	CIS
LAUNCH DATE:	17.53 Dec 1992
DRY MASS (KG):	55
MAIN BODY:	Ellipsoid; 0.6 m diameter by 1 m length
MAJOR APPENDAGES:	None
ATTITUDE CONTROL:	None
ENERGY SOURCES:	On-board propellant

EVENT DATA

DATE:	~22 Apr 2005	LOCATION:	Unknown
TIME:	Unknown	ASSESSED CAUSE:	Propulsion
ALTITUDE:	Unknown		

PRE-EVENT ELEMENTS

EPOCH:	05112.46798568	MEAN ANOMALY:	2.1228
RIGHT ASCENSION:	268.8209	MEAN MOTION:	3.90056983
INCLINATION:	46.7270	MEAN MOTION DOT/2:	0.00018984
ECCENTRICITY:	0.6142562	MEAN MOTION DOT DOT/6:	0.00000003
ARG. OF PERIGEE:	348.3165	BSTAR:	0.0006277

DEBRIS CLOUD DATA

MAXIMUM ΔP:	Unknown
MAXIMUM ΔI:	Unknown

COMMENTS

This event marks the 32nd known breakup of a Proton Blok DM SOZ ullage motor since 1984. This object had a perigee near 200 km. The event happened sometime between 1100 GMT on April 22, and 1200 GMT on April 23. Only about a dozen pieces were detected by the US SSN.

REFERENCE DOCUMENT

History of Soviet/Russian Satellite Fragmentations-A Joint U.S.-Russian Investigation, N. L. Johnson, *et al.,* Kaman Sciences Corporation, October 1995.

"Recent Satellite Breakups," The Orbital Debris Quarterly News, NASA JSC, July 2005.
Available online at https://orbitaldebris.jsc.nasa.gov/quarterly-news/pdfs/ODQNv9i3.pdf.

Insufficient data to construct a Gabbard diagram.

SATELLITE DATA

TYPE:	Payload
OWNER:	CIS
LAUNCH DATE:	22.50 Dec 1992
DRY MASS (KG):	6000
MAIN BODY:	Cylinder; 2.4 m diameter by 7 m length
MAJOR APPENDAGES:	Solar panels
ATTITUDE CONTROL:	Active, 3-axis
ENERGY SOURCES:	On-board propellants, explosive charge

EVENT DATA

DATE:	18 Feb 1993	LOCATION:	55N, 157E (dsc)
TIME:	1856 GMT	ASSESSED CAUSE:	Deliberate
ALTITUDE:	220 km		

PRE-EVENT ELEMENTS

EPOCH:	93040.89217375	MEAN ANOMALY:	244.5776
RIGHT ASCENSION:	125.1196	MEAN MOTION:	16.07940666
INCLINATION:	64.8919	MEAN MOTION DOT/2:	0.00301303
ECCENTRICITY:	0.0039285	MEAN MOTION DOT DOT/6:	0.000049705
ARG. OF PERIGEE:	115.8892	BSTAR:	0.00032572

DEBRIS CLOUD DATA

MAXIMUM ΔP:	> 2.3 min
MAXIMUM ΔI:	> 0.5 deg

COMMENTS

Spacecraft was destroyed with a planned detonation. Fourth fragmentation of the Cosmos 2031 subclass. Early elements on only 10 objects (including the parent) available; 21 objects were observed by Fylingdales soon after the event.

REFERENCE DOCUMENTS

Analysis of Fragmentations from December 1992 - February 1993, Technical Report CS93-LKD-010, Teledyne Brown Engineering, Colorado Springs, Colorado, 30 March 1993.

History of Soviet/Russian Satellite Fragmentations-A Joint U.S.-Russian Investigation, N. L. Johnson, et al., Kaman Sciences Corporation, October 1995.

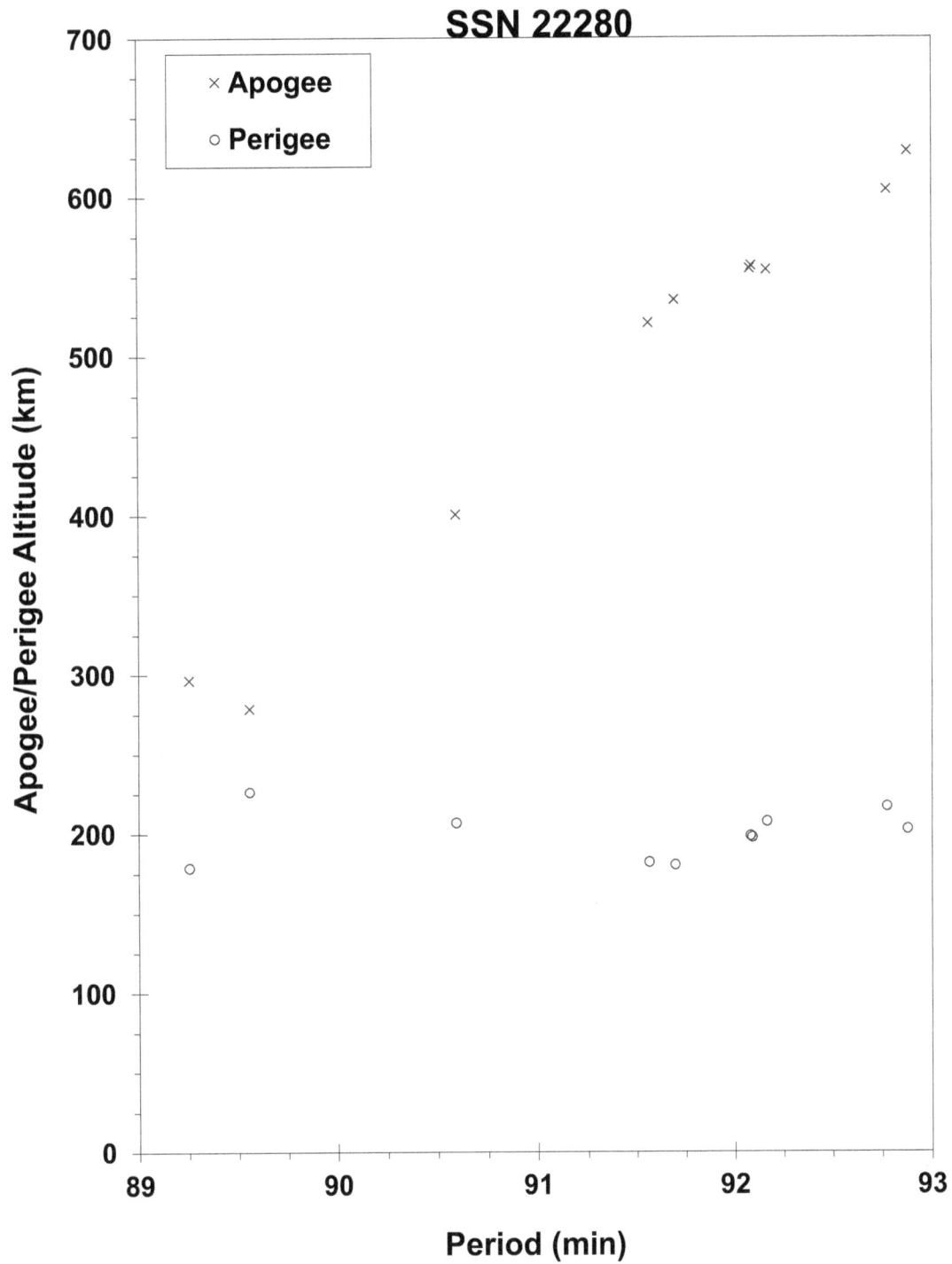

SSN 22280

Legend:
× Apogee
○ Perigee

X-axis: Period (min)
Y-axis: Apogee/Perigee Altitude (km)

Cosmos 2225 debris cloud remnant of 10 objects 4 days after the event as reconstructed from the data provided by Naval Space Surveillance System in a 22 February 1993 Satellite Support message.

COSMOS 2227 R/B   1992-093B   22285

SATELLITE DATA

TYPE:	Zenit Second Stage
OWNER:	CIS
LAUNCH DATE:	25.25 Dec 1992
DRY MASS (KG):	8300
MAIN BODY:	Cylinder; 3.9 m diameter by 12 m length
MAJOR APPENDAGES:	None
ATTITUDE CONTROL:	None at time of the event.
ENERGY SOURCES:	On-board propellants (~900 kg), explosive charge (?)

EVENT DATA (1)

DATE:	26 Dec 1992	LOCATION:	63 N, 60 E (asc)
TIME:	0738 GMT	ASSESSED CAUSE:	Propulsion
ALTITUDE:	830 km		

Note: NAVSPASUR could not correlate with 22285, but 22285 was closest object.

EVENT DATA (2)

DATE:	26 Dec 1992	LOCATION:	44 N, 168 E (asc)
TIME:	2249 GMT	ASSESSED CAUSE:	Unknown
ALTITUDE:	830 km		

EVENT DATA (3)

DATE:	26 Dec 1992	LOCATION:	52 N, 63 E (dsc)
TIME:	2310 GMT	ASSESSED CAUSE:	Unknown
ALTITUDE:	835 km		

EVENT DATA (4)

DATE:	30 Dec 1992	LOCATION:	22 S, 172 E (dsc)
TIME:	0903 GMT	ASSESSED CAUSE:	Propulsion
ALTITUDE:	~ 830 km		

PRE-EVENT ELEMENTS TO EVENT 1

EPOCH:	92361.30431818	MEAN ANOMALY:	289.8749
RIGHT ASCENSION:	227.4354	MEAN MOTION:	14.1258288
INCLINATION:	71.0274	MEAN MOTION DOT/2:	-0.00061925
ECCENTRICITY:	0.0005311	MEAN MOTION DOT DOT/6:	0.0
ARG. OF PERIGEE:	71.7543	BSTAR:	-0.034134

DEBRIS CLOUD DATA

MAXIMUM ΔP:	4.3 min *
MAXIMUM ΔI:	1.4 deg *

* Based upon comparison to pre-event elements,
includes all four events, based upon cataloged elements only.

COMMENTS

Four separate events were reported by NAVSPASUR. The first observed event was accompanied by an initial 18 objects but could not be correlated with the rocket body element set. The rocket body was the closest object to the BLAST point. The second event followed 15 hours later with 96 objects. The third event followed the second by less than 20 minutes and was based upon 51 pieces. The fourth event was accompanied by 3 objects. Element data on 164 objects has been combined into a single Gabbard Diagram.

NAVSPASUR initially generated 164 element sets on the combined debris from these 4 events. On 24 April 1995 object 22366, 1992-093BF, fragmented liberating 1 associated piece.

REFERENCE DOCUMENTS

Cosmos 2227 Rocket Body Fragmentation Event, E. L. Jenkins, et. al., NAVSPASUR, Dahlgren, VA.

Analysis of Fragmentations From December 1992 - February 1993, TBE Technical Report CS93-LKD-010, Teledyne Brown Engineering, Colorado Springs, 30 March 1993.

History of Soviet/Russian Satellite Fragmentations-A Joint U.S.-Russian Investigation, N. L. Johnson, et al., Kaman Sciences Corporation, October 1995.

"The Recent Fragmentations of LEO Upper Stages," G. Chernyavskiy, et al., 45th IAF Congress, October 1994.

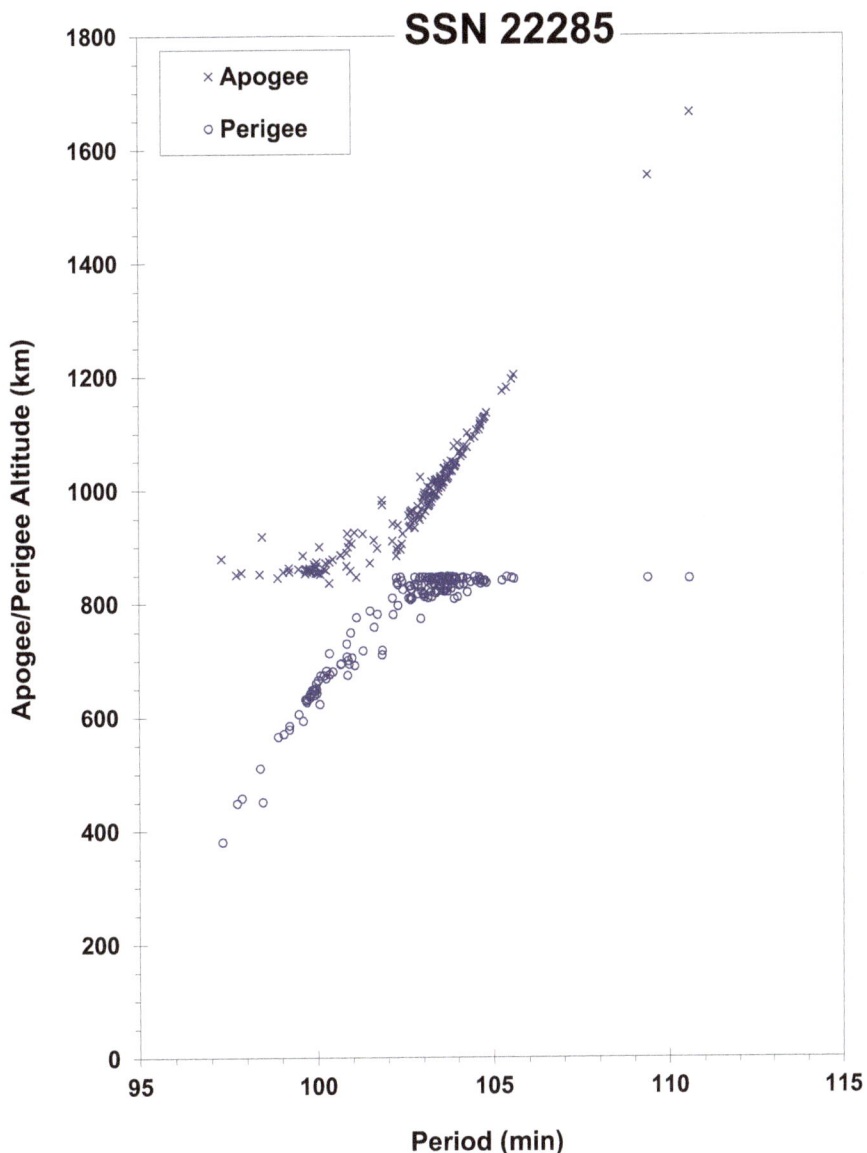

Naval Space Surveillance System generated 164 initial element sets on the four events that are plotted on the single Gabbard diagram above.

COSMOS 2237 R/B   **1993-016B**   **22566**

SATELLITE DATA

TYPE:	Zenit Second Stage
OWNER:	CIS
LAUNCH DATE:	26.10 Mar 1993
DRY MASS (KG):	8300
MAIN BODY:	Cylinder; 3.9 m diameter by 12 m length
MAJOR APPENDAGES:	None
ATTITUDE CONTROL:	None
ENERGY SOURCES:	On-board propellants (~900 kg), explosive charge (?)

EVENT DATA

DATE:	28 Mar 1993	LOCATION:	70N, 37E (dsc)
TIME:	0716 GMT	ASSESSED CAUSE:	Propulsion
ALTITUDE:	840 km		

PRE-EVENT ELEMENTS

EPOCH:	93088.27687915	MEAN ANOMALY:	84.1791
RIGHT ASCENSION:	258.8192	MEAN MOTION:	14.14093359
INCLINATION:	70.9947	MEAN MOTION DOT/2:	0.00255882
ECCENTRICITY:	0.0006748	MEAN MOTION DOT DOT/6:	0.0
ARG. OF PERIGEE:	275.8565	BSTAR:	0.12879

DEBRIS CLOUD DATA

MAXIMUM ΔP:	3.5 min
MAXIMUM ΔI:	0.1 deg

COMMENTS

At least 26 initial element sets were generated on this event by NAVSPASUR. The BLAST point was calculated from 12 objects.

REFERENCE DOCUMENTS

The Fragmentation of the Cosmos 2237 & 2243 Rocket Bodies, D. J. Nauer, TBE Technical Report CS93-LKD-016, Teledyne Brown Engineering, Colorado Springs, 15 June 1993.

History of Soviet/Russian Satellite Fragmentations-A Joint U.S.-Russian Investigation, N. L. Johnson, *et al.,* Kaman Sciences Corporation, October 1995.

"The Recent Fragmentations of LEO Upper Stages," G. Chernyavskiy, *et al.,* 45th IAF Congress, October 1994.

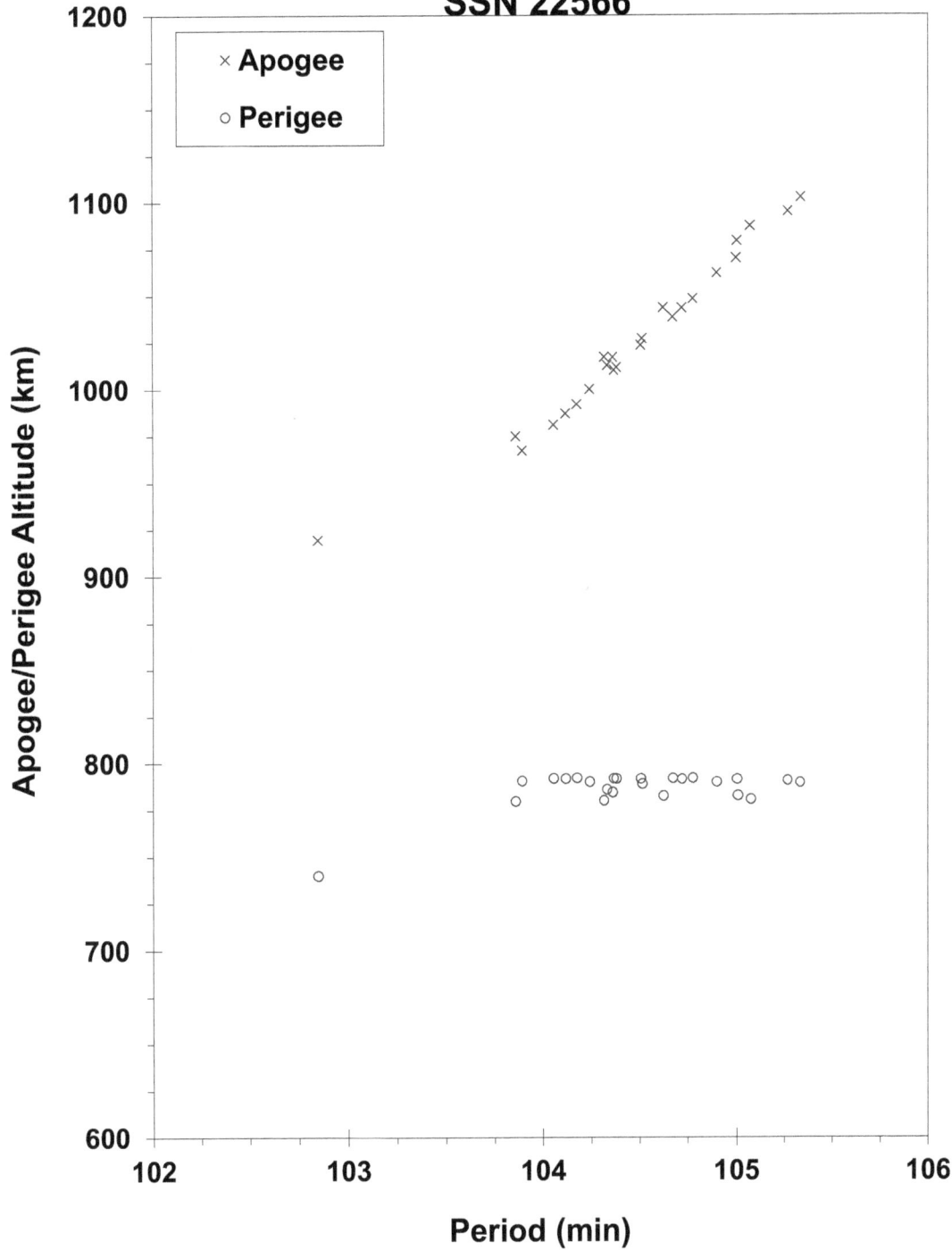

SSN 22566

Naval Space Surveillance System generated 24 initial element sets that are plotted on the Gabbard diagram above.

COSMOS 2238   **1993-018A**   **22585**

SATELLITE DATA

TYPE:	Payload
OWNER:	CIS
LAUNCH DATE:	30.50 Mar 1993
DRY MASS (KG):	3000
MAIN BODY:	Cylinder; 1.3 m diameter by 17 m length
MAJOR APPENDAGES:	Solar panels
ATTITUDE CONTROL:	Active, 3-axis
ENERGY SOURCES:	On-board propellants, explosive charge (?)

EVENT DATA

DATE:	1 Dec 1994	LOCATION:	6.5 S, 243.0 E (dsc)
TIME:	1111 GMT	ASSESSED CAUSE:	Unknown
ALTITUDE:	275 km		

PRE-EVENT ELEMENTS

EPOCH:	94335.21831221	MEAN ANOMALY:	119.6648
RIGHT ASCENSION:	124.7826	MEAN MOTION:	16.06466469
INCLINATION:	65.0063	MEAN MOTION DOT/2:	0.00787680
ECCENTRICITY:	0.0069696	MEAN MOTION DOT DOT/6:	0.0000095760
ARG. OF PERIGEE:	239.7651	BSTAR:	0.00073936

DEBRIS CLOUD DATA

MAXIMUM ΔP:	Unknown
MAXIMUM ΔI:	Unknown

COMMENTS

Cosmos 2238 was a member of the Cosmos 699-type and the first such spacecraft to breakup for 7 years. As many as 26 debris were detected; however, additional debris probably decayed before they were observed. Cosmos is the 17th fragmentation of a Cosmos 699 class payload.

Insufficient data to construct a Gabbard diagram.

SATELLITE DATA

TYPE:	Payload
OWNER:	CIS
LAUNCH DATE:	27.44 Apr 1993
DRY MASS (KG):	5700
MAIN BODY:	Cone-cylinder; 2.7 m diameter by 6.3 m length
MAJOR APPENDAGES:	Solar panels
ATTITUDE CONTROL:	Active, 3-axis
ENERGY SOURCES:	On-board propellants, explosive charge

EVENT DATA

DATE:	27 Apr 1993	LOCATION:	61N, 81E (asc)
TIME:	1044 GMT	ASSESSED CAUSE:	Deliberate
ALTITUDE:	200 km		

PRE-EVENT ELEMENTS *

EPOCH:	93119.28633059	MEAN ANOMALY:	283.6524
RIGHT ASCENSION:	51.8515	MEAN MOTION:	16.26199828
INCLINATION:	70.3602	MEAN MOTION DOT/2:	0.02823100
ECCENTRICITY:	0.0032877	MEAN MOTION DOT DOT/6:	0.000019668
ARG. OF PERIGEE:	76.8057	BSTAR:	0.00077017

* Note: Element Set 1 not generated until ~2 days after the event.

DEBRIS CLOUD DATA

MAXIMUM ΔP:	7.1 min
MAXIMUM ΔI:	0.4 deg

COMMENTS

Although this event was originally thought to be a fragmentation of the Soyuz final stage rocket body, it was actually the payload that fragmented. This event occurred near orbital insertion into the operational orbit. It is unclear whether the payload was attached at the time of the event. The payload malfunctioned and self-destructed. There were 25 initial element sets available after launch. NAVSPASUR reported tracking approximately 27 objects on 30 April 1993 and detected as many as 20 more unknowns. The final official piece count associated with this event was 172 objects. Due to the very low altitude, most objects decayed from this cloud within 2 weeks of launch. No cataloged element sets were released until almost 2 days after the event.

REFERENCE DOCUMENTS

The Fragmentation of the Cosmos 2237 & 2243 Rocket Bodies, D. J. Nauer, TBE Technical Report CS93-LKD-016, Teledyne Brown Engineering, Colorado Springs, 15 June 1993.

History of Soviet/Russian Satellite Fragmentations-A Joint U.S.-Russian Investigation, N. L. Johnson, et al., Kaman Sciences Corporation, October 1995.

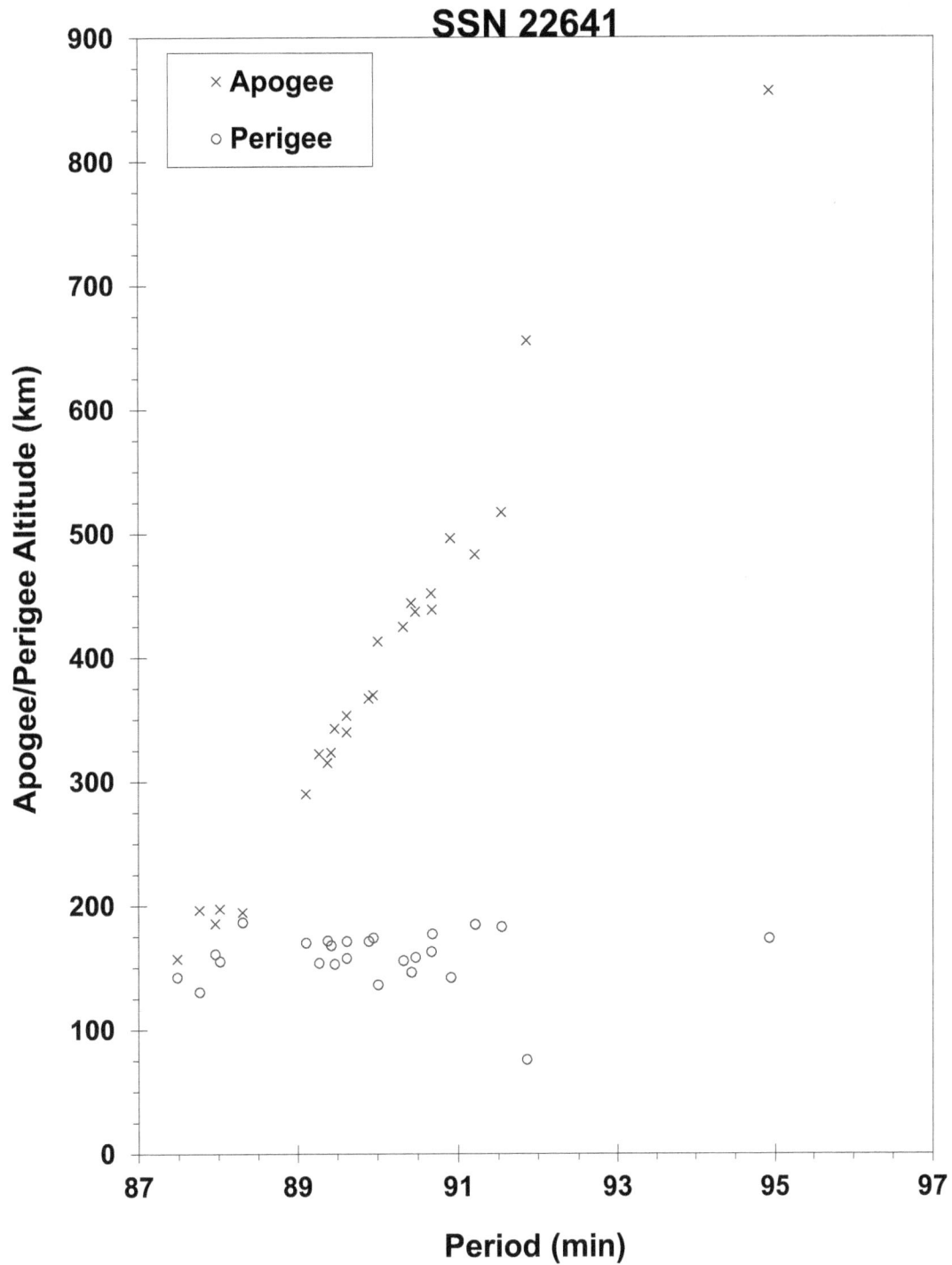

SSN 22641

Apogee/Perigee Altitude (km) vs Period (min)

Naval Space Surveillance System generated 25 initial element sets on the Cosmos 2243 fragmentation that are plotted on the Gabbard diagram above.

COSMOS 2251   1993-036A   22675

SATELLITE DATA

TYPE:	Payload
OWNER:	CIS
LAUNCH DATE:	16 Jun 1993
DRY MASS (KG):	900
MAIN BODY:	cylinder; 2.035 m diameter x 2 m length
MAJOR APPENDAGES:	gravity gradient boom
ATTITUDE CONTROL:	gravity gradient (passive)
ENERGY SOURCES:	pressurized vessel; chemical batteries

EVENT DATA

DATE:	10 Feb 2009	LOCATION:	72.5N, 97.9E (asc)
TIME:	1656 GMT	ASSESSED CAUSE:	Collision
ALTITUDE:	788.597 km		

PRE-EVENT ELEMENTS

EPOCH:	09040.49834364	MEAN ANOMALY:	261.5952
RIGHT ASCENSION:	19.4646	MEAN MOTION:	14.31135643
INCLINATION:	74.0355	MEAN MOTION DOT/2:	-0.00000001
ECCENTRICITY:	0.0016027	MEAN MOTION DOT DOT/6:	0.0
ARG. OF PERIGEE:	98.7014	BSTAR:	0.0000095251

DEBRIS CLOUD DATA

MAXIMUM ΔP:	9.4 min
MAXIMUM ΔI:	24.7 deg

COMMENTS

The spacecraft, a member of the Soviet/Russian Strela-2M low-altitude communications satellite constellation, utilized the versatile KAUR-1 bus (*Kosmicheskiy Apparat Universalnogo Ryada-1*, (Космический Аппарат Универсального Ряда), which can be translated as Spacecraft Bus from the Standardized Line-1). This bus consists of a central pressurized cylinder for the payload, an outer cylinder mounting solar cells (the Solar Battery), and a gravity gradient boom for passive stabilization. The spacecraft was derelict at the time of the event. Cosmos 2251 collided with Iridium 33 [q.v.] in the first, accidental collision of large, intact resident space objects.

REFERENCE DOCUMENT

"Satellite Collision Leaves Significant Debris Clouds," The Orbital Debris Quarterly News, NASA JSC, April 2009. Available online at https://orbitaldebris.jsc.nasa.gov/quarterly-news/ODQNv13i2.pdf.

The COSMOS 2251 debris cloud, including 1648 fragments cataloged up to 5 years after the event.

COSMOS 2259   1993-045A   22716

SATELLITE DATA

TYPE:	Payload
OWNER:	CIS
LAUNCH DATE:	14.69 Jul 1993
DRY MASS (KG):	5700
MAIN BODY:	Cone-cylinder; 2.7 m diameter by 6.3 m length
MAJOR APPENDAGES:	Solar panels
ATTITUDE CONTROL:	Active, 3-axis
ENERGY SOURCES:	On-board propellants, explosive charge

EVENT DATA

DATE:	25 Jul 1993	LOCATION:	Unknown
TIME:	Unknown	ASSESSED CAUSE:	Deliberate
ALTITUDE:	Unknown		

PRE-EVENT ELEMENTS

EPOCH:	93205.96411483	MEAN ANOMALY:	292.3177
RIGHT ASCENSION:	134.4696	MEAN MOTION:	16.09525981
INCLINATION:	67.1310	MEAN MOTION DOT/2:	0.00638090
ECCENTRICITY:	0.0113387	MEAN MOTION DOT DOT/6:	0.000023099
ARG. OF PERIGEE:	68.9805	BSTAR:	0.00025239

DEBRIS CLOUD DATA

MAXIMUM ΔP:	Unknown
MAXIMUM ΔI:	Unknown

COMMENTS

Spacecraft was destroyed after a malfunction prevented reentry and landing in the CIS. Event identified by Russian officials during investigation cited below.

REFERENCE DOCUMENT

History of Soviet/Russian Satellite Fragmentations-A Joint U.S.-Russian Investigation, N. L. Johnson, *et al.,* Kaman Sciences Corporation, October 1995.

Insufficient data to construct a Gabbard diagram.

SATELLITE DATA

TYPE:	Payload
OWNER:	CIS
LAUNCH DATE:	7.56 Sep 1993
DRY MASS (KG):	6000
MAIN BODY:	Cylinder; 2.4 m diameter by 7 m length
MAJOR APPENDAGES:	Solar panels
ATTITUDE CONTROL:	Active, 3-axis
ENERGY SOURCES:	On-board propellants, explosive charge

EVENT DATA

DATE:	18 Dec 1993	LOCATION:	65N, 107E (dsc)
TIME:	0711 GMT	ASSESSED CAUSE:	Deliberate
ALTITUDE:	195 km		

PRE-EVENT ELEMENTS

EPOCH:	93352.09835999	MEAN ANOMALY:	294.6647
RIGHT ASCENSION:	209.9170	MEAN MOTION:	16.17608693
INCLINATION:	64.8761	MEAN MOTION DOT/2:	0.00554324
ECCENTRICITY:	0.0065884	MEAN MOTION DOT DOT/6:	0.0
ARG. OF PERIGEE:	66.1310	BSTAR:	0.00022099

DEBRIS CLOUD DATA

MAXIMUM ΔP:	8.7 min *
MAXIMUM ΔI:	0.8 deg *

* Based on uncataloged debris data

COMMENTS

Spacecraft was destroyed with a planned detonation. Fifth fragmentation of this sub-type (Cosmos 2031 subclass). Early elements on 43 objects (including the parent) were collected; at least 179 objects were reported by the NAVSPOC for early passes through the NAVSPASUR fence.

REFERENCE DOCUMENTS

The Fragmentation of Cosmos 2262, Technical Report CS94-LKD-006, Teledyne Brown Engineering, Colorado Springs, Colorado, 31 December 1993.

History of Soviet/Russian Satellite Fragmentations-A Joint U.S.-Russian Investigation, N. L. Johnson, et al., Kaman Sciences Corporation, October 1995.

SSN 22789

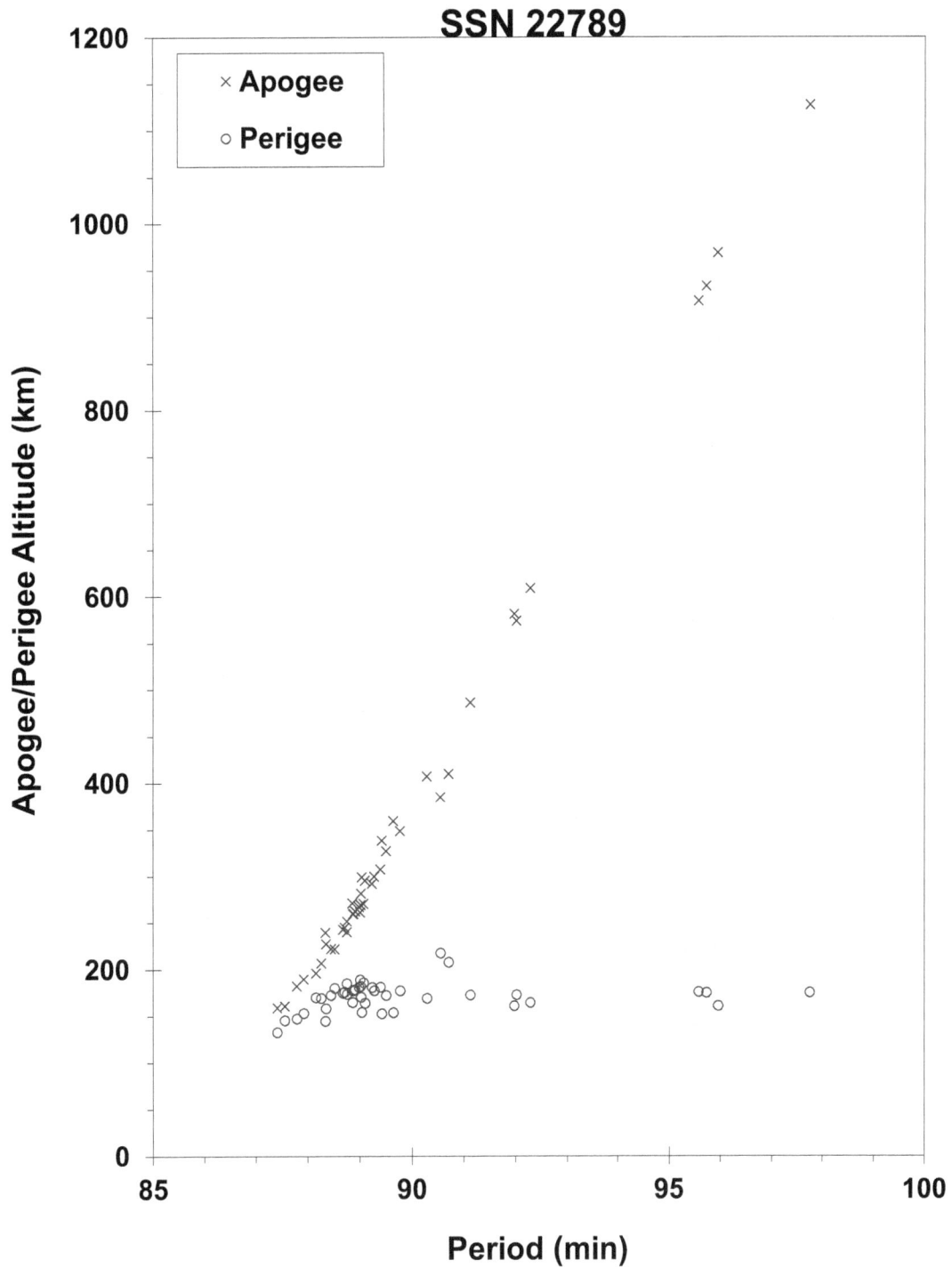

Cosmos 2262 debris cloud based upon 43 early element sets provided by the NAVSPOC.

GORIZONT 29 ULLAGE MOTOR
1993-072E
22925

SATELLITE DATA

TYPE:	Proton Block DM SOZ Ullage Motor
OWNER:	CIS
LAUNCH DATE:	18.58 Nov 1993
DRY MASS (KG):	~55 kg
MAIN BODY:	~0.6 m by 0.6 m by 1.0 m
MAJOR APPENDAGES:	None
ATTITUDE CONTROL:	None
ENERGY SOURCES:	On-board propellants?

EVENT DATA

DATE:	6-7 Sep 2000	LOCATION:	Unknown
TIME:	between 1918–0253 GMT	ASSESSED CAUSE:	Propulsion
ALTITUDE:	Unknown		

PRE-EVENT ELEMENTS

EPOCH:	00250.18110680	MEAN ANOMALY:	305.0033
RIGHT ASCENSION:	135.7916	MEAN MOTION:	6.55809618
INCLINATION:	46.7439	MEAN MOTION DOT/2:	0.00601672
ECCENTRICITY:	0.4592082	MEAN MOTION DOT DOT/6:	0.00000031378
ARG. OF PERIGEE:	109.1361	BSTAR:	0.00059159

DEBRIS CLOUD DATA

MAXIMUM ΔP:	Unknown
MAXIMUM ΔI:	Unknown

COMMENTS

This is the 22nd breakup event for an object of this class, and the first of the year 2000. The breakups are assessed to be caused by residual propellants. Russian officials have been aware of the problem since 1992 and have made design changes, although the date of full implementation is unknown. The environmental consequence of the breakup will be short-lived; the object is in catastrophic decay from a geosynchronous transfer orbit. Latest estimate of the breakup time is between 1918 GMT, 6 September and 0253 GMT, 7 September.

REFERENCE DOCUMENT

"September Breakup is 22nd in Series", <u>The Orbital Debris Quarterly News</u>, NASA JSC, October 2000. Available online at https://orbitaldebris.jsc.nasa.gov/quarterly-news/pdfs/ODQNv5i4.pdf.

Insufficient data to construct a Gabbard diagram.

SATELLITE DATA

TYPE:	Titan II Second Stage
OWNER:	US
LAUNCH DATE:	25.69 Jan 1994
DRY MASS (KG):	2860
MAIN BODY:	Cylinder; 3.048 m diameter x 7.3 m length
MAJOR APPENDAGES:	None
ATTITUDE CONTROL:	None at time of the event.
ENERGY SOURCES:	On-board propellants

EVENT DATA

DATE:	7 Feb 1994	LOCATION:	59S, 234E (dsc)
TIME:	1719 GMT	ASSESSED CAUSE:	Propulsion
ALTITUDE:	260 km		

PRE-EVENT ELEMENTS

EPOCH:	94038.24510489	MEAN ANOMALY:	208.0182
RIGHT ASCENSION:	47.9208	MEAN MOTION:	16.13665058
INCLINATION:	66.9945	MEAN MOTION DOT/2:	0.01050211
ECCENTRICITY:	0.0027030	MEAN MOTION DOT DOT/6:	0.0000059221
ARG. OF PERIGEE:	152.2460	BSTAR:	0.00081413

DEBRIS CLOUD DATA

MAXIMUM ΔP:	5.6 min *
MAXIMUM ΔI:	0.6 deg *

* Based on uncataloged debris data

COMMENTS

First Titan II Second Stage to violently fragment. NAVSPOC reported observing a maximum of 364 objects in the early debris cloud, and the NAVSPOC released 45 element sets. Engineering analysis by the manufacturer (Martin Marietta) indicates no known failure mechanism, although unspent on-board propellants were present.

REFERENCE DOCUMENT

The Fragmentation of the Clementine Rocket Body, TBE Technical Report CS94-LKD-010, Teledyne Brown Engineering, Colorado Springs, Colorado, 31 March 1994.

SSN 22974

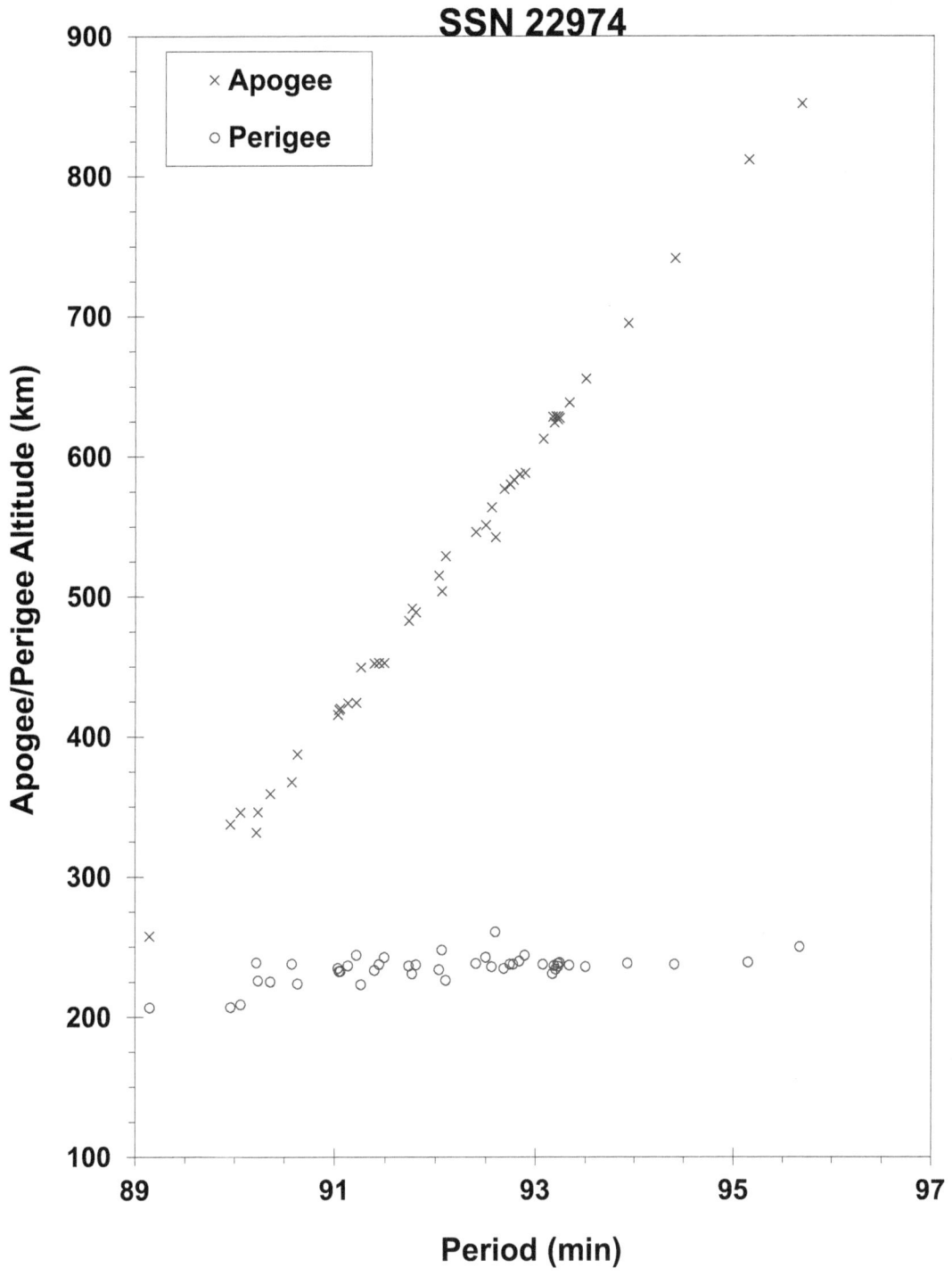

Gabbard diagram of 45 NAVSPOC element sets.

SATELLITE DATA

TYPE:	Pegasus HAPS
OWNER:	USA
LAUNCH DATE:	19.71 May 1994
DRY MASS (KG):	97
MAIN BODY:	Cylinder; 0.97 m diameter by 0.93 m length
MAJOR APPENDAGES:	None
ATTITUDE CONTROL:	None at time of the event.
ENERGY SOURCES:	On-board propellants and pressurants

EVENT DATA

DATE:	3 Jun 1996	LOCATION:	67 S, 56 E (asc)
TIME:	1518 GMT	ASSESSED CAUSE:	Propulsion
ALTITUDE:	625 km		

PRE-EVENT ELEMENTS

EPOCH:	96155.10100506	MEAN ANOMALY:	108.3711
RIGHT ASCENSION:	197.8565	MEAN MOTION:	14.56780581
INCLINATION:	81.9749	MEAN MOTION DOT/2:	0.00000158
ECCENTRICITY:	0.0165742	MEAN MOTION DOT DOT/6:	0
ARG. OF PERIGEE:	249.9583	BSTAR:	0.000025815

DEBRIS CLOUD DATA

MAXIMUM ΔP:	27.9 min
MAXIMUM ΔI:	2.4 deg

COMMENTS

The Pegasus Hydrazine Auxiliary Propulsion Stage (HAPS) was employed for only the second time. It failed to place its payload into the assigned circular orbit and had an estimated 5-8 kg of residual propellant plus propellant for attitude control on board. The fragmentation of the small, 2-year-old upper stage produced a record number (to that time) of more than 750 tracked debris and necessitated the addition of a third place in the piece tag format in two-line element sets. This is about an order of magnitude more than can be expected for the small dry mass of the R/B of 97 kg. Observations suggest that the debris are physically small with a high radar reflectivity. Investigations suggest that a regulator failure led to over pressurization of the propellant tank that in turn ruptured.

REFERENCE DOCUMENT

"Major Satellite Breakup in June," N. Johnson, <u>Orbital Debris Quarterly News</u>, NASA JSC, September 1996, p. 2 and 11. Available online at https://orbitaldebris.jsc.nasa.gov/quarterly-news/pdfs/ODQNV1i2.pdf.

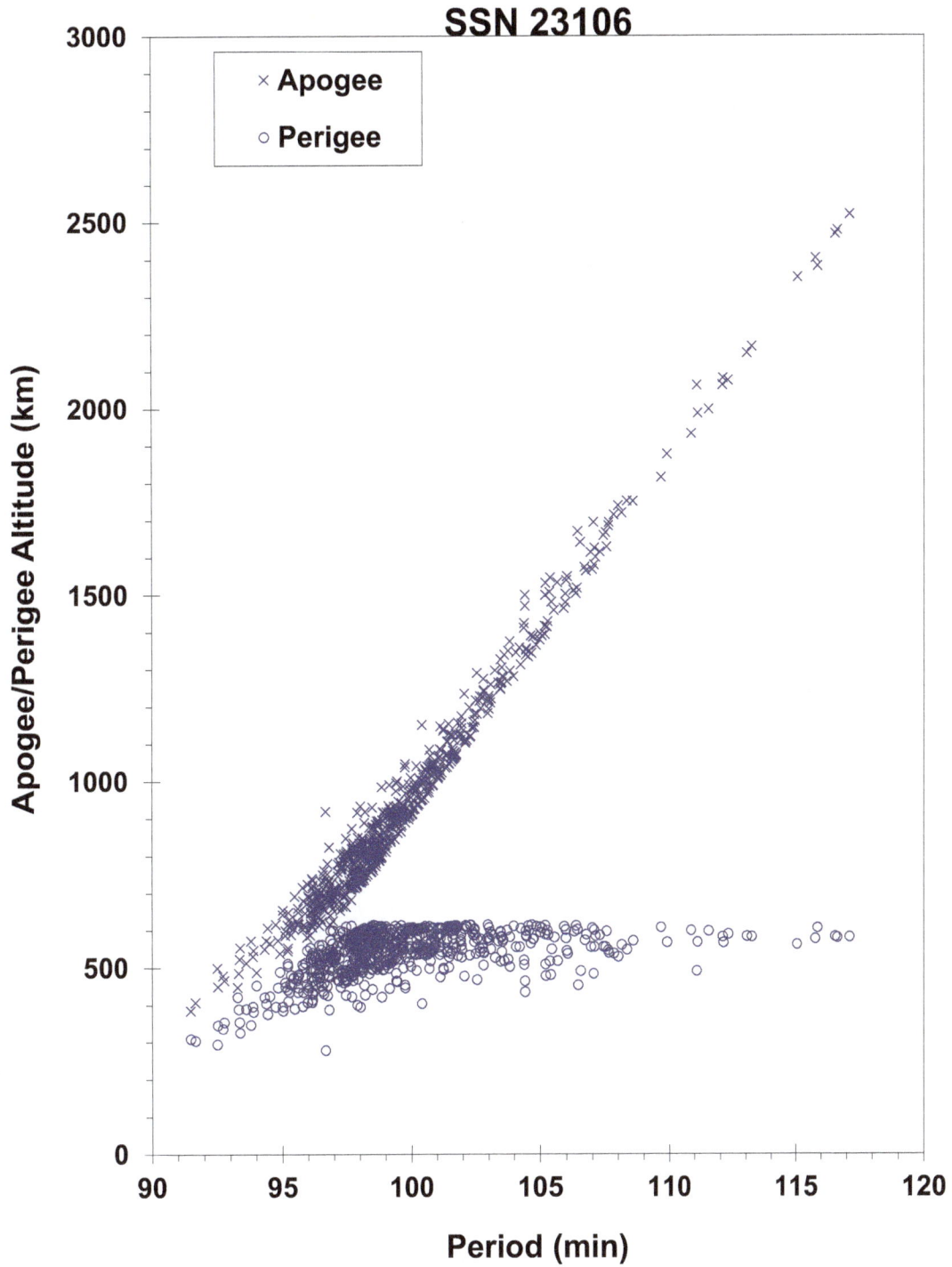

SSN 23106

Step II R/B debris cloud of 713 fragments as of August 29, 1996 as reconstructed from the US SSN database.

SATELLITE DATA

TYPE:	Mission Related Debris
OWNER:	CIS
LAUNCH DATE:	6.99$^+$ Jul 1994
DRY MASS (KG):	55
MAIN BODY:	Ellipsoid; 0.6 m diameter by 1 m length
MAJOR APPENDAGES:	None
ATTITUDE CONTROL:	None
ENERGY SOURCES:	On-board propellants

EVENT DATA

DATE:	Approx. 21 Oct 1995	LOCATION:	Unknown
TIME:	Unknown	ASSESSED CAUSE:	Propulsion
ALTITUDE:	Unknown		

PRE-EVENT ELEMENTS

EPOCH:	95293.99530492	MEAN ANOMALY:	2.33725319
RIGHT ASCENSION:	157.0951	MEAN MOTION:	321.8211
INCLINATION:	47.0485	MEAN MOTION DOT/2:	0.00002472
ECCENTRICITY:	0.7223127	MEAN MOTION DOT DOT/6:	0.0000
ARG. OF PERIGEE:	127.9520	BSTAR:	0.0010694

DEBRIS CLOUD DATA

MAXIMUM ΔP:	Unknown
MAXIMUM ΔI:	Unknown

COMMENTS

Parent satellite was one of two small engine units that are routinely released after the first burn of the Proton fourth stage. The nature of these objects was identified by Dr. Boris V. Chernlatiev, Deputy Constructor for the Energiya NPO, in October 1992. The cause of this fragmentation is assumed to be related to the residual hypergolic propellants on board and failure of the membrane separating the fuel and oxidizer. NAVSPASUR observed 114 objects that were associated with this breakup.

REFERENCE DOCUMENTS

The Fragmentation of Proton Debris, D. J. Nauer, TBE Technical Report CS93-LKD-004, Teledyne Brown Engineering, Colorado Springs, 31 December 1992.

Analysis of Fragmentations From December 1992 - February 1993, TBE Technical Report CS93-LKD-010, Teledyne Brown Engineering, Colorado Springs, 30 March 1993.

History of Soviet/Russian Satellite Fragmentations-A Joint U.S.-Russian Investigation, N. L. Johnson, et al., Kaman Sciences Corporation, October 1995.

"Identification and Resolution of an Orbital Debris Problem with the Proton Launch Vehicle," B. V. Cherniatiev, et al., Proceedings of the First European Conference on Space Debris, April 1993.

SSN 23174

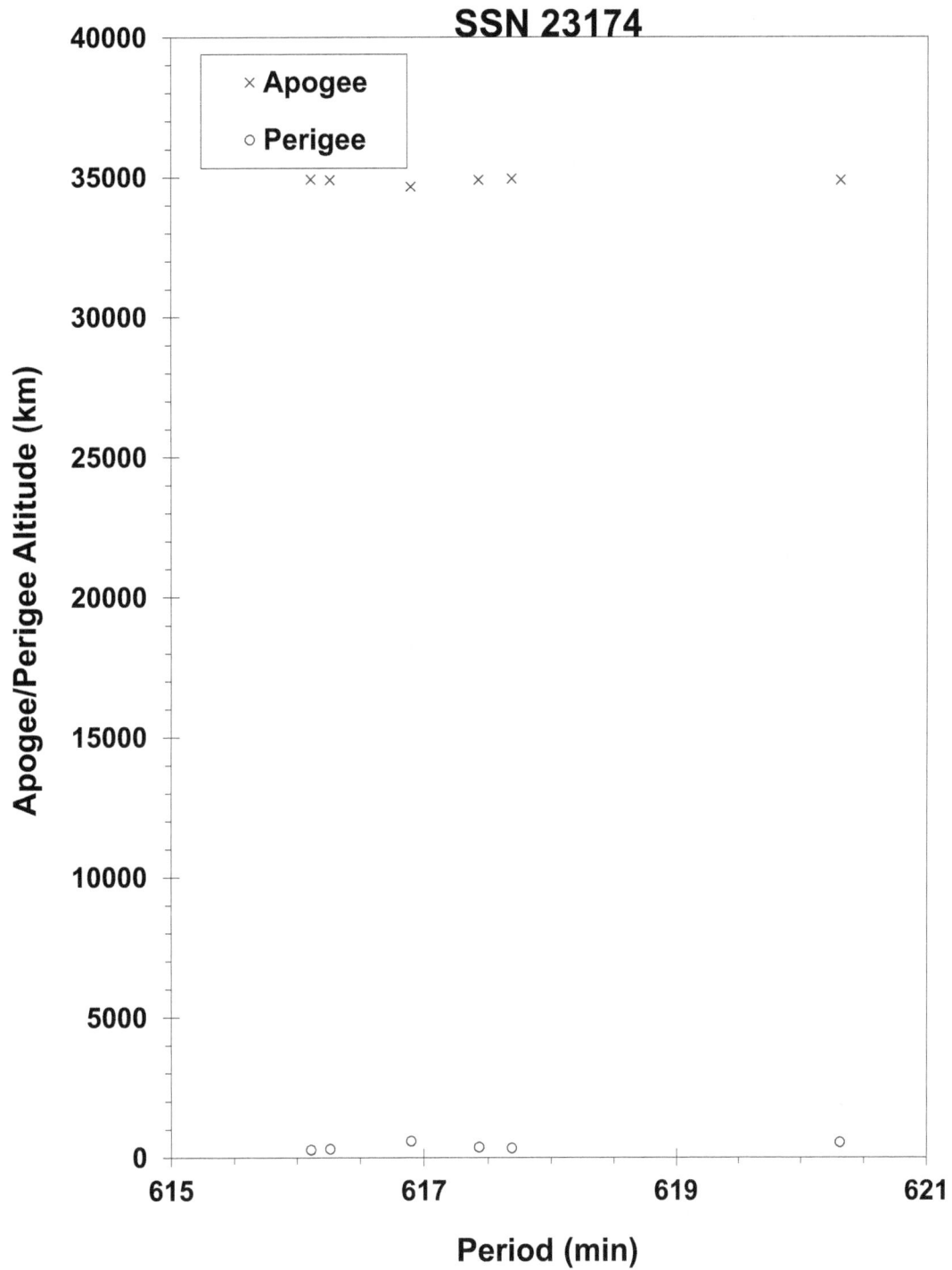

Cosmos 2282 ullage motor debris cloud of 6 fragments assembled by NAVSPOC.

ELEKTRO ULLAGE MOTOR   1994-069E   23338

SATELLITE DATA

TYPE:	Mission Related Debris
OWNER:	CIS
LAUNCH DATE:	31.60 Oct 1994
DRY MASS (KG):	55
MAIN BODY:	Ellipsoid; 0.6 m diameter by 1 m length
MAJOR APPENDAGES:	None
ATTITUDE CONTROL:	None
ENERGY SOURCES:	On-board propellants

EVENT DATA

DATE:	Prior to 0547 GMT, 11 May 1995	LOCATION:	Unknown
TIME:	Unknown	ASSESSED CAUSE:	Propulsion
ALTITUDE:	Unknown		

PRE-EVENT ELEMENTS

EPOCH:	95130.00087914	MEAN ANOMALY:	317.6470
RIGHT ASCENSION:	200.4799	MEAN MOTION:	11.93599702
INCLINATION:	46.9113	MEAN MOTION DOT/2:	0.99999999
ECCENTRICITY:	0.2007574	MEAN MOTION DOT DOT/6:	0.000034693
ARG. OF PERIGEE:	63.6124	BSTAR:	0.021116

DEBRIS CLOUD DATA

MAXIMUM ΔP:	Unknown
MAXIMUM ΔI:	Unknown

COMMENTS

Parent satellite was one of two small engine units that are routinely released after the first burn of the Proton fourth stage. The nature of these objects was identified by Dr. Boris V. Cherniatiev, Deputy Constructor for the Energiya NPO and Mr. Nicholas Johnson of Kaman Sciences, in October 1992. The cause of this fragmentation appears to be related to the residual hypergolic propellants on board and failure of the membrane separating the fuel and oxidizer. NAVSPASUR observed up to 13 objects that were associated with this breakup on 11 May 95. This was the eleventh in a series of fragmentations of this object type.

REFERENCE DOCUMENTS

The Fragmentation of Proton Debris, D. J. Nauer, TBE Technical Report CS93-LKD-004, Teledyne Brown Engineering, Colorado Springs, 31 December 1992.

Analysis of Fragmentations From December 1992 - February 1993, TBE Technical Report CS93-LKD-010, Teledyne Brown Engineering, Colorado Springs, 30 March 1993.

Identification and Resolution of an Orbital Debris Problem with the Proton Launch Vehicle, B. V. Cherniatiev,, et al., First European Conference on Space Debris, 5-7 April 1993.

History of Soviet/Russian Satellite Fragmentations-A Joint U.S.-Russian Investigation, N. L. Johnson, et al., Kaman Sciences Corporation, October 1995.

Insufficient data to construct a Gabbard diagram.

SATELLITE DATA

TYPE:	payload
OWNER:	CIS
LAUNCH DATE:	4 Nov. 1994
DRY MASS (KG):	1900 kg
MAIN BODY:	Meteor-3 cylindrical bus: 1.4 m diameter
MAJOR APPENDAGES:	solar arrays (2), antennas
ATTITUDE CONTROL:	three-axis stabilization
ENERGY SOURCES:	Unknown

EVENT DATA

DATE:	27 Aug. 2020	LOCATION:	Unknown
TIME:	1000 GMT ± 8 hours	ASSESSED CAUSE:	Unknown
ALTITUDE:	Unknown		

PRE-EVENT ELEMENTS

EPOCH:	20239.22001899	MEAN ANOMALY:	283.3416
RIGHT ASCENSION:	287.1592	MEAN MOTION:	14.79443205
INCLINATION:	97.9244	MEAN MOTION DOT/2:	-0.00000019
ECCENTRICITY:	0.0000925	MEAN MOTION DOT DOT/6:	0.0
ARG. OF PERIGEE:	76.7899	BSTAR:	0.0000043823

DEBRIS CLOUD DATA

MAXIMUM ΔP:	3.1 min
MAXIMUM ΔI:	0.8 deg

COMMENTS

This spacecraft is the third of the Resurs-O1 class Earth-observing spacecraft. The spacecraft was observed to exhibit multiple shedding events, in which numerous objects separated from the spacecraft and relatively low velocities.

REFERENCE DOCUMENTS

"Two New Breakups with One Resulting in an ISS Maneuver," The Orbital Debris Quarterly News, NASA JSC, November 2020. Available online at https://www.orbitaldebris.jsc.nasa.gov/quarterly-news/pdfs/odqnv24i4.pdf.

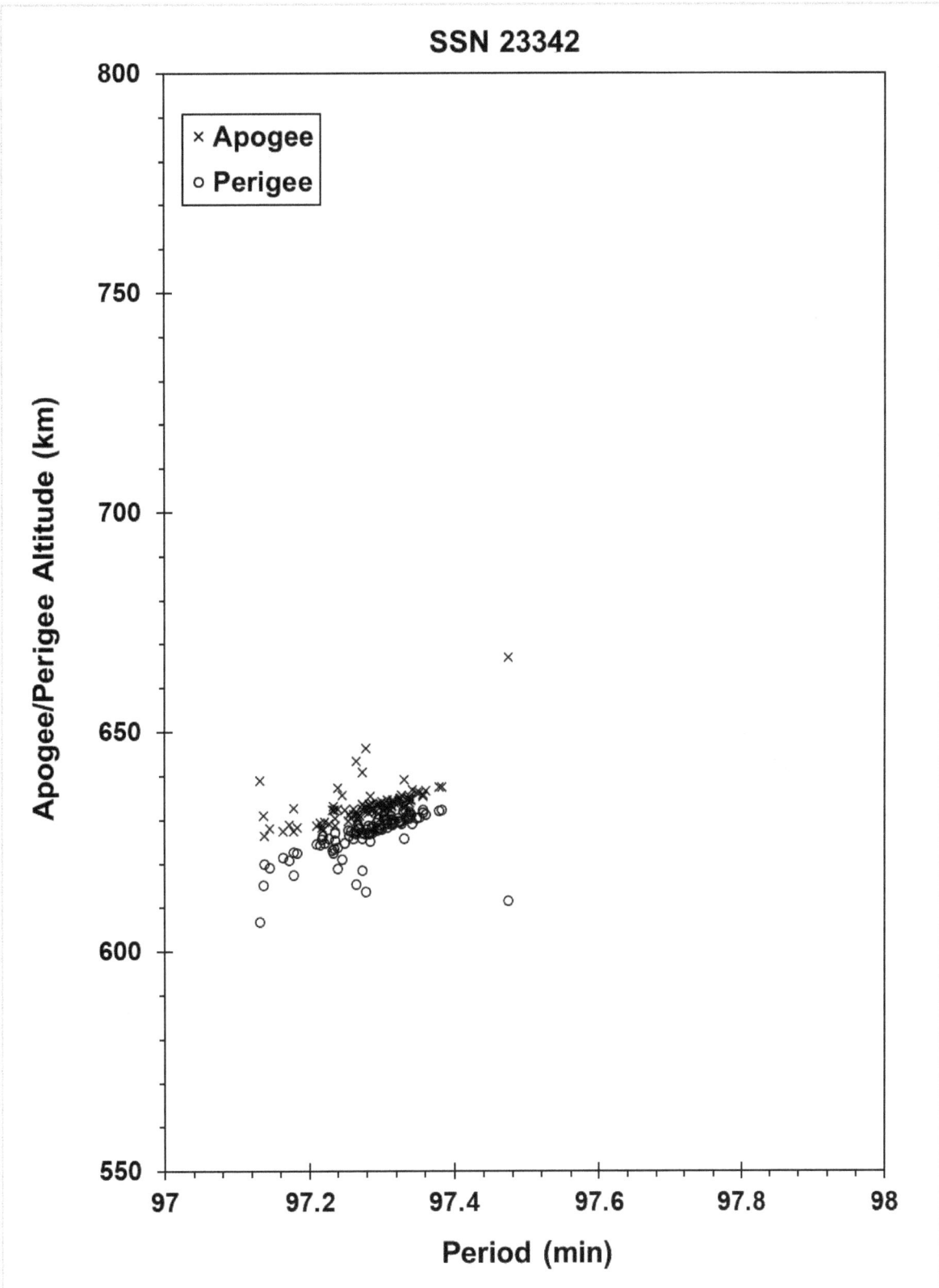

RESURS O1 No. 3 debris cloud, composed of parent spacecraft and 96 additional objects cataloged between 30 and 250 days after the event.

SATELLITE DATA

TYPE:	Mission Related Debris
OWNER:	CIS
LAUNCH DATE:	20 Nov 1994
DRY MASS (KG):	56.0
MAIN BODY:	ellipsoid; 0.6 m diameter x 1 m length
MAJOR APPENDAGES:	none
ATTITUDE CONTROL:	none
ENERGY SOURCES:	on-board propellants

EVENT DATA

DATE:	7 June 2014	LOCATION:	unknown
TIME:	unknown	ASSESSED CAUSE:	Propulsion
ALTITUDE:	unknown		

PRE-EVENT ELEMENTS

EPOCH:	14156.48285780	MEAN ANOMALY:	3.5477
RIGHT ASCENSION:	76.9596	MEAN MOTION:	4.25680746
INCLINATION:	64.9844	MEAN MOTION DOT/2:	0.00004224
ECCENTRICITY:	0.5768591	MEAN MOTION DOT DOT/6:	0.00000
ARG. OF PERIGEE:	344.0988	BSTAR:	0.050698

DEBRIS CLOUD DATA

MAXIMUM ΔP:	unknown
MAXIMUM ΔI:	unknown

COMMENTS

This event was the 41st known fragmentation of Blok-DM ullage motors. The breakups are believed to be caused by residual propellants. Russian officials have been aware of the problem since 1992 and have made design changes, although the date of full implementation is unknown. Although at least 15 small debris were observed by the SSN, none have been officially cataloged.

REFERENCE DOCUMENTS

History of Soviet/Russian Satellite Fragmentations: A Joint U.S.-Russian Investigation. N.L. Johnson et al. Kaman Sciences Corporation, October 2005.

"Flurry of Small Breakups in First Half of 2014", The Orbital Debris Quarterly News, NASA JSC, July 2014. Available online at http://orbitaldebris.jsc.nasa.gov/newsletter/pdfs/ODQNv18i3.pdf.

Insufficient data to construct a Gabbard diagram.

SATELLITE DATA

TYPE:	Rokot Third Stage
OWNER:	CIS
LAUNCH DATE:	26.13 Dec 1994
DRY MASS (KG):	1000
MAIN BODY:	Cylinder; 2.4 m diameter by 2.8 m length
MAJOR APPENDAGES:	None
ATTITUDE CONTROL:	Unknown
ENERGY SOURCES:	On-board propellants

EVENT DATA

DATE:	26 Dec 1994	LOCATION:	51.6S, 307E (asc)
TIME:	0627 GMT	ASSESSED CAUSE:	Unknown
ALTITUDE:	1880 km		

PRE-EVENT ELEMENTS

EPOCH:	94361.79150546	MEAN ANOMALY:	66.1014
RIGHT ASCENSION:	172.1572	MEAN MOTION:	11.27113018
INCLINATION:	64.8297	MEAN MOTION DOT/2:	-0.00000043
ECCENTRICITY:	0.0188748	MEAN MOTION DOT DOT/6:	0.00000
ARG. OF PERIGEE:	292.0126	BSTAR:	0.000000

DEBRIS CLOUD DATA

MAXIMUM ΔP:	4.5 min
MAXIMUM ΔI:	0.2 deg

COMMENTS

Parent satellite was the Rokot third stage. The Rokot is an SS-19 ICBM based vehicle with the third stage being the *Briz-K*, retired after this launch for the improved *Briz-KM* stage. All three stages are fueled with UDMH/N204. NAVSPASUR observed 34 objects that were associated with this breakup, of which 26 entered the public satellite catalog.

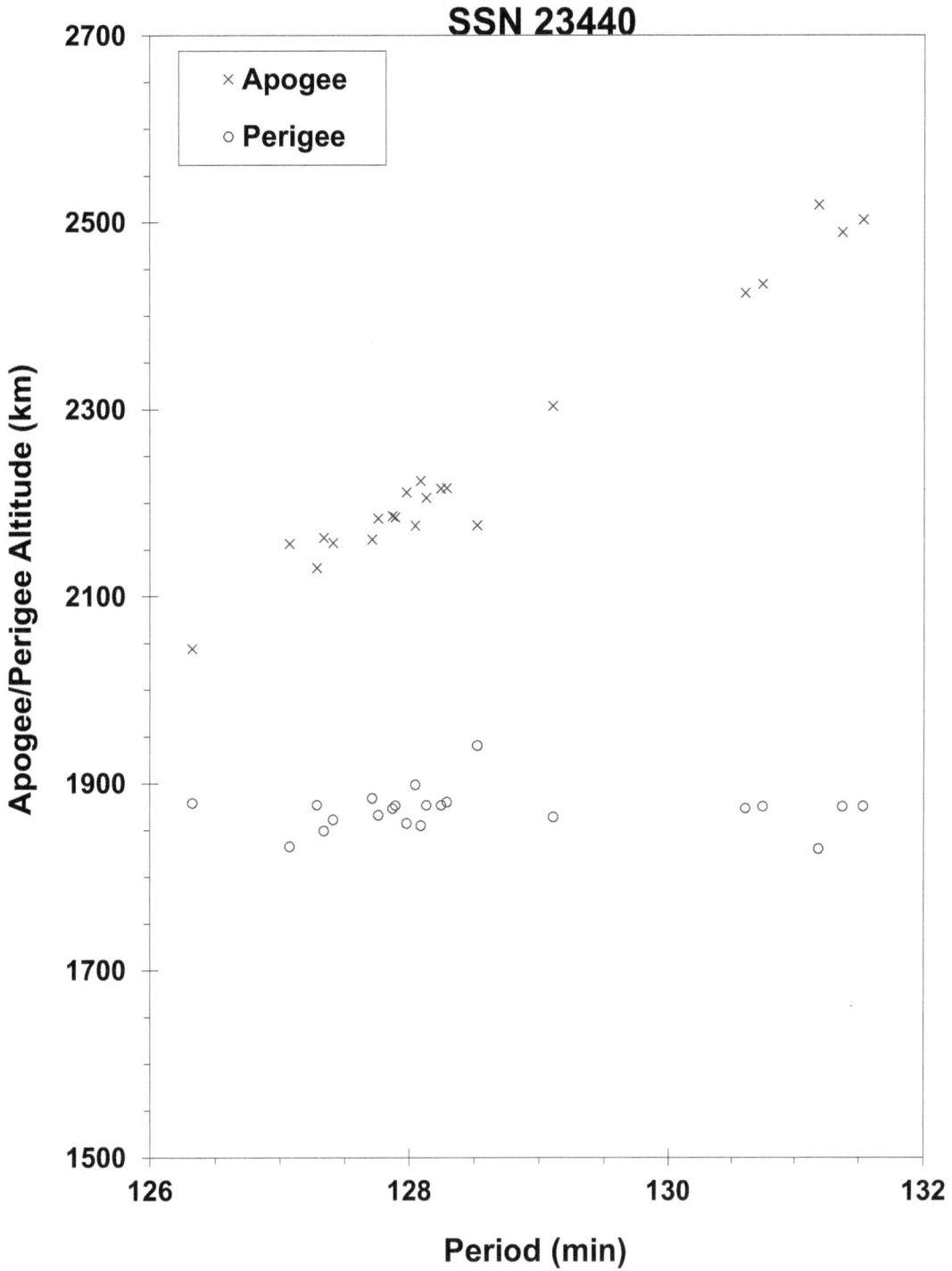

SSN 23440

Gabbard diagram for RS-15 R/B debris cloud of 23 fragments as reconstructed from the US SSN database.

SATELLITE DATA

TYPE:	Payload
OWNER:	US
LAUNCH DATE:	24 Mar 1995
DRY MASS (KG):	767
MAIN BODY:	cylinder; 1.0 m diameter x 3.6 m length
MAJOR APPENDAGES:	solar panel
ATTITUDE CONTROL:	three-axis
ENERGY SOURCES:	on-board propellants and chemical batteries

EVENT DATA

DATE:	3.736 Feb 2015	LOCATION:	40.8S, 7.5E
TIME:	1740 GMT	ASSESSED CAUSE:	Battery
ALTITUDE:	860.7 km		

PRE-EVENT ELEMENTS - unavailable

EPOCH:	MEAN ANOMALY:	
RIGHT ASCENSION:	MEAN MOTION:	
INCLINATION:	MEAN MOTION DOT/2:	
ECCENTRICITY:	MEAN MOTION DOT DOT/6:	
ARG. OF PERIGEE:	BSTAR:	

DEBRIS CLOUD DATA

MAXIMUM ΔP:	unknown	min
MAXIMUM ΔI:	unknown	deg

COMMENTS

This is the second known breakup of a Defense Meteorological Satellite Program (DMSP) Block 5D-2 spacecraft, the first being DMSP 5D-2 F11 (USA 73). USA 109 was active when the event occurred and supporting telemetry was available for the inquiry. Analysis indicates that the event was due to an explosion of one of the Ni-Cd batteries.

REFERENCE DOCUMENT

"Recent Breakup of a DMSP Satellite," The Orbital Debris Quarterly News, NASA JSC, April 2015.
Available online at https://orbitaldebris.jsc.nasa.gov/quarterly-news/pdfs/ODQNv19i2.pdf.

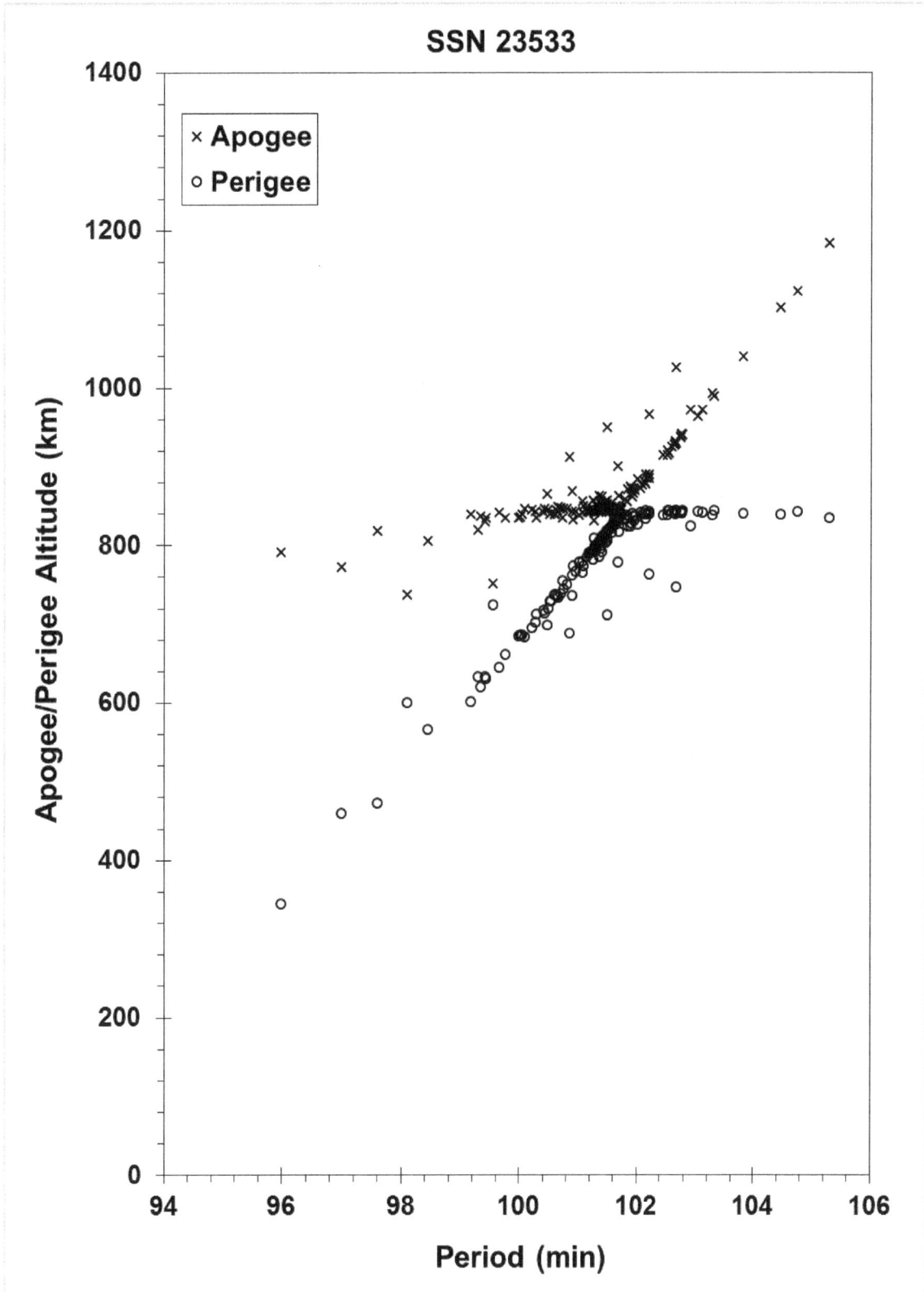

USA 109 debris cloud. Debris cataloged between 22 and 160 days after event.

COSMOS 2313   1995-028A   23596

SATELLITE DATA

TYPE:	Payload
OWNER:	CIS
LAUNCH DATE:	8.20 Jun 1995
DRY MASS (KG):	3000
MAIN BODY:	Cylinder; 1.3 m diameter by 17 m length
MAJOR APPENDAGES:	Solar arrays
ATTITUDE CONTROL:	Active, 3-axis
ENERGY SOURCES:	On-board propellants, explosive charge?

EVENT DATA

DATE:	26 June 1997	LOCATION:	44 N, 173 E (asc.)
TIME:	0257 GMT	ASSESSED CAUSE:	Unknown
ALTITUDE:	285 km		

PRE-EVENT ELEMENTS

EPOCH:	97176.10173599	MEAN ANOMALY:	124.6445
RIGHT ASCENSION:	342.0749	MEAN MOTION:	16.02369895
INCLINATION:	65.0221	MEAN MOTION DOT/2:	0.00306537
ECCENTRICITY:	0.0084335	MEAN MOTION DOT DOT/6:	0.0000069339
ARG. OF PERIGEE:	234.6794	BSTAR:	0.00033322

DEBRIS CLOUD DATA

MAXIMUM ΔP:	Unknown
MAXIMUM ΔI:	Unknown

COMMENTS

Cosmos 2313 was the second spacecraft of its type to breakup since November 1987. Prior to the current event 17 spacecraft of this class (Cosmos 699) have experienced breakups in low Earth orbit. In the 1980's procedures were introduced to deplete remaining propellants at the end of mission, reducing orbital lifetime at the same time. Cosmos 2313 performed such a maneuver during 22-23 April 1997 and was close to reentry at the time of the event. Earlier spacecraft breakups resulted in up to 150 or more trackable debris. The cause of the event may well not be propellant related, but by reducing the orbital lifetime recent vehicles have decayed before the trigger mechanism could activate. At least 90 debris were detected after this event.

REFERENCE DOCUMENT

"Three Satellite Breakups During May-June," The Orbital Debris Quarterly News, NASA JSC, July 1997, p. 2. Available online at https://orbitaldebris.jsc.nasa.gov/quarterly-news/pdfs/ODQNv2i3.pdf.

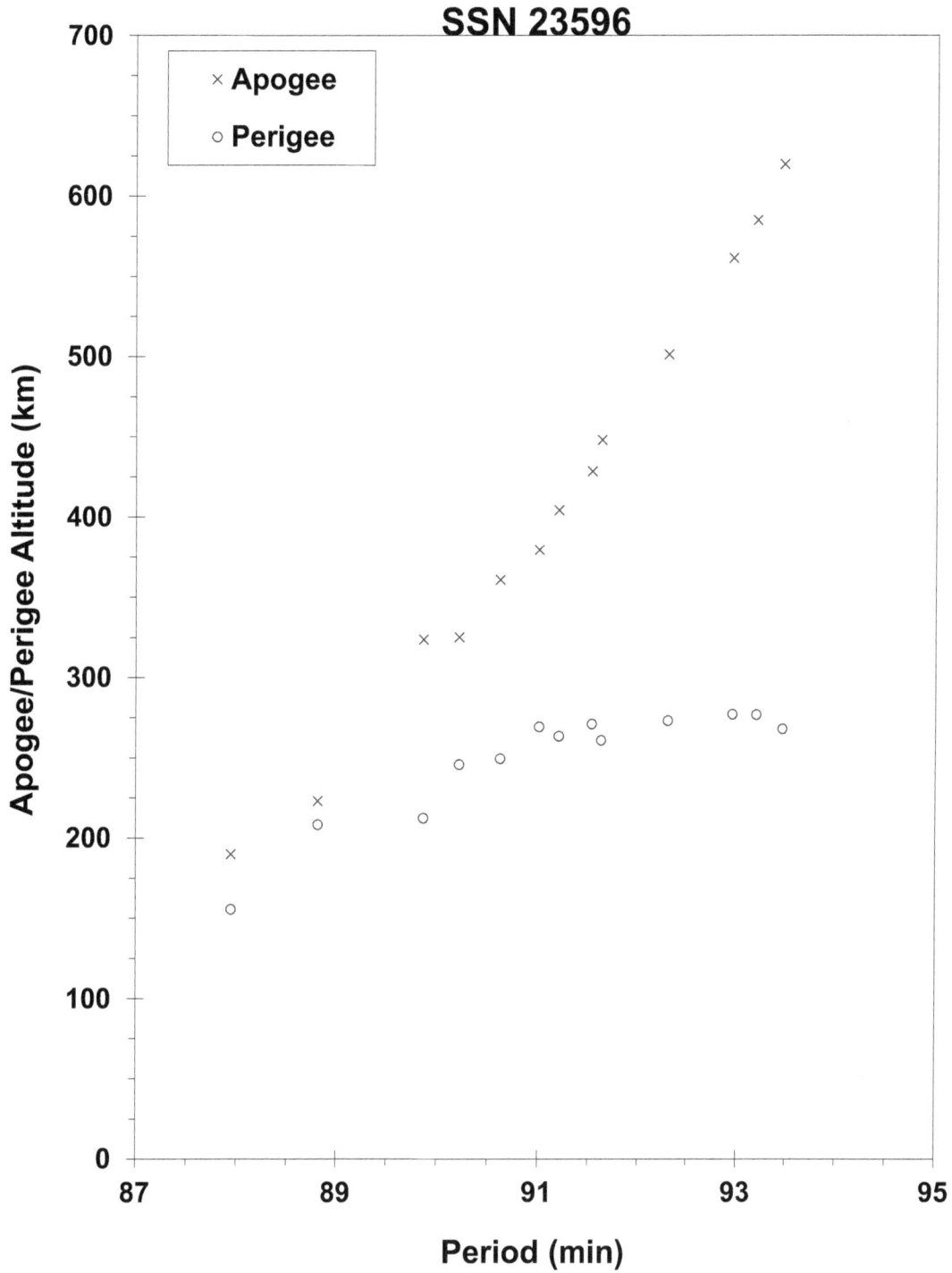

Cosmos 2313 debris cloud of 13 fragments 1 day to 2 weeks after the event as reconstructed from the US SSN database.

SATELLITE DATA

TYPE:	Payload
OWNER:	France
LAUNCH DATE:	7.68 Jul 1995
DRY MASS (KG):	50
MAIN BODY:	Box; 0.6 m by 0.3 m by 0.3 m
MAJOR APPENDAGES:	6 m long gravity-gradient boom; solar panels
ATTITUDE CONTROL:	Gravity-gradient stabilized
ENERGY SOURCES:	Battery

EVENT DATA

DATE:	24 Jul 1996	LOCATION:	38 S, 60 E (asc)
TIME:	0948 GMT	ASSESSED CAUSE:	Collision
ALTITUDE:	685 km		

PRE-EVENT ELEMENTS

EPOCH:	96205.39273562	MEAN ANOMALY:	292.8048
RIGHT ASCENSION:	141.7519	MEAN MOTION:	14.67264268
INCLINATION:	98.1025	MEAN MOTION DOT/2:	0.00000083
ECCENTRICITY:	0.0008991	MEAN MOTION DOT DOT/6:	0
ARG. OF PERIGEE:	67.4104	BSTAR:	0.000023247

DEBRIS CLOUD DATA

MAXIMUM ΔP:	unknown min
MAXIMUM ΔI:	unknown deg

COMMENTS

The incident marked the first time that two objects in the U.S. satellite catalog are known to have accidentally run into one another. The CERISE spacecraft (Satellite Number 23606, International Designator 1995-033B) is a microsatellite of British design. The other participant in the encounter was Satellite Number 18208 (International Designator 1986-019RF), which was generated in November 1986, when ESA's SPOT 1 rocket body broke up into nearly 500 tracked debris. The orbit of this fragmentation debris at the time of the collision was 660 km by 680 km at an inclination of 98.45 degrees. The collision, which occurred with a relative velocity of 14.8 km/s, produced only a single piece of debris large enough to be tracked, *i.e.*, the upper portion of the gravity-gradient boom. Analysis of the manufacturer of the spacecraft bus, Surrey Satellite Technology Ctd. at the University of Surrey, United Kingdom suggested that the 6 m, gravity-gradient boom had been severed at 3.1-3.2 meter from its base.

Using USAF Space Command's COMBO (Computation of Miss Between Orbits) program, a close approach of less than 1 km between Satellite 23606 and Satellite 18208 was determined by NASA JSC to have taken place at 0948 GMT on 24 July over the southern Indian Ocean. Naval Space Operations Center (NAVSPOC) at Dahlgren, Virginia, replicated the NASA findings and, using direct observational data and special perturbation theory, was able to refine the miss distance uncertainty to within 137 m. In addition, NAVSPOC identified a minor perturbation in the orbit of Satellite 18208 that occurred about the time of the event.

REFERENCE DOCUMENTS

"First Natural Collision of Cataloged Earth Satellites," N. Johnson, <u>The Orbital Debris Quarterly News</u>, NASA JSC, September 1996, p. 1. Available online at https://orbitaldebris.jsc.nasa.gov/quarterly-news/pdfs/ODQNV1i2.pdf.

"Collision of CERISE with Space Debris," F. Alby*, et al.*, <u>Proceedings of the Second European Conference on Space Debris</u>, SP-393, p. 589-596.

"First 'Confirmed' Natural Collision Between Two Cataloged Satellites," T. Payne, <u>Proceedings of the Second European Conference on Space Debris</u>, SP-393, p. 597-600.

"Predicting Conjunctions with Trackable Space Debris: Some Recent Experiences," E. L. Jenkins and P. W. Schumacher, Jr., AAS 97-014, 20th Annual AAS Guidance and Control Conference, February 1997.

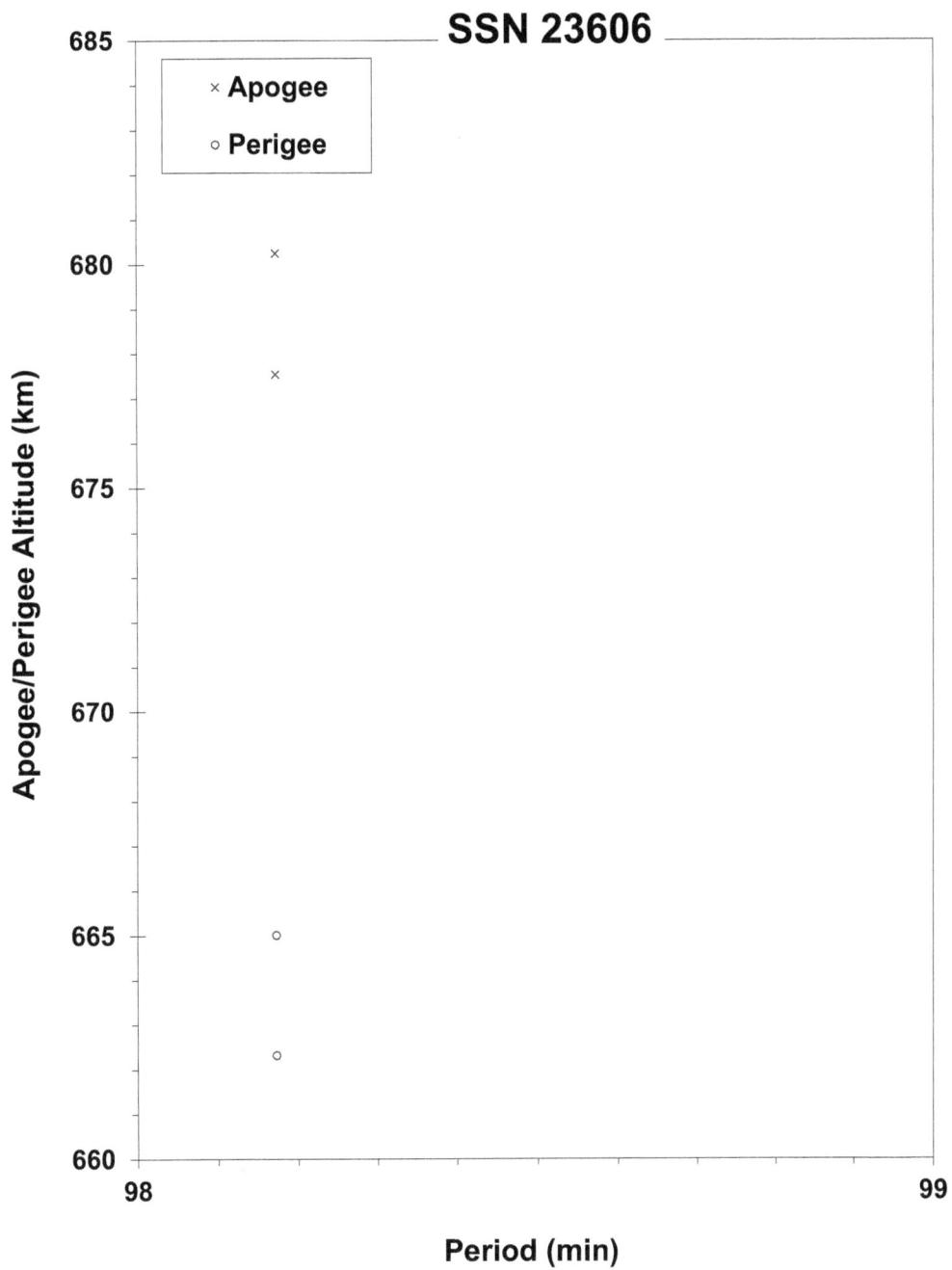

CERISE debris cloud of 2 fragments 4 days after the event as reconstructed from the US SSN database.

SATELLITE DATA

TYPE:	Proton Block DM SOZ Ullage Motor
OWNER:	CIS
LAUNCH DATE:	24.66 Jul 1995
DRY MASS (KG):	~55 kg
MAIN BODY:	~0.6 m by 0.6 m by 1.0 m
MAJOR APPENDAGES:	None
ATTITUDE CONTROL:	None
ENERGY SOURCES:	On-board propellants?

EVENT DATA

DATE:	21 Nov 2000	LOCATION:	Unknown
TIME:	Unknown	ASSESSED CAUSE:	Propulsion
ALTITUDE:	Unknown		

PRE-EVENT ELEMENTS

EPOCH:	00324.99357911	MEAN ANOMALY:	90.3648
RIGHT ASCENSION:	200.0539	MEAN MOTION:	4.50149006
INCLINATION:	64.4375	MEAN MOTION DOT/2:	0.00164632
ECCENTRICITY:	0.5787543	MEAN MOTION DOT DOT/6:	0.00000030156
ARG. OF PERIGEE:	213.7574	BSTAR:	0.00048999

DEBRIS CLOUD DATA

MAXIMUM ΔP:	Unknown
MAXIMUM ΔI:	Unknown

COMMENTS

This is the 4th breakup of 2000 and the 23rd breakup of a Proton SOZ motor (see Orbital Debris Quarterly Newsletter V, Issue 4, p. 2.) This object is associated with the 24 July 1995 launch (1995-037) of the Cosmos 2316-2318 satellites. These members of the GLONASS series are equivalent to GPS/Navstar satellites and reside in middle Earth orbit. This object was one of two pieces left in the transfer orbit and is assessed to be one of the SOZ ullage/orientation motor units. As of November 21, 2000, this object had been on orbit 5 years and 121 days.

REFERENCE DOCUMENT

"SOZ Ullage Motor Breakup," The Orbital Debris Quarterly News, NASA JSC, January 2001. Available online at https://orbitaldebris.jsc.nasa.gov/quarterly-news/pdfs/odqnv6i1.pdf.

Insufficient data to construct a Gabbard diagram.

SATELLITE DATA

TYPE:	Proton Blok DM Fourth Stage
OWNER:	CIS
LAUNCH DATE:	19.36 Feb 1996
DRY MASS (KG):	3400 (?)
MAIN BODY:	Cylinder; 3.7 m diameter by 6.3 m length
MAJOR APPENDAGES:	None
ATTITUDE CONTROL:	Active, 3-axis
ENERGY SOURCES:	On-board propellants, pressurants, and batteries

EVENT DATA

DATE:	19 Feb 1996	LOCATION:	0.2 N, 88.8 E (dsc)
TIME:	1459 GMT	ASSESSED CAUSE:	Propulsion
ALTITUDE:	36511 km		

POST-EVENT ELEMENTS

EPOCH:	96058.46760248	MEAN ANOMALY:	359.9314
RIGHT ASCENSION:	280.4138	MEAN MOTION:	2.23172282
INCLINATION:	48.7	MEAN MOTION DOT/2:	0.0002158
ECCENTRICITY:	0.7321111	MEAN MOTION DOT DOT/6:	0.0000
ARG. OF PERIGEE:	1.7779	BSTAR:	0.00068491

DEBRIS CLOUD DATA

MAXIMUM ΔP:	Unknown
MAXIMUM ΔI:	Unknown

COMMENTS

The first burn of this stage was successful and indicates that 23797 was in a transfer orbit. However, prior to the first pass through the NAVSPOC fence, 23797 fragmented. Twenty (20) pieces were observed during this first pass. During a subsequent pass, 196 pieces were observed that were associated with the upper stage. Stage apparently broke up after main engine restart for GEO apogee maneuver.

REFERENCE DOCUMENT

"Satellite Fragmentations in 1996", N. Johnson, The Orbital Debris Quarterly News, NASA JSC, January 1997, p. 1. Available online at https://orbitaldebris.jsc.nasa.gov/quarterly-news/pdfs/ODQNv2i1.pdf.

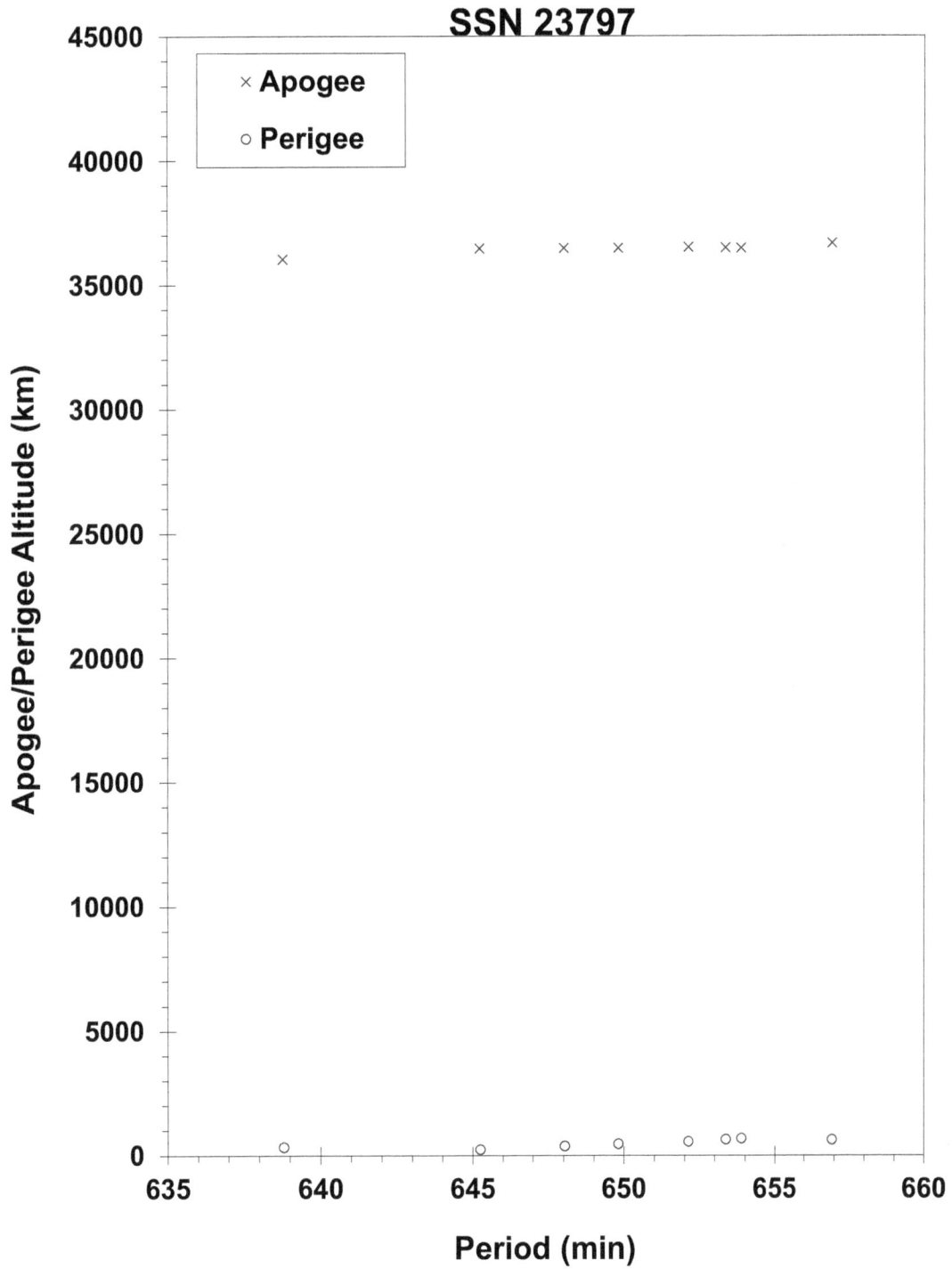

SSN 23797

× Apogee

○ Perigee

Apogee/Perigee Altitude (km)

Period (min)

Gabbard diagram from Raduga 33 from NAVSPOC elements.

435

GORIZONT 32 ULLAGE MOTOR 1996-034F 23887

SATELLITE DATA

TYPE:	Mission Related Debris
OWNER:	CIS
LAUNCH DATE:	25.09 May 1996
DRY MASS (KG):	55
MAIN BODY:	Ellipsoid; 0.6 m diameter by 1 m length
MAJOR APPENDAGES:	None
ATTITUDE CONTROL:	Unknown
ENERGY SOURCES:	On-board Propellants

EVENT DATA

DATE:	~13 Dec 1999	LOCATION:	Unknown
TIME:	Unknown	ASSESSED CAUSE:	Propulsion
ALTITUDE:	Unknown		

PRE-EVENT ELEMENTS

EPOCH:	99347.02294368	MEAN ANOMALY:	314.8549
RIGHT ASCENSION:	194.3249	MEAN MOTION:	9.75630550
INCLINATION:	46.4558	MEAN MOTION DOT/2:	0.00969995
ECCENTRICITY:	0.2950283	MEAN MOTION DOT DOT/6:	0.0000015400
ARG. OF PERIGEE:	75.9037	BSTAR:	0.00055450

CATALOGED DEBRIS CLOUD DATA

MAXIMUM ΔP:	Unknown
MAXIMUM ΔI:	Unknown

COMMENTS

This is 21st breakup event for an object of this class, and the third in 1999. The breakups are believed to be caused by residual propellants. Russian officials have been aware of the problem since 1992 and have made design changes, although the date of full implementation is unknown. This is the most recently launched object to breakup (age = ~3.5 years). The environmental consequence of the breakup was short-lived; the object was in catastrophic decay from a geosynchronous transfer orbit.

REFERENCE DOCUMENTS

The Fragmentation of Proton Debris, D. J. Nauer, TBE Technical Report CS93-LKD-004, Teledyne Brown Engineering, Colorado Springs, 31 December 1992.

Analysis of Fragmentations From December 1992 - February 1993, TBE Technical Report CS93-LKD-010, Teledyne Brown Engineering, Colorado Springs, 30 March 1993.

Insufficient data to construct a Gabbard diagram.

SATELLITE DATA

TYPE:	Payload
OWNER:	CIS
LAUNCH DATE:	15.51 May 1997
DRY MASS (KG):	6000
MAIN BODY:	Cylinder; 2.4 m diameter by 7 m length
MAJOR APPENDAGES:	Solar arrays
ATTITUDE CONTROL:	Active, 3-axis
ENERGY SOURCES:	On-board propellants, explosive charge

EVENT DATA

DATE:	16 Sep 1997	LOCATION:	58.2 N, 157.5 E (asc.)
TIME:	2208 GMT	ASSESSED CAUSE:	Deliberate
ALTITUDE:	230 km		

PRE-EVENT ELEMENTS

EPOCH:	97258.16080604	MEAN ANOMALY:	247.0345
RIGHT ASCENSION:	1.1478	MEAN MOTION:	16.06645410
INCLINATION:	64.8485	MEAN MOTION DOT/2:	0.00206295
ECCENTRICITY:	0.0048612	MEAN MOTION DOT DOT/6:	0.000026376
ARG. OF PERIGEE:	113.5945	BSTAR:	0.00022999

DEBRIS CLOUD DATA

MAXIMUM ΔP:	7.3 min
MAXIMUM ΔI:	0.9 deg

COMMENTS

Cosmos 2343 was the sixth of the Cosmos 2031 class of spacecraft that debuted in 1989 but was not flown since 1993. In all five previous missions (1989-1993), the spacecraft was deliberately exploded at the end of mission. Previous missions of this type include Cosmos 2031, Cosmos 2101, Cosmos 2163, Cosmos 2225, and Cosmos 2262. All such events have occurred over Eastern Russia. This event, as with three of the previous events, occurred over the Kamchatka Peninsula. Highest previous piece count for large debris for this class of vehicle was 180, although more were probably created. Due to the low altitude of the breakup, the debris were short-lived.

REFERENCE DOCUMENT

"International LEO Spacecraft Breakup in September," N. Johnson, The Orbital Debris Quarterly News, NASA JSC, October 1997, p. 2. Available online at https://orbitaldebris.jsc.nasa.gov/quarterly-news/pdfs/ODQNv2i4.pdf.

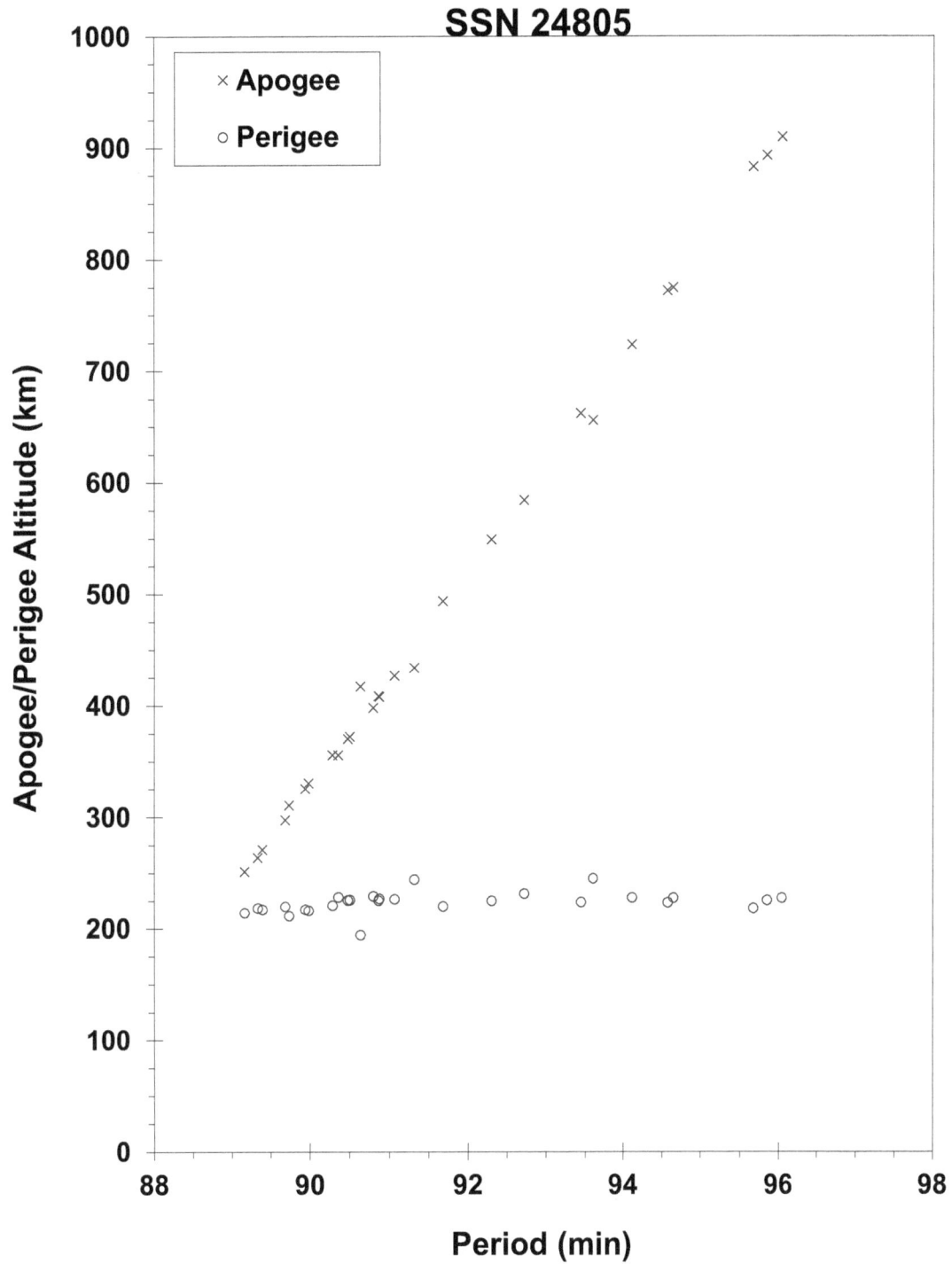

SSN 24805

Apogee/Perigee Altitude (km) vs Period (min)

Legend:
- × Apogee
- ○ Perigee

Cosmos 2343 debris cloud of 28 fragments as reconstructed from the US SSN database.

SATELLITE DATA

TYPE:	Payload
OWNER:	USA
LAUNCH DATE:	14 Sep 1997
DRY MASS (KG):	556
MAIN BODY:	3.6 m long triangular prism
MAJOR APPENDAGES:	single-axis solar panels (2); communication antennas (3)
ATTITUDE CONTROL:	three-axis stabilized
ENERGY SOURCES:	on-board hydrazine monopropellant; NiH_2 batteries

EVENT DATA

DATE:	10 Feb 2009	LOCATION:	72.5N, 97.9E (asc)
TIME:	1656 GMT	ASSESSED CAUSE:	Collision
ALTITUDE:	788.597 km		

PRE-EVENT ELEMENTS

EPOCH:	09040.36587205	MEAN ANOMALY:	274.2112
RIGHT ASCENSION:	121.8779	MEAN MOTION:	14.34219614
INCLINATION:	86.3996	MEAN MOTION DOT/2:	0.00000121
ECCENTRICITY:	0.0002273	MEAN MOTION DOT DOT/6:	0.0
ARG. OF PERIGEE:	85.9342	BSTAR:	0.000036194

DEBRIS CLOUD DATA

MAXIMUM ΔP:	8.0 min
MAXIMUM ΔI:	0.8 deg

COMMENTS

Iridium 33 was a member of the first-generation Iridium low-altitude communication satellite constellation; it utilized a Lockheed Martin LM700A bus and was active at the time of the event. Iridium 33 collided with Cosmos 2251 [q.v.] in the first, accidental collision of large, intact resident space objects.

REFERENCE DOCUMENT

"Satellite Collision Leaves Significant Debris Clouds," The Orbital Debris Quarterly News, NASA JSC, April 2009. Available online at https://orbitaldebris.jsc.nasa.gov/quarterly-news/ODQNv13i2.pdf.

SSN 24946

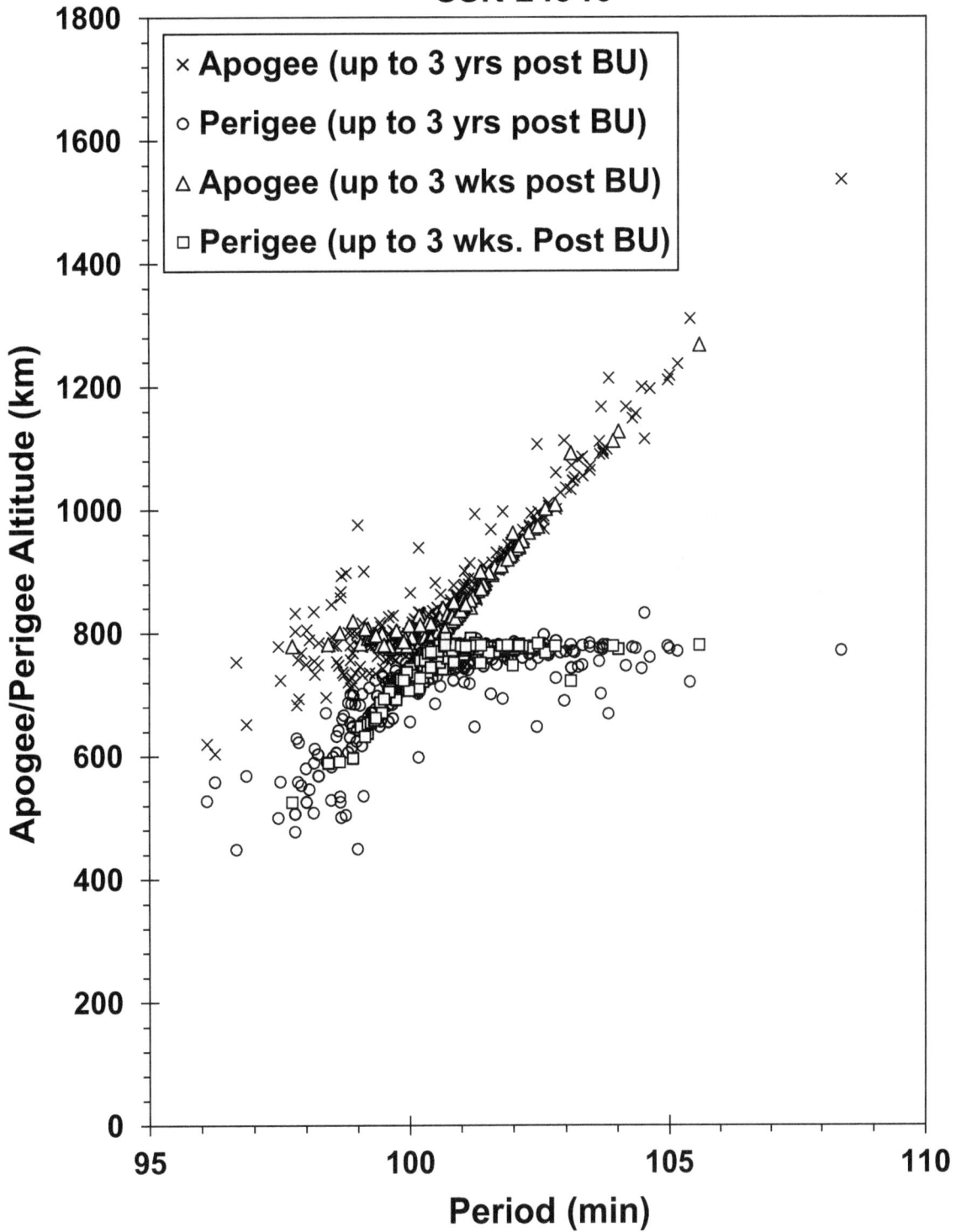

The IRIDIUM 33 debris cloud, including 621 fragments cataloged up to 3 years after the event.

SATELLITE DATA

TYPE:	Mission Related Debris
OWNER:	CIS
LAUNCH DATE:	12.71 Nov 1997
DRY MASS (KG):	55
MAIN BODY:	Ellipsoid; 0.6 m diameter by 1 m length
MAJOR APPENDAGES:	None
ATTITUDE CONTROL:	None
ENERGY SOURCES:	On-board propellant

EVENT DATA

DATE:	14 Feb 2007	LOCATION:	Unknown
TIME:	Unknown	ASSESSED CAUSE:	Propulsion
ALTITUDE:	Unknown		

PRE-EVENT ELEMENTS

EPOCH:	07044.95683864	MEAN ANOMALY:	35.6246
RIGHT ASCENSION:	14.2124	MEAN MOTION:	5.48131368
INCLINATION:	46.649	MEAN MOTION DOT/2:	0.00001445
ECCENTRICITY:	0.5113669	MEAN MOTION DOT DOT/6:	0.0
ARG. OF PERIGEE:	267.6957	BSTAR:	0.00013146

DEBRIS CLOUD DATA

MAXIMUM ΔP:	Unknown
MAXIMUM ΔI:	Unknown

COMMENTS

The object was in a decaying geosynchronous transfer orbit; this event marks the 36th known breakup of a Proton Blok DM SOZ ullage motor since 1984. By accident, an observer in Finland captured about 20 of the debris with two CCD cameras less than 24 hours after the event. The debris would have been too faint to be seen with the naked eye. An estimated 60 debris were detected by the US SSN.

REFERENCE DOCUMENTS

History of Soviet/Russian Satellite Fragmentations-A Joint U.S.-Russian Investigation, N. L. Johnson, *et al.,* Kaman Sciences Corporation, October 1995.

"Four Satellite Breakups in February Add to Debris Population," The Orbital Debris Quarterly News, NASA JSC, April 2007. Available online at https://orbitaldebris.jsc.nasa.gov/quarterly-news/pdfs/ODQNv11i2.pdf.

Insufficient data to construct a Gabbard diagram.

SATELLITE DATA

TYPE:	Payload
OWNER:	CIS
LAUNCH DATE:	9.30 Dec 1997
DRY MASS (KG):	3000
MAIN BODY:	1.3 m diameter by 17 m length, plus solar arrays
MAJOR APPENDAGES:	Solar arrays
ATTITUDE CONTROL:	Active, 3-axis
ENERGY SOURCES:	On-board propellants, explosive charge?

EVENT DATA

DATE:	22 Nov 1999	LOCATION:	31.6N, 4.3E (dsc)
TIME:	0440 GMT	ASSESSED CAUSE:	Unknown
ALTITUDE:	370 km		

PRE-EVENT ELEMENTS

EPOCH:	99325.85267585	MEAN ANOMALY:	85.1293
RIGHT ASCENSION:	332.8746	MEAN MOTION:	15.83563975
INCLINATION:	65.0115	MEAN MOTION DOT/2:	0.00295116
ECCENTRICITY:	0.0134056	MEAN MOTION DOT DOT/6:	0.000036131
ARG. OF PERIGEE:	273.4567	BSTAR:	0.00065869

DEBRIS CLOUD DATA

MAXIMUM ΔP:	2.834 min
MAXIMUM ΔI:	0.22 deg

COMMENTS

Cosmos 2347 was the 19th spacecraft of this type known to have experienced a major fragmentation. Such events were common prior to 1988, but only three breakups have occurred during the past 12 years: Cosmos 2347, Cosmos 2238 (1 Dec 1994), and Cosmos 2313 (26 June 1997). In this case Cosmos 2347 had performed a standard end-of-mission maneuver on 19 November 1999, a little more than 2 days before the breakup. Extensive analyses of these events have been conducted, although the cause is still unknown in the open literature.

A second breakup of Cosmos 2347 was discovered on 10 December when the spacecraft's orbit had decayed to 175 km by 250 km. Three dozen new debris were detected after the second event, but the very low altitude made it difficult to assess accurately the number of large debris. Prior spacecraft (especially Cosmos 1220, 1260, and 1306) also experienced multiple fragmentations.

REFERENCE DOCUMENT

"Satellite Breakups Increase in Last Quarter of 1999", The Orbital Debris Quarterly News, NASA JSC, January 2000. Available online at https://orbitaldebris.jsc.nasa.gov/quarterly-news/pdfs/ODQNv5i1.pdf.

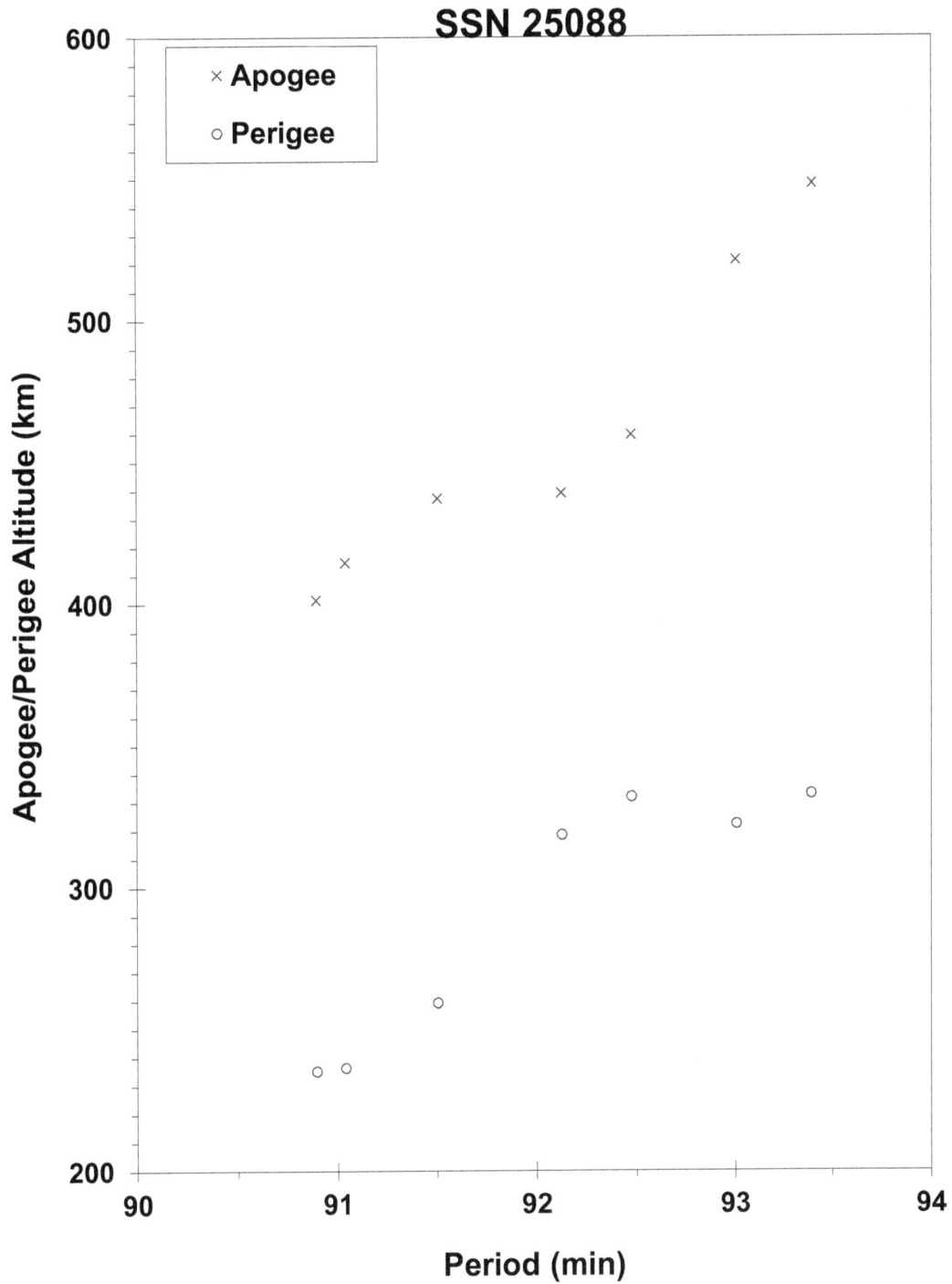

Cosmos 2347 debris cloud of 7 fragments within 1 day of the event as reconstructed from the US SSN database.

ORBCOMM FM5   **1997-084F**   **25117**

SATELLITE DATA

TYPE:	Payload
OWNER:	USA
LAUNCH DATE:	23.80 Dec. 1997
DRY MASS (KG):	40
MAIN BODY:	3.6 m long triangular prism
MAJOR APPENDAGES:	single-axis solar panels (2); communication antenna integrated with gravity-gradient (GG) boom
ATTITUDE CONTROL:	torque rods (believed inactive at event time) & GG boom stabilized
ENERGY SOURCES:	on-board GN2 monopropellant; batteries

EVENT DATA

DATE:	18 Nov. 2021	LOCATION:	Unknown
TIME:	0600-0700 GMT	ASSESSED CAUSE:	Unknown
ALTITUDE:	Unknown		

PRE-EVENT ELEMENTS

EPOCH:	21321.80858342	MEAN ANOMALY:	222.4292
RIGHT ASCENSION:	123.6310	MEAN MOTION:	14.39292417
INCLINATION:	45.0157	MEAN MOTION DOT/2:	0.00000232
ECCENTRICITY:	0.0003400	MEAN MOTION DOT DOT/6:	0.0
ARG. OF PERIGEE:	137.6820	BSTAR:	0.00011961

DEBRIS CLOUD DATA

MAXIMUM ΔP:	1.9 min
MAXIMUM ΔI:	0.3 deg

COMMENTS

ORBCOMM FM5 was a member of the first generation ORBCOMM low-altitude communication satellite constellation; it utilized a Northrup Grumman (Orbital Sciences Corp.) Microstar bus. This is the second ORBCOMM first generation vehicle to fragment.

REFERENCE DOCUMENT

 "Two Minor Breakup Events in Fourth Quarter of 2021", The Orbital Debris Quarterly News, NASA JSC, March 2022. Available online at https://orbitaldebris.jsc.nasa.gov/quarterly-news/ODQNv26i1.pdf.

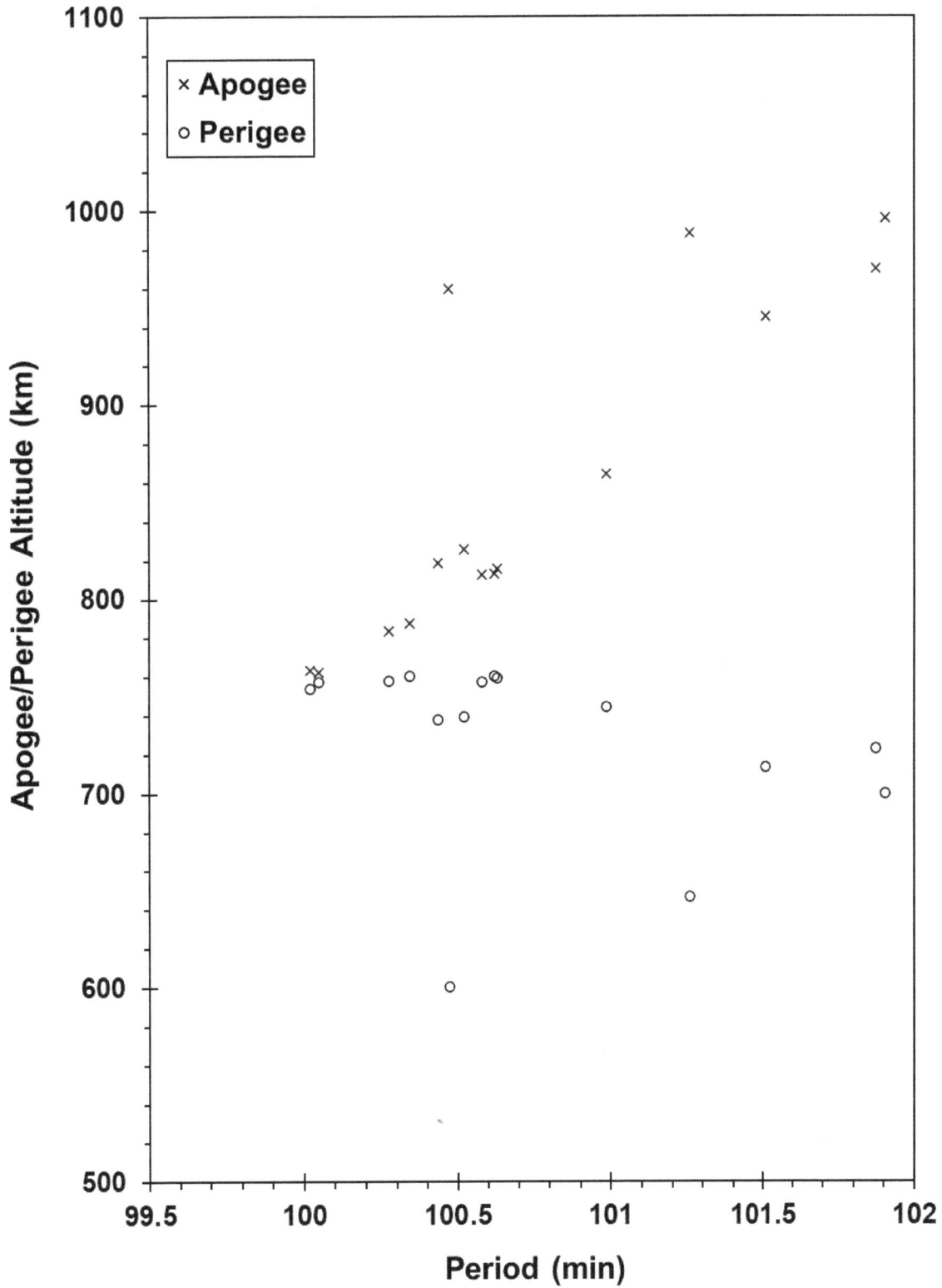

The ORBCOMM FM5 debris cloud, including 15 fragments cataloged over two months after the event.

SATELLITE DATA

TYPE:	Proton Blok DM Fourth Stage
OWNER:	CIS
LAUNCH DATE:	24.97 Dec 1997
DRY MASS (KG):	3400 (?)
MAIN BODY:	Cylinder; 3.7 m diameter by 6.3 m length
MAJOR APPENDAGES:	None
ATTITUDE CONTROL:	Active, 3-axis
ENERGY SOURCES:	On-board propellants

EVENT DATA

DATE:	25 Dec 1997	LOCATION:	0.3 S, 91.2 E (dsc)
TIME:	0550 GMT	ASSESSED CAUSE:	Propulsion
ALTITUDE:	35995 km		

POST-EVENT ELEMENTS

EPOCH:	97359.90803672	MEAN ANOMALY:	359.8589
RIGHT ASCENSION:	92.0594	MEAN MOTION:	2.26580509
INCLINATION:	51.4499	MEAN MOTION DOT/2:	-0.00000507
ECCENTRICITY:	0.7304004	MEAN MOTION DOT DOT/6:	0
ARG. OF PERIGEE:	1.0181	BSTAR:	0

DEBRIS CLOUD DATA

MAXIMUM ΔP:	Unknown
MAXIMUM ΔI:	Unknown

COMMENTS

The commercial Asiasat 3 spacecraft was launched by a Proton launch vehicle at 2319 GMT, 24 December 1997. The fourth stage completed its first burn successfully about 0035 GMT, 25 December, placing the R/B-S/C combination into a GTO. When the vehicle reached its first apogee, the main engine was restarted but shutdown within 1 second, apparently due to a catastrophic failure. The US Space Surveillance Network detected less than 10 objects, and by 9 January only 1-2 were still being observed. The fragmentation is similar to the breakup of the Raduga 33 upper stage on 19 Feb 1996. In that case, nearly 200 debris were detected by the SSN.

REFERENCE DOCUMENT

"Recent Satellite Fragmentation Investigations," N. Johnson, The Orbital Debris Quarterly News, January 1998, p. 3. Available online at https://orbitaldebris.jsc.nasa.gov/quarterly-news/pdfs/ODQNv3i1.pdf.

Insufficient data to construct a Gabbard diagram.

SATELLITE DATA

TYPE:	H-II Second Stage
OWNER:	Japan
LAUNCH DATE:	21.33 Feb 1998
DRY MASS (KG):	3000
MAIN BODY:	Spheroid + cylinder + cone; 4 m diameter by 10.1 m length
MAJOR APPENDAGES:	None
ATTITUDE CONTROL:	Active, 3-axis
ENERGY SOURCES:	On-board propellants

EVENT DATA

DATE:	21 Feb 1998	LOCATION:	Unknown
TIME:	Unknown	ASSESSED CAUSE:	Propulsion
ALTITUDE:	Unknown		

POST-EVENT ELEMENTS

EPOCH:	98054.59975400	MEAN ANOMALY:	162.0601
RIGHT ASCENSION:	294.3031	MEAN MOTION:	13.51967368
INCLINATION:	30.0458	MEAN MOTION DOT/2:	0.0002873
ECCENTRICITY:	0.1097485	MEAN MOTION DOT DOT/6:	-0.000003104
ARG. OF PERIGEE:	194.5714	BSTAR:	0.00029603

DEBRIS CLOUD DATA

MAXIMUM ΔP:	Unknown
MAXIMUM ΔI:	Unknown

COMMENTS

According to a NASDA report, a welding failure caused the LE-5A engine shutdown that stranded the Japanese COMETS satellite (1998-011A) in an elliptical orbit. The failure occurred 47 seconds into an apogee-raising maneuver. The report determined from telemetry data that hot combustion gases managed to penetrate special welding, called brazing, between nickel alloy cooling tubes in the lowest part of the combustion chamber near the top of the engine's nozzle skirt. Burning through the tubes, combustion gases quickly caused a fire, which triggered the engine shutdown. The report concludes the accident was caused by a manufacturing flaw and not a fundamental design problem. At least three dozen debris were detected by optical sensors in Hawaii.

REFERENCE DOCUMENTS

Kallender, P., "LE-5A Shutdown Blamed on Welding." Space News, 23-29 March 1998, p. 6.

"The Upper Stage Breakups in One Week Top February Debris Activity," The Orbital Debris Quarterly News, NASA JSC, April 1998, p. 1. Available online at https://orbitaldebris.jsc.nasa.gov/quarterly-news/pdfs/ODQNv3i2.pdf.

Insufficient data to construct a Gabbard diagram.

SATELLITE DATA

TYPE:	Payload
OWNER:	USA
LAUNCH DATE:	02 Aug. 1998
DRY MASS (KG):	40
MAIN BODY:	3.6 m long triangular prism
MAJOR APPENDAGES:	single-axis solar panels (2); communication antenna integrated with gravity-gradient (GG) boom
ATTITUDE CONTROL:	torque rods (believed inactive at event time) & GG boom stabilized
ENERGY SOURCES:	on-board GN2 monopropellant; batteries

EVENT DATA

DATE:	22 Dec. 2018	LOCATION:	Unknown
TIME:	~ 0712 GMT	ASSESSED CAUSE:	Unknown
ALTITUDE:	Unknown		

PRE-EVENT ELEMENTS

EPOCH:	20355.54764710	MEAN ANOMALY:	21.6240
RIGHT ASCENSION:	350.1557	MEAN MOTION:	14.33299236
INCLINATION:	44.9970	MEAN MOTION DOT/2:	0.00000068
ECCENTRICITY:	0.0003887	MEAN MOTION DOT DOT/6:	0.0
ARG. OF PERIGEE:	338.4442	BSTAR:	0.000075113

DEBRIS CLOUD DATA

MAXIMUM ΔP:	4.8 min
MAXIMUM ΔI:	0.5 deg

COMMENTS

ORBCOMM FM16 was a member of the first generation ORBCOMM low-altitude communication satellite constellation; it utilized a Northrup Grumman (Orbital Sciences Corp.) Microstar bus. It was the first ORBCOMM vehicle to fragment.

REFERENCE DOCUMENT

"2018 Ends with Breakup of an ORBCOMM Constellation Spacecraft", The Orbital Debris Quarterly News, NASA JSC, May 2019. Available online at https://orbitaldebris.jsc.nasa.gov/quarterly-news/ODQNv23i1.pdf.

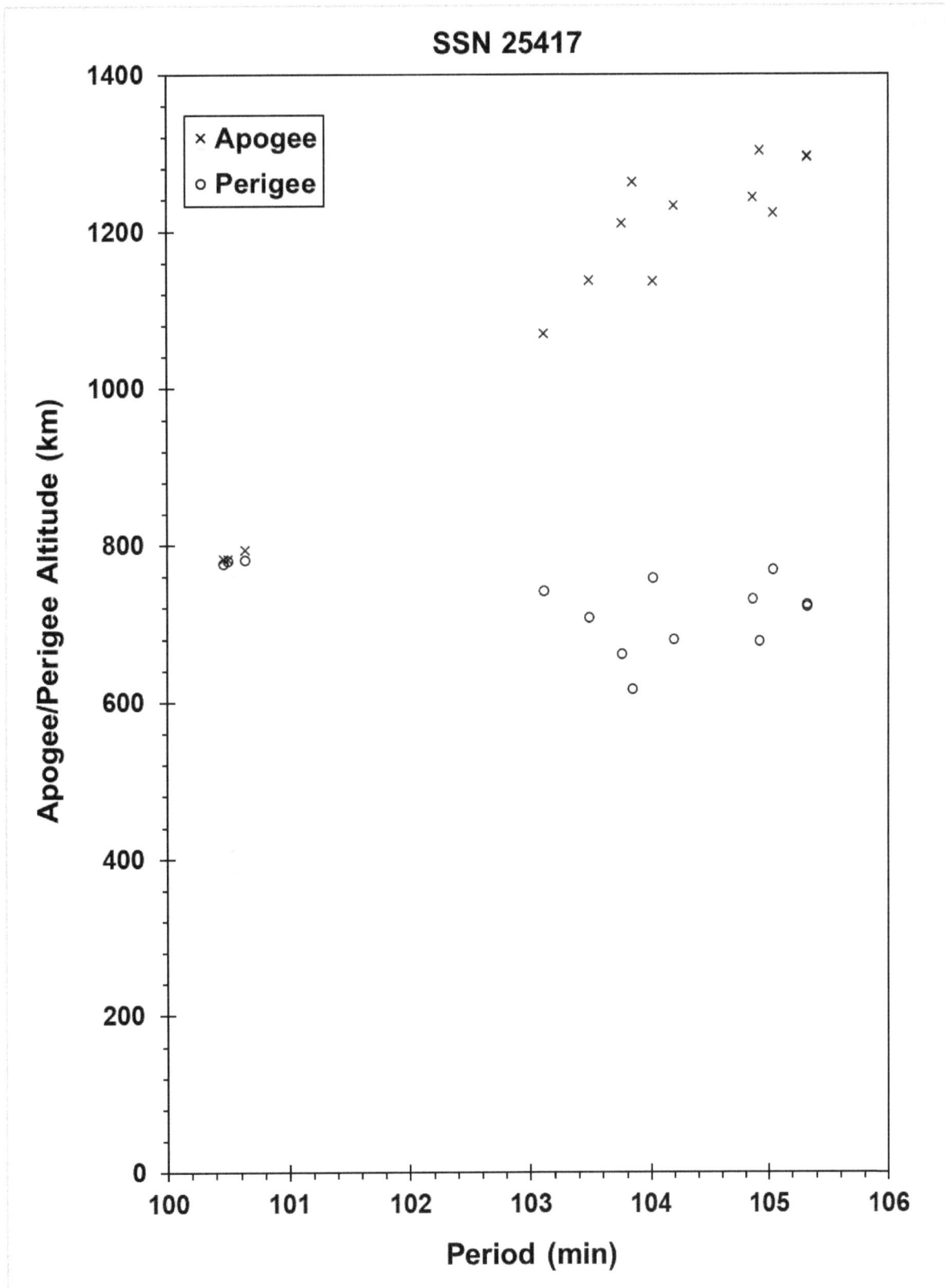

The ORBCOMM FM16 debris cloud, including 13 fragments cataloged up to one month after the event.

ARGOS/ORSTED/SUNSAT R/B   **1999-008D**   **25637**

SATELLITE DATA

TYPE:	Delta II Second Stage Rocket Body
OWNER:	US
LAUNCH DATE:	23 Feb 1999
DRY MASS (KG):	921
MAIN BODY:	2.4 m diameter x 5 m long right circular cylinder
MAJOR APPENDAGES:	none
ATTITUDE CONTROL:	inactive 3-axis
ENERGY SOURCES:	none at time of event

EVENT DATA

DATE:	28 April 2014	LOCATION:	unknown
TIME:	unknown	ASSESSED CAUSE:	Unknown
ALTITUDE:	unknown		

PRE-EVENT ELEMENTS

EPOCH:	14117.05656630	MEAN ANOMALY:	143.8191
RIGHT ASCENSION:	271.0573	MEAN MOTION:	14.45733299
INCLINATION:	96.4621	MEAN MOTION DOT/2:	0.00000245
ECCENTRICITY:	0.0143136	MEAN MOTION DOT DOT/6:	0.00000
ARG. OF PERIGEE:	329.6124	BSTAR:	0.00069149

DEBRIS CLOUD DATA

MAXIMUM ΔP:	5.2 minutes
MAXIMUM ΔI:	0.2°

COMMENTS

The cause of this event, occurring over 15 years after launch, is unknown. Seven fragments entered the catalog, but all decayed from orbit by 16 May 2014. Since 1982, Delta II second stage rocket bodies have been fully passivated, implying that a small MMOD impact may have caused this event.

REFERENCE DOCUMENT

"Flurry of Small Breakups in First Half of 2014", The Orbital Debris Quarterly News, NASA JSC, July 2014. Available online at https://orbitaldebris.jsc.nasa.gov/quarterly-news/pdfs/ODQNv18i3.pdf.

SSN 25637

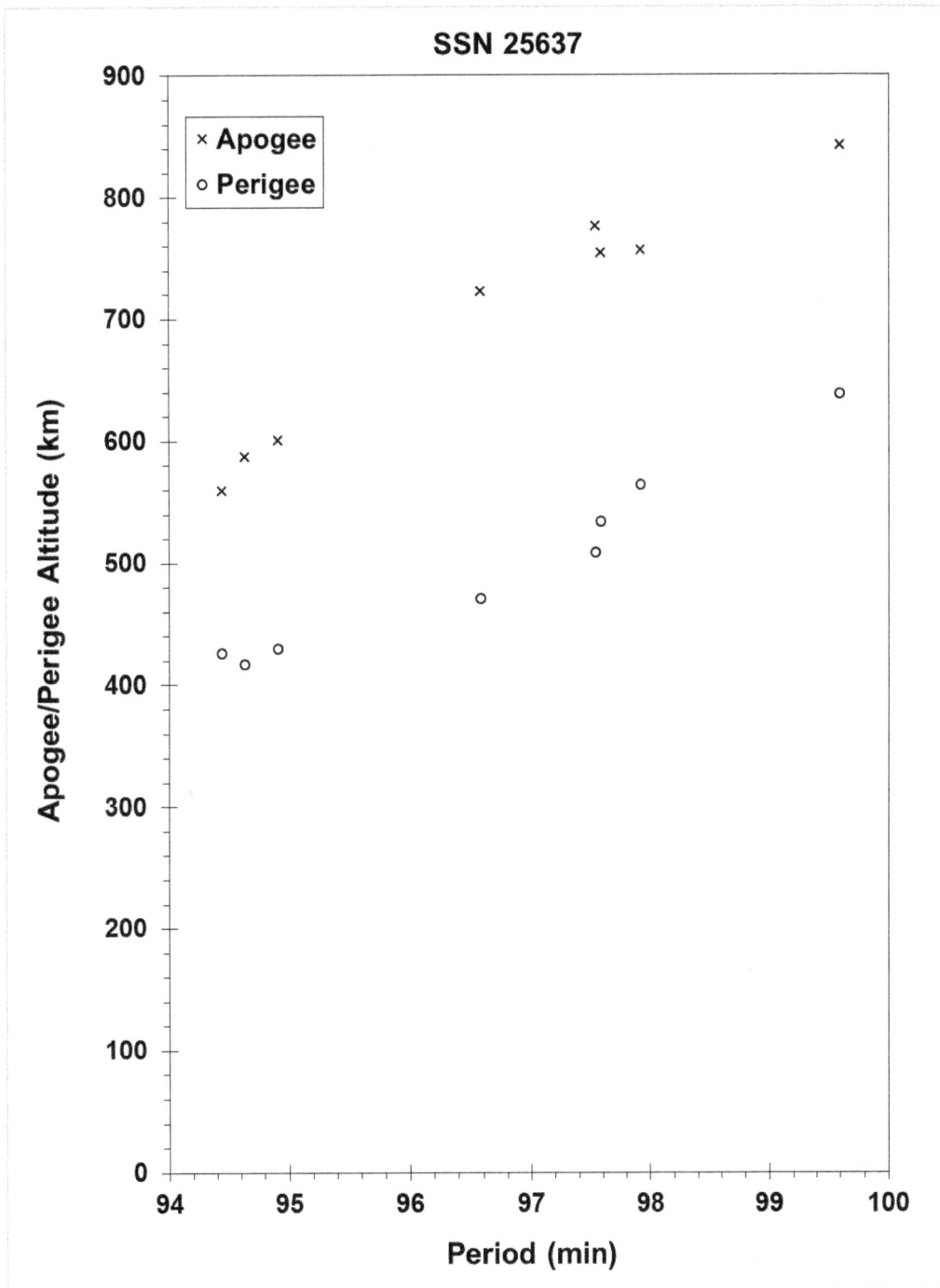

ARGOS/ORSTED/SUNSAT R/B debris cloud.

SATELLITE DATA

TYPE:	Payload
OWNER:	PRC
LAUNCH DATE:	10.06 May 1999
DRY MASS (KG):	950
MAIN BODY:	Box; 1.5 m by 1.5 m by 1.5 m
MAJOR APPENDAGES:	Solar Panels, 1.5 m by 4 m
ATTITUDE CONTROL:	Active, 3-axis
ENERGY SOURCES:	On-board propellants

EVENT DATA

DATE:	11 Jan 2007	LOCATION:	35N, 100E (asc)
TIME:	2226 GMT	ASSESSED CAUSE:	Deliberate (HVI)
ALTITUDE:	860 km		

PRE-EVENT ELEMENTS

EPOCH:	07011.90621003	MEAN ANOMALY:	94.0215
RIGHT ASCENSION:	1.7411	MEAN MOTION:	14.11820274
INCLINATION:	98.6464	MEAN MOTION DOT/2:	0.00000180
ECCENTRICITY:	0.0013513	MEAN MOTION DOT DOT/6:	0.0
ARG. OF PERIGEE:	266.0357	BSTAR:	0.00012153

DEBRIS CLOUD DATA

MAXIMUM ΔP:	33.4 min
MAXIMUM ΔI:	5.0 deg

COMMENTS

The debris cloud created by this anti-satellite test represents the worst contamination of low Earth orbit in history. More than half the identified debris were thrown into orbits exceeding a mean altitude of 850 km, meaning that much of the 10 cm and larger debris will be in orbit for decades or centuries.

REFERENCE DOCUMENT

"Chinese Anti-satellite Test Creates Most Severe Orbital Debris Cloud in History," The Orbital Debris Quarterly News, NASA JSC, April 2007.
Available online at https://orbitaldebris.jsc.nasa.gov/quarterly-news/pdfs/ODQNv11i2.pdf.

SSN 25730

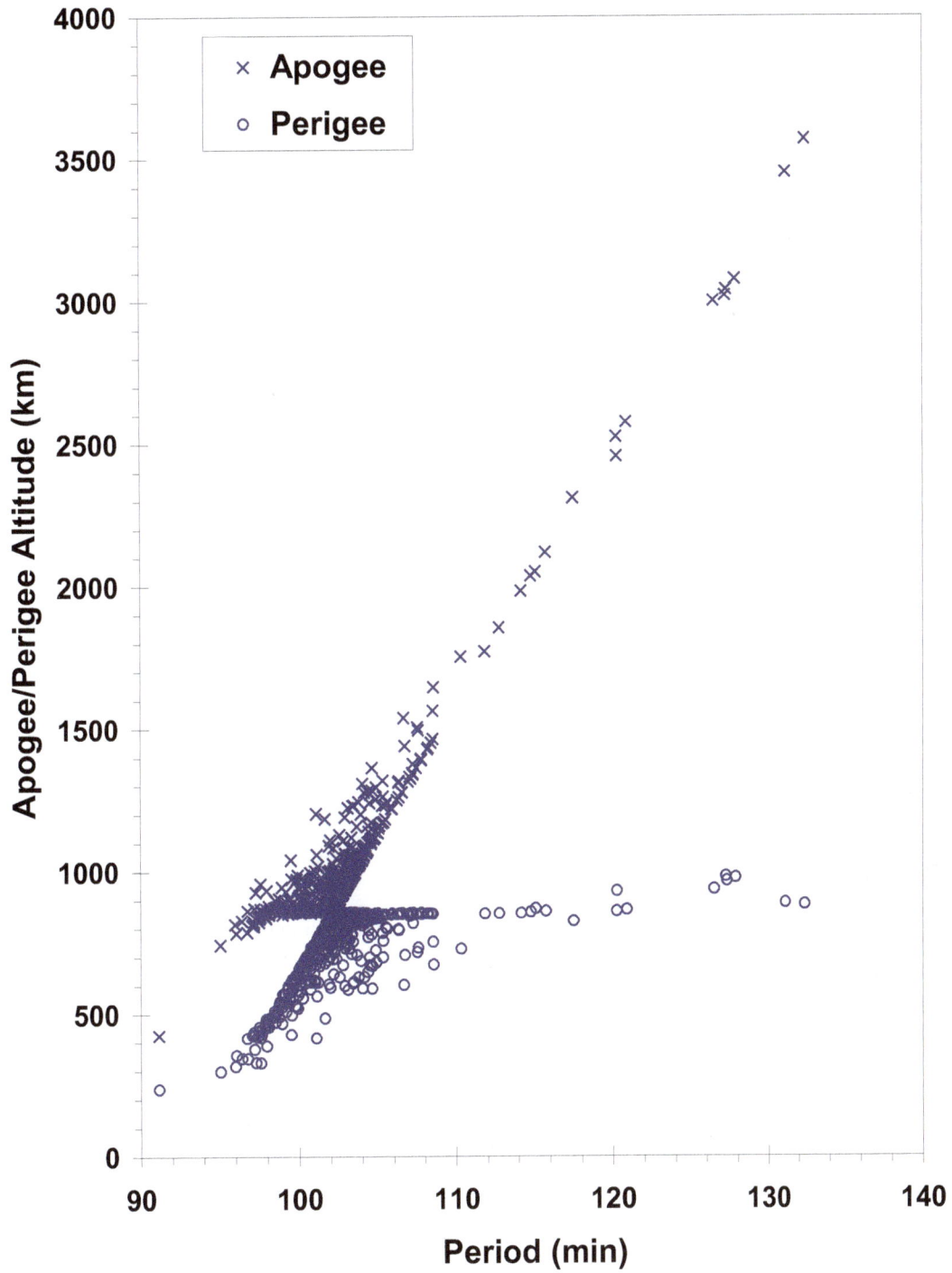

Fengyun 1C debris cloud remnant of 2000+ cataloged fragments 6 months after the event as reconstructed from the US SSN database.

SATELLITE DATA

TYPE:	Payload
OWNER:	PRC/Brazil
LAUNCH DATE:	14.14 Oct 1999
DRY MASS (KG):	1450
MAIN BODY:	Box: 1.8 m by 2.0 m by 2.2 m
MAJOR APPENDAGES:	6.3 m x 2.6 m Solar Panels
ATTITUDE CONTROL:	None at time of event
ENERGY SOURCES:	On-board propellant?

EVENT DATA

DATE:	18 Feb 2007	LOCATION:	35N, 128E (asc)
TIME:	1107 GMT	ASSESSED CAUSE:	Unknown
ALTITUDE:	780 km		

PRE-EVENT ELEMENTS

EPOCH:	07049.17726620	MEAN ANOMALY:	250.9413
RIGHT ASCENSION:	88.9135	MEAN MOTION:	14.34483847
INCLINATION:	98.2175	MEAN MOTION DOT/2:	-0.00000185
ECCENTRICITY:	0.0007520	MEAN MOTION DOT DOT/6:	0.0
ARG. OF PERIGEE:	109.2997	BSTAR:	-0.000051172

DEBRIS CLOUD DATA

MAXIMUM ΔP: 3.4 min
MAXIMUM ΔI: 0.3 deg

COMMENTS

The spacecraft exceeded its expected lifetime and had been moved to a retirement orbit in August 2003. It is unclear whether or not it was entirely passivated. The rocket body associated with this payload (1999-057C, 25942) experienced an unrelated, but significant fragmentation six months after launch.

REFERENCE DOCUMENT

"Four Satellite Breakups in February Add to Debris Population," The Orbital Debris Quarterly News, NASA JSC, April 2007. Available online at https://orbitaldebris.jsc.nasa.gov/quarterly-news/pdfs/ODQNv11i2.pdf.

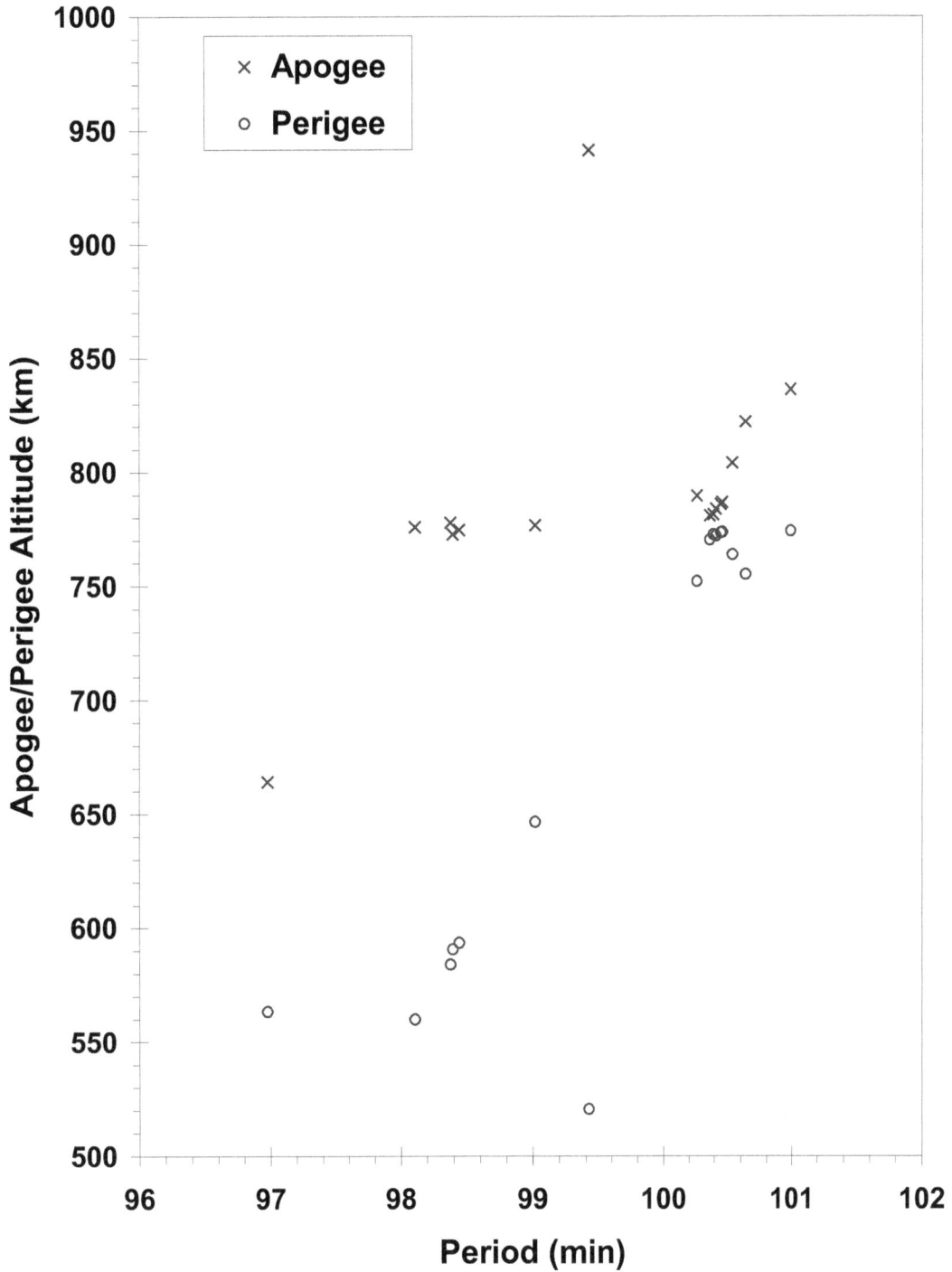

CBERS 1 debris cloud of 16 cataloged fragments 2 weeks after the event as reconstructed from the US SSN database.

SATELLITE DATA

TYPE:	Long March 4 third stage
OWNER:	PRC
LAUNCH DATE:	14.14 Oct 1999
DRY MASS (KG):	1700
MAIN BODY:	Cylinder nozzle; 2.9 m diameter by ~5 m length
MAJOR APPENDAGES:	None
ATTITUDE CONTROL:	None
ENERGY SOURCES:	On-board propellants

EVENT DATA

DATE:	11 Mar 2000	LOCATION:	51.2S, 311.5W
TIME:	1304 GMT	ASSESSED CAUSE:	Propulsion
ALTITUDE:	741 km		

PRE-EVENT ELEMENTS

EPOCH:	00070.18630445	MEAN ANOMALY:	46.4837
RIGHT ASCENSION:	146.5592	MEAN MOTION:	14.46869143
INCLINATION:	98.5376	MEAN MOTION DOT/2:	0.00001547
ECCENTRICITY:	0.0012358	MEAN MOTION DOT DOT/6:	0.0
ARG. OF PERIGEE:	313.5308	BSTAR:	0.00043864

DEBRIS CLOUD DATA

MAXIMUM ΔP:	10.985 min
MAXIMUM ΔI:	0.99 deg

COMMENTS

This is the second Long March 4 to breakup in only four missions. The first breakup (flight 2) occurred on 4 Oct 1990, 1 month after launch. Long March 4 missions did not resume until 1999, when two more were flown. This breakup involved the second 1999 mission (flight 4) and occurred 5 months after launch. This event has created more trackable debris than the 1990 breakup, with more than 300 pieces tracked by the SSN. Chinese officials were aware of the international concern following the 1990 breakup and had pledged to adopt countermeasures before the 1999 missions. Passivation of this vehicle was attempted.

REFERENCE DOCUMENT

"Analyzing the Cause of LM-4 (A)'s Upper Stage's Disintegration and the Countermeasures", W. X. Zang and S. Y. Liao, 5th International Conference of Pacific Basin Societies, 6-9 Jun 1993, Shanghai.

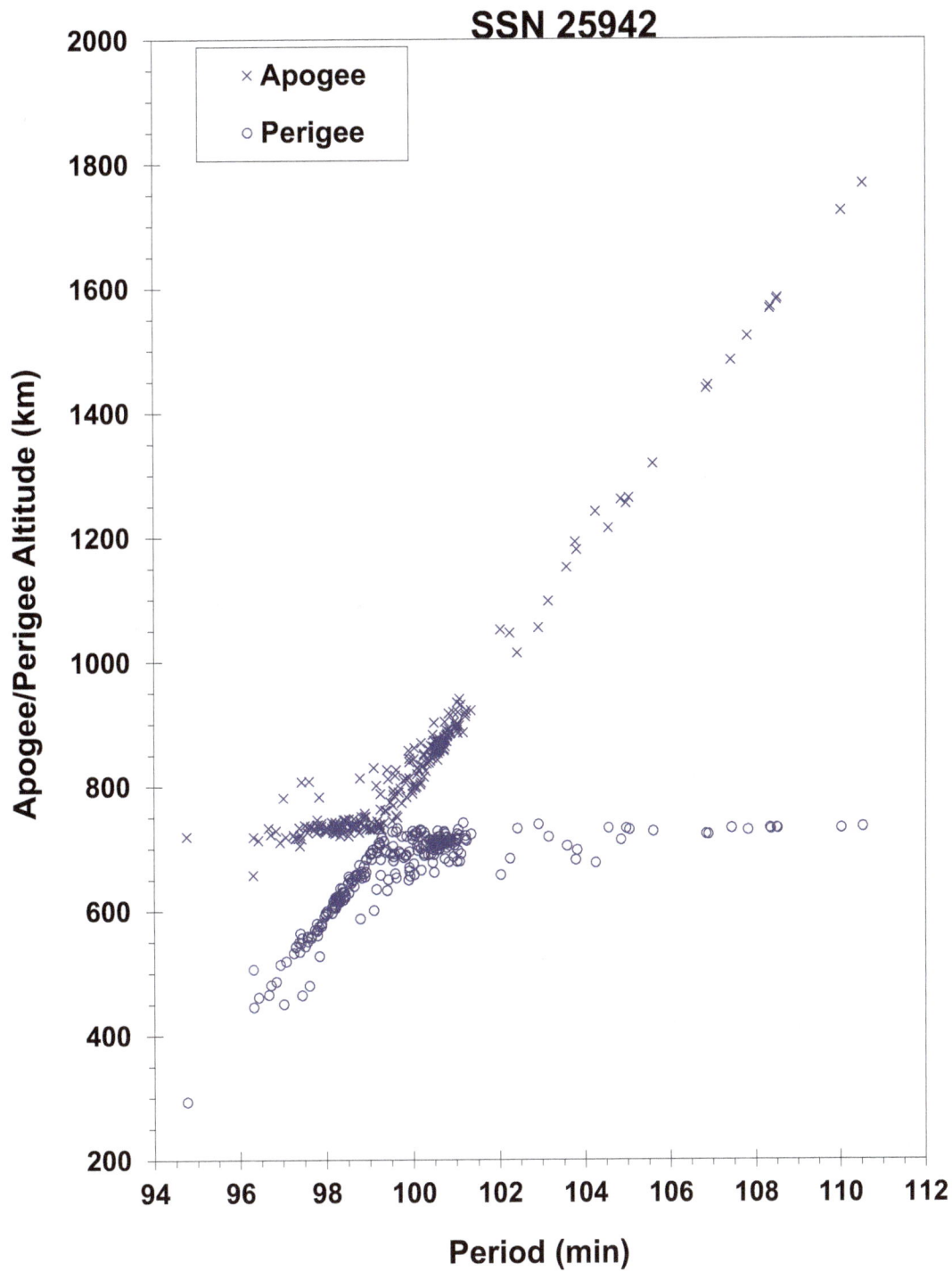

CBERS-1 / SACI-1 Long March 4 third stage debris cloud of 280 fragments within 1 day of the event as reconstructed from the US SSN database.

SATELLITE DATA

TYPE:	Payload
OWNER:	CIS
LAUNCH DATE:	26.33 Dec 1999
DRY MASS (KG):	~3000
MAIN BODY:	Cylinder; 1.3 m diameter by 17 m length, plus solar arrays
MAJOR APPENDAGES:	Solar Arrays
ATTITUDE CONTROL:	Active 3-axis
ENERGY SOURCES:	On-board propellant, explosive charge

EVENT DATA

DATE:	21 Nov 2001	LOCATION:	38.3 S, 342.3 E (dsc)
TIME:	1414Z	ASSESSED CAUSE:	Unknown
ALTITUDE:	410 km		

PRE-EVENT ELEMENTS

EPOCH:	03325.57054648	MEAN ANOMALY:	199.8631
RIGHT ASCENSION:	55.0233	MEAN MOTION:	15.51939724
INCLINATION:	65.0021	MEAN MOTION DOT/2:	0.00131711
ECCENTRICITY:	0.0008788	MEAN MOTION DOT DOT/6:	0.0
ARG. OF PERIGEE:	257.3641	BSTAR:	0.0021441

DEBRIS CLOUD DATA

MAXIMUM ΔP:	10.62 min*
MAXIMUM ΔI:	1.28 deg*

* Based on uncataloged debris data

COMMENTS

Cosmos 2367 was the 20th spacecraft of this type (Cosmos 699 class) known to have experienced a major fragmentation. The previous spacecraft in this series was Cosmos 2347, which experienced two fragmentations, one each in Nov and Dec of 1999. Cosmos 2367 was still in its operational orbit at the time of the event. Over 100 pieces were detected by the SSN 1 week after the breakup. Based upon other observations, the actual number of pieces probably exceeded 300. Although some debris were thrown into orbits with apogees above 1000 km, in general the debris were short-lived.

REFERENCE DOCUMENT

"Two Major Satellite Breakups Near End of 2001", The Orbital Debris Quarterly News, NASA JSC, January 2002. Available online at https://orbitaldebris.jsc.nasa.gov/quarterly-news/pdfs/ODQNv7i1.pdf.

SSN 26040

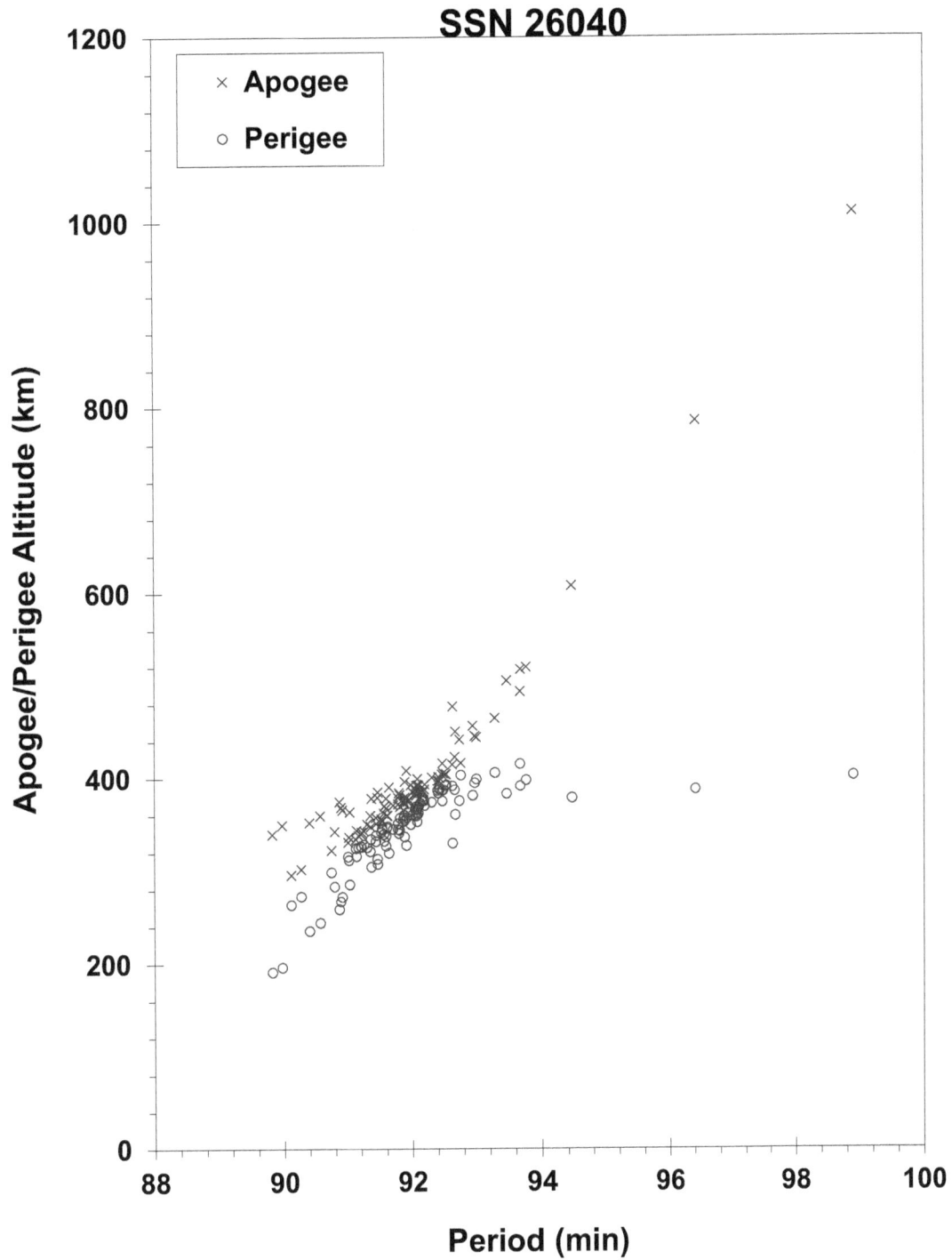

Cosmos 2367 debris cloud of 103 fragments 1 week after the breakup as reconstructed from the US SSN database.

SATELLITE DATA

TYPE:	Mission Related Debris
OWNER:	CIS
LAUNCH DATE:	4.99 Jul 2000
DRY MASS (KG):	55
MAIN BODY:	Ellipsoid; 0.6 m diameter by 1 m length
MAJOR APPENDAGES:	None
ATTITUDE CONTROL:	None
ENERGY SOURCES:	On-board propellant

EVENT DATA

DATE:	~1 Sep 2006	LOCATION:	Unknown
TIME:	Unknown	ASSESSED CAUSE:	Propulsion
ALTITUDE:	Unknown		

PRE-EVENT ELEMENTS

EPOCH:	06244.59746638	MEAN ANOMALY:	38.1280
RIGHT ASCENSION:	18.3906	MEAN MOTION:	3.86574836
INCLINATION:	46.8834	MEAN MOTION DOT/2:	0.00005467
ECCENTRICITY:	0.6151900	MEAN MOTION DOT DOT/6:	0.0
ARG. OF PERIGEE:	248.4110	BSTAR:	0.00034737

DEBRIS CLOUD DATA

MAXIMUM ΔP:	98.8 min
MAXIMUM ΔI:	0.2 deg

COMMENTS

The object was in a decaying geosynchronous transfer orbit; this event marks the 35th known breakup of a Proton Blok DM SOZ ullage motor since 1984. Only a handful of debris was detected from this event.

REFERENCE DOCUMENTS

History of Soviet/Russian Satellite Fragmentations-A Joint U.S.-Russian Investigation, N. L. Johnson, *et al.,* Kaman Sciences Corporation, October 1995.

"Three More Satellites Involved in Fragmentations," The Orbital Debris Quarterly News, NASA JSC, October 2006. Available online at https://orbitaldebris.jsc.nasa.gov/quarterly-news/pdfs/ODQNv10i4.pdf.

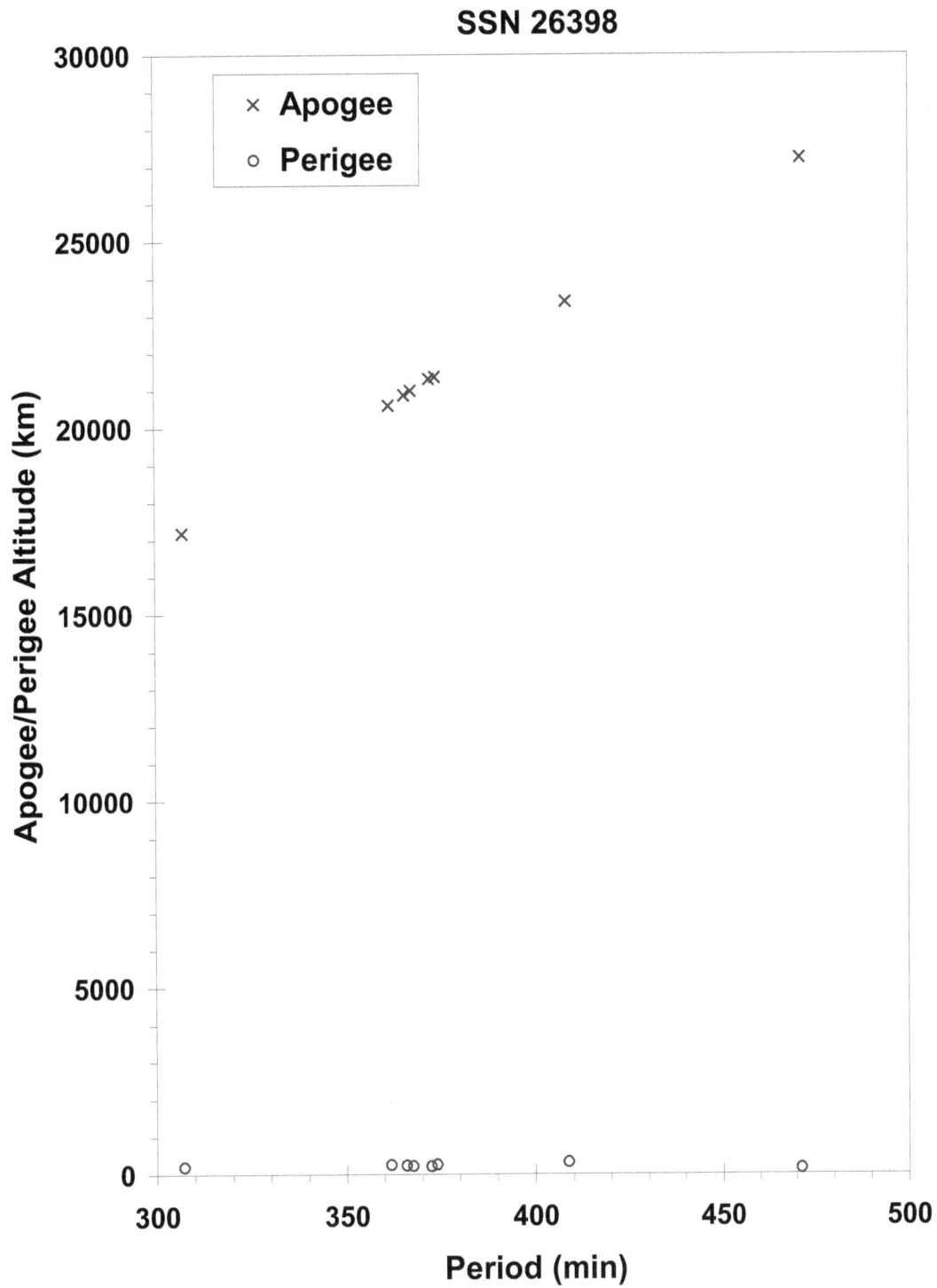

SSN 26398

Cosmos 2371 SOZ motor debris cloud of 7 cataloged fragments plus the parent a day after the event. The debris pieces were never cataloged.

SATELLITE DATA

TYPE:	Payload
OWNER:	US
LAUNCH DATE:	21 Sep 2000
DRY MASS (KG):	1475
MAIN BODY:	box; 1.88 m x 1.88 m x 4.2 m long
MAJOR APPENDAGES:	solar panel
ATTITUDE CONTROL:	none at time of event
ENERGY SOURCES:	on-board propellants and batteries

EVENT DATA

DATE:	25 Nov 2015	LOCATION:	77.5 N, 48.8 E (dsc)
TIME:	0950 GMT	ASSESSED CAUSE:	Battery
ALTITUDE:	855 km		

PRE-EVENT ELEMENTS

EPOCH:	15329.31881564	MEAN ANOMALY:	226.6499
RIGHT ASCENSION:	34.9954	MEAN MOTION:	14.13117297
INCLINATION:	98.9249	MEAN MOTION DOT/2:	0.00000058
ECCENTRICITY:	0.0011308	MEAN MOTION DOT DOT/6:	0.00000
ARG. OF PERIGEE:	133.5614	BSTAR:	0.00054879

DEBRIS CLOUD DATA

MAXIMUM ΔP:	TBD min
MAXIMUM ΔI:	TBD deg

COMMENTS

This event was the second known breakup of a NOAA-series spacecraft, the other being NOAA-8. NOAA-16 performed its nominal mission until placed in backup status in 2005. An on-board anomaly ended communication on 6 June 2014 and the spacecraft was decommissioned on 9 June 2014. The event is assessed as likely being a battery explosion.

REFERENCE DOCUMENTS

"Recent NOAA_16 Satellite Breakup", The Orbital Debris Quarterly News, NASA JSC, April 2016. Available online at http://orbitaldebris.jsc.nasa.gov/newsletter/pdfs/ODQNv20i1-2.pdf.

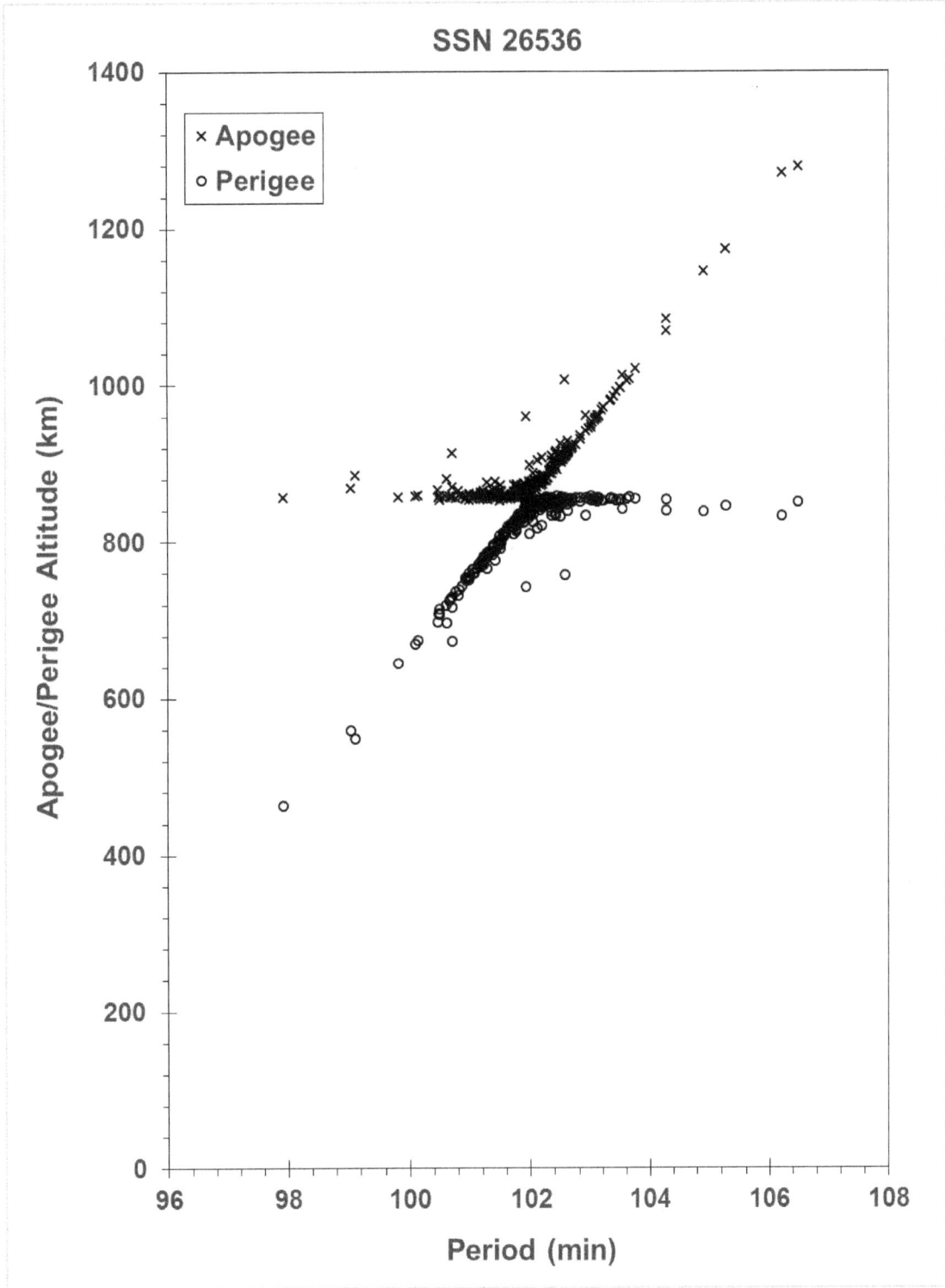

SSN 26536

NOAA 16 debris cloud; debris cataloged between approximately one week and 123 days after event.

467

SATELLITE DATA

TYPE:	PSLV Final Stage
OWNER:	India
LAUNCH DATE:	22.20 Oct 2001
DRY MASS (KG):	~900
MAIN BODY:	Cylinder; 2.8 m diameter by 2.6 m length
MAJOR APPENDAGES:	None
ATTITUDE CONTROL:	None
ENERGY SOURCES:	On-board hypergolic propellants

EVENT DATA

DATE:	19 Dec 2001	LOCATION:	25 S, 340 E (asc)
TIME:	~1140Z	ASSESSED CAUSE:	Propulsion
ALTITUDE:	670 km		

PRE-EVENT ELEMENTS

EPOCH:	01352.90695581	MEAN ANOMALY:	316.4909
RIGHT ASCENSION:	65.6004	MEAN MOTION:	14.85657962
INCLINATION:	97.9010	MEAN MOTION DOT/2:	-0.00000443
ECCENTRICITY:	0.0088752	MEAN MOTION DOT DOT/6:	0.0
ARG. OF PERIGEE:	44.3375	BSTAR:	-0.000041058

DEBRIS CLOUD DATA

MAXIMUM ΔP:	9.86 min*
MAXIMUM ΔI:	3.06 deg*

                     * Based on uncataloged debris data

COMMENTS

This is the first known breakup associated with the Polar Satellite Launch Vehicle (PSLV) fourth stage. While 332 fragments were initially detected by the SSN, 326 debris were cataloged. The vehicle employed hypergolic propellants that were not passivated after payload delivery. Some of the debris could remain in orbit for several years or longer.

REFERENCE DOCUMENT

"Two Major Satellite Breakups Near End of 2001", <u>The Orbital Debris Quarterly News</u>, NASA JSC, January 2002. Available online at https://orbitaldebris.jsc.nasa.gov/quarterly-news/pdfs/ODQNv7i1.pdf.

SSN 26960

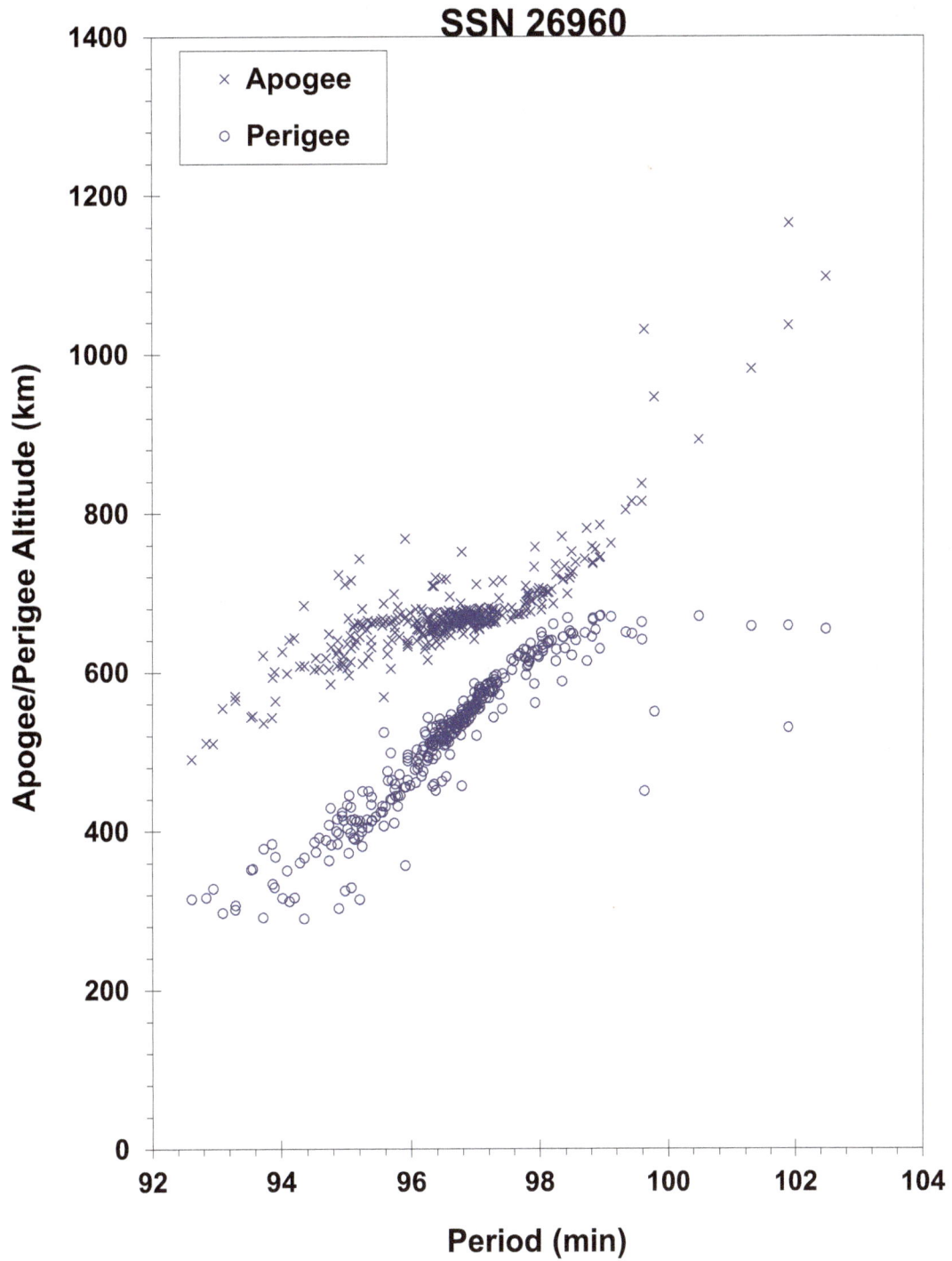

TES R/B debris cloud of 332 fragments 1 week after the breakup as reconstructed from the
US SSN database.

SATELLITE DATA

TYPE:	Payload
OWNER:	CIS
LAUNCH DATE:	21.17 Dec 2001
DRY MASS (KG):	3000
MAIN BODY:	Cylinder; 1.3 m diameter by 17 m length, plus solar arrays
MAJOR APPENDAGES:	Solar Arrays
ATTITUDE CONTROL:	Active 3-axis
ENERGY SOURCES:	On-board propellant; explosive charge

EVENT DATA

DATE:	28 Feb 2004	LOCATION:	26S, 100E (dsc)
TIME:	~1930 GMT	ASSESSED CAUSE:	Unknown
ALTITUDE:	265 km		

PRE-EVENT ELEMENTS

EPOCH:	04059.20843210	MEAN ANOMALY:	95.7196
RIGHT ASCENSION:	358.8049	MEAN MOTION:	15.87352021
INCLINATION:	64.9763	MEAN MOTION DOT/2:	0.00416036
ECCENTRICITY:	0.0131275	MEAN MOTION DOT DOT/6:	0.000069430
ARG. OF PERIGEE:	262.9022	BSTAR:	0.00074756

DEBRIS CLOUD DATA

MAXIMUM ΔP:	3.4 min
MAXIMUM ΔI:	0.3 deg

COMMENTS

Cosmos 2383 was the 21st spacecraft of this type (Cosmos 699 class) known to have experienced a major fragmentation. The previous spacecraft in this series was Cosmos 2367, which experienced a fragmentation at an altitude just above the International Space Station in 2001. This event also produced debris crossing the ISS orbit altitude. Approximately 50 debris were detected by the US SSN. Fortunately, all the cataloged debris from this fragmentation were short-lived. Of the 48 spacecraft of this type, 21 have experienced fragmentation events.

REFERENCE DOCUMENT

"Fragmentation of Cosmos 2383", The Orbital Debris Quarterly News, NASA JSC, April 2004.
Available online at https://orbitaldebris.jsc.nasa.gov/quarterly-news/pdfs/ODQNv8i2.pdf.

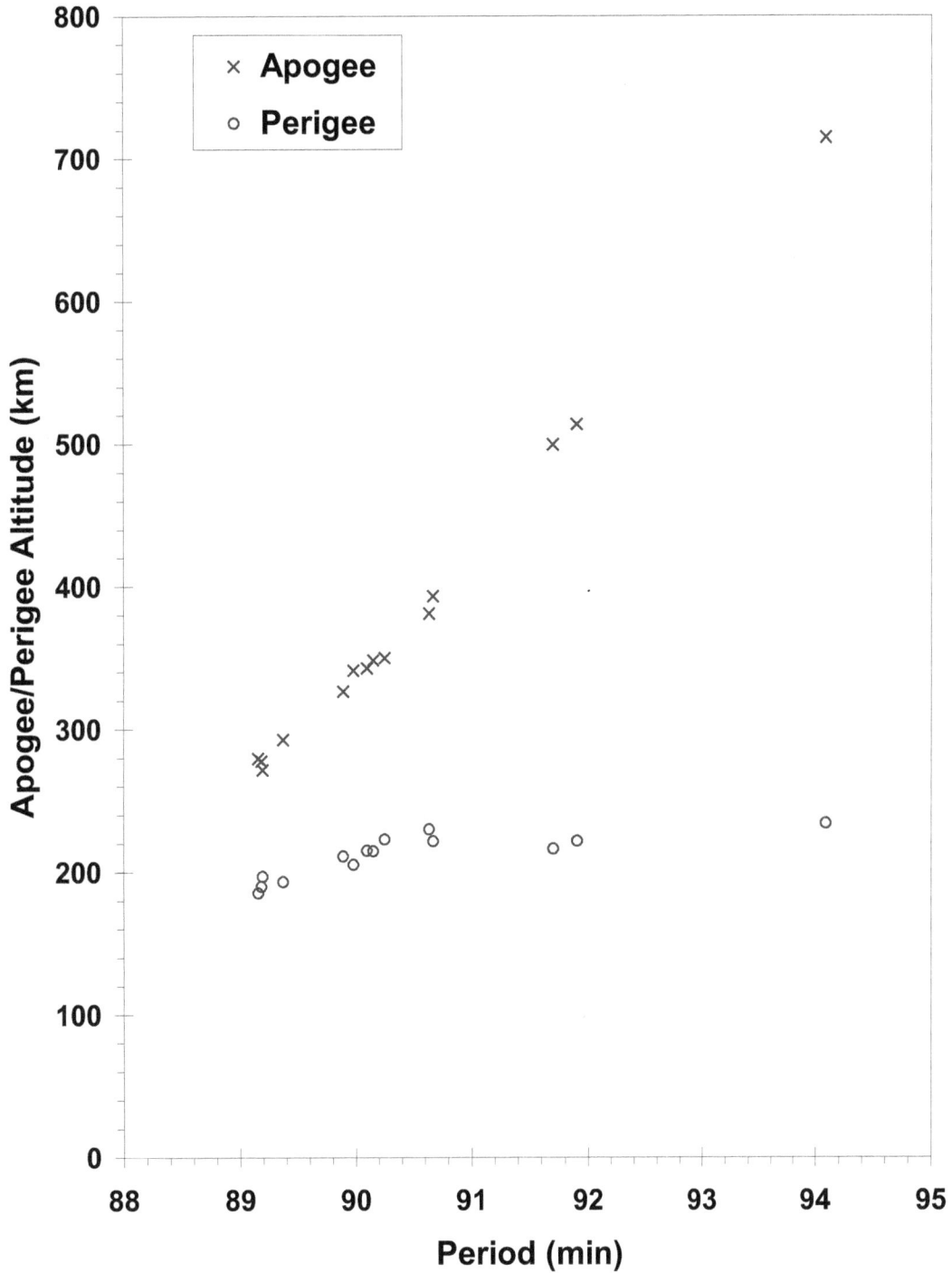

SSN 27053

Cosmos 2383 debris cloud of 14 cataloged fragments a few days after the event as reconstructed from the US SSN database.

SATELLITE DATA

TYPE:	Payload
OWNER:	US
LAUNCH DATE:	24 June 2002
DRY MASS (KG):	1475
MAIN BODY:	box; 1.88 m x 1.88 m x 4.2 m long
MAJOR APPENDAGES:	solar panel
ATTITUDE CONTROL:	none at time of event
ENERGY SOURCES:	on-board propellants and batteries

EVENT DATA

DATE:	10 Mar. 2021	LOCATION:	5.99 N, 281.47 E (asc)
TIME:	0711 GMT	ASSESSED CAUSE:	Unknown
ALTITUDE:	802.2 km		

PRE-EVENT ELEMENTS

EPOCH:	21068.56331182	MEAN ANOMALY:	180.8320
RIGHT ASCENSION:	15.7814	MEAN MOTION:	14.25091535
INCLINATION:	98.6181	MEAN MOTION DOT/2:	0.00000049
ECCENTRICITY:	0.0011552	MEAN MOTION DOT DOT/6:	0.00000
ARG. OF PERIGEE:	179.2878	BSTAR:	0.000039171

DEBRIS CLOUD DATA

MAXIMUM ΔP:	5.0 min
MAXIMUM ΔI:	0.4 deg

COMMENTS

This event was the third known breakup of a NOAA-series spacecraft, the others being NOAA-8 and -16; those events were assessed as likely being battery explosions. NOAA-17 became operational on 15 October 2002 and was decommissioned on 10 April 2013.

REFERENCE DOCUMENTS

"Two On-orbit Breakup Events in March," <u>The Orbital Debris Quarterly News</u>, NASA JSC, June 2021. Available online at http://orbitaldebris.jsc.nasa.gov/newsletter/pdfs/ODQNv25i2.pdf.

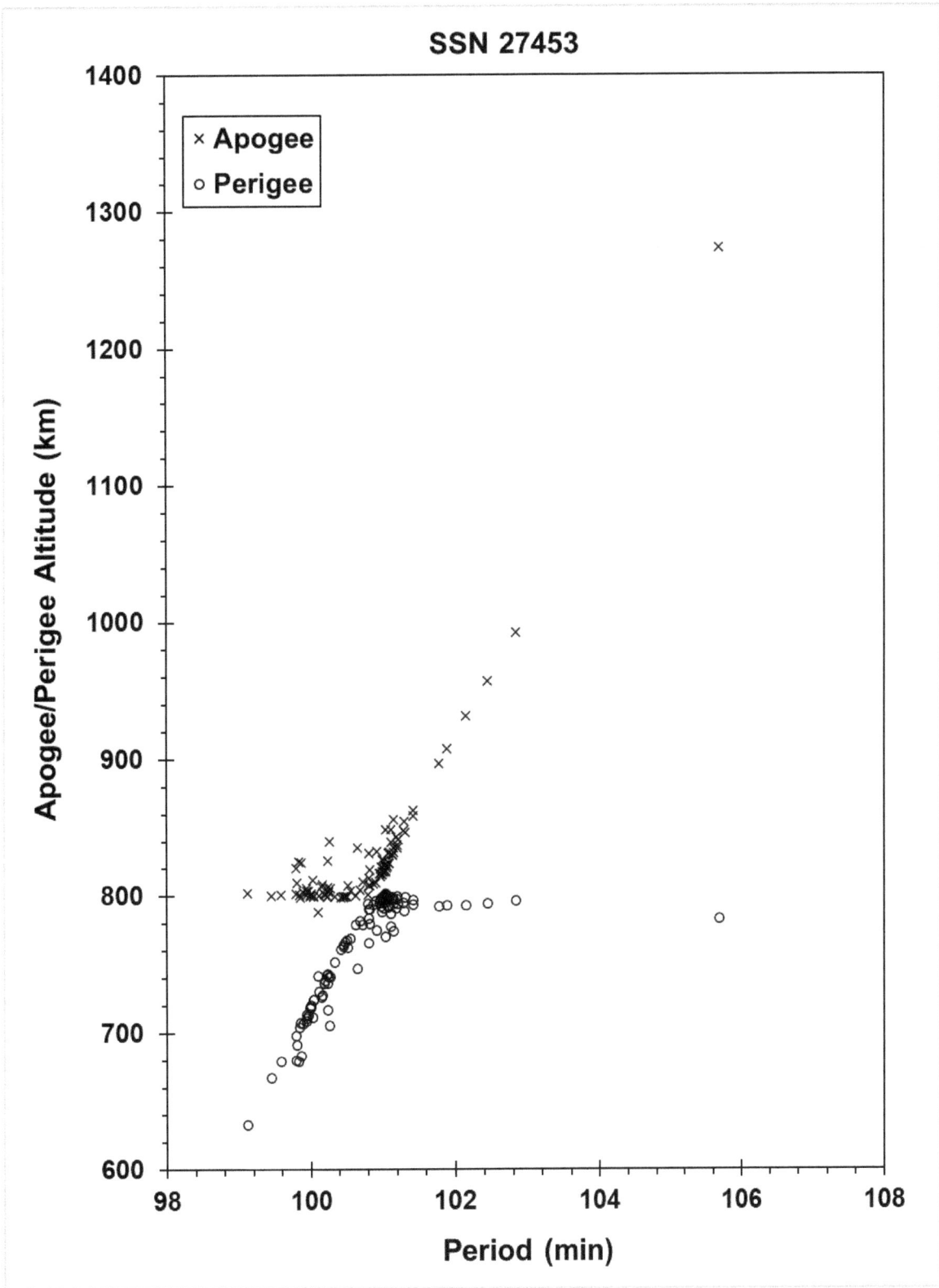

SSN 27453

NOAA 17 debris cloud; debris cataloged up to two months after the event.

COSMOS 2392 ULLAGE MOTOR   **2002-037E**   **27474**

SATELLITE DATA

TYPE:	Mission Related Debris
OWNER:	CIS
LAUNCH DATE:	25.63 Jul 2002
DRY MASS (KG):	55
MAIN BODY:	Ellipsoid; 0.6 m diameter by 1 m length
MAJOR APPENDAGES:	None
ATTITUDE CONTROL:	None
ENERGY SOURCES:	On-board propellant

EVENT DATA

DATE:	1 Jun 2005	LOCATION:	Unknown
TIME:	Unknown	ASSESSED CAUSE:	Propulsion
ALTITUDE:	Unknown		

PRE-EVENT ELEMENTS

EPOCH:	05151.71140009	MEAN ANOMALY:	222.2107
RIGHT ASCENSION:	143.5118	MEAN MOTION:	15.06786995
INCLINATION:	63.6569	MEAN MOTION DOT/2:	0.00075759
ECCENTRICITY:	0.0418207	MEAN MOTION DOT DOT/6:	0.000007275
ARG. OF PERIGEE:	140.9987	BSTAR:	0.00057187

DEBRIS CLOUD DATA

MAXIMUM ΔP:	5.2 min
MAXIMUM ΔI:	1.0 deg

COMMENTS

This event marks the 33rd known breakup of a Proton Block DM SOZ ullage motor since 1984. The motor experienced a second event a month later about 29 June. About 40 new debris were seen after each event.

REFERENCE DOCUMENT

History of Soviet/Russian Satellite Fragmentations-A Joint U.S.-Russian Investigation, N. L. Johnson, et al., Kaman Sciences Corporation, October 1995.

"Recent Satellite Breakups," The Orbital Debris Quarterly News, NASA JSC, July 2005.
Available online at https://orbitaldebris.jsc.nasa.gov/quarterly-news/pdfs/ODQNv9i3.pdf.

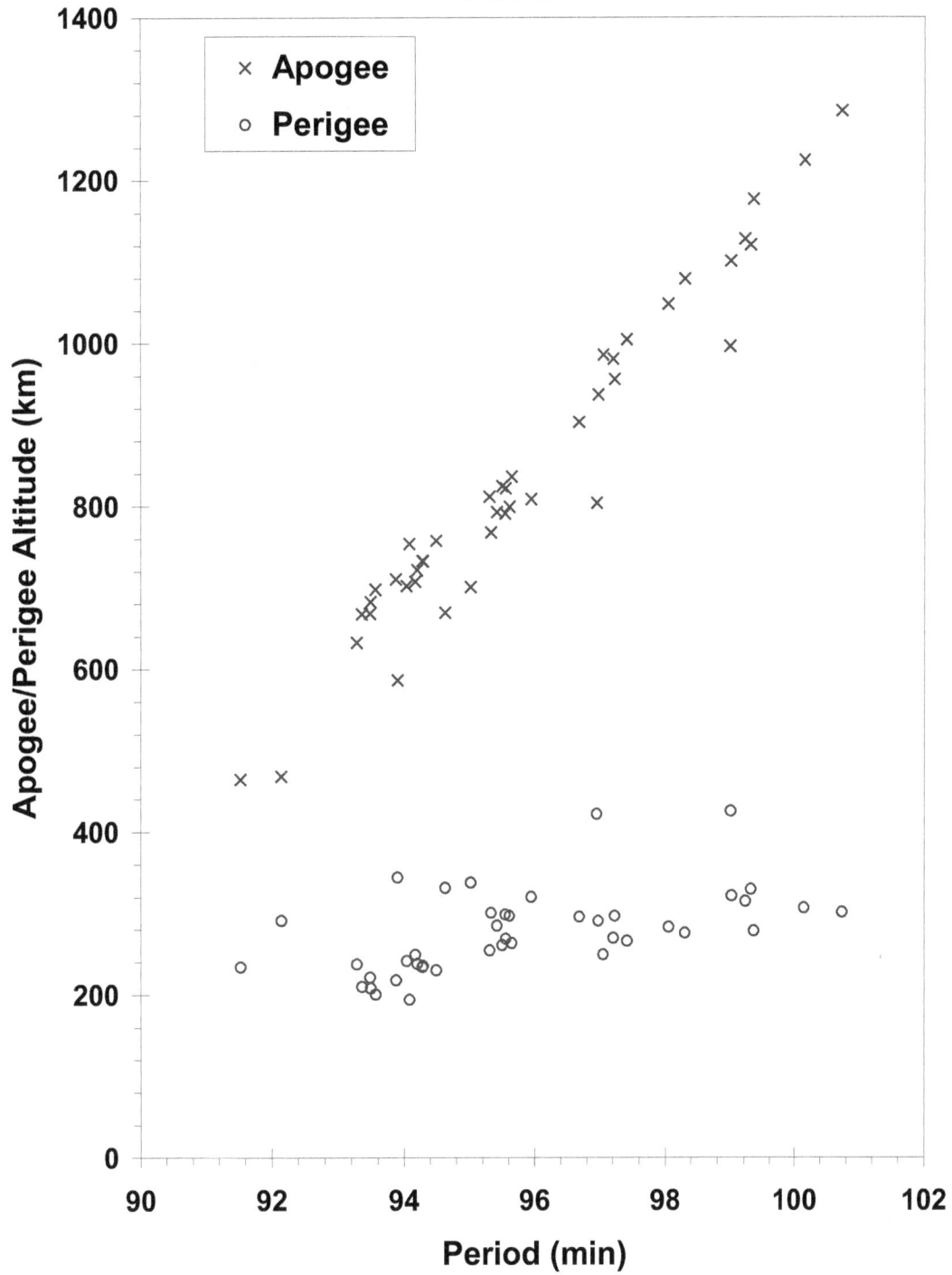

SSN 27474

SOZ motor debris cloud in July 2005 after the second breakup event.

SATELLITE DATA

TYPE:	Mission Related Debris
OWNER:	CIS
LAUNCH DATE:	25.63 Jul 2002
DRY MASS (KG):	55
MAIN BODY:	Ellipsoid; 0.6 m diameter by 1 m length
MAJOR APPENDAGES:	None
ATTITUDE CONTROL:	None
ENERGY SOURCES:	On-board propellant

EVENT DATA

DATE:	29 Oct 2004	LOCATION:	Unknown
TIME:	Unknown	ASSESSED CAUSE:	Propulsion
ALTITUDE:	Unknown		

PRE-EVENT ELEMENTS

EPOCH:	04302.83463691	MEAN ANOMALY:	199.8896
RIGHT ASCENSION:	56.1075	MEAN MOTION:	15.09294101
INCLINATION:	63.6401	MEAN MOTION DOT/2:	0.00146672
ECCENTRICITY:	0.0436823	MEAN MOTION DOT DOT/6:	0.000010335
ARG. OF PERIGEE:	161.8395	BSTAR:	0.00073391

DEBRIS CLOUD DATA

MAXIMUM ΔP:	Unknown
MAXIMUM ΔI:	Unknown

COMMENTS

This event marks the 31st known breakup of a Proton Blok DM SOZ ullage motor since 1984. More than 60 debris were detected by the Naval electronic fence. No debris were in orbit long enough to be cataloged.

REFERENCE DOCUMENTS

History of Soviet/Russian Satellite Fragmentations-A Joint U.S.-Russian Investigation, N. L. Johnson, et al., Kaman Sciences Corporation, October 1995.

"Recent Satellite Breakups," The Orbital Debris Quarterly News, NASA JSC, January 2005.
Available online at https://orbitaldebris.jsc.nasa.gov/quarterly-news/pdfs/ODQNv9i1.pdf.

Insufficient data to construct a Gabbard diagram.

SATELLITE DATA

TYPE:	Payload
OWNER:	CIS
LAUNCH DATE:	12.60 Aug 2003
DRY MASS (KG):	~6000
MAIN BODY:	Cylinder; 2.4 m diameter by 7 m length
MAJOR APPENDAGES:	Solar Arrays
ATTITUDE CONTROL:	Active, 3-axis
ENERGY SOURCES:	On-board propellants, explosive charge

EVENT DATA

DATE:	9 Dec 2003	LOCATION:	64.8 N, 135.4 E (asc)
TIME:	0129Z	ASSESSED CAUSE:	Deliberate
ALTITUDE:	189.33 km		

PRE-EVENT ELEMENTS

EPOCH:	03342.92270571	MEAN ANOMALY:	296.9639
RIGHT ASCENSION:	136.8172	MEAN MOTION:	16.22926227
INCLINATION:	64.9062	MEAN MOTION DOT/2:	0.01025110
ECCENTRICITY:	0.0055948	MEAN MOTION DOT DOT/6:	0.0000073532
ARG. OF PERIGEE:	63.7269	BSTAR:	0.00028689

DEBRIS CLOUD DATA

MAXIMUM ΔP:	7.34 min*
MAXIMUM ΔI:	0.08 deg*

* Based on uncataloged debris data

COMMENTS

Cosmos 2399 was the seventh of the Cosmos 2031 class of spacecraft that debuted in 1989 but was not flown since 1997. In all six previous missions (1989-1997), the spacecraft was deliberately exploded at the end of mission. Previous missions of this type include Cosmos 2031, Cosmos 2101, Cosmos 2163, Cosmos 2225, Cosmos 2262, and Cosmos 2343. All such events have occurred over Eastern Russia. Highest previous piece count for large debris for this class of vehicle was 180, although more were probably created. Approximately 22 debris were detected by the SSN. Due to the low altitude of the breakup, the debris were short-lived.

REFERENCE DOCUMENT

"Satellite Fragmentations in 2003", The Orbital Debris Quarterly News, NASA JSC, January 2004. Available online at https://orbitaldebris.jsc.nasa.gov/quarterly-news/pdfs/ODQNv8i1.pdf.

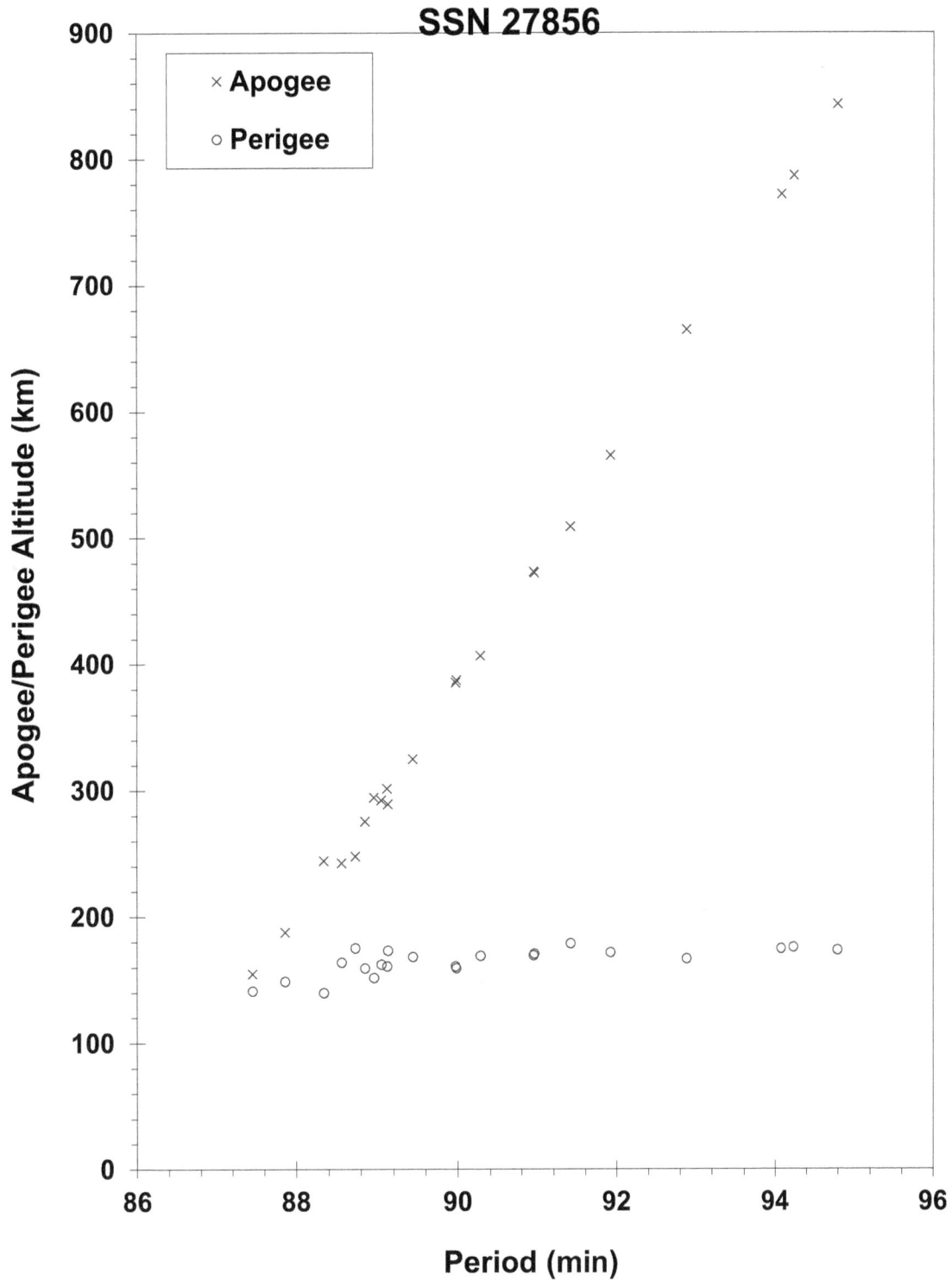

Cosmos 2399 debris cloud of 22 fragments 1 day after the breakup as reconstructed from the US SSN database.

SATELLITE DATA

TYPE:	Mission Related Debris
OWNER:	CIS
LAUNCH DATE:	25 Dec 2005
DRY MASS (KG):	55
MAIN BODY:	ellipsoid; 0.6 m diameter x 1 m length
MAJOR APPENDAGES:	none
ATTITUDE CONTROL:	none
ENERGY SOURCES:	on-board propellants

EVENT DATA

DATE:	24 August 2018	LOCATION:	41.60N, 68.92E (asc)
TIME:	2159 GMT	ASSESSED CAUSE:	Propulsion
ALTITUDE:	4090.1 km		

PRE-EVENT ELEMENTS

EPOCH:	18236.43334214	MEAN ANOMALY:	9.3617
RIGHT ASCENSION:	347.9875	MEAN MOTION:	4.24138578
INCLINATION:	65.2797	MEAN MOTION DOT/2:	-.00000082
ECCENTRICITY:	0. 5700851	MEAN MOTION DOT DOT/6:	0.00000
ARG. OF PERIGEE:	320.8341	BSTAR:	+0.000053164

DEBRIS CLOUD DATA

MAXIMUM ΔP:	5 min
MAXIMUM ΔI:	0.1 deg

COMMENTS

This event was the 49th known fragmentation of Blok-DM ullage motors. The breakups are believed to be caused by residual propellants. Russian officials have been aware of the problem since 1992 and have made design changes, although the date of full implementation is unknown. Approximately 20 small debris were observed by the SSN, but due to difficulties in cataloging fragments in elliptical and deep-space orbits many more fragments could be present.

REFERENCE DOCUMENTS

History of Soviet/Russian Satellite Fragmentations: A Joint U.S.-Russian Investigation. N.L. Johnson *et al*. Kaman Sciences Corporation, October 2005.

"A SOZ Unit Breaks Up in August 2018", The Orbital Debris Quarterly News, NASA JSC, November 2018. Available online at https://www.orbitaldebris.jsc.nasa.gov/quarterly-news/pdfs/odqnv22i4.pdf.

SSN 28920

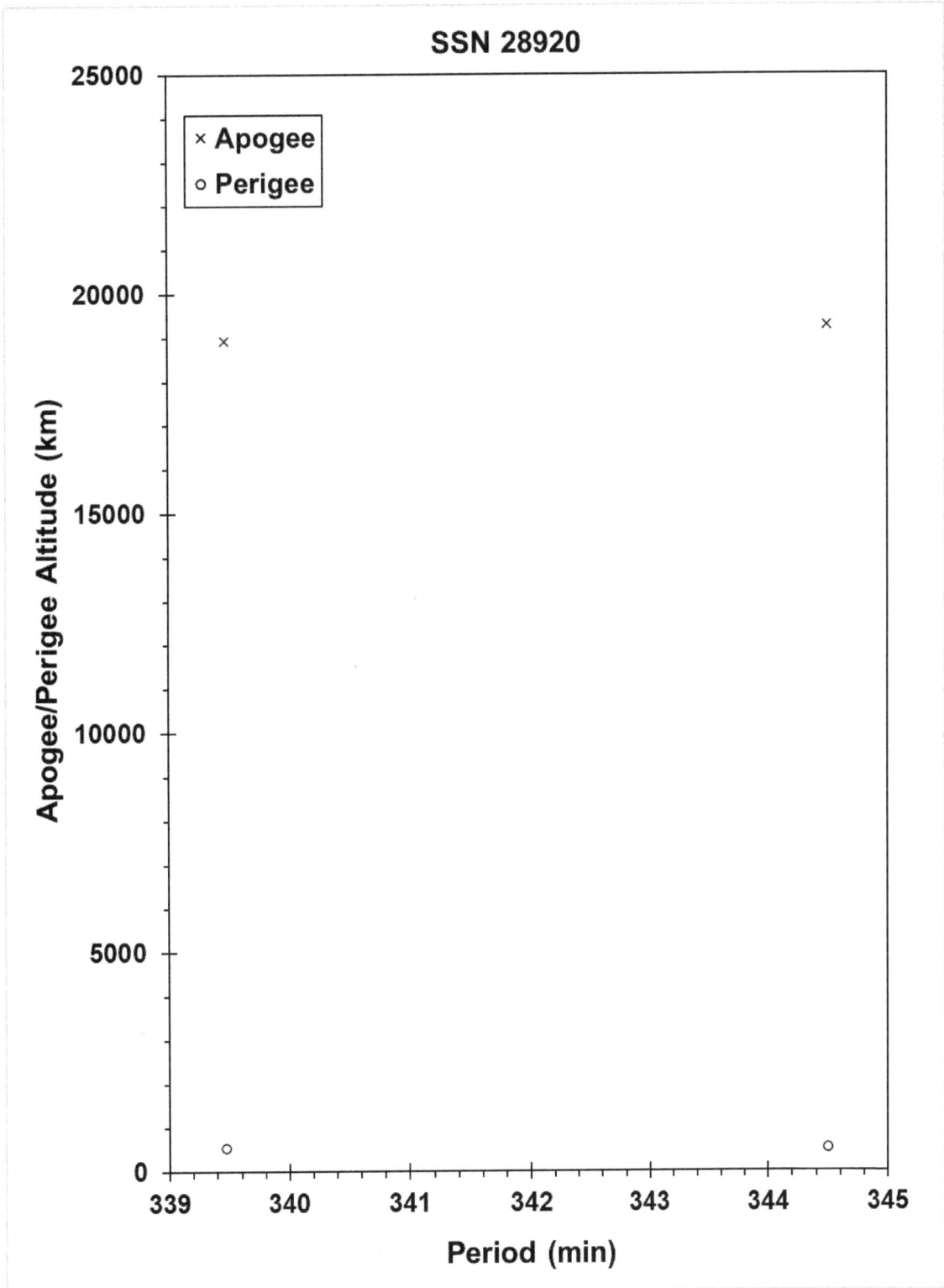

The COSMOS ullage motor debris cloud includes the parent body and one debris fragment
cataloged 131 days post-event.

SATELLITE DATA

TYPE:	Rocket Body
OWNER:	Japan
LAUNCH DATE:	24.06 Jan 2006
DRY MASS (KG):	~3000
MAIN BODY:	Cylinder; 4.0 m diameter by 10.6 m length
MAJOR APPENDAGES:	None
ATTITUDE CONTROL:	None at time of event
ENERGY SOURCES:	On-board propellant

FIRST EVENT DATA

DATE:	8 Aug 2006	LOCATION:	5N, 131E (asc)
TIME:	1407 GMT	ASSESSED CAUSE:	Unknown
ALTITUDE:	685 km		

SECOND EVENT DATA

DATE:	27 Aug 2006	LOCATION:	76S, 135E (asc)
TIME:	1618 GMT	ASSESSED CAUSE:	Unknown
ALTITUDE:	695 km		

PRE-EVENT ELEMENTS

EPOCH:	06220.18259253	MEAN ANOMALY:	147.5075
RIGHT ASCENSION:	300.3416	MEAN MOTION:	14.8204486
INCLINATION:	98.1944	MEAN MOTION DOT/2:	0.00000143
ECCENTRICITY:	0.0106899	MEAN MOTION DOT DOT/6:	0.0
ARG. OF PERIGEE:	211.9623	BSTAR:	0.000023737

DEBRIS CLOUD DATA

MAXIMUM ΔP:	1.0 min
MAXIMUM ΔI:	0.0 deg

COMMENTS

The first event shed four pieces of debris; the second event shed more than 15 pieces. The parent experienced little if any change in orbit and the debris from both events were ejected with relatively low delta velocity from the parent, making a propulsion explosion unlikely for the cause of breakup. All the debris appeared to have high area-to-mass ratios, causing the ejected debris to decay within six months of the events. The parent body remains in orbit as of August 2007. Another H-IIA second stage (2006-037B) experienced two minor fragmentation events a few months later. Possible source of the debris is light-weight insulation material.

REFERENCE DOCUMENTS

"Three More Satellites Involved in Fragmentations," The Orbital Debris Quarterly News, NASA JSC, October 2006. Available online at https://orbitaldebris.jsc.nasa.gov/quarterly-news/pdfs/ODQNv10i4.pdf.

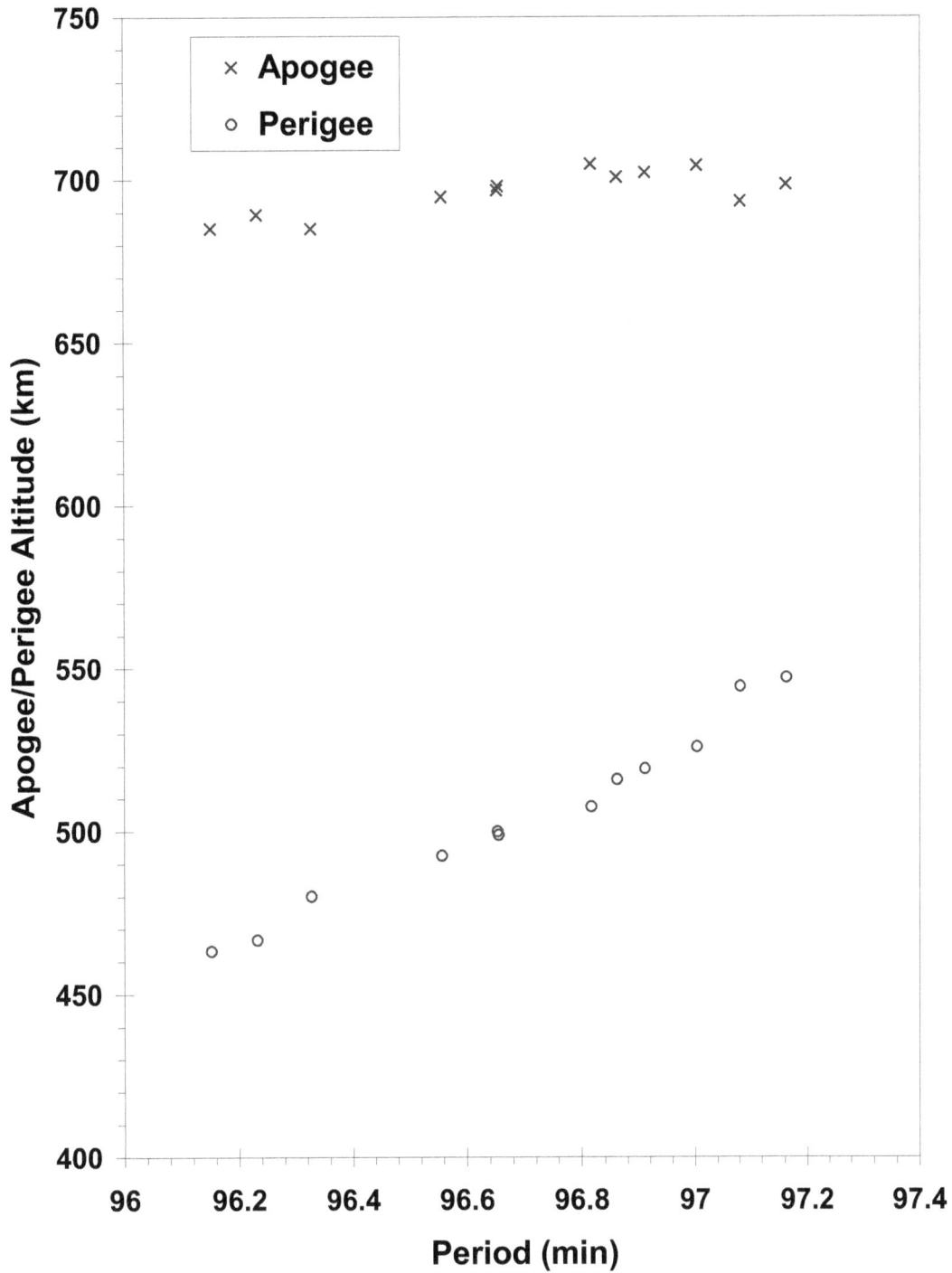

SSN 28932

The ALOS R/B debris cloud from the second event, a few days after the event. The parent body is the piece with the highest perigee, at the right of the graph.

ARABSAT 4 BRIZ-M R/B   **2006-006B**   **28944**

SATELLITE DATA

TYPE:	Rocket Body
OWNER:	CIS
LAUNCH DATE:	28.84 Feb 2006
DRY MASS (KG):	2600
MAIN BODY:	Cylinder; 2.5 m diameter by 2.6 m length
MAJOR APPENDAGES:	None
ATTITUDE CONTROL:	None
ENERGY SOURCES:	On-board propellants

EVENT DATA

DATE:	19 Feb 2007	LOCATION:	31S, 135E (asc)
TIME:	1721 GMT	ASSESSED CAUSE:	Propulsion
ALTITUDE:	7640 km		

PRE-EVENT ELEMENTS

EPOCH:	07050.57138199	MEAN ANOMALY:	134.5177
RIGHT ASCENSION:	213.0506	MEAN MOTION:	5.25304422
INCLINATION:	51.4995	MEAN MOTION DOT/2:	0.00000185
ECCENTRICITY:	0.5083307	MEAN MOTION DOT DOT/6:	0.0
ARG. OF PERIGEE:	197.8403	BSTAR:	0.00029198

DEBRIS CLOUD DATA

MAXIMUM ΔP:	48.6 min
MAXIMUM ΔI:	2.6 deg

COMMENTS

This upper stage malfunctioned, stranding itself and its payload in an eccentric orbit. The cause of the breakup is assessed to be related to the ~8 metric tons of unused propellants. The breakup occurred over Southern Australia, and was captured by three amateur astronomers, which showed expansion of a faint cloud around the exploded fourth stage. Over a thousand objects 10 cm or larger were estimated by the US SSN.

REFERENCE DOCUMENT

"Two More Incidents Add to Growing Space Debris," Space News, February 26, 2007.

"Four Satellite Breakups in February Add to Debris Population," The Orbital Debris Quarterly News, NASA JSC, April 2007. Available online at https://orbitaldebris.jsc.nasa.gov/quarterly-news/pdfs/ODQNv11i2.pdf.

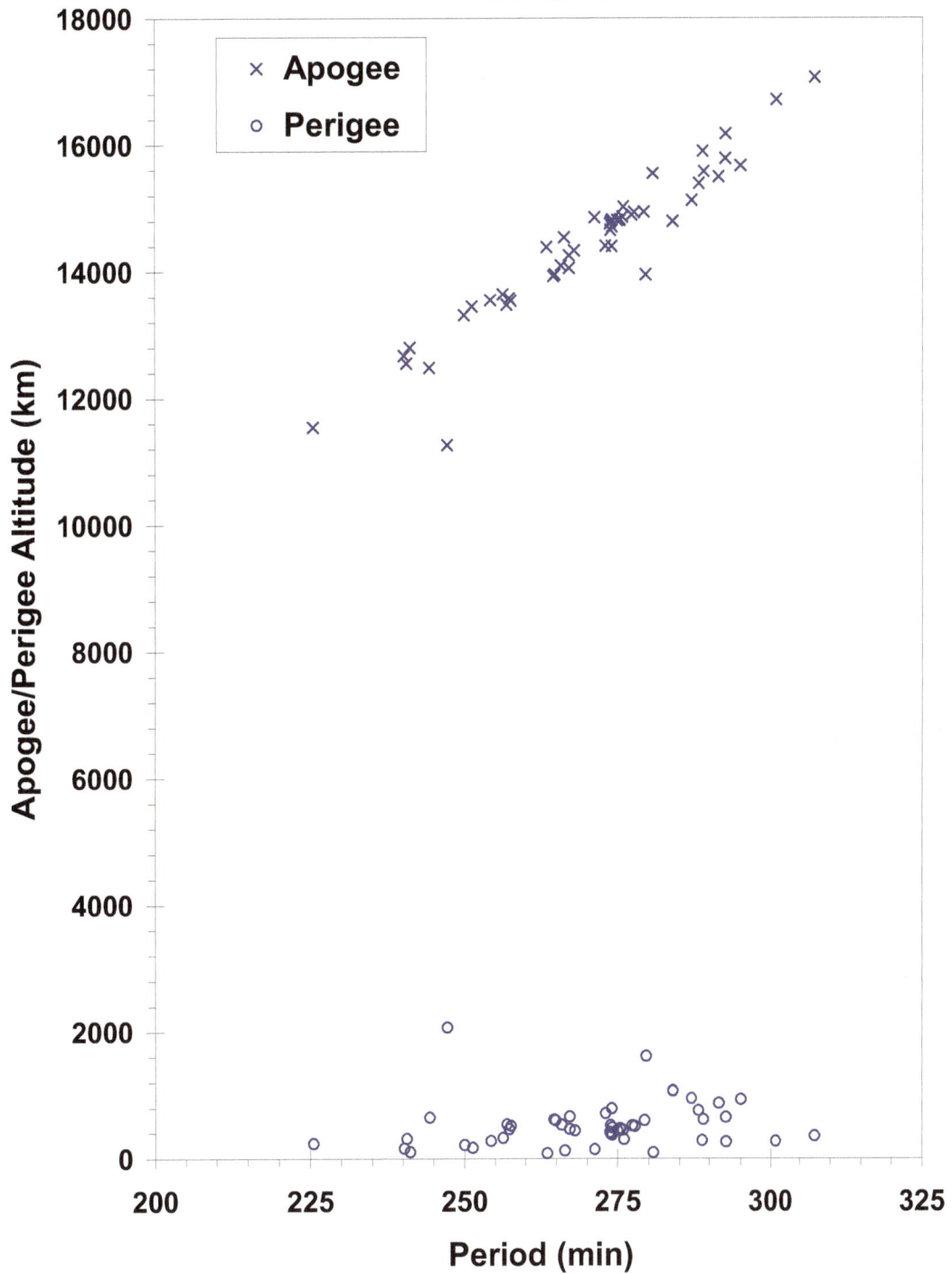

The *Briz-M* R/B debris cloud of uncataloged fragments 10 days after the event.

SATELLITE DATA

TYPE:	Payload
OWNER:	PRC
LAUNCH DATE:	26 Apr 2006
DRY MASS (KG):	2721.0
MAIN BODY:	unknown
MAJOR APPENDAGES:	unknown
ATTITUDE CONTROL:	unknown
ENERGY SOURCES:	unknown

EVENT DATA

DATE:	4 Feb 2010	LOCATION:	56.3N, 140.7E (asc)
TIME:	0649 GMT	ASSESSED CAUSE:	Unknown
ALTITUDE:	636.516 km		

PRE-EVENT ELEMENTS

EPOCH:	10034.53017668	MEAN ANOMALY:	264.9880
RIGHT ASCENSION:	28.5835	MEAN MOTION:	14.80351885
INCLINATION:	97.8566	MEAN MOTION DOT/2:	-0.00000264
ECCENTRICITY:	0.0001956	MEAN MOTION DOT DOT/6:	0.0
ARG. OF PERIGEE:	95.1578	BSTAR:	-0.000027865

DEBRIS CLOUD DATA

MAXIMUM ΔP:	5.4 min
MAXIMUM ΔI:	0.4 deg

COMMENTS

REFERENCE DOCUMENT

"Old and New Satellite Breakups Identified," The Orbital Debris Quarterly News, NASA JSC, April 2010. Available online at https://orbitaldebris.jsc.nasa.gov/quarterly-news/ODQNv14i2.pdf.

SSN 29092

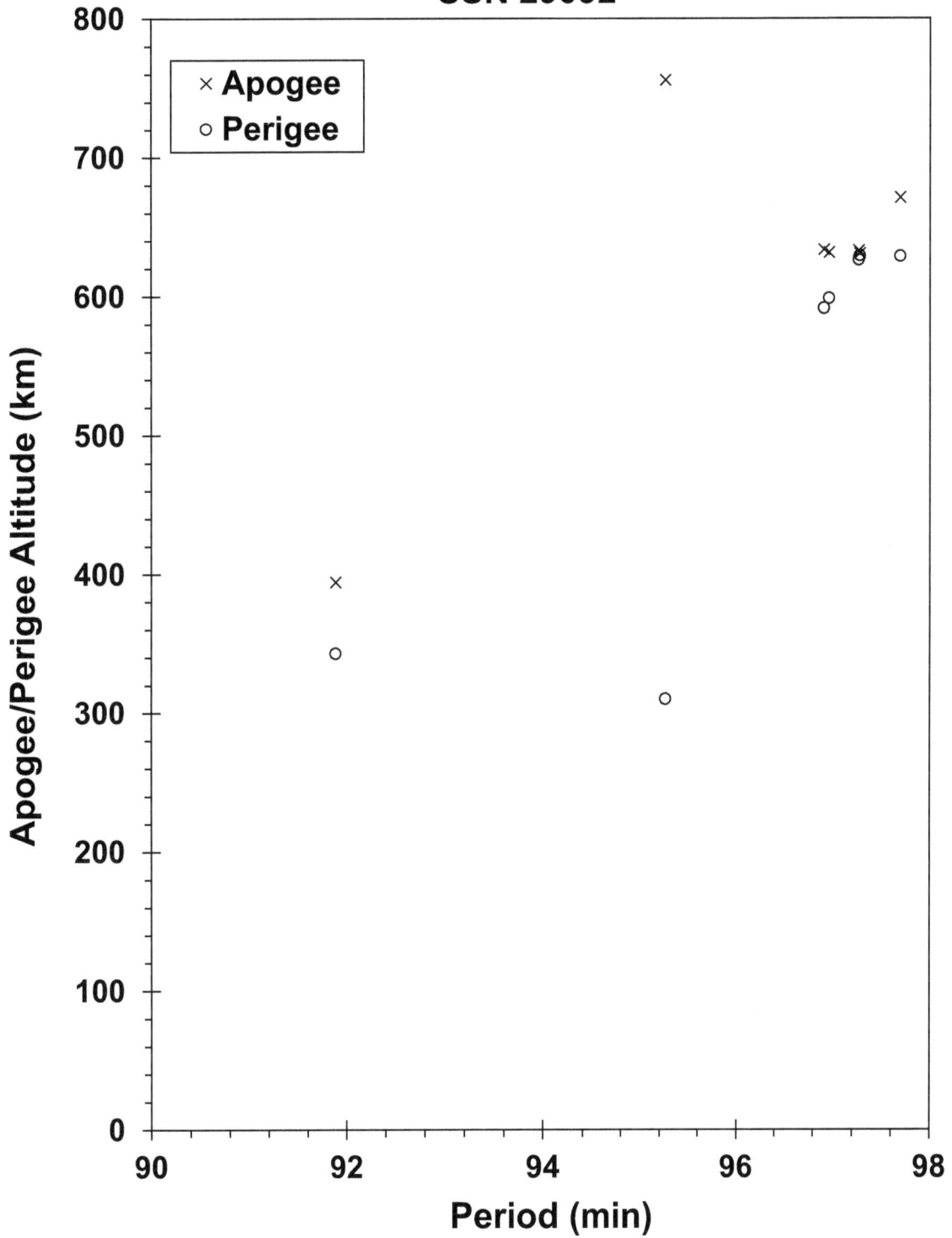

Seven cataloged fragments 7 weeks after the event.

SATELLITE DATA

TYPE:	Payload
OWNER:	CIS
LAUNCH DATE:	25 Jun 2006
DRY MASS (KG):	3200
MAIN BODY:	1.7 m diameter x 17 m long cylinder
MAJOR APPENDAGES:	solar panels
ATTITUDE CONTROL:	active, three-axis
ENERGY SOURCES:	on-board propellants, explosive charge (?)

EVENT DATA

DATE:	14 Mar 2008	LOCATION:	52.247S, 24.130E (dsc)
TIME:	0429 GMT	ASSESSED CAUSE:	Unknown
ALTITUDE:	420.242 km		

PRE-EVENT ELEMENTS

EPOCH:	08073.77012377	MEAN ANOMALY:	145.3615
RIGHT ASCENSION:	48.2539	MEAN MOTION:	15.52978992
INCLINATION:	65.0437	MEAN MOTION DOT/2:	0.00011420
ECCENTRICITY:	0.0014765	MEAN MOTION DOT DOT/6:	0.0
ARG. OF PERIGEE:	285.7968	BSTAR:	0.00018556

DEBRIS CLOUD DATA

MAXIMUM ΔP:	8.1 min
MAXIMUM ΔI:	1.4 deg

COMMENTS

Cosmos 2421 was the last of the Soviet/Russian series of Electronic Ocean Reconnaissance Satellite (EORSAT) spacecraft. As with other spacecraft of the Cosmos 699 class, the high area-to-mass ratio of the majority of debris produced in this event resulted in a short residence on-orbit and no long-term environmental hazard.

REFERENCE DOCUMENT

"Satellite Breakups During First Quarter of 2008", The Orbital Debris Quarterly News, NASA JSC, April 2008. Available online at https://orbitaldebris.jsc.nasa.gov/quarterly-news/ODQNv12i2.pdf.

SSN 29247

The COSMOS 2421 debris cloud, including 509 fragments cataloged up to 48 weeks after the event.

SATELLITE DATA

TYPE:	Rocket Body
OWNER:	Japan
LAUNCH DATE:	11.19 Sep 2006
DRY MASS (KG):	~3000
MAIN BODY:	Cylinder; 4.0 m diameter by 10.6 m length
MAJOR APPENDAGES:	None
ATTITUDE CONTROL:	None at time of event
ENERGY SOURCES:	On-board propellant

EVENT DATA

DATE:	28 Dec 2006	LOCATION:	58S, 131E (asc)
TIME:	1729 GMT	ASSESSED CAUSE:	Unknown
ALTITUDE:	470 km		

PRE-EVENT ELEMENTS

EPOCH:	06361.50058695	MEAN ANOMALY:	347.5081
RIGHT ASCENSION:	117.7926	MEAN MOTION:	15.35084918
INCLINATION:	97.2357	MEAN MOTION DOT/2:	0.00002838
ECCENTRICITY:	0.0043549	MEAN MOTION DOT DOT/6:	0.0
ARG. OF PERIGEE:	12.7250	BSTAR:	0.000083212

DEBRIS CLOUD DATA

MAXIMUM ΔP:	2.6 min
MAXIMUM ΔI:	0.2 deg

COMMENTS

This was the second fragmentation of an H-IIA second stage in 2006 (see 2006-002B). No debris were cataloged from the first event, but at least 20 new objects were detected. However, the rocket body experienced a second fragmentation event on 25 July 2007, releasing at least 15 new debris.

REFERENCE DOCUMENT

"Significant Increase in Satellite Breakups During 2006", The Orbital Debris Quarterly News, NASA JSC, January 2007. Available online at https://orbitaldebris.jsc.nasa.gov/quarterly-news/pdfs/ODQNv11i1.pdf.

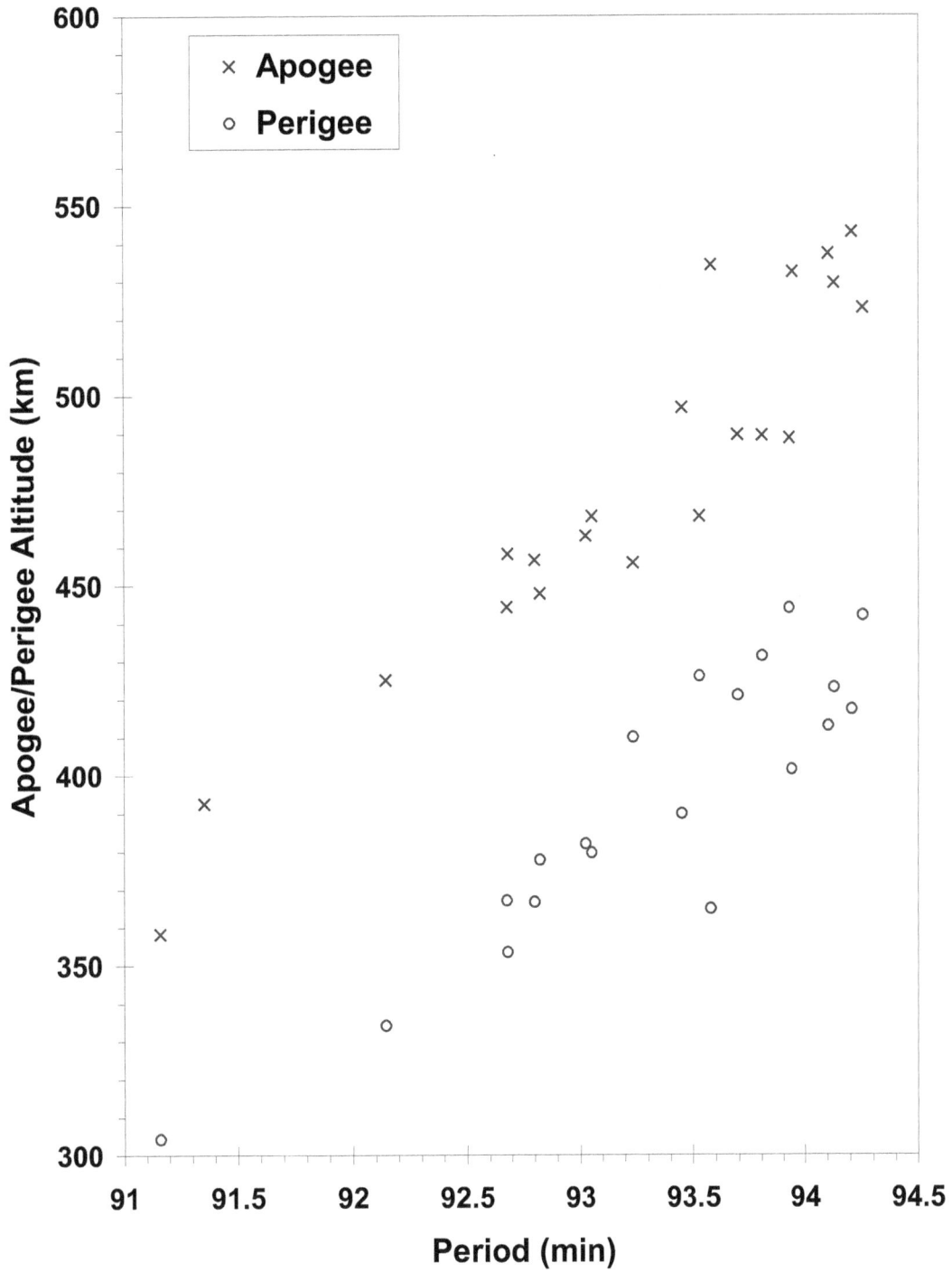

SSN 29394

H-IIA debris cloud 2 weeks after the first event. The debris were never cataloged.

SATELLITE DATA

TYPE:	Payload
OWNER:	CIS
LAUNCH DATE:	14.57 Sep 2006
DRY MASS (KG):	~6000
MAIN BODY:	Cylinder; 2.4 m diameter by 7 m length
MAJOR APPENDAGES:	Solar Arrays
ATTITUDE CONTROL:	Active, 3-axis
ENERGY SOURCES:	On-board propellants, explosive charge

EVENT DATA

DATE:	17 Nov 2006	LOCATION:	Unknown
TIME:	~1800 GMT	ASSESSED CAUSE:	Deliberate
ALTITUDE:	210 km		

PRE-EVENT ELEMENTS

EPOCH:	06321.75318443	MEAN ANOMALY:	8.8408
RIGHT ASCENSION:	358.1498	MEAN MOTION:	16.11335386
INCLINATION:	64.8900	MEAN MOTION DOT/2:	0.00257180
ECCENTRICITY:	0.0061777	MEAN MOTION DOT DOT/6:	0.0000071463
ARG. OF PERIGEE:	106.6782	BSTAR:	0.00018281

DEBRIS CLOUD DATA

MAXIMUM ΔP:	5.8 min
MAXIMUM ΔI:	0.5 deg

COMMENTS

Cosmos 2423 was the eighth of the Cosmos 2031 class of spacecraft, which debuted in 1989. In all seven previous missions, the spacecraft was deliberately exploded at the end of mission. Cosmos 2423 had the shortest lifetime of any of this class of spacecraft, two months instead of a usual four months. Because these spacecraft are deliberately exploded at a low altitude, the lifetime of the trackable debris cloud is usually measured in days.

REFERENCE DOCUMENTS

"Significant Increase in Satellite Breakups During 2006", The Orbital Debris Quarterly News, NASA JSC, January 2007. Available online at https://orbitaldebris.jsc.nasa.gov/quarterly-news/pdfs/ODQNv11i1.pdf.

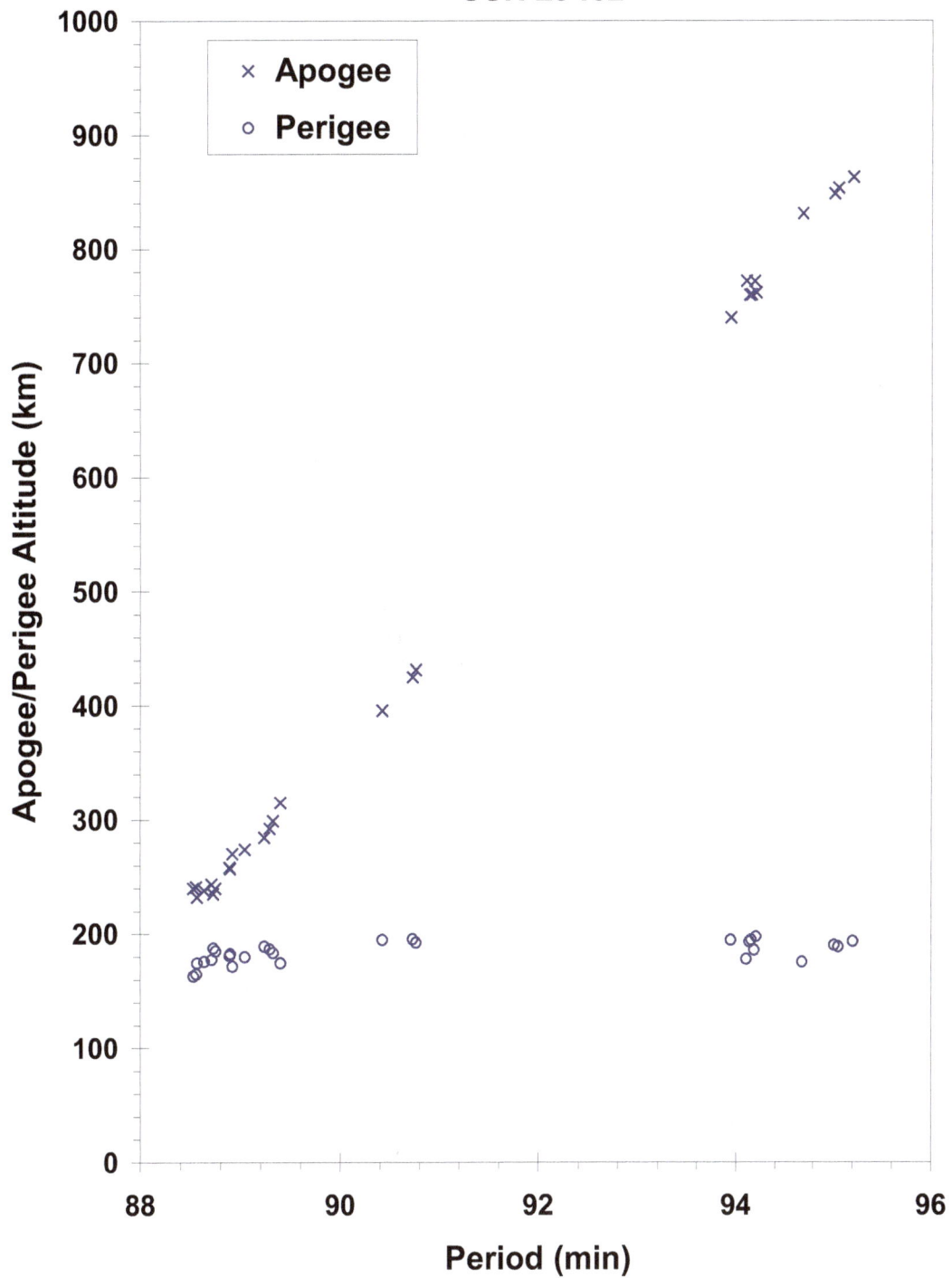

Cosmos 2423 debris cloud of 28 fragments a few days after the event as reconstructed from the US SSN database.

SATELLITE DATA

TYPE:	Rocket Body
OWNER:	US
LAUNCH DATE:	4.58 Nov 2006
DRY MASS (KG):	2850
MAIN BODY:	Cylinder; 4.0 m diameter by 12 m length
MAJOR APPENDAGES:	None
ATTITUDE CONTROL:	None at time of event
ENERGY SOURCES:	On-board propellant

EVENT DATA

DATE:	4 Nov 2006	LOCATION:	Unknown
TIME:	Unknown	ASSESSED CAUSE:	Unknown
ALTITUDE:	Unknown		

PRE-EVENT ELEMENTS

EPOCH:	06308.62553010	MEAN ANOMALY:	119.1776
RIGHT ASCENSION:	307.4245	MEAN MOTION:	14.13676442
INCLINATION:	98.7885	MEAN MOTION DOT/2:	-0.00000044
ECCENTRICITY:	0.0022197	MEAN MOTION DOT DOT/6:	0.0
ARG. OF PERIGEE:	240.7178	BSTAR:	0.0

DEBRIS CLOUD DATA

MAXIMUM ΔP:	9.5 min
MAXIMUM ΔI:	0.3 deg

COMMENTS

This was the first major event associated with a Delta IV second stage. Debris appeared to separate from the parent body in a retrograde direction soon after orbit insertion. Over 60 pieces were eventually cataloged from this event. The rocket body did not see any performance degradation and was reentered directly after payload delivery. The cause of the debris release was investigated to ensure any countermeasures can be implemented for future Delta IV missions.

REFERENCE DOCUMENTS

"Significant Increase in Satellite Breakups During 2006", The Orbital Debris Quarterly News, NASA JSC, January 2007. Available online at https://orbitaldebris.jsc.nasa.gov/quarterly-news/pdfs/ODQNv11i1.pdf.

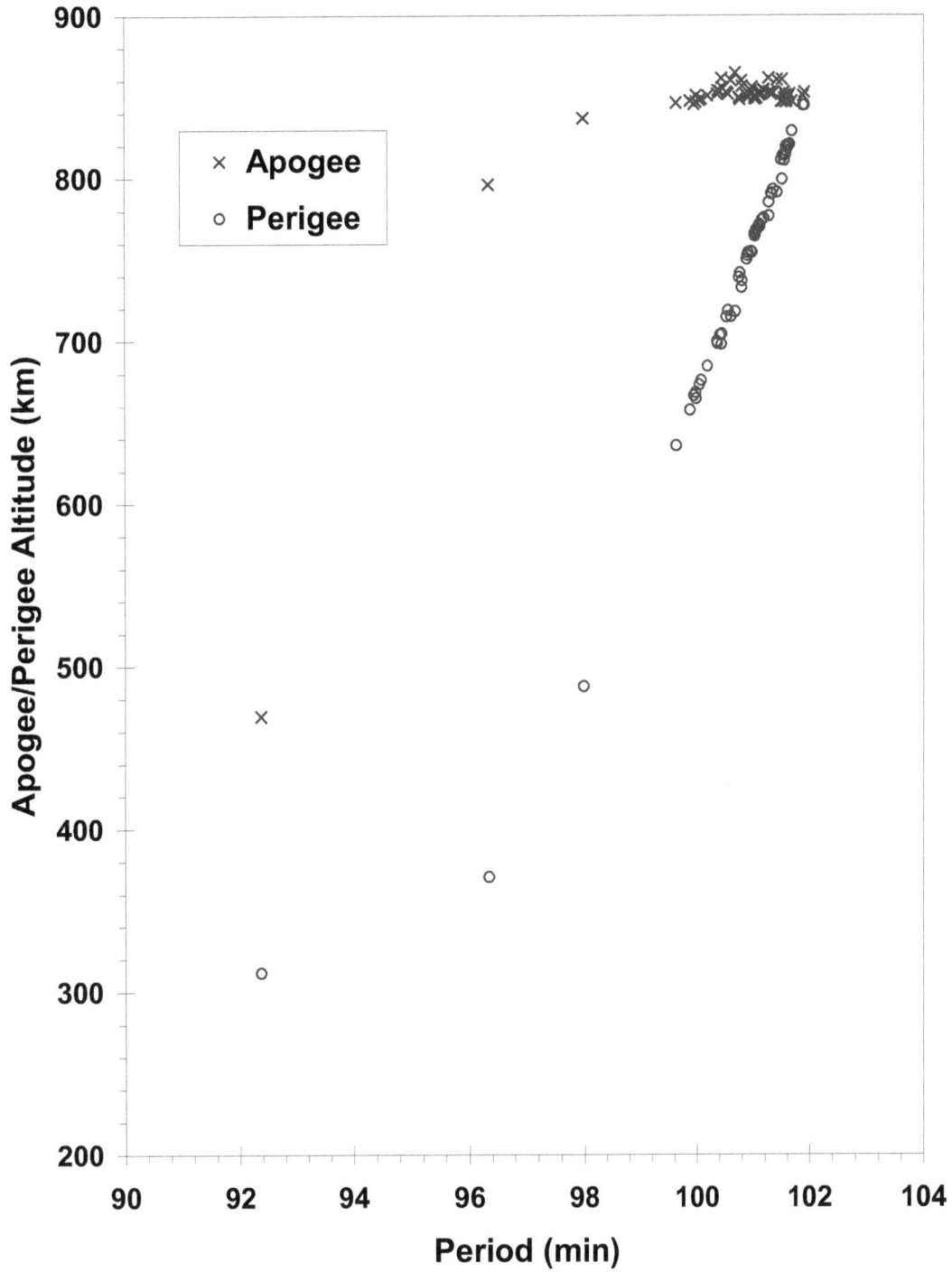

SSN 29523

Delta 4 debris cloud of 62 fragments less than two weeks after the event.

SATELLITE DATA

TYPE:	Payload
OWNER:	USA
LAUNCH DATE:	14 Dec 2006
DRY MASS (KG):	unknown
MAIN BODY:	unknown
MAJOR APPENDAGES:	unknown
ATTITUDE CONTROL:	unknown
ENERGY SOURCES:	unknown

EVENT DATA

DATE:	21 Feb 2008	LOCATION:	8.813N, 189.023E (asc)
TIME:	0326 GMT	ASSESSED CAUSE:	Deliberate (HVI)
ALTITUDE:	247.769 km		

PRE-EVENT ELEMENTS - unavailable

EPOCH:		MEAN ANOMALY:	
RIGHT ASCENSION:		MEAN MOTION:	
INCLINATION:		MEAN MOTION DOT/2:	
ECCENTRICITY:		MEAN MOTION DOT DOT/6:	
ARG. OF PERIGEE:		BSTAR:	

DEBRIS CLOUD DATA

MAXIMUM ΔP:	unknown min
MAXIMUM ΔI:	unknown deg

COMMENTS

This spacecraft was intentionally destroyed in order to minimize the potential ground hazard should it reenter with its fuel tank intact. NASA's standard reentry hazard analysis tool indicated a high probability that frozen/slush hydrazine monopropellant would survive intact to the ground, presenting a significant chemical hazard. The intercept was accomplished by an SM-3 missile launched by the US Navy Aegis cruiser USS *Lake Erie*.

REFERENCE DOCUMENT

"Satellite Breakups During First Quarter of 2008", The Orbital Debris Quarterly News, NASA JSC, April 2008. Available online at https://orbitaldebris.jsc.nasa.gov/quarterly-news/ODQNv12i2.pdf.

SSN 29651

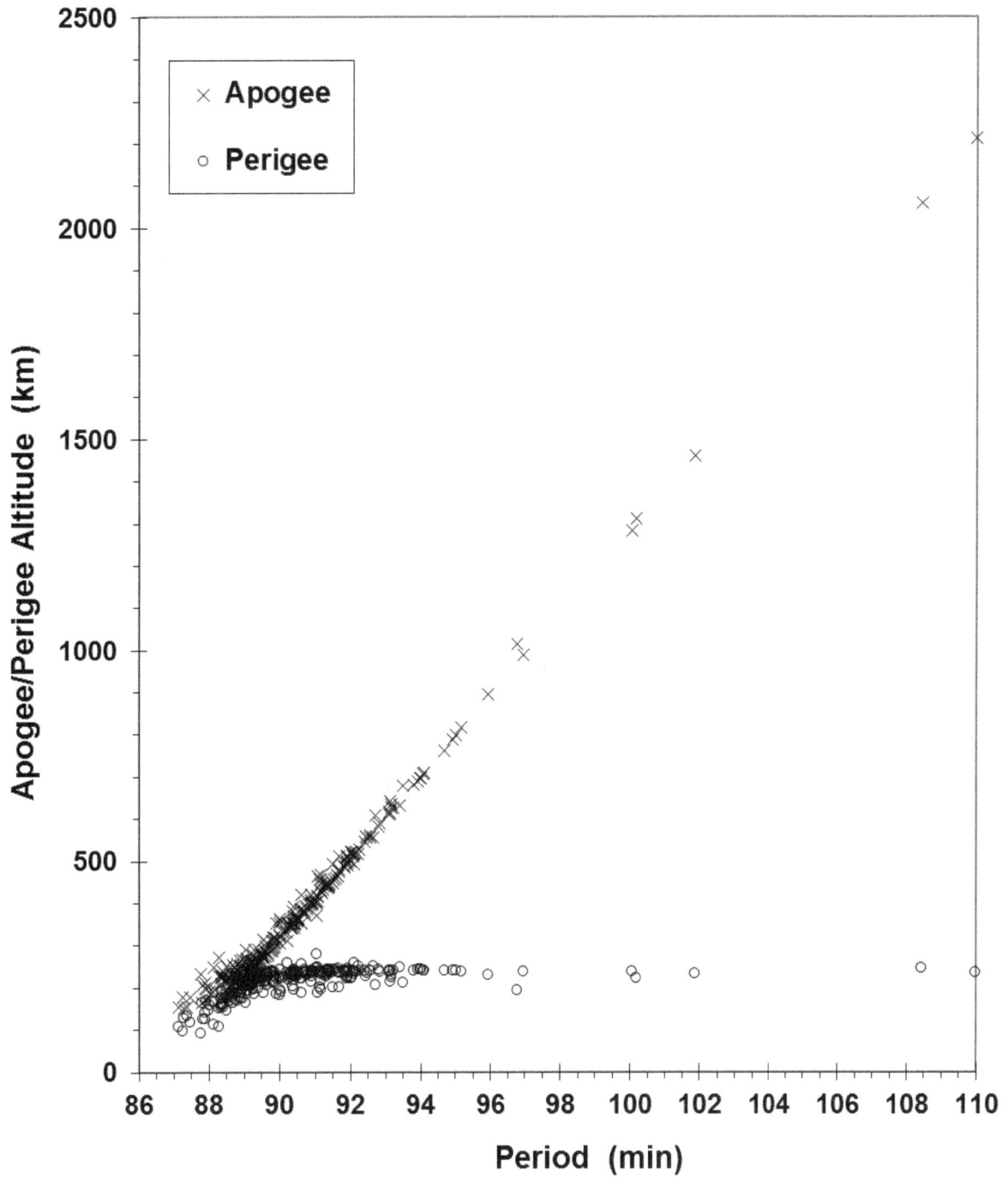

The USA 193 debris cloud, including 174 fragments cataloged up to 1 month after the event.

COSMOS 2424-2426 ULLAGE MOTOR   2006-062G   29680

SATELLITE DATA

TYPE:	Mission Related Debris
OWNER:	CIS
LAUNCH DATE:	25 Dec 2006
DRY MASS (KG):	55
MAIN BODY:	ellipsoid; 0.6 m diameter x 1 m length
MAJOR APPENDAGES:	none
ATTITUDE CONTROL:	none
ENERGY SOURCES:	on-board propellants

EVENT DATA

DATE:	27 July 2016	LOCATION:	15.38S, 124.1E (asc)
TIME:	0119 GMT	ASSESSED CAUSE:	Propulsion
ALTITUDE:	18551.3 km		

PRE-EVENT ELEMENTS

EPOCH:	16208.24667120	MEAN ANOMALY:	356.6314
RIGHT ASCENSION:	261.9075	MEAN MOTION:	4.23611820
INCLINATION:	64.7769	MEAN MOTION DOT/2:	+0.00000237
ECCENTRICITY:	0.5783296	MEAN MOTION DOT DOT/6:	0.00000
ARG. OF PERIGEE:	27.2587	BSTAR:	+0.00030726

DEBRIS CLOUD DATA

MAXIMUM ΔP:	10 min
MAXIMUM ΔI:	0.7 deg

COMMENTS

This event was the 46th known fragmentation of Blok-DM ullage motors. The breakups are believed to be caused by residual propellants. Russian officials have been aware of the problem since 1992 and have made design changes, although the date of full implementation is unknown. Multiple small debris were observed by the SSN, but due to difficulties in cataloging fragments in elliptical and deep-space orbits many more fragments could be present.

REFERENCE DOCUMENTS

History of Soviet/Russian Satellite Fragmentations: A Joint U.S.-Russian Investigation. N.L. Johnson *et al*. Kaman Sciences Corporation, October 2005.

"New SOZ Breakup in July 2016", The Orbital Debris Quarterly News, NASA JSC, October 2016. Available online at https://www.orbitaldebris.jsc.nasa.gov/quarterly-news/pdfs/odqnv20i4.pdf.

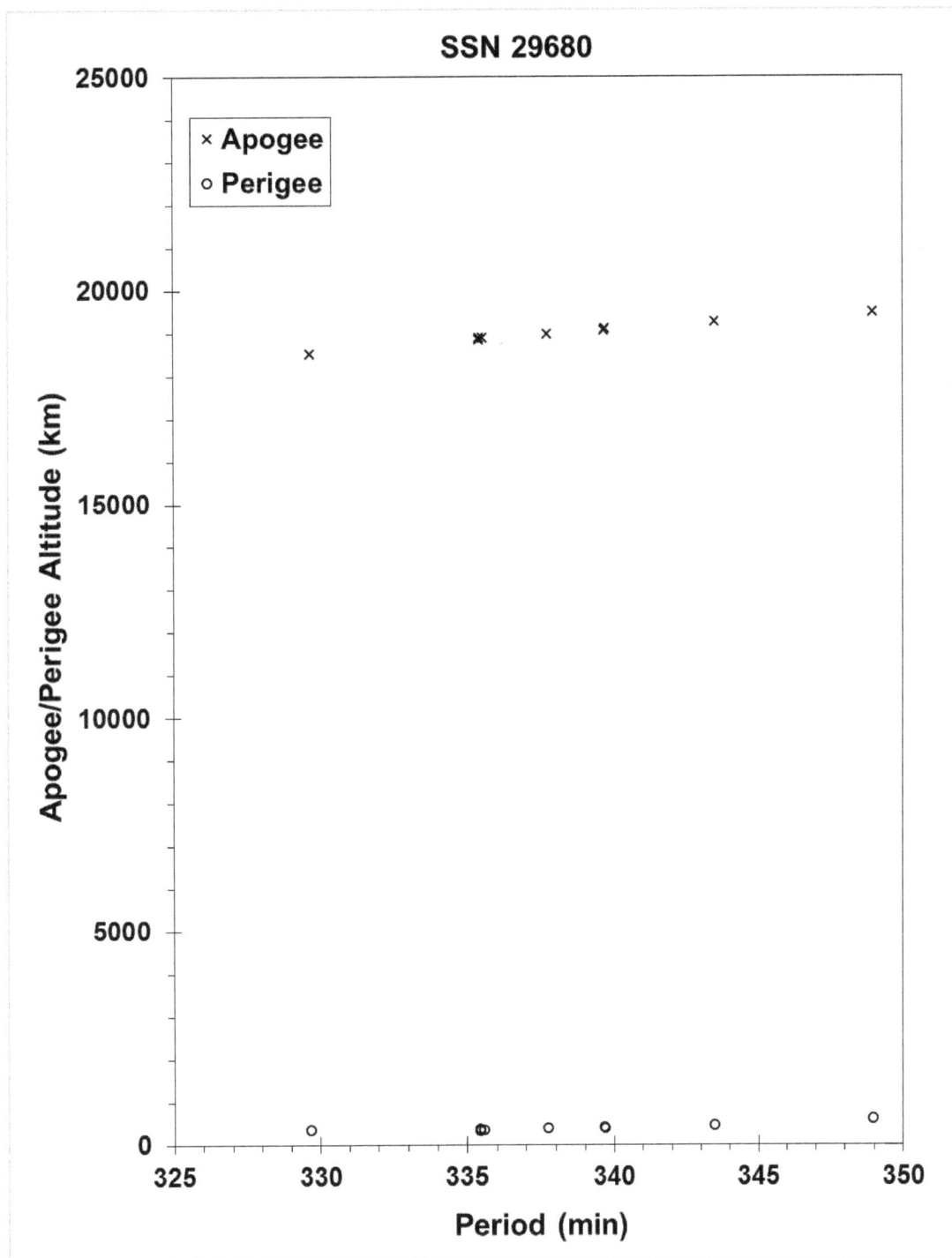

SSN 29680

The COSMOS ullage motor debris cloud includes the parent body and eight debris fragments cataloged up to 75 days post-event.

COSMOS 2424-2426 ULLAGE MOTOR   2006-062H   29682

SATELLITE DATA

TYPE:	Mission Related Debris
OWNER:	CIS
LAUNCH DATE:	25 Dec 2006
DRY MASS (KG):	55
MAIN BODY:	ellipsoid; 0.6 m diameter x 1 m length
MAJOR APPENDAGES:	none
ATTITUDE CONTROL:	none
ENERGY SOURCES:	on-board propellants

EVENT DATA

DATE:	23 Oct. 2019	LOCATION:	7.05S, 149.37E (dsc)
TIME:	1200 GMT	ASSESSED CAUSE:	Propulsion
ALTITUDE:	18562.5 km		

PRE-EVENT ELEMENTS

EPOCH:	19296.36239282	MEAN ANOMALY:	0.5707
RIGHT ASCENSION:	177.6633	MEAN MOTION:	4.24195467
INCLINATION:	64.4587	MEAN MOTION DOT/2:	+0.00001135
ECCENTRICITY:	0. 5860854	MEAN MOTION DOT DOT/6:	0.00000
ARG. OF PERIGEE:	357.7058	BSTAR:	+0.00030496

DEBRIS CLOUD DATA

MAXIMUM ΔP:	1 min
MAXIMUM ΔI:	0.9 deg

COMMENTS

This event was the 51st known fragmentation of Blok-DM ullage motors. The breakups are believed to be caused by residual propellants. Russian officials have been aware of the problem since 1992 and have made design changes, although the date of full implementation is unknown. Ten small debris were initially observed by the SSN, but due to difficulties in cataloging fragments in elliptical and deep-space orbits many more fragments could be present.

REFERENCE DOCUMENTS

History of Soviet/Russian Satellite Fragmentations: A Joint U.S.-Russian Investigation. N.L. Johnson *et al*. Kaman Sciences Corporation, October 2005.

"Fifty-first SOZ Unit Breaks Up," The Orbital Debris Quarterly News, NASA JSC, February 2020. Available online at https://www.orbitaldebris.jsc.nasa.gov/quarterly-news/pdfs/odqnv24i1.pdf.

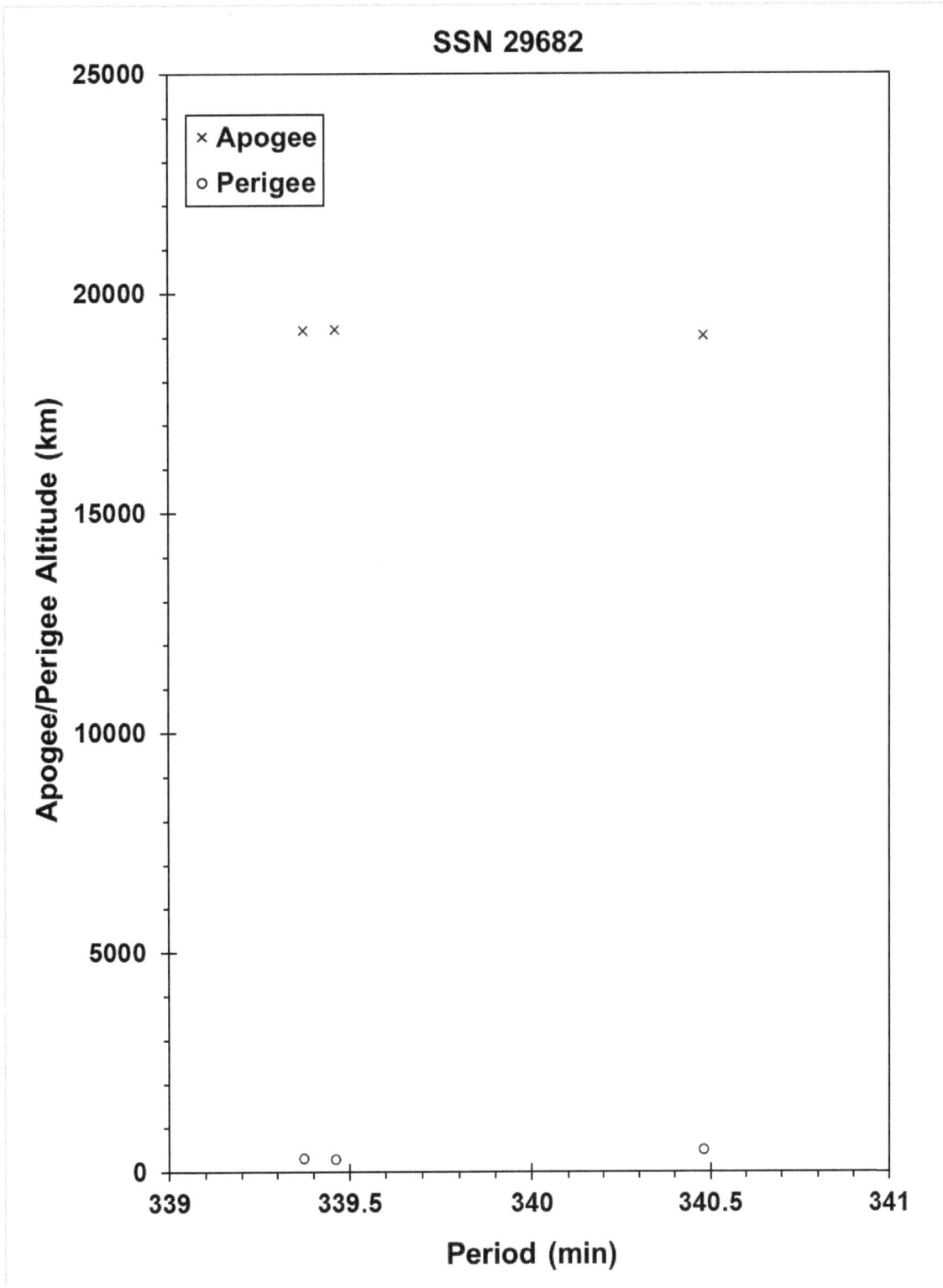

SSN 29682

The COSMOS ullage motor debris cloud includes the parent body and two debris fragments cataloged up to 74 days post-event.

BEIDOU 1D R/B   **2007-003B**   **30324**

SATELLITE DATA

TYPE:	Rocket Body
OWNER:	PRC
LAUNCH DATE:	02 Feb 2007
DRY MASS (KG):	3062
MAIN BODY:	Cylinder; 3.0 m diameter x 12.4 m long
MAJOR APPENDAGES:	none
ATTITUDE CONTROL:	status unknown at time of event
ENERGY SOURCES:	on-board propellants

EVENT DATA

DATE:	02 Feb 2007	LOCATION:	Unknown
TIME:	Unknown	ASSESSED CAUSE:	Unknown
ALTITUDE:	Unknown		

PRE-EVENT ELEMENTS

EPOCH:	07033.76400703	MEAN ANOMALY:	43.007
RIGHT ASCENSION:	9.4841	MEAN MOTION:	1.91693848
INCLINATION:	25.0136	MEAN MOTION DOT/2:	-0.00001383
ECCENTRICITY:	0.7597957	MEAN MOTION DOT DOT/6:	0.0000011914
ARG. OF PERIGEE:	179.8053	BSTAR:	0.0

DEBRIS CLOUD DATA

MAXIMUM ΔP:	5.5 min
MAXIMUM ΔI:	0.2 deg

COMMENTS

In the 14th edition of this work, this event was identified as being associated with the payload. To quote: "The spacecraft experienced problems soon after insertion into a geosynchronous transfer orbit. In March 2007, the spacecraft was recovered and maneuvered into GEO. As many as 100 debris were detected by the US SSN." This event was more recently identified as being associated with the CZ-3A R/B. This is now recognized as being the first known CZ-3 third stage fragmentation event.

REFERENCE DOCUMENT

"Four Satellite Breakups in February Add to Debris Population," The Orbital Debris Quarterly News, NASA JSC, January 2007. Available online at https://orbitaldebris.jsc.nasa.gov/quarterly-news/pdfs/ODQNv11i2.pdf.

SSN 30323

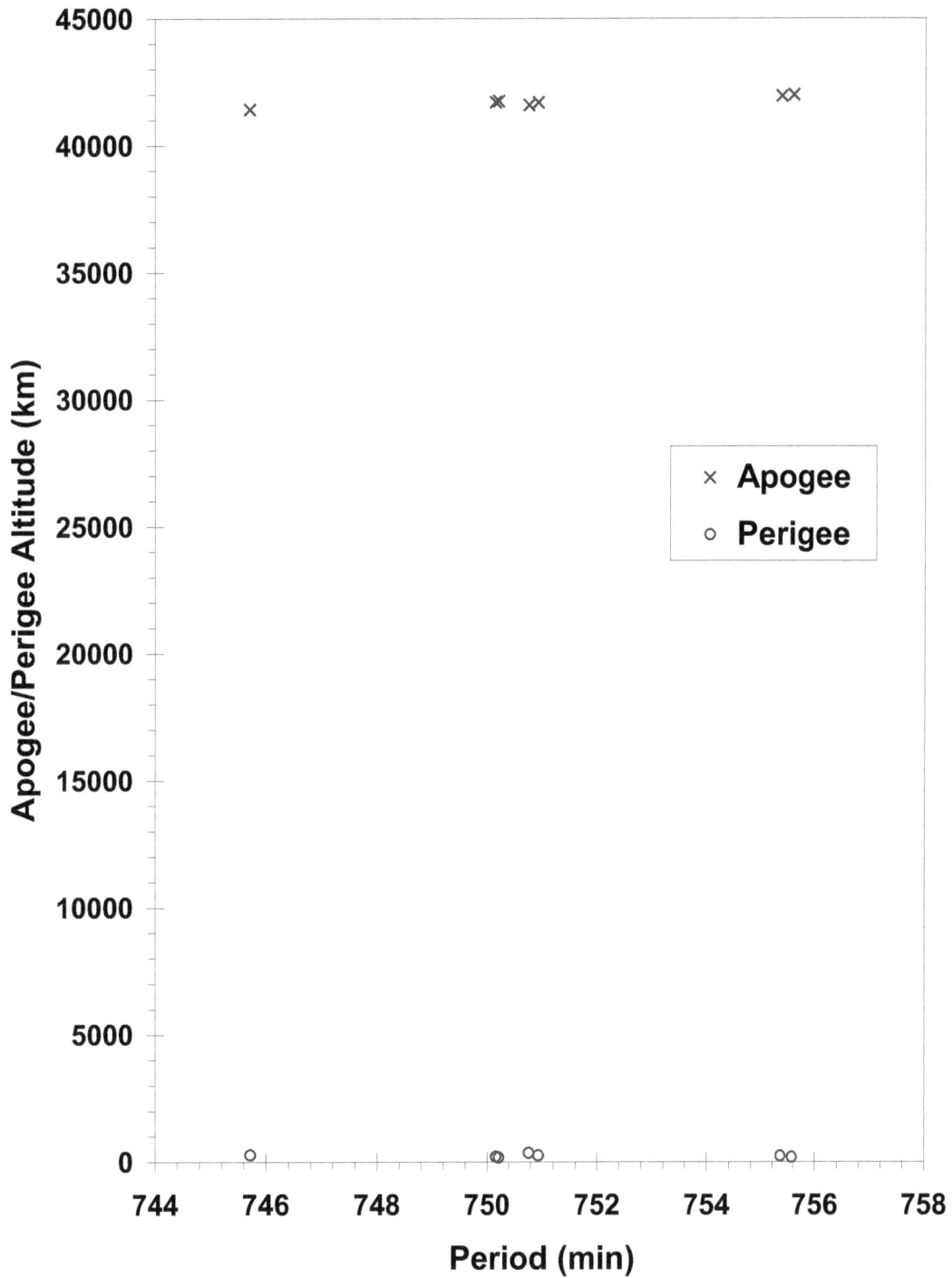

Identified Beidou R/B debris cloud 2 weeks after the event as reconstructed from the US SSN database.

SATELLITE DATA

TYPE:	Mission Related Debris
OWNER:	Japan
LAUNCH DATE:	24 Feb 2007
DRY MASS (KG):	unknown
MAIN BODY:	truncated cone; 4.070 m diameter tapering to < 1.98 m diameter x ~ 1.5 m length
MAJOR APPENDAGES:	none
ATTITUDE CONTROL:	none
ENERGY SOURCES:	unknown

EVENT DATA

DATE:	23 Dec 2010	LOCATION:	Unknown
TIME:	1600 GMT	ASSESSED CAUSE:	Unknown
ALTITUDE:	Unknown		

PRE-EVENT ELEMENTS

EPOCH:	10357.22196763	MEAN ANOMALY:	297.1353
RIGHT ASCENSION:	114.8768	MEAN MOTION:	15.43699236
INCLINATION:	97.3356	MEAN MOTION DOT/2:	0.00018749
ECCENTRICITY:	0.0005569	MEAN MOTION DOT DOT/6:	0.0
ARG. OF PERIGEE:	63.1040	BSTAR:	0.0

DEBRIS CLOUD DATA

MAXIMUM ΔP:	0.9 min
MAXIMUM ΔI:	0.0 deg

COMMENTS

Parent object is the upper cap section of the three-piece Kawasaki Heavy Industries, Ltd. 4/4D-LC dual payload fairing's lower payload encapsulation system. Six objects were initially observed and three were officially cataloged. These objects displayed large area-to-mass ratios (0.38-0.85 m2/kg) and decayed rapidly from orbit. Subsequently an additional 11 pieces entered the catalog. All have decayed.

REFERENCE DOCUMENT

"New Satellite Fragmentations Add to Debris Population," The Orbital Debris Quarterly News, NASA JSC, January 2011. Available online at https://orbitaldebris.jsc.nasa.gov/quarterly-news/pdfs/ODQNv15i1.pdf.

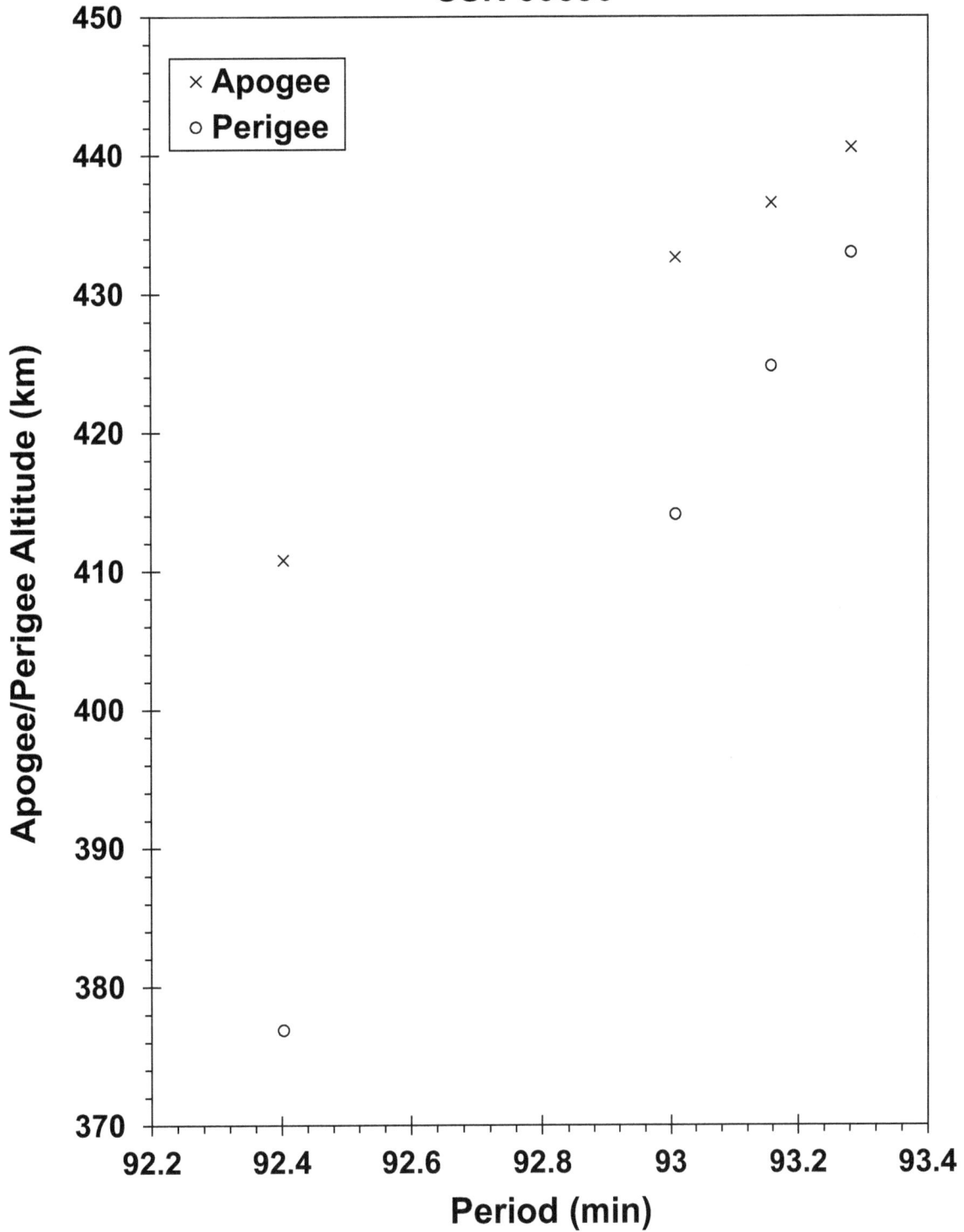

Four HII-A debris fragments cataloged within 5 days of the event.

SATELLITE DATA

TYPE:	Payload
OWNER:	CIS
LAUNCH DATE:	26 Oct 2007
DRY MASS (KG):	3250
MAIN BODY:	stepped cylinder
MAJOR APPENDAGES:	solar panels & antennas
ATTITUDE CONTROL:	gravity gradient (passive)
ENERGY SOURCES:	unknown at time of event

EVENT DATA

DATE:	10 May 2014	LOCATION:	TBD
TIME:	2024 GMT	ASSESSED CAUSE:	Unknown
ALTITUDE:	TBD km		

PRE-EVENT ELEMENTS

EPOCH:	14129.36499359	MEAN ANOMALY:	236.8999
RIGHT ASCENSION:	270.8988	MEAN MOTION:	14.12320352
INCLINATION:	70.9495	MEAN MOTION DOT/2:	0.00000475
ECCENTRICITY:	0.0010837	MEAN MOTION DOT DOT/6:	0.00000
ARG. OF PERIGEE:	200.7382	BSTAR:	0.0028345

DEBRIS CLOUD DATA

MAXIMUM ΔP:	TBD min
MAXIMUM ΔI:	TBD deg

COMMENTS

Cosmos 2428 is believed to be the last Tselina-2 class spacecraft flown. Approximately 17 fragments have been observed, but only nine have been officially cataloged. Of these, eight have decayed from orbit by 23 September 2014. The remaining fragment is in an orbit similar to four pieces of previously cataloged SL-16 launch vehicle debris and may not be related to this event.

REFERENCE DOCUMENTS

"Flurry of Small Breakups in First Half of 2014", The Orbital Debris Quarterly News, NASA JSC, July 2014. Available online at https://orbitaldebris.jsc.nasa.gov/quarterly-news/pdfs/ODQNv18i3.pdf.

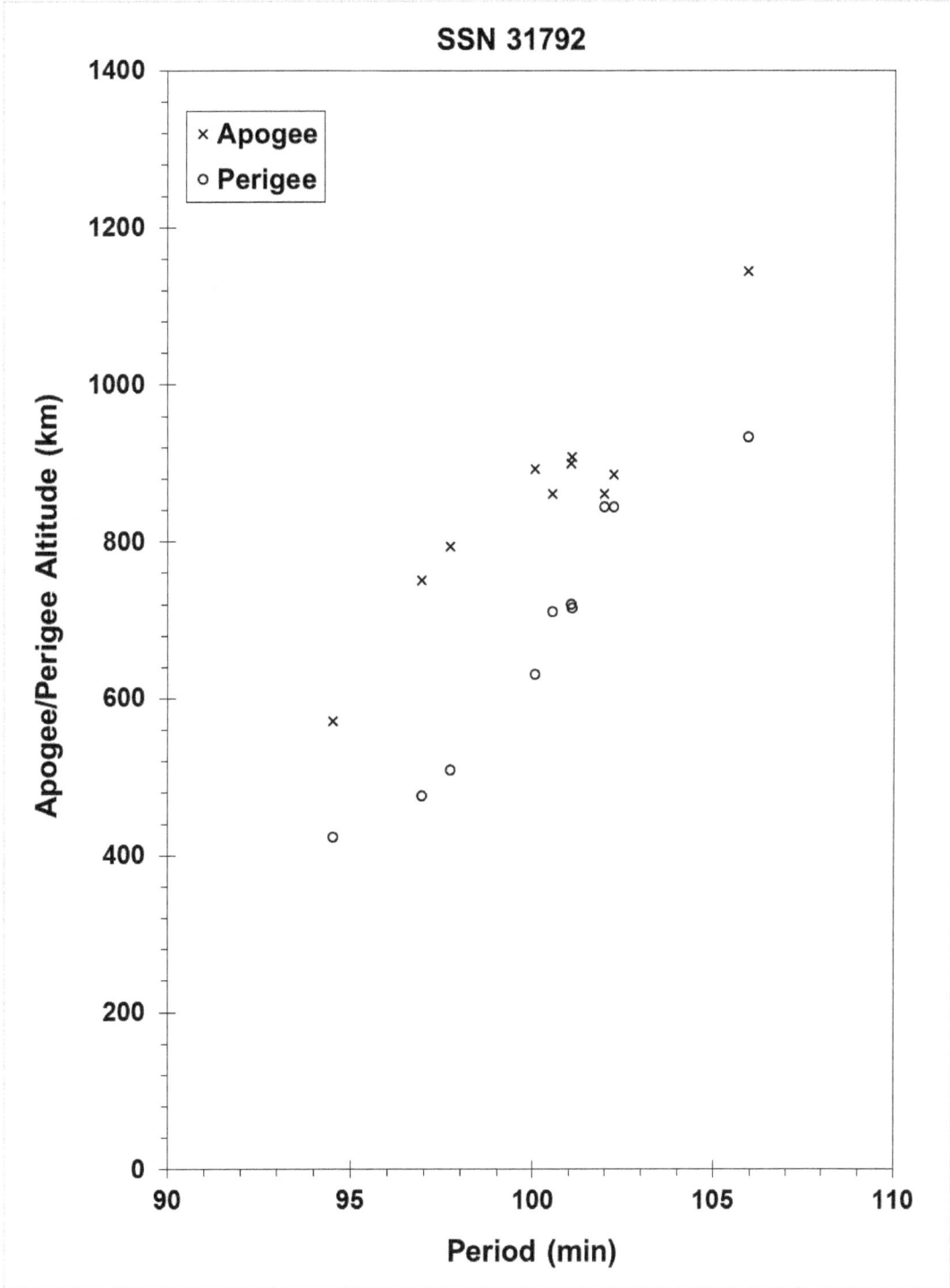

SSN 31792

Cosmos 2428 debris cloud cataloged approximately one week after event.

SATELLITE DATA

TYPE:	Mission Related Debris
OWNER:	CIS
LAUNCH DATE:	26 Oct 2007
DRY MASS (KG):	56
MAIN BODY:	ellipsoid; 0.6 m diameter x 1 m length
MAJOR APPENDAGES:	none
ATTITUDE CONTROL:	none
ENERGY SOURCES:	on-board propellants

EVENT DATA

DATE:	13 Aug 2014	LOCATION:	30.7S, 201.4E (dsc)
TIME:	1340 GMT	ASSESSED CAUSE:	Propulsion
ALTITUDE:	6877.5 km		

PRE-EVENT ELEMENTS

EPOCH:	14223.45723491	MEAN ANOMALY:	338.9903
RIGHT ASCENSION:	24.9798	MEAN MOTION:	4.23372123
INCLINATION:	65.3851	MEAN MOTION DOT/2:	-0.00001815
ECCENTRICITY:	0.5593625	MEAN MOTION DOT DOT/6:	0.0
ARG. OF PERIGEE:	72.9539	BSTAR:	-0.19915

DEBRIS CLOUD DATA

MAXIMUM ΔP:	0 min
MAXIMUM ΔI:	0 deg

COMMENTS

This is the 43rd known breakup of a Proton Blok-DM SOZ ullage unit since 1984. The breakups are believed to be caused by residual propellants. Russian officials have been aware of the problem since 1992 and have made design changes, although the date of full implementation is unknown. Approximately 70 fragments have been observed, but none have been officially cataloged. Due to the difficulty in observing and cataloging debris in deep-space, highly elliptical orbits, this event may have created a much larger debris ensemble.

REFERENCE DOCUMENTS

History of Soviet/Russian Satellite Fragmentations: A Joint U.S.-Russian Investigation. N.L. Johnson et al. Kaman Sciences Corporation, October 2005.

"Three Additional Breakups Mar 2014", The Orbital Debris Quarterly News, NASA JSC, October 2014. Available online at https://orbitaldebris.jsc.nasa.gov/quarterly-news/pdfs/ODQNv18i4.pdf.

Insufficient data to construct a Gabbard diagram.

SATELLITE DATA

TYPE:	Rocket Body
OWNER:	USA
LAUNCH DATE:	11 Nov 2007
DRY MASS (KG):	3,490.0
MAIN BODY:	stepped cylinder: 5 m (max) diameter x 13.7 m length
MAJOR APPENDAGES:	none
ATTITUDE CONTROL:	active; 3 axis RCS
ENERGY SOURCES:	on-board propellants

EVENT DATA

DATE:	11 Nov 2007	LOCATION:	Unknown
TIME:	Unknown	ASSESSED CAUSE:	Unknown
ALTITUDE:	Unknown		

PRE-EVENT ELEMENTS - unavailable

EPOCH:	MEAN ANOMALY:
RIGHT ASCENSION:	MEAN MOTION:
INCLINATION:	MEAN MOTION DOT/2:
ECCENTRICITY:	MEAN MOTION DOT DOT/6:
ARG. OF PERIGEE:	BSTAR:

DEBRIS CLOUD DATA

MAXIMUM ΔP:	unknown min
MAXIMUM ΔI:	unknown deg

COMMENTS

First operational launch of a Delta IV Heavy (Delta 9250H), though second launch of the 5 m-diameter cryogenic second stage. The flight plan called for the second stage to insert itself and the payload into three distinct orbits: a LEO parking orbit; a geosynchronous transfer orbit; and a near geosynchronous final orbit. At least two dozen debris appear to have been released in the parking orbit of approximately 1575 km x 220 km. The debris release event had no apparent effect on subsequent orbit maneuvers, and the payload was successfully delivered to the planned final orbit.

REFERENCE DOCUMENT

"Two Minor Fragmentations End Worst Debris Year Ever," The Orbital Debris Quarterly News, NASA JSC, January 2008. Available online at http://orbitaldebris.jsc.nasa.gov/newsletter/pdfs/ODQNv12i1.pdf.

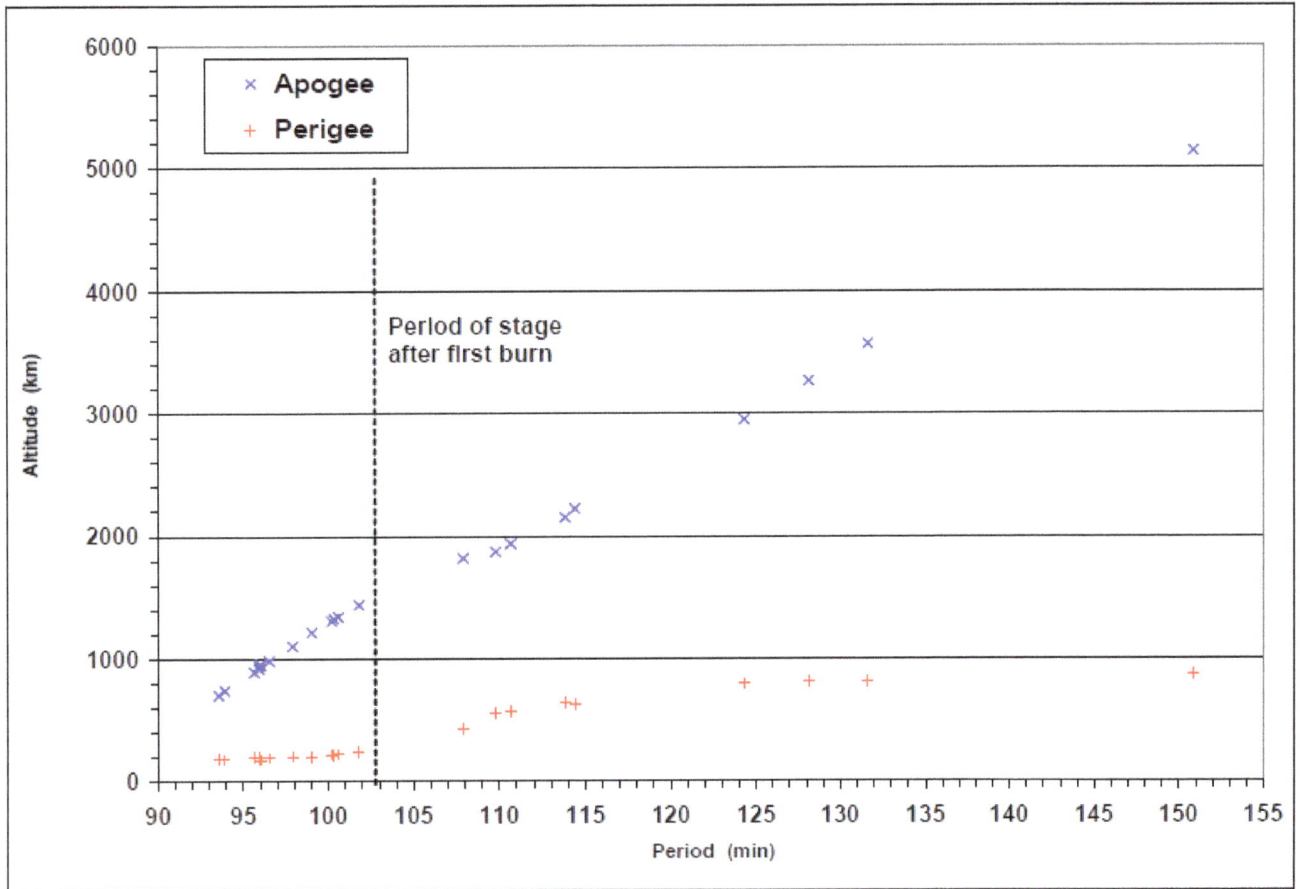

The Delta IV second stage R/B debris cloud of two dozen uncatalogued fragments were identified soon after launch.

SATELLITE DATA

TYPE:	Mission Related Debris
OWNER:	CIS
LAUNCH DATE:	25.81 Dec 2007
DRY MASS (KG):	56.0
MAIN BODY:	ellipsoid; 0.6 m diameter x 1 m length
MAJOR APPENDAGES:	none
ATTITUDE CONTROL:	none
ENERGY SOURCES:	on-board propellants

EVENT DATA

DATE:	15 Apr. 2022	LOCATION:	Unknown
TIME:	0520-2033 GMT	ASSESSED CAUSE:	Propulsion
ALTITUDE:	Unknown		

PRE-EVENT ELEMENTS

EPOCH:	22102.86767241	MEAN ANOMALY:	4.3732
RIGHT ASCENSION:	318.2399	MEAN MOTION:	4.24611595
INCLINATION:	64.8361	MEAN MOTION DOT/2:	0.00003985
ECCENTRICITY:	0.5799705	MEAN MOTION DOT DOT/6:	0.0
ARG. OF PERIGEE:	340.4310	BSTAR:	0.0031307

DEBRIS CLOUD DATA

MAXIMUM ΔP:	0.8 min
MAXIMUM ΔI:	0.4 deg

COMMENTS

This event was the 53rd known fragmentation of Blok-DM ullage motors. The breakups are believed to be caused by residual propellants. Russian officials have been aware of the problem since 1992 and have made design changes, although the date of full implementation is unknown.

This SOZ unit appeared to issue forth a fragment on or about 31 August 2021, which did not enter the Satellite Catalog.

REFERENCE DOCUMENTS

History of Soviet/Russian Satellite Fragmentations: A Joint U.S.-Russian Investigation. N.L. Johnson et al. Kaman Sciences Corporation, October 2005.

"Recent Fragmentation of an Ullage Motor," The Orbital Debris Quarterly News, NASA JSC, June 2022.
Available online at http://orbitaldebris.jsc.nasa.gov/newsletter/pdfs/ODQNv26i2.pdf.

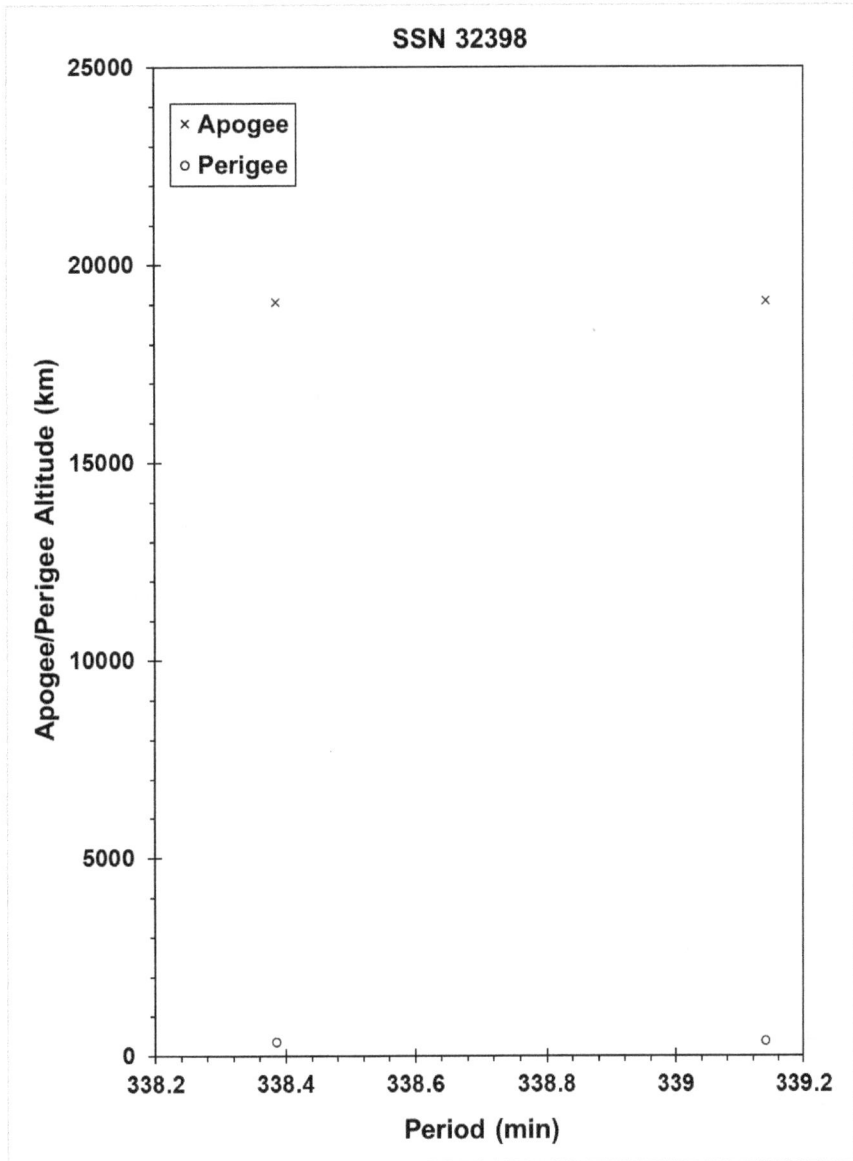

SSN 32398

The COSMOS ullage motor debris cloud includes the parent body and one debris fragment.

SATELLITE DATA

TYPE:	Mission Related Debris
OWNER:	CIS
LAUNCH DATE:	25.81 Dec 2007
DRY MASS (KG):	56.0
MAIN BODY:	ellipsoid; 0.6 m diameter x 1 m length
MAJOR APPENDAGES:	none
ATTITUDE CONTROL:	none
ENERGY SOURCES:	on-board propellants

EVENT DATA

DATE:	18 Aug 2011	LOCATION:	Unknown
TIME:	Unknown	ASSESSED CAUSE:	Propulsion
ALTITUDE:	Unknown		

PRE-EVENT ELEMENTS

EPOCH:	11230.12943848	MEAN ANOMALY:	279.9120
RIGHT ASCENSION:	286.1835	MEAN MOTION:	4.23764448
INCLINATION:	65.0463	MEAN MOTION DOT/2:	0.00000228
ECCENTRICITY:	0.5712685	MEAN MOTION DOT DOT/6:	0.0
ARG. OF PERIGEE:	140.2548	BSTAR:	0.0

DEBRIS CLOUD DATA

MAXIMUM ΔP:	0.0 min
MAXIMUM ΔI:	0.0 deg

COMMENTS

This event was the 38th known fragmentation of Blok-DM ullage motors. The breakups are believed to be caused by residual propellants. Russian officials have been aware of the problem since 1992 and have made design changes, although the date of full implementation is unknown. Although some small debris were observed by the SSN, none have been officially cataloged.

REFERENCE DOCUMENTS

History of Soviet/Russian Satellite Fragmentations: A Joint U.S.-Russian Investigation. N.L. Johnson et al. Kaman Sciences Corporation, October 2005.

"Only a Few Minor Satellite Breakups in 2011", The Orbital Debris Quarterly News, NASA JSC, January 2012. Available online at http://orbitaldebris.jsc.nasa.gov/newsletter/pdfs/ODQNv16i1.pdf.

Insufficient data to construct a Gabbard diagram.

SATELLITE DATA

TYPE:	Rocket Body
OWNER:	CIS
LAUNCH DATE:	14 Mar 2008
DRY MASS (KG):	2510
MAIN BODY:	Cylinder + toroid; 4.1 m diameter x 2.65 m length
MAJOR APPENDAGES:	none
ATTITUDE CONTROL:	none at time of event
ENERGY SOURCES:	on-board propellants

EVENT DATA

DATE:	13 Oct 2010	LOCATION:	33.768N, 120.670E (asc)
TIME:	0553 GMT	ASSESSED CAUSE:	Propulsion
ALTITUDE:	10,799.1 km		

PRE-EVENT ELEMENTS

EPOCH:	10285.89612768	MEAN ANOMALY:	15.0680
RIGHT ASCENSION:	195.2356	MEAN MOTION:	3.07372467
INCLINATION:	48.9364	MEAN MOTION DOT/2:	0.00000252
ECCENTRICITY:	0.6485876	MEAN MOTION DOT DOT/6:	0.0
ARG. OF PERIGEE:	287.1455	BSTAR:	0.0025757

DEBRIS CLOUD DATA

MAXIMUM ΔP:	189.6 min
MAXIMUM ΔI:	2.2 deg

COMMENTS

Event occurred 31 months after launch. The dry mass quoted above includes that of the *Briz-M* core stage and the toroidal Auxiliary Propulsion Tank (APT)—sometimes referred to as the Additional Fuel Tank (AFT); the malfunction stranding the stage in its elliptical orbit occurred prior to APT jettison. An estimated five metric tons of propellant was aboard.

REFERENCE DOCUMENT

"New Satellite Fragmentations Add to Debris Population," The Orbital Debris Quarterly News, NASA JSC, January 2011. Available online at https://orbitaldebris.jsc.nasa.gov/quarterly-news/pdfs/ODQNv15i1.pdf.

SSN 32709

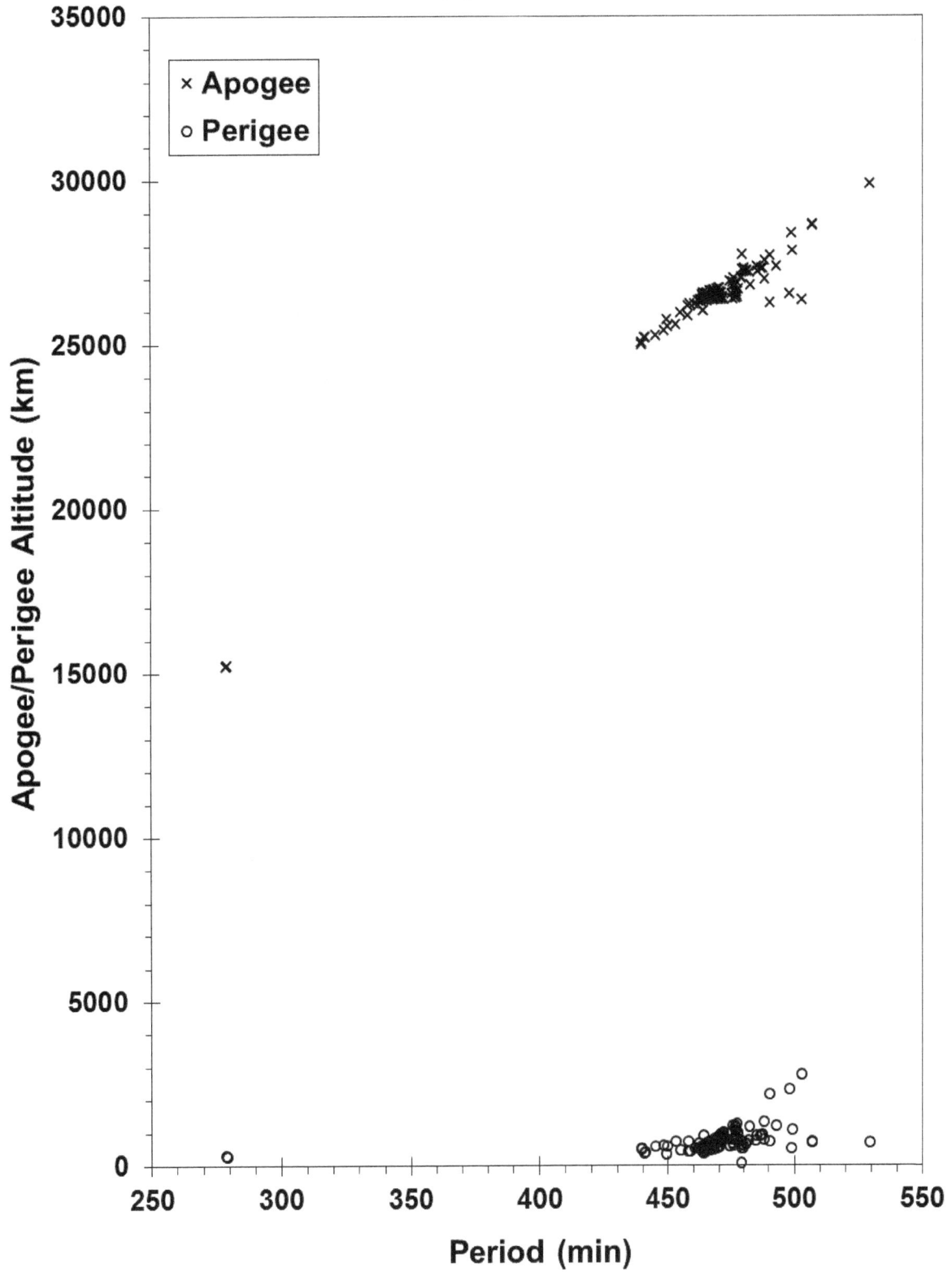

The *Briz-M* R/B debris cloud of 116 fragments cataloged up to 3.6 years after the event.

COSMOS 2442-44 ULLAGE MOTOR   2008-046H   33385

SATELLITE DATA

TYPE:	Mission Related Debris
OWNER:	CIS
LAUNCH DATE:	25 Sep 2008
DRY MASS (KG):	56
MAIN BODY:	ellipsoid; 0.6 m diameter x 1 m length
MAJOR APPENDAGES:	none
ATTITUDE CONTROL:	none
ENERGY SOURCES:	on-board propellants

EVENT DATA

DATE:	20 May 2014	LOCATION:	62.4S, 148.5E (asc)
TIME:	0032 GMT	ASSESSED CAUSE:	Propulsion
ALTITUDE:	16,288.3 km		

PRE-EVENT ELEMENTS

EPOCH:	14135.66838656	MEAN ANOMALY:	322.5973
RIGHT ASCENSION:	154.3825	MEAN MOTION:	4.23142676
INCLINATION:	65.3219	MEAN MOTION DOT/2:	-0.00000189
ECCENTRICITY:	0.5517601	MEAN MOTION DOT DOT/6:	0.00000
ARG. OF PERIGEE:	101.3049	BSTAR:	0.00000

DEBRIS CLOUD DATA

MAXIMUM ΔP:	12.7 min
MAXIMUM ΔI:	0.3 deg

COMMENTS

This is the 40th known breakup of a Proton Blok-DM SOZ ullage unit since 1984. The breakups are believed to be caused by residual propellants. Russian officials have been aware of the problem since 1992 and have made design changes, although the date of full implementation is unknown. Due to the difficulty in observing and cataloging debris in deep-space, highly elliptical orbits, a debris ensemble much larger than that cataloged may have been created by this event.

REFERENCE DOCUMENT

History of Soviet/Russian Satellite Fragmentations: A Joint U.S.-Russian Investigation. N.L. Johnson et al. Kaman Sciences Corporation, October 2005.

"Flurry of Small Breakups in First Half of 2014", The Orbital Debris Quarterly News, NASA JSC, July 2014.
Available online at https://orbitaldebris.jsc.nasa.gov/quarterly-news/pdfs/ODQNv18i3.pdf.

SSN 33385

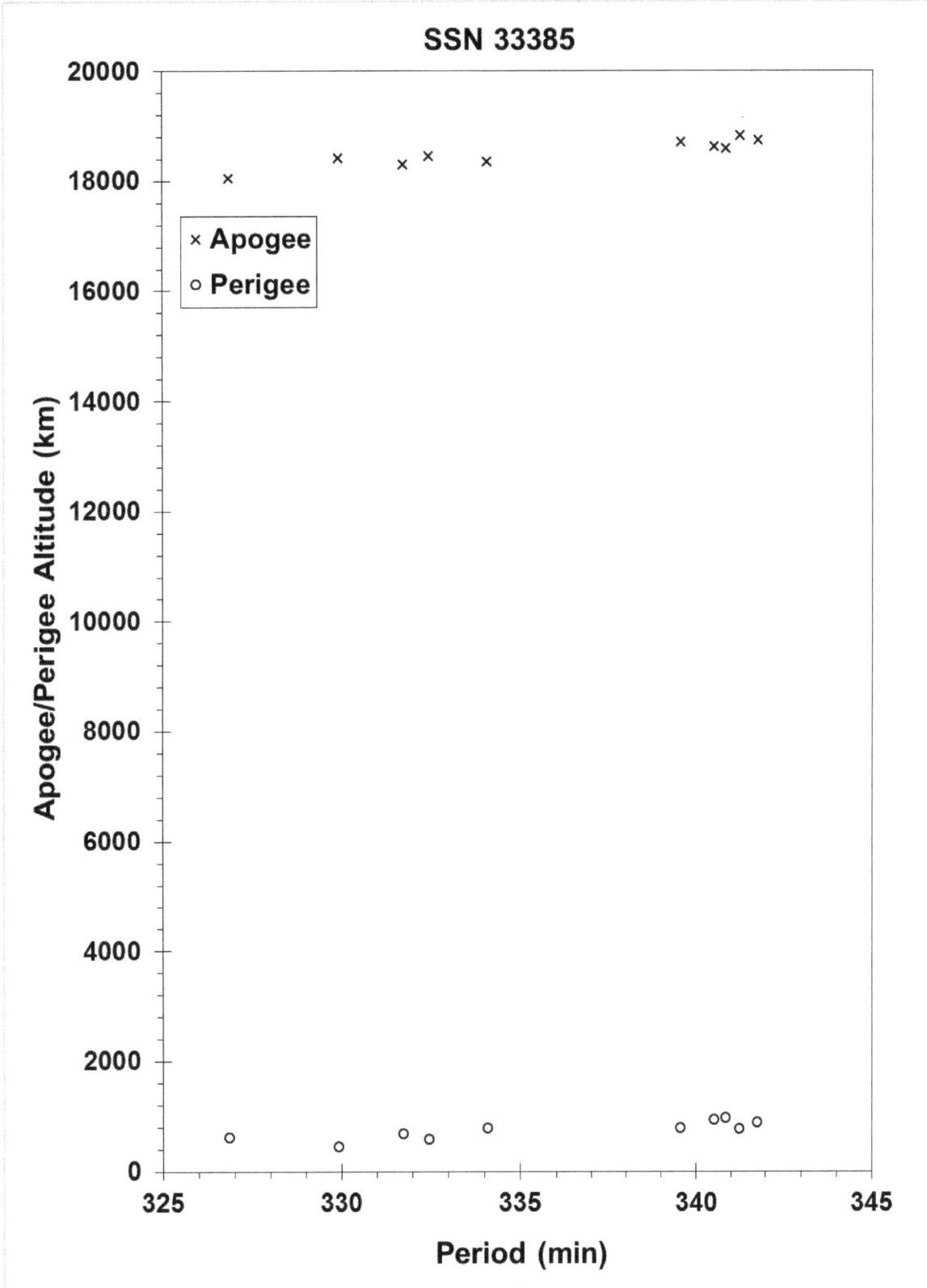

The COSMOS ullage motor debris cloud includes the parent body and 10 debris fragments
(this plot omits 2008-046S, SSN#40121).

SATELLITE DATA

TYPE:	Mission Related Debris
OWNER:	CIS
LAUNCH DATE:	25 Dec 2008
DRY MASS (KG):	56.0
MAIN BODY:	ellipsoid; 0.6 m diameter x 1 m length
MAJOR APPENDAGES:	none
ATTITUDE CONTROL:	none
ENERGY SOURCES:	on-board propellants

EVENT DATA

DATE:	26 Mar 2016	LOCATION:	46.91S, 285.50E (asc)
TIME:	1212 GMT	ASSESSED CAUSE:	Propulsion
ALTITUDE:	18075.34 km		

PRE-EVENT ELEMENTS

EPOCH:	16084.51547544	MEAN ANOMALY:	341.8531
RIGHT ASCENSION:	57.2039	MEAN MOTION:	4.23930172
INCLINATION:	65.3219	MEAN MOTION DOT/2:	+0.00000377
ECCENTRICITY:	0.5630529	MEAN MOTION DOT DOT/6:	0.00000
ARG. OF PERIGEE:	66.1764	BSTAR:	+0.0029410

DEBRIS CLOUD DATA

MAXIMUM ΔP:	11.4 min
MAXIMUM ΔI:	0.2 deg

COMMENTS

This event was the 44th known fragmentation of Blok-DM ullage motors. The breakups are believed to be caused by residual propellants. Russian officials have been aware of the problem since 1992 and have made design changes, although the date of full implementation is unknown. Twenty-one pieces were initially observed. Due to the difficulty of cataloging debris in elliptical and deep space orbits there may be many more debris fragments than this number or those cataloged.

REFERENCE DOCUMENTS

History of Soviet/Russian Satellite Fragmentations: A Joint U.S.-Russian Investigation. N.L. Johnson *et al*. Kaman Sciences Corporation, October 2005.

"Russian SOZ Unit Breakup Up in March," The Orbital Debris Quarterly News, NASA JSC, April 2016.
Available online at https://www.orbitaldebris.jsc.nasa.gov/quarterly-news/pdfs/odqnv20i1-2.pdf.

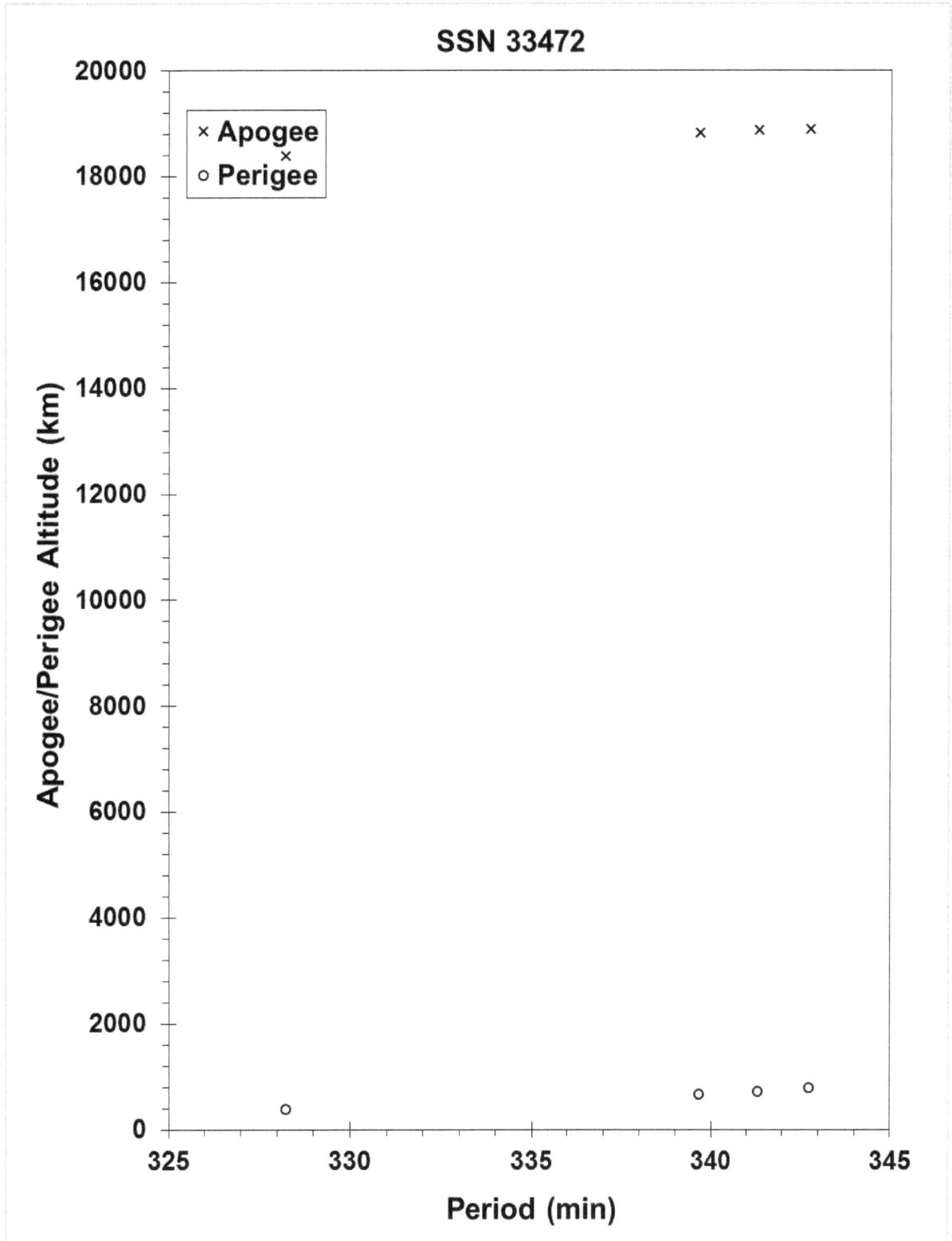

SSN 33472

The COSMOS ullage motor debris cloud includes the parent body and three debris fragments.

SATELLITE DATA

TYPE:	Mission Related Debris
OWNER:	CIS
LAUNCH DATE:	25 Dec 2008
DRY MASS (KG):	56.0
MAIN BODY:	ellipsoid; 0.6 m diameter x 1 m length
MAJOR APPENDAGES:	none
ATTITUDE CONTROL:	none
ENERGY SOURCES:	on-board propellants

EVENT DATA

DATE:	1 June 2016	LOCATION:	32.86S, 339.90E (dsc)
TIME:	0925 GMT	ASSESSED CAUSE:	Propulsion
ALTITUDE:	7148.03 km		

PRE-EVENT ELEMENTS

EPOCH:	16152.95219263	MEAN ANOMALY:	5.7190
RIGHT ASCENSION:	28.8315	MEAN MOTION:	4.23951640
INCLINATION:	65.2630	MEAN MOTION DOT/2:	+0.00005748
ECCENTRICITY:	0.5609396	MEAN MOTION DOT DOT/6:	0.00000
ARG. OF PERIGEE:	72.5352	BSTAR:	+0.053896

DEBRIS CLOUD DATA

MAXIMUM ΔP:	8.9 min
MAXIMUM ΔI:	0.4 deg

COMMENTS

This event was the 45th known fragmentation of Blok-DM ullage motors. The breakups are believed to be caused by residual propellants. Russian officials have been aware of the problem since 1992 and have made design changes, although the date of full implementation is unknown. Due to the difficulty of cataloging debris in elliptical and deep space orbits there may be many more debris fragments than those cataloged.

REFERENCE DOCUMENTS

History of Soviet/Russian Satellite Fragmentations: A Joint U.S.-Russian Investigation. N.L. Johnson *et al*. Kaman Sciences Corporation, October 2005.

"Two Additional Russian Breakups in 2016", The Orbital Debris Quarterly News, NASA JSC, July 2016. Available online at https://www.orbitaldebris.jsc.nasa.gov/quarterly-news/pdfs/odqnv20i3.pdf.

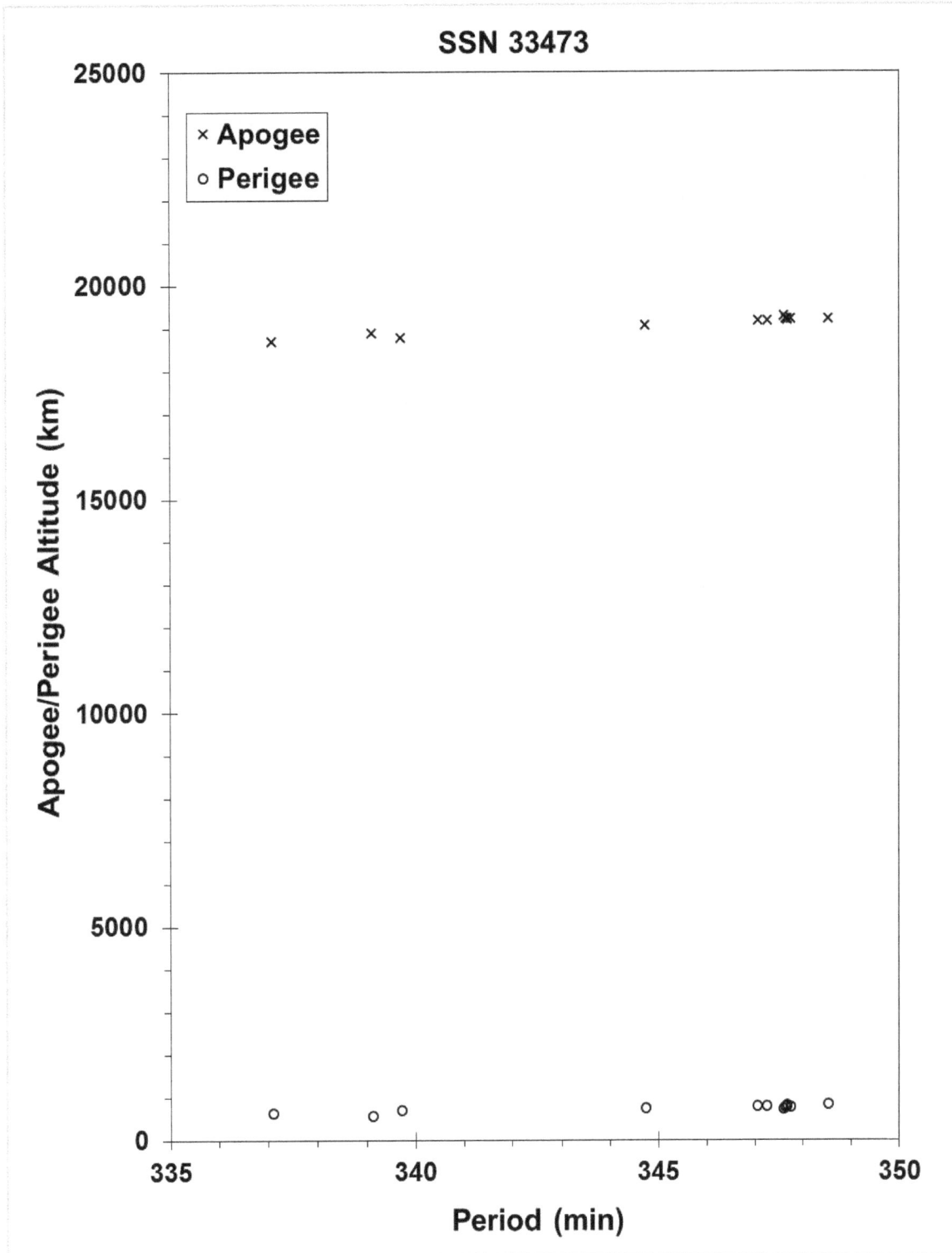

SSN 33473

The COSMOS ullage motor debris cloud includes the parent body and 11 debris fragments cataloged up to 37 days post-event.

SATELLITE DATA

TYPE:	payload
OWNER:	PRC
LAUNCH DATE:	14 Apr. 2009
DRY MASS (KG):	1100
MAIN BODY:	DFH-3 bus derivative; box 2.2 m long x 1.72 m wide x 2.0 m high
MAJOR APPENDAGES:	solar arrays and antennas
ATTITUDE CONTROL:	three-axis stabilized
ENERGY SOURCES:	propellant, batteries, reaction wheels

EVENT DATA

DATE:	29 June 2016	LOCATION:	unknown
TIME:	unknown	ASSESSED CAUSE:	unknown
ALTITUDE:	unknown		

PRE-EVENT ELEMENTS

EPOCH:	16180.12570702	MEAN ANOMALY:	164.3864
RIGHT ASCENSION:	61.3828	MEAN MOTION:	1.00365151
INCLINATION:	4.7106	MEAN MOTION DOT/2:	-0.00000325
ECCENTRICITY:	0.0089308	MEAN MOTION DOT DOT/6:	0.00000
ARG. OF PERIGEE:	195.1803	BSTAR:	0.00000

DEBRIS CLOUD DATA

MAXIMUM ΔP:	unknown
MAXIMUM ΔI:	unknown

COMMENTS

This spacecraft was the first launch of the People's Republic of China (PRC) BeiDou 2nd generation regional navigation satellites in the Compass Navigation Satellite System, and is sometimes labeled as "Beidou-2 G2" to indicate 2nd generation, 2nd Geosynchronous (GEO) spacecraft (the –G1 was launched in 2010). This designation ("G") identifies it as being apart from the PRC's middle Earth orbit (MEO, "M") or inclined GEO ("IG") spacecraft constellations. The spacecraft uses the Chinese Academy of Spacecraft Technology (CAST) Dong Fang Hong 3 (DFH-3) communication satellite-heritage bus with the specialized navigational payload.

The spacecraft fragmented into at least five pieces, though no fragments have entered the catalog to accompany the parent body. Due to the difficulty in cataloging fragments in deep-space orbits there may be many more fragments resident at or near GEO.

REFERENCE DOCUMENTS

"BeiDou G2 Spacecraft Fragments in Geosynchronous Orbit", The Orbital Debris Quarterly News, NASA JSC, October 2016. Available online at https://www.orbitaldebris.jsc.nasa.gov/quarterly-news/pdfs/odqnv20i4.pdf.

Insufficient data to construct a Gabbard diagram.

SATELLITE DATA

TYPE:	Mission Related Debris
OWNER:	CIS
LAUNCH DATE:	14 Dec. 2009
DRY MASS (KG):	56.0
MAIN BODY:	ellipsoid; 0.6 m diameter x 1 m length
MAJOR APPENDAGES:	none
ATTITUDE CONTROL:	none
ENERGY SOURCES:	on-board propellants

EVENT DATA

DATE:	21-23 Dec. 2019	LOCATION:	Unknown
TIME:	Unknown	ASSESSED CAUSE:	Propulsion
ALTITUDE:	Unknown		

PRE-EVENT ELEMENTS

EPOCH:	19354.40630414	MEAN ANOMALY:	353.6318
RIGHT ASCENSION:	40.1421	MEAN MOTION:	4.24206914
INCLINATION:	64.6921	MEAN MOTION DOT/2:	-.00000077
ECCENTRICITY:	0.5733441	MEAN MOTION DOT DOT/6:	0.00000
ARG. OF PERIGEE:	28.1846	BSTAR:	0.00013771

DEBRIS CLOUD DATA

MAXIMUM ΔP:	19.8 min
MAXIMUM ΔI:	0.4 deg

COMMENTS

This event was the 52nd known fragmentation of Blok-DM ullage motors. The breakups are believed to be caused by residual propellants. Russian officials have been aware of the problem since 1992 and have made design changes, although the date of full implementation is unknown. Although at least 25 small debris were observed by the SSN, though only seven have been officially cataloged.

REFERENCE DOCUMENTS

History of Soviet/Russian Satellite Fragmentations: A Joint U.S.-Russian Investigation. N.L. Johnson *et al.* Kaman Sciences Corporation, October 2005.

"Three Recent Breakup Events," The Orbital Debris Quarterly News, NASA JSC, April 2020.
Available online at https://www.orbitaldebris.jsc.nasa.gov/quarterly-news/pdfs/odqnv24i2.pdf.

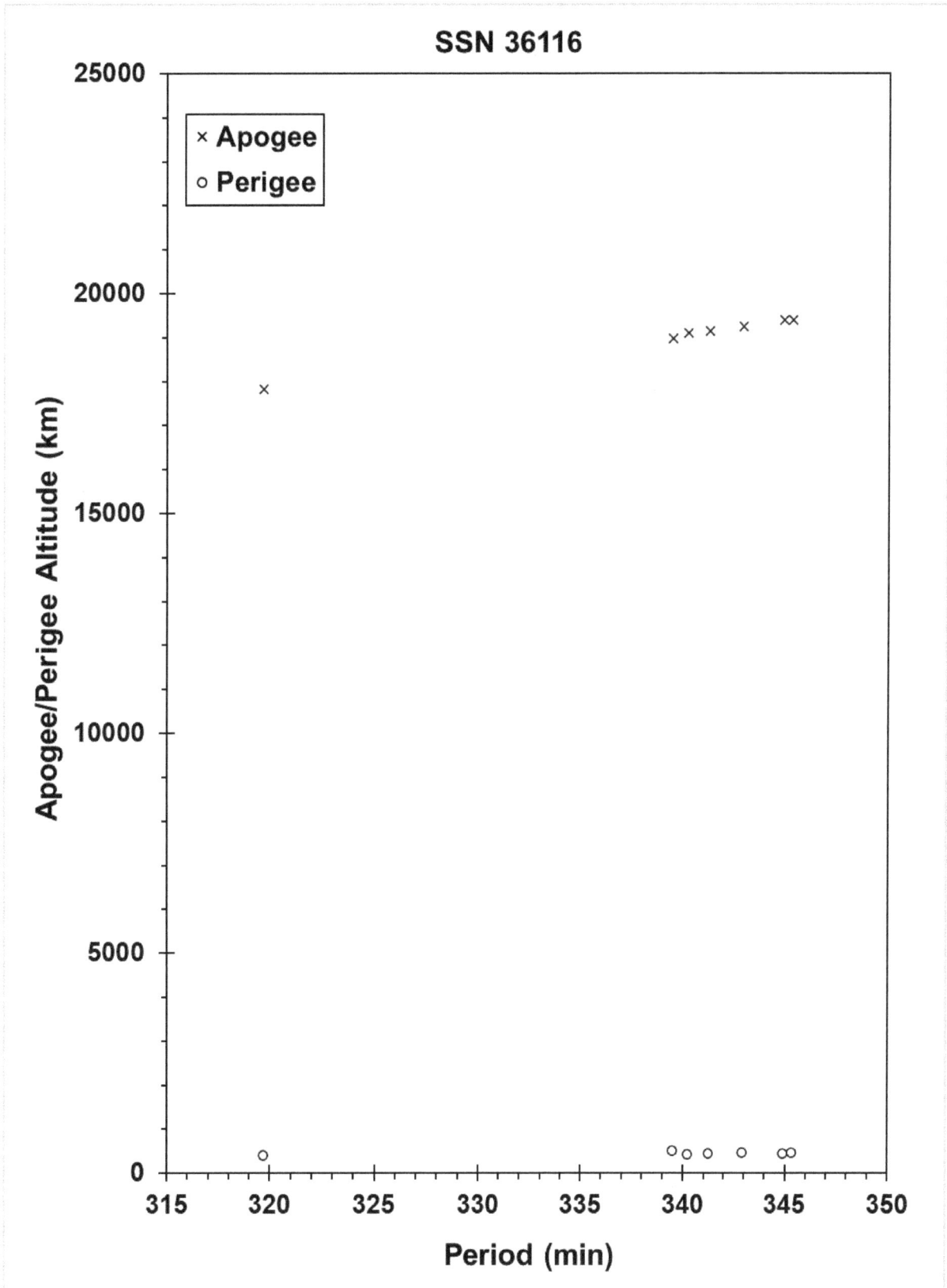

SSN 36116

The COSMOS ullage motor debris cloud includes the parent body and 6 debris fragments cataloged up to 59 days post-event.

SATELLITE DATA

TYPE:	Mission Related Debris
OWNER:	CIS
LAUNCH DATE:	01 Mar 2010
DRY MASS (KG):	56.0
MAIN BODY:	ellipsoid; 0.6 m diameter x 1 m length
MAJOR APPENDAGES:	none
ATTITUDE CONTROL:	none
ENERGY SOURCES:	on-board propellants

EVENT DATA

DATE:	9 July 2014	LOCATION:	33.96N, 41.3E (asc)
TIME:	0828 GMT	ASSESSED CAUSE:	Propulsion
ALTITUDE:	755.2 km		

PRE-EVENT ELEMENTS

EPOCH:	14188.41970654	MEAN ANOMALY:	296.6908
RIGHT ASCENSION:	294.1777	MEAN MOTION:	4.23407390
INCLINATION:	65.1872	MEAN MOTION DOT/2:	-0.00000214
ECCENTRICITY:	0.5623059	MEAN MOTION DOT DOT/6:	0.00000
ARG. OF PERIGEE:	128.6500	BSTAR:	-0.013117

DEBRIS CLOUD DATA

MAXIMUM ΔP:	16.4 min
MAXIMUM ΔI:	0.6 deg

COMMENTS

This event was the 42nd known fragmentation of Blok-DM ullage motors. The breakups are believed to be caused by residual propellants. Russian officials have been aware of the problem since 1992 and have made design changes, although the date of full implementation is unknown. Although at least 16 small debris were observed by the SSN, none have been officially cataloged.

REFERENCE DOCUMENTS

History of Soviet/Russian Satellite Fragmentations: A Joint U.S.-Russian Investigation. N.L. Johnson *et al.* Kaman Sciences Corporation, October 2005.

"Three Additional Breakups Mar 2014", The Orbital Debris Quarterly News, NASA JSC, October 2014. Available online at https://www.orbitaldebris.jsc.nasa.gov/quarterly-news/pdfs/odqnv18i4.pdf.

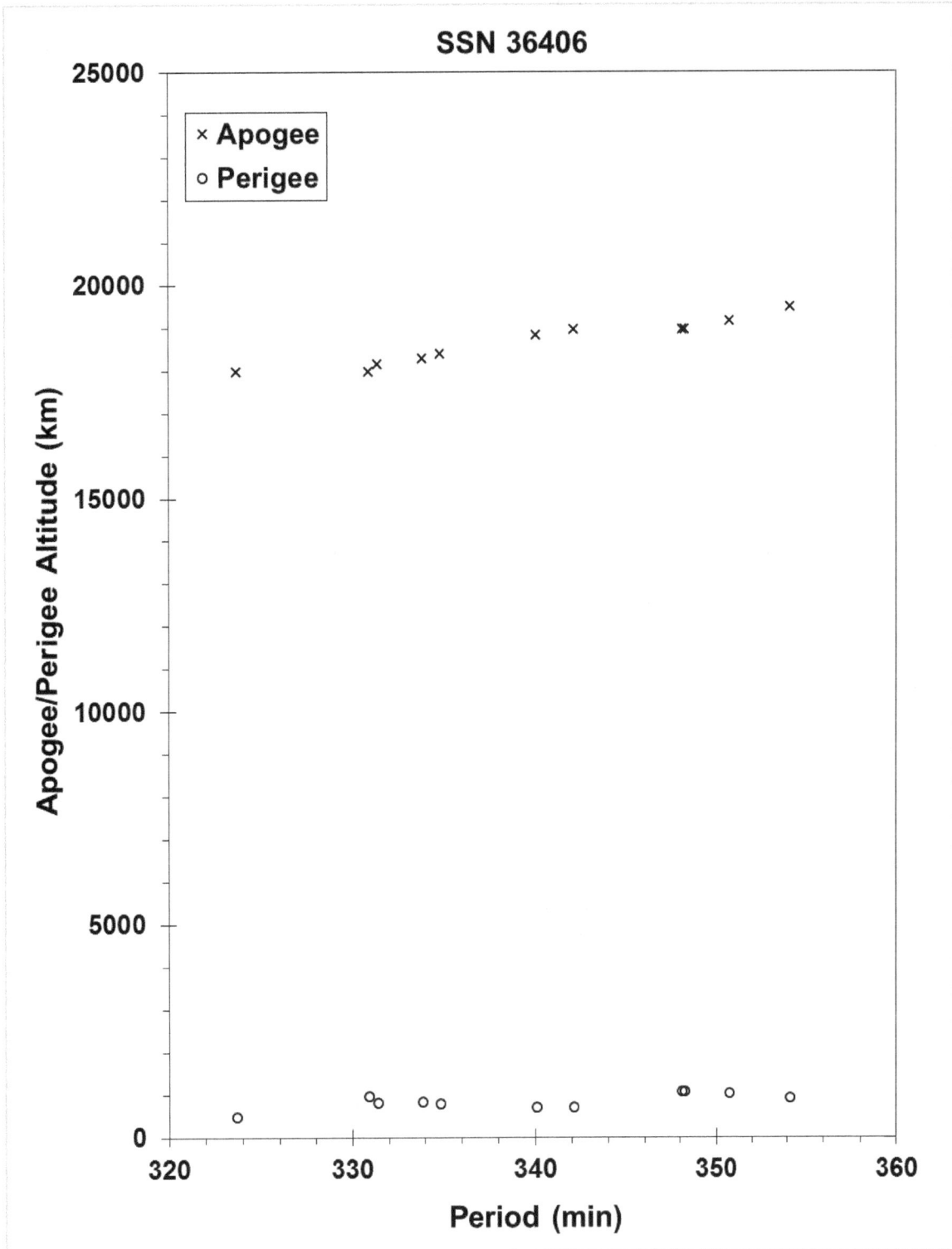

SSN 36406

The COSMOS ullage motor debris cloud includes the parent body and 10 debris fragments cataloged up to two years post-event.

SATELLITE DATA

TYPE:	Mission Related Debris
OWNER:	CIS
LAUNCH DATE:	01 Mar 2010
DRY MASS (KG):	56.0
MAIN BODY:	ellipsoid; 0.6 m diameter x 1 m length
MAJOR APPENDAGES:	none
ATTITUDE CONTROL:	none
ENERGY SOURCES:	on-board propellants

EVENT DATA

DATE:	22 May 2018	LOCATION:	8.59S, 90.0E (asc)
TIME:	0206 GMT	ASSESSED CAUSE:	Propulsion
ALTITUDE:	13745.8 km		

PRE-EVENT ELEMENTS

EPOCH:	18141.55091406	MEAN ANOMALY:	348.7596
RIGHT ASCENSION:	177.3536	MEAN MOTION:	4.23280497
INCLINATION:	65.0720	MEAN MOTION DOT/2:	-0.00001771
ECCENTRICITY:	0.5676424	MEAN MOTION DOT DOT/6:	0.00000
ARG. OF PERIGEE:	45.9342	BSTAR:	-0.0089882

DEBRIS CLOUD DATA

MAXIMUM ΔP:	23.7 min
MAXIMUM ΔI:	1 deg

COMMENTS

This event was the 48th known fragmentation of Blok-DM ullage motors. The breakups are believed to be caused by residual propellants. Russian officials have been aware of the problem since 1992 and have made design changes, although the date of full implementation is unknown. Although at least 60 small debris were observed by the SSN, many more fragments could be present due to difficulties in cataloging fragments in elliptical and deep-space orbits.

REFERENCE DOCUMENTS

History of Soviet/Russian Satellite Fragmentations: A Joint U.S.-Russian Investigation. N.L. Johnson *et al*. Kaman Sciences Corporation, October 2005.

"A SOZ Unit Breakup Predicted and Observed in May 2018", The Orbital Debris Quarterly News, NASA JSC, September 2018. Available online at https://www.orbitaldebris.jsc.nasa.gov/quarterly-news/pdfs/odqnv22i3.pdf.

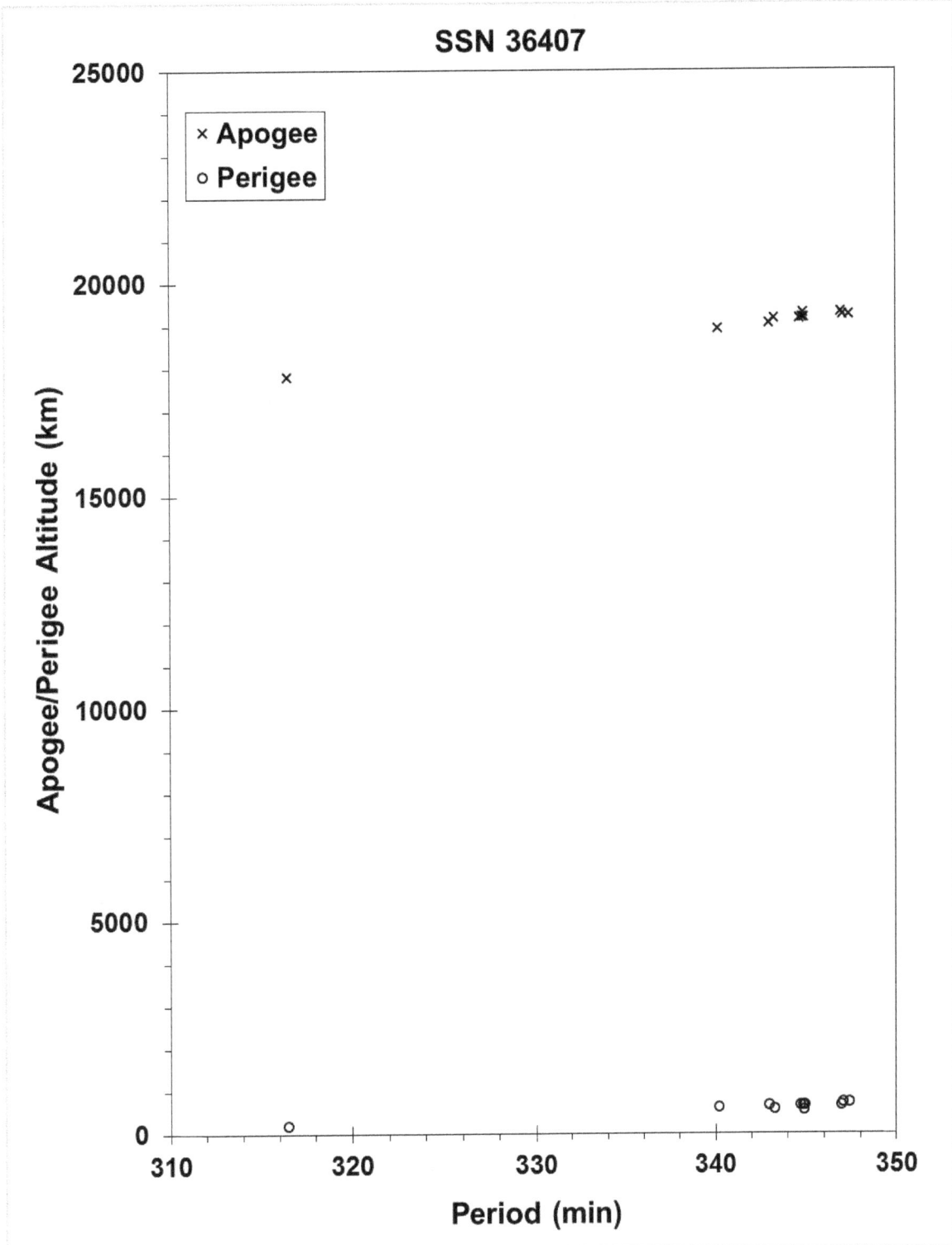

SSN 36407

The COSMOS ullage motor debris cloud includes the parent body and 10 debris fragments cataloged up to 64 days post-event.

COSMOS 2464-2466 ULLAGE MOTOR   **2010-041G**   **37143**

SATELLITE DATA

TYPE:	Mission Related Debris
OWNER:	CIS
LAUNCH DATE:	2 Sep 2010
DRY MASS (KG):	55
MAIN BODY:	ellipsoid; 0.6 m diameter x 1 m length
MAJOR APPENDAGES:	none
ATTITUDE CONTROL:	none
ENERGY SOURCES:	on-board propellants

EVENT DATA

DATE:	03 Sept. 2017	LOCATION:	54.96S, 115.3E (dsc)
TIME:	0237 GMT	ASSESSED CAUSE:	Propulsion
ALTITUDE:	14887.2 km		

PRE-EVENT ELEMENTS

EPOCH:	17245.45043553	MEAN ANOMALY:	333.2513
RIGHT ASCENSION:	178.2175	MEAN MOTION:	4.25073663
INCLINATION:	65.2305	MEAN MOTION DOT/2:	+0.00009868
ECCENTRICITY:	0.5568611	MEAN MOTION DOT DOT/6:	0.00000
ARG. OF PERIGEE:	84.7390	BSTAR:	+0.12084

DEBRIS CLOUD DATA

MAXIMUM ΔP:	19.2 min
MAXIMUM ΔI:	0.6 deg

COMMENTS

This event was the 47th known fragmentation of Blok-DM ullage motors. The breakups are believed to be caused by residual propellants. Russian officials have been aware of the problem since 1992 and have made design changes, although the date of full implementation is unknown. Multiple small debris were observed by the SSN, but due to difficulties in cataloging fragments in elliptical and deep-space orbits many more fragments could be present.

REFERENCE DOCUMENTS

History of Soviet/Russian Satellite Fragmentations: A Joint U.S.-Russian Investigation. N.L. Johnson *et al*. Kaman Sciences Corporation, October 2005.

"Latest SOZ Breakup Occurs in September 2017", The Orbital Debris Quarterly News, NASA JSC, November 2017. Available online at https://www.orbitaldebris.jsc.nasa.gov/quarterly-news/pdfs/odqnv21i4.pdf.

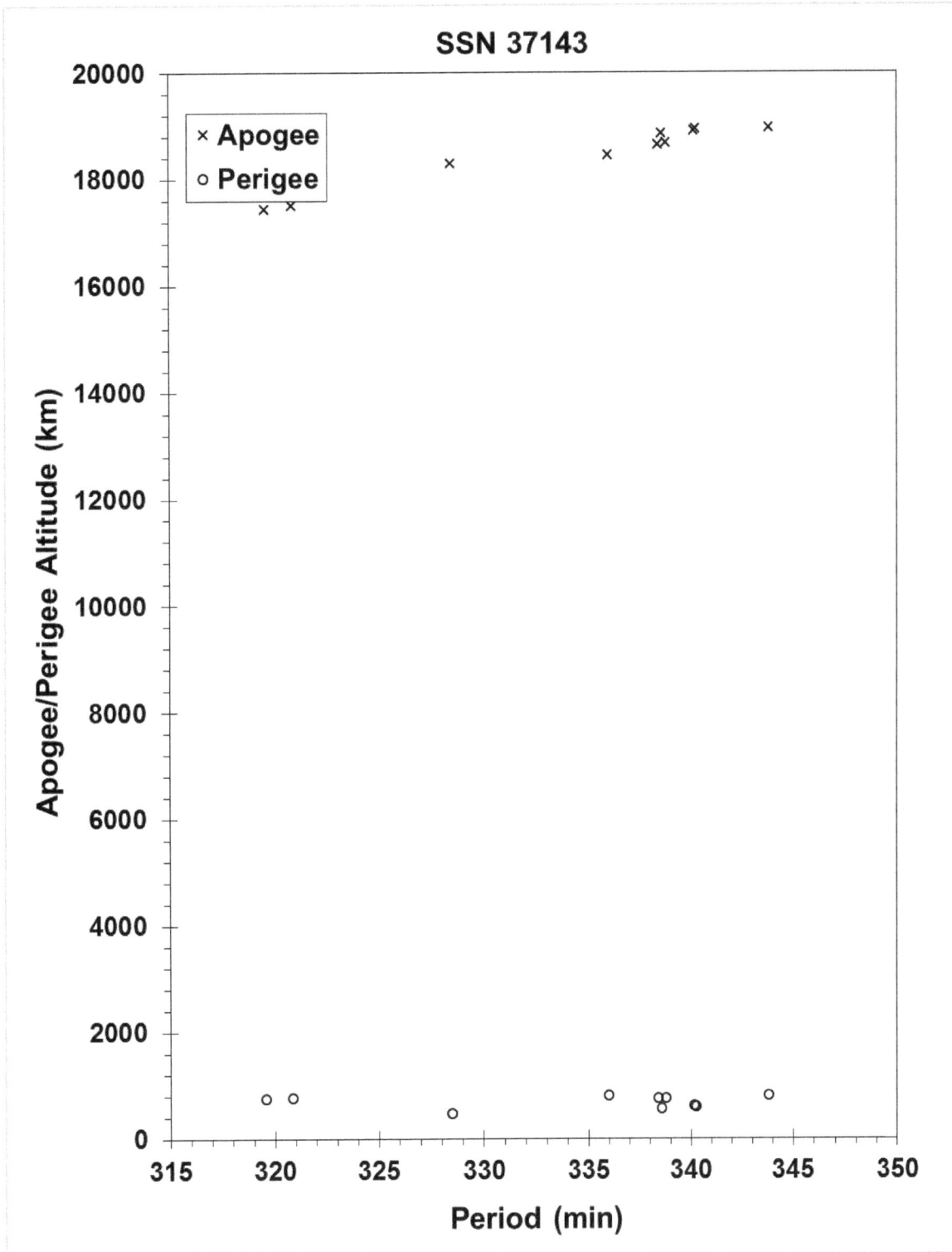

The COSMOS ullage motor debris cloud includes the parent body and nine debris fragments cataloged up to 45 days post-event.

SATELLITE DATA

TYPE:	Mission Related Debris
OWNER:	CIS
LAUNCH DATE:	2 Sep 2010
DRY MASS (KG):	55
MAIN BODY:	ellipsoid; 0.6 m diameter x 1 m length
MAJOR APPENDAGES:	none
ATTITUDE CONTROL:	none
ENERGY SOURCES:	on-board propellants

EVENT DATA

DATE:	12 August 2019	LOCATION:	24.46S, 316.99E (asc)
TIME:	1358 GMT	ASSESSED CAUSE:	Propulsion
ALTITUDE:	18220.1 km		

PRE-EVENT ELEMENTS

EPOCH:	19224.48092398	MEAN ANOMALY:	350.7247
RIGHT ASCENSION:	295.0100	MEAN MOTION:	4.24858480
INCLINATION:	64.9546	MEAN MOTION DOT/2:	-.00000030
ECCENTRICITY:	0. 5694965	MEAN MOTION DOT DOT/6:	0.00000
ARG. OF PERIGEE:	38.9302	BSTAR:	-0.0000055210

DEBRIS CLOUD DATA

MAXIMUM ΔP:	19.7 min
MAXIMUM ΔI:	0.9 deg

COMMENTS

This event was the 50th known fragmentation of Blok-DM ullage motors. The breakups are believed to be caused by residual propellants. Russian officials have been aware of the problem since 1992 and have made design changes, although the date of full implementation is unknown. Multiple small debris were observed by the SSN, but due to difficulties in cataloging fragments in elliptical and deep-space orbits many more fragments could be present.

REFERENCE DOCUMENTS

History of Soviet/Russian Satellite Fragmentations: A Joint U.S.-Russian Investigation. N.L. Johnson *et al*. Kaman Sciences Corporation, October 2005.

"Three Recent Rocket Body Breakups," The Orbital Debris Quarterly News, NASA JSC, November 2019. Available online at https://www.orbitaldebris.jsc.nasa.gov/quarterly-news/pdfs/odqnv23i4.pdf.

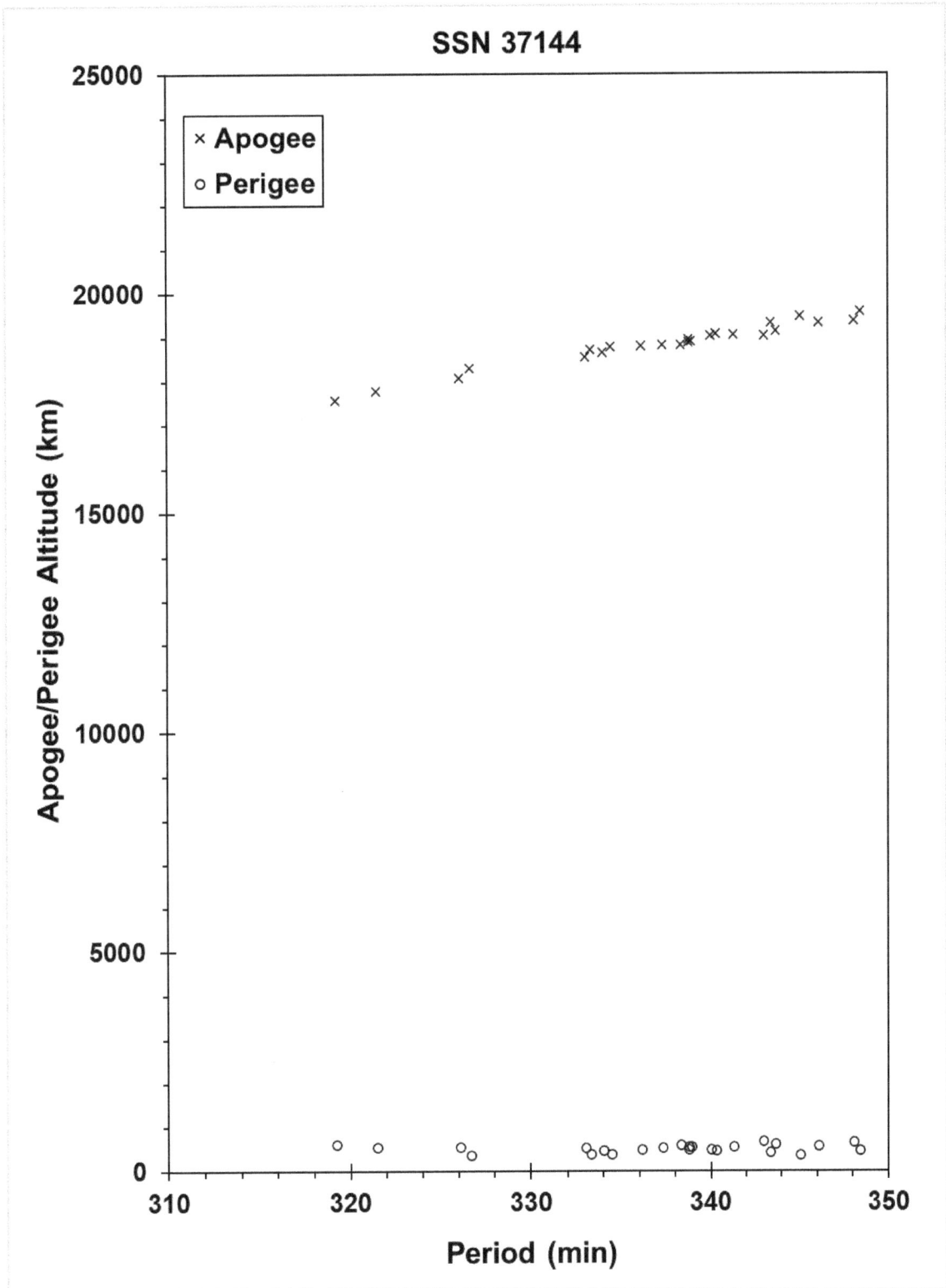

SSN 37144

The COSMOS ullage motor debris cloud includes the parent body and 23 debris fragments cataloged up to 240 days post-event.

SATELLITE DATA

TYPE:	Rocket Body
OWNER:	PRC
LAUNCH DATE:	4 Sep 2010
DRY MASS (KG):	3062
MAIN BODY:	Cylinder; 3.0 m diameter x 12.4 m long
MAJOR APPENDAGES:	none
ATTITUDE CONTROL:	status unknown at time of event
ENERGY SOURCES:	on-board propellants

EVENT DATA

DATE:	~4 Sep 2010	LOCATION:	Unknown
TIME:	Unknown	ASSESSED CAUSE:	Unknown
ALTITUDE:	Unknown		

PRE-EVENT ELEMENTS

EPOCH:	10247.43584147	MEAN ANOMALY:	183.2812
RIGHT ASCENSION:	224.5993	MEAN MOTION:	1.91160163
INCLINATION:	25.2307	MEAN MOTION DOT/2:	0.00001931
ECCENTRICITY:	0.7603361	MEAN MOTION DOT DOT/6:	-0.11660e-5
ARG. OF PERIGEE:	179.3215	BSTAR:	0.0001

DEBRIS CLOUD DATA

MAXIMUM ΔP:	74.8 min
MAXIMUM ΔI:	3.2 deg

COMMENTS

Detected by software.

This was the second fragmentation of a CZ-3 third stage. Like the prior event, this event occurred within hours of launch and separation of the payload. Cataloging was delayed by high eccentricity of orbit.

SSN 37151

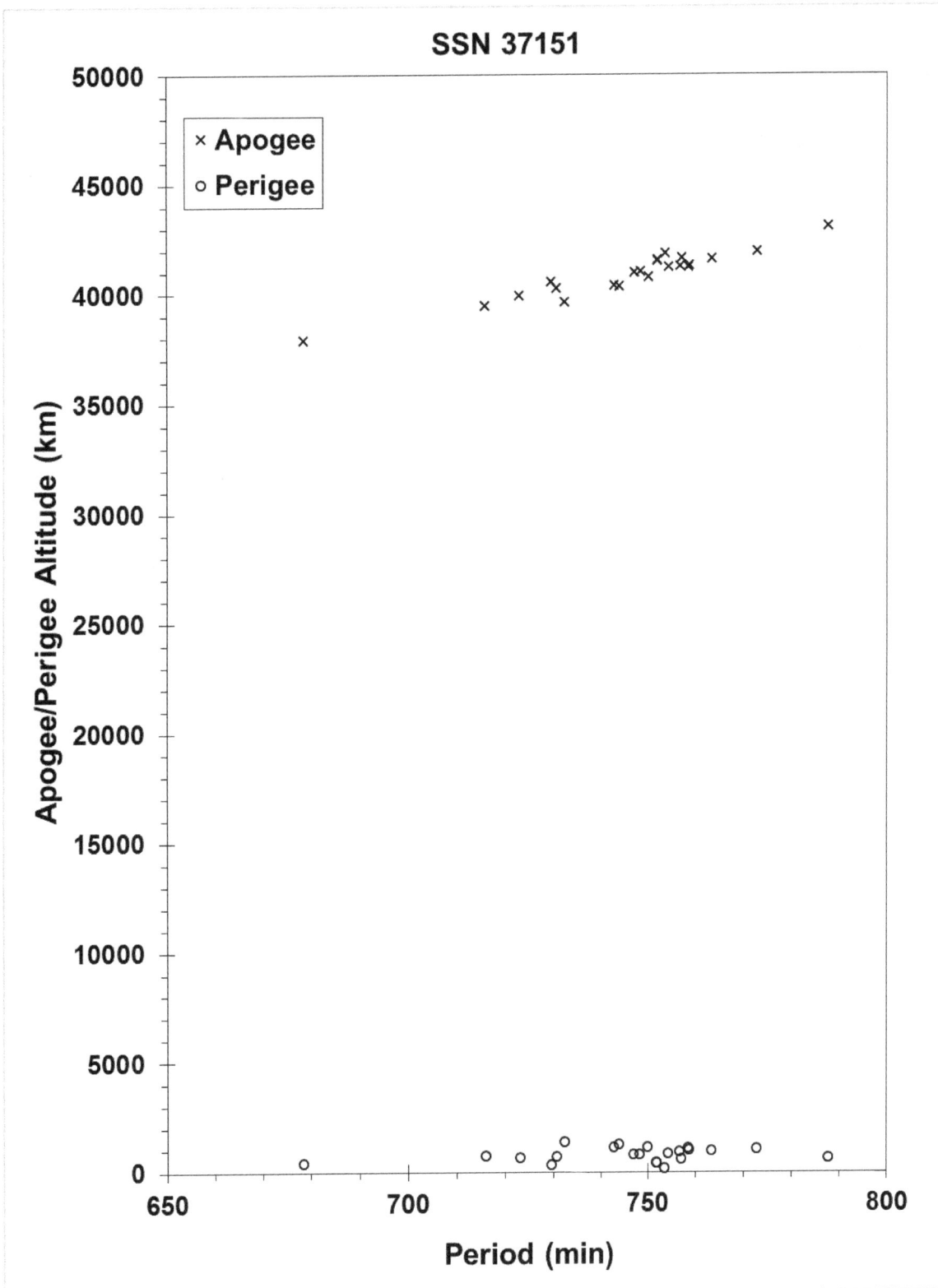

The Chinasat 6A CZ-3B R/B debris cloud of 22 fragments cataloged up to 5 years after the event.

SATELLITE DATA

TYPE:	Rocket Body
OWNER:	PRC
LAUNCH DATE:	1 Nov 2010
DRY MASS (KG):	3062
MAIN BODY:	Cylinder; 3.0 m diameter x 12.4 m long
MAJOR APPENDAGES:	none
ATTITUDE CONTROL:	status unknown at time of event
ENERGY SOURCES:	on-board propellants

EVENT DATA

DATE:	1 Nov 2010	LOCATION:	4.407S, 328.867E (asc)
TIME:	1731 GMT	ASSESSED CAUSE:	Unknown
ALTITUDE:	33,072.2 km		

PRE-EVENT ELEMENTS

EPOCH:	10305.30000000	MEAN ANOMALY:	132.3340
RIGHT ASCENSION:	284.6090	MEAN MOTION:	2.28504000
INCLINATION:	20.4700	MEAN MOTION DOT/2:	0.00000669
ECCENTRICITY:	0.7313927	MEAN MOTION DOT DOT/6:	-0.74391e-6
ARG. OF PERIGEE:	179.8990	BSTAR:	0.0

DEBRIS CLOUD DATA

MAXIMUM ΔP:	446.6 min
MAXIMUM ΔI:	3.1 deg

COMMENTS

This was the third known fragmentation of a CZ-3 third stage. Like the prior event, this event occurred within hours of launch and separation of the payload.

REFERENCE DOCUMENT

"New Satellite Fragmentations Add to Debris Population," The Orbital Debris Quarterly News, NASA JSC, January 2011. Available online at http://orbitaldebris.jsc.nasa.gov/newsletter/pdfs/ODQNv15i1.pdf.

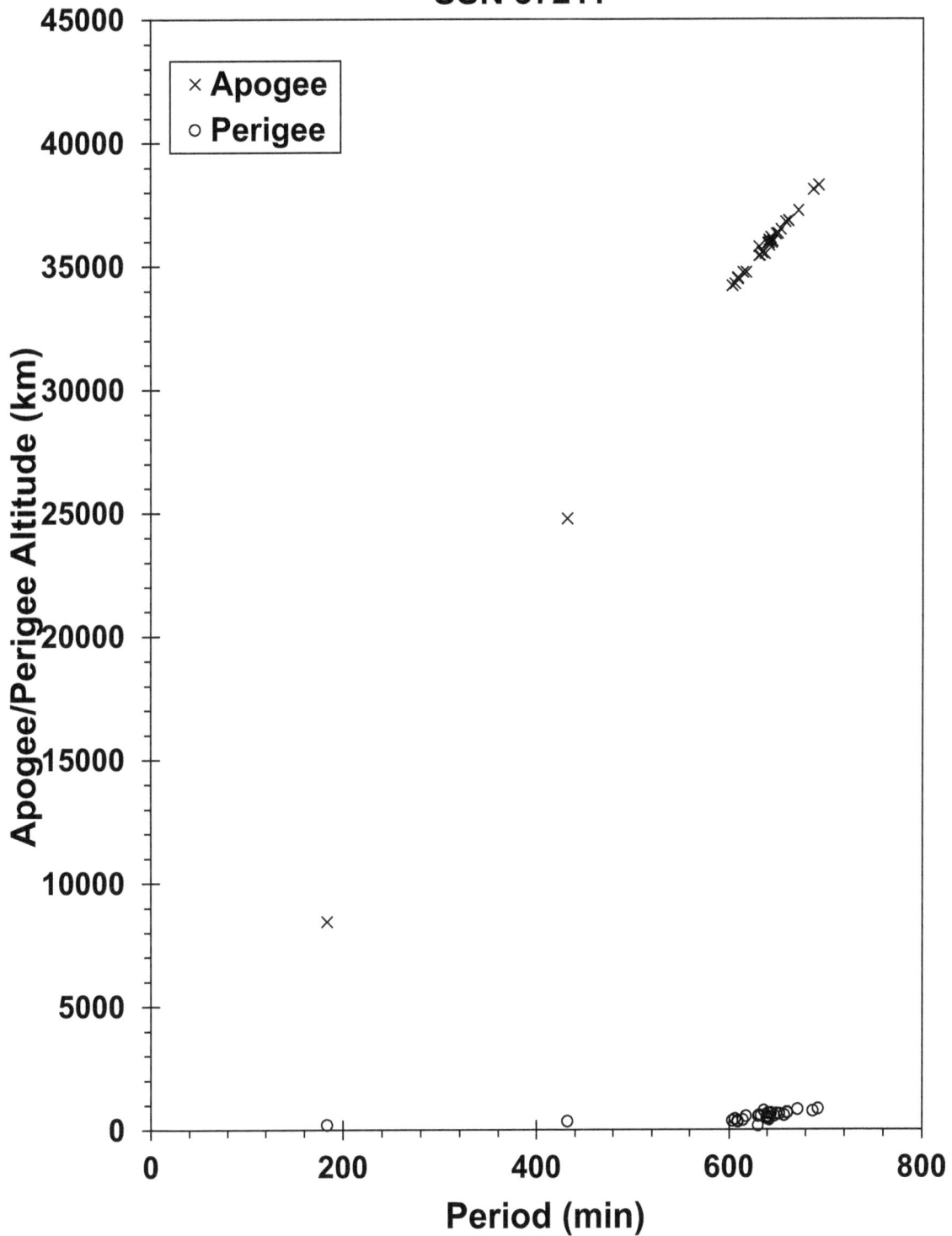

SSN 37211

The CZ-3C R/B debris cloud of 30 fragments cataloged up to 4 years after the event.

SATELLITE DATA

TYPE:	Mission Related Debris
OWNER:	CIS
LAUNCH DATE:	21 Sept. 2000
DRY MASS (KG):	375
MAIN BODY:	toroidal fuel tank; 0.821 m tank diameter x 3.44 m overall diameter
MAJOR APPENDAGES:	none
ATTITUDE CONTROL:	none
ENERGY SOURCES:	unknown

FIRST EVENT DATA (1)

DATE:	3-4 Aug 2015	LOCATION:	unknown
TIME:	unknown	ASSESSED CAUSE:	unknown
ALTITUDE:	unknown		

PRE-EVENT ELEMENTS (1)

EPOCH:	15214.44003403	MEAN ANOMALY:	298.4148
RIGHT ASCENSION:	218.9517	MEAN MOTION:	11.24281384
INCLINATION:	51.4462	MEAN MOTION DOT/2:	0.00000437
ECCENTRICITY:	0.1913343	MEAN MOTION DOT DOT/6:	0.00000
ARG. OF PERIGEE:	234.4912	BSTAR:	0.0011788

SECOND EVENT DATA (2)

DATE:	8 May 2020	LOCATION:	unknown
TIME:	0402-0551 GMT	ASSESSED CAUSE:	unknown
ALTITUDE:	unknown		

PRE-EVENT ELEMENTS (2)

EPOCH:	20129.14401989	MEAN ANOMALY:	6.7150
RIGHT ASCENSION:	94.5637	MEAN MOTION:	11.29195849
INCLINATION:	51.4509	MEAN MOTION DOT/2:	0.00000429
ECCENTRICITY:	0.1896822	MEAN MOTION DOT DOT/6:	0.00000
ARG. OF PERIGEE:	350.0613	BSTAR:	0.00011268

DEBRIS CLOUD DATA (2)

MAXIMUM ΔP:	36.9 min
MAXIMUM ΔI:	1.6 deg

COMMENTS

The parent body is assumed to be the jettisonable auxiliary fuel tank discarded by the Fregat-SB upper stage. Twenty-four debris were observed to be associated with the first (2015) event, but none have entered the SSN catalog to date.

REFERENCE DOCUMENTS

(1) "Fragmentation of *Fregat* Upper Stage Debris," The Orbital Debris Quarterly News, NASA JSC, April 2016.
Available online at http://orbitaldebris.jsc.nasa.gov/newsletter/pdfs/ODQNv20i1-2.pdf.
(2) "Second Fragmentation of *Fregat* Upper Stage Debris," The Orbital Debris Quarterly News, NASA JSC, August 2020.
Available online at http://orbitaldebris.jsc.nasa.gov/newsletter/pdfs/ODQNv24i3.pdf.

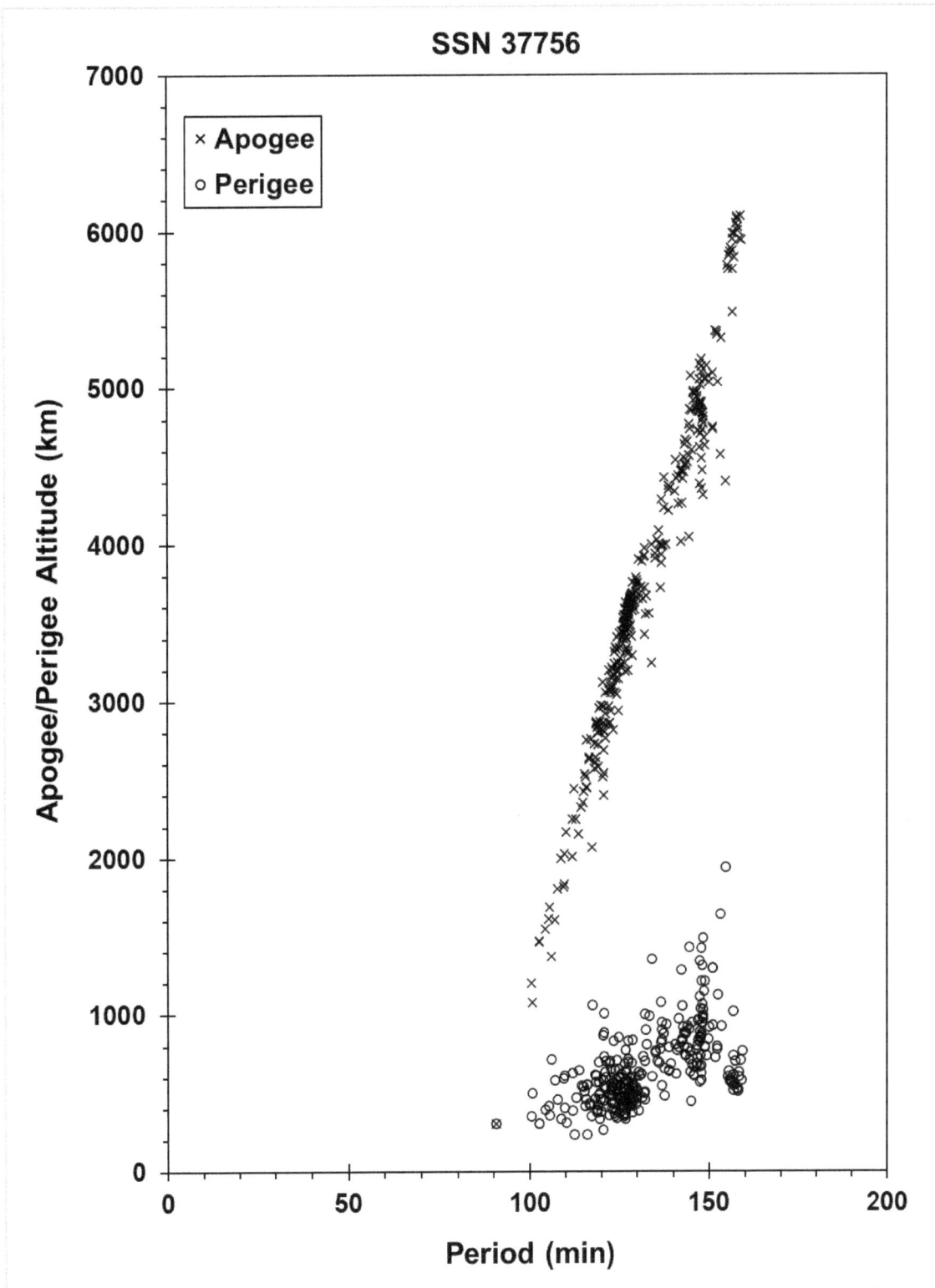

SSN 37756

SL-23 DEB cloud for the second (2020) breakup event. Many of the 324 fragments entered the catalog withing two months of the second event.

NIGCOMSAT 1R R/B   **2011-077B**   **38015**

SATELLITE DATA

TYPE:	Rocket Body
OWNER:	PRC
LAUNCH DATE:	19 Dec 2011
DRY MASS (KG):	3062
MAIN BODY:	Cylinder; 3.0 m diameter x 12.4 m length
MAJOR APPENDAGES:	none
ATTITUDE CONTROL:	status unknown at time of event
ENERGY SOURCES:	on-board propellants

EVENT DATA

DATE:	21 Dec 2011	LOCATION:	Unknown
TIME:	Unknown	ASSESSED CAUSE:	Unknown
ALTITUDE:	Unknown		

PRE-EVENT ELEMENTS

EPOCH:	11355.16500251	MEAN ANOMALY:	285.0544
RIGHT ASCENSION:	335.3965	MEAN MOTION:	1.91939558
INCLINATION:	24.3335	MEAN MOTION DOT/2:	-0.00001209
ECCENTRICITY:	0.7584148	MEAN MOTION DOT DOT/6:	0.0
ARG. OF PERIGEE:	179.8362	BSTAR:	-0.00031775

DEBRIS CLOUD DATA

MAXIMUM ΔP:	91.3 min
MAXIMUM ΔI:	2.4 deg

COMMENTS

This was the fourth fragmentation of a CZ-3 third stage. Like prior events, this event occurred within two days of launch and separation of the payload.

REFERENCE DOCUMENT

"Only a few minor satellite breakups in 2011", The Orbital Debris Quarterly News, NASA JSC, January 2012. Available online at http://orbitaldebris.jsc.nasa.gov/newsletter/pdfs/ODQNv16i1.pdf.

SSN 38015

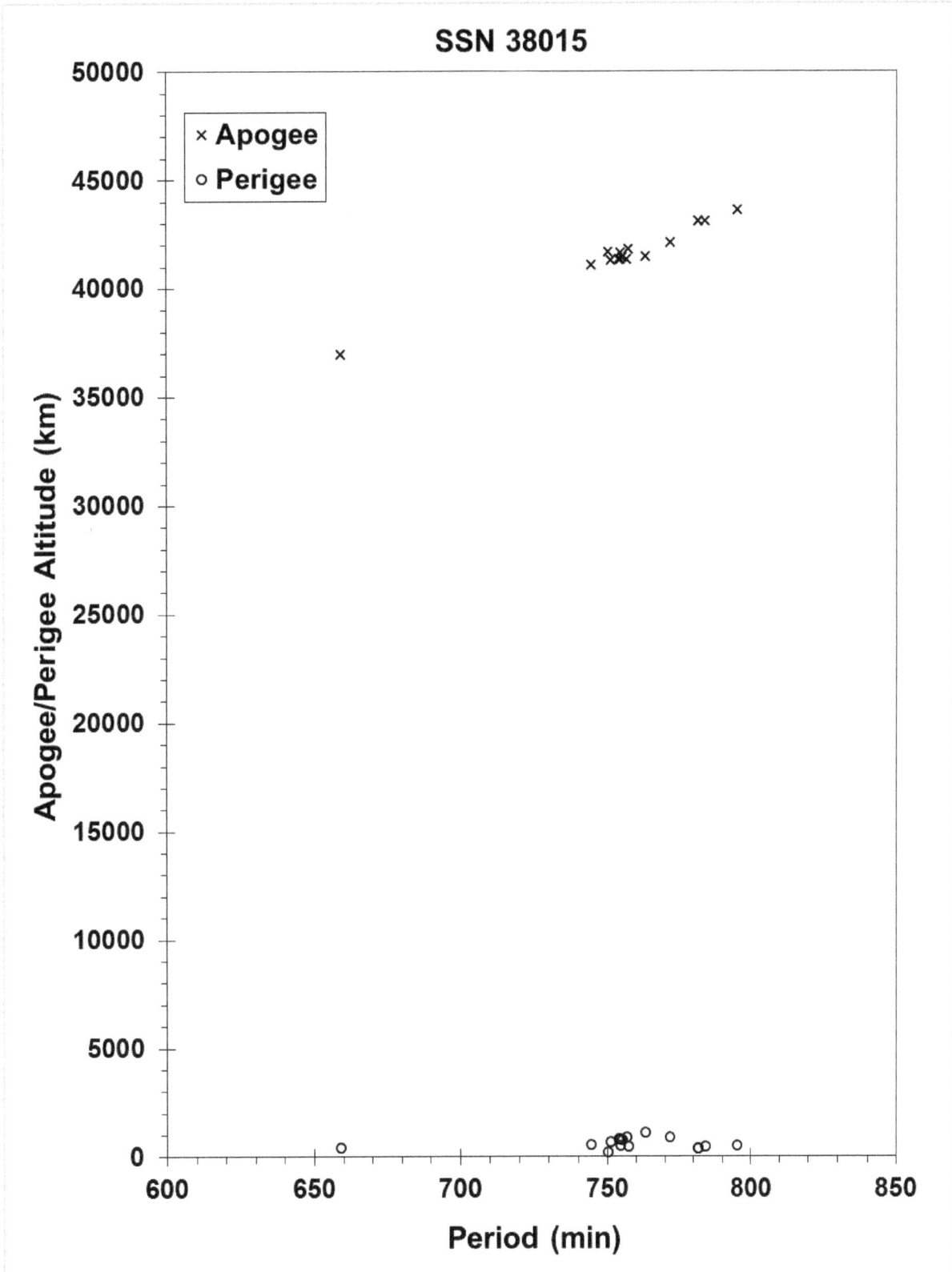

The CZ-3C R/B debris cloud of 16 fragments cataloged up to 2.7 years after the event.

SATELLITE DATA

TYPE:	Rocket Body
OWNER:	PRC
LAUNCH DATE:	24 Feb 2012
DRY MASS (KG):	3062
MAIN BODY:	Cylinder; 3.0 m diameter x 12.4 m length
MAJOR APPENDAGES:	none
ATTITUDE CONTROL:	status unknown at time of event
ENERGY SOURCES:	on-board propellants

EVENT DATA

DATE:	26 Feb 2012	LOCATION:	Unknown
TIME:	Unknown	ASSESSED CAUSE:	Unknown
ALTITUDE:	Unknown		

PRE-EVENT ELEMENTS

EPOCH:	12056.56279906	MEAN ANOMALY:	353.3665
RIGHT ASCENSION:	35.2298	MEAN MOTION:	2.27377203
INCLINATION:	20.6965	MEAN MOTION DOT/2:	0.00007641
ECCENTRICITY:	0.7327583	MEAN MOTION DOT DOT/6:	0.0000012139
ARG. OF PERIGEE:	180.1005	BSTAR:	0.00010000

DEBRIS CLOUD DATA

MAXIMUM ΔP:	56.1 min
MAXIMUM ΔI:	2.6 deg

COMMENTS

This was the fifth fragmentation of a CZ-3 third stage. Like its preceding event, this event occurred within two days of launch and separation of the Beidou G5 spacecraft.

REFERENCE DOCUMENT

"Chinese Rocket Body Explosions Continue," The Orbital Debris Quarterly News, NASA JSC, April 2012. Available online at https://orbitaldebris.jsc.nasa.gov/quarterly-news/pdfs/ODQNv16i2.pdf.

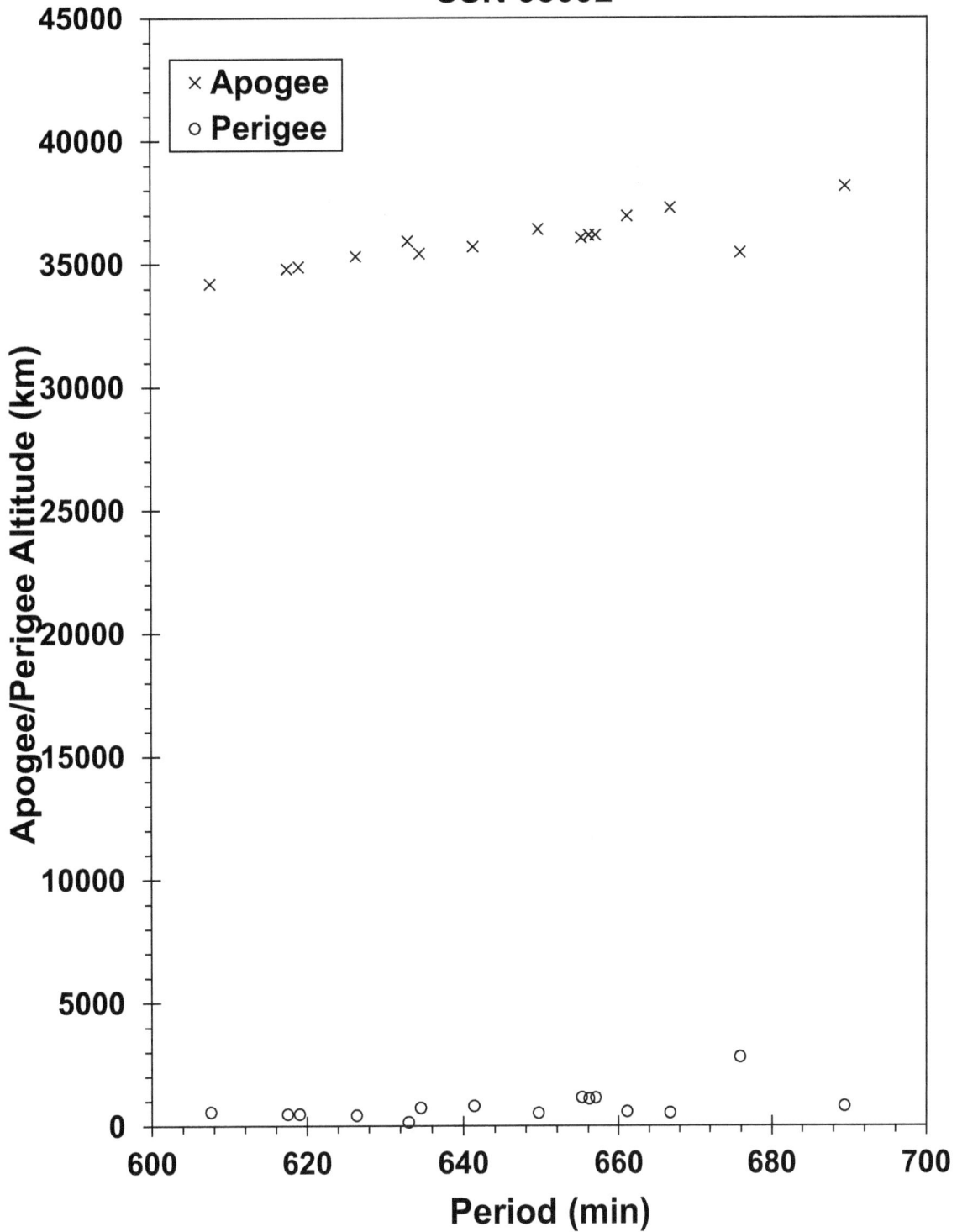

SSN 38092

The CZ-3C R/B debris cloud of 15 fragments cataloged up to 2.6 years after the event.

SATELLITE DATA

TYPE:	*Briz-M* Upper Stage Core
OWNER:	CIS
LAUNCH DATE:	17 May 2012
DRY MASS (KG):	1220
MAIN BODY:	cylinder 2.49 m diameter x 2.65 m length
MAJOR APPENDAGES:	none
ATTITUDE CONTROL:	none at time of event
ENERGY SOURCES:	unknown

EVENT DATA

DATE:	23 Dec 2015	LOCATION:	11.9N, 178E (dsc)
TIME:	1600 GMT	ASSESSED CAUSE:	Unknown
ALTITUDE:	24,310 km		

PRE-EVENT ELEMENTS

EPOCH:	15356.29062075	MEAN ANOMALY:	128.0880177
RIGHT ASCENSION:	68.3779	MEAN MOTION:	1.76901584
INCLINATION:	12.0183	MEAN MOTION DOT/2:	-0.00000143
ECCENTRICITY:	0.4187335	MEAN MOTION DOT DOT/6:	0.00000
ARG. OF PERIGEE:	203.9714	BSTAR:	0.00000

DEBRIS CLOUD DATA

MAXIMUM ΔP:	13.9 min
MAXIMUM ΔI:	0.09 deg

COMMENTS

Eight debris (in addition to the parent object) have been officially cataloged. Debris in deep-space orbits are difficult for the SSN to track and catalog. Hundreds of additional fragments could be on-orbit.

REFERENCE DOCUMENTS

"Briz-M Core Stage Fragments in Elliptical Orbit," The Orbital Debris Quarterly News, NASA JSC, April 2016. Available online at http://orbitaldebris.jsc.nasa.gov/newsletter/pdfs/ODQNv20i1-2.pdf.

SSN 38343

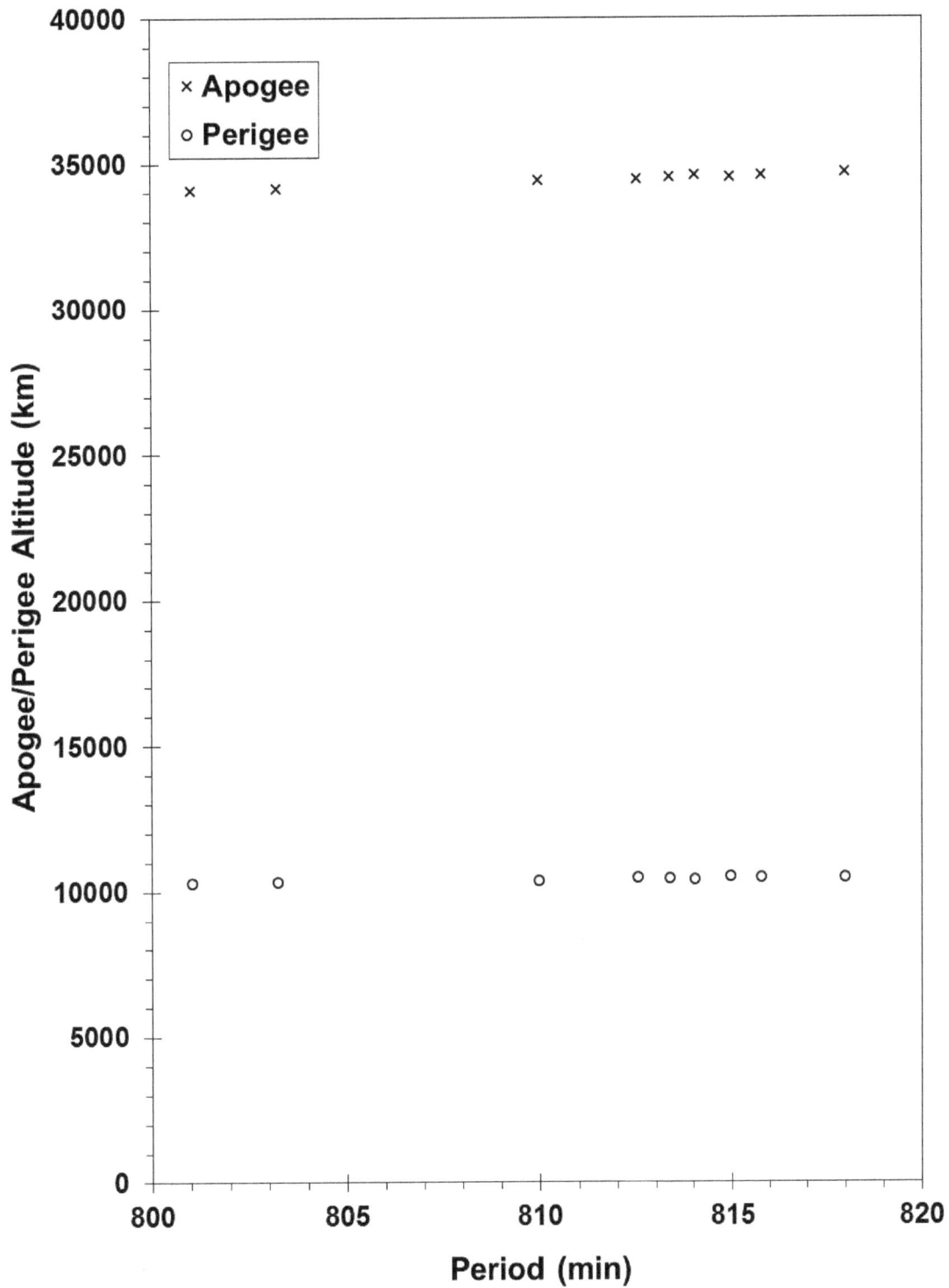

The *Briz-M* debris cloud; debris cataloged approximately 39 days after the event.

SATELLITE DATA

TYPE:	Rocket Body
OWNER:	CIS
LAUNCH DATE:	6 August 2012
DRY MASS (KG):	2510
MAIN BODY:	Cylinder + toroid; 4.1 m diameter x 2.65 m length
MAJOR APPENDAGES:	none
ATTITUDE CONTROL:	none at the time of the event
ENERGY SOURCES:	on-board propellants

EVENT DATA

DATE:	16 Oct 2012	LOCATION:	32.447N, 335.711E (asc)
TIME:	1631 GMT	ASSESSED CAUSE:	Propulsion
ALTITUDE:	273.061 km		

PRE-EVENT ELEMENTS

EPOCH:	12289.67349823	MEAN ANOMALY:	259.1600
RIGHT ASCENSION:	103.4658	MEAN MOTION:	10.14046751
INCLINATION:	49.8913	MEAN MOTION DOT/2:	0.00003146
ECCENTRICITY:	0.2633226	MEAN MOTION DOT DOT/6:	0.0000019595
ARG. OF PERIGEE:	127.8077	BSTAR:	0.00011788

DEBRIS CLOUD DATA

MAXIMUM ΔP:	38.9 min
MAXIMUM ΔI:	0.8 deg

COMMENTS

The dry mass quoted above includes that of the *Briz-M* core stage and the toroidal Auxiliary Propulsion Tank (APT)—sometimes referred to as the Additional Fuel Tank (AFT); the malfunction stranding the stage in its elliptical orbit occurred prior to APT jettison. The event occurred almost 70 days after launch. An estimated 10 metric tons of propellant was aboard.

REFERENCE DOCUMENT

"Upper Stage Explosion Places LEO Satellites at Risk," The Orbital Debris Quarterly News, NASA JSC, January 2013. Available online at http://orbitaldebris.jsc.nasa.gov/newsletter/pdfs/ODQNv17i1.pdf.

"New Russian Launch Failure Raises Breakup Concern," The Orbital Debris Quarterly News, NASA JSC, October 2012. Available online at http://orbitaldebris.jsc.nasa.gov/newsletter/pdfs/ODQNv16i4.pdf.

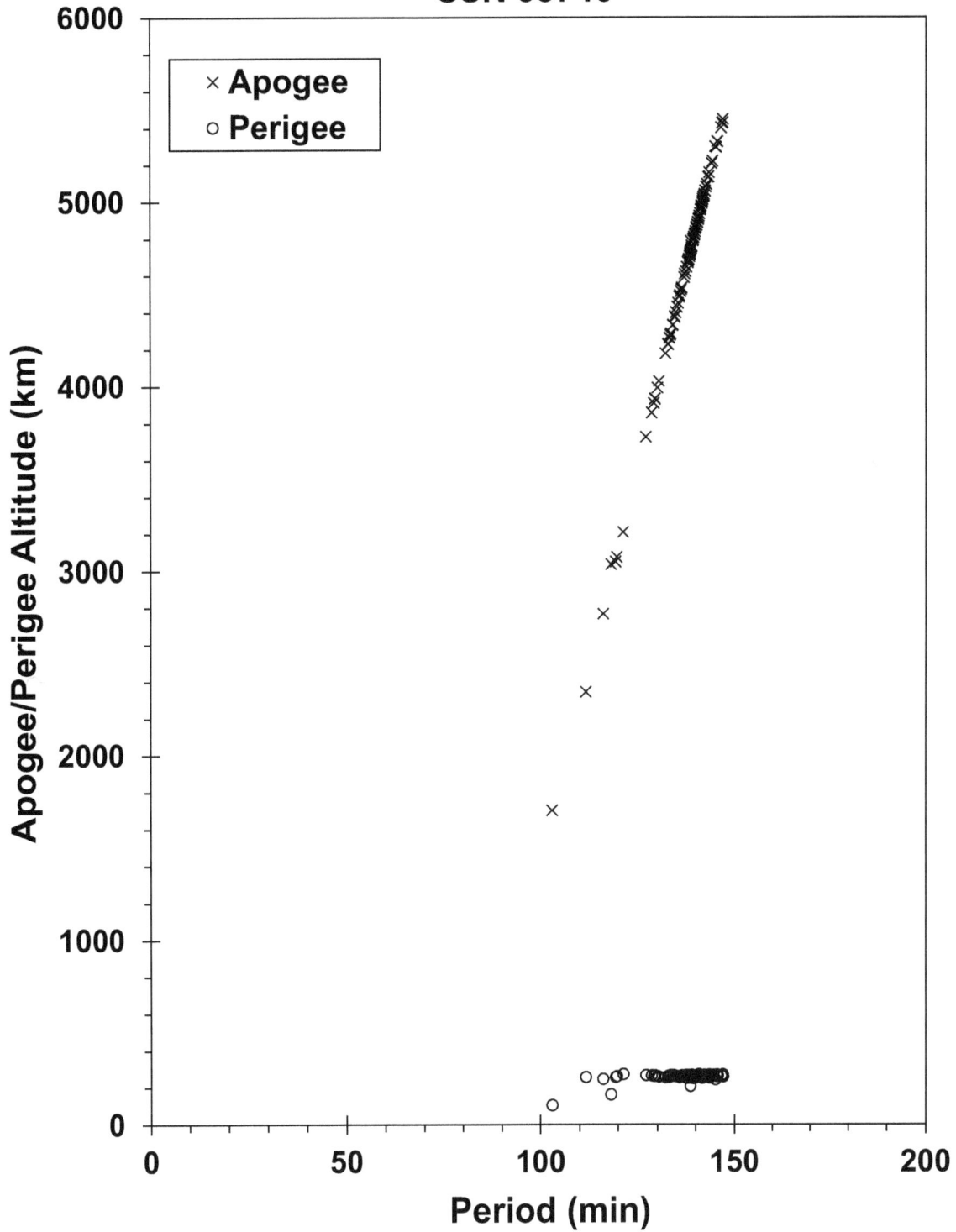

SSN 38746

The *Briz-M* R/B debris cloud of 113 fragments cataloged up to 3 weeks after the event.

CASSIOPE R/B   **2013-055G**   **39271**

SATELLITE DATA

TYPE:	Rocket Body
OWNER:	USA
LAUNCH DATE:	29 Sep 2013
DRY MASS (KG):	3900 kg
MAIN BODY:	cylinder; 3.66 m diameter x ~ 14 m length
MAJOR APPENDAGES:	none
ATTITUDE CONTROL:	active, 3 axis RCS
ENERGY SOURCES:	on-board propellants

EVENT DATA

DATE:	29 Sep 2013	LOCATION:	78.026S, 8.302E (asc)
TIME:	1641 GMT	ASSESSED CAUSE:	Unknown
ALTITUDE:	1204.12 km		

PRE-EVENT ELEMENTS

EPOCH:	13272.73704787	MEAN ANOMALY:	327.9211
RIGHT ASCENSION:	315.1956	MEAN MOTION:	13.93734602
INCLINATION:	80.9876	MEAN MOTION DOT/2:	0.00000033
ECCENTRICITY:	0.0693087	MEAN MOTION DOT DOT/6:	0.0
ARG. OF PERIGEE:	153.5198	BSTAR:	0.0 (unavailable)

DEBRIS CLOUD DATA

MAXIMUM ΔP:	2.1 min
MAXIMUM ΔI:	0.16 deg

COMMENTS

The elements listed above are from the first published TLE; the epoch time is approximately one hour and 21 seconds after the debris-producing event. This was the first flight of the Falcon 9 v1.1 second stage. The second stage engine restart, intended to cast the vehicle into its disposal orbit, failed.

REFERENCE DOCUMENT

"Upper Stage of New Falcon 9 Rocket Did Not Explode After Launch, SpaceX Says", http://www.space.com/23038-spacex-falcon-9-rocket-explosion-rumors.html, retrieved 23 April 2015.

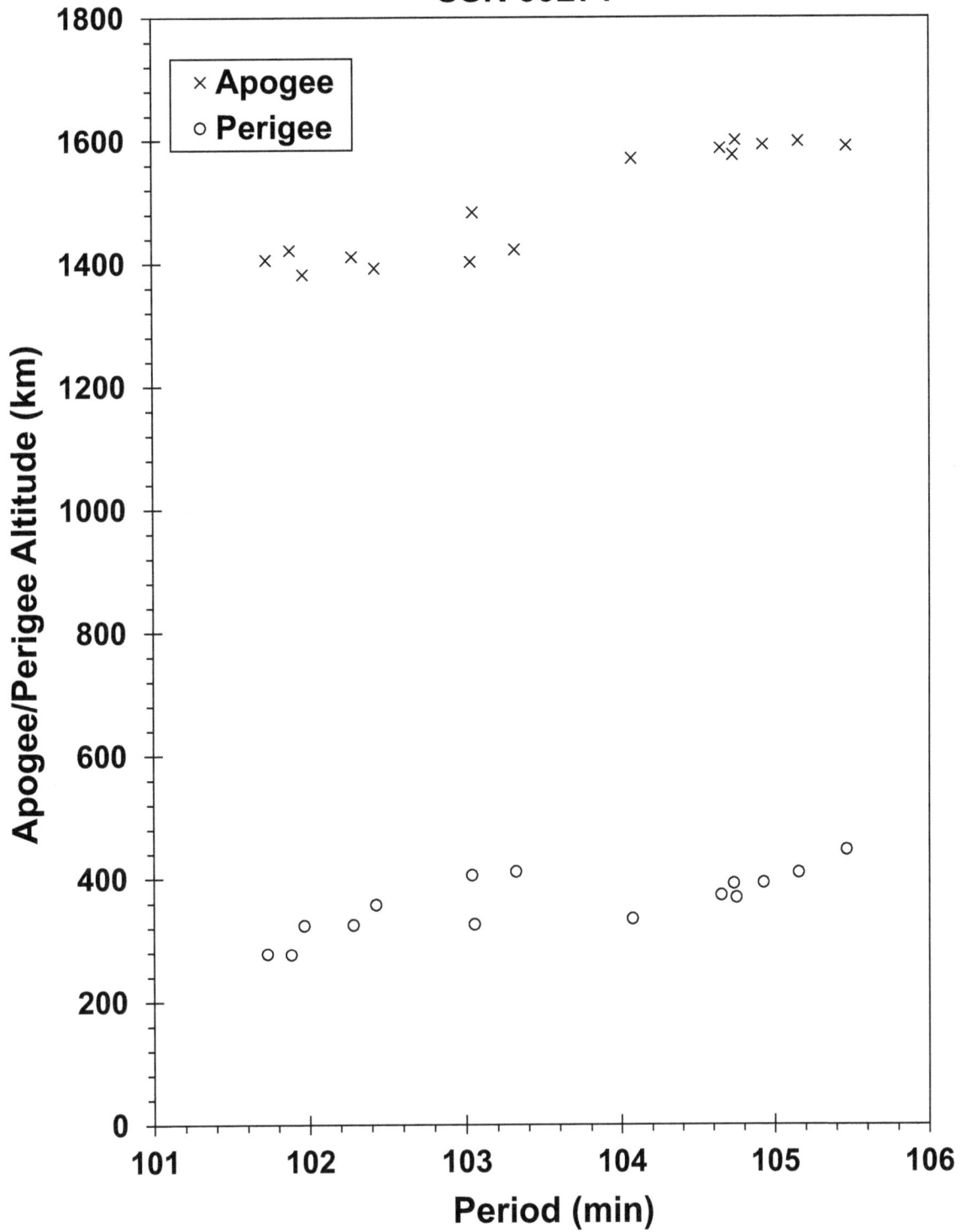

SSN 39271

The Falcon 9 R/B debris cloud of 15 fragments cataloged within 2 weeks of the event.

YAOGAN 19 R/B   **2013-065B**   **39411**

SATELLITE DATA

TYPE:	Long March 4C third stage
OWNER:	PRC
LAUNCH DATE:	20 Nov. 2013
DRY MASS (KG):	1700
MAIN BODY:	Cylinder nozzle; 2.9 m diameter by ~5 m length
MAJOR APPENDAGES:	None
ATTITUDE CONTROL:	None
ENERGY SOURCES:	On-board propellants

EVENT DATA

DATE:	17 Aug. 2018	LOCATION:	55.24N, 272.09E (dsc)
TIME:	0846 GMT	ASSESSED CAUSE:	Propulsion
ALTITUDE:	1192.5 km		

PRE-EVENT ELEMENTS

EPOCH:	18228.90645880	MEAN ANOMALY:	155.0427
RIGHT ASCENSION:	24.2178	MEAN MOTION:	13.43492155
INCLINATION:	100.5034	MEAN MOTION DOT/2:	-0.00000033
ECCENTRICITY:	0.0140154	MEAN MOTION DOT DOT/6:	0.0
ARG. OF PERIGEE:	204.4022	BSTAR:	0.000025679

DEBRIS CLOUD DATA

MAXIMUM ΔP:	4.2 min
MAXIMUM ΔI:	0.4 deg

COMMENTS

This is the third Long March 4 to breakup, prior breakups being of the 4A and 4B models. Passivation of this vehicle was attempted.

REFERENCE DOCUMENTS

"CZ-4C Upper Stage Fragments in August", The Orbital Debris Quarterly News, NASA JSC, September 2018. Available online at http://orbitaldebris.jsc.nasa.gov/newsletter/pdfs/ODQNv22i3.pdf.

"Analyzing the Cause of LM-4 (A)'s Upper Stage's Disintegration and the Countermeasures", W. X. Zang and S. Y. Liao, 5th International Conference of Pacific Basin Societies, 6-9 Jun 1993, Shanghai.

"Passivation Investigation and Engineering Applications for Orbital Stage of LM-4B/-4C Launch Vehicle," Yanfeng, G., W. Yijin, and F. Hongtuan, Space Debris Research, Special 2013, Chinese National Space Agency (2013), pp. 3-6.

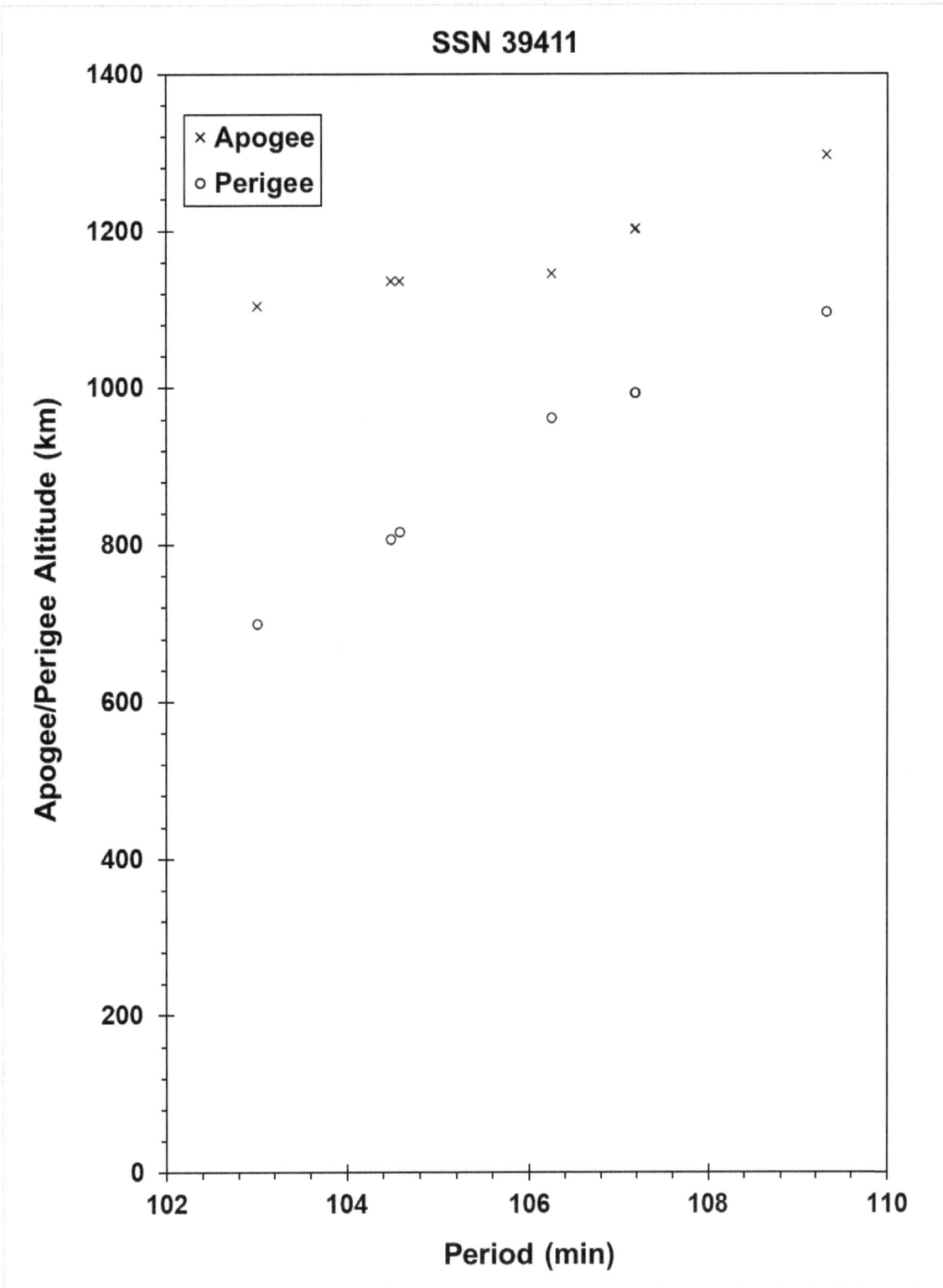

YAOGAN 19 Long March 4C third stage debris cloud of six fragments within (generally) one month of the event as reconstructed from the US SSN database.

SATELLITE DATA

TYPE:	payload
OWNER:	CIS
LAUNCH DATE:	25 Dec. 2013
DRY MASS (KG):	Unknown
MAIN BODY:	Unknown
MAJOR APPENDAGES:	Unknown
ATTITUDE CONTROL:	Unknown
ENERGY SOURCES:	Unknown

EVENT DATA

DATE:	23 Dec. 2019	LOCATION:	31.95 S, 102.4 E
TIME:	1302 GMT	ASSESSED CAUSE:	Unknown
ALTITUDE:	1494.1 km		

PRE-EVENT ELEMENTS

EPOCH:	19356.69052255	MEAN ANOMALY:	135.7280
RIGHT ASCENSION:	205.5631	MEAN MOTION:	12.42682561
INCLINATION:	82.4845	MEAN MOTION DOT/2:	0.00000033
ECCENTRICITY:	0.0016799	MEAN MOTION DOT DOT/6:	0.0
ARG. OF PERIGEE:	224.2458	BSTAR:	0.00016848

DEBRIS CLOUD DATA

MAXIMUM ΔP:	14.4 min
MAXIMUM ΔI:	1.3 deg

COMMENTS

This spacecraft is believed to be of the CIS' first generation "Satellite Inspector" class spacecraft. Other members of the class are believed to be Cosmos 2499 [q.v.] and Cosmos 2504. Cosmos 2491 carried the Radio Sputnik 46 (RS-46) amateur radio payload.

REFERENCE DOCUMENTS

"Russian Satellite Fragments," The Orbital Debris Quarterly News, NASA JSC, February 2020. Available online at https://www.orbitaldebris.jsc.nasa.gov/quarterly-news/pdfs/odqnv24i1.pdf.

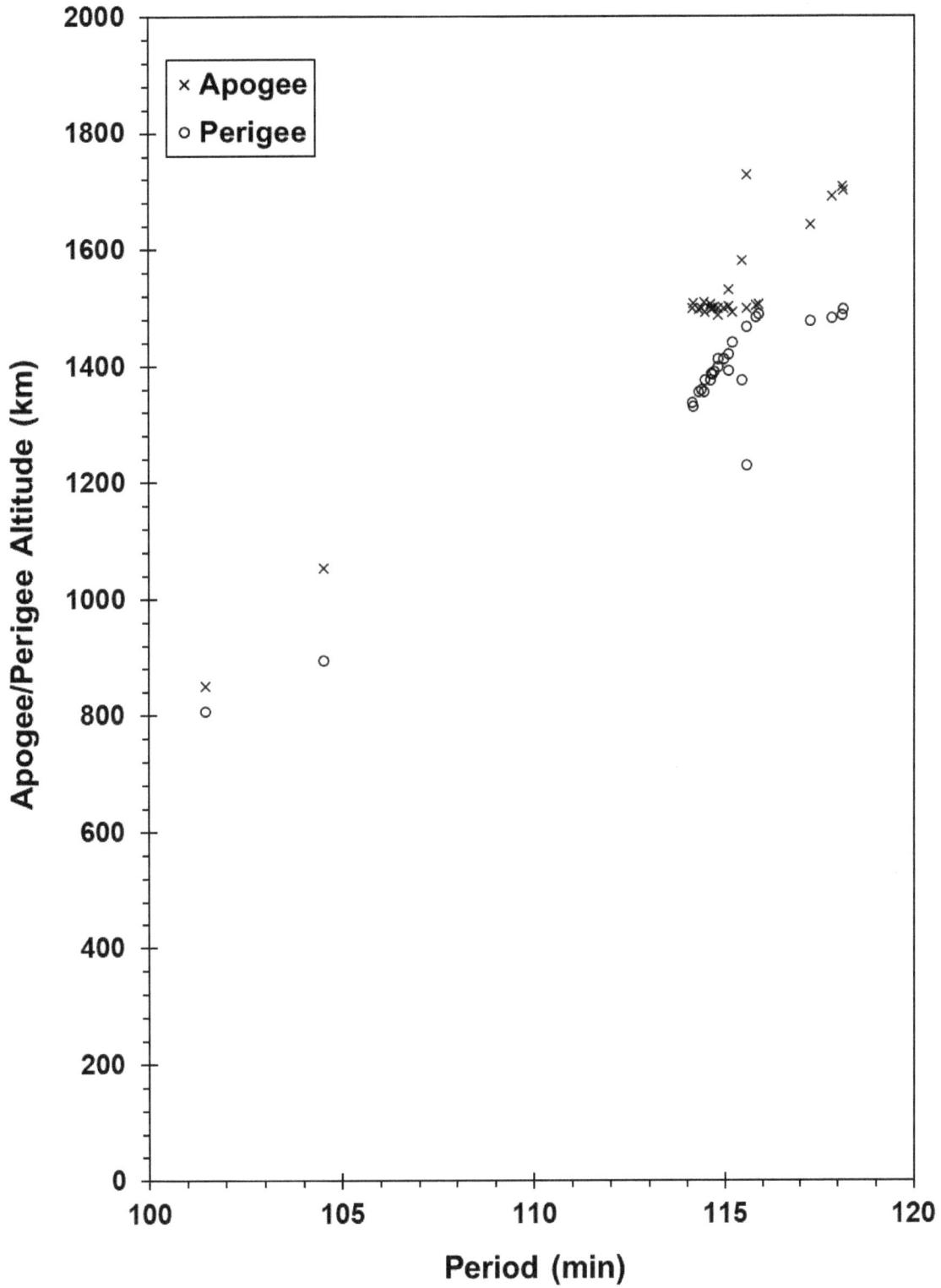

SSN 39497

COSMOS 2491 debris cloud composed of parent spacecraft and 26 additional objects cataloged between two weeks and a year after the event. One object is likely erroneously assigned to this cloud.

SENTINEL 1A   **2014-016A**   **39634**

SATELLITE DATA

TYPE:	payload
OWNER:	ESA
LAUNCH DATE:	3 Apr 2014
DRY MASS (KG):	2170
MAIN BODY:	cubical 3.9 m x 2.6 m x 2.5 m spacecraft envelope
MAJOR APPENDAGES:	dual solar arrays and synthetic aperture radar antenna
ATTITUDE CONTROL:	three-axis stabilized
ENERGY SOURCES:	monopropellant, batteries, gyroscope, reaction wheels

EVENT DATA

DATE:	23 Aug 2016	LOCATION:	72.03S, 39.59E (dsc)
TIME:	1707 GMT	ASSESSED CAUSE:	Collision, accidental
ALTITUDE:	723.20 km		

PRE-EVENT ELEMENTS

EPOCH:	16235.56155967	MEAN ANOMALY:	283.0743
RIGHT ASCENSION:	241.4752	MEAN MOTION:	14.59198146
INCLINATION:	98.1816	MEAN MOTION DOT/2:	+0.00000017
ECCENTRICITY:	0.0001365	MEAN MOTION DOT DOT/6:	0.00000
ARG. OF PERIGEE:	77.0610	BSTAR:	+0.000013361

DEBRIS CLOUD DATA

MAXIMUM ΔP:	2.9 min
MAXIMUM ΔI:	0.1 deg

COMMENTS

This event is categorized as an accidental collision with a small particle. Normally, an event of this nature would have been categorized as an anomalous event; however, spacecraft attitude control was upset by the solar array impact and on-board cameras were able to document the before/after state of the penetrated solar array. While either small, un-cataloged debris or a micrometeoroid could have been this event's projectile, the source of the resulting tracked, and cataloged debris was unambiguous.

REFERENCE DOCUMENT

Collecte Localisation Satellites, "Sentinel-1A Debris Collision August 2016," MPC-S1 (18 October 2016). Retrieved 28 September 2018 from https://sentinel.esa.int/documents/247904/2142675/Sentinel-1A_Debris_Collision_August_2016_MPC.pdf.

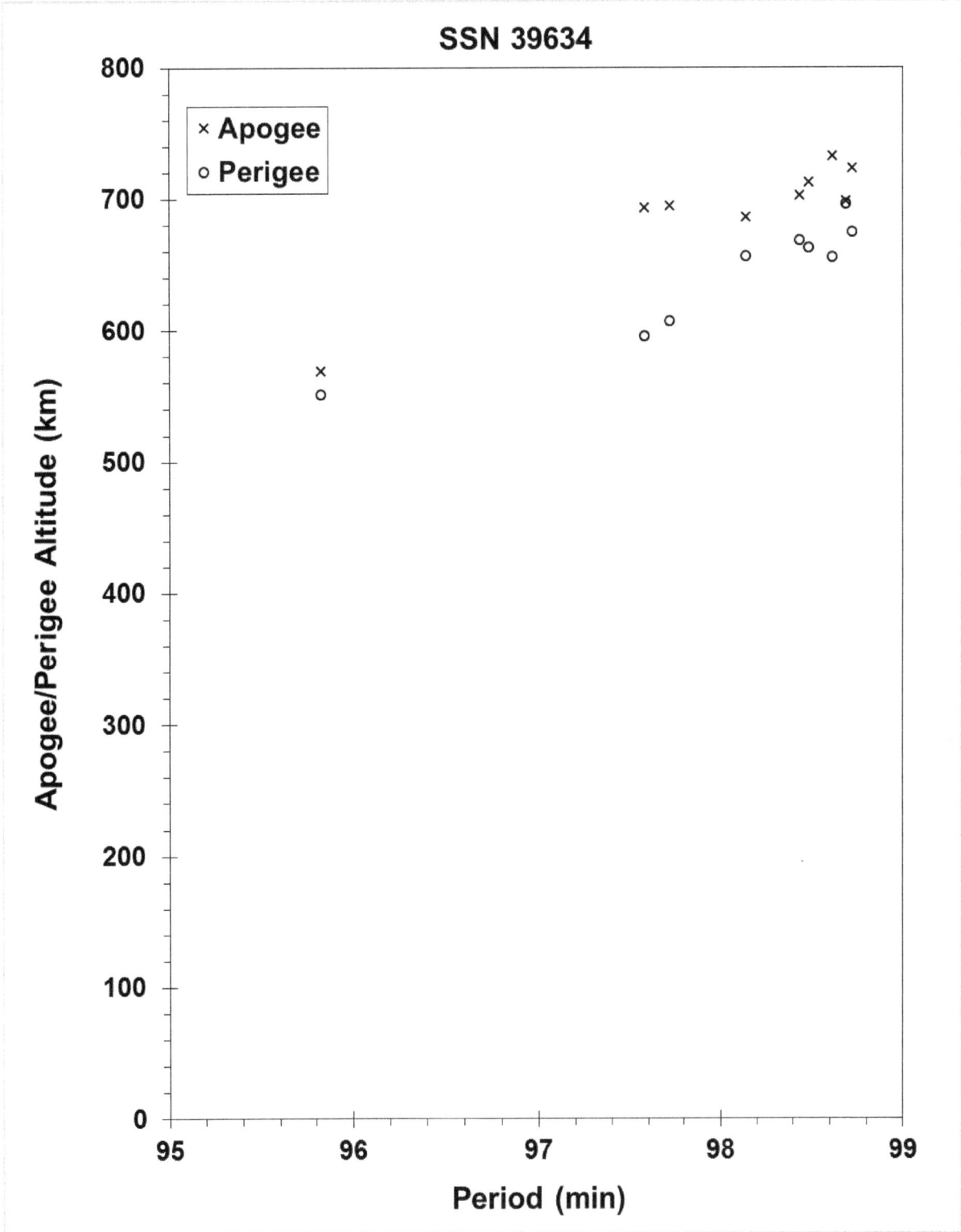

The Sentinel 1A debris cloud; debris cataloged within approximately 18 days of the event.

SATELLITE DATA

TYPE:	payload
OWNER:	CIS
LAUNCH DATE:	23.23 May 2014
DRY MASS (KG):	Unknown
MAIN BODY:	Unknown
MAJOR APPENDAGES:	Unknown
ATTITUDE CONTROL:	Unknown
ENERGY SOURCES:	Unknown

EVENT DATA

DATE:	23 Oct. 2021	LOCATION:	38.31 N, 186.24 E
TIME:	1814 GMT	ASSESSED CAUSE:	Unknown
ALTITUDE:	1405.3 km		

PRE-EVENT ELEMENTS

EPOCH:	21296.65192163	MEAN ANOMALY:	334.1036
RIGHT ASCENSION:	318.1879	MEAN MOTION:	12.82931495
INCLINATION:	82.4413	MEAN MOTION DOT/2:	0.00000027
ECCENTRICITY:	0.0231563	MEAN MOTION DOT DOT/6:	0.0
ARG. OF PERIGEE:	27.1975	BSTAR:	0.000058494

DEBRIS CLOUD DATA

MAXIMUM ΔP:	1.3 min
MAXIMUM ΔI:	0.1 deg

COMMENTS

This spacecraft is believed to be of the CIS' first generation "Satellite Inspector" class spacecraft. Other members are believed to be Cosmos 2491 [q.v.] and Cosmos 2504. Like the former spacecraft, Cosmos 2499 carried an amateur radio payload, Radio Sputnik 47 (RS-47).

REFERENCE DOCUMENTS

"Two Minor Breakup Events in Fourth Quarter of 2021," The Orbital Debris Quarterly News, NASA JSC, March 2022. Available online at https://www.orbitaldebris.jsc.nasa.gov/quarterly-news/pdfs/odqnv26i1.pdf.

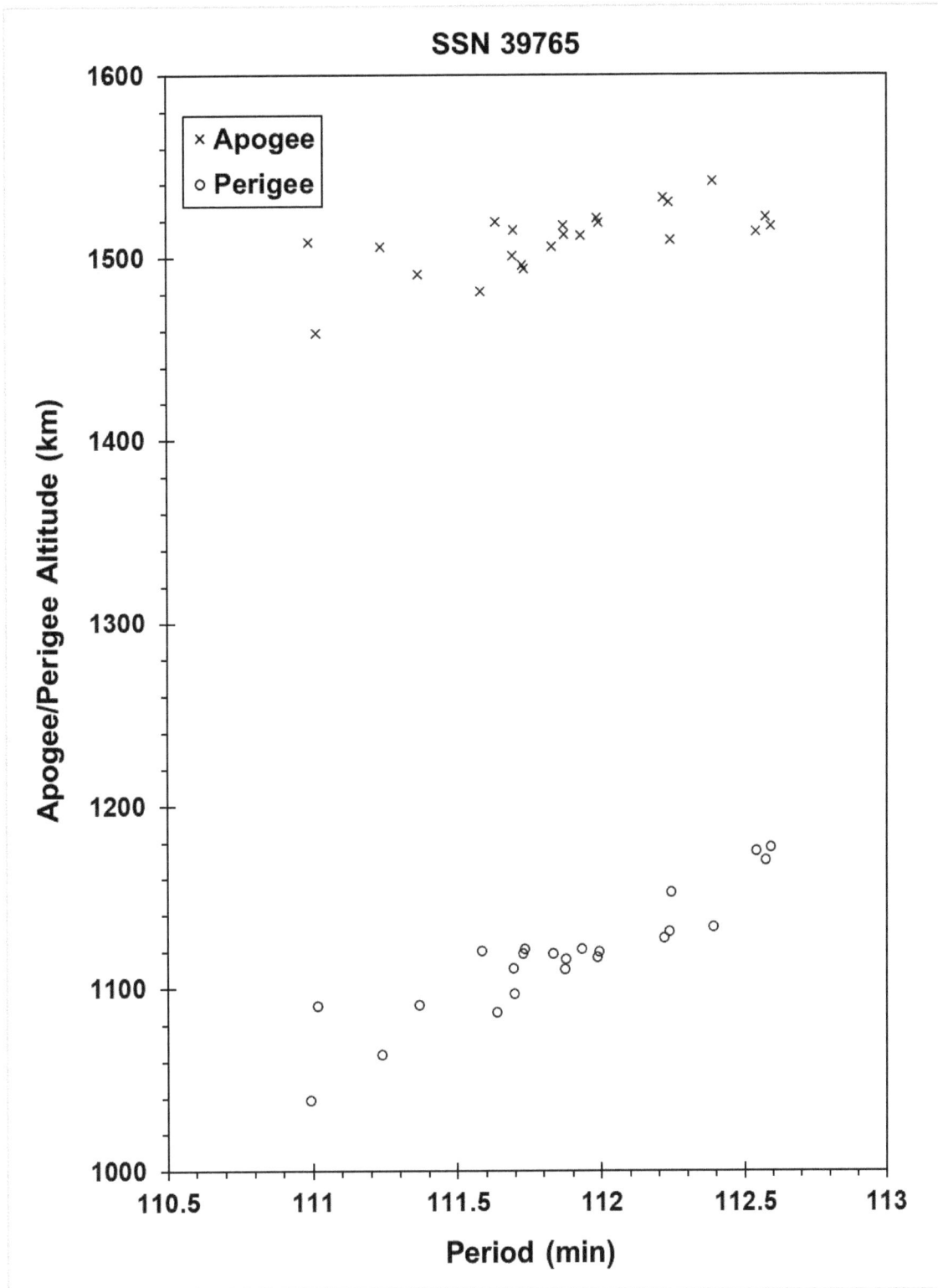

COSMOS 2499 debris cloud, composed of parent spacecraft and 22 additional objects cataloged up to 72 days after the event.

USA 257 R/B   **2014-055B**   **40209**

SATELLITE DATA

TYPE:	Centaur V (Single Engine Centaur) Rocket Body
OWNER:	USA
LAUNCH DATE:	17 Sept. 2014
DRY MASS (KG):	2243
MAIN BODY:	Cylinder; 3.048 m diameter x 12.68 m long
MAJOR APPENDAGES:	none
ATTITUDE CONTROL:	status unknown at time of event
ENERGY SOURCES:	status unknown at time of event

EVENT DATA

DATE:	31 Aug. 2018	LOCATION:	Unknown
TIME:	0200 GMT	ASSESSED CAUSE:	Unknown
ALTITUDE:	Unknown		

POST-EVENT ELEMENTS

EPOCH:	19024.70582927	MEAN ANOMALY:	6.8426
RIGHT ASCENSION:	35.1040	MEAN MOTION:	1.85507267
INCLINATION:	22.2682	MEAN MOTION DOT/2:	-.00000179
ECCENTRICITY:	0.4821670	MEAN MOTION DOT DOT/6:	0.0
ARG. OF PERIGEE:	78.4468	BSTAR:	0.0

DEBRIS CLOUD DATA*

MAXIMUM ΔP: 82.5 min
MAXIMUM ΔI: 0.9 deg
*descriptive data derived from debris entering public catalog within 150 days of event

COMMENTS

The apparent discrepancy between the known even time and lack of pre-event elements is because elements of the Centaur upper stage were not publicly available until well after the event. This is the first acknowledged Centaur V upper stage fragmentation event. The cause of the event is unknown.

REFERENCE DOCUMENTS

"Atlas V Upper Stage Fragmentation Event Concludes an Active August 2018", The Orbital Debris Quarterly News, NASA JSC, November 2018. Available online at https://orbitaldebris.jsc.nasa.gov/quarterly-news/pdfs/ODQNv22i4.pdf.

"Centaur V Upper Stage Fragmentation Update," The Orbital Debris Quarterly News, NASA JSC, May 2019. Available online at https://orbitaldebris.jsc.nasa.gov/quarterly-news/pdfs/ODQNv23i1.pdf.

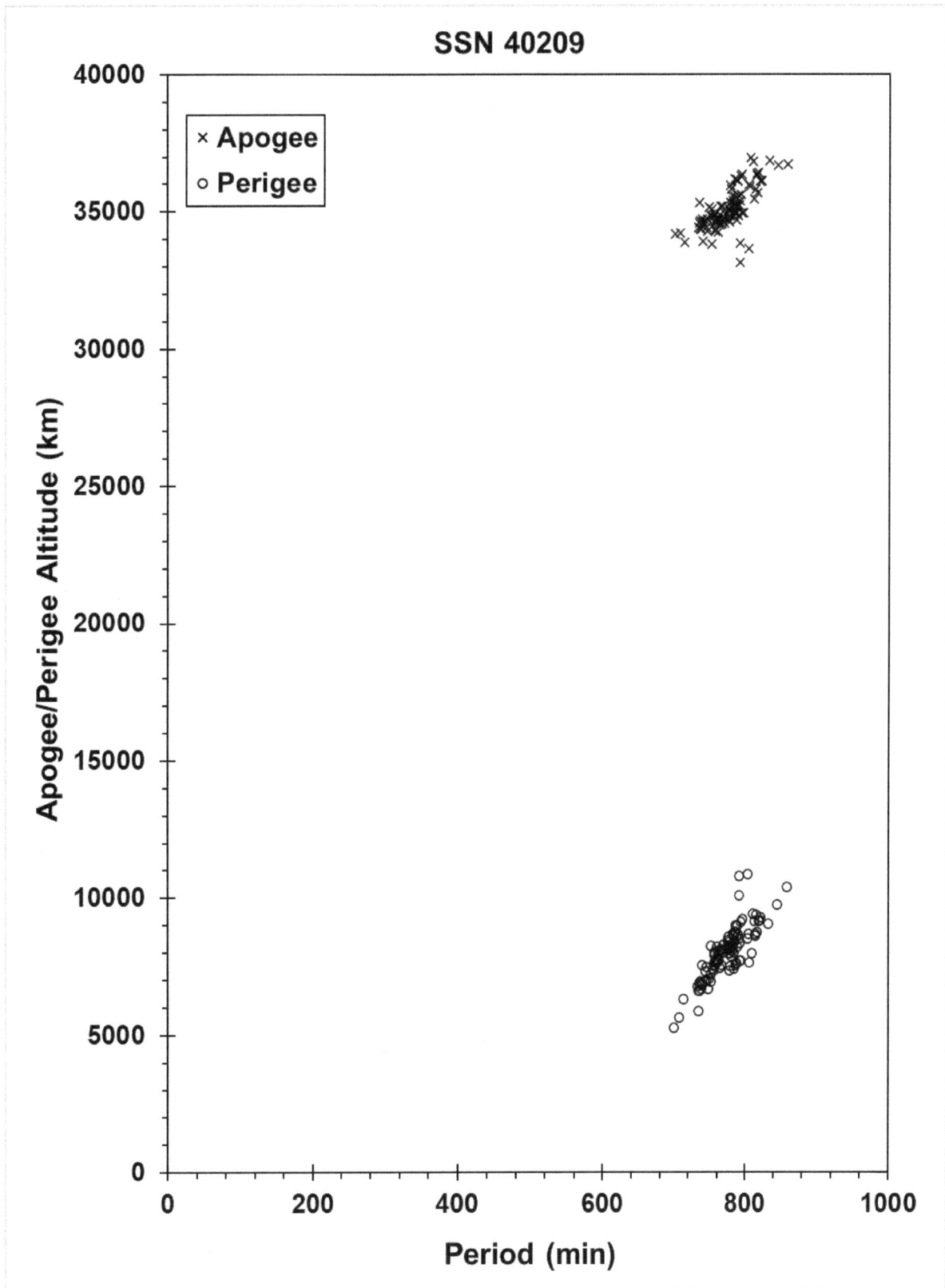

SSN 40209

Identified Centaur V R/B debris cloud; 103 debris objects cataloged up to two years after the event as reconstructed from the US SSN database.

SATELLITE DATA

TYPE:	Soyuz-U (SL-4) Third Stage Rocket Body
OWNER:	CIS
LAUNCH DATE:	28 April 2015
DRY MASS (KG):	2400
MAIN BODY:	2.66 m diameter x 8.1 m long cylinder
MAJOR APPENDAGES:	none
ATTITUDE CONTROL:	three-axis
ENERGY SOURCES:	on-board propellants

EVENT DATA

DATE:	28 April 2015	LOCATION:	TBD
TIME:	07:18:35 GMT	ASSESSED CAUSE:	Unknown
ALTITUDE:	TBD km		

PRE-EVENT ELEMENTS

EPOCH:	TBD	MEAN ANOMALY:	TBD
RIGHT ASCENSION:	TBD	MEAN MOTION:	TBD
INCLINATION:	TBD	MEAN MOTION DOT/2:	TBD
ECCENTRICITY:	TBD	MEAN MOTION DOT DOT/6:	TBD
ARG. OF PERIGEE:	TBD	BSTAR:	TBD

DEBRIS CLOUD DATA

MAXIMUM ΔP:	unknown
MAXIMUM ΔI:	unknown

COMMENTS

In addition to the parent body, 20 debris have been cataloged. Due to the low altitude of the event, all had decayed from orbit by 7 May 2015.

REFERENCE DOCUMENTS

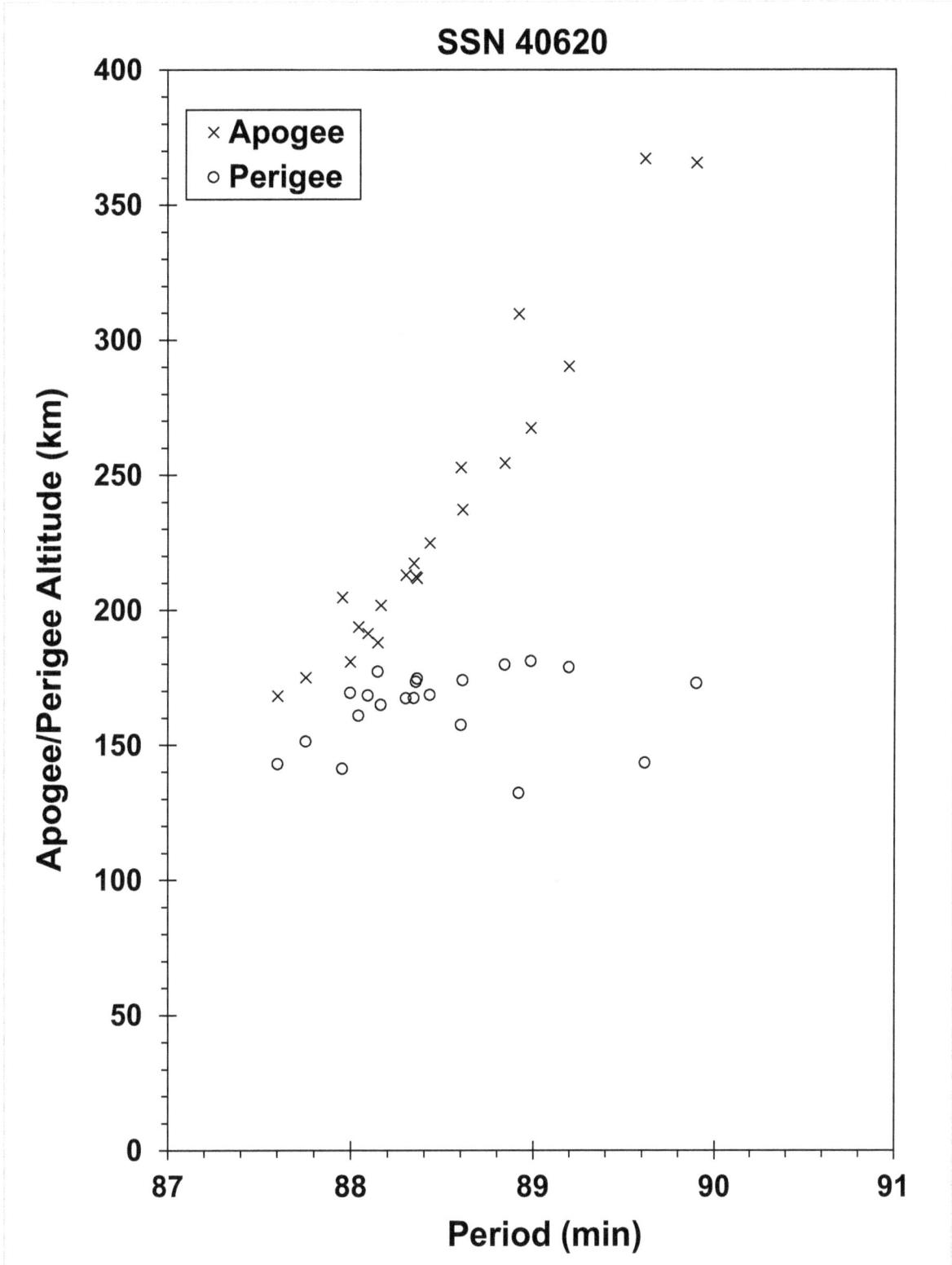

The SL-4 R/B debris cloud cataloged between 28 April and 1 May 2015.

COSMOS 2513 R/B   **2015-075B**   **41122**

SATELLITE DATA

TYPE:	*Briz-M* Upper Stage Core
OWNER:	CIS
LAUNCH DATE:	13 Dec 2015
DRY MASS (KG):	1220
MAIN BODY:	cylinder 2.49 m diameter x 2.65 m long
MAJOR APPENDAGES:	none
ATTITUDE CONTROL:	none at time of event
ENERGY SOURCES:	unknown

EVENT DATA

DATE:	16 Jan 2016	LOCATION:	0.18S, 223.0E (dsc)
TIME:	0350 GMT	ASSESSED CAUSE:	Unknown
ALTITUDE:	34,880.09 km		

PRE-EVENT ELEMENTS

EPOCH:	16015.10700804	MEAN ANOMALY:	221.1059
RIGHT ASCENSION:	135.1430	MEAN MOTION:	1.04652118
INCLINATION:	0.1737	MEAN MOTION DOT/2:	-0.00000109
ECCENTRICITY:	0.0286832	MEAN MOTION DOT DOT/6:	0.00000
ARG. OF PERIGEE:	5.8561	BSTAR:	0.00000

DEBRIS CLOUD DATA

MAXIMUM ΔP:	33.1 min
MAXIMUM ΔI:	0.7 deg

COMMENTS

Seven debris (including the parent object) have been officially cataloged. Debris in deep-space orbits are difficult for the SSN to track and catalog. Hundreds of additional fragments could be on-orbit.

REFERENCE DOCUMENTS

"Briz-M Core Stage Fragments Near Geosynchronous Orbit," The Orbital Debris Quarterly News, NASA JSC, April 2016. Available online at https://orbitaldebris.jsc.nasa.gov/quarterly-news/pdfs/odqnv20i1-2.pdf.

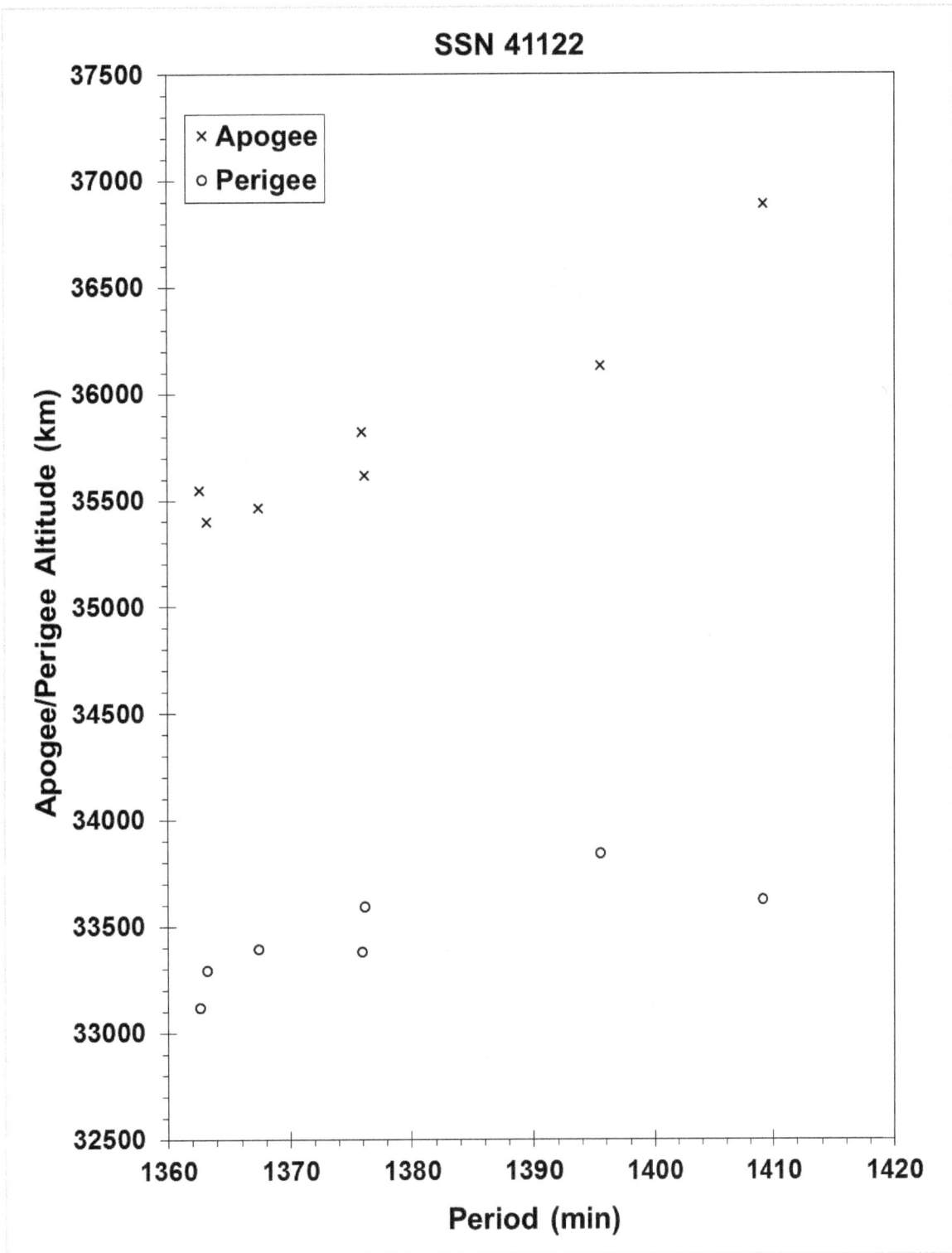

The *Briz-M* debris cloud; debris cataloged up to 127 days after the event.

ASTRO H (HITOMI)   **2016-012A**   **41337**

SATELLITE DATA

TYPE:	payload
OWNER:	Japan
LAUNCH DATE:	17 Feb 2016
DRY MASS (KG):	< 2700
MAIN BODY:	cylinder 1.8 m diameter x 7.9 m long (13.6 m length overall after EOB erection)
MAJOR APPENDAGES:	Solar Array Paddles (SAP) and Extensible Optical Bench (EOB)
ATTITUDE CONTROL:	three-axis stabilized
ENERGY SOURCES:	propellant, batteries, reaction wheels

EVENT DATA

DATE:	26 Mar 2016	LOCATION:	20.82N, 267.36E (asc)
TIME:	0142 GMT	ASSESSED CAUSE:	Propulsion
ALTITUDE:	573.92 km		

PRE-EVENT ELEMENTS

EPOCH:	16084.46380907	MEAN ANOMALY:	86.8693
RIGHT ASCENSION:	87.9716	MEAN MOTION:	14.98194942
INCLINATION:	31.0065	MEAN MOTION DOT/2:	+0.00001057
ECCENTRICITY:	0.0011434	MEAN MOTION DOT DOT/6:	0.00000
ARG. OF PERIGEE:	273.0633	BSTAR:	+0.00008

DEBRIS CLOUD DATA

MAXIMUM ΔP:	0.47 min
MAXIMUM ΔI:	0.02 deg

COMMENTS

This event is categorized as a propulsion-induced structural failure.

REFERENCE DOCUMENTS

"ASTRO-H Spacecraft Fragments During Payload Check-out Operations," The Orbital Debris Quarterly News, NASA JSC, October 2016. Available online at https://www.orbitaldebris.jsc.nasa.gov/quarterly-news/pdfs/odqnv20i4.pdf.

JAXA, "Hitomi Experience Report: Investigation of Anomalies Affecting the X-ray Astronomy Satellite 'Hitomi' (ASTRO-H)," 8 June 2016 edition. Retrieved 1 October 2016 from
http://global.jaxa.jp/projects/sat/astro_h/topics.html?utm_source=dlvr.it&utm_medium=twitter#topics7815

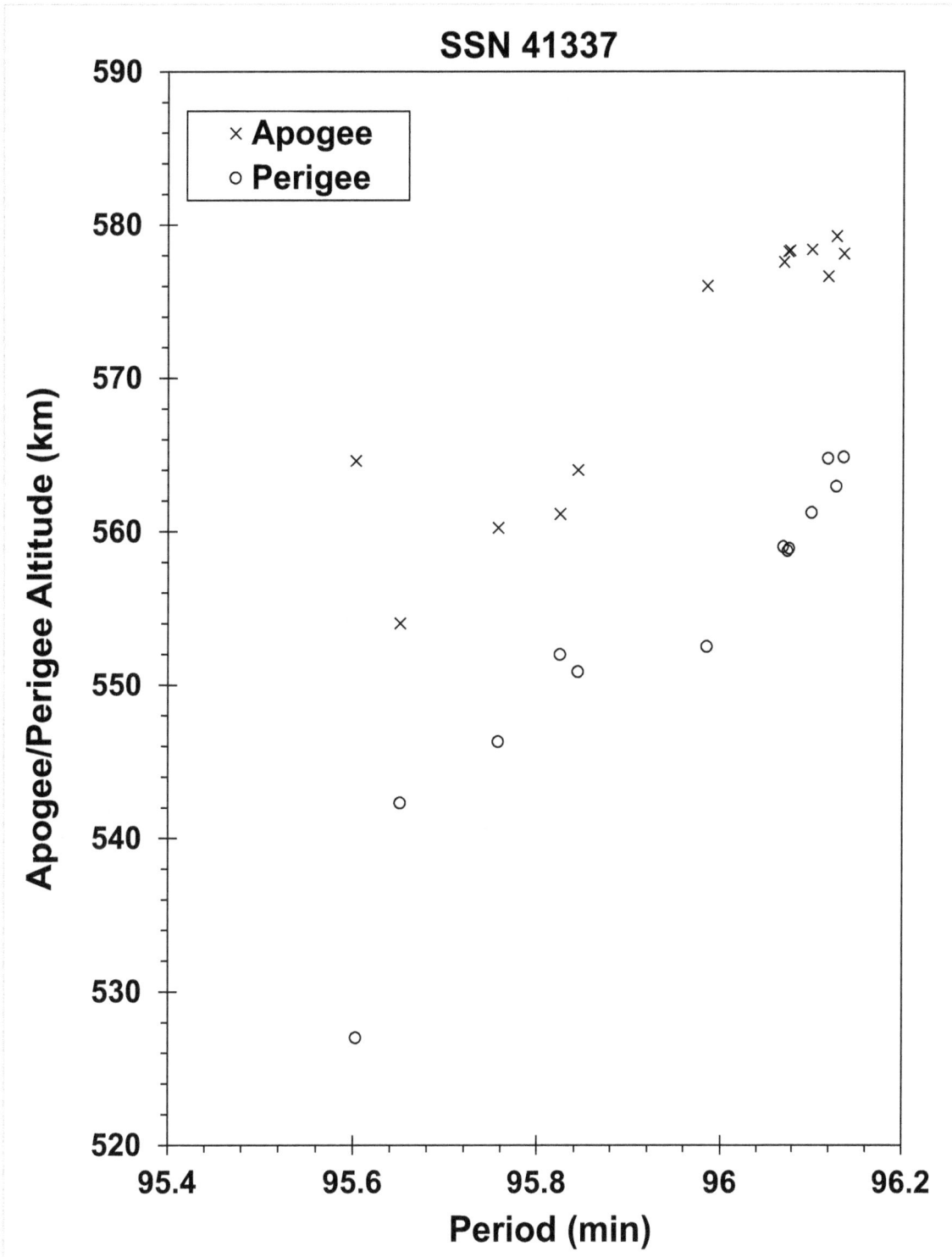

The Hitomi debris cloud; debris cataloged up to approximately one year after the event.

SATELLITE DATA

TYPE:	Mission Related Debris
OWNER:	CIS
LAUNCH DATE:	26 Dec 2017
DRY MASS (KG):	375
MAIN BODY:	toroidal fuel tank; 0.821 m tank diameter x 3.44 m overall diameter
MAJOR APPENDAGES:	none
ATTITUDE CONTROL:	none
ENERGY SOURCES:	unknown

EVENT DATA

DATE:	12 Feb 2018	LOCATION:	38.06N, 117.94E (asc)
TIME:	0957 GMT	ASSESSED CAUSE:	unknown
ALTITUDE:	396.48 km		

PRE-EVENT ELEMENTS

EPOCH:	18041.75579119	MEAN ANOMALY:	287.5127
RIGHT ASCENSION:	273.8260	MEAN MOTION:	10.99960807
INCLINATION:	50.4229	MEAN MOTION DOT/2:	0.00005800
ECCENTRICITY:	0.2212829	MEAN MOTION DOT DOT/6:	0.0000034339
ARG. OF PERIGEE:	98.0587	BSTAR:	0.00020634

DEBRIS CLOUD DATA

MAXIMUM ΔP:	26.3 min
MAXIMUM ΔI:	0.3 deg

COMMENTS

The parent body is assumed to be the jettisonable auxiliary fuel tank discarded by the Fregat-SB upper stage.

REFERENCE DOCUMENT

"Fragmentation of *Fregat-SB* Upper Stage Debris," The Orbital Debris Quarterly News, NASA JSC, May 2018. Available online at https://www.orbitaldebris.jsc.nasa.gov/quarterly-news/pdfs/odqnv22i2.pdf.

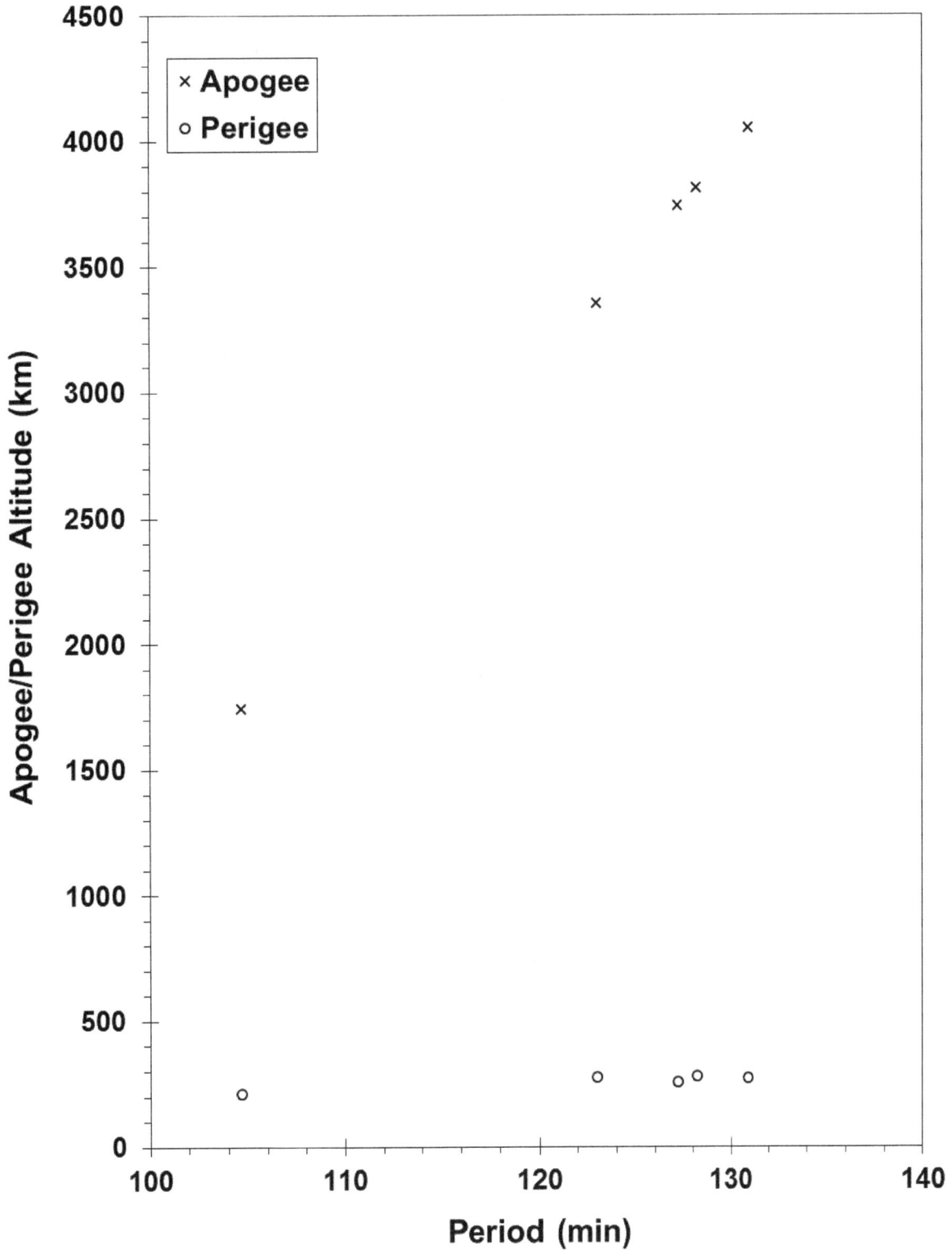

SSN 43089

The Fregat-SB Auxiliary Propellant Tank debris cloud composed of parent body and four fragments; debris cataloged up to 11 days after the event.

SATELLITE DATA

TYPE:	Centaur V (Single Engine Centaur) Rocket Body
OWNER:	USA
LAUNCH DATE:	17 Oct. 2018
DRY MASS (KG):	2243
MAIN BODY:	Cylinder; 3.048 m diameter x 12.68 m long
MAJOR APPENDAGES:	none
ATTITUDE CONTROL:	status unknown at time of event
ENERGY SOURCES:	status unknown at time of event

EVENT DATA

DATE:	06 Apr. 2019	LOCATION:	12.07N, 257.39E (dsc)
TIME:	1859 GMT	ASSESSED CAUSE:	Unknown
ALTITUDE:	21122.1 km		

POST-EVENT ELEMENTS

EPOCH:	19096.51224855	MEAN ANOMALY:	115.9623
RIGHT ASCENSION:	284.3574	MEAN MOTION:	1.83449283
INCLINATION:	12.0165	MEAN MOTION DOT/2:	-.00000066
ECCENTRICITY:	0.4712344	MEAN MOTION DOT DOT/6:	0.0
ARG. OF PERIGEE:	207.7825	BSTAR:	0.0

DEBRIS CLOUD DATA

MAXIMUM ΔP:	184.1 min
MAXIMUM ΔI:	3.4 deg

COMMENTS

This is the second acknowledged Centaur V upper stage fragmentation event. The cause of the event is unknown.

REFERENCE DOCUMENT

"Two Breakup Events Reported," The Orbital Debris Quarterly News, NASA JSC, August 2019. Available online at https://orbitaldebris.jsc.nasa.gov/quarterly-news/pdfs/ODQNv23i3.pdf.

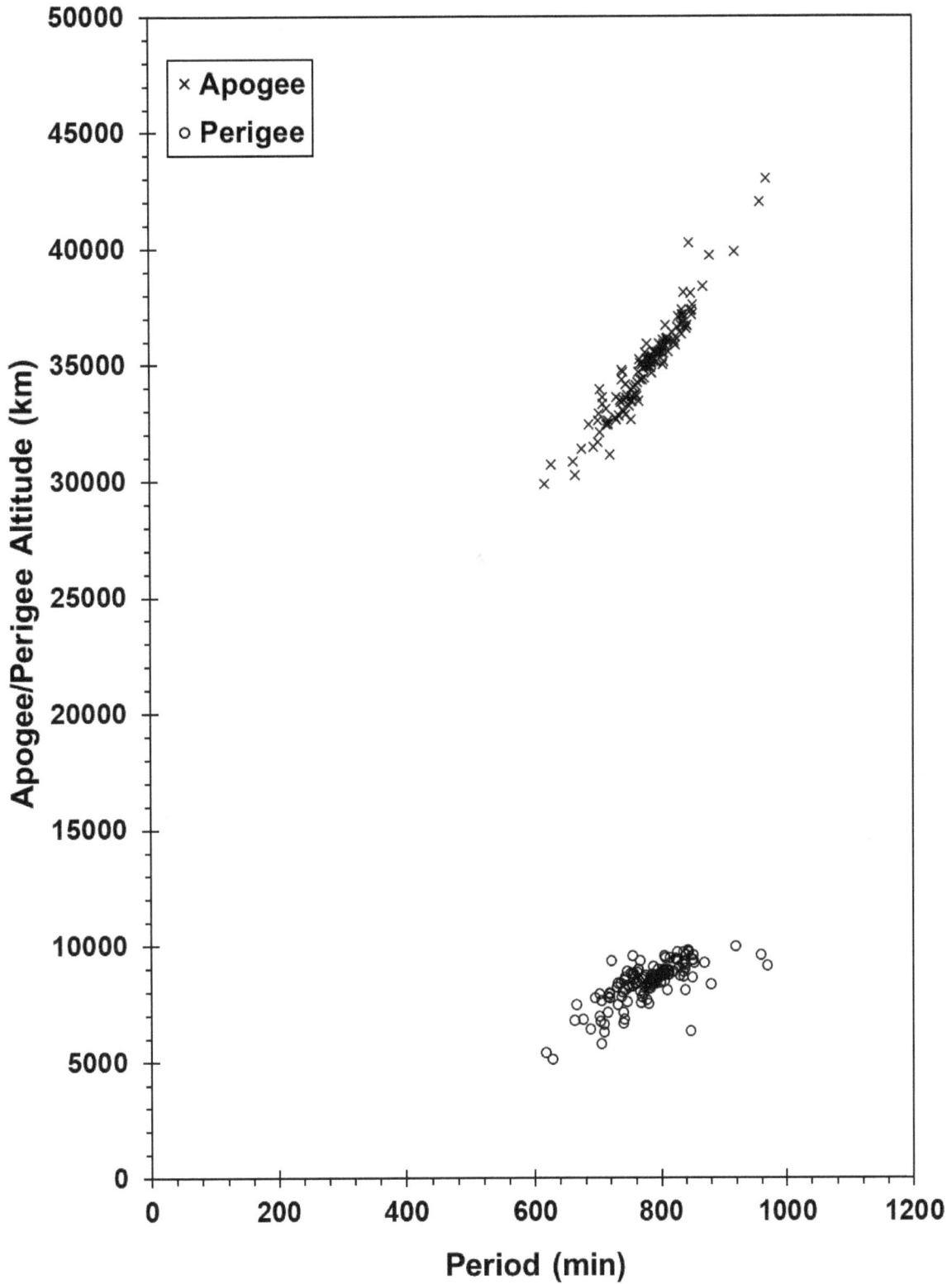

SSN 43652

Identified Centaur V R/B debris cloud; 134 debris objects cataloged up to two years after the event as reconstructed from the US SSN database.

SATELLITE DATA

TYPE:	Mission Related Debris (Payload Encapsulation System [PES])
OWNER:	Japan
LAUNCH DATE:	29 Oct. 2018
DRY MASS (KG):	unknown, but estimated to be more than 100 kg
MAIN BODY:	right circular cylinder split lengthwise; 4.070 m cylinder diameter ~ 4.44 m length
MAJOR APPENDAGES:	none
ATTITUDE CONTROL:	none
ENERGY SOURCES:	unknown

EVENT DATA

DATE:	12 July 2020	LOCATION:	28.03 N, 224.609 E (asc)
TIME:	1000 GMT	ASSESSED CAUSE:	Unknown
ALTITUDE:	595.79 km		

PRE-EVENT ELEMENTS

EPOCH:	20193.87359992	MEAN ANOMALY:	308.9927
RIGHT ASCENSION:	309.0367	MEAN MOTION:	14.88389968
INCLINATION:	97.8854	MEAN MOTION DOT/2:	+0.00000462
ECCENTRICITY:	0.0016456	MEAN MOTION DOT DOT/6:	0.0
ARG. OF PERIGEE:	51.2758	BSTAR:	0.000055838

DEBRIS CLOUD DATA

MAXIMUM ΔP:	3.9 min
MAXIMUM ΔI:	0.7 deg

COMMENTS

Parent object is likely one of the split cylinder lower sections of the three-piece Kawasaki Heavy Industries, Ltd. 4/4D-LC dual payload fairing's lower payload encapsulation system.

REFERENCE DOCUMENT

"Two New Breakups with One Resulting in an ISS Maneuver," The Orbital Debris Quarterly News, NASA JSC, November 2020. Available online at https://orbitaldebris.jsc.nasa.gov/quarterly-news/pdfs/ODQNv24i4.pdf.

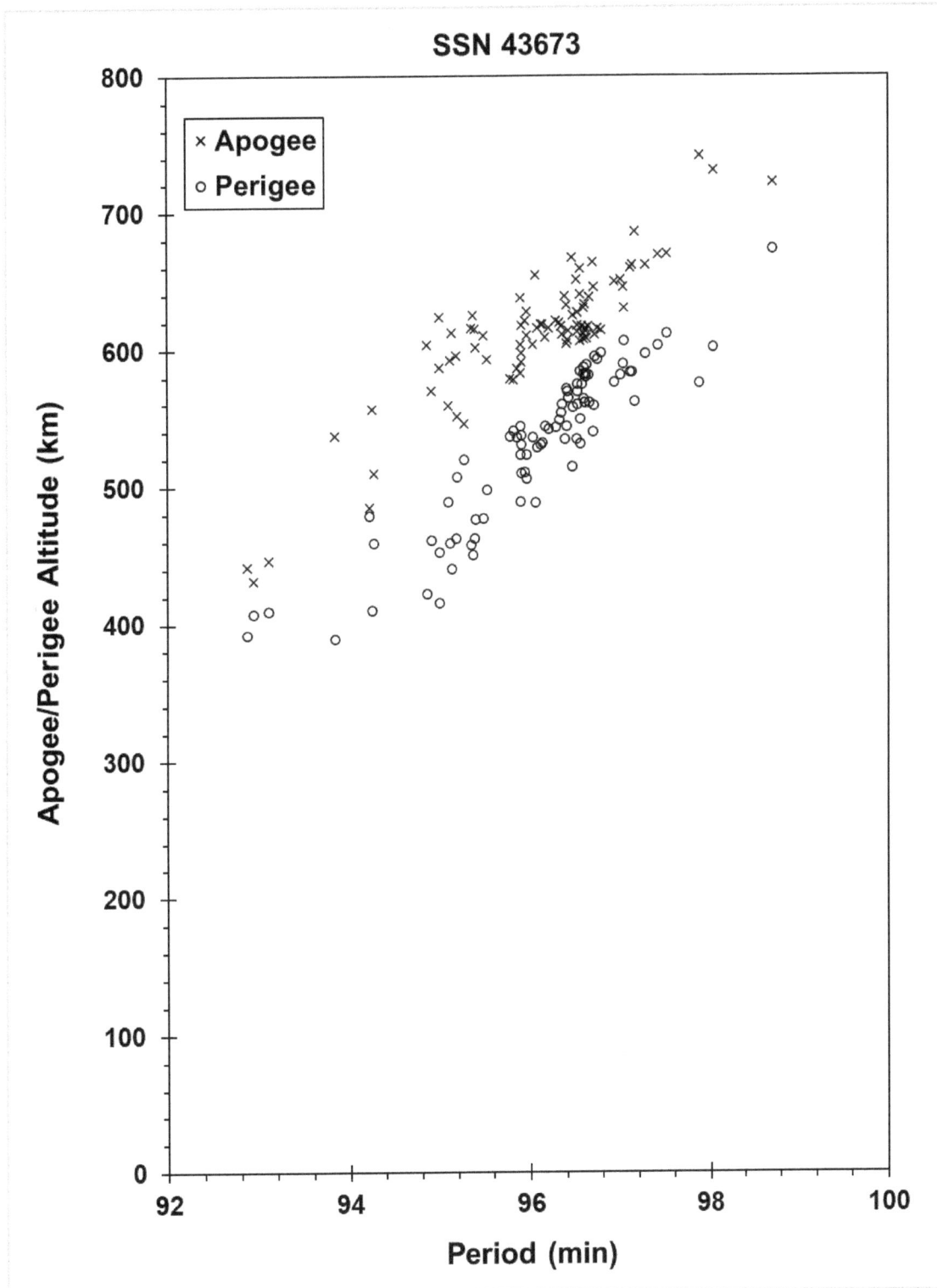

SSN 43673

The GOSAT 2 mission-related debris cloud was composed of the parent object and 87 additional fragments, as reconstructed from the US SSN catalog.

SATELLITE DATA

TYPE:	payload
OWNER:	India
LAUNCH DATE:	24 Jan. 2019
DRY MASS (KG):	740
MAIN BODY:	cubical box
MAJOR APPENDAGES:	Unknown
ATTITUDE CONTROL:	Unknown
ENERGY SOURCES:	Unknown

EVENT DATA

DATE:	27 Mar. 2019	LOCATION:	10.25 N, 89.02 E
TIME:	0540 GMT	ASSESSED CAUSE:	Deliberate Collision
ALTITUDE:	284.5 km		

PRE-EVENT ELEMENTS

EPOCH:	19086.19975398	MEAN ANOMALY:	301.5020
RIGHT ASCENSION:	359.5987	MEAN MOTION:	16.01262446
INCLINATION:	96.6372	MEAN MOTION DOT/2:	0.00067582
ECCENTRICITY:	0.0015984	MEAN MOTION DOT DOT/6:	0.0000037637
ARG. OF PERIGEE:	218.8134	BSTAR:	0.00012331

DEBRIS CLOUD DATA

MAXIMUM ΔP:	20.9 min
MAXIMUM ΔI:	4.4 deg

COMMENTS

This spacecraft was the target of the first Indian test of a ground-based direct ascent anti-satellite (ASAT) system. Over 400 objects were observed initially but tracking, correlation, and cataloging were compromised by the low altitude at which the test was conducted.

REFERENCE DOCUMENTS

"Two Breakup Events Reported," The Orbital Debris Quarterly News, NASA JSC, August 2019. Available online at https://www.orbitaldebris.jsc.nasa.gov/quarterly-news/pdfs/odqnv23i3.pdf.

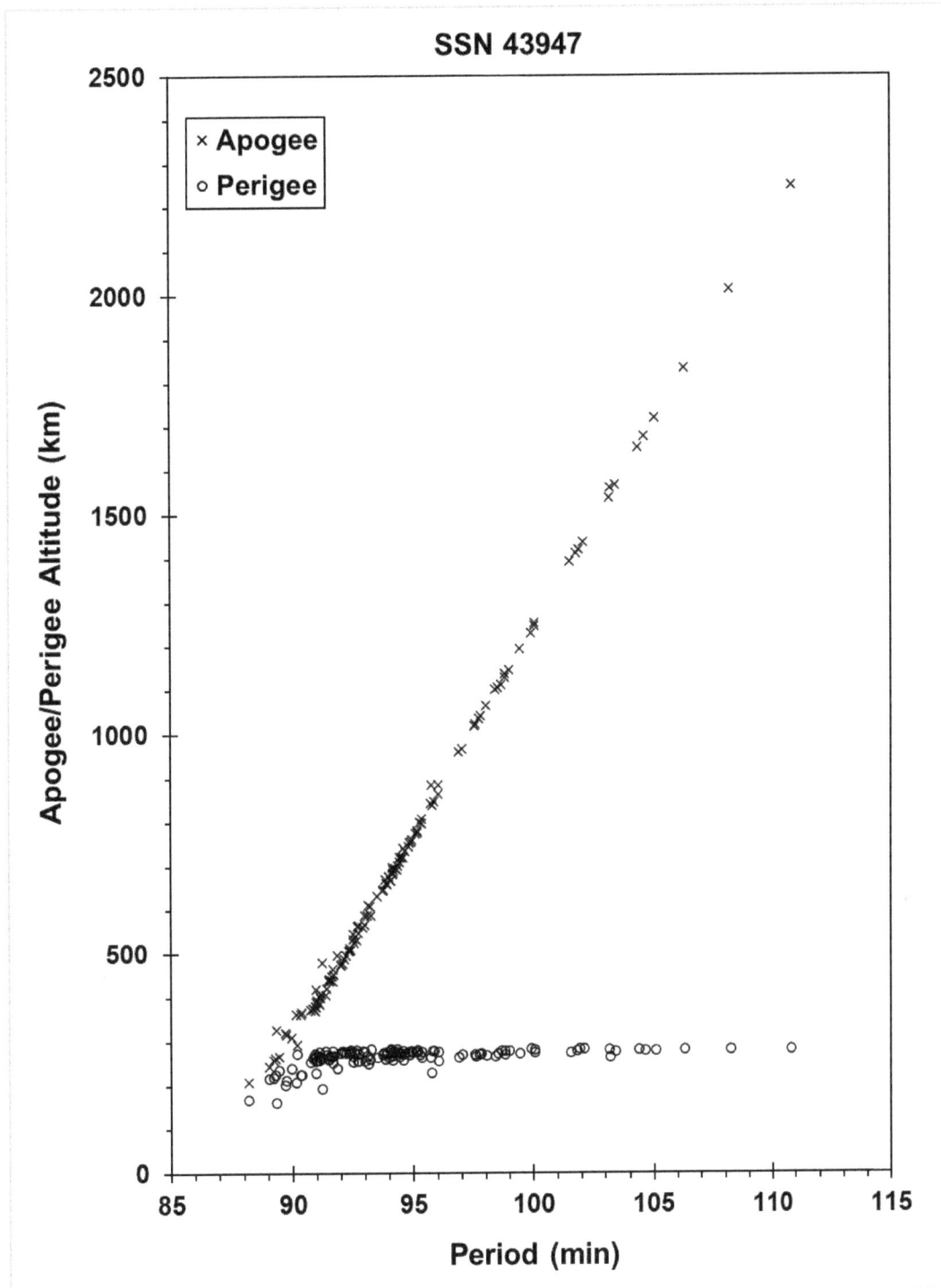

MICROSAT-R ASAT test debris cloud, composed of parent spacecraft and 128 additional objects cataloged between a week and over a year after the event.

SATELLITE DATA

TYPE:	payload
OWNER:	CIS
LAUNCH DATE:	10 July 2019
DRY MASS (KG):	Unknown
MAIN BODY:	Unknown
MAJOR APPENDAGES:	Unknown
ATTITUDE CONTROL:	Unknown
ENERGY SOURCES:	Unknown

EVENT DATA

DATE:	9 Jan. 2020	LOCATION:	80.52 N, 221.58 E
TIME:	2200 GMT	ASSESSED CAUSE:	Unknown
ALTITUDE:	618.9 km		

PRE-EVENT ELEMENTS

EPOCH:	20009.89888825	MEAN ANOMALY:	307.8833
RIGHT ASCENSION:	64.7016	MEAN MOTION:	14.85897343
INCLINATION:	97.9012	MEAN MOTION DOT/2:	0.00000064
ECCENTRICITY:	0.0010215	MEAN MOTION DOT DOT/6:	0.0
ARG. OF PERIGEE:	52.3307	BSTAR:	0.000013659

DEBRIS CLOUD DATA

MAXIMUM ΔP:	3.6 min
MAXIMUM ΔI:	1.1 deg

COMMENTS

This spacecraft is believed to be of the CIS' "Satellite Inspector" class spacecraft.

REFERENCE DOCUMENTS

"Three Recent Breakup Events," The Orbital Debris Quarterly News, NASA JSC, April 2020.   Available online at https://www.orbitaldebris.jsc.nasa.gov/quarterly-news/pdfs/odqnv24i2.pdf.

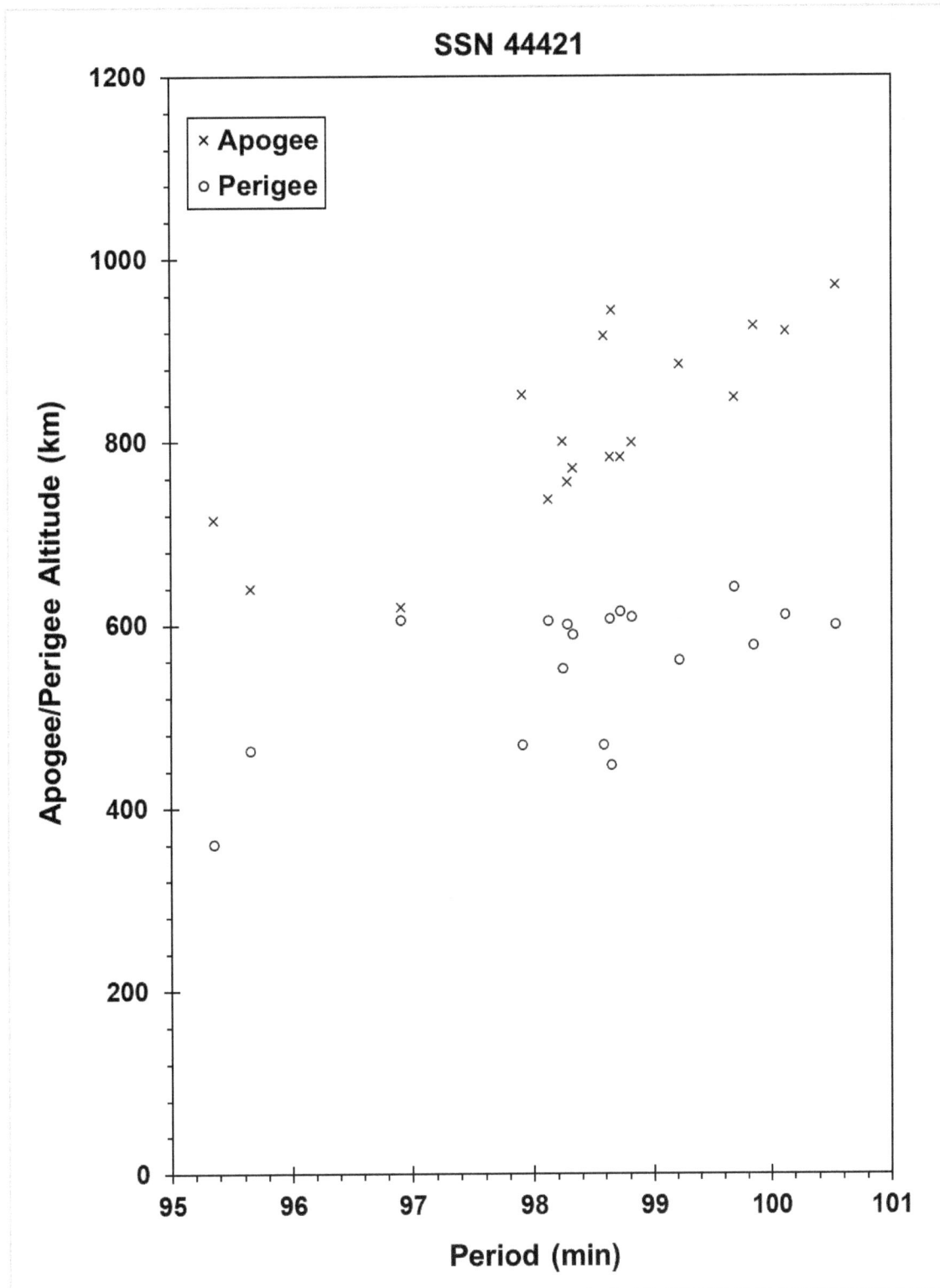

COSMOS 2535 debris cloud, composed of parent spacecraft and 17 additional objects cataloged up to a week after the event.

SATELLITE DATA

TYPE:	payload
OWNER:	PRC
LAUNCH DATE:	25 Sept. 2019
DRY MASS (KG):	~ 1000 kg
MAIN BODY:	CAST2000 bus?
MAJOR APPENDAGES:	solar arrays (2)
ATTITUDE CONTROL:	Unknown
ENERGY SOURCES:	batteries; other sources unknown

EVENT DATA

DATE:	18 Mar. 2021	LOCATION:	71.99 N, 21.19 E (dsc)
TIME:	0741 GMT	ASSESSED CAUSE:	Unknown
ALTITUDE:	792.4 km		

PRE-EVENT ELEMENTS

EPOCH:	21076.81075416	MEAN ANOMALY:	249.3088
RIGHT ASCENSION:	104.6815	MEAN MOTION:	14.32749022
INCLINATION:	98.5409	MEAN MOTION DOT/2:	0.00000177
ECCENTRICITY:	0.0003504	MEAN MOTION DOT DOT/6:	0.0
ARG. OF PERIGEE:	110.8469	BSTAR:	0.000080885

DEBRIS CLOUD DATA

MAXIMUM ΔP:	4.0 min
MAXIMUM ΔI:	0.4 deg

COMMENTS

YunHai 1-class spacecraft are meteorological spacecraft, and YunHai 1-02 likely was operational at the time of the event and afterwards. The cause of the breakup was an accidental collision with SL-16 debris (international designator 1996-051Q, SSN 48078).

REFERENCE DOCUMENTS

"Two On-orbit Breakup Events in March," The Orbital Debris Quarterly News, NASA JSC, June 2021. Available online at https://www.orbitaldebris.jsc.nasa.gov/quarterly-news/pdfs/odqnv25i2.pdf.

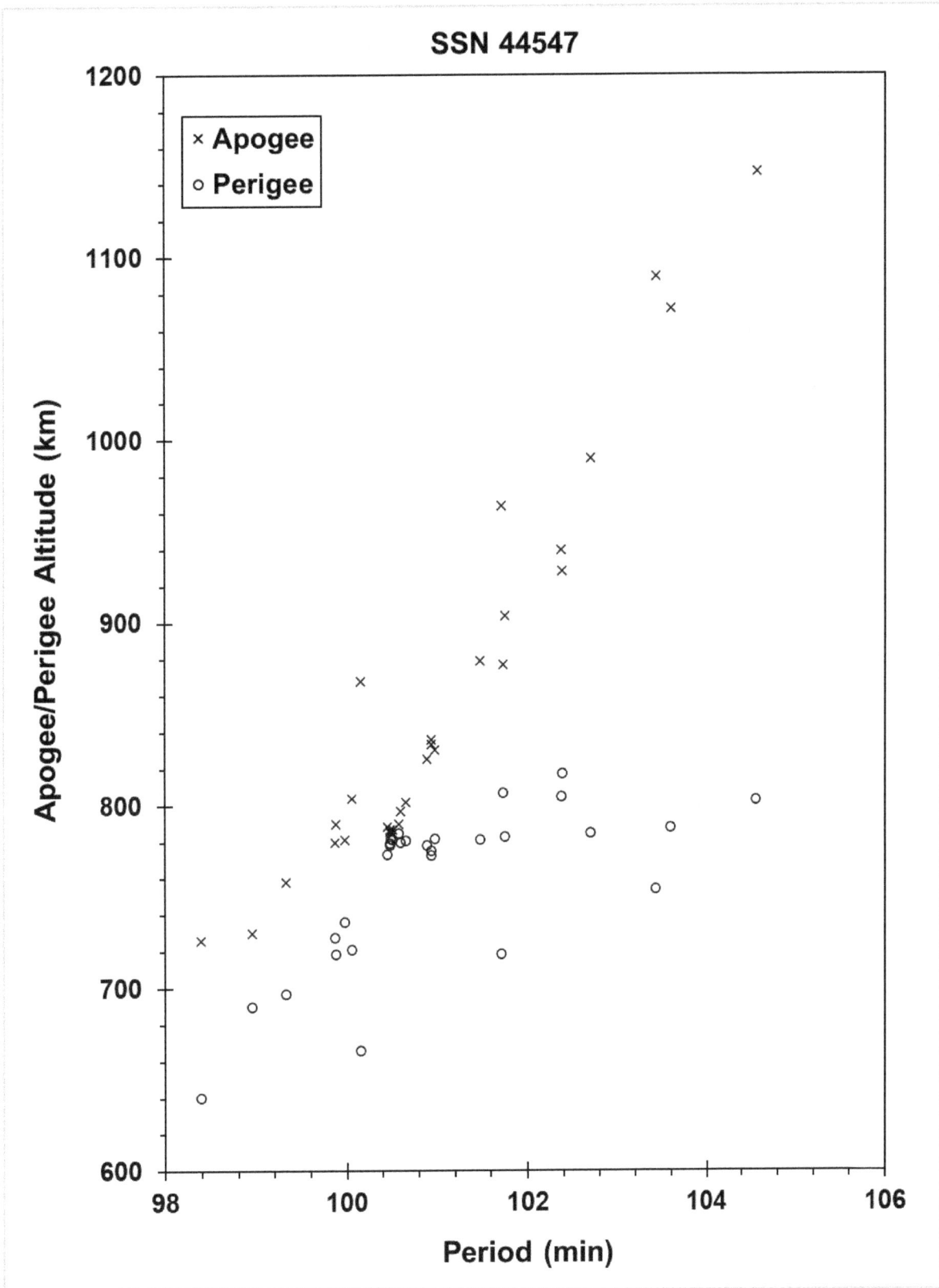

SSN 44547

YunHai 1-02 debris cloud, composed of parent spacecraft and 29 additional objects cataloged up to approximately two months after the event.

SATELLITE DATA

TYPE:	Minotaur IV fourth stage Rocket Body/Guidance Control Assembly
OWNER:	USA
LAUNCH DATE:	15.57 Jul. 2020
DRY MASS (KG):	176
MAIN BODY:	Cylinder; 1.0 m diameter x 1.8 m long
MAJOR APPENDAGES:	none
ATTITUDE CONTROL:	none at time of event
ENERGY SOURCES:	batteries, flight termination system ordnance, cold gas attitude control system

EVENT DATA

DATE:	26 Nov. 2021	LOCATION:	39.81 N, 17.80 E (dsc)
TIME:	2245 GMT	ASSESSED CAUSE:	Unknown
ALTITUDE:	572.1 km		

POST-EVENT ELEMENTS

EPOCH:	21338.26483458	MEAN ANOMALY:	179.1420
RIGHT ASCENSION:	250.4848	MEAN MOTION:	14.97771267
INCLINATION:	53.9810	MEAN MOTION DOT/2:	0.00000097
ECCENTRICITY:	0.0009167	MEAN MOTION DOT DOT/6:	0.0
ARG. OF PERIGEE:	180.9566	BSTAR:	0.000030077

DEBRIS CLOUD DATA*

MAXIMUM ΔP:	Unknown
MAXIMUM ΔI:	Unknown

*unknown, as pre-event elements are not available.

COMMENTS

The apparent discrepancy between the known even time and lack of pre-event elements is because elements of the Minotaur IV upper stage were not publicly available until after the event. The cause of the event is unknown.

REFERENCE DOCUMENT

"Additional Minor Breakup Identified in Fourth Quarter 2021," The Orbital Debris Quarterly News, NASA JSC, June 2022. Available online at https://orbitaldebris.jsc.nasa.gov/quarterly-news/pdfs/ODQNv26i2.pdf.

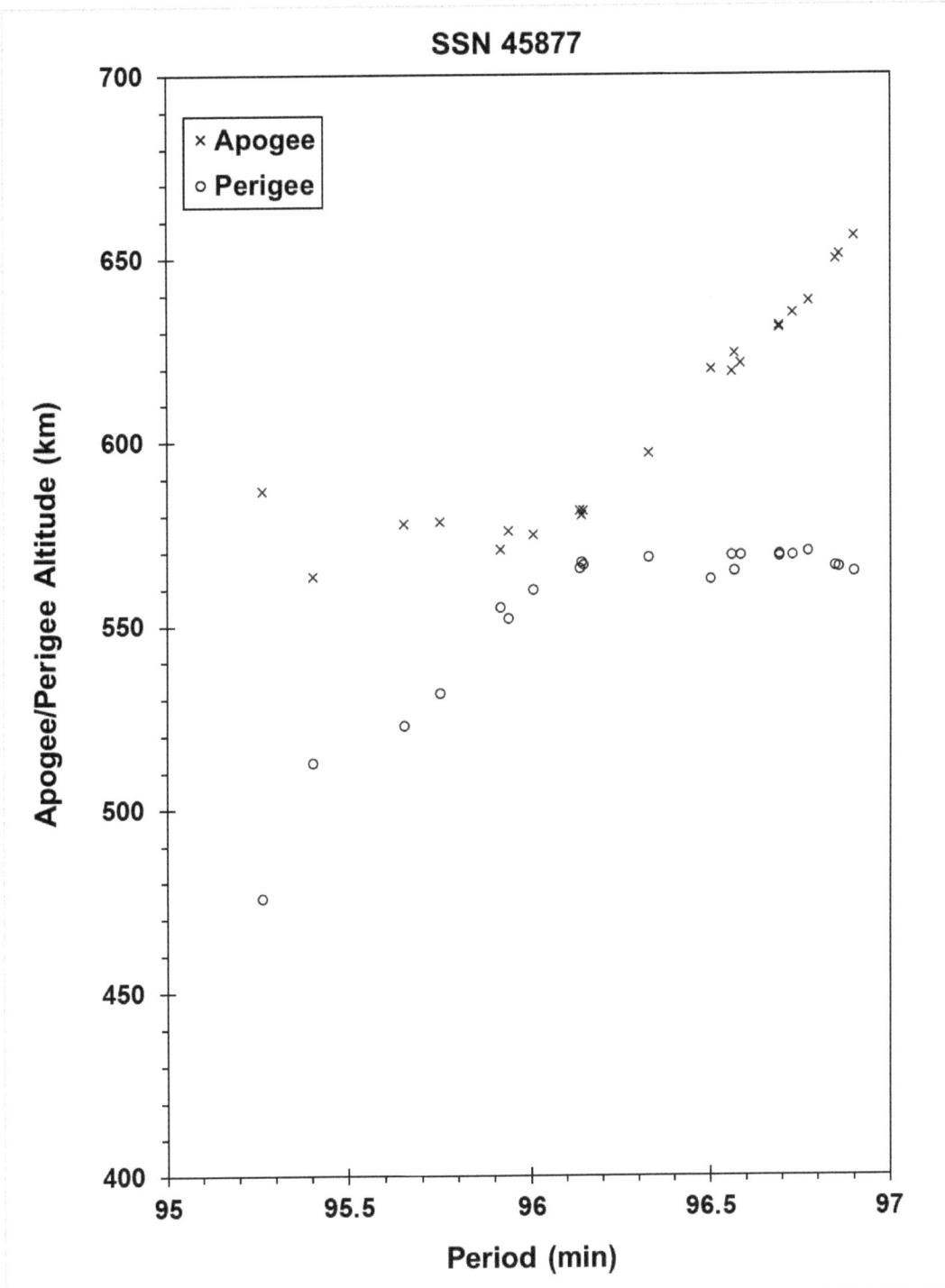

SSN 45877

Minotaur IV R/B debris cloud; parent body and 21 debris objects cataloged up to 110 days after the event as reconstructed from the US SSN database.

This Page Intentionally Left Blank

3.0   SATELLITE ANOMALOUS EVENTS

This section describes the identified anomalous events compiled throughout the years of the U.S. Satellite Catalog and orbital debris analysis associated with this volume. No exhaustive search for anomalous events has yet been conducted, although the following compilation should represent the most significant events noted thus far.

3.1   Background and Status

As defined in the introduction of this volume, an anomalous event is the unplanned separation, usually at low velocity, of one or more detectable objects from a satellite that remains essentially intact. The assessment that the configuration of the parent satellite has not changed significantly is to a degree subjective and is often based on indirect parameters and not on detailed imagery.

Anomalous events can be caused by material deterioration of items such as thermal blankets, protective shields, and solar panels and also by impacts of small debris, either natural or man-made. Other satellite deteriorations, *e.g.*, paint debonding, are known to take place, but are undetectable with the sensors of the US SSN. Interestingly, 40 of the 87 satellites in this section are attributable to the U.S. with another (JASON 1) being a joint U.S.-French spacecraft; 27 are of CIS or predecessor state origin; 4 are French; 3 are Canadian, Japanese, or belong to the PRC; 2 each are of ESA origin or Indian; and 1 each are from Indonesia and a commercial organization headquartered in Luxembourg, SES. Of the 87 satellites, 62 are payloads, 24 are rocket bodies, and 1 is mission-related debris. These events are summarized in Tables 3.1 and 3.2.

Because of the usually low velocity of debris ejection and the potential delay in detecting debris liberated in small numbers, the accuracy of the calculated time of separation is often degraded. Hence, only the month and year of each event are provided, although in some cases the time of the event has been narrowed to a shorter interval. As in the previous section, orbital altitudes are cited to the nearest 5 km based on a mean Earth radius and on the last element set prior to the assessed event date.

Anomalous event debris often exhibit unusually high decay rates, which are indicative of high area-to-mass ratios. This feature, coupled with the normal small size of the debris, hinders official tracking and cataloging. Consequently, some debris are observed but are lost or decay before being assigned a permanent catalog number. The numbers of cataloged debris listed in this section are only from the anomalous events and do not include normal mission related debris identified with the particular launch nor the parent itself.

Historically, anomalous events have often been confused with satellite breakups and have not been the subject of separate, extensive analyses. The list of events in this section is known to be incomplete. Several other satellites have been tentatively tagged as sources of anomalous events. Moreover, preliminary satellite catalog surveys suggest that additional anomalous events have occurred but remain unrecognized as such. Table 3.2 suggests a potential correlation of anomalous events with high solar activity. This section will be updated as future studies warrant.

For additional information on anomalous events, see [3].

TABLE 3.1 HISTORY OF SATELLITE ANOMALOUS EVENTS BY LAUNCH DATE

INTERNATIONAL DESIGNATOR	US SATELLITE NUMBER	LAUNCH DATE	FIRST EVENT DATE	KNOWN EVENTS	CATALOGED DEBRIS	IN-ORBIT DEBRIS	APOGEE (KM)	PERIGEE (KM)	INCLINATION (DEG)
1959-007A	20	18-Sep-59	14-Feb-06	1	2	0	3310	510	33.4
1963-049B	704	5-Dec-63	9/10-Jan-98	1	2	2	1110	1060	90.1
1964-006A	746	30-Jan-64	prior to 1-Jan-80	1	24	1	6900	416	60.9
1964-026A	801	4-Jun-64	Dec-80	4	4	0	930	845	90.5
1964-053B	877	28-Aug-64	Nov-90	1	1	1	775	655	65.1
1965-016A	1271	9-Mar-65	Nov-80	1	2	2	935	900	70.1
1965-027A	1314	3-Apr-65	1-Nov-79	7	158	158	1320	1270	90.3
1965-048A	1420	24-Jun-65	Aug-80	4	9	0	1135	1025	89.9
1965-098A	1804	29-Nov-65	Aug-90	2	2	0	2705	505	79.8
1965-101B	1815	6-Dec-65	Nov-80	1	2	0	660	655	75.8
1965-109A	1864	22-Dec-65	21-Mar-03	2	7	1	1065	895	89.1
1966-005A	1952	28-Jan-66	30-Nov-96	4	2	1	895	855	89.8
1966-024A	2119	26-Mar-66	Apr-80	1	7	2	1115	890	89.9
1966-040A	2173	15-May-66	Jul-81	Many	68	68	3710	3660	100.4
1966-077A	2403	19-Aug-66	Nov-97	5	25	25	1175	1095	89.7
1967-034A	2754	14-Apr-67	Mar-91	1	5	4	1065	1035	90.1
1967-048A	2807	18-May-67	Sep-92	4	4	2	1090	1035	89.6
1967-092A	2965	25-Sep-67	Apr-81	4	7	0	1110	1060	89.3
1968-019B	3151	14-Mar-68	Nov-90	1	0	0	515	450	81.2
1969-009A	3669	30-Jan-69	24-May-07	1	2	2	3455	580	88.5
1970-067A	4507	27-Aug-70	7-Mar-98	1	1	0	1205	945	90.0
1971-003B	4850	20-Jan-71	Jun-87	1	1	0	665	535	81.2
1972-049B	6080	30-Jun-72	Sep-89	1	1	1	935	860	81.2
1972-102A	6319	21-Dec-72	21-Apr-02	1	1	0	1380	1340	74.0
1975-027B	7735	9-Apr-75	Mar-78	1	3	2	845	835	115.0
1978-014A	10664	4-Feb-78	Jan-88	2	2	0	4220	760	65.0
1978-064A	10967	27-Jun-78	Jul-83	>12	20	2	780	780	108.0
1978-094A	11055	10-Oct-78	Feb-93	1	1	0	435	435	81.2
1978-096A	11060	13-Oct-78	Sep-87	2	5	0	855	835	99.0
1978-098B	11081	24-Oct-78	May-81	2	1	0	955	935	99.3
1979-057A	11416	27-Jun-79	Sep-92	2	3	0	805	790	98.7
1979-095A	11605	31-Oct-79	prior to 1-Jan-05	Multiple	90	62	881	862	81.2
1981-043A	12456	14-May-81	Mar-04	1	21	16	895	825	81.3
1981-043B	12457	14-May-81	Oct-96	1	1	1	920	825	81.3
1981-054E	12519	9-Jun-81	Jul-98	1	0	0	33415	85	62.0
1981-059A	12553	23-Jun-81	26-Jul-93	2	6	1	835	830	98.9
1982-102B	13618	19-Oct-82	Early-09	1	1	1	1000	955	83.0
1983-022A	13923	28-Mar-83	Early-05	2	4	2	810	790	98.7
1985-066A	15935	3-Aug-85	Nov-86	2	4	4	1255	1000	89.9
1985-090B	16111	3-Oct-85	5-May-02	1	1	1	565	510	97.7
1986-073A	16969	17-Sep-86	Jun-07	2	5	0	810	790	98.7
1987-011A	17369	1-Feb-87	4-Jul-08	1	53	52	800	775	65.0

TABLE 3.1 HISTORY OF SATELLITE ANOMALOUS EVENTS BY LAUNCH DATE (CONT'D)

INTERNATIONAL DESIGNATOR	US SATELLITE NUMBER	LAUNCH DATE	FIRST EVENT DATE	KNOWN EVENTS	CATALOGED DEBRIS	IN-ORBIT DEBRIS	APOGEE (KM)	PERIGEE (KM)	INCLINATION (DEG)
1987-020A	17535	20-Feb-87	Apr-May 97	3	3	3	1525	1480	73.6
1987-060A	18187	10-Jul-87	21-Mar to 4-Apr-14	1	42	42	800	775	65.0
1988-005A	18820	30-Jan-88	2000-2001	2	2	2	960	936	82.5
1988-032B	19046	20-Apr-88	30-Jul-96	2	2	2	655	585	97.6
1988-089A	19531	24-Sep-88	24-Nov-10	1	2	2	850	835	98.8
1989-089A	20322	18-Nov-89	Mar-93	12	76	0	885	870	99.0
1990-017B	20509	27-Feb-90	22-Jun-05	1	2	2	1015	950	83.0
1990-037B	20580	24-Apr-90	5-Aug-03	1	1	0	575	570	28.5
1991-032A	21263	14-May-91	2-Oct-11	1	4	4	815	800	98.7
1991-039A	21397	4-Jun-91	12-Oct-98	1	1	0	655	620	82.5
1991-050A	21574	17-Jul-91	7-Jul-02	7	8	8	800	750	98.5
1991-050E	21578	17-Jul-91	22-Aug-03	3	4	0	730	730	98.1
1991-050F	21610	17-Jul-91	Apr-01	1	1	0	770	770	98.2
1991-063B	21701	12-Sep-91	10-Nov-07	3	5	0	485	355	57.0
1993-014A	22561	25-Mar-93	4-Mar-98	1	2	2	970	685	75.8
1993-014B	22562	25-Mar-93	Late-02	Multiple	53	16	920	680	75.8
1994-057A	23233	29-Aug-94	23-Oct-16	1	4	2	848	832	99.1
1994-074A	23342	4-Nov-94	2006	Multiple	36	9	660	633	97.9
1994-077B	23405	24-Nov-94	Jun-98	2?	1	0	845	845	71.0
1994-089A	23455	30-Dec-94	Jul-12	1	1	1	860	850	98.8
1995-021A	23560	21-Apr-95	24-Jul-03	Multiple	6	4	787	785	98.6
1995-021B	23561	21-Apr-95	prior to 1-Jan-18	1	1	1	772	762	98.8
1995-041B	23640	5-Aug-95	6-Dec-95	1	1	0	1375	935	26.7
1995-059B	23711	4-Nov-95	30-Jan-96	1	2	0	1495	935	100.6
1996-017A	23827	21-Mar-96	Oct-00	1	1	0	822	820	98.6
1997-082C	25106	20-Dec-97	7-Jun-14	1	10	5	781	778	86.4
1999-035A	25791	24-Jun-99	6-Jun-04	1	9	0	760	745	25.0
1999-040C	25868	23-Jul-99	2001	2	11	11	13499	279	28.5
1999-042A	25880	12-Aug-99	25-Aug-17	1	0	0	35793	35781	0.0112
1999-051A	25919	24-Sep-99	19-Mar-01	1	1	0	680	680	98.2
2001-055A	26997	7-Dec-01	16-Mar-02	1	2	2	1345	1330	66.0
2002-005A	27372	11-Feb-02	30-Nov-14	1	4	1	781	777	86.4
2002-022A	27424	04-May-02	Apr-19	1?	4	2	703	702	98.21
2003-024A	27820	6-Jun-03	17-Jun-17	1	0	0	35798	35774	0.0174
2004-049A	28492	18-Dec-04	26-Jul-21	1	7	6	682*	646*	98.09*
2009-049G	35871	17-Sep-09	22-Jan-13	1	1	1	825	815	98.6
2009-055A	35946	8-Oct-09	18-Jul-16	1	16	10	768	767	98.5
2011-043A	37781	15-Aug-11	6/7-Jul-14	2	3	0	965	965	99.0
2012-017A	38248	26-Apr-12	30-Sep-16	1	1	0	544	538	97.6
2015-020E	40556	31-Mar-15	29-Apr-15	1	6	6	1342	1339	82.5
2018-084E	43675	29-Oct-18	6-Feb-19	1	4	0	629	594	97.9

TABLE 3.1 HISTORY OF SATELLITE ANOMALOUS EVENTS BY LAUNCH DATE (CONT'D)

INTERNATIONAL DESIGNATOR	US SATELLITE NUMBER	LAUNCH DATE	FIRST EVENT DATE	KNOWN EVENTS	CATALOGED DEBRIS	IN-ORBIT DEBRIS	APOGEE (KM)	PERIGEE (KM)	INCLINATION (DEG)
2018-084L	43682	29-Oct-18	prior to Sep-19	1	1	1	590	493	98.8
2021-092C	52260	15-Oct-21	15-Apr-22	1	4	4	378	368	41.5
2021-133A	50505	27-Dec-21	28-Dec-21	1	3	0	206	183	63.38
2022-031C	52152	29-Mar-22	29-Mar-22	1	0	0	594	556	97.8

587

TABLE 3.2 HISTORY OF SATELLITE ANOMALOUS EVENTS BY EVENT DATE

INTERNATIONAL DESIGNATOR	US SATELLITE NUMBER	LAUNCH DATE	FIRST EVENT DATE	KNOWN EVENTS	CATALOGED DEBRIS	IN-ORBIT DEBRIS	APOGEE (KM)	PERIGEE (KM)	INCLINATION (DEG)
1975-027B	7735	9-Apr-75	Mar-78	1	3	2	845	835	115.0
1965-027A	1314	3-Apr-65	1-Nov-79	7	158	158	1320	1270	90.3
1964-006A	746	30-Jan-64	prior to 1-Jan-80	1	24	1	6900	416	60.9
1966-005A	1952	28-Jan-66	Apr-80	4	7	1	1205	855	89.8
1965-048A	1420	24-Jun-65	Aug-80	4	9	0	1135	1025	89.9
1965-016A	1271	9-Mar-65	Nov-80	1	2	2	935	900	70.1
1964-026A	801	4-Jun-64	Dec-80	4	4	0	930	845	90.5
1967-092A	2965	25-Sep-67	Apr-81	4	7	0	1110	1035	89.3
1978-098B	11081	24-Oct-78	May-81	2	1	0	955	935	99.3
1966-024A	2119	26-Mar-66	Jul-81	1	2	2	1115	890	89.9
1978-064A	10967	27-Jun-78	Jul-83	>12	20	2	780	780	108.0
1985-066A	15935	3-Aug-85	Nov-86	2	4	4	1255	1000	89.9
1971-003B	4850	20-Jan-71	Jun-87	1	1	0	665	535	81.2
1978-096A	11060	13-Oct-78	Sep-87	2	5	0	855	835	99.0
1978-014A	10664	4-Feb-78	Jan-88	2	2	0	4220	760	65.0
1972-049B	6080	30-Jun-72	Sep-89	1	1	0	935	860	81.2
1965-098A	1804	29-Nov-65	Aug-90	2	2	0	2705	505	79.8
1964-053B	877	28-Aug-64	Nov-90	1	1	1	775	655	65.1
1968-019B	3151	14-Mar-68	Nov-90	1	0	0	515	450	81.2
1966-077A	2403	19-Aug-66	Mar-91	5	25	25	3710	3660	89.7
1967-034A	2754	14-Apr-67	Sep-92	1	5	4	1065	1035	90.1
1979-057A	11416	27-Jun-79	Sep-92	2	3	0	805	790	98.7
1978-094A	11055	10-Oct-78	Feb-93	1	1	0	435	435	81.2
1989-089A	20322	18-Nov-89	Mar-93	12	76	0	885	870	99.0
1981-059A	12553	23-Jun-81	26-Jul-93	2	6	1	835	830	98.9
1967-048A	2807	18-May-67	Feb-95	1	4	2	1090	1060	89.6
1995-041B	23640	5-Aug-95	6-Dec-95	1	1	0	1375	935	26.7
1995-059B	23711	4-Nov-95	30-Jan-96	1	2	0	1495	935	100.6
1988-032B	19046	20-Apr-88	30-Jul-96	2	2	2	655	585	97.6
1981-043B	12457	14-May-81	Oct-96	1	1	1	920	825	81.3
1965-109A	1864	22-Dec-65	30-Nov-96	2	2	1	1065	895	89.1
1987-020A	17535	20-Feb-87	Apr-May 97	3	3	3	1525	1480	73.6
1966-040A	2173	15-May-66	Nov-97	Many	68	68	1175	1095	100.4
1963-049B	704	5-Dec-63	9/10-Jan-98	1	2	2	1110	1060	90.1
1993-014A	22561	25-Mar-93	4-Mar-98	1	2	2	970	685	75.8
1970-067A	4507	27-Aug-70	7-Mar-98	1	1	0	1205	945	90.0
1994-077B	23405	24-Nov-94	Jun-98	2?	1	0	845	845	71.0
1981-054E	12519	9-Jun-81	Jul-98	1	0	0	33415	85	62.0
1991-039A	21397	4-Jun-91	12-Oct-98	1	1	0	655	620	82.5
1988-005A	18820	30-Jan-88	2000-2001	2	2	2	960	936	82.5
1996-017A	23827	21-Mar-96	Oct-00	1	1	0	822	820	98.6
1999-040C	25868	23-Jul-99	2001	2	11	11	13499	279	28.5
1999-051A	25919	24-Sep-99	19-Mar-01	1	1	0	680	680	98.2
1991-050F	21610	17-Jul-91	Apr-01	1	1	0	770	770	98.2

TABLE 3.2 HISTORY OF SATELLITE ANOMALOUS EVENTS BY EVENT DATE (CONT'D)

INTERNATIONAL DESIGNATOR	US SATELLITE NUMBER	LAUNCH DATE	FIRST EVENT DATE	KNOWN EVENTS	CATALOGED DEBRIS	IN-ORBIT DEBRIS	APOGEE (KM)	PERIGEE (KM)	INCLINATION (DEG)
2001-055A	26997	7-Dec-01	16-Mar-02	1	2	2	1345	1330	66.0
1972-102A	6319	21-Dec-72	21-Apr-02	1	1	0	1380	1340	74.0
1985-090B	16111	3-Oct-85	5-May-02	1	1	1	565	510	97.7
1991-050A	21574	7-Jul-91	7-Jul-02	7	8	8	800	750	98.5
1993-014B	22562	25-Mar-93	Late-02	Multiple	53	16	920	680	75.8
1965-101B	1815	6-Dec-65	21-Mar-03	1	2	0	660	655	75.8
1995-021A	23560	21-Apr-95	24-Jul-03	Multiple	6	4	787	785	98.6
1990-037B	20580	24-Apr-90	5-Aug-03	1	1	0	575	570	28.5
1991-050E	21578	17-Jul-91	22-Aug-03	3	4	0	730	730	98.1
1981-043A	12456	14-May-81	Mar-04	1	21	16	895	825	81.3
1999-035A	25791	24-Jun-99	6-Jun-04	1	9	0	760	745	25.0
1979-095A	11605	31-Oct-79	prior to 1-Jan-05	Multiple	90	62	881	862	81.2
1983-022A	13923	28-Mar-83	Early-05	2	4	2	810	790	98.7
1990-017B	20509	27-Feb-90	22-Jun-05	1	2	2	1015	950	83.0
1994-074A	23342	4-Nov-94	2006	Multiple	36	9	660	633	97.9
1959-007A	20	18-Sep-59	14-Feb-06	1	2	0	3310	510	33.4
1969-009A	3669	30-Jan-69	24-May-07	1	2	2	3455	580	88.5
1986-073A	16969	17-Sep-86	Jun-07	2	5	5	810	790	98.7
1991-063B	21701	12-Sep-91	10-Nov-07	3	5	0	485	355	57.0
1987-011A	17369	1-Feb-87	4-Jul-08	1	53	52	800	775	65.0
1982-102B	13618	19-Oct-82	Early-09	1	1	1	1000	955	83.0
1988-089A	19531	24-Sep-88	24-Nov-10	1	2	2	850	835	98.8
1991-032A	21263	14-May-91	2-Oct-11	1	4	4	815	800	98.7
1994-089A	23455	30-Dec-94	Jul-12	1	1	1	860	850	98.8
2009-049G	35871	17-Sep-09	22-Jan-13	1	1	1	825	815	98.6
1987-060A	18187	10-Jul-87	21-Mar to 4-Apr-14	1	42	42	800	775	65.0
1997-082C	25106	20-Dec-97	7-Jun-14	1	10	5	781	778	86.4
2011-043A	37781	15-Aug-11	6/7-Jul-14	2	3	0	965	965	99.0
2002-005A	27372	11-Feb-02	30-Nov-14	1	4	1	781	777	86.4
2015-020E	40556	31-Mar-15	29-Apr-15	1	6	6	1342	1339	82.5
2009-055A	35946	8-Oct-09	18-Jul-16	1	1	1	768	767	98.5
2012-017A	38248	26-Apr-12	30-Sep-16	1	1	0	544	538	97.6
1994-057A	23233	29-Aug-94	23-Oct-16	1	4	2	848	832	99.1
2003-024A	27820	6-Jun-03	17-Jun-17	1	0	0	35798	35774	0.0174
1999-042A	25880	12-Apr-99	25-Aug-17	1	0	0	35793	35781	0.0112
1995-021B	23561	21-Apr-95	prior to 1-Jan-18	1	1	1	772	762	98.8
2018-084E	43675	29-Oct-18	6-Feb-19	1	4	4	629	594	97.9
2002-022A	27424	04-May-02	Apr-19	1?	4	2	703	702	98.21
2018-084L	43682	29-Oct-18	prior to Sep-19	1	1	1	590	493	98.8
2004-049A	28492	18-Dec-04	26-Jul-21	1	7	6	682*	646*	98.09*
2021-133A	50505	27-Dec-21	28-Dec-21	1	3	0	206	183	63.38
2022-031C	52152	29-Mar-22	29-Mar-22	1	0	0	594	556	97.8
2021-092C	52260	15-Oct-21	15-Apr-22	1	4	4	378	368	41.5

This Page Intentionally Left Blank

3.2 Identified Satellite Anomalous Events

Much like section 2.2 above, this section identifies particulars for the limited number of anomalous events thus far cataloged. There is no Gabbard Diagram included with these events, and each page often refers to multiple events. The first known date of the first anomalous event is categorized for each satellite. Where possible the best estimate of the cause and potential failure are noted in the comments section.

REFERENCE DOCUMENT

3. Johnson, N. L. "Environmentally induced Debris Sources," Advances in Space Research 34, pp. 993-999, (2004).

SATELLITE DATA

TYPE:	Payload
OWNER:	US
LAUNCH DATE:	18 Sep 1959
DRY MASS (KG):	45
MAIN BODY:	50.8 cm sphere with third stage attached
MAJOR APPENDAGES:	66 cm boom(s)
ATTITUDE CONTROL:	spin stabilized

EVENT DATA

KNOWN EVENTS:	1
FIRST DATE:	14 Feb 2006

APOGEE	PERIGEE	PERIOD	INCLINATION
3310	510	125.14 min	33.4 deg

COMMENTS

At the time of the event, Vanguard 3 was the 5th oldest object in orbit. Two objects released, although the second object was not cataloged until May 2007.

REFERENCE DOCUMENT

"First Satellite Breakups of 2006", The Orbital Debris Quarterly News, NASA JSC, July 2006.
Available online at https://orbitaldebris.jsc.nasa.gov/quarterly-news/pdfs/ODQNv10i3.pdf.

"Detection of Debris from Chinese ASAT Test Increases; One Minor Fragmentation Event in Second Quarter of 2007", The Orbital Debris Quarterly News, NASA JSC, July 2007.
Available online at https://orbitaldebris.jsc.nasa.gov/quarterly-news/pdfs/ODQNv11i3.pdf.

SATELLITE DATA

TYPE:	Payload
OWNER:	US
LAUNCH DATE:	5 December 1963
DRY MASS (KG):	75
MAIN BODY:	Octagon; 0.46 m diameter by 0.5 m length
MAJOR APPENDAGES:	Boom
ATTITUDE CONTROL:	None at the time of event

EVENT DATA

KNOWN EVENTS:	1
FIRST DATE:	9-10 January 1998

APOGEE	PERIGEE	PERIOD	INCLINATION
1110 km	1060 km	106.98 min	90.1 deg

COMMENTS

One of several Transit-class spacecraft involved in anomalous events. Spacecraft may have experienced earlier anomalous events in 1960s and 1970s. Only one object associated with January 1998 event.

SATELLITE DATA

TYPE:	payload
OWNER:	CIS
LAUNCH DATE:	30 Jan 1964
DRY MASS (KG):	329
MAIN BODY:	Cylinder with hemispherical ends; 0.75 m diameter by 1.3 m length
MAJOR APPENDAGES:	Six solar arrays, antennas
ATTITUDE CONTROL:	None at the time of event

EVENT DATA

KNOWN EVENTS: 1

FIRST DATE: 1982-1984

APOGEE	PERIGEE	PERIOD	INCLINATION
6825.3 km	404.7 km	165.7 min	60.8 deg

COMMENTS

"Event Data" is for epoch of 1 January 1982, given the uncertainty in actual time of separation event(s). It is likely that four of the 24 anomalous debris are associated with other space missions, including the sole piece on orbit as of this edition.

SATELLITE DATA

TYPE:	Payload
OWNER:	US
LAUNCH DATE:	4 June 1964
DRY MASS (KG):	60
MAIN BODY:	Octagonal cylinder; 0.5 m by 0.4 m
MAJOR APPENDAGES:	4 solar panels; gravity-gradient boom
ATTITUDE CONTROL:	Gravity-gradient boom

EVENT DATA

KNOWN EVENTS:	4
FIRST DATE:	December 1980

APOGEE	PERIGEE	PERIOD	INCLINATION
930 km	845 km	102.7 min	90.5 deg

COMMENTS

Second event observed July 1982. The third event occurred in May 1994. Fourth event date not determined but also close to May 1994. First fragment decayed rapidly; the second decayed more slowly. Two latest pieces not cataloged as of publication date. One of several known Transits involved in anomalous events.

SATELLITE DATA

TYPE:	Vostok Final Stage
OWNER:	CIS
LAUNCH DATE:	28 August 1964
DRY MASS (KG):	1440
MAIN BODY:	Cylinder; 2.6 m diameter by 3.8 m length
MAJOR APPENDAGES:	None
ATTITUDE CONTROL:	None at time of event

EVENT DATA

KNOWN EVENTS:	1
FIRST DATE:	Late-1990

APOGEE	PERIGEE	PERIOD	INCLINATION
775 km	655 km	99.1 min	65.1 deg

COMMENTS

Cosmos 44 was the first prototype spacecraft of the Meteor 1 program. This is one of several Vostok final stages associated with this old program to shed a piece of debris since 1987.

SATELLITE DATA

TYPE:	Payload
OWNER:	US
LAUNCH DATE:	9 May 1965
DRY MASS (KG):	40
MAIN BODY:	Sphere
MAJOR APPENDAGES:	Unknown
ATTITUDE CONTROL:	Unknown

EVENT DATA

KNOWN EVENTS: 1

FIRST DATE: November 1980

APOGEE	PERIGEE	PERIOD	INCLINATION
935 km	900 km	103.4 min	70.1 deg

COMMENTS

No other events observed.

SATELLITE DATA

TYPE:	Payload (attached to Agena D upper stage)
OWNER:	US
LAUNCH DATE:	3 April 1965
DRY MASS (KG):	2500 (approx.)
MAIN BODY:	Cylinder-cone; 1.5 m by 11.6 m
MAJOR APPENDAGES:	None
ATTITUDE CONTROL:	None at time of event

EVENT DATA

KNOWN EVENTS:	7
FIRST DATE:	November 1979

APOGEE	PERIGEE	PERIOD	INCLINATION
1320 km	1270 km	111.5 min	90.3 deg

COMMENTS

Six additional events observed: Dec 1980, Aug 1981, Mar 1983, Aug 1983, Nov 1983, and Jan 1985. Decay rates of all debris are nominal for this altitude. One debris was administratively decayed in February 1989.

REFERENCE DOCUMENTS

Investigation of Certain Anomalies Associated with Object 1314, A US Nuclear Powered Satellite, G. T. DeVere, Technical Memorandum 85-S-001, Headquarters NORAD/ADCOM, DCS/Plans, March 1985 (Appendix TM-85-001A, Secret).

Environmentally-Induced Debris Sources, N. L. Johnson, NASA Lyndon B. Johnson Space Center, Second World Space Congress, 2002.

SATELLITE DATA

TYPE:	Payload
OWNER:	US
LAUNCH DATE:	24 June 1965
DRY MASS (KG):	60
MAIN BODY:	Octagonal cylinder; 0.5 m by 0.4 m
MAJOR APPENDAGES:	4 solar panels; gravity-gradient boom
ATTITUDE CONTROL:	Gravity-gradient

EVENT DATA

KNOWN EVENTS:	Multiple
FIRST DATE:	August 1980

APOGEE	PERIGEE	PERIOD	INCLINATION
1135 km	1025 km	106.8 min	89.9 deg

COMMENTS

Three additional events observed: one 2 days after the initial event, one in June 1981, and the most recent in late 1999. All debris appear very small. One of several known Transits involved in anomalous events.

REFERENCE DOCUMENT

Environmentally-Induced Debris Sources, N. L. Johnson, NASA Lyndon B. Johnson Space Center, Second World Space Congress, 2002.

SATELLITE DATA

TYPE:	Payload
OWNER:	Canada
LAUNCH DATE:	29 November 1965
DRY MASS (KG):	146
MAIN BODY:	Oblate spheroid; 0.86 m high by 1.1 m diameter
MAJOR APPENDAGES:	73 m- and 22.8 m-long dipole antennas
ATTITUDE CONTROL:	spin-stabilized

EVENT DATA

KNOWN EVENTS:	2
FIRST DATE:	August 1990

APOGEE	PERIGEE	PERIOD	INCLINATION
2742 km	506 km	118.7 min	79.8 deg

COMMENTS

First event in August 1990 (SSN 20833); second separation event 25 July 1996 (SSN 25058). Both objects displayed a high area-to-mass ratio resulting in relatively rapid decays on 11 February 1992 and 15 December 1999, respectively.

REFERENCE DOCUMENT

Corliss, W.R., <u>Scientific Satellites</u>. NASA SP-133 (1967).

SATELLITE DATA

TYPE:	Rocket Body
OWNER:	US
LAUNCH DATE:	6 December 1965
DRY MASS (KG):	26
MAIN BODY:	Cylinder, 0.64 m diameter by 2.53 m length
MAJOR APPENDAGES:	None
ATTITUDE CONTROL:	None at the time of the event

EVENT DATA

KNOWN EVENTS:	1
FIRST DATE:	21 March 2003

APOGEE	PERIGEE	PERIOD	INCLINATION
660 km	655 km	97.89 min	75.8 deg

COMMENTS

There was only one piece cataloged from the relatively small Scout R/B stage.

SATELLITE DATA

TYPE:	Payload
OWNER:	US
LAUNCH DATE:	22 December 1965
DRY MASS (KG):	60
MAIN BODY:	Octagon; 0.5 m diameter by 0.4 m length
MAJOR APPENDAGES:	4 vanes
ATTITUDE CONTROL:	None at time of event

EVENT DATA

KNOWN EVENTS:	2
FIRST DATE:	30 November 1996

APOGEE	PERIGEE	PERIOD	INCLINATION
1065 km	895 km	104.66 min	89.1 deg

COMMENTS

One of several Transit-class spacecraft involved in anomalous events. Two debris objects (one cataloged and one not cataloged) were being tracked in 1997.

SATELLITE DATA

TYPE:	Payload
OWNER:	US
LAUNCH DATE:	28 January 1966
DRY MASS (KG):	60
MAIN BODY:	Octagonal cylinder; 0.5 m by 0.4 m
MAJOR APPENDAGES:	4 solar panels; gravity-gradient boom
ATTITUDE CONTROL:	Gravity-gradient

EVENT DATA

KNOWN EVENTS:	3
FIRST DATE:	April 1980

APOGEE	PERIGEE	PERIOD	INCLINATION
1205 km	855 km	105.8 min	89.8 deg

COMMENTS

Two additional events observed: Sep 1980 and Jul 1983. Last event may have originated with a piece of debris from earlier event. One of several known Transits involved in anomalous events.

SATELLITE DATA

TYPE:	Payload
OWNER:	US
LAUNCH DATE:	26 March 1966
DRY MASS (KG):	60
MAIN BODY:	Octagonal cylinder; 0.5 m by 0.4 m
MAJOR APPENDAGES:	4 solar panels; gravity-gradient boom
ATTITUDE CONTROL:	Gravity-gradient

EVENT DATA

KNOWN EVENTS:	1
FIRST DATE:	July 1981

APOGEE	PERIGEE	PERIOD	INCLINATION
1115 km	890 km	105.1 min	89.9 deg

COMMENTS

No other events observed. One of several known Transits involved in anomalous events.

SATELLITE DATA

TYPE:	Payload
OWNER:	US
LAUNCH DATE:	15 May 1966
DRY MASS (KG):	414
MAIN BODY:	Conical skeleton; 1.45 m diameter by 3.0 m length
MAJOR APPENDAGES:	2 Paddles
ATTITUDE CONTROL:	None at time of event

EVENT DATA

KNOWN EVENTS:	Multiple
FIRST DATE:	November 1997

APOGEE	PERIGEE	PERIOD	INCLINATION
1175 km	1095 km	108.03 min	100.4 deg

COMMENTS

A single piece of debris was detected on 16 November 1997. Separation may have occurred about 1 November. Numerous debris were released from the late 1990's to 2001. See cited reference below.

REFERENCE DOCUMENT

Environmentally-Induced Debris Sources, N. L. Johnson, NASA Lyndon B. Johnson Space Center, Second World Space Congress, 2002.

SATELLITE DATA

TYPE:	Agena D Stage
OWNER:	US
LAUNCH DATE:	19 August 1966
DRY MASS (KG):	600
MAIN BODY:	Cylinder; 1.5 m diameter by 8 m length
MAJOR APPENDAGES:	None
ATTITUDE CONTROL:	None at time of events

EVENT DATA

KNOWN EVENTS:	4
FIRST DATE:	March 1991

APOGEE	PERIGEE	PERIOD	INCLINATION
3710 km	3660 km	167.5 min	89.7 deg

COMMENTS

Second, third, and fourth events observed on 16 June 1992, 23 June 1992, and 1 November 1995, respectively. Additional events may have occurred.

SATELLITE DATA

TYPE:	Payload
OWNER:	US
LAUNCH DATE:	14 April 1967
DRY MASS (KG):	60
MAIN BODY:	Octagonal cylinder; 0.5 m by 0.4 m
MAJOR APPENDAGES:	4 solar panels; gravity-gradient boom
ATTITUDE CONTROL:	Gravity-gradient

EVENT DATA

KNOWN EVENTS:	1
FIRST DATE:	September 1992

APOGEE	PERIGEE	PERIOD	INCLINATION
1065 km	1035 km	106.2	90.1

COMMENTS

Event most likely happened around 28 September 1992 based on element data near the event time and analysis using the COMBO algorithm in the SATRAK Astrodynamics Toolkit. One of several Transit-class satellites involved in anomalous events.

REFERENCE DOCUMENT

Environmentally-Induced Debris Sources, N. L. Johnson, NASA Lyndon B. Johnson Space Center, Second World Space Congress, 2002.

SATELLITE DATA

TYPE:	Payload
OWNER:	US
LAUNCH DATE:	18 May 1967
DRY MASS (KG):	60
MAIN BODY:	Octagonal cylinder; 0.5 m by 0.4 m
MAJOR APPENDAGES:	4 solar panels; gravity-gradient boom
ATTITUDE CONTROL:	Gravity-gradient boom

EVENT DATA

KNOWN EVENTS:	1
DATE:	11/12 February 1995

APOGEE	PERIGEE	PERIOD	INCLINATION
1090 km	1060 km	106.12 min	89.6 deg

COMMENTS

One piece of debris liberated. One of several Transit-class satellites involved in anomalous events.

SATELLITE DATA

TYPE:	Payload
OWNER:	US
LAUNCH DATE:	25 September 1967
DRY MASS (KG):	60
MAIN BODY:	Octagonal cylinder; 0.5 m by 0.4 m
MAJOR APPENDAGES:	4 solar panels; gravity-gradient boom
ATTITUDE CONTROL:	Gravity-gradient

EVENT DATA

KNOWN EVENTS:	4
FIRST DATE:	April 1981

APOGEE	PERIGEE	PERIOD	INCLINATION
1110 km	1035 km	106.7 min	89.3 deg

COMMENTS

Second event observed in August 1986. One of several known Transits involved in anomalous events.

SATELLITE DATA

TYPE:	Vostok Final Stage
OWNER:	CIS
LAUNCH DATE:	14 March 1968
DRY MASS (KG):	1440
MAIN BODY:	Cylinder; 2.6 m diameter by 3.8 m length
MAJOR APPENDAGES:	None
ATTITUDE CONTROL:	None at the time of event

EVENT DATA

KNOWN EVENTS:	1
FIRST DATE:	Late-1990

APOGEE	PERIGEE	PERIOD	INCLINATION
515 km	450 km	94.3 min	81.2 deg

COMMENTS

Cosmos 206 was a prototype spacecraft of the Meteor 1 program. This is one of several Vostok final stages to shed a piece of debris since 1987. One piece of debris was released, although never officially cataloged.

SATELLITE DATA

TYPE:	Payload
OWNER:	Canada
LAUNCH DATE:	30 January 1969
DRY MASS (KG):	240
MAIN BODY:	Cylinder; 1.27 m diameter x 1.07 m length
MAJOR APPENDAGES:	Several antennae
ATTITUDE CONTROL:	None at the time of event

EVENT DATA

KNOWN EVENTS:	1
FIRST DATE:	24.82 May 2007

APOGEE	PERIGEE	PERIOD	INCLINATION
3455 km	580 km	127.57 min	88.5 deg

COMMENTS

One piece of debris cataloged. Altitude at the time of breakup was approximately 2940 km.

REFERENCE DOCUMENT

"Two Minor Satellite Fragmentations Identified in the Third Quarter," The Orbital Debris Quarterly News, NASA JSC, October 2007. Available online at: https://orbitaldebris.jsc.nasa.gov/quarterly-news/pdfs/ODQNv11i4.pdf

SATELLITE DATA

TYPE:	Payload
OWNER:	US
LAUNCH DATE:	27 August 1970
DRY MASS (KG):	60
MAIN BODY:	Octagon; 0.5 m diameter by 0.4 m length
MAJOR APPENDAGES:	4 solar panels; gravity-gradient boom
ATTITUDE CONTROL:	None at time of event

EVENT DATA

KNOWN EVENTS:	1
FIRST DATE:	7 March 1998

APOGEE	PERIGEE	PERIOD	INCLINATION
1205 km	945 km	106.75 min	90.0 deg

COMMENTS

One of several Transit-class satellites involved in anomalous events.

SATELLITE DATA

TYPE:	Vostok Final Stage
OWNER:	CIS
LAUNCH DATE:	20 January 1971
DRY MASS (KG):	1440
MAIN BODY:	Cylinder; 2.6 m diameter by 3.8 m length
MAJOR APPENDAGES:	None
ATTITUDE CONTROL:	None at time of the event.

EVENT DATA

KNOWN EVENTS:	1
FIRST DATE:	June 1987

APOGEE	PERIGEE	PERIOD	INCLINATION
665 km	535 km	96.7 min	81.2 deg

COMMENTS

No other events observed. One of several Vostok final stages to be involved in anomalous events.

SATELLITE DATA

TYPE:	Vostok Final Stage
OWNER:	CIS
LAUNCH DATE:	30 June 1972
DRY MASS (KG):	1440
MAIN BODY:	Cylinder; 2.6 m diameter by 3.8 m length
MAJOR APPENDAGES:	None
ATTITUDE CONTROL:	None at time of the event.

EVENT DATA

KNOWN EVENTS:	1
FIRST DATE:	September 1989

APOGEE	PERIGEE	PERIOD	INCLINATION
935 km	860 km	102.9 min	81.2 deg

COMMENTS

No other events observed. One of several Vostok final stages involved in anomalous events.

SATELLITE DATA

TYPE:	Payload
OWNER:	CIS
LAUNCH DATE:	21 December 1972
DRY MASS (KG):	600
MAIN BODY:	Unknown
MAJOR APPENDAGES:	Unknown
ATTITUDE CONTROL:	None at the time of the event

EVENT DATA

KNOWN EVENTS:	1
FIRST DATE:	April 2002

APOGEE	PERIGEE	PERIOD	INCLINATION
1380 km	1340 km	112.9 min	74.0 deg

COMMENTS

One piece of debris cataloged. It was concluded that because of the deduced debris ejecta velocity, Cosmos 539 was apparently struck by a small meteoroid or man-made object.

REFERENCE DOCUMENTS

Environmentally-Induced Debris Sources, N. L. Johnson, NASA Lyndon B. Johnson Space Center, Second World Space Congress, 2002.

"A New Collision in Space?" The Orbital Debris Quarterly News, NASA JSC, July 2002. Available online at https://orbitaldebris.jsc.nasa.gov/quarterly-news/pdfs/ODQNv7i3.pdf

SATELLITE DATA

TYPE:	Delta Second Stage (2410)
OWNER:	US
LAUNCH DATE:	9 April 1975
DRY MASS (KG):	900
MAIN BODY:	Cylinder-nozzle; 2.4 m diameter by 8 m length
MAJOR APPENDAGES:	None
ATTITUDE CONTROL:	None at time of the event.

EVENT DATA

KNOWN EVENTS:	1
FIRST DATE:	March 1978

APOGEE	PERIGEE	PERIOD	INCLINATION
845 km	835 km	101.7 min	115.0 deg

COMMENTS

Only one event noted with three fragments cataloged 12 March 1978. There was repeated mis-tagging of objects during 1978 among rocket body and debris. One fragment lost in 1978 and administratively decayed in 1983. This event may be related to series of major Delta second stage breakups.

KYOKKOH 1 (EXOS-A)   1978-014A   10664

SATELLITE DATA

TYPE:	Payload
OWNER:	Japan
LAUNCH DATE:	4 February 1978
DRY MASS (KG):	103
MAIN BODY:	Octagonal cylinder; 0.95 m by 0.8 m
MAJOR APPENDAGES:	3 small booms
ATTITUDE CONTROL:	Unknown

EVENT DATA

KNOWN EVENTS:	2
FIRST DATE:	January 1988

APOGEE	PERIGEE	PERIOD	INCLINATION
4219 km	760 km	134.0 min	65.0 deg

COMMENTS

First event (object 18816) may have occurred much earlier than the January 1988 date. Catalog actions taken at the end of 1988. The second event occurred in June 1992. Object 22008 led a short life, decaying on 2 August 1992.

SATELLITE DATA

TYPE:	Payload (attached to Agena R/B)
OWNER:	US
LAUNCH DATE:	27 June 1978
DRY MASS (KG):	2300
MAIN BODY:	Cylinder; 1.5 m diameter by 21 m length
MAJOR APPENDAGES:	2 solar panels; 1 antenna panel; miscellaneous booms
ATTITUDE CONTROL:	None at time of event

EVENT DATA

KNOWN EVENTS:	Multiple
FIRST DATE:	July 1983

APOGEE	PERIGEE	PERIOD	INCLINATION
780 km	780 km	100.5 min	108.0 deg

COMMENTS

On average, one piece of debris is released per year, but sometimes in groups. Most debris experience very rapid decay for this altitude. Last known release was in 2007.

REFERENCE DOCUMENTS

"Environmentally-Induced Debris Sources," N.L. Johnson. Journal of Advances in Space Research, Vol. 34, Issue 5, 2004, pp. 993-999.

"Detection of Debris from Chinese ASAT Test Increases; One Minor Fragmentation Event in Second Quarter of 2007", The Orbital Debris Quarterly News, NASA JSC, July 2007.
Available online at https://orbitaldebris.jsc.nasa.gov/quarterly-news/pdfs/ODQNv11i3.pdf.

COSMOS 1043   **1978-094A**   **11055**

SATELLITE DATA

TYPE:	Payload
OWNER:	CIS
LAUNCH DATE:	10 October 1978
DRY MASS (KG):	2200 (est.)
MAIN BODY:	Cylinder; dimensions ~1.5 m diameter by 5 m length
MAJOR APPENDAGES:	Solar panels; payload panels; gravity-gradient boom
ATTITUDE CONTROL:	None at time of the event.

EVENT DATA

KNOWN EVENTS:	1
FIRST DATE:	February 1993

APOGEE	PERIGEE	PERIOD	INCLINATION
437 km	435 km	94.9 min	81.2 deg

COMMENTS

No other satellite of this type has experienced an anomalous event. The piece was cataloged on 28 Feb 93 and decayed on 11 Mar 93. Given prior cataloging practices, other spacecraft could have experienced similar events that went unrecorded.

SATELLITE DATA

TYPE:	Payload
OWNER:	US
LAUNCH DATE:	13 October 1978
DRY MASS (KG):	725
MAIN BODY:	Cylinder; 1.9 m diameter by 3.7 m length
MAJOR APPENDAGES:	1 solar panel
ATTITUDE CONTROL:	None at time of the event

EVENT DATA

KNOWN EVENTS:	2
FIRST DATE:	September 1987

APOGEE	PERIGEE	PERIOD	INCLINATION
855 km	835 km	101.9 min	99.0 deg

COMMENTS

Both fragments from the first event decayed rapidly during winter of 1988-89. A second event associated with 1978-096A occurred on 23 Feb 96 liberating 1 piece.

SATELLITE DATA

TYPE:	Delta Second Stage (2910)
OWNER:	US
LAUNCH DATE:	24 October 1978
DRY MASS (KG):	900
MAIN BODY:	Cylinder-nozzle; 2.4 m diameter by 8 m length
MAJOR APPENDAGES:	None
ATTITUDE CONTROL:	None at time of the event.

EVENT DATA

KNOWN EVENTS: 2

FIRST DATE: May 1981

APOGEE	PERIGEE	PERIOD	INCLINATION
955 km	935 km	104.0 min	99.3 deg

COMMENTS

Second anomalous event apparently occurred about January 1987. A more prolific event in December 1981 is tentatively categorized as a satellite breakup (see Section 2). The cataloged debris section above refers only to the new fragment observed after the second anomalous event and does not include the Delta second stage that is accounted for in the tables of Section 2. These events may be related to the series of major Delta second stage breakups.

SATELLITE DATA

TYPE:	Payload
OWNER:	US
LAUNCH DATE:	27 June 1979
DRY MASS (KG):	723
MAIN BODY:	Rectangular; 3.71 m by 1.88 m
MAJOR APPENDAGES:	Solar panels
ATTITUDE CONTROL:	3-axis reaction control

EVENT DATA

KNOWN EVENTS:	2
FIRST DATE:	September 1992

APOGEE	PERIGEE	PERIOD	INCLINATION
810 km	795 km	100.8 min	98.68 deg

COMMENTS

One piece of debris cataloged from the first event. A second event took place in June 1995 with one piece of debris liberated, but none cataloged.

REFERENCE DOCUMENT

Environmentally-Induced Debris Sources, N. L. Johnson, NASA Lyndon B. Johnson Space Center, Second World Space Congress, 2002.

SATELLITE DATA

TYPE:	Payload
OWNER:	CIS
LAUNCH DATE:	31 Oct 1979
DRY MASS (KG):	2750
MAIN BODY:	Cylinder
MAJOR APPENDAGES:	Large Solar Arrays
ATTITUDE CONTROL:	None at the time of event

EVENT DATA

KNOWN EVENTS:	Multiple
FIRST DATE:	prior to 1 Jan. 2005; orbital data for 1 Jan. 2014.

APOGEE	PERIGEE	PERIOD	INCLINATION
880.7 km	862 km	102.35 min	81.2 deg

COMMENTS

Multiple anomalous debris from multiple individual events; the origination date of the pieces is not conclusive. The first events occurred in the 1990s and the last event occurred as recently as late 2013 to early 2014.

SATELLITE DATA

TYPE:	Payload
OWNER:	CIS
LAUNCH DATE:	14 May 1981
DRY MASS (KG):	2750
MAIN BODY:	Cylinder
MAJOR APPENDAGES:	Large Solar Arrays
ATTITUDE CONTROL:	None at the time of event

EVENT DATA

KNOWN EVENTS:	1
FIRST DATE:	March 2004

APOGEE	PERIGEE	PERIOD	INCLINATION
895 km	825 km	102.15 min	81.3 deg

COMMENTS

Eight pieces of debris cataloged, may have been from two individual events, the origination date of the pieces is not conclusive. The rocket body associated with the launch of this spacecraft experienced an anomalous event over seven years earlier. The events are unrelated.

SATELLITE DATA

TYPE:	Vostok Final Stage
OWNER:	CIS
LAUNCH DATE:	14 May 1981
DRY MASS (KG):	1440
MAIN BODY:	Cylinder; 2.6 m diameter by 3.8 m length
MAJOR APPENDAGES:	None
ATTITUDE CONTROL:	None at time of the event.

EVENT DATA

KNOWN EVENTS:	1
FIRST DATE:	October 1996

APOGEE	PERIGEE	PERIOD	INCLINATION
920 km	825 km	102.41 min	81.3 deg

COMMENTS

One of several Vostok upper stages involved in anomalous events.

SATELLITE DATA

TYPE:	Molniya Final Stage
OWNER:	CIS
LAUNCH DATE:	9 Jun 1981
DRY MASS (KG):	1100
MAIN BODY:	Cylinder; 2.7 m diameter by 3 m length
MAJOR APPENDAGES:	None
ATTITUDE CONTROL:	None

EVENT DATA

KNOWN EVENTS:	1
FIRST DATE:	July 1998

APOGEE	PERIGEE	PERIOD	INCLINATION
33415 km	85 km	583.42 min	62.0 deg

COMMENTS

No debris was cataloged from this event.

SATELLITE DATA

TYPE:	Payload
OWNER:	US
LAUNCH DATE:	23 June 1981
DRY MASS (KG):	723
MAIN BODY:	Rectangular; 3.71 m by 1.88 m
MAJOR APPENDAGES:	Solar panels
ATTITUDE CONTROL:	None at the time of event

EVENT DATA

KNOWN EVENTS:	2
FIRST DATE:	26.5 July 1993

APOGEE	PERIGEE	PERIOD	INCLINATION
835 km	830 km	101.6 min	98.9 deg

COMMENTS

Two objects were detected by the NAVSPOC and subsequently entered in the US SSN catalog. A piece separation analysis by the NAVSPOC identified the precise time these objects separated from the parent. It is unclear whether this event is a small breakup or whether other explanations such as spacecraft degradation could explain this separation. Unless other evidence is uncovered, this event will be classified as an anomalous event. The NOAA 7 payload was inactive for 3 years prior to this event. In 1997 a second, more curious event occurred. Three new debris appeared simultaneously with a discrete decrease in the orbital period of NOAA 7 of approximately 1 second.

REFERENCE DOCUMENT

Environmentally-Induced Debris Sources, N. L. Johnson, NASA Lyndon B. Johnson Space Center, Second World Space Congress, 2002.

COSMOS 1417 R/B   **1982-102B**   **13618**

SATELLITE DATA

TYPE:	Cosmos 3M/SL-8 Second Stage
OWNER:	CIS
LAUNCH DATE:	19 October 1982
DRY MASS (KG):	1435
MAIN BODY:	Cylinder; 2.4 m diameter by 6 m length
MAJOR APPENDAGES:	None
ATTITUDE CONTROL:	None

EVENT DATA

KNOWN EVENTS:	1
FIRST DATE:	early 2009

APOGEE	PERIGEE	PERIOD	INCLINATION
1000 km	955 km	104.6 min	83.0 deg

COMMENTS

One piece of debris was cataloged. The object resides in an orbit similar to its parent.

SATELLITE DATA

TYPE:	Payload
OWNER:	US
LAUNCH DATE:	28.66 March 1983
DRY MASS (KG):	740
MAIN BODY:	Rectangular; 3.71 m by 1.88 m
MAJOR APPENDAGES:	Solar panels
ATTITUDE CONTROL:	None at the time of event

EVENT DATA

KNOWN EVENTS:	2
FIRST DATE:	2004

APOGEE	PERIGEE	PERIOD	INCLINATION
813 km	790 km	100.9 min	98.7 deg

COMMENTS

Four objects, liberated in two unique events, have been cataloged. The first event occurred in in 2004, and the last in 2011 (three debris). All objects displayed very high area-to-mass ratios and have decayed from orbit. Unless other evidence is uncovered, these events will be classified as anomalous events. The NOAA 8 payload employed the Advanced TIROS-N bus and was declared lost on 29 December 1985 following a battery thermal runaway. The battery was destroyed, and that event is described in Section 2.

SATELLITE DATA

TYPE:	Payload
OWNER:	US
LAUNCH DATE:	3 August 1985
DRY MASS (KG):	60
MAIN BODY:	Octagonal cylinder; 0.5 m by 0.4 m
MAJOR APPENDAGES:	4 solar panels; gravity-gradient boom
ATTITUDE CONTROL:	Gravity-gradient boom

EVENT DATA

KNOWN EVENTS:	1
FIRST DATE:	27 December 1991

APOGEE	PERIGEE	PERIOD	INCLINATION
1255 km	1000 km	107.8 min	89.9 deg

COMMENTS

Other debris pieces are associated with this dual payload launch. The most recent event identified (SCC 21878) apparently originated from Oscar 30 on 27 December 1991, when a portion of the gravity-gradient boom broke off. One of several Transit-class satellites involved in anomalous events.

REFERENCE DOCUMENT

"The Transit System," L. Lee Pryor, AIAA Paper 92-1708, Applied Physics Laboratory, 1992.

SATELLITE DATA

TYPE:	Vostok Final Stage
OWNER:	CIS
LAUNCH DATE:	3 October 1985
DRY MASS (KG):	1440
MAIN BODY:	Cylinder; 2.6 m diameter by 3.8 m length
MAJOR APPENDAGES:	None
ATTITUDE CONTROL:	None at the time of the event.

EVENT DATA

KNOWN EVENTS:	1
FIRST DATE:	May 2002

APOGEE	PERIGEE	PERIOD	INCLINATION
565 km	510 km	95.4 min	97.7 deg

COMMENTS

One of several Vostok stages involved in anomalous events.

SATELLITE DATA

TYPE:	Payload
OWNER:	US
LAUNCH DATE:	17.66 September 1986
DRY MASS (KG):	740
MAIN BODY:	Rectangular; 3.71 m by 1.88 m
MAJOR APPENDAGES:	Solar panels
ATTITUDE CONTROL:	None at the time of event

EVENT DATA

KNOWN EVENTS:	3
FIRST DATE:	April 2000

APOGEE	PERIGEE	PERIOD	INCLINATION
816 km	799 km	101.0 min	98.6 deg

COMMENTS

Four objects, liberated in three unique events, have been cataloged. The first event occurred in April 2000, the second in 2007, and the last in 2010. All objects displayed very high area-to-mass ratios and have decayed from orbit. Unless other evidence is uncovered, this event will be classified as an anomalous event. The NOAA 10 payload employed the Advanced TIROS-N bus, was placed into a standby status on 17 September 1991 and had been decommissioned on 30 August 2001.

SATELLITE DATA

TYPE:	Payload
OWNER:	CIS
LAUNCH DATE:	23.86 April 1987
DRY MASS (KG):	2500
MAIN BODY:	cylinder-cone 1.4 m diameter x 9.3 m long
MAJOR APPENDAGES:	~ 3 m beam extending from base of cylindrical section
ATTITUDE CONTROL:	inactive at time of event
ENERGY SOURCES:	inactive nuclear power source; other sources speculative

EVENT DATA

KNOWN EVENTS: 1

FIRST DATE: 04 Jul 2008

APOGEE	PERIGEE	PERIOD	INCLINATION
803	774	100.62 min	65.0 deg

COMMENTS

Cosmos 1818 was the first of two Plazma-A spacecraft, the second being Cosmos 1867. These spacecraft used a thermionic nuclear power reactor in lieu of the thermoelectric reactors aboard the predecessor Radar Ocean Reconnaissance (RORSAT) spacecraft series. However, like RORSATs, the Plazma-A spacecraft used sodium-potassium (NaK) as a coolant. Cosmos 1818's operational lifetime is estimated to be approximately five months. The event occurred after approximately 21.4 years on orbit. Radar observations indicate signatures consistent with small, metallic spheres, most likely NaK coolant droplets.

REFERENCE DOCUMENT

"New Debris Seen from Decommissioned Satellite with Nuclear Power Source," The Orbital Debris Quarterly News, NASA JSC, January 2009. Available online at https://orbitaldebris.jsc.nasa.gov/quarterly-news/pdfs/ODQNv13i1.pdf.

SATELLITE DATA

TYPE:	Payload
OWNER:	CIS
LAUNCH DATE:	20.20 Feb 1987
DRY MASS (KG):	1500
MAIN BODY:	Cylinder; 2.4 m diameter by 4 m length
MAJOR APPENDAGES:	Gravity-gradient boom; 10 small solar panels
ATTITUDE CONTROL:	Gravity gradient

EVENT DATA

KNOWN EVENTS:	3
FIRST DATE:	Apr-May 1997

APOGEE	PERIGEE	PERIOD	INCLINATION
1525 km	1480 km	116.0 min	73.6 deg

COMMENTS

Cosmos 1823 appears to have experienced three separate anomalous events, two in 1997 and one in 1999. Because Cosmos 1823 suffered a serious fragmentation in December 1987, the anomalous debris pieces may have been loosely attached to the spacecraft, then separated after continued exposure to the space environment or change in attitude of the spacecraft remnant.

REFERENCE DOCUMENTS

History of Soviet/Russian Satellite Fragmentations-A Joint U.S.-Russian Investigation, N. L. Johnson, *et al.,* Kaman Sciences Corporation, October 1995.

Environmentally-Induced Debris Sources, N. L. Johnson, NASA Lyndon B. Johnson Space Center, Second World Space Congress, 2002.

SATELLITE DATA

TYPE:	Payload
OWNER:	CIS
LAUNCH DATE:	10.65 July 1987
DRY MASS (KG):	2500
MAIN BODY:	cylinder-cone 1.4 m diameter x 9.3 m long
MAJOR APPENDAGES:	~ 3 m beam extending from base of cylindrical section
ATTITUDE CONTROL:	inactive at time of event
ENERGY SOURCES:	inactive nuclear power source; other sources speculative

EVENT DATA

KNOWN EVENTS:	1
FIRST DATE:	21 March to 4 April 2014

APOGEE	PERIGEE	PERIOD	INCLINATION
802	776	100.63 min	65.0 deg

COMMENTS

Cosmos 1867 was the second of two Plazma-A spacecraft, the first being Cosmos 1818. These spacecraft used a thermionic nuclear power reactor in lieu of the thermoelectric reactors aboard the predecessor Radar Ocean Reconnaissance (RORSAT) spacecraft series. However, like RORSATs, the Plazma-A spacecraft used sodium-potassium (NaK) as a coolant. Cosmos 1867's operational lifetime is estimated to be approximately five months. The event occurred after approximately 26.9 years on orbit and was an extended event. Radar observations indicate signatures consistent with small, metallic spheres, most likely NaK coolant droplets.

SATELLITE DATA

TYPE:	Payload
OWNER:	CIS
LAUNCH DATE:	30 January 1988
DRY MASS (KG):	2750
MAIN BODY:	Cylinder
MAJOR APPENDAGES:	Large Solar Arrays
ATTITUDE CONTROL:	None at the time of event

EVENT DATA

KNOWN EVENTS:	2
FIRST DATE:	2000-2001

APOGEE	PERIGEE	PERIOD	INCLINATION
960	936	104 min	82.5 deg

COMMENTS

One debris object separated from the parent spacecraft in '00-'01-time frame, the second in the '02-'03-time frame. The debris were tracked under various analyst satellite numbers until being cataloged in 2012 as SSN 38540 and 38543.

COSMOS 1939 R/B   **1988-032B**   **19046**

SATELLITE DATA

TYPE:	Vostok Final Stage
OWNER:	CIS
LAUNCH DATE:	20 April 1988
DRY MASS (KG):	1440
MAIN BODY:	Cylinder; 2.6 m diameter by 3.8 m length
MAJOR APPENDAGES:	None
ATTITUDE CONTROL:	None at the time of the event.

EVENT DATA

KNOWN EVENTS:	2
FIRST DATE:	30 July 1996

APOGEE	PERIGEE	PERIOD	INCLINATION
655 km	585 km	97.14 min	97.6 deg

COMMENTS

One of several Vostok final stages involved in anomalous events.

SATELLITE DATA

TYPE:	Payload
OWNER:	US
LAUNCH DATE:	24.42 September 1988
DRY MASS (KG):	740
MAIN BODY:	Rectangular; 3.71 m by 1.88 m
MAJOR APPENDAGES:	Solar panels
ATTITUDE CONTROL:	None at the time of event

EVENT DATA

KNOWN EVENTS:	1
FIRST DATE:	24 November 2010

APOGEE	PERIGEE	PERIOD	INCLINATION
816 km	799 km	101.0 min	98.6 deg

COMMENTS

Two low area-to-mass ratio objects have been cataloged. Unless other evidence is uncovered, this event will be classified as an anomalous event. The NOAA 11 payload employed the Advanced TIROS-N bus and was decommissioned on 16 June 2004. This event occurred approximately 6.4 years after decommissioning.

SATELLITE DATA

TYPE:	Payload
OWNER:	US
LAUNCH DATE:	18.61 November 1989
DRY MASS (KG):	2265
MAIN BODY:	Cylinder; with protective shield, 4.0 m diameter by 5.8 m length
MAJOR APPENDAGES:	3 - 8.6 m solar arrays
ATTITUDE CONTROL:	Spin stabilized; gyroscopes

EVENT DATA

KNOWN EVENTS:	Multiple
FIRST DATE:	January 1993

APOGEE	PERIGEE	PERIOD	INCLINATION
885 km	870 km	102.5 min	99.0 deg

COMMENTS

At least 12 separate event dates have been calculated by the NAVSPOC, and other events are certain to have occurred. Through December 1993 the satellite remained active, and the cause of the separations could be determined. No degradation of satellite performance was reported by the satellite operators.

REFERENCE DOCUMENT

Environmentally-Induced Debris Sources, N. L. Johnson, NASA Lyndon B. Johnson Space Center, Second World Space Congress, 2002.

SATELLITE DATA

TYPE:	Rocket Body
OWNER:	CIS
LAUNCH DATE:	27 February 1990
DRY MASS (KG):	1434
MAIN BODY:	Cylinder; 2.4 m diameter x 6.0 m length
MAJOR APPENDAGES:	None
ATTITUDE CONTROL:	None at the time of the event

EVENT DATA

KNOWN EVENTS:	1
FIRST DATE:	22 June 2005

APOGEE	PERIGEE	PERIOD	INCLINATION
1015	950	104.65 min	83.0 deg

COMMENTS

A piece was detected in a more eccentric and higher inclination orbit, indicating that this event may have been a collision with a small, uncataloged object or meteoroid.

REFERENCE DOCUMENT

"Recent Satellite Breakups," The Orbital Debris Quarterly News, NASA JSC, July 2005.
Available online at https://orbitaldebris.jsc.nasa.gov/quarterly-news/pdfs/ODQNv9i3.pdf.

SATELLITE DATA

TYPE:	Payload
OWNER:	US
LAUNCH DATE:	24 April 1990
DRY MASS (KG):	10863
MAIN BODY:	Cylinder
MAJOR APPENDAGES:	Two Solar Array Panels
ATTITUDE CONTROL:	CMG controlled

EVENT DATA

KNOWN EVENTS:	1
FIRST DATE:	August 2003

APOGEE	PERIGEE	PERIOD	INCLINATION
575 km	570 km	96.1 min	28.5 deg

COMMENTS

The Hubble Space Telescope (HST) debris decayed rapidly after the event.

SATELLITE DATA

TYPE:	Payload
OWNER:	US
LAUNCH DATE:	14.66 May 1991
DRY MASS (KG):	740
MAIN BODY:	Rectangular; 3.71 m by 1.88 m
MAJOR APPENDAGES:	Solar panels
ATTITUDE CONTROL:	None at the time of event

EVENT DATA

KNOWN EVENTS:	1
FIRST DATE:	2 October 2011

APOGEE	PERIGEE	PERIOD	INCLINATION
815 km	800 km	100 min	98.7 deg

COMMENTS

Four low area-to-mass ratio objects have been cataloged. Unless other evidence is uncovered, this event will be classified as an anomalous event. The NOAA 12 (NOAA D pre-launch) payload employed the last TIROS-N bus, No. 5, and was decommissioned on 10 August 2007. This event occurred approximately 4.1 years after decommissioning.

SATELLITE DATA

TYPE:	Payload
OWNER:	CIS
LAUNCH DATE:	4 June 1991
DRY MASS (KG):	1922
MAIN BODY:	Cylinder; 1.4-0.8 m diameter by 3.5 m length
MAJOR APPENDAGES:	Solar arrays, payload trays, radar antenna
ATTITUDE CONTROL:	Gravity-gradient

EVENT DATA

KNOWN EVENTS:	1
DATE:	12 October 1998

APOGEE	PERIGEE	PERIOD	INCLINATION
665 km	620 km	97.5 min	82.5 deg

COMMENTS

First event for this type object. No other events observed.

REFERENCE DOCUMENT

Environmentally-Induced Debris Sources, N. L. Johnson, NASA Lyndon B. Johnson Space Center, Second World Space Congress, 2002.

SATELLITE DATA

TYPE:	Payload
OWNER:	ESA
LAUNCH DATE:	17 July 1991
DRY MASS (KG):	2084
MAIN BODY:	SPOT Mk. 1 bus. Cubical box. Approximately 2 m x 2 m x 3 m long payload support module; 11.8 m length overall when antennas and solar panel deployed.
MAJOR APPENDAGES:	Solar Array Panel, multiple instruments and antennas
ATTITUDE CONTROL:	CMG/monopropellant thrusters; inactive at time of event

EVENT DATA

KNOWN EVENTS:	7
FIRST DATE:	7 July 2002

APOGEE	PERIGEE	PERIOD	INCLINATION
801.6 km	750.8 km	100.3 min	98.5 deg

COMMENTS

Seven debris objects were cataloged from seven unique production events, the most recent being in 2017. The ERS-1 mission ended on 10 March 2000 due to the failure of the attitude control system, approximately 2.3 years before the initial release event.

SATELLITE DATA

TYPE:	Payload
OWNER:	France
LAUNCH DATE:	17 July 1991
DRY MASS (KG):	26
MAIN BODY:	Cube; 360 mm per side
MAJOR APPENDAGES:	Several deployable 5 m long antennae
ATTITUDE CONTROL:	None at the time of events

EVENT DATA

KNOWN EVENTS: 3

FIRST DATE: August 2003

APOGEE	PERIGEE	PERIOD	INCLINATION
730 km	730 km	99.4 min	98.1 deg

COMMENTS

This French "Microsat" was no longer active at the time of the events. Some objects may be a piece broken off from any of the long antennae. Follow on events occurred on 17 April 2005 and 15 October 2006. Four debris totals have been cataloged from the parent object.

SATELLITE DATA

TYPE:	Ariane 40 Rocket Body
OWNER:	France
LAUNCH DATE:	17 July 1991
DRY MASS (KG):	1720
MAIN BODY:	Cylinder; 2.6 m diameter by 10 m length
MAJOR APPENDAGES:	None
ATTITUDE CONTROL:	None

EVENT DATA

KNOWN EVENTS:	1
FIRST DATE:	April 2001

APOGEE	PERIGEE	PERIOD	INCLINATION
770 km	770 km	100.2 min	98.2 deg

COMMENTS

One piece of debris cataloged. Parent object was in a sun-synchronous orbit at the time of the event. First occurrence of an anomalous event with an Ariane R/B.

REFERENCE DOCUMENT

Environmentally-Induced Debris Sources, N. L. Johnson, NASA Lyndon B. Johnson Space Center, Second World Space Congress, 2002.

SATELLITE DATA

TYPE:	Payload
OWNER:	US
LAUNCH DATE:	12 September 1991
DRY MASS (KG):	10863
MAIN BODY:	Box; approximately 4.6 m x 4.6 m x 10.7 m long
MAJOR APPENDAGES:	Solar Array Panel, multiple instruments and antennas
ATTITUDE CONTROL:	CMG/torque rod controlled; inactive at time of event

EVENT DATA

KNOWN EVENTS:	3
FIRST DATE:	10 November 2007

APOGEE	PERIGEE	PERIOD	INCLINATION
483 km	353 km	92.9 min	57 deg

COMMENTS

Five debris objects were cataloged, and all debris decayed rapidly after their production events. UARS was decommissioned on 14 December 2005, approximately 1.9 years before this series of events.

SATELLITE DATA

TYPE:	Payload
OWNER:	CIS
LAUNCH DATE:	25 March 1993
DRY MASS (KG):	260
MAIN BODY:	Two cylinders; < 1.5 m diameter
MAJOR APPENDAGES:	Solar panels; gravity-gradient boom
ATTITUDE CONTROL:	Gravity-gradient

EVENT DATA

KNOWN EVENTS:	Multiple
FIRST DATE:	4 March 1998

APOGEE	PERIGEE	PERIOD	INCLINATION
970 km	685 km	101.43 min	75.8 deg

COMMENTS

EKA 1 is a test payload prior to launches of small communications satellites. First orbital launch of Start-1 booster derived from SS-20/SS-25 missiles.

SATELLITE DATA

TYPE:	Rocket Body
OWNER:	CIS
LAUNCH DATE:	25 March 1993
DRY MASS (KG):	200
MAIN BODY:	Cylinder; 1.4 m diameter x 2.5 m length
MAJOR APPENDAGES:	None
ATTITUDE CONTROL:	None at the time of the event

EVENT DATA

KNOWN EVENTS:	Multiple
FIRST DATE:	Late 2002

APOGEE	PERIGEE	PERIOD	INCLINATION
920 km	680 km	100.85 min	75.8 deg

COMMENTS

Dozens of pieces have been cataloged from this rocket body, starting in late 2002 through 2005. Exact time and date of the event(s) is unknown. The payload associated with this rocket body, START-1, also experienced an anomalous event in March of 1998, the events are unrelated.

SATELLITE DATA

TYPE:	Payload
OWNER:	US
LAUNCH DATE:	29 Aug. 1994
DRY MASS (KG):	767
MAIN BODY:	TIROS N bus. Cylindrical; approximately 1.88 m diameter x 3.71 m long
MAJOR APPENDAGES:	Solar Array Panel, multiple instruments and antennas
ATTITUDE CONTROL:	reaction wheels/monopropellant thrusters; inactive at time of event

EVENT DATA

KNOWN EVENTS:	1
FIRST DATE:	23 Oct. 2016

APOGEE	PERIGEE	PERIOD	INCLINATION
848 km	832 km	101.7 min	99.1 deg

COMMENTS

Four objects were cataloged from a single production event, prior cataloged debris being mission-related. Two-line elements are not available for the payload after 2013 but are for the four anomalous debris; payload elements are taken from the 11 October 2016 public satellite catalog. This event may be similar in nature to the breakup events suffered by USA 29 (DMSP 5D-2 F9, SSN 18822), USA 73 (DMSP 5D-2 F11, SSN 21798), and USA 109 (DMSP 5D-2 F13, SSN 23533).

SATELLITE DATA

TYPE:	payload
OWNER:	CIS
LAUNCH DATE:	4 Nov. 1994
DRY MASS (KG):	1900 kg
MAIN BODY:	Meteor-3 cylindrical bus: 1.4 m diameter
MAJOR APPENDAGES:	solar arrays (2), antennas
ATTITUDE CONTROL:	three-axis stabilization; inactive at event times

EVENT DATA

KNOWN EVENTS: 7?

FIRST DATE: early 2006?

APOGEE*	PERIGEE	PERIOD	INCLINATION
642 km	641 km	97.52 min	97.9 deg

*parent body, 1 Jan. 2006

COMMENTS

This spacecraft is the third of the Resurs-O1 class Earth-observing spacecraft. The spacecraft was observed to exhibit multiple shedding events, in which numerous objects separated from the spacecraft and relatively low velocities.

This event may be similar to other Meteor-bus shedding events [*q.v.*] described in this Chapter.

.

SATELLITE DATA

TYPE:	Zenit Second Stage
OWNER:	CIS
LAUNCH DATE:	24 November 1994
DRY MASS (KG):	8300
MAIN BODY:	Cylinder; 3.9 m diameter by 12 m length
MAJOR APPENDAGES:	None
ATTITUDE CONTROL:	None

EVENT DATA

KNOWN EVENTS:	2?
FIRST DATE:	June 1998

APOGEE	PERIGEE	PERIOD	INCLINATION
845 km	845 km	101.82 min	71.0 deg

COMMENTS

One piece of debris was cataloged.

REFERENCE DOCUMENTS

History of Soviet/Russian Satellite Fragmentations-A Joint U.S.-Russian Investigation, N. L. Johnson, *et al.,* Kaman Sciences Corporation, October 1995.

Environmentally-Induced Debris Sources, N. L. Johnson, NASA Lyndon B. Johnson Space Center, Second World Space Congress, 2002.

SATELLITE DATA

TYPE:	Payload
OWNER:	US
LAUNCH DATE:	30 December 1994
DRY MASS (KG):	740
MAIN BODY:	Rectangular; 3.71 m by 1.88 m
MAJOR APPENDAGES:	Solar panels
ATTITUDE CONTROL:	None at the time of event

EVENT DATA

KNOWN EVENTS:	1
FIRST DATE:	July 2012

APOGEE	PERIGEE	PERIOD	INCLINATION
860 km	850 km	101 min	98.8 deg

COMMENTS

A single low area-to-mass ratio object has been cataloged. Unless other evidence is uncovered, this event will be classified as an anomalous event. The NOAA 14 payload employed the Advanced TIROS-N bus and was decommissioned on 23 May 2007.

SATELLITE DATA

TYPE:	Payload
OWNER:	ESA
LAUNCH DATE:	21 Apr. 1995
DRY MASS (KG):	2216
MAIN BODY:	SPOT Mk. 1 bus. Cubical box. Approximately 2 m x 2 m x 3 m long payload support module; 11.8 m length overall when antennas and solar panel deployed.
MAJOR APPENDAGES:	Solar Array Panel, multiple instruments and antennas
ATTITUDE CONTROL:	CMG/monopropellant thrusters; inactive at time of event

EVENT DATA

KNOWN EVENTS:	4
FIRST DATE:	24 July 2003

APOGEE	PERIGEE	PERIOD	INCLINATION
786.6 km	784.8 km	100.5 min	98.6 deg

COMMENTS

Six debris objects were cataloged from at least four production events, the most recent being in 2011. The ERS-2 spacecraft was decommissioned on 5 Sept. 2011 when it was passivated in accordance with debris mitigation guidelines.

SATELLITE DATA

TYPE:	Ariane 40+ Rocket Body
OWNER:	France
LAUNCH DATE:	21 Apr. 1995
DRY MASS (KG):	1720
MAIN BODY:	Cylinder; 2.6 m diameter by 10 m length
MAJOR APPENDAGES:	None
ATTITUDE CONTROL:	None

EVENT DATA

KNOWN EVENTS:	1
FIRST DATE:	27 July 2015

APOGEE	PERIGEE	PERIOD	INCLINATION
773.4 km	762.3 km	100.2 min	98.7 deg

COMMENTS

One piece of debris (SSN 43351) cataloged. Parent object was in a sun-synchronous orbit at the time of the event.

KOREASAT 1 R/B   **1995-041B**   **23640**

SATELLITE DATA

TYPE:	Delta Second Stage
OWNER:	US
LAUNCH DATE:	5 August 1995
DRY MASS (KG):	900
MAIN BODY:	Cylinder; 2.4 m diameter by 8 m length
MAJOR APPENDAGES:	None
ATTITUDE CONTROL:	None at time of the event.

EVENT DATA

KNOWN EVENTS:	1
DATE:	6 December 1995

APOGEE	PERIGEE	PERIOD	INCLINATION
1375 km	935 km	108.5 min	26.7 deg

COMMENTS

One piece was liberated.

RADARSAT R/B   **1995-059B**   **23711**

SATELLITE DATA

TYPE:	Delta Second Stage
OWNER:	US
LAUNCH DATE:	5 November 1995
DRY MASS (KG):	900
MAIN BODY:	Cylinder; 2.4 m diameter by 8 m length
MAJOR APPENDAGES:	None
ATTITUDE CONTROL:	None

EVENT DATA

KNOWN EVENTS:	1
DATE:	30 January 1996

APOGEE	PERIGEE	PERIOD	INCLINATION
1495 km	935 km	109.7 min	100.6 deg

COMMENTS

One piece was liberated.

SATELLITE DATA

TYPE:	Payload
OWNER:	India
LAUNCH DATE:	21 March 1996
DRY MASS (KG):	838
MAIN BODY:	Cubical box; 1.6 m x 1.6 m by 1.2 m high
MAJOR APPENDAGES:	Solar panels
ATTITUDE CONTROL:	three-axis stabilized; reaction wheels, torque rods, and monopropellant reaction control system

EVENT DATA

KNOWN EVENTS:	1
FIRST DATE:	October 2000

APOGEE	PERIGEE	PERIOD	INCLINATION
821.9 km	820.3 km	101.3 min	98.6 deg

COMMENTS

A single relatively high area-to-mass ratio object has been cataloged. "Event Data" epoch is 30 October 2000. Unless other evidence is uncovered, this event will be classified as an anomalous event. The payload was operational at the time of separation and was decommissioned in January 2006.

SATELLITE DATA

TYPE:	Payload
OWNER:	US
LAUNCH DATE:	20 December 1997
DRY MASS (KG):	556
MAIN BODY:	Triangular prism, 3.6 m long
MAJOR APPENDAGES:	Two solar arrays, three antennas
ATTITUDE CONTROL:	Three-axis stabilized

EVENT DATA

KNOWN EVENTS:	1
FIRST DATE:	7 June 2014

APOGEE	PERIGEE	PERIOD	INCLINATION
781 km	778 km	100.4 min	86.4 deg

COMMENTS

Ten pieces of debris were cataloged from this event. Three of the cataloged debris had reentered within 10 months indicating relatively large area-to-mass ratios. The event did not affect vehicle operations or performance.

REFERENCE DOCUMENT

"Iridium Anomalous Debris Events," The Orbital Debris Quarterly News, NASA JSC, January 2015. Available online at https://orbitaldebris.jsc.nasa.gov/quarterly-news/pdfs/odqnv19i1.pdf.

SATELLITE DATA

TYPE:	Payload
OWNER:	US
LAUNCH DATE:	24 June 1999
DRY MASS (KG):	1360
MAIN BODY:	Box; 1.3 m by 0.9 m by 0.9 m
MAJOR APPENDAGES:	Two 3.5 m^2 solar Arrays
ATTITUDE CONTROL:	Three-axis stabilized

EVENT DATA

KNOWN EVENTS:	1
FIRST DATE:	6 June 2004

APOGEE	PERIGEE	PERIOD	INCLINATION
760 km	745 km	99.90 min	25.0 deg

COMMENTS

Eight pieces of debris were cataloged from this event. An additional piece was detected but never cataloged. The event might have been coincidental with a "safe mode" entry around 5 June 2004, which cause the closing and reopening of several sensor doors. Five of the cataloged debris had reentered within 8 months indicating higher than normal area-to-mass ratios. The event did not affect vehicle operations or performance.

REFERENCE DOCUMENT

"FUSE Satellite Releases Unexpected Debris," The Orbital Debris Quarterly News, NASA JSC, July 2004. Available online at https://orbitaldebris.jsc.nasa.gov/quarterly-news/pdfs/ODQNv8i3.pdf.

CXO IUS R/B (1)   **1999-040C**   **25868**

SATELLITE DATA

TYPE:	IUS first stage
OWNER:	US
LAUNCH DATE:	23 July 1999
DRY MASS (KG):	1123 (burnout mass)
MAIN BODY:	Cylinder; 2.97 m diameter by 5.2 m length
MAJOR APPENDAGES:	None
ATTITUDE CONTROL:	None at time of events

EVENT DATA*

KNOWN EVENTS: 2

DATE: 2001

APOGEE	PERIGEE	PERIOD	INCLINATION
13498.9 km	278.8 km	253.5 min	28.5 deg

COMMENTS

Three pieces were liberated in the 2001 event and eight additional pieces separated in 2011. These are international designator piece tags E-G and H-Q respectively.

*Event orbit data is for 1 Jan. 2001

TELKOM 1   **1999-042A**   **25880**

SATELLITE DATA

TYPE:	Payload
OWNER:	Indonesia
LAUNCH DATE:	12 Aug. 1999
DRY MASS (KG):	1640
MAIN BODY:	Lockheed-Martin A2100A bus. Cubical box; 1.8 m x 1.8 m by 3.7 m high
MAJOR APPENDAGES:	Solar panels
ATTITUDE CONTROL:	three-axis stabilized; reaction wheels and monopropellant reaction control system

EVENT DATA

KNOWN EVENTS: 1

FIRST DATE: 25 Aug. 2017

APOGEE	PERIGEE	PERIOD	INCLINATION
35793 km	35781 km	1436.1 min	0.01 deg

COMMENTS

This event occurred after approximately 18.1 years on-orbit, three years past the nominal operational lifetime. The payload was operational at the time of the event. No debris have been cataloged as of this edition.

REFERENCE DOCUMENT

"Two Anomalous Events in GEO," The Orbital Debris Quarterly News, NASA JSC, February 2018. Available online at https://www.orbitaldebris.jsc.nasa.gov/quarterly-news/pdfs/odqnv22i1.pdf.

SATELLITE DATA

TYPE:	Payload
OWNER:	US
LAUNCH DATE:	24.76 September 1999
DRY MASS (KG):	735
MAIN BODY:	Box; 1.8 m by 1.8 m by 1.6 m
MAJOR APPENDAGES:	3 solar panels
ATTITUDE CONTROL:	3 axis stabilization

EVENT DATA

KNOWN EVENTS:	1
DATE:	19 March 2001

APOGEE	PERIGEE	PERIOD	INCLINATION
680 km	678 km	98.3 min	98.2 deg

COMMENTS

One piece was liberated. A very high ballistic coefficient resulted in the anomalous debris object reentering on 11 April 2001.

SATELLITE DATA

TYPE:	Payload
OWNER:	US-France joint mission
LAUNCH DATE:	7 Dec. 2001
DRY MASS (KG):	472
MAIN BODY:	box; 0.95 m x 0.95 m x 2.2 m high
MAJOR APPENDAGES:	Solar panels
ATTITUDE CONTROL:	three-axis stabilized; reaction wheels and monopropellant reaction control system

EVENT DATA

KNOWN EVENTS:	1
FIRST DATE:	16 Mar. 2002

APOGEE	PERIGEE	PERIOD	INCLINATION
1344.7 km	1332.2 km	112.4 min	66.0 deg

COMMENTS

On 16 March 2002 spacecraft controllers noted an attitude upset of the payload. A detailed study of the spacecraft's perturbations allowed analysts to infer that the left solar array had been impacted by a small particle (micrometeoroid or orbital debris unknown). Two anomalous debris were observed and associated with this payload, although the objects did not enter the public catalog until 2009 (SSN 35414) and 2011 (SSN 37379). A second anomaly was experienced in 2005 but there is no evidence that this event liberated debris.

REFERENCE DOCUMENTS

"New Evidence of Particle Impact on Jason-1 Spacecraft", The Orbital Debris Quarterly News, NASA JSC, July 2011. Available online at https://www.orbitaldebris.jsc.nasa.gov/quarterly-news/pdfs/odqnv15i3.pdf.

SATELLITE DATA

TYPE:	Payload
OWNER:	US
LAUNCH DATE:	11 February 2002
DRY MASS (KG):	556
MAIN BODY:	Triangular prism, 3.6 m long
MAJOR APPENDAGES:	Two solar arrays, three antennas
ATTITUDE CONTROL:	Three-axis stabilized

EVENT DATA

KNOWN EVENTS:	1
FIRST DATE:	30 November 2014

APOGEE	PERIGEE	PERIOD	INCLINATION
781 km	777 km	100.4 min	86.4 deg

COMMENTS

Four pieces of debris were cataloged from this event. Two of the cataloged debris had reentered within a year indicating relatively large area-to-mass ratios. The event did not affect vehicle operations or performance.

REFERENCE DOCUMENT

"Iridium Anomalous Debris Events," The Orbital Debris Quarterly News, NASA JSC, January 2015. Available online at https://orbitaldebris.jsc.nasa.gov/quarterly-news/pdfs/odqnv19i1.pdf.

SATELLITE DATA

TYPE:	Payload
OWNER:	US
LAUNCH DATE:	04 May 2002
DRY MASS (KG):	2832
MAIN BODY:	Rectangular box, 2.5 x 4.8 x 8.0 m
MAJOR APPENDAGES:	solar array, multiple antennas
ATTITUDE CONTROL:	Three-axis stabilized; active at event time

EVENT DATA

KNOWN EVENTS:	1
FIRST DATE:	29 April 2019

APOGEE	PERIGEE	PERIOD	INCLINATION
705 km	703 km	98.8 min	98.2 deg

COMMENTS

Five pieces of debris were observed from this event, but none have entered the catalog. One of the cataloged debris had reentered in early December 2019 indicating a relatively large area-to-mass ratio. The event did not affect vehicle operations or performance.

SATELLITE DATA

TYPE:	Payload
OWNER:	SES (formerly Société Européenne des Satellites; Luxembourg)
LAUNCH DATE:	6 June 2003
DRY MASS (KG):	1663
MAIN BODY:	Thales Alenia Space Spacebus-3000B3 bus. Cubical box; 3.2 m x 2.4 m by 4 m high
MAJOR APPENDAGES:	Solar panels
ATTITUDE CONTROL:	three-axis stabilized; reaction wheels and bipropellant reaction control system

EVENT DATA

KNOWN EVENTS:	1
FIRST DATE:	17 June 2017

APOGEE	PERIGEE	PERIOD	INCLINATION
35798 km	35774 km	1436.1 min	0.02 deg

COMMENTS

This event occurred after approximately 14 years on-orbit. The payload was operational at the time of the event and was subsequently boosted to the so-called GEO "graveyard orbit." No debris have been cataloged as of this edition.

REFERENCE DOCUMENT

"Two Anomalous Events in GEO," The Orbital Debris Quarterly News, NASA JSC, February 2018.
Available online at https://www.orbitaldebris.jsc.nasa.gov/quarterly-news/pdfs/odqnv22i1.pdf.

SATELLITE DATA

TYPE: Payload

OWNER: Délégation Générale pour l'Armament (DGA, France, Ministry of Defense; space segment managed by CNES)

LAUNCH DATE: 18 Dec. 2004

DRY MASS (KG):   4200 (launch mass); dry mass Unknown

MAIN BODY: SPOT Mk. 3 bus. Multiple cubical boxes with planar solar array panel.   Approximately 3.7 m x 3.4 m x 6.0 m long payload module; longer when solar panel deployed.

MAJOR APPENDAGES: Solar Array Panel

ATTITUDE CONTROL: 3 axis stabilized; assumed active at time of event and until approximately 4 March 2022

EVENT DATA

KNOWN EVENTS: Unknown

FIRST DATE: 26 July 2021

APOGEE	PERIGEE	PERIOD	INCLINATION
601.7 km	579.5 km	96.5 min	98.2 deg

*All orbital data is from the post-event final (uncontrolled?) orbit, epoch 4 March 2022.

COMMENTS

Seven debris objects have been cataloged from an unknown number of production events. The HELIOS 2A spacecraft appears to have experienced one or more events which may have been energetic in nature, but this is speculative based on the unknown orbital behavior of the payload between 2012 and 2022. Orbital data again became available on 1 March 2022 and the payload appears to have engaged in one or more orbit-lowering maneuvers, perhaps to comply with debris mitigation guidelines, culminating on or about 4 March in an orbit with a semimajor axis altitude of approximately 590 km. Unfortunately, given the lack of orbital data pre- and post-event, this poorly understood event must remain an anomalous event at this time.

SATELLITE DATA

TYPE:	Payload
OWNER:	CIS
LAUNCH DATE:	17 September 2009
DRY MASS (KG):	7.53
MAIN BODY:	Nested spheres; 0.17 m diameter
MAJOR APPENDAGES:	none
ATTITUDE CONTROL:	none

EVENT DATA

KNOWN EVENTS:	1
FIRST DATE:	22 January 2013

APOGEE	PERIGEE	PERIOD	INCLINATION
825 km	818 km	101.3 min	98.6 deg

COMMENTS

A single piece of debris was produced by this event. There remains uncertainty about the exact nature of this event, so it is characterized as an anomalous event at this time.

REFERENCE DOCUMENT

"Small Satellite Possibly Hit by Even Smaller Object," The Orbital Debris Quarterly News, NASA JSC, April 2013. Available online at https://orbitaldebris.jsc.nasa.gov/quarterly-news/pdfs/odqnv17i2.pdf.

SATELLITE DATA

TYPE:	Payload
OWNER:	DigitalGlobe (US)
LAUNCH DATE:	8 Oct. 2009
DRY MASS (KG):	2385
MAIN BODY:	Cylinder; 2.5 m diameter x 4.3 m high
MAJOR APPENDAGES:	Solar panels
ATTITUDE CONTROL:	three-axis stabilized; CMGs and monopropellant reaction control system

EVENT DATA

KNOWN EVENTS:	1
FIRST DATE:	18 July 2016

APOGEE	PERIGEE	PERIOD	INCLINATION
767.9 km	767.0 km	100.2 min	98.5 deg

COMMENTS

Nine anomalous debris were initially observed, but sixteen have been cataloged as of this edition. An initial analysis indicated a separation velocity of approximately 3 m/s for the longest-period debris object, and debris were observed with a maximum change, with respect to the parent body, in period of 0.8 minutes and inclination 0.02 deg. Debris were found at semimajor axes both larger and smaller than the parent body, indicating this event was not a simple shedding event. The spacecraft remained operational after the event.

REFERENCE DOCUMENTS

"WorldView 2 Spacecraft Fragments in July 2016", <u>The Orbital Debris Quarterly News</u>, NASA JSC, October 2016. Available online at https://www.orbitaldebris.jsc.nasa.gov/quarterly-news/pdfs/odqnv20i4.pdf.

HAIYANG 2A   **2011-043A**   **37781**

SATELLITE DATA

TYPE:	Payload
OWNER:	PRC
LAUNCH DATE:	15 August 2011
DRY MASS (KG):	1500?
MAIN BODY:	Box, 3 m x 2 m x 2 m?
MAJOR APPENDAGES:	one solar array
ATTITUDE CONTROL:	Unknown at time of event

EVENT DATA

KNOWN EVENTS: 2

FIRST DATE: 6-7 July 2014

APOGEE	PERIGEE	PERIOD	INCLINATION
969 km	967 km	100.4 min	99.4 deg

COMMENTS

Four pieces of debris were produced from these events, of which three entered the catalog. All three of the cataloged debris had reentered by 15 December 2014 indicating relatively large area-to-mass ratios. This spacecraft uses the CAST 968 bus; however, actual physical parameters are highly uncertain.

REFERENCE DOCUMENT

"Three Additional Breakups Mar 2014", The Orbital Debris Quarterly News, NASA JSC, October 2014. Available online at https://orbitaldebris.jsc.nasa.gov/quarterly-news/pdfs/odqnv18i4.pdf.

SATELLITE DATA

TYPE:	Payload
OWNER:	India
LAUNCH DATE:	26 Apr. 2012
DRY MASS (KG):	1858 at launch (wet mass)
MAIN BODY:	Cubical box; 2 m x 2 m by 5 m high
MAJOR APPENDAGES:	Solar panels
ATTITUDE CONTROL:	three-axis stabilized; reaction wheels, gyros, and monopropellant reaction control system

EVENT DATA

KNOWN EVENTS:	1
FIRST DATE:	30 Sept. 2016

APOGEE	PERIGEE	PERIOD	INCLINATION
543.7 km	538 km	95.4 min	97.6 deg

COMMENTS

Over 12 anomalous debris were initially observed, but only one (SSN 41797) has been cataloged as of this edition. All known debris have decayed.

REFERENCE DOCUMENTS

"Indian RISAT-1 Spacecraft Experiences Possible Fragmentation", The Orbital Debris Quarterly News, NASA JSC, October 2016. Available online at https://www.orbitaldebris.jsc.nasa.gov/quarterly-news/pdfs/odqnv20i4.pdf.

"Indian RISAT-1 Spacecraft Fragments in Late September - Update", The Orbital Debris Quarterly News, NASA JSC, February 2017. Available online at https://www.orbitaldebris.jsc.nasa.gov/quarterly-news/pdfs/odqnv21i1.pdf.

SATELLITE DATA

TYPE:	rocket body
OWNER:	CIS
LAUNCH DATE:	31 Mar. 2015
DRY MASS (KG):	1220
MAIN BODY:	Cylinder; 2.5 m diameter by 2.6 m length
MAJOR APPENDAGES:	None
ATTITUDE CONTROL:	None at time of event

EVENT DATA

KNOWN EVENTS:	1
DATE:	29 Apr. 2015

APOGEE	PERIGEE	PERIOD	INCLINATION
1342.5 km	1339.2 km	112.5 min	82.5 deg

SATELLITE DATA

TYPE:	mission related debris
OWNER:	Japan
LAUNCH DATE:	29 Oct. 2018
DRY MASS (KG):	unknown
MAIN BODY:	truncated cone; 4.070 m diameter tapering to < 1.98 m diameter x ~ 1.5 m length
MAJOR APPENDAGES:	None
ATTITUDE CONTROL:	None

EVENT DATA

KNOWN EVENTS:	1
DATE:	6-7 Feb. 2019

APOGEE	PERIGEE	PERIOD	INCLINATION
616.8 km	592.9 km	96.8 min	97.9 deg

COMMENTS

Parent object is believed to be the upper cap section of the three-piece Kawasaki Heavy Industries, Ltd. 4/4D-LC dual payload fairing's lower payload encapsulation system. Five objects were initially observed and four were officially cataloged. All have decayed.

Readers should note that the reference document cited was developed shortly after the event. The conclusions reached were predicated upon incomplete data. As the record matured over time, the actual events were identified as anomalous events attributable to the Payload Encapsulation System and the rocket body.

REFERENCE DOCUMENT

"2019 Fragmentations Commence with Japanese Rocket Body Breakup", The Orbital Debris Quarterly News, NASA JSC, May 2019. Available online at https://www.orbitaldebris.jsc.nasa.gov/quarterly-news/pdfs/odqnv23i1.pdf.

SATELLITE DATA

TYPE:	rocket body
OWNER:	Japan
LAUNCH DATE:	29 Oct. 2018
DRY MASS (KG):	3100
MAIN BODY:	Cylinder; 4 m diameter by 10.5 m length
MAJOR APPENDAGES:	None
ATTITUDE CONTROL:	None at time of event

EVENT DATA

KNOWN EVENTS:	1
DATE:	prior to Sept. 2019*

APOGEE	PERIGEE	PERIOD	INCLINATION
591.7 km	494.1 km	95.5 min	98.8 deg

*elements at epoch 6 February 2019

COMMENTS

Seven objects were initially observed, and one was officially cataloged; this entered the catalog well after the event itself.

Readers should note that the reference document cited was developed shortly after the event. The conclusions reached were predicated upon incomplete data. As the record matured over time, the actual events were identified as anomalous events attributable to the Payload Encapsulation System and the rocket body.

REFERENCE DOCUMENT

"2019 Fragmentations Commence with Japanese Rocket Body Breakup", The Orbital Debris Quarterly News, NASA JSC, May 2019. Available online at https://www.orbitaldebris.jsc.nasa.gov/quarterly-news/pdfs/odqnv23i1.pdf.

SZ-13 MODULE   **2021-092C**   **52260**

SATELLITE DATA

TYPE:	Shenzhou spacecraft's Orbital Module
OWNER:	PRC
LAUNCH DATE:	15 Oct. 2021
DRY MASS (KG):	1500 kg
MAIN BODY:	Cylinder: 2.25 m diameter by 2.80 m length
MAJOR APPENDAGES:	None
ATTITUDE CONTROL:	Unknown at time of event

EVENT DATA*

KNOWN EVENTS:	1
DATE:	15 April 2022

APOGEE	PERIGEE	PERIOD	INCLINATION
378 km	368.0 km	92.0 min	41.5 deg

*orbital elements based on first post-event TLE, 21 April 2022

COMMENTS

Ten to 15 objects were initially observed, and four were officially cataloged; all had decayed within two months following the event.

The Shenzhou crewed spacecraft, like the Soyuz crewed spacecraft, is composed of an orbital module, a reentry module, and a service module. The spacecraft, at end of mission, separates into these three modules and the crew return to Earth in the reentry module. This event is believed to have occurred at the time of separation; originally attributed to the (intact) "SZ-13" it was later associated with the orbital module, which remained on orbit. The production mechanism and identities of the observed debris are unknown hence this event's classification as anomalous.

SATELLITE DATA

TYPE:	rocket body with payload mass simulator
OWNER:	CIS
LAUNCH DATE:	27 Dec. 2021
DRY MASS (KG):	5180 kg (estimated), with up to 18,700 kg propellant
MAIN BODY:	Cylinder + conical mass simulator; 3.7 m diameter by ~ 10 m length
MAJOR APPENDAGES:	None
ATTITUDE CONTROL:	Unknown at time of event

EVENT DATA

KNOWN EVENTS:	1
DATE:	28 December 2021, 1836 GMT

APOGEE	PERIGEE	PERIOD	INCLINATION
200.8 km	178.8 km	88.3 min	63.4 deg

COMMENTS

Three objects were initially observed, and all were officially cataloged; all had decayed within one week following the event.

The Persey (*Perseus;* also transliterated as *Persei*) upper stage is in actuality derived from the venerable Blok-DM family, being a Blok DM-03 configured to the Angara A5 space launch vehicle. The Blok DM-03 differs from other marks in having extended propellant tanks. The DM-03 has flown previously on the Proton space launch vehicle and this was the first flight on the Angara A5. The payload has also been referred to as the IPN-1 payload.

The intended destination for the IPM-3 mass simulator was geostationary orbit. The upper stage engines failed to fire for the geosynchronous transfer orbit GTO injection burn and the stack was stranded in low Earth orbit. The production mechanism and identities of the observed debris are unknown hence this event's classification as anomalous.

SATELLITE DATA

TYPE:	rocket body
OWNER:	PRC
LAUNCH DATE:	29 Mar. 2022
DRY MASS (KG):	5800 kg
MAIN BODY:	Cylinder; 3.35 m diameter by ~ 9 m length
MAJOR APPENDAGES:	None
ATTITUDE CONTROL:	three-axis control upon ascent

POST-EVENT DATA

KNOWN EVENTS: 1

DATE: 29 March 2022, approximately 0950 GMT

APOGEE	PERIGEE	PERIOD	INCLINATION
593.9 km	555.7 km	96.1 min	97.8 deg

COMMENTS

Of the 10 to 20 objects initially observed, none were officially cataloged; all had decayed within one week following the event.

The reader should note that:
(a) this poorly characterized event evidently occurred during the ascent of the second stage--post event elements cited above are after an apparently successful separation of the two payloads and launch vehicle end-of-mission; and
(b) the CZ-6A space launch vehicle is dissimilar from the CZ-6 small launch vehicle. The vehicle is also referred to as the "Chang Zheng 6 Gao", or "Improved CZ-6" and is not apparently a derivative of the CZ-6 in the usual sense. Configuration is four strap-on solid boosters (stage 0) and liquid first and second core stages based upon launch imagery. Physical specifications cited above are estimates for the CZ-7A medium launch vehicle second stage.

4.0 OTHER SATELLITES ASSOCIATED WITH FRAGMENTATIONS

4.1 Aerodynamic Associations with Fragmentation Events

Aerodynamic breakups are associated with the breakup of an RSO caused by interacting with Earth's atmosphere. Forty-two such events have occurred between 1994 and May 2022. Because of the orbit elements of the parent object at the time of fragmentation, only seven of these events showed any cataloged debris and all parent objects reentered within one year of the event (most reentered within a few days). It is understood that only a fraction of these fragmentations can be detected because of the short remaining life of the parent and debris created. These events have no impact to the mid- or long-term debris environment, and therefore, it was deemed more appropriate to separate these from the fragmentations in Chapter 2. The parent object for these aerodynamic events shall not be considered fragmentation debris when discussing object categorization. Seven of these events produced cataloged debris other than the parent, and these debris objects represent the difference between the decayed fragmentation debris count in Table 1.3.2 and the decayed fragmentation debris count in Tables 2.1 and 2.2.

The following missions, listed by international designator in Table 4.1-1 and by event date in 4.1-2, have been determined to be solely related to aerodynamic effects at the time of reentry and therefore did not contribute to the orbital environment. Note that the tables strictly interpret cataloging; in the case of there being no cataloged debris, there were one to multiple objects observed that decayed prior to entry into the public satellite catalog.

TABLE 4.1-1 HISTORY OF SATELLITE AERODYNAMIC EVENTS BY LAUNCH DATE

SATELLITE NAME	INTERNATIONAL DESIGNATOR	US SATELLITE NUMBER	LAUNCH DATE	BREAKUP DATE	REENTRY DATE	DEBRIS CATALOGED	APOGEE (KM)	PERIGEE (KM)	INCLINATION (DEG)	ADDITIONAL INFORMATION
ELEKTRON 1/2 R/B	1964-006D	751	30-Jan-64	13-Feb-98	15-Feb-98	1	56315	90	56.2	VOSTOK FINAL STAGE
COSMOS 41	1964-049E	898	22-Aug-64	Apr-04	7-May-04	0	~35750	~115	64.5	
COSMOS 862 DEB	1976-105F	9889	22-Oct-76	29-Mar-14	29-May-14	2	14990	110	62.0	
COSMOS 1030	1978-083A	11015	6-Sep-78	14-Aug-04	17-Aug-04	1	~4560	~95	61.9	
COSMOS 1172	1980-028A	11758	12-Apr-80	23-Dec-97	26-Dec-97	1	5125	75	61.8	
COSMOS 1188	1980-050A	11844	14-Jun-80	23-May-13	24-May-13	0	1745	100	62.2	
MOLNIYA 3-16	1981-054A	12512	9-Jun-81	5-Feb-98	10-Feb-98	0	7670	85	62.1	
MOLNIYA 3-16 R/B	1981-054E	12519	9-Jun-81	28-Jul-98	30-Apr-99	0	33415	85	62.0	MOLNIYA FINAL STAGE
COSMOS 41 DEB	1964-049F	13091	22-Aug-64	30-Dec-02	31-Dec-02	0	1200	85	64.4	
MOLNIYA 3-19	1982-083A	13432	27-Aug-82	13-Jan-02	13-Jan-02	0	2075	95	62.3	
COSMOS 1658	1985-045A	15808	11-Jun-85	12-Nov-05	12-Nov-05	0	1730	80	62.1	
MOLNIYA 3-26	1985-091A	16112	3-Oct-85	21-Feb-01	22-Feb-01	0	5690	80	62.6	
MOLNIYA 1-66 R/B	1985-103D	16223	28-Oct-85	13-Jan-03	13-Jan-03	0	~1600	~120	62.4	MOLNIYA FINAL STAGE
COSMOS 1701	1985-105A	16235	9-Nov-85	29-Apr-01	11-May-01	0	25570	85	62.9	
COSMOS 1849	1987-048A	18083	4-Jun-87	27-Jan-03	4-Feb-03	0	7450	95	62.1	
COSMOS 1966	1988-076A	19445	30-Aug-88	~02-Nov-05	10-Nov-05	0	11535	90	62.9	
MOLNIYA 3-35	1989-043A	20052	8-Jun-89	14-Dec-01	14-Dec-01	0	593	65	61.9	
MOLNIYA 3-36	1989-094A	20338	28-Nov-89	19-May-00	20-May-00	0	1795	80	63.4	
MOLNIYA 3-36 R/B	1989-094B	20339	28-Nov-89	28-Jun-00	4-Jul-00	0	7145	75	63.6	MOLNIYA FINAL STAGE
MOLNIYA 1-77	1990-039A	20583	26-Apr-90	24-Feb-05	25-Feb-05	0	1710	75	62.0	
MOLNIYA 3-38 R/B	1990-052D	20649	13-Jun-90	~Sep-06	13-Sep-06	0	37710	130	62.4	MOLNIYA FINAL STAGE
COSMOS 2105	1990-099A	20941	20-Nov-90	16-Jan-08	21-Jan-08	0	2470	65	62.6	
MOLNIYA 1-82	1991-053A	21630	1-Aug-91	8-Oct-04	9-Oct-04	0	1510	75	61.7	
COSMOS 2176	1992-003A	21847	24-Jan-92	16-Jan-12	17-Jan-12	0	2555	75	62.1	
MOLNIYA 1-83 R/B	1992-011D	21900	4-Mar-92	26-Sep-06	26-Sep-06	0	1090	70	62.0	MOLNIYA FINAL STAGE
MOLNIYA 1-84	1992-050A	22068	6-Aug-92	3-Apr-08	4-Apr-08	0	2600	80	61.5	
MOLNIYA 3-44	1993-025A	22633	21-Apr-93	25-Jan-04	25-Jan-04	0	~1000	~90	63.4	

TABLE 4.1-1 HISTORY OF SATELLITE AERODYNAMIC EVENTS BY LAUNCH DATE (CONT'D)

SATELLITE NAME	INTERNATIONAL DESIGNATOR	US SATELLITE NUMBER	LAUNCH DATE	BREAKUP DATE	REENTRY DATE	DEBRIS CATALOGED	APOGEE (KM)	PERIGEE (KM)	INCLINATION (DEG)	ADDITIONAL INFORMATION
MOLNIYA 3-46 R/B	1994-051D	23214	23-Aug-94	17-Feb-08	19-Feb-08	0	5530	115	62.3	
ETS-VI R/B	1994-056B	23231	28-Aug-94	31-Mar-95	2-Apr-95	0	4840	100	28.6	H-II SECOND STAGE
MOLNIYA 3-48	1996-060A	24640	24-Oct-96	13-Oct-07	18-Oct-07	0	7825	100	63.4	
MOLNIYA 3-52	2001-050A	26970	25-Oct-01	5-Dec-11	6-Dec-11	0	2745	85	63.9	
HELLAS SAT-2 R/B	2003-020B	27812	13-May-03	11-Dec-04	12-Dec-04	0	10300	90	17.5	ATLAS V
MOLNIYA 1-93	2004-005A	28163	18-Feb-04	15-Apr-16	16-Apr-16	0	2415	77	62.9	
MERIDIAN 1 R/B	2006-061B	29669	24-Dec-06	2-Feb-22	10-Mar-22	0	6293	102	61.1	SL-4 R/B
USA 195 R/B	2007-046B	32259	11-Oct-07	21-Mar-08	21-Mar-08	0	59015	95	19.2	ATLAS V
CHANG'E R/B	2007-051B	32274	24-Oct-07	27-Jan-08	28-Jan-08	0	6035	80	30.7	CZ-3A FINAL STAGE
CTDRS R/B	2008-019B	32780	25-Apr-08	5-Mar-11	5-Mar-11	1	1065	110	17.9	CZ-3C FINAL STAGE
BRIZ-M TANK	2009-042C	35698	11-Aug-09	21-Jun-10	22-Jun-10	88	1490	90	48.4	
USA 230 R/B	2011-019B	37482	7-May-11	17-Aug-11	17-Aug-11	1	2285	95	21.0	ATLAS V
BEIDOU IGSO4 R/B	2011-038B	37764	26-Jul-11	2-Mar-12	3-Mar-12	0	1580	95	54.4	CZ-3A FINAL STAGE
BREEZE-M DEB (TANK)	2014-064C	40279	21-Oct-14	17-Jun-15	18-Jun-15	0	4690	100	48.6	Briz-M APT
GAOFEN 13 R/B	2020-071B	46611	11-Oct-20	2-Jan-21	2-Jan-21	0	900	150	28.2	CZ-3B R/B

TABLE 4.1-2 HISTORY OF SATELLITE AERODYNAMIC EVENTS BY EVENT DATE

SATELLITE NAME	INTERNATIONAL DESIGNATOR	US SATELLITE NUMBER	LAUNCH DATE	BREAKUP DATE	REENTRY DATE	DEBRIS CATALOGED	APOGEE (KM)	PERIGEE (KM)	INCLINATION (DEG)	ADDITIONAL INFORMATION
ETS-VI R/B	1994-056B	23231	28-Aug-94	31-Mar-95	2-Apr-95	0	4840	100	28.6	H-II SECOND STAGE
COSMOS 1172	1980-028A	11758	12-Apr-80	23-Dec-97	26-Dec-97	1	5125	75	61.8	
MOLNIYA 3-16	1981-054A	12512	9-Jun-81	5-Feb-98	10-Feb-98	0	7670	85	62.1	
ELEKTRON 1/2 R/B	1964-006D	751	30-Jan-64	13-Feb-98	15-Feb-98	1	56315	90	56.2	VOSTOK FINAL STAGE
MOLNIYA 3-16 R/B	1981-054E	12519	9-Jun-81	28-Jul-98	30-Apr-99	0	33415	85	62.0	MOLNIYA FINAL STAGE
MOLNIYA 3-36	1989-094A	20338	28-Nov-89	19-May-00	20-May-00	0	1795	80	63.4	
MOLNIYA 3-36 R/B	1989-094B	20339	28-Nov-89	28-Jun-00	4-Jul-00	0	7145	75	63.6	MOLNIYA FINAL STAGE
MOLNIYA 3-26	1985-091A	16112	3-Oct-85	21-Feb-01	22-Feb-01	0	5690	80	62.6	
COSMOS 1701	1985-105A	16235	9-Nov-85	29-Apr-01	11-May-01	0	25570	85	62.9	
MOLNIYA 3-35	1989-043A	20052	8-Jun-89	14-Dec-01	14-Dec-01	0	593	65	61.9	
MOLNIYA 3-19	1982-083A	13432	27-Aug-82	13-Jan-02	13-Jan-02	0	2075	95	62.3	
COSMOS 41 DEB	1964-049F	13091	22-Aug-64	30-Dec-02	31-Dec-02	0	1200	85	64.4	
MOLNIYA 1-66 R/B	1985-103D	16223	28-Oct-85	13-Jan-03	13-Jan-03	0	~1600	~120	62.4	MOLNIYA FINAL STAGE
COSMOS 1849	1987-048A	18083	4-Jun-87	27-Jan-03	4-Feb-03	0	7450	95	62.1	
MOLNIYA 3-44	1993-025A	22633	21-Apr-93	25-Jan-04	25-Jan-04	0	~1000	~90	63.4	
COSMOS 41	1964-049E	898	22-Aug-64	Apr-04	7-May-04	0	~35750	~115	64.5	
COSMOS 1030	1978-083A	11015	6-Sep-78	14-Aug-04	17-Aug-04	1	~4560	~95	61.9	
MOLNIYA 1-82	1991-053A	21630	1-Aug-91	8-Oct-04	9-Oct-04	0	1510	75	61.7	
HELLAS SAT-2 R/B	2003-020B	27812	13-May-03	11-Dec-04	12-Dec-04	0	10300	90	17.5	ATLAS V
MOLNIYA 1-77	1990-039A	20583	26-Apr-90	24-Feb-05	25-Feb-05	0	1710	75	62.0	
COSMOS 1966	1988-076A	19445	30-Aug-88	~02-Nov-05	10-Nov-05	0	11535	90	62.9	
COSMOS 1658	1985-045A	15808	11-Jun-85	12-Nov-05	12-Nov-05	0	1730	80	62.1	

TABLE 4.1-2 HISTORY OF SATELLITE AERODYNAMIC EVENTS BY EVENT DATE (CONT'D)

SATELLITE NAME	INTERNATIONAL DESIGNATOR	US SATELLITE NUMBER	LAUNCH DATE	BREAKUP DATE	REENTRY DATE	DEBRIS CATALOGED	APOGEE (KM)	PERIGEE (KM)	INCLINATION (DEG)	ADDITIONAL INFORMATION
MOLNIYA 3-38 R/B	1990-052D	20649	13-Jun-90	~Sep-06	13-Sep-06	0	37710	130	62.4	MOLNIYA FINAL STAGE
MOLNIYA 1-83 R/B	1992-011D	21900	4-Mar-92	26-Sep-06	26-Sep-06	0	1090	70	62.0	MOLNIYA FINAL STAGE
MOLNIYA 3-48	1996-060A	24640	24-Oct-96	13-Oct-07	18-Oct-07	0	7825	100	63.4	
COSMOS 2105	1990-099A	20941	20-Nov-90	16-Jan-08	21-Jan-08	0	2470	65	62.6	
CHANG'E R/B	2007-051B	32274	24-Oct-07	27-Jan-08	28-Jan-08	0	6035	80	30.7	CZ-3A FINAL STAGE
MOLNIYA 3-46 R/B	1994-051D	23214	23-Aug-94	17-Feb-08	19-Feb-08	0	5530	115	62.3	
USA 195 R/B	2007-046B	32259	11-Oct-07	21-Mar-08	21-Mar-08	0	59015	95	19.2	ATLAS V
MOLNIYA 1-84	1992-050A	22068	6-Aug-92	3-Apr-08	4-Apr-08	0	2600	80	61.5	
BRIZ-M TANK	2009-042C	35698	11-Aug-09	21-Jun-10	22-Jun-10	88	1490	90	48.4	
CTDRS R/B	2008-019B	32780	25-Apr-08	5-Mar-11	5-Mar-11	1	1065	110	17.9	CZ-3C FINAL STAGE
USA 230 R/B	2011-019B	37482	7-May-11	17-Aug-11	17-Aug-11	1	2285	95	21.0	ATLAS V
MOLNIYA 3-52	2001-050A	26970	25-Oct-01	5-Dec-11	6-Dec-11	0	2745	85	63.9	
COSMOS 2176	1992-003A	21847	24-Jan-92	16-Jan-12	17-Jan-12	0	2555	75	62.1	
BEIDOU IGSO4 R/B	2011-038B	37764	26-Jul-11	2-Mar-12	3-Mar-12	0	1580	95	54.4	CZ-3A FINAL STAGE
COSMOS 1188	1980-050A	11844	14-Jun-80	23-May-13	24-May-13	0	1745	100	62.2	
COSMOS 862 DEB	1976-105F	9889	22-Oct-76	29-Mar-14	29-May-14	2	14990	110	62.0	
BREEZE-M DEB (TANK)	2014-064C	40279	21-Oct-14	17-Jun-15	18-Jun-15	0	4690	100	48.6	Briz-M APT
MOLNIYA 1-93	2004-005A	28163	18-Feb-04	15-Apr-16	16-Apr-16	0	2415	77	62.9	
GAOFEN 13 R/B	2020-071B	46611	11-Oct-20	2-Jan-21	2-Jan-21	0	900	150	28.2	CZ-3B R/B
MERIDIAN 1 R/B	2006-061B	29669	24-Dec-06	2-Feb-22	10-Mar-22	0	6293	102	61.1	SL-4 R/B

4.2　Reactor Coolant Debris

The Soviet Union conducted the Radar Ocean Reconnaissance satellite (RORSAT) program with developmental and operational flights spanning 1965 (Cosmos 102) to 1988 (Cosmos 1932). Starting with Cosmos 198 (1967), large operational debris objects were left in their operational orbit while the nuclear reactor power source was boosted to a higher storage (or "graveyard") orbit. Beginning with Cosmos 1176 (1980), RORSATs routinely ejected their reactor core. This practice resulted in the venting of all or part of the reactor's sodium-potassium (NaK) liquid metal coolant. Although this population is largely undetectable by its sensors, some coolant of sufficient size to be detected by at least some sensors has been cataloged by the SSN. Table 4.2 lists the coolant objects cataloged as of 03 May 2022.

TABLE 4.2:　CATALOGED RORSAT PROGRAM NAK COOLANT DEBRIS

International Designator	Common Name	Total Coolant Cataloged	Coolant On-Orbit
1976-103	COSMOS 860	3	3
1980-034	COSMOS 1176	2	2
1981-021	COSMOS 1249	17	17
1981-037	COSMOS 1266	1	1
1982-099	COSMOS 1412	1	1
1984-069	COSMOS 1579	31	31
1987-011	COSMOS 1818*	53	52
1987-060	COSMOS 1867*	42	42

*not operational RORSAT program element; discussed in Section 3.

4.3　Spurious Associations with Fragmentation Events

Satellite fragmentation lists compiled by other organizations, in particular by the National Security Council and the U.S. Naval Space Command Space Surveillance System (NAVSPASUR), were carefully reviewed during the preparation of the fourth edition of the History of On-orbit Satellite Fragmentations. Due to the frequent exchange of information within the small orbital debris and space operations community and the long period during which satellite fragmentation lists have been maintained, no current list is completely independent from all others.

These reviews also revealed the need to define better the terms "satellite breakup" and "anomalous event" as discussed in Section 1.0. Many breakup lists have historically included entries related to normal launch and mission activities that resulted in numbers of debris in excess of the handful usually observed on these occasions. Some researchers have been misled by tracking difficulties and cataloging procedures that may cause late cataloging or misidentification of debris, superficially giving the appearance of fragmentations. A higher than average number of debris alone is not sufficient to assume a satellite fragmentation. Such pitfalls can generally be avoided by conducting analyses with complete satellite element set data rather than the limited orbital data available in the U.S. Satellite Catalog.

The following space missions, listed by international designator, have been examined in detail and have failed to qualify as either a satellite breakup or an anomalous event, as set forth in Section 1.0. The source of debris associated with nearly all of these flights is of a mission-related nature. Bolded items indicate the alleged source of the debris when unique or appropriate. The Common Name prefix C indicates a Cosmos-series satellite.

TABLE 4.3: SPURIOUS ASSOCIATION WITH FRAGMENTATIONS BY LAUNCH DATE

Int'l Des.	Common Name	S/C	R/B	Total Debris	Debris On-orbit
1963-014	FTV 1169	Payload	Agena B spacecraft	147	45
1965-073	C 86-90	Payload	**Cosmos 3**	5	5
1965-112	C 103	Payload	**Cosmos 3**	13	0
1967-001	INTELSAT 2-F2	INTELSAT II	**Delta 1 R/B (2): FW-4**	17	1
1967-011	**Diademe 1**	Payload	**Diamant**	13	0
1967-014	**Diademe 2**	Payload	**Diamant**	12	3
1967-024	**C 149**	DS-MO	Cosmos 2	16	0
1967-086	**C 176**	DS-P1-Yu	Cosmos 2	9	0
1968-117	C 261	DS-U2-GK	**Cosmos 2**	22	0
1969-021	C 269	*Tselina*-O	**Cosmos 3**	21	0
1970-005	C 320	DS-MO	**Cosmos 2**	5	0
1970-033	**C 334**	DS-P1-Yu	Cosmos 2	3	0
1970-065	C 359	Venera	**Molniya**	2	0
1972-078	**C 523**	DS-P1-Yu	Cosmos 2	10	0
1973-027	Skylab 1	Skylab workshop	**Saturn V**	22	0
1973-075	**C 601**	DS-P1-Yu	Cosmos 2	12	0
1974-074	**C 686**	DS-P1-Yu	Cosmos 2	18	0
1974-104	**Salyut 4**	Crewed station	Proton	17	0
1976-012	C 801	DS-P1-I	**Cosmos 2**	15	0
1976-037	**C 816**	Romb	Cosmos 3	23	0
1976-057	Salyut 5	Crewed station	Proton	8	0
1976-124	**C 885**	*Romb*	Cosmos 3	17	0
1977-042	**C 913**	*Romb*	Cosmos 3	20	0
1977-097	**Salyut 6**	Crewed station	Proton	104	0

Int'l Des.	Common Name	S/C	R/B	Total Debris	Debris On-orbit
1977-111	C 965	*Romb*	Cosmos 3	25	0
1978-043	C 1004	*Zenit*-2M	**Soyuz**	5	0
1978-120	C 1065	Romb	Cosmos 3	6	0
1979-008	C 1074	*Soyuz* T test vehicle	Soyuz	5	0
1979-063	C 1112	Romb	Cosmos 3	24	0
1980-047	C 1186	*Romb*	Cosmos 3	25	0
1980-067	C 1204	*Romb*	Cosmos 3	22	0
1980-083	C 1215	Payload	**Cosmos 3**	2	0
1981-093	SJ-2/-2A/-2B	Payloads	**CZ-2B**	6	0
1981-097	C 1311	Romb	Cosmos 3	24	0
1982-006	**OPS 2849**	Payload	Titan 3B Agena	4	0
1982-007	C 1335	Romb	Cosmos 3	22	0
1982-033	**Salyut 7**	Crewed station	Proton	197	0
1982-034	C 1351	*Romb*	Cosmos 3	24	0
1982-076	C 1397	*Romb*	Cosmos 3	22	0
1983-034	C 1453	*Romb*	Cosmos 3	22	0
1983-049	C 1465	*Romb*	Cosmos 3	8	0
1983-091	C 1494	*Romb*	Cosmos 3	25	0
1983-101	C 1501	*Romb*	Cosmos 3	24	0
1984-008	STTW-T1	Payload	**CZ-3**	2	0
1984-104	C 1601	Romb	Cosmos 3	28	0
1985-021	GEOSAT	Payload	**Atlas 41E (OIS R/B)**	5	3
1985-050	C 1662	Romb	Cosmos 3	27	0
1985-075	C 1677	RORSAT	Tsyklon	2	0
1985-097	C 1697	*Tselina*-2	**Zenit**	4	4
1986-017	Mir	Crewed station	Proton	323	0
1986-024	C 1736	RORSAT	Tsyklon	26	0
1986-030	C 1741	Payload	**Cosmos 3**	2	2
1986-052	C 1763	Payload	**Cosmos 3**	4	3
1986-067	C 1776	Romb	Cosmos 3	28	0
1986-101	C 1809	Payload	**Tsyklon**	9	9
1988-019	C 1932	RORSAT	Tsyklon	3	2

Int'l Des.	Common Name	S/C	R/B	Total Debris	Debris On-orbit
1988-065	**C 1960**	Romb	Cosmos 3	28	0
1988-067	FSW-1 2	Payload	**CZ-2C**	5	0
1988-113	**C 1985**	Duga-K	Tsyklon	36	0
1989-012	C 2002	*Romb*	**Cosmos 3**	10	0
1989-100	**C 2053**	*Duga-K*	Tsyklon	37	0
1990-012	C 2059	*Romb*	**Cosmos 3**	10	0
1990-038	**C 2075**	*Romb*	Cosmos 3	14	0
1990-104	**C 2106**	*Romb*	Tsyklon	28	0
1995-008	**C 2306**	*Romb*	Cosmos 3	23	0
1998-067	**Zarya**	International Space Station (ISS)	Proton	129	8

Note: the debris count associated with 1998-067, the ISS, does not include payloads deployed from the ISS by the U.S., the CIS) partners, or participant countries and non-governmental organizations. Of the 18 additional objects known by "OBJECT NN"-series generic names associated with the ISS, 6 were on-orbit as of 03 May 2022.

For more information on these events, the references below.

REFERENCE DOCUMENTS

4. Johnson, N.L. and Nauer, D.J. "History of On-orbit Satellite Fragmentations (4th Edition), Teledyne Brown Engineering, CS90-TR-JSC-002, Colorado Springs, CO, USA, 1990.
5. Pedersen, K.S., Kunsberg, P., and Binns, Jr., J.H. "Report on Orbital Debris by Interagency Group (Space) for National Security Council," NASA, Washington, D.C., 1989.
6. 99th Congress of the U.S., Senate Committee on Commerce, Science, and Transportation. "Soviet Space Programs: 1976-80, Part 3," Senate Print 98-235 Part 3, U.S. Government Printing Office, Washington, D.C., May 1985.

This Page Intentionally Left Blank

5.0 SATELLITES NOT ASSOCIATED WITH BREAKUPS

The table below identifies specific SSN numbers of objects that possess the same International Designator year and number but are not associated with the indicated event. For example, 1961-015C was an Ablestar rocket body that broke up. The mission deployed two objects (Transit 4A and Solrad 3/Injun 1) not associated with the rocket body explosion. Those two objects are not counted in the 1961-015 totals, though they are associated with the 1961-015 international designator.

Occasionally, it is not obvious whether an object should be included in a fragmentation event. In those cases, historical research and historical satellite catalogs may reveal whether an object should be included in the count. This list represents the best summary of excluded objects. The parent object is always considered a fragment. Aerodynamic breakups are included in this list if they produced cataloged fragmentation other than the parent object.

The list is formatted as follows: The international designator and number of excluded debris in parenthesis are followed by the SSN numbers, which are not debris. A blank line separates years.

International Designator

1961-015 (2) -   116 117

1962-057 (0) -

1963-047 (0) -

1964-006 (28) -   746 748 750 751 14427 14428 15786 16544 16545 16546 16547 16548 18589 18686 19010 19173 19990 19991 19992 19993 19994 19995 19996 19997 19998 20101 20224 21621
1964-070 (1) -   920

1965-012 (1) -   1095
1965-020 (3) -   1267 1268 1269
1965-082 (1) -   1641
1965-088 (23)   1707 1708 1740 1741 1784 1785 1786 1787 1788 1789 1790 1791 1792 1793 1794 1795 1796 1797 1798 1799 1800 1801 1802
1965-108 (4) -   1870 1902 1941 13912

1966-012 (2) -   2012 2014
1966-046 (3) -   2186 2189 2190
1966-056 (3) -   2255 2256 2511
1966-059 (1) -   2291
1966-088 (1) –   2438
1966-101 (0) –

1967-116 (1) -   3048

1968-003 (1) -   3096
1968-025 (1) -   3170
1968-081 (5) -   3428 3429 3430 3431 5999
1968-090 (0) -
1968-091 (1) -   3505
1968-097 (0) –
1968-114 (1) -   3615

```
1969-013 (1) -    3691
1969-029 (1) -    3835
1969-064 (1) -    4051
1969-082 (10) -   4111 4132 4166 4168 4237 4247 4256 4257 4259 4295

1970-025 (2) -    4362 4363
1970-089 (1) -    4597
1970-091 (0) -

1971-015 (1) -    4965
1971-106 (4) -    5650 5664 5665 5672

1972-058 (1) -    6126

1973-017 (1) -    6398
1973-021 (4) -    6434 6436 6442 6443
1973-086 (1) -    6920

1974-015 (1) -    7218
1974-089 (3) -    7529 7530 7531
1974-103 (1) -    7588

1975-004 (1) -    7615
1975-052 (2) -    7924 7965
1975-080 (1) -    8192
1975-102 (1) -    8417

1976-023 (9) -    8746  8747  8748  8749  8750  8752  8753  8832 13753
1976-063 (1) -    8933
1976-067 (2) -    9013 9016
1976-072 (1) -    9048
1976-077 (1) -    9057
1976-105 (3) -    9496 9497 9506
1976-120 (2) -    9604 9605
1976-123 (4) -    9623 9624 9639 9640
1976-126 (3) -    9643 9644 9645

1977-027 (3) -    9912 9913 9921
1977-047 (3) -    10060 10066 10089
1977-065 (3) -    10143 10145 10156
1977-068 (3) -    10151 10152 10167
1977-092 (6) -    10366 10367 10368 10408 10484 11571
1977-121 (1) -    10532

1978-026 (2) -    10702 10703
1978-083 (3) -    11016 11017 11076
1978-098 (2) -    11080 18605
1978-100 (4) -    11084 11085 11086 11177

1979-017 (3) -    11279 11291 11322
1979-033 (2) -    11334 11367
1979-058 (3) -    11418 11423 11555
1979-077 (3) -    11512 11513 11550
1979-087 (4) -    11562 17939 18966 18971
1979-101 (2) -    11636 11637
1979-104 (3) -    11645 24754 25098
```

```
1980-021 (1) -    11730
1980-028 (5) -    11758 11759 11760 11761 11762
1980-030 (1) -    11766
1980-050 (3) -    11845 11846 11847
1980-057 (3) -    11872 11873 11888
1980-085 (3) -    12033 12034 12035
1980-089 (1) -    12055

1981-016 (4) -    12304 12305 12306 12311
1981-028 (1) -    12365
1981-031 (3) -    12377 12378 12384
1981-053 (1) -    12508
1981-058 (3) -    12548 12549 12561
1981-071 (3) -    12629 12630 12680
1981-072 (1) -    12632
1981-088 (5) -    12818 12819 12820 12821 12822
1981-089 (1) -    12829
1981-108 (3) -    12934 12935 12940

1982-025 (1) -    13114
1982-029 (3) -    13125 13126 13169
1982-038 (1) -    13151
1982-055 (2) -    13260 13261
1982-088 (1) -    13509
1982-092 (2) -    13553 42060
1982-093 (5) -    13555 13556 13583 13584 14115
1982-115 (4) -    13685 13686 13692 13693

1983-020 (3) -    13901 13903 20413
1983-022 (8) -    13924 14477 16502 16503 28604 38839 38840 38841
1983-038 (6) -    14036 14037 14038 14041 14042 14043
1983-044 (1) -    14065
1983-070 (3) -    14183 14184 14191
1983-075 (5) -    14208 14209 14229 14631 14928
1983-127 (7) -    14590 14591 14592 14593 14594 14595 14607

1984-011 (6) -    14681 14688 14689 14692 14695 14696
1984-083 (1) -    15168
1984-106 (6) -    15333 15334 15335 15336 15337 17358
1984-114 (2) -    15385 15386

1985-030 (1) -    15654
1985-037 (7) -    15697 15698 15699 15700 15701 15702 15715
1985-039 (1) -    15735
1985-042 (5) -    15755 15770 15771 15772 15774
1985-082 (1) -    16055
1985-094 (6) -    16138 16140 16141 16142 16143 16144
1985-108 (1) -    16262
1985-118 (10) -   16396 16397 16398 16399 16403 16404 16405 16406 16407 16445
1985-121 (5) -    16434 16435 16436 16437 16438

1986-019 (3) -    16613 16614 16616
1986-059 (1) -    16896
1986-069 (4) -    16946 16947 16948 16949
```

1987-004 (1) -   17298
1987-020 (4) -   17536 26111 26601 26982
1987-059 (2) -   18185 18186
1987-062 (1) -   18215
1987-068 (1) -   18312
1987-078 (3) -   18350 18351 18353
1987-079 (6) -   18355 18356 18357 18358 18359 18360
1987-108 (1) -   18714
1987-109 (5) -   18715 18716 18717 18718 18722

1988-005 (3) -   18821 38540 38543
1988-006 (5) -   18845 18846 18855 18956 18984
1988-007 (1) -   18824
1988-023 (1) -   18986
1988-040 (1) -   19121
1988-085 (6) -   19501 19502 19503 19504 19505 21751
1988-109 (3) -   19687 19688 19690

1989-001 (6) -   19749 19750 19751 19752 19753 19754
1989-004 (5) -   19765 19766 19767 19768 19776
1989-006 (1) -   19772
1989-039 (7) -   20024 20025 20026 20027 20028 20044 20082
1989-052 (5) -   20107 20108 20109 20110 20115
1989-054 (1) -   20125
1989-056 (2) -   20137 20138
1989-089 (79) -   20322 20324 20328 22625 22683 22695 22747 22748 22749 22750 22751 22752 22753 22754
                  22755 22756 22757 22758 22759 22760 22761 22762 22763 22764 22765 22766 22767 22768
                  22769 22770 22771 22772 22773 22774 22775 22776 22820 22852 22853 22854 22855 22856
                  22857 22858 22972 23053 23054 23055 23056 23057 23058 23059 23060 23061 23062 23063
                  23064 23065 23066 23067 23068 23069 23070 23071 23072 23073 23074 23075 23076 23077
                  23078 23079 23080 23081 23082 23083 23084 23085 23086
1989-100 (38) -   20389 20397 20398 20408 20467 20468 20515 20522 20531 20532 20637 20640 20802 20803
                  20821 20822 20823 20911 21020 21021 21022 21023 21042 21043 21064 21205 21206 21207
                  21537 21540 21767 21768 21769 21770 21771 21772 21773 21774
1989-101 (6) -   20391 20392 20393 20394 20400 21648

1990-045 (5) -   20619 20620 20621 20622 20623
1990-081 (7) -   20788 20789 20790 20792 20793 20797 20798
1990-087 (1) -   20829
1990-102 (5) -   20953 20954 20955 20958 21046
1990-105 (1) -   20978
1990-110 (6) -   21006 21007 21008 21009 21010 21011

1991-003 (3) -   21055 21056 21058
1991-009 (8) -   21100 21101 21102 21103 21104 21105 21106 21107
1991-010 (5) -   21111 21112 21113 21122 21129
1991-015 (4) -   21139 21140 21142 21904
1991-025 (5) -   21216 21217 21218 21219 21221
1991-056 (1) -   21655
1991-068 (6) -   21728 21729 21730 21731 21732 21733
1991-071 (1) -   21742
1991-075 (1) -   21765
1991-082 (4)   21800 21801 21825 21836

1992-021 (3) -   21939 21940 21942
1992-041 (8) -   22027 22028 22033 27484 27485 27486 27487 27675

```
1992-047 (6) -    22056 22057 22058 22059 22060 22061
1992-082 (5) -    22245 22246 22247 22248 22249
1992-088 (5) -    22269 22270 22271 22272 22273
1992-091 (1) -    22281
1992-093 (5) -    22284 22290 22291 22292 22293

1993-016 (3) -    22565 22575 22576
1993-018 (1) -    22586
1993-028 (1) -    22642
1993-036 (3) -    22676 23007 25028
1993-045 (1) -    22717
1993-057 (2) -    22790 22953
1993-072 (5) -    22907 22908 22909 22910 22926

1994-004 (2) -    22973 22987
1994-029 (1) -    23105
1994-038 (5) -    23168 23169 23170 23171 23172
1994-069 (5) -    23327 23328 23329 23330 23339
1994-074 (41) -   23343 23344 23345 23346 23347 29297 29298 31580 33281 35263 35264 35265
                  35266 35267 36848 36849 36850 36851 36852 36853 39047 39048 39049 39050
                  39051 39516 39517 39611 40650 40669 40670 40671 43347 43348 43349 43350
                  44577 44578 44579 44580 44581
1994-076 (7) -    23396 23397 23398 23399 23400 23401 23403
1994-085 (1) -    23439

1995-015 (4) -    23534 23535 23594 23595
1995-028 (1) -    23597
1995-033 (3) -    23605 23607 23608
1995-037 (9) -    23620 23621 23622 23623 23624 23625 23626 23627 23630

1996-010 (5) -    23794 23795 23796 23824 24736
1996-034 (5) -    23880 23881 23882 23883 23886

1997-024 (1) -    24806
1997-051 (7) -    24944 24945 24947 24948 24949 24950 24951
1997-070 (5) -    25045 25046 25047 25048 25053
1997-079 (1) -    25089
1997-084 (9) -    25112 25113 25114 25115 25116 25118 25119 25120 25121
1997-086 (3) -    25126 25127 25128

1998-011 (1) -    25175

1999-008 (3) -    25634 25635 25636
1999-025 (5) -    25731 25732 25733 37580 37581
1999-057 (1) -    25941
1999-072 (1) -    26041

2000-036 (5) -    26394 26395 26396 26397 26399
2000-055 (1) -    27477

2001-049 (3) -    26957 26958 26959
2001-057 (1) -    27054

2002-032 (1) -    38555
2002-037 (7) -    27470 27471 27472 27473 27476 27494 43372
```

2003-035 (6) -   27857 28084 28085 28086 28087 28088

2005-050 (6) -   28915 28916 28917 28918 28919 28921

2006-002 (1) -   28931
2006-006 (1) -   28943
2006-015 (5) -   29093 29536 29537 29538 29539
2006-026 (1) -   29248
2006-037 (4) -   29393 29395 29396 29493
2006-039 (2) -   29397 29403
2006-050 (5) -   29522 29524 29525 29600 29637
2006-057 (1) -   29652
2006-062 (6) -   29670 29671 29672 29673 29674 29675

2007-003 (2) -   30323 30479
2007-005 (14) -   30586 30587 30588 30589 30591 30651 31105 31106 31107 31108
                  31109 31110
                  31111 31112
2007-029 (7) -   31793 31794 31795 31796 31799 43374 43375
2007-052 (6) -   32275 32276 32277 32278 32279 32281
2007-054 (1) -   32287
2007-065 (9) -   32393 32394 32395 32396 32397 32400 32401 32402 32403

2008-011 (1) -   32708
2008-019 (2) -   32779 32780
2008-046 (7) -   33378 33379 33380 33381 33382 33383 33384
2008-067 (6) -   33466 33467 33468 33469 33470 33471

2009-018 (1) -   34780
2009-042 (2) -   35696 35697
2009-070 (10) -   36111 36112 36113 36114 36115 36117 36118 36126 36127 36128

2010-007 (9) -   36400 36401 36402 36403 36404 36405 36408 36409 36410
2010-041 (10) -   37137 37138 37139 37140 37141 37142 37145 37146 37147 37149
2010-042 (1) -   37150
2010-057 (1) -   37210

        2011-019 (2) -37481 37482
2011-037 (7) -   37755 37757 37758 37759 37760 37761 46736
2011-077 (1) -   38014

2012-008 (2) -   38091 38095
2012-026 (2) -   38342 38344
2012-044 (2) -   38744 38745

2013-055 (11) -   39265 39266 39267 39268 39269 39270 39280 39290 39292 39293
                  39369
2013-076 (4) -   39483 39484 39485 39486

2014-016 (1) -   39634
2014-028 (4) -   39761 39762 39763 39764
2014-055 (1) -   40208

2015-024 (1) -   40619
2015-075 (2) -   41121 41123

2016-012 (5) -   41338 41339 41340 41341 41342

2017-086 (7) -   43087 43088 43090 43091 43092 43093 43094

2018-079 (1) -   43651
2018-084 (15) -   43671 43672 43674 43675 43676 43677 43678 43679 43681 43682 44051 44052
                  44053 44054 44620

2019-006 (1) -   43948
2019-039 (13) -   44422 44423 44424 44425 44621 44630 44631 44632 44633 44640
                  44641 44711 44712
2019-063 (4) -   44548 44549 47974 47975

2020-046 (4) -   45873 45874 45875 45876

This Page Intentionally Left Blank

6.0   SATELLITES NOT ASSOCIATED WITH ANOMALOUS EVENTS

The table below identifies specific SSN numbers of objects that possess the same International Designator year and number but are not associated with the indicated anomalous event. This list represents the best summary of excluded objects. The parent object is not considered a fragment.

The list is formatted as follows: The international designator and number of excluded debris in parentheses are followed by the SSN numbers that are not debris.

International Designator

1959-007 (1) -   20

1963-049 (12) -   703 704 705 706 715 753 2432 2620 2930 4586 6182 6283

1964-006 (5) -   746 748 750 751 25278
1964-026 (5) -   801 805 806 809 2986
1964-053 (2) -   876 877

1965-016 (9) -   1208 1244 1245 1271 1272 1291 1292 1293 1310
1965-027 (3) -   1314 1315 1316
1965-048 (6) -   1420 1425 1428 1435 2701 3592
1965-098 (9) -   1804 1806 1807 1808 1944 1948 1951 2092 2153 20833
1965-101 (4) -   1814 1815 1934 1935
1965-109 (5) -   1864 1865 2086 2226 2353

1966-005 (6) -   1952 1953 2140 2141 2889 2989
1966-024 (3) -   2119 2120 3590
1966-040 (2) -   2173 2174
1966-077 (3) -   2403 2411 2412

1967-034 (6) -   2754 2755 2777 2778 6718 7670
1967-048 (4) -   2807 2811 17723 19222
1967-092 (4) -   2965 2967 2994 3122

1968-019 (2) -   3150 3151

1969-009 (2) -   3669 3670

1970-067 (5) -   4507 4515 5036 5447 6372

1971-003 (2) -   4849 4850

1972-049 (2) -   6079 6080
1972-102 (2) -   6319 6320

1975-027 (2) -   7734 7735

1978-014 (6) -   10664 10665 12329 12330 12331 12406
1978-064 (1) -   10967
1978-094 (2) -   11055 11056
1978-096 (3) -   11060 11061 11062
1978-098 (2) -   11080 11081

1979-057 (3) -   11416 11419 11634
1979-095 (2) -   11605 11608

1981-043 (3) -   12456 12457 15769
1981-054 (5) -   12512 12513 12514 12515 12519
1981-059 (3) -   12553 12559 12560

1982-102 (3) -   13617 13618 37865

1983-022 (9) -   13923 13924 14477 16442 16443 16444 16502 16503 16504

1985-066 (6) -   15935 15936 15938 15950 15951 16020
1985-090 (2) -   16110 16111

1986-073 (4) -   16969 16982 16983 26303

1987-011 (4) -   17369 17370 17399 17400
1987-020 (2) -   17535 17536 (there are over 100 pieces of fragmentation as well)
1987-060 (4) -   18187 18188 18191 18524

1988-005 (45) -   18821 28988 28999 28990 28991 28992 28993 28994 28995 29066 29067 29299
                  29300 29301 29302 31397 31398 31399 31400 31401 31402 35364 35365 35366
                  35367 35368 35369 35370 35371 35372 35373 38281 38282 38283 38284 38285
                  38286 38287 38330 38539 38541 38542 38544 38655 38656
1988-032 (2) -   19045 19046
1988-089 (3) -   19531 19532 19534

1989-089 (2) -   20322 20323

1990-017 (2) -   20508 20509
1990-037 (3) -   20579 20580 22920

1991-032 (3) -   21263 21267 21298
1991-039 (3) -   21397 21398 21842
1991-050 (6) -   21574 21575 21576 21577 21578 21610
1991-063 (2) -   21700 21701

```
1993-014 (5) -      22561 22562 22567 22568 22599

1994-057 (5) -      23233 23234 23235 12150 23277
1994-074 (102) -    23342 23343 23344 23345 23346 23347 46508 46509 46510 46511
                    46512 46513 46514 46515 46516 46873 46874 46875 46876 46877
                    46878 46879 46880 46881 46882 46883 46884 46885 46886 46887
                    46888 46889 46890 46891 46892 46893 46894 46895 46896 46897
                    46898 46899 46900 46901 46902 46903 47198 47199 47200 47201
                    47204 47205 47206 47207 47208 47209 47210 47211 47213 47214
                    47215 47216 47217 47218 47219 47220 47221 47222 47223 47224
                    47225 47323 47324 47325 47326 47327 47328 47329 48493 48494
                    48495 48496 48497 48498 48499 48500 48501 48502 48503 48504
                    48505 48506 48507 48508 48509 48510 48511 48512 48513 48514
                    48515 48516
1994-077 (11) -     23404 23405 23406 23407 23408 23409 23410 23417 23418 23419 27760
1994-079 (6) -      23342 23343 23344 23345 23346 23347
1994-089 (3) -      23455 23457 23458

1995-021 (3) -      23560 23561 28066
1995-041 (3) -      23639 23640 23641
1995-059 (2) -      23710 23711

1996-017 (3) –      23827 23828 31403

1997-082 (7) -      25104 25105 25106 25107 25108 25109 25141

1999-035 (2) -      25791 25792
1999-040 (3) -      25866 25867 25869
1999-042 (2) -      25880 25881
1999-051 (3) -      25919 25920 25921

2001-055 (5) -      26997 26998 26999 27000 27497

2002-005 (6) -      27372 27373 27374 27375 27376 27377
2002-022 (1) -      27424

2003-024 (3) -      27820 27821 28998

2004-049 (7) -      28493 28494 28495 28496 28497 28498 28499

2009-049 (8) -      35865 35866 35867 35868 35869 35870 35871 35872
2009-055 (2) -      35946 35947

2011-043 (6) -      37781 37782 38313 38314 38315 38316

2012-017 (2) -      38248 38249

2015-020 (5) -      40552 40553 40554 40555 40556

2018-084 (98) -     43671 43672 43673 43674 43675 43676 43677 43678 43679 43681
                    43682 46181 46182 46183 46184 46185 46186 46187 46188 46189
                    46190 46191 46192 46193 46194 46195 46196 46197 46198 46199
                    46200 46201 46202 46203 46204 46205 46206 46207 46208 46209
                    46210 46211 46212 46213 46214 46215 46216 46217 46218 46219
                    46220 46221 46222 46223 46224 46225 46226 46227 46228 46229
                    46230 46231 46414 46415 46416 46417 46418 46419 46420 46421
```

46422 46423 46424 46425 46426 46427 46428 46473 46474 46475
46476 46477 46598 46599 46600 47187 47188 47189 47190 47191
47192 47193 47194 47195 47196 47197 48551 48552

2021-092 (2) -   49326 49327
2021-133 (0) –

2022-031 (2) -   52150 52151